HOLOGRAPHY & HOMOLOGY OF TRAVERSING FLOWS

HOLOGRAPHY & HOMOLOGY OF TRAVERSING FLOWS

Gabriel Katz
Massachusetts Institute of Technology, USA

World Scientific

NEW JERSEY • LONDON • SINGAPORE • BEIJING • SHANGHAI • HONG KONG • TAIPEI • CHENNAI • TOKYO

Published by

World Scientific Publishing Co. Pte. Ltd.
5 Toh Tuck Link, Singapore 596224
USA office: 27 Warren Street, Suite 401-402, Hackensack, NJ 07601
UK office: 57 Shelton Street, Covent Garden, London WC2H 9HE

British Library Cataloguing-in-Publication Data
A catalogue record for this book is available from the British Library.

Cover image by Gabriel Katz: The Holography Wizard, controlling the traversing flows of "holographic magic", creates a holographic image of a Topological Dragon in between the palms of his hands. The picture is inspired by the Holography Theorems from the book that recreate the interior of a manifold (the dragon) from the residual information on its boundary, generated by the traversing flow (the creative hands of the wizard). In fact, the Holography Theorems require a partition of the boundary into two domains (thus the two hands) and a flow generated correspondence between the two domains.

HOLOGRAPHY & HOMOLOGY OF TRAVERSING FLOWS

ISBN 978-981-98-2057-3 (hardcover)
ISBN 978-981-98-2058-0 (ebook for institutions)
ISBN 978-981-98-2059-7 (ebook for individuals)

For any available supplementary material, please visit
https://www.worldscientific.com/worldscibooks/10.1142/14506#t=suppl

Desk Editor: Tan Rok Ting

Dedication Page

to Uma, Zev, and Oren, and to their rapid time flows

Preface

"The stringent precision attainable in Mathematics has led many authors to a mode of writing which must give the reader the impression of being shut up in a brightly illuminated cell where every detail sticks out with the same dazzling intensity, but without relief. I prefer the open landscape with a clear sky with its depth of perspective where a wealth of sharply defined nearby details gradually fade away towards the horizon."

Hermann Weyl

This monograph, a natural extension of our book [K7], is an account of the author's thoughts about certain class of structures, induced by smooth vector flows on the boundaries of compact smooth manifolds. We are mostly concerned with the vector flows that admit Lyapunov functions; these functions are strictly monotone along each trajectory. We call such flows traversing.

Part I of the book deals with the inverse boundary value problems, generated by traversing vector flows v (in a sense, by generic boundary value problems for ordinary differential equations) on compact smooth manifolds X with a boundary. The word "holography" in the title of the book refers to the process by which the reconstruction of the pair (X, v) is often possible from the flow-generated data, confined to the boundary ∂X. These data are commonly called "boundary data".

Consider a physical model of a liquid flowing trough a given volume X with a smooth boundary. We assume that the flow velocity v does not vanish in X. We think about ∂X as the hypersurface where a multitude of measuring devices are installed. The basic assumption is that their presence and measuring activity does not alter the flow.

Any flow particle which enters the volume is registered, and its next appearance at a point of the boundary ∂X is registered as well. According to the Holography Theorem (see Corollary 3.5), these data allow for a reconstruction of the bulk X and of the un-parametrized dynamics of the flow in it, up to a homeomorphism (often a diffeomorphism) of X which is the identity on its boundary.

Thus in our approach, the effects produced by interactions of the v-trajectories with the boundary ∂X, take the central stage.

We denote by $\partial_1^+ X(v)$ the portion of the boundary ∂X where the vector field v points inside of X, and by $\partial_1^- X(v)$ the portion of the boundary ∂X where v points outside of X.

The causality map $C_v : \partial_1^+ X(v) \to \partial_1^- X(v)$ (also called "the scattering map" in various contexts) takes each point $x \in \partial_1^+ X(v)$ to the next point $C_v(x) \in \partial_1^- X(v)$ along the v-trajectory through x. A particular concave geometry of the boundary ∂X with respect to the flow causes the flow discontinuity, which results in discontinuity of the causality map C_v. These discontinuities are of a fundamental importance for the reconstruction of the topology of the pair (X, v) from appropriate boundary data, including the causality map C_v. In other words, in the book we promote the slogan:

"The topology of (X, v) hides in the discontinuity of the causality map C_v!"

Among other things, this holography colored approach leads to new representations of smooth compact $(n+1)$-manifolds X. These representations describe compact $(n+1)$-manifolds as maps (in general, discontinuous) between two n-dimensional manifolds-screens. Such representations are reminiscent of holographic pictures and thus deserve the name "holography". In particular, any compact $(n+1)$-manifold with spherical boundary admits a representation in the form of a causality map between two n-balls.

We stress that the structures, arising from the interactions of the generic traversing vector flows on X with the boundary ∂X, are surprisingly universal, although intricate. Locally, they are described in terms of the Morin's maps [Morin] and their singularities.

We often restrict the combinatorial patterns ω that characterize the tangency to ∂X patterns of the trajectories of smooth traversing vector flows v on X.

In this book, we are preoccupied with the following basic question:

"What are the constraints imposed on the topology of X by such combinatorial restrictions?"

Chapter 1 serves as a gentle introduction to the rest of the book. In the chapter, we consider traversing flows on surfaces. In this simple $2D$-landscape, we present few key general results from the subsequent chapters.

Chapter 2 contains the necessary trivia about vector fields on smooth manifolds with boundary, presented along the lines from [K7].

For a given traversing pair (X, v), in order to inject some dynamics into our settings, we employ an involution $\tau : \partial X \to \partial X$ (mimicking the elastic reflections from the billiard boundary) to convert the scattering map C_v to the proto-billiard map $B_{v,\tau} =_{\mathsf{def}} \tau \circ C_v$.

In Chapter 3, we study the dynamics of the $(B_{v,\tau})$-iterations, in particular, the various implications of their ergodicity. These investigations are based on the measure preserving property of the proto-billiard map $B_{v,\tau}$. They revolve around the Santaló formula [S]. In particular, we study the harmonizing metrics for the pair (X, v) [Ca], [FKL], [K14].

We also devote significant space and time to the holographic considerations of contact structures on compact odd dimensional manifolds with a boundary.

Let g be a Riemannian metric on a connected compact Riemannian manifold M with boundary.

In Chapter 4, we consider the tangent spherical fibration $\{S_gM \to M\}$ of M. The metric g generates the geodesic vector field v^g on S_gM. If v^g admits a Lyapunov function, we call the metric g "non-trapping". Assuming that the metrics g are non-trapping, we apply the same holographic approach to the study of inverse scattering problems for the scattering/causality maps C_{v^g} and to the dynamics of the billiard maps $B_{v^g,\tau}$.

In Part II of the book, we study the rich combinatorial structures that reflect the tangency patterns $\{\omega\}$ of generic traversing flows on compact manifolds X to their boundaries ∂X. This leads us to a topological study of spaces $\mathcal{P}_d^{\mathbf{c}\Theta}$ of degree d real monic univariate polynomials, whose real divisors avoid an a priori given closed poset Θ of ω-patterns (see [KSW1], [KSW2]).

In Chapter 5, we are concerned with immersions of curves ("doodles") and surfaces ("blobs") in a ruled page, shaped as the products $\mathbb{R} \times [0, 1]$ or $\mathbb{R} \times S^1$. This chapter is a somewhat involved introduction into the subjects of Chapters 7 and 8, but without more complex combinatorics that characterizes tangency patterns on high-dimensional "ruled pages".

In Chapter 6, in terms of the given poset Θ, we describe the (co)homology and the fundamental groups of the polynomial spaces $\mathcal{P}_d^{\mathbf{c}\Theta}$. Their points represent real univariate monic polynomials whose real divisors avoid the closed poset Θ. The problem of computing the (co)homology of $\mathcal{P}_d^{\mathbf{c}\Theta}$ is effectively reduced to combinatorics of the poset Θ.

Let M and Y be compact smooth n-manifolds.

In Chapter 7, we show that the polynomial spaces $\mathcal{P}_d^{\mathbf{c}\Theta}$ play the role of universal objects in our study of immersions and embeddings $\beta : M \to \mathbb{R} \times Y$. We require that the tangency patterns of these immersions and embeddings to the fibers of the obvious projection $\pi : \mathbb{R} \times Y \to Y$ avoid a given closed poset Θ.

We fix a compact smooth manifold Y, a poset Θ, and a natural number d. Then we consider immersions $\beta : M \to \mathbb{R} \times Y$ whose normal line bundles are trivial. We organize the β's that avoid the tangency patterns $\omega \in \Theta$ and such that the cardinality of the fibers $\{(\pi \circ \beta)^{-1}(y)\}_{y \in Y}$ is $\leq d$ into some equivalence classes. We call this equivalence relation "$\beta_0 \sim \beta_1$" between two immersions a "quasitopy". Quasitopies of immersions is a crossover between their pseudo-isotopies and cobordisms. We denote by $\mathcal{QT}_d^{\mathsf{imm}}(Y; \mathbf{c}\Theta)$ the set of quasitopy classes of immersions.

Switching from immersions to embeddings, we introduce a similar set $\mathcal{QT}_d^{\mathsf{emb}}(Y; \mathbf{c}\Theta)$ of quasitopies of embeddings. In Chapter 7, utilizing our understanding of the polynomial spaces $\mathcal{P}_d^{\mathbf{c}\Theta}$, we compute $\mathcal{QT}_d^{\mathsf{emb}}(Y; \mathbf{c}\Theta)$ in homotopy-theoretical terms.

In Chapter 8, the spaces $\mathcal{P}_d^{\mathbf{c}\Theta}$ also play the role of universal objects in our study of Θ-restricted traversing flows v on X such that (X, v) can be "encapsulated" in a, so called, convex pseudo-envelope $(\hat{X}, \hat{v})$. Informally speaking, a convex pseudo-envelope $(\hat{X}, \hat{v})$ provides a "virtual global section" of the v-flow.

The study of traversing flows on manifolds with boundaries leads organically to many encounters with Singularity Theory, Combinatorics, Differential Topology, Differential and Contact Geometry, and Dynamical Systems. It is this diversity of connections that has motivated the writing of this monograph in the first place.

Throughout our journey, we try to keep the interplay between the boundary value problems for families of ordinary differential equations and the topology of the underlying manifolds as the unifying theme.

We presume that our reader is familiar with basic classical tools and techniques of Differential and Algebraic Topology, Differential Geometry, and Dynamical Systems. However, we strive to describe these tools in the book when we can.

We are not hesitant to mark the borders of our understanding and progress with multiple conjectures in hope that our readers will be able to progress beyond these borders.

Few more words about the organization of the book. It is divided in two parts: the first one is under the banner of Holography of traversing flows, and the second one is under the banner of their Homology.

The book consists of 8 chapters. Each chapter starts with a description of its purpose and main results.

As much as we could, we tried to provide our reader, a researcher or a graduate student, with independent entrances to different chapters and sections. As a result, the book has certain redundancy in its design: some constructions and notions that first appear in more special settings, later reappear in their full generality.

The chart below indicates pathways through the text:

$$\begin{array}{ccccccc} \mathit{Chapter}\,1 \Longrightarrow \mathit{Chapter}\,2 \Longrightarrow \mathit{Chapter}\,3 \Longrightarrow \mathit{Chapter}\,4 \\ \Downarrow \\ \mathit{Chapter}\,5 \rightsquigarrow \mathit{Chapter}\,6 \Longrightarrow \mathit{Chapter}\,7 \Longrightarrow \mathit{Chapter}\,8 \end{array}$$

Acknowledgments

"The most important question anyone can ask is: What myth am I living?"
Carl Jung

While working on this book, I enjoyed hospitality of the Mathematics Department of the Massachusetts Institute of Technology. It is my pleasant duty to acknowledge the stimulating environment the MIT provided for my research.

My deep gratitude goes to my coauthors Boris Shapiro, Volkmar Welker, John Etnyre, and Leonid Bunimovich. I am also grateful to Christofer Croke, Yakov Eliashberg, Pavel Etingof, Vladimir Matveev, Serge Tabachnikov, Gunther Uhlmann, and Shmuel Weinberger for very enlightening conversations about various topics of the book.

I am thankful to D. Rajesh Babu and Tan Rok Ting for the valuable technical support.

Most importantly, I would not be able to complete this project without the wise and steady support of my wife Adela Katz. She managed to keep the chaos of life at the doorsteps of our home. I am happy to thank Adela for her sacrifice and remarkable patience.

Contents

PART 1

Holography of Traversing Flows

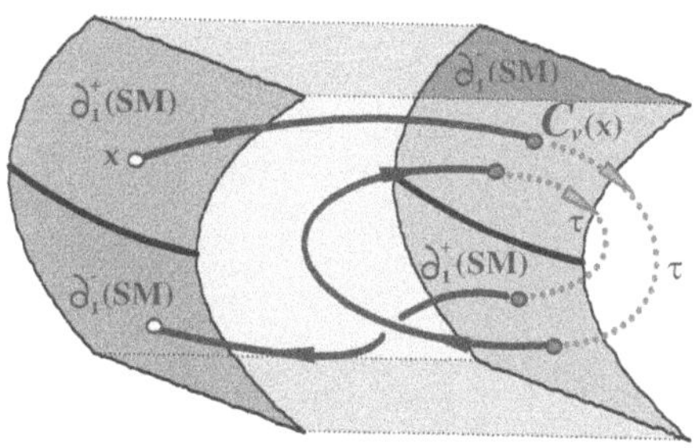

Chapter 1

Flows in 2D-land, their Holography, and Combinatorics

"I call our world Flatland, not because we call it so, but to make its nature clearer to you, my happy readers, who are privileged to live in Space."

Mr. Square, cited by Edvin A. Abbot

1.1 In this chapter...

This chapter is an informal introduction into the philosophy and flavor of Part I of the book. Most of the results from Chapter 1 may be found in [K4] and [K7].

We use the relatively simple environment of the gradient-like non-vanishing vector flows on smooth compact surfaces with boundary to illustrate the main ideas and assertions from Chapters 2-4, as they manifest themselves in $2D$. The complete proofs of these results will be delivered in Chapters 2-4, unless the results are specific to dimension two.

The remarkable convergence of topological, geometrical, and analytical approaches to the study of closed surfaces is widely recognized by the practitioners for more than century and a half. We will exhibit a similar convergence of different investigative approaches to vector flows on compact connected surfaces *with boundary*.

1.2 On Morse Theory without critical points on surfaces with boundary and beyond

The Morse theory of smooth functions and their gradient flows on compact manifolds X *with boundary* has been studied (under general boundary conditions) in the pioneering papers of Morse [Mo] and of Morse and Van Schaack [MS].

The Morse Theory on manifolds with boundary can be viewed as a very special instance of the Morse Theory on stratified spaces, where the two strata ∂X and X form the stratification. The latter theory was developed by Goresky and MacPherson in [GM]-[GM2]. See also [Ha], [La] for constructions of the Morse-Smale differential complex on manifolds with conners.

If a Morse function $f : X \to \mathbb{R}$ is assumed to be constant on the boundary ∂X, then its gradient-like field $\mathsf{grad} f$ is transversal to the boundary. Then the *relative* topology of the pair $(X, \partial X)$ is classically captured in terms of the gradient flow [Mi3] in terms of interactions of the f-critical points via the flow lines that connect them.

In [K7] as well as in this book, we offer a different philosophy for the Morse-like Theory on compact connected manifolds X *with boundary.*

We consider vector fields v on X that admit a smooth real-valued function $f : X \to \mathbb{R}$ such that $df(v) > 0$ on X. We call such f a Lyapunov function for v.

We propose to focus on the geometry of the boundary in relation to the v-flows that admit Lyapunov functions.

For surfaces, *typically*, the v-trajectories are interacting with the boundary ∂X in very particular and stable ways: they are either transversal to ∂X, or are simply tangent to it in a concave or convex fashion (see Fig. 1.1). Of course, there are vector fields v for which some trajectories will be, say, cubically tangent to the boundary. However, a majority of vector fields v avoids such cubic tangencies. We will see that this "curvy" geometry of the v-flow in connection to the boundary is the crucial ingredient for reconstructions of the surface/manifold X in terms of the appropriate boundary data.

Crucially, in the vicinity of a *concave* tangency point, the v-flow is *discontinuous* in the same way the gradient flow of a Morse function is discontinuous in the vicinity of its critical point: under the flow, initially close points become distant in time. In our setting, the divergence of initially close points occurs due to very different travel times available to them; unlike the infinite travel time for the gradient flows of the classical Morse theory on *closed* surfaces, in the case of the non-singular gradient flows on surfaces with boundary, every point exits the surface in finite time. As a result, the surface is *not* flow-invariant! Again, these discontinuities of the flow reflect the topology of the surface. Let us clarify this point.

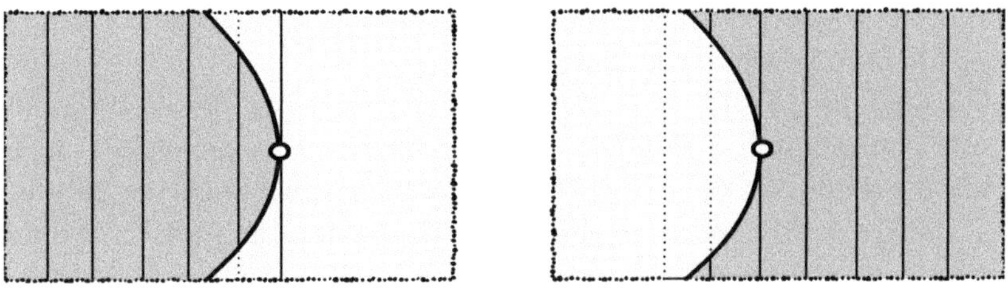

Fig. 1.1 A boundary generic vector field v in the vicinity of a "convex point" from ∂X (on the left) and a "concave point" (on the right). The surface X is the shaded region, and the v-trajectories are shown as vertical lines.

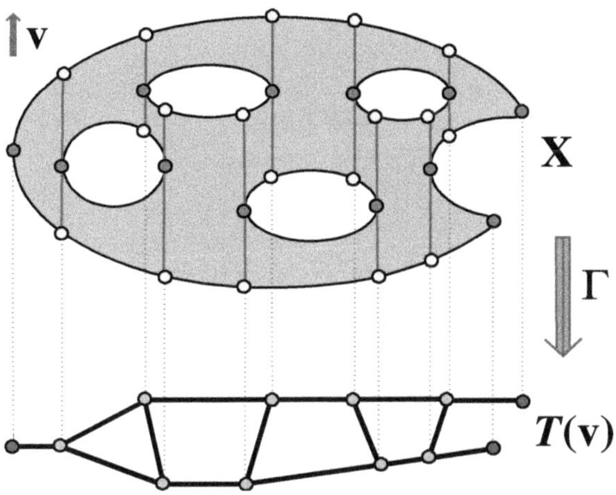

Fig. 1.2 The quotient map $\Gamma : X \to \mathcal{T}(v)$ from a surface X, a disk with 4 holes, to the space $\mathcal{T}(v)$ of trajectories of a traversally generic (vertical) vector field v. The trajectory space $\mathcal{T}(v)$ is a graph whose vertices are of valencies 1 and 3.

Fig. 1.2 shows a non-vanishing gradient flow v on a surface $X \subset \mathbb{R}^2$, topologically a disk with 4 holes. The Lyapunov function $f : X \to \mathbb{R}$ for v is the vertical coordinate in $\mathbb{R}^2$. Each v-trajectory is either a closed segment, or a singleton.

By collapsing each trajectory to a point, we create the quotient space of trajectories $\mathcal{T}(v)$. Since the flow trajectories are closed segments or singletons, $\mathcal{T}(v)$ is a "decent space", a finite graph with vertices of valency 1 or 3 only. The vertices of valency 3 correspond to the points on ∂X where the boundary is concave with respect to the flow, and the univalent vertices to the points on ∂X where the flow is convex.

The obvious map $\Gamma : X \to \mathcal{T}(v)$ cellular. Moreover, because the fibers of Γ are contractable, Γ is a *homotopy equivalence.* In particular, the fundamental groups $\pi_1(X)$ and $\pi_1(\mathcal{T}(v))$ are isomorphic with the help of Γ. Thus, the trajectory spaces of generic vector fields v that admit Lyapunov functions on connected surfaces X with boundary deliver 1-dimensional *homotopy-theoretical models* of X (see Corollary 2.2 for the multidimensional analogues of this fact).

1.3 Vector fields and Morse stratifications on surfaces

Let v be a smooth vector field on a compact connected surface X with boundary, such that $v \neq 0$ along the boundary ∂X. Following [Mo], we consider the closed locus $\partial_1^+ X(v) \subset \partial X$, where the field points inside X or is tangent to ∂X, and the closed locus $\partial_1^- X(v) \subset \partial X$, where v points outside X or is tangent to ∂X. The intersection

$$\partial_2 X(v) =_{\mathsf{def}} \partial_1^+ X(v) \cap \partial_1^+ X(v)$$

is the finite locus where v is *tangent* to the boundary ∂X. Points $z \in \partial_2 X(v)$ come in two flavors: by definition, $z \in \partial_2^+ X(v)$ when $v(z)$ points inside of the locus $\partial_1^+ X(v)$, otherwise $z \in \partial_2^- X(v)$.

• To achieve some uniformity of notations, let us put $\partial_0^+ X =_{\mathsf{def}} X$ and $\partial_1 X =_{\mathsf{def}} \partial X$.

We denote by $T(X)$ the tangent bundle of a surface X and by $T(\partial X)$ the tangent bundle of its boundary ∂X.

Definition 1.1. We say that a vector field v on a compact surface X is boundary generic if:

- $v|_{\partial X} \neq 0$,
- $v|_{\partial X}$, viewed as a section of the normal 1-dimensional (quotient) bundle $n_1 =_{\mathsf{def}} \big(T(X)|_{\partial X}\big)/T(\partial X)$, is transversal to its zero section (at the points of the locus $\partial_2 X(v)$). $\diamondsuit$

In particular, for a boundary generic v, the loci $\partial_1^\pm X(v)$ are finite unions of closed intervals and circles, residing in $\partial_1 X$; the loci $\partial_2^\pm X(v)$ are finite unions of points, residing in $\partial_1 X$ (see Fig. 1.2).

• We denote by $\mathcal{B}^\dagger(X)$ the space of all boundary generic vector fields, equipped with the C^∞-topology, on a compact surface X.

Definition 1.2. If a vector field v admits a smooth Lypunov function f so that $df(v) > 0$, we call v traversing. $\diamondsuit$

1.4 Concavity and complexity of vector flows in 2D

Definition 1.3. We say that a boundary generic vector field v is boundary convex if $\partial_2^+ X(v) = \emptyset$. We say that a boundary generic v is boundary concave if $\partial_2^- X(v) = \emptyset$ (see Fig. 1.1). $\diamondsuit$

The existence of a boundary convex field puts severe restrictions on the topology of the surface.

Lemma 1.1. *([K4]) If a compact connected surface X with boundary admits a* boundary convex *traversing (see Definition 1.2) vector field v, then X is either a disk D^2, or an annulus An^2.* $\diamondsuit$

In contrast, the boundary *concave* traversing vector fields are plentiful. For example, consider the radial vector field v on an annulus $\mathsf{An}^2 \subset \mathbb{R}^2$. Delete from An^2 any number of convex disks and restrict v to the resulting 2-disk with holes. This construction implies that any disk with holes admits a non-vanishing boundary concave gradient-like vector field v.

Many other surfaces admit concave fields. For example, consider a Morse function $f : Y \to \mathbb{R}$ on a closed surface Y and its gradient field v. Then removing from Y small convex (in the local Morse coordinates) balls, centered on the f-critical points, produces a *boundary concave* non-vanishing gradient field on X. In particular, if Y is a sphere with g handles, then one can find a Morse function with $2g + 2$ critical points. Therefore, the surface X, obtained from Y by removing $2g + 2$ balls, admits a non-vanishing concave gradient-like vector field v.

The following theorem summarizes the concave-convex dichotomy for traversing flows on surfaces.

Theorem 1.1. *([K4]) Any compact connected surface X with boundary, but the disk and the Möbius band, admit a traversing vector field v with only concave tangencies to the boundary ∂X.*

The disk and the Möbius band do not admit such a vector field.

Any compact connected surface X that admits a traversing vector field with convex tangencies to ∂X is a disk or an annulus.

The annulus is the only surface that admits both convex and concave traversing vector fields. $\diamondsuit$

1.5 On spaces of vector fields

It turns out (see [K1]) that the following definition is equivalent to Definition 1.2.

Definition 1.4. We say that a non-vanishing vector field v on a compact surface X is traversing if all its trajectories are closed segments or singletons. (It is easy to see that the ends of these segments, as well as the singletons, must reside in the boundary ∂X.) $\diamondsuit$

Each trajectory γ of a traversing vector field v must reach the boundary both in positive and negative times: otherwise γ is not homeomorphic to a closed interval.

• We denote by $\mathcal{V}_{\mathsf{trav}}(X)$ the space (in the C^∞-topology) of all traversing vector fields on a given compact surface X.

• We denote by $\mathcal{V}_{\mathsf{grad}}(X)$ the space (in the C^∞-topology) of all gradient-like vector fields on X, and by $\mathcal{V}_{\neq 0}(X)$ the space of all non-vanishing fields on X.

The next lemma claims that v is traversal if and only if it is non-vanishing and gradient-like.

Lemma 1.2. *([K1]) For any compact connected surface X with boundary,*

$$\mathcal{V}_{\mathsf{trav}}(X) = \mathcal{V}_{\mathsf{grad}}(X) \cap \mathcal{V}_{\neq 0}(X).$$

$\diamondsuit$

All surfaces X and vector fields v we consider are smooth. We can add an external collar to X to form a diffeomorphic surface $\hat{X} \supset X$ and then to extend v to a smooth vector field $\hat{v}$ on $\hat{X}$. Let $\hat{\gamma}$ be a $\hat{v}$-trajectory (or rather its germ) through a point $x \in \partial X$. We can talk about the order of tangency of two smooth curves, $\hat{\gamma}$ and ∂X, at a point $x \in \hat{\gamma} \cap \partial X$ in $\hat{X}$ (see Definition 1.6 below). We say that the tangency of $\hat{\gamma}$ to ∂X is simple if its order of tangency is 2. When the two curves are transversal at x we say that the order of tangency is 1. In fact, these notions depend only on (X, v) and not on the extension $(\hat{X}, \hat{v})$.

Definition 1.5. A traversing vector field v on a compact surface X is called traversally generic, if the following properties hold:

(1) if a trajectory γ is tangent to the boundary ∂X, then the tangency is simple,

(2) no v-trajectory γ contains more then one simple point of tangency to ∂X. $\diamondsuit$

In particular, a traversally generic vector field v is boundary generic.

• We denote by $\mathcal{V}^{\ddagger}(X)$ the space of all traversally generic vector fields on a compact surface X.

As the name suggests, the traversally generic vector fields are typical among all traversing fields. Furthermore, a smooth perturbation of any traversally generic field is traversally generic. This is the content of the next theorem. Its validation requires an involved argument, which even in dimension two resists a significant simplification.

Theorem 1.2. *([K2]) For any compact connected surface X with boundary, the space $\mathcal{V}^{\ddagger}(X)$ traversally generic vector fields is open and dense in the space $\mathcal{V}_{\mathsf{trav}}(X) = \mathcal{V}_{\mathsf{grad}}(X) \cap \mathcal{V}_{\neq 0}(X)$.* ♢

1.6 Combinatorics of tangency for traversing flows in 2D

We pick an extension $\hat{X}$ of a given compact surface X by adding an external collar to X. Let $\hat{v}$ be an extension of a given field v into $\hat{X}$. We pick also an auxiliary smooth function $z : \hat{X} \to \mathbb{R}$ such that:

(1.1)

- 0 is a regular value of z,
- $z^{-1}(0) = \partial X$,
- $z^{-1}((-\infty, 0]) = X$,

Definition 1.6. Let $\hat{\gamma}$ be a $\hat{v}$-trajectory through a point $x \in \partial X$. We say that $\hat{\gamma}$ has the order/multiplicity of tangency k to ∂X at x, if

$$\mathcal{L}_{\hat{v}}^{\{j\}}(z) = 0 \text{ for all } j < k, \text{ and } \mathcal{L}_{\hat{v}}^{\{k\}}(z) \neq 0 \text{ at } x.$$

Here $\mathcal{L}_{\hat{v}}^{\{j\}}(z)$ denotes the jth iterated $\hat{v}$-directional derivative of the function z as in (1.1). ♢

Given a traversally generic vector field v on a compact connected surface X, we will attach the combinatorial pattern (11) to a typical v-trajectory $\gamma \subset X$ that corresponds to the edges of the trajectory graph $\mathcal{T}(v)$, the pattern (121) to the trajectories that correspond to the trivalent vertices of $\mathcal{T}(v)$, and the pattern (2) to the univalent vertices (see Fig. 1.2). The numbers 1 and 2 in these patterns reflect the order of tangency (see Definition 1.6) of the curves $\hat{\gamma}$ and ∂X at the points of $\gamma \cap \partial X$. On a given compact surface X, for *traversally generic* vector fields v no other patterns (say, like (1221) or (13)) occur. In $2D$, this conclusion follows from Definition 1.5.

The lemma below is another way to state this fact. Its proof, relying on the Malgrange Preparation Theorem [Mal], can be found in [K2] or [K7].

Lemma 1.3. *Let v be a traversally generic vector field on a compact connected surface X. Extend (X, v) to an enveloping pair $(\hat{X}, \hat{v})$.*

In the vicinity of each v-trajectory γ, there exist special local coordinates (u, x) on $\hat{X}$ and a real polynomial $P(u, x)$ of degree 2 or 4 such that:

- *each $\hat{v}$-trajectory is given by the equation $\{x = const\}$,*
- *the boundary ∂X is given by the polynomial equation $\{P(u, x) = 0\}$,*
- *X is given by the polynomial inequality $\{P(u, x) \leq 0\}$.*

The polynomial $P(u, x)$ takes the following canonical forms:

(1) $u(u-1)$, *which corresponds to the combinatorial pattern* (11),
(2) $u^2 - x$, *which corresponds to the combinatorial pattern* (2),
(3) $u\big((u-1)^2 + x\big)(u-2)$, *which corresponds to the pattern* (121). ◇

To summarize, at $\partial_2 X(v)$ the order of tangency is 2; the trajectories through $\partial_2^+ X(v)$ have the combinatorial tangency pattern (121), and through $\partial_2^- X(v)$ the combinatorial tangency pattern (2). The rest of the trajectories have the generic pattern (11).

We denote by $\mathbf{\Omega}^\bullet_{|\sim|' \leq 1}$ the *partially ordered set* whose elements are $(11), (2), (121)$ and the order is defined by $(11) \succ (2)$ and $(11) \succ (121)$. This combinatorics does not look impressive. However, in higher dimensions, traversally generic vector fields on $(n+1)$-dimensional manifolds with boundary generate rich and interesting partially ordered finite sets ("posets" for short) $\mathbf{\Omega}^\bullet_{|\sim|' \leq n}$ of combinatorial tangency patters. In fact, the poset $\mathbf{\Omega}^\bullet_{|\sim|' \leq n}$ is *universal* in each dimension [K3]. We will discuss its properties in Chapter 2.

1.7 Holography of traversing flows on surfaces

Consider a compact connected surface X with boundary and a traversing vector field v on it. As a default, we assume that all the structures are smooth.

We start with an "existential" question:

"Are there some flow-generated structures on the boundary ∂X that will allow for (perhaps, even partial) reconstruction of X and the flow?"

We call such desirable structures holographic since the information, recorded on 1-dimensional (in general, $(\dim X - 1)$-dimensional) screen(s), makes it possible to restore the topology of X and the dynamics of the 2-dimensional (in general, $\dim X$-dimensional) flow on it.

Let us describe one particular holographic structure on ∂X that will dominate the landscape. For each point $x \in \partial_1^+ X$, consider the v-trajectory γ through x. Let $C_v(x) \in \partial X$ be the first along γ point, where γ exits X or touches its boundary. The existence of $C_v(x)$ is guaranteed by the property of v being traversing. The correspondence $x \rightsquigarrow C_v(x)$ gives rise to a well-defined map

$$C_v : \partial_1^+ X(v) \to \partial_1^- X(v).$$

We call C_v the causality map (see Theorem 3.7 for a justification of the name).

Note also that $C_v(x) = x$ if and only if $x \in \partial_2^- X(v)$.

Alternatively, one can think of C_v as determining a partial order "$x \prec C_v(x)$" among the points of the boundary ∂X.

The map C_v is discontinuous; in what follows, we assign the most important role to the discontinuities of C_v (see Fig. 1.3 depicting C_v).

The word "causality" in the name of C_v is motivated by the pivotal special case, described in Example 1.1 below.

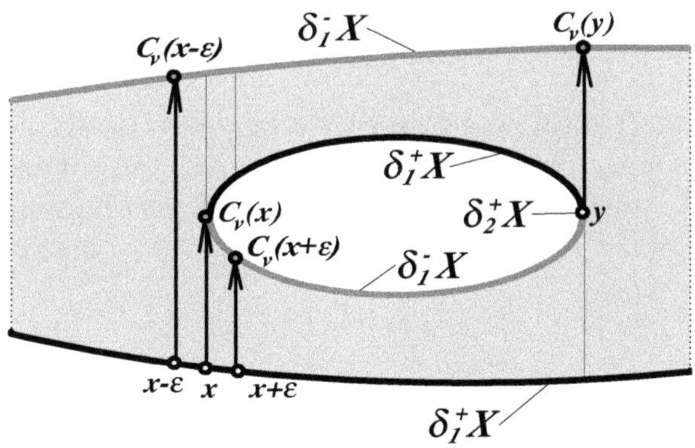

Fig. 1.3 An example of the causality map $C_v : \partial_1^+ X(v) \to \partial_1^- X(v)$. Note the discontinuity of C_v in the vicinity of point x.

Example 1.1. Let $w = w(\theta, t)$ be a smooth time-dependent (t stands for time) vector field on the circle S^1, equipped with the angular coordinate θ.

It gives rise to a vector field $v = (w, 1)$ on the cylinder $S^1 \times \mathbb{R}$. We think about the factor S^1 as space and about the factor $\mathbb{R}$ as time. Therefore, we call $S^1 \times \mathbb{R}$ the space of events. Note that v admits the obvious Lyapunov global time function $T : S^1 \times \mathbb{R} \to \mathbb{R}$.

Let us pick a compact surface $X \subset S^1 \times \mathbb{R}$ with a smooth boundary ∂X. We call X the event capsule, and its boundary ∂X the event horizon.

Since v is traversing in X, the map C_v is well-defined. Then the map $C_v : \partial_1^+ X(v) \to \partial_1^- X(v)$ indeed gives rise to the causality relation on the event horizon: the correspondence C_v reflects the *evolution* of an event z into the event $C_v(z)$. $\diamondsuit$

Let $\mathcal{C}(\partial_2^+ X(v))$ denote the union of v-trajectories through the points of the concavity locus $\partial_2^+ X(v)$.

The causality map is *discontinuous* at the points of the intersection $\mathcal{C}(\partial_2^+ X(v)) \cap \partial_1^+ X(v)$ (see Fig. 1.3). On the positive side, in $2D$, the discontinuities of the causality map C_v are not too bad: the map has the "left" and the "right" limits.

Trajectories of a traversing boundary generic vector field v, viewed as *un*parametrized v-oriented curves, produce an oriented 1-dimensional (singular) foliation $\mathcal{F}(v)$ on the surface X. More accurately, if $(\hat{X}, \hat{v})$ is an open traversing extension of a traversing pair (X, v), then the $\hat{v}$-flow defines an oriented foliation $\mathcal{F}(\hat{v})$ in $\hat{X}$. We denote its restriction to X by $\mathcal{F}(v)$. Thus, some of the leaves of $\mathcal{F}(v)$ may be singletons. The rest are closed segments.

Theorem 1.3. (Conjugate Causal Holography in 2D).
Let (X_1, v_1) and (X_2, v_2) be two compact connected surfaces with boundaries, carrying traversally generic vector fields v_1 and v_2, respectively. Assume that there is a diffeomorphism $\Phi^\partial : \partial X_1 \to \partial X_2$ which conjugates the two causality maps:

$$C_{v_2} \circ \Phi^\partial = \Phi^\partial \circ C_{v_1}.$$

Then Φ^∂ extends to a diffeomorphism $\Phi : X_1 \to X_2$ which maps the oriented foliation $\mathcal{F}(v_1)$ to the oriented foliation $\mathcal{F}(v_2)$.

Proof. We will only sketch the argument here. Theorem 3.3, the main result of Chapter 3, contains a fully developed proof of the multidimensional analogue of Theorem 1.3. Moreover, the assumption that the vector fields are traversally generic can be weakened: the traversing and boundary generic assumptions are sufficient for the validity of the claim.

Since the maps C_{v_1} and C_{v_2} are Φ^∂-conjugate, the diffeomorphism Φ^∂ induces a well-defined continuous map $\Phi_{\mathcal{T}} : \mathcal{T}(v_1) \to \mathcal{T}(v_2)$ of the trajectory spaces. Moreover, $\Phi_{\mathcal{T}}$ preserves the stratifications of the two trajectory spaces/graphs by the combinatorial types of trajectories. That is, the trivalent vertices of $\mathcal{T}(v_1)$ are mapped to the trivalent vertices of $\mathcal{T}(v_2)$, the univalent vertices are mapped to univalent vertices, and the interior of the edges to the interior of the edges.

Then we pick a smooth function $f_2 : X_2 \to \mathbb{R}$ such that $df_2(v_2) > 0$. With the help of Φ^∂, we pull-back the function $f_2|_{\partial X_2}$ to get a smooth function $f_1^\partial : \partial X_1 \to \mathbb{R}$ such that $f_1^\partial(z) < f_1^\partial(C_{v_1}(z))$ for all $z \in \partial_1^+ X_1(v_1)$.

Then we argue that f_1^∂ extends to a smooth function $f_1 : X_1 \to \mathbb{R}$ such that $df_1(v_1) > 0$ (see Lemma 3.3 for the validation of this claim in any dimension).

We use this new function f_1 to embed X_1 in the product $\mathcal{T}(v_1) \times \mathbb{R}$ by the formula

$$\alpha_{(v_1,\, f_1)}(z) = ([\gamma_z], f_1(z)),$$

where γ_z, the v_1-trajectory through a typical point z, is viewed as the point $[\gamma_z] = \Gamma_1(z)$ of the graph $\mathcal{T}(v_1)$. Similarly, we use f_2 to embed, via the map $\alpha_{(v_2,\, f_2)}$, the surface X_2 in the product $\mathcal{T}(v_2) \times \mathbb{R}$.

Finally, we employ $\Phi_{\mathcal{T}}$, f_1, and f_2 to construct a map

$$\hat{\Phi} : \mathcal{T}(v_1) \times \mathbb{R} \to \mathcal{T}(v_2) \times \mathbb{R}$$

by the formula

$$\hat{\Phi}([\gamma], t) = \big(\Phi_{\mathcal{T}}([\gamma]),\, f_1(f_2^{-1}(t))\big),$$

where t belongs to the f_2-image of the trajectory $\Gamma_2^{-1}\big(\Phi_{\mathcal{T}}(\gamma)\big)$.

Crudely, the restriction of the map $\hat{\Phi}$ to $\alpha_{v_1, f_1}(X_1) \subset \mathcal{T}(v_1) \times \mathbb{R}$ is the desired diffeomorphism $\Phi : X_1 \to X_2$.

Note that, in general, $\Phi_*(v_1)$ is not v_2; so the parametrizations of the trajectories are not respected by the diffeomorphism Φ, but the oriented 1-foliations $\mathcal{F}(v_1)$ and $\mathcal{F}(v_2)$ are. □

Theorem 1.3 implies the following claim.

Corollary 1.1. (Causal Holography in 2D). *Let X be a compact connected surface with boundary, and v a smooth traversally generic vector field on it.*

Then the knowledge of the causality map $C_v : \partial_1^+ X(v) \to \partial_1^- X(v)$ is sufficient for a reconstruction of the pair $(X, \mathcal{F}(v))$, up to a diffeomorphism that is the identity on ∂X. ◇

In fact, we can relax the hypotheses in Theorem 1.3 and Corollary 1.1 by replacing traversally generic vector fields with just traversing and boundary generic ones.

The world "holography" is present in the name of Theorem 1.3 and Corollary 1.1 since the surface X and the qualitative $2D$-dynamics of the v-flow on it are recorded on two 1-dimensional screens, $\partial_1^+ X(v)$ and $\partial_1^- X(v)$.

Example 1.2. Let v be a traversally generic field on a connected surface X whose boundary ∂X is a *single* loop. Then the boundary ∂X is divided into q disjoint arcs $a_1, \ldots, a_q$ that form $\partial_1^+ X(v)$ and q complementary arcs $b_1, \ldots, b_q$ that form $\partial_1^- X(v)$.

The causality map

$$C_v : \coprod_{i=1}^{q} a_i \longrightarrow \coprod_{i=1}^{q} b_i$$

can be represented by its graph $G(C_v) \subset \coprod_{i,j=1}^q (a_i \times b_j)$. The map C_v (the curve $G(C_v)$) is discontinuous at exactly $c^+(v)$ points that belong to the intersection set $\mathcal{C}(\partial_2^+ X(v)) \cap \partial_1^+ X(v) \subset \coprod_{i=1}^q a_i$. Here $\mathcal{C}(\partial_2^+ X(v))$ denotes the set of trajectories through the points of $\partial_2^+ X(v)$. There the map C_v has distinct left and right limits.

According to the Corollary 1.1, the curve $G(C_v) \subset \coprod_{i,j=1}^q (a_i \times b_j)$ determines X and the un-parametrized the v-flow, up to a diffeomorphism $\Phi : X \to X$ which is the identity on ∂X. Note that the number q alone is not sufficient even to determine the genus of the surface X. ◇

Revisiting Example 1.1, we get the following interpretation of Corollary 1.1 in terms of the event horizon and event capsule, introduced in the example.

Corollary 1.2. *For any smooth time-dependent vector field $w(\theta, t)$ on the circle S^1 and an event capsule $X \subset S^1 \times \mathbb{R}$, the causality relation on the event horizon $\partial X \subset S^1 \times \mathbb{R}$ is sufficient for a reconstruction of the event capsule X and the un-parametrized dynamics of the $v(\theta, t) =_{\mathsf{def}} (w(\theta, t), 1)$-flow, up to a diffeomorphism of X which is the identity on ∂X.* ◇

In order to interpret the causality map as a version of the Poincaré return map of a dynamic system, we need to leave the Flatland and to sail briefly into the $3D$ world.

Let us recall informally the notion of geodesic flow. Consider a compact surface X with a Riemannian metric g. We denote by SX the space of unit tangent vectors on X. Given a point $x \in X$ and a unit tangent vector $v \in T_xX$, we take the germ of the unique geodesic curve $\gamma : (-\epsilon, \epsilon) \to X$ through x such that $\gamma(0) = x$ and $\dot{\gamma}(0) = v$. Then the pair $(\gamma(t), \dot{\gamma}(t))$ is a path in SX; its tangent vector $v^g = (\dot{\gamma}, \ddot{\gamma})$ belongs to the tangent space $T_{(x,v)}(SX)$. The vector field v^g on SX is called geodesic, and the flow on SX it generates is called the geodesic flow.

The metrics g for which the geodesic vector fields v^g are traversing are called non-trapping. Such metrics form an open set in the space of all Riemannian metrics.

The theory of billiards on connected Riemmanian surfaces (X, g) with boundary benefits from applying a $3D$-version of the Causal Holography Theorem 1.3 to the geodesic flow on the 3-dimensional manifold SX. In addition to geodesic billiards, these applications include the classic inverse geodesic scattering problems. We devote Chapter 4 to these applications.

Let us outline some of these applications. When ∂X is connected (i.e., is a circle), then $\partial_1(SX)$ is a 2-torus. If the metric g on X is non-trapping, then the causality (scattering) map C_{v^g} acts from the annulus $\partial_1^+(SX)$ to the annulus $\partial_1^-(SX)$. Both annuli can be identified via a diffeomorphism $\tau : \partial_1^-(SX) \to \partial_1^+(SX)$ that reflects each tangent vector $w \in T_xX$ at a point $x \in \partial X$ with respect to the boundary ∂X; that is, $\tau((x, v)) = (x, v^\bullet)$, where $x \in \partial X$ and $v^\bullet$ is the image of $v \in T_xX$ under the flip of T_xX with respect to the line $T_x(\partial X)$. The composition $B_{v^g} =_{\mathsf{def}} \tau \circ C_{v^g}$, called the billiard map, is a self-map of the annulus $\partial_1^+(SX)$. The iterations of the billiard map B_{v^g} give rise to an interesting dynamic on the annulus, a focal point of many investigations in the theory of billiards [Tab].

In this context, a challenging and well-studied question can be posed: for which Riemannian metrics g on X, the knowledge of the scattering map $C_{v^g} : \partial_1^+(SX) \to \partial_1^-(SX)$ (equivalently, of the billiard map) is sufficient for a reconstruction of the non-trapping metric g, say, up to a scalar? Up to an action of diffeomorphisms $\Phi : X \to X$? We will offer some answers to questions of this kind in Chapter 4.

We have completed our description of vector flows in $2D$-lands and are ready to find out how things flow in the lands of "higher dimensions".

Chapter 2

Trivia About Vector Fields on Manifolds with Boundary

"They've all done things, often beautiful things, in a context that was already set out before them, which they had no inclination to disturb. Without being aware of it, they've remained prisoners of those invisible and despotic circles which delimit the universe of a certain milieu in a given era. To have broken these bounds they would have had to rediscover in themselves that capability which was their birth-right, as it was mine: the capacity to be alone."

Alexander Grothendieck

2.1 In this chapter...

We start with a review of some classes of vector fields on compact connected smooth manifolds with boundary and properties of these classes, which will be essential for the rest of the book. The relevant definitions and facts are borrowed from the monograph [K7], the prequel of this book. The reader interested in validation of these properties can consult with [K1]–[K4].

2.2 Traversing, boundary generic, and traversally generic vector fields

Let X be a compact connected smooth $(n+1)$-dimensional manifold with boundary.

Definition 2.1. A non-vanishing vector field v on X is called traversing if each v-trajectory is homeomorphic to ether a closed interval with its ends residing in the boundary ∂X, or to a singleton, also residing in ∂X.

We denote by $\mathcal{V}_{\mathsf{trav}}(X)$ the space of all traversing vector fields on X. $\diamondsuit$

We denote by $\mathcal{F}(v)$ the singular oriented foliation on X, generated by the v-trajectories. Such foliations, under the name "short foliations" were studied in [EG].

The following lemma (see [K7], Lemma 5.6 and Corollary 5.8) will be employed on many occasions.

Lemma 2.1. *([K1]). Any smooth traversing vector field v on a compact connected manifold X with boundary is of the* gradient type*, i.e., there exists a smooth Lyapunov function $f : X \to \mathbb{R}$ such that $df(v) > 0$ in X.*

Moreover, the converse is true: on such a manifold X, any vector field v that admits a Lyapunov function is traversing. $\diamondsuit$

Any smooth vector field v on X, which does not vanish along the boundary $\partial_1 X =_{\mathsf{def}} \partial X$, gives rise to a partition $\partial_1^+ X(v) \cup \partial_1^- X(v)$ of the boundary $\partial_1 X$ into two sets: the locus $\partial_1^+ X(v)$, where the field is directed inward of X or is tangent to $\partial_1 X$, and the locus $\partial_1^- X(v)$, where it is directed outward of X or is tangent to $\partial_1 X$.

We assume that $v|_{\partial_1 X}$, viewed as a section of the quotient line bundle $T(X)/T(\partial_1 X)$ over $\partial_1 X$, is transversal to its zero section. This assumption implies that both sets, $\partial_1^+ X(v)$ and $\partial_1^- X(v)$, are compact manifolds that share a common boundary $\partial_2 X(v) =_{\mathsf{def}} \partial(\partial_1^+ X(v)) = \partial(\partial_1^- X(v))$. Evidently, $\partial_2 X(v)$ is the locus where v is *tangent* to the boundary $\partial_1 X$.

In his groundbreaking work [Mo], M. Morse made an important observation: for a generic vector field v, the tangent locus $\partial_2 X(v)$ inherits a similar structure in relation to $\partial_1^+ X(v)$, as $\partial_1 X$ has in relation to X. That is, v gives rise to a partition $\partial_2^+ X(v) \cup \partial_2^- X(v)$ of $\partial_2 X(v)$ into two sets: the locus $\partial_2^+ X(v)$, where the vector field v is directed inward of $\partial_1^+ X(v)$ or is tangent to $\partial_2 X(v)$, and the locus $\partial_2^- X(v)$, where it is directed outward of $\partial_1^+ X(v)$ or is tangent to $\partial_2 X(v)$. Again, we assume that the restriction $v|_{\partial_2 X(v)}$, viewed as a section of the quotient line bundle $T(\partial_1 X)/T(\partial_2 X(v))$ over $\partial_2 X(v)$, is transversal to its zero section.

For generic fields, this structure replicates itself: the cuspidal locus $\partial_3 X(v)$ is defined as the locus where v is tangent to $\partial_2 X(v)$. The locus $\partial_3 X(v)$ is divided into two manifolds, $\partial_3^+ X(v)$ and $\partial_3^- X(v)$. In $\partial_3^+ X(v)$, the field is directed inward of $\partial_2^+ X(v)$ or is tangent to its boundary, in $\partial_3^- X(v)$, outward of $\partial_2^+ X(v)$ or is tangent to its boundary. We can repeat this construction until we reach the zero-dimensional locus $\partial_{n+1} X(v) = \partial_{n+1}^+ X(v) \cup \partial_{n+1}^- X(v)$.

To achieve some uniformity in our notations, put $\partial_0^+ X =_{\mathsf{def}} X$ and $\partial_1 X =_{\mathsf{def}} \partial X$.

Thus a generic vector field v on X gives rise to two stratifications:

$$\partial X =_{\mathsf{def}} \partial_1 X \supset \partial_2 X(v) \supset \cdots \supset \partial_{n+1} X(v),$$
$$X =_{\mathsf{def}} \partial_0^+ X \supset \partial_1^+ X(v) \supset \partial_2^+ X(v) \supset \cdots \supset \partial_{n+1}^+ X(v), \tag{2.1}$$

the first one by closed submanifolds, the second one—by compact ones. Here $\dim(\partial_j X(v)) = \dim(\partial_j^+ X(v)) = n + 1 - j$.

• We will often use the notation "$\partial_j^\pm X$" instead of "$\partial_j^\pm X(v)$" when the vector field v is fixed or its choice is obvious.

These considerations motivate a more formal definition.

Definition 2.2. Let X be a compact connected smooth $(n+1)$-dimensional manifold with boundary $\partial X \neq \emptyset$, and v a smooth vector field on X.

We say that v is boundary generic if the restriction $v|_{\partial X}$ does not vanish and produces a filtrations of X as in (2.1). Its strata $\{\partial_j^+ X \subset \partial_j X\}_{1 \leq j \leq n+1}$ are defined inductively in j as follows:

- $\partial_0 X =_{\mathsf{def}} \partial X$, $\partial_1 X =_{\mathsf{def}} \partial X$ (hence $\partial_0 X$ and $\partial_1 X$—the base of induction—do not depend on v).
- for each $k \in [1, j]$, the v-generated stratum $\partial_k X$ is a closed smooth submanifold of $\partial_{k-1} X$,
- the field v, viewed as section of the quotient 1-bundle

$$T_k^\nu =_{\mathsf{def}} T(\partial_{k-1} X)/T(\partial_k X) \longrightarrow \partial_k X,$$

  is transversal to the zero section of the bundle $T_k^\nu \to \partial_k X$ for all $k \leq j$.
- the stratum $\partial_{j+1} X$ is the zero set of the section $v \in T_j^\nu$.
- the stratum $\partial_{j+1}^+ X \subset \partial_{j+1} X$ is the closure of the locus where v points inside $\partial_j^+ X$.

• Let $\mathcal{B}^\dagger(X)$ denote the space of boundary generic vector fields on X.

$\diamondsuit$

By the second bullet of Theorem 6.6 in [K7], the smooth topological type of the stratification $\{\partial_j X(v)\}_j$ is stable under perturbations of v within the space $\mathcal{B}^\dagger(X)$ of boundary generic fields. The same argument shows that the strata $\{\partial_j^+ X(v)\}_j$ are topologically stable as well.

Definition 2.3. We say that a boundary generic vector field v is convex if $\partial_2^+ X(v) = \emptyset$. When $\partial_2^- X(v) = \emptyset$, we say that the vector field v is concave.

$\diamondsuit$

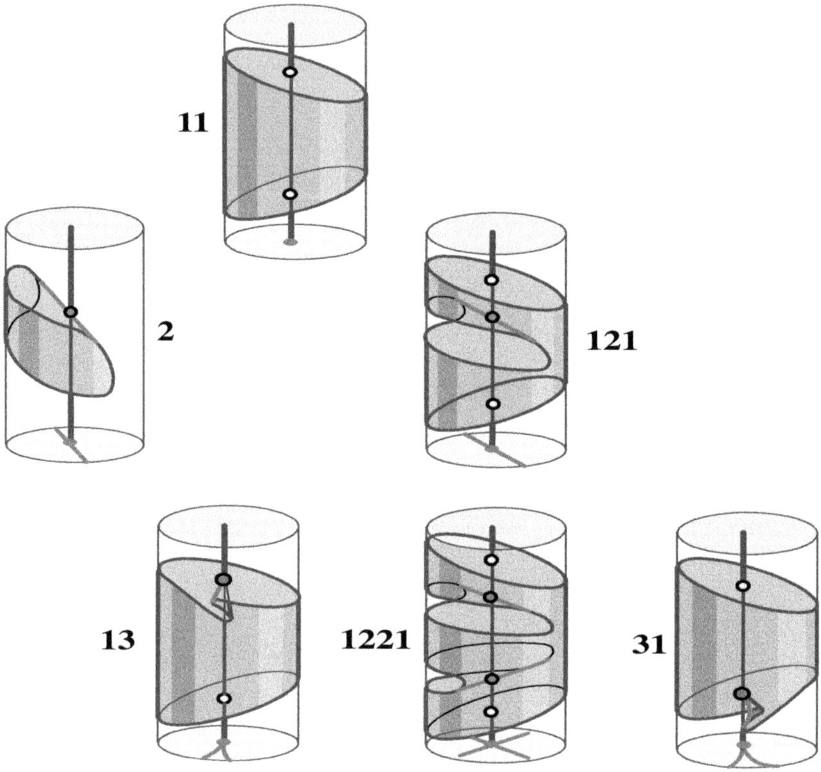

Fig. 2.1 Six semi-local patterns of traversally generic vector fields (the vertical flow in the cylinders) on 3-folds, labeled by the six combinatorial types $\omega \in \{(11), (2), (121), (3,1), (1221), (13)\}$ of the core trajectories.

Note that convexity or concavity of v implies that the locus $\partial_3 X(v) = \emptyset$.

As a default, we assume that the field v on X extends to a non-vanishing vector field $\hat{v}$ on some open manifold $\hat{X}$ which properly contains X. Throughout the book, we treat the extension $(\hat{X}, \hat{v})$ as a germ that contains (X, v). One may think of $\hat{X}$ as being obtained from X by attaching an external collar to X along $\partial_1 X$. In fact, the treatment of (X, v) will not depend on the germ of extension $(\hat{X}, \hat{v})$, but many constructions are simplified by introducing it.

The trajectories γ of a boundary generic vector field v on X interact with the boundary $\partial_1 X$ so that each point $a \in \gamma \cap \partial_1 X$ acquires a multiplicity

$m(a) \in \mathbb{N}$, the order of tangency of γ to $\partial_1 X$ at a. We associate a divisor

$$D_\gamma = \sum_{a \in \gamma \cap \partial_1 X} m(a) \cdot a$$

with each v-trajectory γ. In fact, for any boundary generic v, $m(a) \leq \dim(X)$ and the support of D_γ is finite ([K2], Lemma 3.1).

Therefore, we may also associate with each v-trajectory γ a finite ordered sequence

$$\omega(\gamma) = (\omega_1, \omega_2, \dots, \omega_q)$$

of multiplicities $\{\omega_i\}_i$. The multiplicity ω_i is the order of tangency between the curve γ and the hypersurface $\partial_1 X$ at the ith point of the finite set $\gamma \cap \partial_1 X$. The linear order in $\gamma \cap \partial_1 X$ is determined by v.

Such sequences $(\omega_1, \omega_2, \dots, \omega_q)$ form an universal poset $(\mathbf{\Omega}, \succ)$. The partial order "$\succ$" in $\mathbf{\Omega}$ is defined in terms of two types of elementary operations: merges $\{\mathsf{M}_i\}_i$ and inserts $\{\mathsf{I}_i\}_i$ (see [K3], [K7], and Chapter 6 for details).

The operation M_i merges a pair of adjacent components ω_i, ω_{i+1} of $\omega = (\omega_1, \dots, \omega_i, \omega_{i+1}, \dots, \omega_q)$ into a single component $\tilde{\omega}_i = \omega_i + \omega_{i+1}$, thus forming a new shorter sequence $\mathsf{M}_i(\omega) = (\omega_1, \dots, \tilde{\omega}_i, \dots, \omega_q)$.

The operation I_i either inserts the number 2 in-between ω_i and ω_{i+1}, thus forming a new longer sequence $\mathsf{I}_i(\omega) = (\dots, \omega_i, 2, \omega_{i+1}, \dots)$, or, in the case of I_0, appends 2 before the sequence ω, or, in the case of I_q, appends 2 after the sequence ω.

Therefore, the merge operation $\mathsf{M}_j : \mathbf{\Omega} \to \mathbf{\Omega}$ sends the composition $\omega = (\omega_1, \dots, \omega_\ell)$ to the composition

$$\mathsf{M}_j(\omega) = (M_j(\omega)_1, \dots, M_j(\omega)_{\ell-1}),$$

where, for any $j \geq \ell$, one has $\mathsf{M}_j(\omega) = \omega$, and for $1 \leq j < \ell$,

$$\begin{cases} \mathsf{M}_j(\omega)_i &= \omega_i \text{ if } i < j, \\ \mathsf{M}_j(\omega)_j &= \omega_j + \omega_{j+1}, \\ \mathsf{M}_j(\omega)_i &= \omega_{i+1} \text{ if } i+1 < j \leq \ell - 1. \end{cases} \tag{2.2}$$

Similarly, we introduce the insert operation $\mathsf{I}_j : \mathbf{\Omega} \to \mathbf{\Omega}$ that sends $\omega = (\omega_1, \dots, \omega_\ell)$ to the composition $\mathsf{I}_j(\omega) = (I_j(\omega)_1, \dots, I_j(\omega)_{\ell+1})$, where for any $j > \ell + 1$, one has $\mathsf{I}_j(\omega) = \omega$, and for $1 \leq j \leq \ell + 1$,

$$\begin{cases} \mathsf{I}_j(\omega)_i &= \omega_i \text{ if } i < j, \\ \mathsf{I}_j(\omega)_j &= 2, \\ \mathsf{I}_j(\omega)_i &= \omega_{i-1} \text{ if } j \leq i \leq \ell + 1. \end{cases} \tag{2.3}$$

Example 2.1. For $\omega = (12243)$, we have:

$$\mathsf{M}_1(\omega) = (\tilde{3}243), \quad \mathsf{M}_2(\omega) = (1\tilde{4}43), \quad \mathsf{M}_3(\omega) = (12\tilde{6}3), \quad \mathsf{M}_4(\omega) = (122\tilde{7}),$$

$$\mathsf{I}_0(\omega) = (\tilde{2}12243), \quad \mathsf{I}_1(\omega) = (1\tilde{2}2243), \quad \mathsf{I}_2(\omega) = (12\tilde{2}243),$$

$$\mathsf{I}_3(\omega) = (122\tilde{2}43), \quad \mathsf{I}_4(\omega) = ((1224\tilde{2}3), \quad \mathsf{I}_5(\omega) = (12243\tilde{2}). \quad \diamondsuit$$

Definition 2.4. We say that $\omega \succ \omega'$, if one can produce ω' from ω by applying a sequence of elementary merge and insert operations as in (2.2) and (2.3). $\diamondsuit$

For each trajectory γ of a boundary generic and traversing v, we introduce two important quantities:

$$m(\gamma) =_{\mathsf{def}} \sum_{a \in \gamma \cap \partial_1 X} m(a), \quad \text{and} \quad m'(\gamma) =_{\mathsf{def}} \sum_{a \in \gamma \cap \partial_1 X} (m(a) - 1), \tag{2.4}$$

the $\mathsf{multiplicity}$ and the $\mathsf{reduced\ multiplicity}$ of the trajectory γ.

Similarly, for a sequence $\omega = (\omega_1, \omega_2, \ldots, \omega_q)$ of natural numbers, we introduce its norm and its $\mathsf{reduced\ norm}$ by the formulas:

$$|\omega| =_{\mathsf{def}} \sum_i \omega_i \quad \text{and} \quad |\omega|' =_{\mathsf{def}} \sum_i (\omega_i - 1). \tag{2.5}$$

Note that q, the cardinality of the $\mathsf{support}$ of ω, is equal to $|\omega| - |\omega|'$.

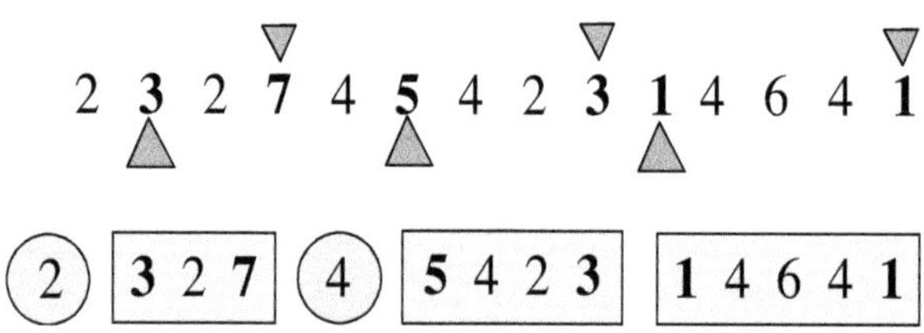

Fig. 2.2 An example of the deconstruction $\omega \Rightarrow \Xi(\omega)$. The strings are boxed, the atoms are circled.

For boundary generic and traversing vector fields v, the combinatorial types $\omega = (\omega_1, \ldots, \omega_q)$ of v-trajectories are somewhat restrictive: for $q > 1$, the first and the last entries of $\omega \in \mathbf{\Omega}^\bullet$ are $\mathsf{odd\ positive\ integers}$, the rest of the entries are even. When $q = 1$, $\omega = (\omega_q)$ must be even. These combinatorial types form $\mathsf{another\ universal\ poset}$ $\mathbf{\Omega}^\bullet$. As a set, $\mathbf{\Omega}^\bullet$ is the closure $\langle(11)\rangle$ of the element $\omega = (11) \in \mathbf{\Omega}$ under the relation $\prec$; however, the order relations in $\mathbf{\Omega}^\bullet$ are different from the subposet $\langle(11)\rangle \subset \mathbf{\Omega}$ (see Fig. 2.3 for an example of the partial order $\prec_\bullet$ in $\mathbf{\Omega}^\bullet$).

The partial order in the set $\mathbf{\Omega}^\bullet$, matches the changing geometry of trajectories for traversally generic vector fields. For a more accurate description, we turn to few auxiliary combinatorial constructions.

Given any $\omega \in \mathbf{\Omega}$, we would like to "chop it" into a number of "strings" and "atoms": each string belongs to $\mathbf{\Omega}^\bullet$ and starts and ends with an odd number, while each atom has a singleton for its support and takes there an even value (see Fig. 2.2). Prior to defining this canonical deconstruction $\Xi(\omega)$ of ω, we need to introduce some notations.

Let $\omega \in \mathbf{\Omega}$ be such that $|\omega| \equiv 0\ (2)$. Denote by $\sup_{\mathsf{odd}}(\omega)$ the points $l \in \mathbb{N}$ in the support of ω such that $\omega(l) \equiv 1\ (2)$, and by $\sup_{\mathsf{ev}}(\omega)$ the points l in the support of ω such that $\omega(l) \equiv 0\ (2)$. In the geometrical context of traversing flows, the number $|\sup_{\mathsf{odd}}(\omega)|$ is even. We count the elements of $\sup_{\mathsf{odd}}(\omega)$ as they appear in the list; some of them acquire odd numerals, others acquire even ones. A string in $A \subset \sup(\omega)$ is formed by all the elements that are bounded on the left by an element of $i \in \sup_{\mathsf{odd}}(\omega)$ with an odd numeral and on the right by the next element $j \in \sup_{\mathsf{odd}}(\omega)$ (it has an even numeral attached to it). In other words, a string $A := [i, j]$ is formed by such i, j and all the elements from $\sup_{\mathsf{ev}}(\omega)$ that lie in-between. The points from $\sup_{\mathsf{ev}}(\omega)$ that do not belong to any string are called "atoms". We can compute the values of ω at the elements of each string A as well as at each atomic support. This gives rise to a unique ordered sequence $\Xi(\omega)$ of strings, interrupted by a number of atoms.

Definition 2.5. We define the partial order "$\prec_\bullet$" in $\mathbf{\Omega}^\bullet$ as follows: for $\omega_1, \omega_2 \in \mathbf{\Omega}^\bullet$, we write "$\omega_1 \prec_\bullet \omega_2$" if ω_2 occurs as a string or an atom from the list $\Xi(\omega)$ for some $\omega \succ \omega_1$, where $\omega \in \mathbf{\Omega}$ and the last ordering is considered in the poset $(\mathbf{\Omega}, \succ)$. ◇

For a boundary generic and traversing vector field v, the trajectory space $\mathcal{T}(v)$ is naturally stratified by subspaces, labeled with the elements $\omega \in \mathbf{\Omega}^\bullet$.

We consider also an important subclass of traversing and boundary generic vector fields v, which we call traversally generic (see Definition 2.6 below and Definition 3.2 from [K2]). Such fields admit v-flow-adjusted local coordinate systems, in which the boundary is given by a quite special polynomial equation (see formula (2.11)), and all the trajectories are parallel to the preferred coordinate axis (see [K2], Lemma 3.4). More importantly,

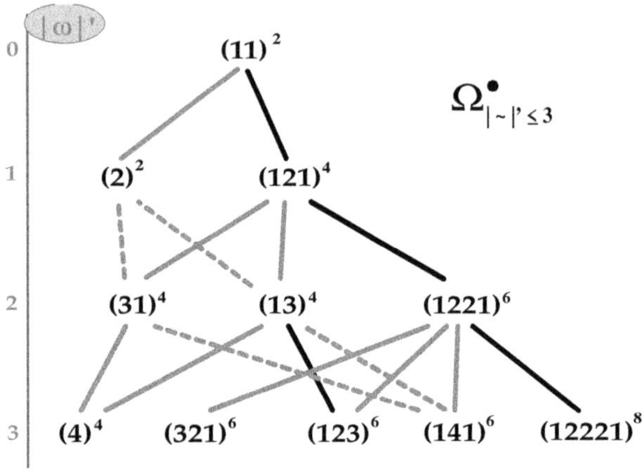

Fig. 2.3 The poset $\mathbf{\Omega}^\bullet_{|\sim|'\leq 3}$ consists of atoms and strings whose reduced norm ≤ 3. The height function in the graph is given by the reduced norm $|\omega|'$. The bold dark lines indicate the insert operations, the lighter bold lines the merge operations. The dotted lines represent the order relations that are *not directly* induced from the ambient poset $\mathbf{\Omega}_{\langle 8]} \supset \mathbf{\Omega}^\bullet_{|\sim|'\leq 3}$. The upper index next to each ω shows the value of the norm $|\omega|$.

for such vector fields, the smooth topological type of the flow in the vicinity of each trajectory of a combinatorial type ω is *determined* only by the combinatorial pattern ω!

Let us give a more accurate definition of a traversally generic vector field. Given a boundary generic and traversing vector field v, for each trajectory γ, consider the finite set $\gamma \cap \partial_1 X = \{a_i\}_i$ and the collection of tangent spaces $\{T_{a_i}(\partial_{j_i} X^\circ)\}_i$ to the pure strata $\{\partial_{j_i} X^\circ\}_i$. Note that each point a_i belongs to a unique stratum with the *maximal possible* index $j_i =_{\mathsf{def}} j(a_i)$. In fact, each space $T_{a_i}(\partial_{j_i} X^\circ)$ is transversal to the curve γ [K2].

Let S be a local transversal section of the $\hat{v}$-flow at a point $a_\star \in \gamma$, and let $\mathsf{T}_\star$ be the space, tangent to the section S at $a_\star$. Each space $T_{a_i}(\partial_j X^\circ)$, with the help of the $\hat{v}$-flow, determines a vector subspace $\mathsf{T}_i = \mathsf{T}_i(\gamma)$ in $\mathsf{T}_\star$. It is the image of the tangent space $T_{a_i}(\partial_j X^\circ)$ under the composition of two maps:

(1) the differential of the v-flow-generated diffeomorphism that maps a_i to $a_\star$, and

(2) the linear projection $T_{a_\star}(X) \to \mathsf{T}_\star$ whose kernel is generated by $v(a_\star)$.

The configuration $\{\mathsf{T}_i\}$ of affine subspaces $\mathsf{T}_i \subset \mathsf{T}_\star$ is called generic (or stable) when all the multiple intersections of spaces from the configuration have the least possible dimensions, consistent with the dimensions of $\{\mathsf{T}_i\}$. In other words,

$$\mathrm{codim}(\bigcap_s \mathsf{T}_{i_s}, \mathsf{T}_\star) = \sum_s \mathrm{codim}(\mathsf{T}_{i_s}, \mathsf{T}_\star)$$

for any subcollection $\{\mathsf{T}_{i_s}\}$ of spaces from the list $\{\mathsf{T}_i\}$.

Consider the case when $\{\mathsf{T}_i\}$ are *vector* subspaces of $\mathsf{T}_\star$. If we interpret each subspace T_i as the kernel of a linear epimorphism $\Phi_i : \mathsf{T}_\star \to \mathbb{R}^{n_i}$, then the property of $\{\mathsf{T}_i\}$ being generic can be reformulated as the property of the direct product map $\prod_i \Phi_i : \mathsf{T}_\star \to \prod_i \mathbb{R}^{n_i}$ being an epimorphism. In particular, for a generic configuration of affine subspaces, if a point belongs to several T_i's, then the sum of their codimensions n_i does not exceed the dimension of the ambient space $\mathsf{T}_\star$.

The definition below resembles and is inspired by the "Condition NC" imposed on, so called, Boardman maps between smooth manifolds (see [GG], page 157, for the relevant definitions). In fact, for a generic traversing vector fields v, the v-flow delivers germs of Boardman maps $p(v,\gamma) : \partial_1 X \to \mathbb{R}^n$, available in the vicinity of each trajectory γ [K2]. Here $\mathbb{R}^n$ is identified with a transversal section S of the flow in the vicinity of γ.

Definition 2.6. A traversing field v on X is called traversally generic if:

- the field is boundary generic in the sense of Definition 2.2,
- for each v-trajectory $\gamma \subset X$, the collection of subspaces $\{\mathsf{T}_i(\gamma)\}_i$ is generic in $\mathsf{T}_\star$; that is, the obvious quotient map $\mathsf{T}_\star \to \prod_i \big(\mathsf{T}_\star / \mathsf{T}_i(\gamma)\big)$ is surjective.

- We denote by $\mathcal{V}^\ddagger(X)$ the space of all traversally generic fields on X.

◇

Fig. 2.2 depicts semi-local models of *traversally generic* vector fields in $3D$.

Remark 2.1. The second bullet in Definition 2.6 implies the inequality

$$\sum_i \mathrm{codim}(\mathsf{T}_i(\gamma), \mathsf{T}_\star) \leq \dim(\mathsf{T}_\star) = n.$$

In other words, for traversally generic vector fields, the reduced multiplicity of each trajectory γ satisfies the inequality

$$m'(\gamma) = \sum_i (j_i - 1) \leq n. \tag{2.6}$$

Evidently, the property of the configuration $\{\mathsf{T}_i(\gamma)\}_i$ being generic in $\mathsf{T}_\star$ does not depend on the choice of the point $a_\star \in \gamma$ and on the smooth transversal flow section S at $a_\star$. $\diamondsuit$

All sufficiently close in the C^∞-topology vector fields to a traversally generic field will remain traversally generic. Moreover, Theorem 3.5 from [K2] or Theorem 6.7 from [K7] can be stated as follows.

Theorem 2.1. • *The space $\mathcal{B}^\dagger(X)$ of boundary generic vector fields is open and dense in the space of all smooth vector fields on X that do not vanish along the boundary ∂X.*

• *The space $\mathcal{V}^\dagger(X)$ of traversing and boundary generic vector fields is open and dense in the space $\mathcal{V}_{\mathsf{trav}}(X)$ of all traversing vector fields on X.*

• *The space $\mathcal{V}^\ddagger(X)$ of traversally generic vector fields is open and dense in the space $\mathcal{V}_{\mathsf{trav}}(X)$ of smooth traversing vector fields.* $\diamondsuit$

These properties of the spaces $\mathcal{V}^\dagger(X)$ and $\mathcal{V}^\ddagger(X)$ will be of great importance for our endeavor.

We denote by $\mathbf{\Omega}^\bullet_{|\sim|' \leq k}$ the subposet of $\mathbf{\Omega}^\bullet$, defined by the constraint $|\omega|' \leq k$.

For *traversally generic* vector fields v on a $(n+1)$-dimensional X, the trajectory space $\mathcal{T}(v)$ is stratified by subspaces, labeled with the elements of the poset $\mathbf{\Omega}^\bullet_{|\sim|' \leq n} \subset \mathbf{\Omega}^\bullet$, defined by the constraint $|\omega|' \leq n$. This poset depends only on $\dim(X) = n+1$ (see [K3] for the properties of $\mathbf{\Omega}^\bullet_{|\sim|' \leq n}$).

Let us revisit the Morse stratum $\partial_j X =_{\mathsf{def}} \partial_j X(v)$, the locus of points $a \in \partial_1 X$ such that the multiplicity of the v-trajectory γ_a through a at a is greater than or equal to j. This locus has an alternative description in terms of an auxiliary smooth function $z : \hat{X} \to \mathbb{R}$, satisfying the following three properties:

$$\begin{aligned} &\bullet \quad 0 \text{ is a regular value of } z, \\ &\bullet \quad z^{-1}(0) = \partial X, \\ &\bullet \quad z^{-1}((-\infty, 0]) = X. \end{aligned} \tag{2.7}$$

In terms of z, the locus $\partial_j X$ is defined by the equations:

$$\{z = 0,\ \mathcal{L}_v z = 0,\ \dots,\ \mathcal{L}_v^{(j-1)} z = 0\}, \tag{2.8}$$

where $\mathcal{L}_v^{(k)}$ stands for the k-th iteration of the Lie derivative operator $\mathcal{L}_v$ in the direction of v (see [K2], [K7]).

The pure stratum $\partial_j X^\circ \subset \partial_j X$ is defined by the additional constraint $\mathcal{L}_v^{(j)} z \neq 0$. The locus $\partial_j X$ is the union of two loci:

(1) $\partial_j^+ X$, defined by the constraint $\mathcal{L}_v^{(j)} z \geq 0$, and

(2) $\partial_j^- X$, defined by the constraint $\mathcal{L}_v^{(j)} z \leq 0$.

The two loci, $\partial_j^+ X$ and $\partial_j^- X$, share a common boundary $\partial_{j+1} X$.

The validation of the following lemma is on the level of definitions.

Lemma 2.2. *A vector field v on a compact connected smooth $(n+1)$-manifold X with boundary is boundary generic if and only if, for each $j \in [1, n+1]$, the differential j-form*

$$\Xi_j(z, v) =_{\mathsf{def}} dz \wedge \mathcal{L}_v(dz) \wedge \ldots (\mathcal{L}_v)^{j-1}(dz), \tag{2.9}$$

a section of the bundle $\bigwedge^j T^*X$, does not vanish along the locus $\partial_j X(v)$. $\diamondsuit$

The next lemma is proven in [Morin] and is based on the Malgrange Preparation Lemma [Mal].

Lemma 2.3. *Let v be a boundary generic vector field on a $(n+1)$-dimensional smooth manifold X with boundary. Let a v-trajectory $\gamma_\star$ be tangent to $\partial_1 X$ at a point $b \in \gamma_\star \cap \partial_1 X$ with the order of tangency $j \in [1, n+1]$.*

Then, in the vicinity of b in X, there exists a system of smooth coordinates $\{u, \vec{x}, \vec{y}\} = \{u, x_0, \ldots, x_{j-2}, y_1, \ldots y_{n-j+1}\}$ such that:

- *the boundary $\partial_1 X$ is given by the equation*
$$P(u, \vec{x}) =_{\mathsf{def}} u^j + \sum_{\ell=0}^{j-2} x_\ell \, u^\ell = 0, \tag{2.10}$$
and X by the inequality $P(u, \vec{x}) \leq 0$,
- *each v-trajectory is produced by freezing the coordinates $\{\vec{x}, \vec{y}\}$, subject to the constraint $P(u, \vec{x}) \leq 0$.* $\diamondsuit$

Lemma 2.3 implies the following lemma (see [K7] or [K2]).

Lemma 2.4. *Let X be a $(n+1)$-dimensional compact connected smooth manifold X with boundary, and v a traversing boundary generic vector field on X. Let γ be a v-trajectory of a combinatorial type ω.*

Then there is a $\hat{v}$-adjusted neighborhood $U \subset \hat{X}$ of γ and a system of coordinates $(u, \vec{x}) : U \to \mathbb{R} \times \mathbb{R}^n$ such that, for a sufficiently small $\epsilon > 0$, X is given by the inequalities $\{P(u, \vec{x}) \leq 0, \; \|\vec{x}\| < \epsilon\}$, where

$$P(u, \vec{x}) =_{\mathsf{def}} u^{|\omega|} + \sum_{\ell=0}^{|\omega|-1} \phi_\ell(\vec{x}) \, u^\ell,$$

and $\{\phi_\ell(\vec{x})\}_\ell$ *are smooth functions. The real divisor of* $P(u, \vec{0})$ *has the combinatorial type* ω. *Each* $\hat{v}$*-trajectory in* U *is given by freezing the coordinate* $\vec{x} \in \mathbb{R}^n$, *subject to the constraint* $P(u, \vec{x}) \leq 0$. ◇

Let v be a traversing and boundary generic vector field. For each v-trajectory γ and each point $a_i \in \gamma \cap \partial_1 X$ of multiplicity $j_i =_{\mathsf{def}} j(a_i)$, we consider the differential form $\Xi_j(z, v)|_{a_i} \in \bigwedge^{j_i} T^*_{a_i} X$ from (2.9) and spread it via the v-flow along γ. We denote by $\tilde{\Xi}_{j_i}(z, \gamma)$ the resulting section (the j_i-form) of the bundle $\bigwedge^{j_i} T^*X\,\big|_\gamma$.

Lemma 2.5. *A traversing and boundary generic vector field* v *on a smooth compact* $(n+1)$*-manifold* X *with boundary is traversally generic if and only if, for each trajectory* γ, *the* $m(\gamma)$*-dimensional differential form*

$$\tilde{\Xi}(z, \gamma) =_{\mathsf{def}} \wedge_{i=1}^{s}\ \tilde{\Xi}_{j_i}(z, \gamma) \in \bigwedge^{|\omega_\gamma|} T^*X\,\big|_\gamma,$$

where $s = \#(\gamma \cap \partial X)$ *and* $|\omega_\gamma| = \sum_{i=1}^{s} j_i$, *does not vanish along* γ. ◇

Given a traversally generic v (see Definition 2.6) on a $(n+1)$-dimensional X, each v-trajectory γ of a combinatorial type $\omega \in \mathbf{\Omega}^\bullet$ has a neighborhood $U \subset \hat{X}$, equipped with a special coordinate system

$$(u, \vec{x}, \vec{y}) : U \to \mathbb{R} \times \mathbb{R}^{|\omega|'} \times \mathbb{R}^{n-|\omega|'}.$$

By [K7], Lemma 6.4, in the vicinity of γ, in these coordinates the boundary $\partial_1 X$ is given by the polynomial equation (compare formula (2.11) below with formula (2.10):

$$P(u, \vec{x}) =_{\mathsf{def}} \prod_{i=1}^{|\omega|-|\omega|'} \Big[(u - \alpha_i)^{\omega_i} + \sum_{\ell=0}^{\omega_i - 1} x_{i,\ell}(u - \alpha_i)^\ell\Big] = 0 \qquad (2.11)$$

of an even degree $|\omega|$ in u. Here $\vec{x} = \{x_{i,l}\}_{i,\ell}$, and the numbers $\{\alpha_i\}_i$ are all distinct real roots of the polynomial $P(u, 0)$, ordered so that $\alpha_i < \alpha_{i+1}$ for all i. At the same time, X is given by the polynomial inequality $\{P(u, \vec{x}) \leq 0\}$. Each v-trajectory in U is produced by freezing all the coordinates $\vec{x}, \vec{y}$, while letting u to vary freely.

In fact, by choosing $\alpha_i = i$, we may rewrite this equation for $\partial_1 X$ in U as

$$\wp_\omega(u, \vec{x}) =_{\mathsf{def}} \prod_{i=1}^{|\omega|-|\omega|'} \Big[(u - i)^{\omega_i} + \sum_{\ell=0}^{\omega_i - 1} x_{i,\ell}(u - i)^\ell\Big] = 0, \qquad (2.12)$$

where $|\omega|' \leq \dim X - 1$, $|\omega| \leq 2\dim X$, and $|\omega| \equiv 0 \mod 2$. Equation (2.12) may be viewed as a *working definition* of a traversally generic vector field [K2].

2.3 Trajectory spaces of traversing flows

In this section, we summarize the content of Section 6.4 from [K7].

Let X be a compact connected smooth $(n+1)$-dimensional manifold with boundary, and v a vector field on X. The points of trajectory space $\mathcal{T}(v)$ are the v-trajectories, and the topology in $\mathcal{T}(v)$ is the quotient topology for the obvious map $\Gamma : X \to \mathcal{T}(v)$.

In general, the space $\mathcal{T}(v)$ is pathological (non-separable). However, for any traversally generic v, $\mathcal{T}(v)$ is a compact CW-complex which is homotopy equivalent to X (Corollary 6.8, [K7]). Conjecturally, this is true for any traversing and boundary generic v.

For a traversing vector field v, $\mathcal{T}(v)$ is homology equivalent to X via the obvious map $\Gamma : X \to \mathcal{T}(v)$ (Theorem 5.1, [K3]).

We denote by $X(v, \omega) \subset X$ the union of v-trajectories whose patterns of tangency to $\partial_1 X$ are of a given combinatorial type $\omega \in \mathbf{\Omega}^\bullet$. We use the notation $X(v, \omega_{\succeq})$ for the closure $\bigcup_{\omega' \preceq \omega} X(v, \omega')$ of $X(v, \omega)$.

For a traversing vector field v, the space $\mathcal{T}(v)$ has an interesting feature: it comes equipped with a vector n-bundle $\tau(\mathcal{T}(v))$ which plays the role of "surrogate tangent bundle".

For a traversally generic v, each pure stratum $\mathcal{T}(v, \omega) \subset \mathcal{T}(v)$, consisting of trajectories of the combinatorial type ω, is an open smooth manifold of dimension $n - |\omega|'$. As such, it has a "conventional" tangent bundle. In particular, the pure strata of maximal dimension n have tangent bundles. It turns out that these "honest" tangent n-bundles extend across the singularities of the space $\mathcal{T}(v)$ to form a "surrogate tangent" n-bundle $\tau(\mathcal{T}(v))$ over $\mathcal{T}(v)$! However, at the singularities, no exponential map, taking a vector from $\tau(\mathcal{T}(v))$ to a point in $\mathcal{T}(v)$, is available; the bundle $\tau(\mathcal{T}(v))$ does not reflect faithfully the local singular geometry of the trajectory space $\mathcal{T}(v)$.

In order to define the dual of the bundle $\tau(\mathcal{T}(v))$ intrinsically, we consider a surrogate of "smooth structure" on the singular space $\mathcal{T}(v)$.

Definition 2.7. Let $\Gamma : X \to \mathcal{T}(v)$ be the projection that takes each point $x \in X$ to the trajectory $\gamma_x \in \mathcal{T}(v)$ that contains x. We say that a function $h : \mathcal{T}(v) \to \mathbb{R}$ is smooth, if the composition $h \circ \Gamma$ is smooth on X.

We denote by $C^\infty(\mathcal{T}(v)) =_{\mathsf{def}} C^\infty(\mathcal{T}(v), \mathbb{R})$ the algebra of all smooth functions on the space $\mathcal{T}(v)$. ◇

Definition 2.8. Let v_1, v_2 be two traversally generic vector fields on manifolds X_1, X_2, respectively.

- A map $\Phi : \mathcal{T}(v_1) \to \mathcal{T}(v_2)$ is called smooth if, for any function h from $C^\infty(\mathcal{T}(v_2))$, its pull-back $\Phi^*(h) \in C^\infty(\mathcal{T}(v_1))$.
- A bijective map $\Phi : \mathcal{T}(v_1) \to \mathcal{T}(v_2)$ is called a diffeomorphism, is both Φ and Φ^{-1} are smooth. $\diamondsuit$

For any traversing vector field v, the algebra $C^\infty(\mathcal{T}(v))$ of smooth functions on the trajectory space $\mathcal{T}(v)$ can be identified with the subalgebra of $C^\infty(X)$, formed by functions $f : X \to \mathbb{R}$ with the property $\{\mathcal{L}_v(f) = df(v) \equiv 0\}$, where $\mathcal{L}_v$ stands for the v-directional derivative (this property is independent of an extension $(\hat{X}, \hat{v}) \supset (X, v)$). Such functions are constant along each trajectory $\gamma \subset X$.

For a traversing v, we consider the algebra $\mathsf{Ker}(\mathcal{L}_v) \approx C^\infty(\mathcal{T}(v))$ of smooth functions on X that are constants along each v-trajectory.

We denote by $C^\infty_\gamma(\mathcal{T}(v))$ the algebra of germs of smooth functions from $C^\infty(\mathcal{T}(v))$ at a given point-trajectory $\gamma \in \mathcal{T}(v)$, by $\mathsf{m}_\gamma(\mathcal{T}(v)) \lhd C^\infty_\gamma(\mathcal{T}(v))$ the maximal ideal, formed by the germs of functions that vanish at the point γ, and by $\mathsf{m}^2_\gamma(\mathcal{T}(v))$ the square of the ideal $\mathsf{m}_\gamma(\mathcal{T}(v))$.

Then the quotients $\mathsf{m}_\gamma(\mathcal{T}(v))/\mathsf{m}^2_\gamma(\mathcal{T}(v))$ are real n-dimensional vector spaces. They form a vector n-bundle $\tau^*(\mathcal{T}(v))$ over $\mathcal{T}(v)$. Its pull-back $\Gamma^*(\tau^*(\mathcal{T}(v)))$ can be identified with the subbundle $\tau^*(v)$ of the cotangent bundle $T^*(X)$, formed by the "horizontal" differential 1-forms α that vanish on v and have the property $\mathcal{L}_v(\alpha) = 0$. The identification is via the correspondence $\Gamma^*(f) \Rightarrow d(\Gamma^*(f))$, where $f \in \mathsf{m}_\gamma(\mathcal{T}(v))$.

Now we *define* the surrogate tangent bundle $\tau(\mathcal{T}(v))$ as the dual bundle of the bundle $\tau^*(\mathcal{T}(v))$.

Let $(11) \in \mathbf{\Omega}^\bullet_{|\sim|' \leq n}$ denote the unique maximal element of the poset; it labels the trajectories that intersect the boundary $\partial_1 X$ only at a pair of distinct points, where they are transversal to the boundary.

Lemma 2.6. *For any traversing vector field v, the tangent bundles to the components of the maximal stratum $\mathcal{T}(v, (11))$ extend to a n-dimensional vector bundle $\tau(\mathcal{T}(v))$ over the trajectory space $\mathcal{T}(v)$.*

Moreover, for a traversally generic vector field v and each element $\omega \in \mathbf{\Omega}^\bullet_{|\sim|' \leq n}$, the tangent bundle of the pure stratum $\mathcal{T}(v, \omega)$ embeds in the restriction $\tau\big(\mathcal{T}(v)\big)\,\big|_{\mathcal{T}(v,\omega)}$ of the tangent bundle $\tau\big(\mathcal{T}(v)\big)$ as a subbundle with a canonically trivialized complement. $\diamondsuit$

The only non-trivial statement of the lemma is the existence of a preferred trivialization in the quotient bundle $\tau(\mathcal{T}(v))\big|_{\mathcal{T}(v,\omega)} / \tau(\mathcal{T}(v, \omega))$. It follows from the last claim of Theorem 2.2 below. Thus, we get the bundle isomorphism

$$\Psi : \tau(\mathcal{T}(v,\omega)) \oplus \underline{\mathbb{R}}^{|\omega|'} \approx \tau(\mathcal{T}(v))\,|_{\mathcal{T}(v,\omega)},$$

canonically defined by v.

By Theorem 6.8 from [K7], the map $\Gamma : X \to \mathcal{T}(v)$ is a homological equivalence for any traversing v. In fact, using that Γ is a quasifibration in the sense of [May], we conclude that Γ is a homotopy equivalence. Therefore, we get the following claim.

Corollary 2.1. *For a traversally generic vector field v on X, the stable characteristic classes of the tangent bundles $\tau(\mathcal{T}(v))$ and $\tau(X)$ coincide via the cohomological isomorphism, induced by the projection $\Gamma : X \to \mathcal{T}(v)$.*

$\diamondsuit$

According to Theorem 6.9 from [K7], for a traversally generic v, the space $\mathcal{T}(v)$ comes equipped with two distinct *intrinsically-defined orientations* of its pure strata $\{\mathcal{T}(v,\omega)\}_\omega$. These orientations depend only on v and the preferred orientation of X.

Let us sketch the constructions that leads to these orientations.

Local transversal sections S of the $\hat{v}$-flow have a well-defined orientation due to the global orientation of X and the preferred orientation of the v-trajectories.

Recall that, for a traversally generic v on a $(n+1)$-dimensional X, each v-trajectory γ of the combinatorial type ω has a neighborhood $U \subset \hat{X}$, equipped with a special coordinate system

$$(u, \vec{x}, \vec{y}) : U \to \mathbb{R} \times \mathbb{R}^{|\omega|'} \times \mathbb{R}^{n-|\omega|'},$$

were $\vec{x} = \{x_{i,\ell}\}_{i,\ell}$. Each v-trajectory in U is produced by freezing all the coordinates $\vec{x}, \vec{y}$, while letting u to vary freely.

In these coordinates, by (2.11), the boundary $\partial_1 X$ is given by the polynomial equation $\{P(u, \vec{x}) = 0\}$ in u of degree $|\omega|$. The numbers $\{\alpha_i\}_i$ in (2.11) are all distinct real roots of the polynomial $P(u, \vec{0})$, ordered so that $\alpha_i < \alpha_{i+1}$ for all i.

We order the coordinates $\{x_{i,\ell}\}_{i,\ell}$ lexicographically: first we order them by the increasing i's; then, for a fixed i, the ordering among $\{x_{i,\ell}\}_\ell$ is defined by the increasing powers ℓ of the binomial $(u - \alpha_i)$ in the formula (2.11). This ordering of $\{x_{i,\ell}\}_{i,\ell}$, together with the orientation of S (induced with the help of v by the orientation of X), gives rise to an orientation of the y-coordinates. The $\vec{y}$-coordinates correspond to the space, tangent to the pure stratum $\mathcal{T}(v,\omega)$ at γ. We call this orientation of the tangent space $\tau_\gamma(\mathcal{T}(v,\omega))$ versal.

By the proof of Theorem 6.9 from [K7], this ordering of $\{x_{i,l}\}_l$ is determined by the *geometry of tangency* and does not depend on a particular choice of the special coordinates $\{x_{i,l}\}_l$.

On the other hand, each manifold $\partial_j X$, $j > 0$, comes equipped with its own preferred orientation. It depends only on the stratification $\{\partial_k^+ X(v)\}_k$ (and not on properties of the field v inside X) and on the preferred orientation of X. Here is the recipe for its construction: the orientation of X, with the help of the inward normals, induces a preferred orientation of $\partial_1 X$, and thus of $\partial_1^\pm X$. In turn, the inward normals to $\partial_2 X = \partial(\partial_1^+ X)$ in $\partial_1^+ X$ produce a preferred orientation of $\partial_2 X$, and thus of $\partial_2^\pm X$; and the process continues: the preferred orientation of $\partial_{j-1} X$, with the help of the inward normal to $\partial_j X$ in $\partial_{j-1}^+ X$, determines a preferred orientation of $\partial_j X$, and hence of $\partial_j^\pm X$.

Therefore, along each trajectory γ, every space T_i, tangent to $\partial_{j_i} X^\circ$ and transversal to γ at the point $a_i \in \gamma \cap \partial_1 X$, is preferably oriented. For a traversally generic v, the $\hat{v}$-flow propagates these spaces T_i's along γ in such a way that they form complementary vector bundles over γ (see Definition 2.6). We order them by the increasing values of i. This ordering, together with the preferred orientations of the T_i's (based on the orientations of the strata $\partial_{j_i}^+ X$), generates a new preferred orientation of the tangent space $\tau_\gamma(\mathcal{T}(v,\omega))$. This preferred orientation may agree or disagree with the versal orientation of the same space (produced with the help of special coordinates in the vicinity of γ). In the first case, we attach the polarity "$\oplus$" to γ, in the second case, the polarity of γ is defined to be "$\ominus$".

Therefore, not only the components of pure strata $\mathcal{T}(v,\omega)$ are canonically oriented open manifolds, but they also come in two flavors: "$\oplus$" and "$\ominus$"!

We summarize these considerations and constructions in the next theorem (Theorem 6.9 from [K7]).

Theorem 2.2. *Let X be an oriented compact $(n+1)$-manifold, and v a traversally generic vector field. Then*

- *each component of any pure stratum $\mathcal{T}(v,\omega)$, where $\omega \in \mathbf{\Omega}^\bullet_{|\sim|' \leq n}$ and $|\omega|' > 0$, acquires two distinct orientations, called* preferred *and* versal. *Switching the orientation of X affects both orientations of $\mathcal{T}(v,\omega)$ by the same factor $(-1)^{|sup(\omega)|}$.*

- *With the help of these two orientations, each component of* $\mathcal{T}(v,\omega)$ *acquires one of the two polarities "*$\oplus$*" and "*$\ominus$*". They do not depend on the orientation of* X. ◊
- *Each manifold* $X(v,\omega)$ *comes equipped with a* v*-induced* normal framing *in* X. *Similarly, the normal* $|\omega|'$*-dimensional bundle*

$$\nu(\mathcal{T}(v,\omega)) =_{\mathsf{def}} \tau(\mathcal{T}(v))\,|_{\mathcal{T}(v,\omega)} / \, \tau(\mathcal{T}(v,\omega))$$

acquires a v*-induced preferred trivialization.*

The preferred orientation of each $\partial_j X$, $j \geq 1$, depends only on $v|_{\partial_1 X}$ and the orientation of X. In particular, the preferred orientation of $\partial_1 X$ depends on the orientation of X only. As we flip the orientation of X, the preferred orientation of each $\partial_j X$ flips as well. Therefore, the preferred orientation of the tangent bundle $\tau(\mathcal{T}(v,\omega))$ changes, as a result of flipping the orientation of X, only when the cardinality of the intersection $\gamma \cap \partial_1 X$—the interger $|\sup(\omega)|$—is odd.

The versal orientation of $\mathcal{T}(v,\omega)$ behaves similarly under the flip of an orientation of X. As a result, the polarity "$\oplus$" or "$\ominus$" of each component of $\mathcal{T}(v,\omega)$ is independent of the orientation of X.

Definition 2.9. Let Z be a closed subset of a smooth manifold M, which decomposes as $Z = \coprod_{\alpha \in \mathcal{S}} Z_\alpha$, where $\mathcal{S}$ a finite poset.

We say that Z is a Whitney space if the following properties hold:

(1) each stratum Z_α locally is a smooth submanifold of M,
(2) take any pair $Z_\alpha \subset \bar{Z}_\beta$ of strata and consider any pair of infinite sequences $\{x_i \in Z_\beta\}_i$, $\{y_i \in Z_\alpha\}_i$, both converging to the same point $y \in Z_\alpha$. In a local coordinate system on M, centered on y, we form the secant lines $\{\ell_i = [x_i, y_i]\}_i$ so that that $\{\ell_i\}_i$ converge to a limiting line ℓ. Also consider a sequence of tangent spaces $\{T_{x_i}(Z_\beta)\}_i$ that converge to a limiting space $\tau \subset T_y(M)$.
Then we require that the limiting line ℓ is contained in the space τ. ◊

Let us state Theorem 6.10 from [K7] (see also [K8], Theorem 3.2).

Theorem 2.3. *For a traversally generic vector field* v *on a compact smooth* $(n+1)$*-dimensional manifold* X, *the* $\mathbf{\Omega}^\bullet_{|\sim|' \leq n}$*-stratified trajectory space* $\mathcal{T}(v)$ *can be given the structure of Whitney space (residing in an Euclidean space).* ◊

The fundamental results of [Gor], [Jo], and [Ve], imply that Whiney spaces Y admit smooth triangulations $\tau : T \to Y$, amenable to their stratifications. The adjective "smooth" here refers to the homeomorphism τ being smooth on the interior of each simplex $\Delta \subset T$. With respect to such triangulations, the strata of Y are subcomplexes. Combining these facts with Theorem 2.3, leads to the following corollary, important for the Holography Theorems in the next section.

Corollary 2.2. *Let X be a $(n+1)$-dimensional compact connected smooth manifold, carrying a traversally generic vector field v. Then the following claims are valid:*

The space of trajectories $\mathcal{T}(v)$ admits the structure of finite cell/simplicial complex.

For each $\omega \in \mathbf{\Omega}^\bullet_{|\sim|' \leq n}$, the stratum $\mathcal{T}(v, \omega_{\succeq})$ is a codimension $|\omega|'$ subcomplex of $\mathcal{T}(v)$.

*With respect to an appropriate cellular/simplicial structure on X, the obvious map $\Gamma : X \to \mathcal{T}(v)$ is cellular/simplicial. It preserves the $(\mathbf{\Omega}^\bullet_{|\sim|' \leq n})$-strata of both spaces. As a result, Γ is a $(\mathbf{\Omega}^\bullet_{|\sim|' \leq n})$-*stratified *homotopy equivalence.* ◇

Remark 2.2. Most probably, $\mathcal{T}(v)$ is a compact CW-complex for any traversing and boundary generic (and not necessary *traversally* generic!) field v. However, for such vector fields, we do not have "open book" algebraic models (as in Fig. 6.4 in [K7]) for their interactions with boundary in the vicinity of a typical trajectory; so we do not know how to extend our arguments to a larger class of vector fields. ◇

Next we introduce one key construction from [K9] (see Fig. 2.4) which will turn to be very useful throughout our investigations.

Lemma 2.7. *For any traversing vector field v on X, there exists an embedding $\alpha : X \subset \mathcal{T}(v) \times \mathbb{R}$. In fact, any smooth Lyapunov function $f : X \to \mathbb{R}$ (i.e., $df(v) > 0$) generates such embedding $\alpha = \alpha(f, v)$ in a canonical fashion.*

For any smooth (in the sense of Definition 2.7) map $\beta : \mathcal{T}(v) \to \mathbb{R}^N$, the composite map

$$A(v, f) : X \xrightarrow{\alpha} \mathcal{T}(v) \times \mathbb{R} \xrightarrow{\beta \times id} \mathbb{R}^N \times \mathbb{R}$$

is smooth.

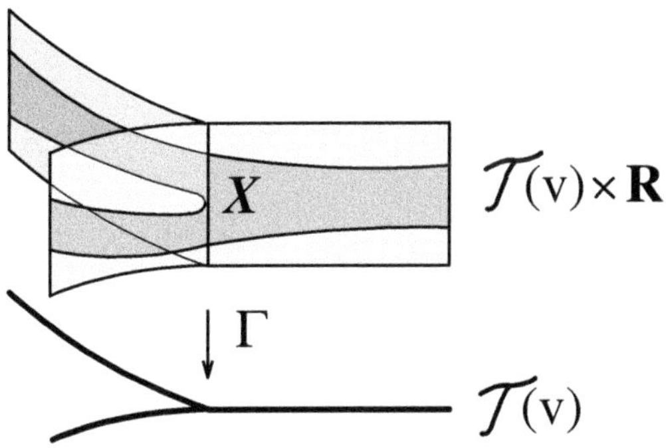

Fig. 2.4 The embedding $\alpha(f, v)$ of X into the product $\mathcal{T}(v) \times \mathbb{R}$.

Any two embeddings $\alpha(f_1, v)$ and $\alpha(f_2, v)$ are isotopic through homeomorphisms, provided that $df_1(v) > 0$, $df_2(v) > 0$.

Proof. Since f is strictly increasing along the v-trajectories, any point $x \in X$ is determined by the v-trajectory γ_x through x and the value $f(x)$. Therefore, x is determined by the point $\gamma_x \times f(x) \in \mathcal{T}(v) \times \mathbb{R}$. By the definition of topology in $\mathcal{T}(v)$, the correspondence $\alpha(f, v) : x \to \gamma_x \times f(x)$ is a continuous map.

In fact, $\alpha(f, v)$ is a smooth map in the spirit of Definition 2.8: more accurately, for any map $\beta : \mathcal{T}(v) \to \mathbb{R}^N$, given by N smooth functions on $\mathcal{T}(v)$, the composite map $A(v, f) : X \to \mathbb{R}^N \times \mathbb{R}$ is smooth. The verification of this fact is on the level of definitions.

For a fixed v, the condition $df(v) > 0$ defines an open convex cone $\mathcal{C}(v)$ in the space $C^\infty(X, \mathbb{R})$. Thus, f_1 and f_2 can be linked by a path in $\mathcal{C}(v)$, which results in $\alpha(f_1, v)$ and $\alpha(f_2, v)$ being homotopic through homeomorphisms. □

Remark 2.3. By examining Fig. 2.4, we notice an interesting phenomenon: the embedding $\alpha : X \subset \mathcal{T}(v) \times \mathbb{R}$ does not extend to an embedding of a larger manifold $\hat{X} \supset X$, where $\hat{X} \setminus X \approx \partial_1 X \times [0, \epsilon)$ and $\epsilon > 0$. In other words, $\alpha(\partial_1 X)$ has no outward "normal field" in the ambient $\mathcal{T}(v) \times \mathbb{R}$; in that sense, the image $\alpha(\partial_1 X)$ is rigid in $\mathcal{T}(v) \times \mathbb{R}$! ◇

Chapter 3

Causal Holography of Traversing Flows

"Mathematics are the result of mysterious powers which no one understands, and which the unconscious recognition of beauty must play an important part. Out of an infinity of designs a mathematician chooses one pattern for beauty's sake and pulls it down to earth."

Marston Morse

3.1 In this chapter...

In this chapter, we are preoccupied with the central for this book and already familiar from the Introduction question:

"For a traversing vector field v on a compact manifold X, what kind of v-generated residual structure on the boundary $\partial_1 X$ allows for a reconstruction of the pair (X, v) up to, say, a homeomorphism, or a piecewise differential homomorphism, or a diffeomorphism?"

If such a structure on the boundary is available, it deserves to be called holographic, since the information about the $(n+1)$-dimensional bulk X and the flow is recorded on a n-dimensional record (or rather on the appropriately linked pair of such records). We owe an apology to the fellow physicists: the name does not suggest a direct connection to the holography principles in the quantum field theory and the dual theories of gravitation.

For a traversing vector field v, with the dream of holography in mind, we introduce the causality map

$$C_v : \partial_1^+ X(v) \to \partial_1^- X(v)$$

that takes any point $x \in \partial_1^+ X(v)$ where the field is directed inwards, to the "next" along the trajectory γ_x point $C_v(x)$, where the field is directed outwards (see Fig. 1.3). We think of $x \in \partial_1 X$ as the cause and of $C_v(x) \in \partial_1 X$ as the effect. Thus, the name "causality map" (see Theorem 3.7 for a more convincing justification of the name).

In general, C_v is a discontinuous map, with a very particular type of discontinuity.

The causality map plays a role somewhat similar to the role played by the classical Poincaré return map: the continuous dynamics on X is reduced to a single map between lower-dimensional slices.

Let v_1 be a traversing and boundary generic vector field on a manifold X_1, and let v_2 be a traversing and boundary generic vector field on a manifold X_2, the dimensions of both manifolds being equal. We denote by $\mathcal{F}(v_i)$ $(i = 1, 2)$ the oriented 1-dimensional (singular) foliation of the manifold X_i, produced by the vector field v_i; hence, the leaves of $\mathcal{F}(v_i)$ are the v_i-trajectories.

Theorem 3.1 and Theorem 3.3—the main results of this chapter—claim, *roughly speaking*, that any piecewise-differential (PD for short) homeomorphism (respectively, diffeomorphism) $\Phi^\partial : \partial_1 X_1 \to \partial_1 X_2$ which commutes with the causality maps C_{v_1} and C_{v_2}, extends to a PD-homeomorphism (respectively, diffeomorphism) $\Phi : X_1 \to X_2$. Moreover, Φ takes each v_1-trajectory to a v_2-trajectory, thus mapping the v_1-oriented 1-dimensional foliation $\mathcal{F}(v_1)$ to the v_2-oriented foliation $\mathcal{F}(v_1)$.

In fact, the causality map C_v allows for a reconstruction of the pair $(X, \mathcal{F}(v))$, up to a PD-homeomorphism (respectively, up to a diffeomorphism).

Therefore, the PD/smooth topology of X and of the unparametrized v-flow are rigid for the given "boundary conditions" $C_v : \partial_1^+ X(v) \to \partial_1^- X(v)$.

For technical reasons, the hypotheses of the two main theorems are slightly different: in Theorem 3.1, the vector fields are assumed to be traversally generic, while in Theorem 3.3 they are only traversing and boundary generic.

This topological rigidity has a number of implication for general dynamical systems (see Theorem 3.7 and Corollary 3.8). We summarize them in the Holographic Causality Principle. Vaguely, it states that the causality relation on a generic event horizon in the space-time of a given dynamic system determines the compact portion of the event space, bounded by the event horizon (we call such portion "space-time capsule"), and the evolution of the system in it, up to a diffeomorphism/homeomorphism of that portion.

In Section 3.4, we introduce the concepts of proto-holographic and holographic structures on closed manifolds Y (see Definitions 3.3-3.5). These definitions postulate the existence of a special cover of Y by local charts which mimic the intersections of a $\hat{v}$-adjusted cylindrical neighborhoods of $\hat{v}$-trajectories in $\hat{X} \supset X$ with the boundary $\partial_1 X = Y$. Here v is a traversally generic vector field on a manifold X such that $\dim X = \dim Y + 1$.

In Theorem 3.8, we show that any traversally generic vector field v on X generates a holographic structure $\mathcal{H}(v)$ on the boundary ∂X, a structure which is cobordant to the empty set (see Definition 3.5). Conversely, any cobordant to the empty set holographic structure $\mathcal{H}$ on $\partial_1 X$ is generated by a traversally generic vector field v on X.

It turns out, that the language of holographic and cobordant to $\emptyset$ structures $\{\mathcal{H}\}$ on $\partial_1 X$ is equivalent to the language of causality maps $\{C_v : \partial_1^+ X(v) \to \partial_1^- X(v)\}$. Thus, $\mathcal{H}(v)$ allows for a reconstruction of the pair $(X, \mathcal{F}(v))$, up to a diffeomorphism.

We complete this section with few remarks about the holographic structures on homotopy spheres and their relation with the underling smooth exotic spheres.

In the following Section 3.5, we study the causality maps as measure-preserving transformations.

Section 3.6 is devoted to properties of, so called, proto-billiard maps and the dynamics they generate.

In Section 3.7 we investigate the v-harmonizing metrics defined for, so called, Calabi vector fields and the minimal foliations we associate with them.

Finally, Section 3.8 is devoted to Contact Holography. It revolves around the problem of recovering contact geometry from the appropriate boundary data. Because of its volume, this section has its own detailed introduction.

3.2 Holography theorems for traversing flows

Let v be a traversing and boundary generic vector field on compact connected smooth $(n+1)$-manifold X. We denote by γ_x the v-trajectory through a point $x \in X$. For a boundary generic and traversing v, the intrsection $\gamma_x \cap \partial_1 X$ is a finite set; moreover, by the compactness of the trajectory space $\mathcal{T}(v)$, we get that $\sup_{\gamma \in \mathcal{T}(v)} \{\#(\gamma \cap \partial_1 X)\} < \infty$.

Consider the causality map

$$C_v : \partial_1^+ X(v) \to \partial_1^- X(v).$$

that takes any point $x \in \partial_1^+ X(v)$ to the next point $y \in \gamma_x \cap \partial_1 X$, the intersection of the trajectory γ_x through x with the boundary $\partial_1 X =_{\mathsf{def}} \partial X$.

• We notice that, for any smooth positive function $\lambda : X \to \mathbb{R}_+$, we have $C_{\lambda \cdot v} = C_v$. Thus, the causality map depends only on the conformal class of a traversing vector field v.

Typically, the map C_v is *discontinuous* and this discontinuity captures some interesting topology of X! The bright spot is that C_v is semi-continuous relative to any nonsingular smooth function $f : X \to \mathbb{R}$ with the property $df(v) > 0$. This semi-continuity of C_v has the following description: for any $x \in \partial_1^+ X$ and $\epsilon > 0$, there is a neighborhood $U_\epsilon(x) \subset \partial_1 X$ such that

$$\begin{aligned} &f(C_v(y)) - f(y) \geq 0, \quad \text{and} \\ &f(C_v(y)) - f(y) > f(C_v(x)) - f(x) - \epsilon \;\text{ for all } y \in U_\epsilon(x). \end{aligned} \tag{3.1}$$

Note that $C_v(x) = x$ exactly when

$$x \in \partial_2^- X(v)^\circ \cup \partial_3^- X(v)^\circ \cup \cdots \cup \partial_{n+1}^- X(v).$$

We may adopt an alternative and more formal view of the map C_v. Note that the traversing v-flow on X defines a structure of a partially ordered set on $\partial_1 X$: we write $x \prec x'$, where $x, x' \in \partial_1 X$, if there is an ascending v-trajectory (not a singleton) that connects x to x'. We denote by $\mathcal{C}^\partial(v)$ this poset $(\partial_1 X, \prec)$. Evidently, $x \preceq x'$ if and only if x' is an image of x under a number of iterations of the causality map C_v, provided that v is boundary generic. Therefore, the poset $(\mathcal{C}^\partial(v), \succ)$ allows for a reconstruction of the causality map C_v for any traversing and boundary generic v.

Equivalently, we can consider the points of $\partial_1 X$ as elements of a category $\mathbf{Cat}^\partial(v)$. For any $x, x' \in \partial_1 X$, we define the set of morphisms $Mor(x, x')$ to be empty if no ascending v-trajectory in X connects x to x'. Otherwise, $Mor(x, x')$ consists of a single element—the portion of an ascending trajectory that starts at x and terminates at x'.

If $Mor(x, x') \neq \emptyset$ and $Mor(x', x'') \neq \emptyset$, then $Mor(x, x'') \neq \emptyset$. This defines the obvious composition map $Mor(x, x') \times Mor(x', x'') \to Mor(x, x'')$.

Remark 3.1. Note that formula (2.12) provides, among other things, for *local models of the causality maps* C_v, generated by traversally generic fields v. In the special coordinates (u, x, y) as in (2.12) and for a sufficiently small $\epsilon > 0$, the map C_v amounts to taking each root of the u-polynomial $\wp(u, x)$, residing in a maximal interval $I(x)$ where $\wp(u, x) \leq 0$, either to the next root residing in $I(x)$, or to itself (when $I(x)$ happens to be a singleton).

By Lemma 6.4 from [K7], this is a (discontinuous) map from the semi-algebraic set $\{\wp(u,x) = 0,\ \|x\| \leq \epsilon\}$ to itself. These observations will form a base for Definitions 3.3-3.5 from Section 3.4 that introduce the notion of a holographic structure on $\partial_1 X$. $\diamondsuit$

We will discuss *two kinds* of holography problems.

• The first kind is described by posing the question:

To what extend given boundary data are sufficient for reconstructing the unknown bulk and the traversing v-flow on it?

This question may be represented symbolically by the diagrams

Holographic Reconstruction Problem

$$(\partial_1 X,\ \succ_v,) \overset{??}{\rightsquigarrow} (X,\ \mathcal{F}(v)), \tag{3.2}$$

$$(\partial_1 X,\ \succ_v,\ f^\partial) \overset{??}{\rightsquigarrow} (X,\ \mathcal{F}(v),\ f), \tag{3.3}$$

where $\succ_v$ denotes the partial order on boundary, defined by the causality map C_v, f is a Lyapunov function for v, f^∂ is its restriction to the boundary $\partial_1 X$, and $\mathcal{F}(v)$ is the oriented 1-dimensional foliation on X, produced by v. The question sign "??" marks the trust of our quest.

Of course, the reconstruction cannot be absolutely rigid, but only up to a homeomorphism $\Phi : X \to X$ of certain kind (say, a PD-homeomorphism, or a smooth diffeomorphism) that respects the boundary data.

• The second kind of holography problems can be stated also as the question:

Given two manifolds, equipped with traversing flows, and a transformation Φ^∂ of their boundaries respecting the relevant boundary data, is it possible to extend Φ^∂ to a transformation Φ of the manifolds that respects the corresponding flows-generated structures in their interior?

These problems may be represented by the commutative diagrams:

• Holographic Extension Problem

$$\begin{array}{ccc} (\partial_1 X_1,\ \succ_{v_1}) & \xrightarrow{\mathsf{inc}} & (X_1,\ \mathcal{F}(v_1)) \\ \downarrow\ \Phi^\partial & & \downarrow ??\ \Phi \\ (\partial_1 X_2,\ \succ_{v_2}) & \xrightarrow{\mathsf{inc}} & (X_2,\ \mathcal{F}(v_2)) \end{array} \tag{3.4}$$

$$\begin{array}{ccc} (\partial_1 X_1,\ \succ_{v_1},\ f_1^\partial) & \xrightarrow{\mathsf{inc}} & (X_1,\ \mathcal{F}(v_1)) \\ \downarrow\ \Phi^\partial & & \downarrow ??\ \Phi \\ (\partial_1 X_2,\ \succ_{v_2},\ f_2^\partial) & \xrightarrow{\mathsf{inc}} & (X_2,\ \mathcal{F}(v_2),\ f_2), \end{array} \tag{3.5}$$

where inc stands for the inclusion.

The two types of holography problems come in a big variety of flavors, depending on the more or less rich boundary data and on the quality of the transformations Φ^∂ and Φ (homeomorphisms, PD-homeomorphisms, Hölder homeomorphisms with some control of the Hölder exponent, and diffeomorphisms of different order of smoothness).

3.2.1 *Piecewise differentiable holography*

Temporarily, we move away from the category smooth maps towards the category of piecewise differentiable ("PD" for short) maps.

Definition 3.1. Let v be a traversing vector field on a compact smooth manifold X.

We say that a triangulation T^∂ of the boundary $\partial_1 X$ is invariant under the causality map $C_v : \partial_1^+ X \to \partial_1^- X$, if the interior of each simplex from T^∂ is mapped homeomorphically by C_v onto the interior of a simplex[1]. $\diamondsuit$

Recall that by Theorem 2.3, for a traversally generic vector field v on a compact connected smooth $(n+1)$-dimensional manifold X with boundary, the $(\mathbf{\Omega}^\bullet_{|\sim|' \leq n})$-stratified trajectory space $\mathcal{T}(v)$ has the structure of a Whitney space (see Definition 2.9). As a result, $\mathcal{T}(v)$ admits a finite triangulation (see [K7], Theorem 6.10, and [K8]). Combining this fact with Corollary 2.2, we get the following claim.

Proposition 3.1. *For a traversally generic vector field v on a compact smooth manifold X, the boundary $\partial_1 X$ admits a C_v-invariant smooth triangulation.*

Proof. For boundary generic fields v, the map $\Gamma : \partial_1 X \to \mathcal{T}(v)$ is finitely ramified and onto. For a traversally generic v, the claim follows from Corollary 2.2 which asserts the existence of a triangulation T of the trajectory space $\mathcal{T}(v)$, consistent with its $\mathbf{\Omega}^\bullet$-stratification. Any such triangulation T of $\mathcal{T}(v)$, with the help of Γ^{-1}, lifts to a triangulation T^∂ of $\partial_1 X$. Indeed, for each $\omega \in \mathbf{\Omega}^\bullet$, by Corollary 7.2 from [K7], the smooth submersion

$$\Gamma :\ \Gamma^{-1}(\mathcal{T}(v, \omega)) \cap \partial_1 X \to \mathcal{T}(v, \omega)$$

is a trivial covering, whose fiber is a finite set. By its very construction, the triangulation T^∂ is C_v-invariant. $\square$

[1] Remember, C_v is typically a discontinuous map!

Remark 3.2. The existence of a triangulation on $\partial_1 X$ by itself does not imply the existence of a triangulation on the trajectory space $\mathcal{T}(v)$: in fact, there are smooth manifolds that can serve as finite covering spaces over topological manifold bases that do not admit *any* triangulation! For example, the standard sphere may be a 2-fold cover space of a non-triangulable fake real projective space (see [CS]). This is why Corollary 3.1 is nontrivial. See [Gor] for general treatment of triangulations on stratified spaces. $\diamondsuit$

In the proofs of the main holography Theorems 3.1 and 3.3, we will rely on the following lemma. Speaking informally, it claims that points $x \in \partial_j^{\pm} X^\circ(v)$ are recognizable by the *localized* combinatorics of tangencies of the v-trajectories in their vicinity.

Lemma 3.1. *For a traversing boundary generic vector field v, the combinatorial type $\omega = \omega(\gamma) \in \mathbf{\Omega}^\bullet$ of each v-trajectory $\gamma \subset X$ can be recovered from the causality map $C_v : \partial_1^+ X \to \partial_1^- X$ in the vicinity of the set $\gamma \cap \partial_1 X$.*

Proof. For each point $y \in \partial_1 X$, its multiplicity $m(y)$ with respect to a boundary generic flow v can be detected by the unique minimal (pure) Morse stratum $\partial_j X^\circ =_{\mathsf{def}} \partial_j X^\circ(v)$, $j = m(y)$, to which y belongs.

On the other hand, it can be also detected in terms of the causality map C_v and its iterations, restricted to the vicinity of y in $\partial_1 X$. Let us justify this key observation. Recall that, for boundary generic vector fields, Lemmas 2.3 and 2.4 and formula (2.12) provide us with models for the divisors $\{D_{\hat\gamma}\}_{\hat\gamma}$, localized to a sufficiently small neighborhood U_y of y in X. Recall that the set $\hat\gamma \cap \partial_1 X \cap U_y$ is the support of $D_{\hat\gamma}|_{U_y}$. We choose the $\hat v$-flow adjusted neighborhood U_y with some care: first we add a narrow collar to X to form a larger open manifold $\hat X \supset X$ and extend the vector field v to a traversing vector field $\hat v$ in $\hat X$ as shown in Fig. 3.2. Then we chose a small smooth transversal section $S \subset \hat X$ of the $\hat v$-flow, which contains y, and consider the union V_y of $\hat v$-trajectories through the points of S. Finally, we let $U_y =_{\mathsf{def}} V_y \cap X$.

In the appropriate coordinates $(u, x) : V_y \to \mathbb{R} \times \mathbb{R}^n$, the boundary $\partial_1 X$ is given by an equation $\{F(u, x) = 0\}$, where a smooth function F has 0 as its regular value. Then the $\hat v$-trajectory $\hat\gamma$ is given by the equation $\{x = 0\}$ and the v-trajectory γ by $\{F(u, x) \leq 0,\ x = 0\}$.

Since v is boundary generic, each point $y = (u_\star, 0) \in \gamma \cap \partial X$ has multiplicity $m(y) \leq \dim(X)$. Thus, $F(u, 0)$ has a zero at $u_\star$ of multiplicity $m(y) \leq \dim(X)$. By the Taylor formula, this implies that, in the vicinity of $u_\star$, any smooth function $g(u)$ that is sufficiently close to $F(u, 0)$ in the C^∞-

topology has finitely many zeros of finite multiplicities that add up to $m(y)$ at most. Moreover, in the vicinity of $u_\star$, $g(u) = P(u)\cdot Q(u)$, where $P(u)$ is a real polynomial of degree $m(y)$ and $Q(u) > 0$ is a smooth function.

Thus, for any trajectory $\gamma' = \{x = x'\}$ in the vicinity of γ (for any x' sufficiently close to 0), the intersection $\gamma' \cap \partial_1 X$ is given by the equation, $\{F(u, x') = 0\}$, and the zero divisor $D_{\gamma'}$, associated with γ', coincides with the zero divisor $D_{\mathbb{R}}(\tilde{P})$ of a real polynomial $\tilde{P}$ of degree $|\omega(\gamma')| = \sum_{y_\star \in \gamma' \cap \partial_1 X} m(y_\star)$, where $\deg(D_{\mathbb{R}}(\tilde{P})) \equiv |\omega(\gamma')| \mod 2$.

Using these algebraic models, the *maximal length of a chain*

$$\{z_1 \rightsquigarrow z_2 = C_v(z_1) \rightsquigarrow z_3 = C_v(z_2) \rightsquigarrow \dots\}$$

in a sufficiently small v-adjusted neighborhood $U_y \subset \partial_1 X$ of y is $\lceil m(y)/2 \rceil$, where $\lceil \sim \rceil$ denotes the integral part of a positive number. Indeed, if $m(y)$ is even, then the maximal number of roots of even multiplicity for a polynomial of degree $m(y)$ is $m(y)/2$, and by (2.10) such polynomials

$$u^{m(y)} + \sum_{j=0}^{m(y)-2} x_j u^j$$

are present in an arbitrary small neighborhood of the polynomial $u^{m(y)}$ in the coefficient space. When $m(y)$ is odd, then the maximal length of a chain $\{z_1 \rightsquigarrow z_2 \rightsquigarrow \dots\}$ in the vicinity of y in $\partial_1 X$ is $(m(y)-1)/2 = \lceil m(y)/2 \rceil$. It corresponds either to the $m(y)$-polynomials with one simple root, followed by the maximal number of multiplicity 2 roots (when y is the first in its localized chain), or to the $m(y)$-polynomials with the maximal number of multiplicity 2 roots, followed by a simple root (when y is the last in its localized chain).

Evidently, the order in which the points $\gamma \cap \partial_1 X$ appear along each trajectory γ is also determined by C_v. Therefore, the combinatorial type $\omega(\gamma) \in \mathbf{\Omega}^\bullet$ of each v-trajectory $\gamma \subset X$ can be recovered from the causality map $C_v : \partial_1^+ X \to \partial_1^- X$ in the vicinity of γ and its partially-defined iterations. Thus, the information encoded in C_v is sufficient for a reconstruction of the $\mathbf{\Omega}^\bullet$-stratified trajectory space $\mathcal{T}(v)$. □

The holography theorem below claims that the poset $\mathcal{C}^\partial(v)$, as well as the category $\mathbf{Cat}^\partial(v)$, allows for a reconstruction of the manifold X, together with the unparameterized v-flow, up to a PD-homeomorphism. Unfortunately, the proof is lengthy, so the reader who is more interested in the smooth version of this theorem and its geometric applications may jump directly to Theorem 3.3.

Theorem 3.1. (The PD-Holography Theorem)
Let X_1, X_2 be two smooth compact connected $(n+1)$-manifolds with boundary, equipped with traversing and boundary generic vector fields v_1, v_2, respectively.

For $i = 1, 2$, let T_i^∂ be a smooth triangulation of the boundary $\partial_1 X_i$, invariant under the causality map C_{v_i} in the sense of Definition 3.1.[2]

Then any PD*-homeomorphism $\Phi^\partial : (\partial_1 X_1, \mathsf{T}_1^\partial) \to (\partial_1 X_2, \mathsf{T}_2^\partial)$, such that*

$$\Phi^\partial \circ C_{v_1} = C_{v_2} \circ \Phi^\partial,$$

extends to a PD*-homeomorphism $\Phi : X_1 \to X_2$ which takes the unparametrized v_1-trajectories to the unparametrized v_2-trajectories so that the v_i-induced orientations of trajectories are preserved.*

Proof. Since each v_i is a traversing vector field, any trajectory reaches the boundary; so the obvious maps $\Gamma_i^\partial : \partial_1 X_i \to \mathcal{T}(v_i)$, $i = 1, 2$, are onto.

Evidently, the fiber of the map Γ_i^∂ consists of the maximal chain of points $\{x_1 \rightsquigarrow x_2 \rightsquigarrow \dots \rightsquigarrow x_q\}$ from $\partial_1 X_i$ such that $C_{v_i}(x_j) = x_{j+1}$ for all $j \in [1, q-1]$. By the definition of C_{v_i}, such chain is exactly the ordered locus $\gamma_{x_1} \cap \partial_1 X_i$. Therefore, using that $\Phi^\partial : \partial_1 X_1 \to \partial_1 X_2$ commutes with the causality maps C_{v_i}, we see that Φ^∂ gives rise to a well-defined continuous map $\Phi^{\mathcal{T}} : \mathcal{T}(v_1) \to \mathcal{T}(v_2)$ of the trajectory spaces.

By Lemma 3.1, the PD-homemorphism $\Phi^\partial : \partial_1 X_1 \to \partial_1 X_2$ that commutes with the causality maps C_{v_1} and C_{v_2} must take any chain

$$\{z_1 \rightsquigarrow z_2 = C_{v_1}(z_1) \rightsquigarrow z_3 = C_{v_1}(z_2) \rightsquigarrow \dots\}$$

in $\partial_1 X_1$ to a similar chain in $\partial_1 X_2$ with the same multiplicity pattern.

Let $i = 1, 2$. Since the triangulation T_i^∂ is invariant under the causality map C_{v_i}, it gives rise to a triangulation T_i of the trajectory space $\mathcal{T}(v_i)$. This triangulation T_i is consistent with its $\mathbf{\Omega}^\bullet$-stratification of $\mathcal{T}(v_i)$. Therefore, Φ^∂ gives rise to the PD-homeomorphism

$$\Phi^{\mathcal{T}} : \mathcal{T}(v_1) \to \mathcal{T}(v_2)$$

which preserves the $\mathbf{\Omega}^\bullet$-stratifications of the trajectory spaces.

Next, we pick two smooth Lyapunov functions $\{f_i : X_i \to \mathbb{R}\}_{i=1,2}$ (thus $df_i(v_i) > 0$) and consider the embeddings

$$\{\alpha_i =_{\mathsf{def}} \alpha_i(f_i, v_i) : X_i \subset \mathcal{T}(v_i) \times \mathbb{R}\}_{i=1,2}$$

[2]We stress that if v_1, v_2 are traversally generic, then, by Proposition 3.1, such triangulations $\mathsf{T}_1^\partial, \mathsf{T}_2^\partial$ do exist.

as in Lemma 2.7 and Fig. 2.4. It is given by the formula

$$\alpha_i(x) =_{\mathsf{def}} (\Gamma_i(x), f_i(x)).$$

We would like to lift the newly constructed PD-homeomorphism $\Phi^{\mathcal{T}} : \mathcal{T}(v_1) \to \mathcal{T}(v_2)$ to a PD-homeomorphism

$$\tilde{\Phi} : \mathcal{T}(v_1) \times \mathbb{R} \to \mathcal{T}(v_2) \times \mathbb{R}$$

that takes $\alpha_1(X_1)$ to $\alpha_2(X_2)$ and has the property $\tilde{\Phi} \circ \alpha_1|_{\partial_1 X_1} = \alpha_2 \circ \Phi^{\partial}$.

• We repeatedly will abuse the notations by denoting the v-trajectories $\gamma \subset X$ and the points $\Gamma(\gamma) \in \mathcal{T}(v)$ in the trajectory space they represent by the same symbol "γ".

Put

$$\tilde{\Phi}(\gamma, t) =_{\mathsf{def}} \big(\Phi^{\mathcal{T}}(\gamma),\ \tilde{\phi}(\gamma)(t)\big), \tag{3.6}$$

where $\gamma \in \mathcal{T}(v_1)$, $t \in \mathbb{R}$, and $\tilde{\phi}(\gamma) : \mathbb{R} \to \mathbb{R}$ is an orientation-preserving PL-homeomorphism *to be constructed.*

Since $\Phi^{\mathcal{T}}$ preserves the $\mathbf{\Omega}^{\bullet}$-stratifications, the combinatorial types $\omega(\gamma)$ and $\omega(\Phi^{\mathcal{T}}(\gamma))$ of the divisors $\mathcal{D}_\gamma$ and $\mathcal{D}_{\Phi^{\mathcal{T}}(\gamma)}$ are the same.

There is a natural choice for an orientation-preserving PL-homeomorphism $\phi : \mathbb{R} \to \mathbb{R}$ that maps the ordered set $D^{(1)}_\gamma =_{\mathsf{def}} \alpha_1(\gamma \cap \partial_1 X_1)$ to the ordered set $D^{(2)}_\gamma =_{\mathsf{def}} \alpha_2(\Phi^{\mathcal{T}}(\gamma) \cap \partial_1 X_2)$.

Let $q = |\sup(\omega)| = |\omega| - |\omega|'$. There is a unique increasing PL-function $\phi : \mathbb{R} \to \mathbb{R}$ that takes the k-th point $x_k \in \alpha_1(\gamma \cap \partial_1 X_1)$ to the k-th point $y_k \in \alpha_2(\Phi^{\mathcal{T}}(\gamma) \cap \partial_1 X_2)$, $1 \leq k \leq q$, such that ϕ is a linear polynomial in each of the intervals

$$(-\infty, x_1], [x_1, x_2], \dots, [x_{q-1}, x_q], [x_q, +\infty),$$

and ϕ has slope 1 in the intervals $(-\infty, x_1) \cup (x_q, +\infty)$. We will denote such ϕ by $\phi_{\vec{x},\vec{y}}$ (see Fig. 3.1, diagram A).

Let $\mathsf{Homeo}^{\mathsf{PL}}_+(\mathbb{R})$ denote the group of orientation-preserving piecewise linear homeomorphisms of the number line $\mathbb{R}$ that are smooth away from a finite set of points. Each element of the group is represented by a monotone PL-function. Therefore, $\mathsf{Homeo}^{\mathsf{PL}}_+(\mathbb{R})$ forms a convex cone in the vector space of all PL-functions on $\mathbb{R}$.

We view $\vec{x} =_{\mathsf{def}} (x_1 < x_2 < \dots < x_q)$ and $\vec{y} =_{\mathsf{def}} (y_1 < y_2 < \dots < y_q)$ as points of $(Sym^q\mathbb{R})^\circ$, the configuration space of q *distinct* and thus ordered points in $\mathbb{R}$.

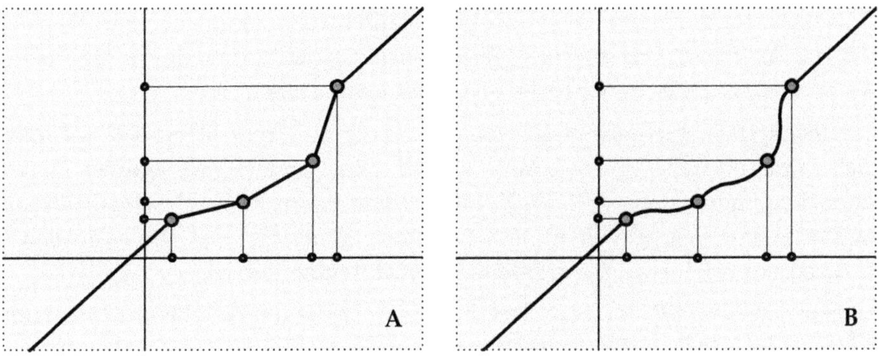

Fig. 3.1 The graphs of PL (diagram A) and smooth (diagram B) interpolating homeomorphisms $\phi_{\vec{x},\vec{y}} : \mathbb{R} \to \mathbb{R}$. In the figures, the vectors $\vec{x}, \vec{y} \in \mathbb{R}^4$.

Evidently, the construction of the function $\phi_{\vec{x},\vec{y}}$ gives rise to a continuous map

$$\phi_q : (Sym^q\mathbb{R})^\circ \times (Sym^q\mathbb{R})^\circ \to \mathsf{Homeo}_+^{\mathsf{PL}}(\mathbb{R}). \tag{3.7}$$

Unfortunately, the map ϕ_d from (3.7) is not "well-behaved" under the merge operations (see formula (2.2)) and especially under insert operations (see formula (2.3)), being applied to of divisors in $\mathbb{R}$. Thus the difficulty shifts towards the problem of matching the maps $\{\phi_q\}$ for different q's of the same parity. Although, thanks to the contractibility of $\mathsf{Homeo}_+^{\mathsf{PL}}(\mathbb{R})$, this problem is solvable, it requires some effort.

For a given finite closed subposet (an ideal) $\Theta \subset \mathbf{\Omega}^\bullet$, we denote by $X_i(v_i, \Theta)$ (where $i = 1, 2$) the union of trajectories in X_i whose combinatorial types belong to Θ. We denote by $\mathcal{T}(v_i, \Theta)$ the Γ_i-image of $X_i(v_i, \Theta)$ under the obvious map $\Gamma_i : X_i \to \mathcal{T}(v_i)$.

The inductive argument, aimed at proving the existence of the extension $\tilde{\Phi}$ as in (3.6), proceeds as follows. We assume that for a finite *closed* subposet $\Theta \subset \mathbf{\Omega}^\bullet$ we already managed to construct a PL-homeomorphism

$$\tilde{\Phi} : X_1(v_1, \Theta) \to X_2(v_2, \Theta)$$

as in (3.6) which covers the homeomorphism $\Phi^{\mathcal{T}} : \mathcal{T}(v_1, \Theta) \to \mathcal{T}(v_2, \Theta)$ and has the property

$$\tilde{\Phi} \circ \alpha_1\big|_{\partial_1 X_1(v_1,\Theta)} = \alpha_2 \circ \Phi^{\partial}\big|_{\partial_1 X_1(v_1,\Theta)}.$$

Let $\omega \in \mathbf{\Omega}^\bullet \setminus \Theta$ be an element such that $\omega_\succ$—the set of elements smaller than ω—is contained in Θ. Then $\omega_\succeq = \{\omega\} \cup \omega_\succ$.

The induction step $\Theta \Rightarrow \omega \cup \Theta$ requires to extend the map $\tilde{\Phi}$ over the set $X_1(v_1, \omega \cup \Theta)$ so that the extension would coincide on the locus $X_1(v_1, \omega \cup \Theta) \cap \partial_1 X_1$ with the given homeomorphism Φ^∂.

Recall that, for traversally generic v_1, v_2, by Proposition 3.1, we may pick triangulations of trajectory spaces $\mathcal{T}(v_i)$, which lift against Γ_i to smooth triangulations T_i^∂ of $\partial_1 X_i$ and are consistent with $\mathbf{\Omega}^\bullet$-stratifications in $\mathcal{T}(v_i)$. For just traversing and boundary generic v_1, v_2, the hypotheses of Theorem 3.1 insure the existence of such triangulations.

For each $\omega \in \mathbf{\Omega}^\bullet$, we consider the closure $\mathcal{T}(v_1, \omega_{\succeq})$ of the pure stratum $\mathcal{T}(v_1, \omega)$ in $\mathcal{T}(v_1)$ and the closed set

$$\mathcal{T}(v_1, \omega_{\succ}) =_{\mathsf{def}} \mathcal{T}(v_1, \omega_{\succeq}) \setminus \mathcal{T}(v_1, \omega).$$

Employing the Lyapunov functions f_1 and f_2 of v_1 and v_2, let

$$D_\gamma^{(1)} =_{\mathsf{def}} f_1(\gamma \cap \partial_1 X_1) \subset \mathbb{R} \quad \text{and} \quad D_\gamma^{(2)} =_{\mathsf{def}} f_2(\Phi^{\mathcal{T}}(\gamma) \cap \partial_1 X_2) \subset \mathbb{R}.$$

Consider the map $\tilde{\Phi} : \mathcal{T}(v_1) \times \mathbb{R} \to \mathcal{T}(v_2) \times \mathbb{R}$ defined by the formula

$$\tilde{\Phi}(\gamma,\, t) =_{\mathsf{def}} \big(\Phi^{\mathcal{T}}(\gamma),\;\; \phi\big(D_\gamma^{(1)},\, D_\gamma^{(2)}\big)(t)\big), \tag{3.8}$$

where $t \in \mathbb{R}$ (here we suppress the dependence of $\phi \in \mathsf{Hom}^{\mathsf{PL}}{}_+(\mathbb{R})$ on the cardinality q of the set $D_\gamma^{(1)}$).

Since, as has been described above, $\Phi^{\mathcal{T}}$ is determined by Φ^∂, we get the desired property $\alpha_2 \circ \Phi^\partial|_{\partial_1 X_1} = \tilde{\Phi} \circ \alpha_1|_{\partial_1 X_1}$. Unfortunately, the map $\tilde{\Phi}$ is *discontinuous*!

Following the induction step $\Theta \Rightarrow \omega \cup \Theta$, we will gradually repair the map $\tilde{\Phi}(\gamma,\, t)$ from (3.8) to insure its continuity. This will be accomplished by *two consecutive corrections.*

Of course, for Θ that is formed by the *minimal* combinatorial types of trajectories from X_1, the formula (3.8) delivers a PL-homomorphism

$$\tilde{\Phi}| : \alpha_1(X_1(v_1, \Theta)) \to \alpha_2(X_2(v_2, \Theta))$$

such that $\alpha_2 \circ \Phi^\partial = \tilde{\Phi}| \circ \alpha_1$ over $\partial_1 X_1 \cap X_1(v_1, \Theta)$. For such γ's, there is no need to modify the homeomorphism $\phi\big(D_\gamma^{(1)},\, D_\gamma^{(2)}\big)$. Thus, we have validated the basis of the induction.

Next, by the inductive assumption, suppose that the desired matching homeomorphisms $\tilde{\phi}(\gamma) \in \mathsf{Homeo}_+^{\mathsf{PL}}(\mathbb{R})$, such that

$$\alpha_2 \circ \Phi(x) = \tilde{\Phi} \circ \alpha_1(x) \;\; \text{and} \;\; \tilde{\Phi}(\gamma_x, t) =_{\mathsf{def}} (\Phi^{\mathcal{T}}(\gamma_x),\, \tilde{\phi}(\gamma_x)(t)),$$

has been constructed for all $x \in X_1(v_1, \omega_{\succ})$.

Let U_ω be a regular neighborhood of the closed simplicial subcomplex $\mathcal{T}(v_i, \omega_\succ)$ in $\mathcal{T}(v_i, \omega_\succeq)$ with the properties as in Lemma 3.2 below. In particular, the PL-retraction $p_\omega : U_\omega \to \mathcal{T}(v, \omega_\succ)$ from that lemma is available.

Put $\mathsf{T}(v_i, \omega) =_{\mathsf{def}} \mathcal{T}(v_i, \omega) \setminus U_\omega$.

• The first correction. Over $\mathsf{T}(v_1, \omega)$, we keep the continuous family of PL-homeomorphisms $\phi\big(D^{(1)}_\gamma, D^{(2)}_\gamma\big)$. In the neighborhood U_ω, for each point $\gamma \in \partial U_\omega$, we linearly interpolate between the function $\phi\big(D^{(1)}_\gamma, D^{(2)}_\gamma\big)$ and previously constructed function $\tilde{\phi}\big(D^{(1)}_{\gamma'}, D^{(2)}_{\gamma'}\big)$. Here the point-trajectory $\gamma' = p_\omega(\gamma)$, where the retraction $p_\omega : U_\omega \to \mathcal{T}(v, \omega_\succ)$ is as in Lemma 3.2. The interpolation uses the product structure in $U_\omega \setminus \mathcal{T}(v_1, \omega_\succ)$, facilitated by the auxiliary PL vector field w as in Lemma 3.2. We connect a typical point $\gamma \in \partial U_\omega$ with the unique point $p_\omega(\gamma) \in \mathcal{T}(v_1, \omega')$, where $\omega' \prec \omega$, by an arc $\{\eta_\gamma(s)\}_{s \in [0,1]} \subset U_\omega$ that integrates the vector field w so that $\eta_\gamma(0) = \gamma$, and $\eta_\gamma(1) = p_\omega(\gamma)$.

We define the new s-family $\varphi_{\eta_\gamma(s)} \in \mathsf{Hom}^{\mathsf{PL}}_+(\mathbb{R})$ by the linear homotopy formula

$$\varphi_{\eta_\gamma(s)} =_{\mathsf{def}} (1-s)\cdot \phi\big(D^{(1)}_\gamma,\, D^{(2)}_\gamma\big) \,+\, s\cdot \tilde{\phi}\big(D^{(1)}_{p_\omega(\gamma)},\, D^{(2)}_{p_\omega(\gamma)}\big). \tag{3.9}$$

Note that $\varphi_{\eta_\gamma(s)}$ is *continuous* in $\gamma \in \partial U_\omega$ and $s \in [0,1]$. Here we rely on the fact that a linear homotopy between monotonically increasing PL-functions on $\mathbb{R}$ is realized through monotonically increasing PL-functions.

For $s \in (0,1)$, the graph of the function $\varphi_{\eta_\gamma(s)}$, acquires singular (i.e., non-smooth) points from both sets, $D^{(1)}_\gamma$ and $D^{(1)}_{p_\omega(\gamma)}$, residing in $\mathbb{R}$. As a result, a second correction is needed.

• The second correction. When $s \in (0,1)$, the map $\varphi_{\eta_\gamma(s)}$ still does not take the set

$$D^{(1)}_{\eta_\gamma(s)} =_{\mathsf{def}} \alpha_1\big(\eta_\gamma(s) \cap \partial_1 X_1\big)$$

onto the set

$$D^{(2)}_{\Phi^{\mathcal{T}}(\eta_\gamma(s))} =_{\mathsf{def}} \alpha_2\big(\Phi^{\mathcal{T}}(\eta_\gamma(s)) \cap \partial_1 X_2\big),$$

(although the orders in which the $(\alpha_2 \circ \Phi^\partial \circ \alpha_1^{-1})$-corresponding elements of two sets, $\varphi_{\eta_\gamma(s)}\big(D^{(1)}_{\eta_\gamma(s)}\big)$ and $D^{(2)}_{\Phi^{\mathcal{T}}(\eta_\gamma(s))}$, appear are similar: by choosing the neighborhood U_ω small enough, we can insure that the distance between these two sets in $\mathbb{R}$ is uniformly small for all w-trajectories η_γ in U_ω.)

To compensate for this failure, consider an additional correction:

$$\xi_{\eta_\gamma(s)} =_{\mathsf{def}} \phi\Big(\varphi_{\eta_\gamma(s)}\big(D^{(1)}_{\eta_\gamma(s)}\big),\; D^{(2)}_{\Phi^{\mathcal{T}}(\eta_\gamma(s))}\Big) \in \mathsf{Homeo}^{\mathsf{PL}}_+(\mathbb{R}), \tag{3.10}$$

where ϕ denotes the canonical PL-homeomorphism of $\mathbb{R}$ as in (3.7). It takes the ordered set $\varphi_{\eta_\gamma(s)}\big(D^{(1)}_{\eta_\gamma(s)}\big)$ to the ordered set $D^{(2)}_{\Phi^{\mathcal{T}}(\eta_\gamma(s))}$.

Now the composition

$$\chi_{\eta_\gamma(s)} =_{\mathsf{def}} \xi_{\eta_\gamma(s)} \circ \varphi_{\eta_\gamma(s)} \in \mathsf{Homeo}^{\mathsf{PL}}_+(\mathbb{R})$$

does the job: it takes takes $D^{(1)}_{\eta_\gamma(s)}$ bijectively to $D^{(2)}_{\Phi^{\mathcal{T}}(\eta_\gamma(s))}$ and is continuous in $\gamma \in \partial U_\omega$ and $s \in [0,1]$. Indeed, chasing the definitions, we get $\xi_{\eta_\gamma(0)} = \mathsf{id} = \xi_{\eta_\gamma(1)}$, so that

$$\chi_{\eta_\gamma(0)} = \phi(D^{(1)}_\gamma, D^{(2)}_\gamma) \text{ and } \chi_{\eta_\gamma(1)} = \tilde{\phi}(D^{(1)}_{p_\omega(\gamma)}, D^{(2)}_{p_\omega(\gamma)}).$$

As $s \to 1$, each point $x(s)$ from the set $\alpha_1(\eta_\gamma(s) \cap \partial_1 X_1)$ approaches a unique point $x(1)$ from the set $\alpha_1(p_\omega(\gamma) \cap \partial_1 X_1)$ (see Lemma 7.5 from [K7]). At the same time, the s-family of intervals $\{\alpha_1(\eta_\gamma(s))\}$ converges to a closed interval $\beta(\eta_\gamma(1))$, bounded by a pair of points from the set $\alpha_1(p_\omega(\gamma))$. By the definition of $\varphi_{\eta_\gamma(s)}$, we get $\lim_{s\to 1} \varphi_{\eta_\gamma(s)} = \varphi_{\eta_\gamma(1)}$. Moreover, since $D^{(1)}_{\eta_\gamma(s)} \subset \mathbb{R}$ and $D^{(2)}_{\Phi^{\mathcal{T}}(\eta_\gamma(s))} \subset \mathbb{R}$ depend continuously on $s \in [0,1]$ and on $\gamma \in \partial U_\omega$, the restrictions of the functions $\xi_{\eta_\gamma(s)}$ to the intervals $\alpha_1(\eta_\gamma(s))$ converge to the restriction of $\xi_{\eta_\gamma(1)} = \mathsf{id}$ to the interval $\beta(\eta_\gamma(1))$. As a result,

$$\lim_{s\to 1} \chi_{\eta_\gamma(s)}\big|_{\beta(\eta_\gamma(1))} = \chi_{\eta_\gamma(1)}\big|_{\beta(\eta_\gamma(1))}.$$

Due to the PL-nature of the field w in U_ω (and thus of the retraction $p_\omega : U_\omega \to \partial\mathcal{T}(v_1,\omega)$), the family maps $\{\chi_{\eta_\gamma(s)}\}_{\gamma\in\partial U_\omega,\, s\in[0,1]}$ helps to extend the piecewise differential homeomorphism

$$\tilde{\Phi} : \alpha_1(X_1(v_1, \omega_\succ)) \to \alpha_2(X_2(v_2, \omega_\succ))$$

to a piecewise differential homeomorphism

$$\tilde{\Phi} : \alpha_1(X_1(v_1, \omega_\succeq)) \to \alpha_2(X_2(v_2, \omega_\succeq)).$$

Thus, we have validated the induction step $\Theta \Rightarrow \omega \cup \Theta$ in the construction of the desired PD-homeomorphism $\Phi =_{\mathsf{def}} \alpha_2^{-1} \circ \tilde{\Phi} \circ \alpha_1 : X_1 \to X_2$. □

Lemma 3.2. *For each $\omega \in \mathbf{\Omega}^\bullet$ and a traversally generic vector field v on X, there exits a regular neighborhood U_ω of the closed subspace $\mathcal{T}(v, \omega_\succ)$ in the space $\mathcal{T}(v, \omega_\succeq)$ and a piecewise-differential vector field w in U_ω, subject to the following properties:*

(1) $w|_{\mathcal{T}(v,\omega_\succ)} = 0$,
(2) w does not vanish in $U_\omega \setminus \mathcal{T}(v, \omega_\succ)$,

(3) w is inward transversal to the boundary ∂U_ω,[3]

(4) the projection $p_\omega : U_\omega \to \mathcal{T}(v, \omega_\succ)$, defined by the continuous w-trajectories, is a PL-map. It helps to identify U_ω with the mapping cylinder of $p_\omega : \partial U_\omega \to \mathcal{T}(v, \omega_\succ)$.

Proof. Using Proposition 3.1, we take a triangulation T of $\mathcal{T}(v)$, consistent with its $\mathbf{\Omega}^\bullet$-stratification. Consider the set of all simplices Δ in the second barycentric subdivision $\beta^2\mathsf{T}$ of $\mathcal{T}(v, \omega_\succeq)$ that share points with $\mathcal{T}(v, \omega_\succ)$. Each such simplex Δ is a join of a maximal simplex $\Delta' \subset \mathcal{T}(v, \omega_\succ)$ and the maximal subsimplex $\Delta'' \subset \Delta$ that does not intersect $\mathcal{T}(v, \omega_\succ)$. For any line segment $[x, y] \subset \Delta$ which connects $x \in \Delta'$ with $y \in \Delta''$, consider a subsegment $[x, \tilde{y}] \subset [x, y]$ where $\tilde{y}$ is the midpoint of $[x, y]$. Denote by $\tilde{\Delta} \subset \Delta$ the polyhedron, formed by all such segments $[x, \tilde{y}]$.

Now define U_ω as the union of all the polyhedra $\tilde{\Delta}$ as above. This choice of U_ω implies that $U_\omega \setminus \mathcal{T}(v, \omega_\succ) \approx \partial U_\omega \times [0, 1)$, where the boundary ∂U_ω is a codimension one PL-submanifold of $\mathcal{T}(v, \omega)$. Indeed, for each simplex Δ as above, consider a standard vector field w_Δ, consistent with the join structure $\mathsf{join}(\Delta', \Delta'')$ in Δ. It points from Δ'' towards Δ', vanishes only on $\Delta' \sqcup \Delta''$, and is transversal to the face $\delta\tilde{\Delta} =_{\mathsf{def}} \tilde{\Delta} \cap \overline{(\Delta \setminus \tilde{\Delta})} \subset \partial U_\omega$. In the barycentric coordinates β_0, β_1 on the segment $[x, y]$, the magnitude of w_Δ is prescribed by the function $(1 - \beta_0)(1 - \beta_1)$.

This local construction of w_Δ gives rise to a global PL-vector field w in U_ω. In particular, w is inward transversal to ∂U_ω, vanishes on $\mathcal{T}(v, \omega_\succ)$, and the w-trajectories $\{\eta_x(t)\}_x$ define a PL-retraction $p_\omega : U_\omega \to \mathcal{T}(v, \omega_\succ)$ via a formula $p_\omega(x) = lim_{t \to \infty} \eta_x(t)$.

The properties of w help to identify U_ω with the mapping cylinder of $p_\omega : \partial U_\omega \to \mathcal{T}(v, \omega_\succ)$. $\square$

Remark 3.3. Recall that, by Whitehead's Theorem [Wh], any smooth manifold admits a *unique* PD/PL-structure, consistent with its differentiable structure. Therefore, different C_v-invariant smooth triangulations $\{\mathsf{T}^\partial\}$ of $\partial_1 X$ all are PD-equivalent, but perhaps not as *C_v-invariant* triangulations! In other words, a common refinement of two C_v-invariant differentiable triangulations of $\partial_1 X$ may be not C_v-invariant.

We conjecture that, in fact, any two smooth C_v-invariant triangulations have a C_v-invariant smooth refinement. In other words, the trajectory space

[3]properties (2) and (3) imply that $U_\omega \setminus \mathcal{T}(v, \omega_\succ)$ is PD-homeomorphic to the product $\partial U_\omega \times [0, 1)$, the product structure being defined by the w-trajectories.

$\mathcal{T}(v)$ admits a *unique* PD-structure, consistent (via the pull-back Γ^{-1}) with the preferred PD-structure on the smooth manifold $\partial_1 X$. $\diamondsuit$

Remark 3.4. For traversally generic vector fields v_1, v_2, *not any* homeomorphism $\Phi^{\mathcal{T}} : \mathcal{T}(v_1) \to \mathcal{T}(v_2)$, which preserves the $\mathbf{\Omega}^\bullet$-stratification of the trajectory spaces, lifts to a homeomorphism $\Phi : X_1 \to X_2$ that maps the v_1-trajectories to the v_2-trajectories so that their field-induced orientations are preserved. Therefore, using the homeomorphisms $\Phi^\partial : \partial X_1 \to \partial X_2$ which commute with the causality maps is essential for the reconstruction of the homeomorphism $\Phi : X_1 \to X_2$ in Theorem 3.1!

Let us illustrate this observation with the most basic example. Consider a 2-dimensional disk X. It admits a traversally generic vector field v with a single v-trajectory γ of the combinatorial type $\omega = (121)$ and tree trajectories with the combinatorial type (2). Let $f : X \to \mathbb{R}$ be a Lyapunov function for v. Then in the vicinity of γ, the space $\mathcal{T}(v)$ is a graph, shaped as the letter Y. The Lyapunov function f breaks the six-fold symmetry of the graph $\mathcal{T}(v) \approx \mathsf{Y}$. With the help of f, the three edges of the graph can be labeled by the three distinct elements $\rho_1, \rho_2, \rho_3 \in Mor((11), (121))$. By definition, ρ_1 sends the first 1 in (11) to the first 1 in (121), and the second one to 2; ρ_2 sends the first 1 in (11) to 2, and the second one to the third 1 in (121); finally, ρ_3 sends the first 1 in (11) to the first 1 in (121), and the second 1 to the third 1 in (121).

Let $\Phi^{\mathcal{T}} : \mathcal{T}(v) \to \mathcal{T}(v)$ be the automorphism which switches the two edges that are labeled by ρ_1 and ρ_2 and keeps the edge labeled by ρ_3 fixed. Evidently, $\Phi^{\mathcal{T}}$ preserves the $\mathbf{\Omega}^\bullet$-stratification of $\mathcal{T}(v)$. We leave to the reader to verify that $\Phi^{\mathcal{T}}$ cannot be lifted to a homeomorphism $\Phi : X \to X$ which preserves the v-induced orientations of the trajectories. $\diamondsuit$

We denote by $\mathsf{Homeo}^{\mathsf{PD}}(Y)$ the group of PD-homeomorphisms of a PD-manifold Y. For a compact PD-manifold X with boundary, there is the obvious restriction group homomorphism

$$\mathcal{R}^\partial : \mathsf{Homeo}^{\mathsf{PD}}(X) \to \mathsf{Homeo}^{\mathsf{PD}}(\partial_1 X).$$

Recall that the coning construction over a sphere delivers an epimorphism $\mathcal{R}^\partial : \mathsf{Homeo}^{\mathsf{PD}}(D^{n+1}) \to \mathsf{Homeo}^{\mathsf{PD}}(S^n)$.

Let $\mathsf{Homeo}_0(\sim)$ denotes the connected component of the group $\mathsf{Homeo}(\sim)$ that contains the identity transformation.

Let us state one beautiful result of L. Chen and K. Mann.

Theorem 3.2. *([ChMa], Theorems 1.6 and 1.7)*

Let X be a connected topological manifold, $\dim X \geq 3$. Suppose that the fundamental group $\pi_1(X)$ has no nontrivial action on the interval (i.e., $\pi_1(X)$ is generated by torsion).

Then the restriction map $\mathcal{R}_0^\partial : \mathsf{Homeo}_0(X) \to \mathsf{Homeo}_0(\partial_1 X)$ has a continuous section if and only if $X = D^{n+1}$ and $\partial_1 X = S^n$.

Moreover, for $n \geq 2$, any section of $\mathcal{R}_0^\partial : \mathsf{Homeo}_0(D^{n+1}) \to \mathsf{Homeo}_0(S^n)$ is conjugate to the standard coning. ◇

In general, the natural induced map

$$\mathcal{R}_*^\partial : \pi_0(\mathsf{Homeo}^{\mathsf{PD}}(X)) \to \pi_0(\mathsf{Homeo}^{\mathsf{PD}}(\partial_1 X))$$

is not necessarily a surjection.

Keeping in mind the claim of Theorem 3.2, consider the subgroup $\mathsf{Homeo}^{\mathsf{PD}}(\partial_1 X, C_v, f^\partial)$ of $\mathsf{Homeo}^{\mathsf{PD}}(\partial_1 X)$ that consists of PD-homeomorphisms that commute with the causality map $C_v : \partial_1^+ X(v) \to \partial_1^- X(v)$ (equivalently, that preserve the partial order structure $\succ$ on $\partial_1 X$, determined by the v-flow) and preserve the restriction $f^\partial : \partial_1 X \to \mathbb{R}$ of a Lyapunov function $f : X \to \mathbb{R}$ to $\partial_1 X$.

The Holography Theorem 3.1 leads to the following corollary, which contrasts with Theorem 3.2.

Corollary 3.1. *Let v be a traversally generic vector field on a compact connected smooth manifold X. Then any* PD*-homeomorphism $\Psi : \partial_1 X \to \partial_1 X$ which commutes with the causality map C_v extends to a* PD*-homeomorphism $\Phi : X \to X$.*

Moreover, over the subgroup $\mathsf{Homeo}^{\mathsf{PD}}(\partial_1 X, C_v, f^\partial) \subset \mathsf{Homeo}^{\mathsf{PD}}(\partial_1 X)$, there exists a continuous group-theoretical section of the natural restriction homomorphism $\mathcal{R}^\partial : \mathsf{Homeo}^{\mathsf{PD}}(X) \to \mathsf{Homeo}^{\mathsf{PD}}(\partial_1 X)$.

Proof. The first claim follows instantly from Theorem 3.1, where we choose $X_1 = X_2 = X$, $v_1 = v_2 = v$, so that $\Psi = \Phi^\partial : \partial_1 X \to \partial_1 X$ extends to $\Phi : X \to X$. We rely here on Proposition 3.1 for the existence of C_v-invariant PD-triangulation T^∂ on $\partial_1 X$.

Let us turn to the second claim of the corollary. Let T^∂ be a PD-triangulation of $\partial_1 X$ that is C_v-invariant. Take a PD-homeomorphism $\Psi : T^\partial \to T^\partial$ which commutes with C_v and such that $\Psi^*(f^\partial) = f^\partial$. Since Ψ commutes with C_v, it induces a PD-homeomorphism $\Psi^{\mathcal{T}} : \mathcal{T}(v) \to \mathcal{T}(v)$.

Recall that the embedding $\alpha : X \to \mathcal{T}(v) \times \mathbb{R}$ employs the Lyapunov function $f : X \to \mathbb{R}$ and the map C_v, while the embedding $\alpha^\partial : \partial_1 X \to \mathcal{T}(v) \times \mathbb{R}$ employs the Lyapunov function $f^\partial : \partial_1 X \to \mathbb{R}$ and the map C_v.

By Theorem 3.1, there exists a PD-homeomorphism $\tilde{\Phi}_\Psi : \mathcal{T}(v) \times \mathbb{R} \to \mathcal{T}(v) \times \mathbb{R}$ that takes $\alpha(\partial_1 X)$ to itself (and thus $\alpha(X)$ to itself) and has the property $\tilde{\Phi}_\Psi \circ \alpha^\partial = \alpha^\partial \circ \Psi_\Psi$. Formula (3.6) describes the nature of the extension $\tilde{\Phi}_\Psi(\gamma, t) = \big(\Psi^{\mathcal{T}}(\gamma),\ \tilde{\phi}(\gamma)(t)\big)$, where $\gamma \in \mathcal{T}(v)$, $t \in \mathbb{R}$, and the PL-homeomorphisms $\{\tilde{\phi}(\gamma) : \mathbb{R} \to \mathbb{R}\}_\gamma$ have been constructed in the proof of Theorem 3.1.

Finally, all these constructions are consistent with the compositions $\Psi \circ \Psi'$ of $\Psi, \Psi' \in \mathsf{Homeo}^{\mathsf{PD}}(\partial_1 X, C_v, f^\partial)$; in other words, $\tilde{\Phi}_{\Psi\circ\Psi'} = \tilde{\Phi}_\Psi \circ \tilde{\Phi}_{\Psi'}$. Therefore, the section $\Psi \to \tilde{\Phi}_\Psi$ delivers a group homomorphism $\mathsf{Homeo}^{\mathsf{PD}}(\partial_1 X, C_v, f^\partial) \to \mathsf{Homeo}^{\mathsf{PD}}(\alpha(X)) \approx \mathsf{Homeo}^{\mathsf{PD}}(X)$. □

3.2.2 *Smooth holography*

Now we are abandoning the PD-category and returning to the smooth settings.

We abbreviate to $C^\infty(X)$ the notation $C^\infty(X, \mathbb{R})$ for the algebra of smooth functions on a manifold X.

Let $\pi_k : J^k(X, \mathbb{R}) \to X$ be the vector bundle of k-jets of smooth maps from X to $\mathbb{R}$. We choose a semi-norm $| \sim |_k$ in the fibers of the k-jet bundle π_k and use it to define a sup-norm $\| \sim \|_k$ for the sections of π_k.

The Whitney topology [W3] in the space $C^\infty(X) = \{h : X \to \mathbb{R}\}$ is defined in terms of a countable family of norms $\{\|jet^k(h)\|_k\}_{k\in\mathbb{N}}$ of such π_k-sections $jet^k(h)$. This topology insures the uniform convergence over the compact subsets in X of functions and their partial derivatives, up to an arbitrary chosen order k.

For a traversing vector field v, let us consider the kernel of the directional derivative operator $\mathcal{L}_v : C^\infty(X) \to C^\infty(X)$. Of cause, the smooth functions from $\ker(\mathcal{L}_v)$ are constant along each v-trajectory γ. Their restrictions to the boundary $\partial_1 X$ generate a subalgebra $C^\infty(\partial_1 X, v)$ of $C^\infty(\partial_1 X)$.

Evidently, functions from $C^\infty(\partial_1 X, v)$ are constant along each discrete C_v-trajectory $\gamma^\partial = \gamma \cap \partial_1 X$ of the causality map. Furthermore, any smooth function $\psi : \partial_1 X \to \mathbb{R}$ that is constant on each finite set γ^∂ gives rise to a unique *continuous* function on X that is constant along each v-trajectory γ. However, such functions ψ may not be automatically *smooth* on X (a priori, they are just Hölderian with some control of the Hölder exponent that depends on the dimension of X only)! This potential complication leads to the following question to which we do not have a satisfactory answer in the general case.

Question 3.1. *For a traversing and boundary generic (alternatively, traversally generic) vector field v on X, is it possible to characterize the image $C^\infty(\partial_1 X, v)$ of the algebra $\mathsf{ker}(\mathcal{L}_v) \subset C^\infty(X)$ in the algebra $C^\infty(\partial_1 X)$ in terms of the causality map C_v and, perhaps, some additional v-generated data, residing in $\partial_1 X$?* $\diamondsuit$

Let $\dim(X) = n+1$ and v be a boundary generic traversing vector field on X. We consider its restriction v_j to the locus $\partial_j X = \partial_j X(v)$. Note that v_j is tangent to $\partial_{j-1} X$ along $\partial_j X$. Thus, the directional derivative $\mathcal{L}_{v_j} = \mathcal{L}_v|_{\partial_j X}$ operates on germs of smooth functions on $\partial_{j-1} X$ at the points of the locus $\partial_j X$.

Let us describe now a good candidate for the subalgebra $C^\infty(\partial_1 X, v)$ in the algebra $C^\infty(\partial X)$. By Lemma 3.1, the map $C_v : \partial_1^+ X(v) \to \partial_1^- X(v)$ allows to recover the Morse stratification $\{\partial_j^\pm X(v)\}_{j>0}$.

We denote by $\mathcal{L}_v^{(j)}$ the j-th iteration of the Lie derivative $\mathcal{L}_v$. Let $\mathsf{M}(v)$ be the subalgebra of smooth functions $\psi : \partial_1 X \to \mathbb{R}$ such that $(\mathcal{L}_v^{(j)}\psi)\big|_{\partial_{j+1} X} = 0$ for all $j \in [1, n]$. By the Leibniz rule, $\mathsf{M}(v)$ is indeed a subalgebra.

Let us denote by $\mathsf{M}(v)^{C_v}$ the subalgebra of functions from $\mathsf{M}(v)$ that are constants on each (finite) C_v-trajectory $\gamma^\partial = \gamma \cap \partial_1 X$.

Conjecture 3.1. *Let v be a traversing and boundary generic vector field on a smooth compact $(n+1)$-manifold X. Then the algebra $C^\infty(\partial_1 X, v)$ coincides with the subalgebra $\mathsf{M}(v)^{C_v} \subset C^\infty(\partial_1 X)$.*

In particular, $C^\infty(\partial_1 X, v)$ can be determined by the causality map C_v and the restriction of v to $\partial_2 X(v)$. $\diamondsuit$

It is easy to check that $C^\infty(\partial_1 X, v) \subset \mathsf{M}(v)^{C_v}$; the challenge is to show that the two algebras coincide.

We are now moving towards proving Theorem 3.3, a smooth analogue of Theorem 3.1 (under somewhat relaxed hypotheses).

Recall that a function f on a closed subset Y of a smooth manifold X is called smooth if it is the restriction to Y of a smooth function, defined in an open neighborhood of Y in X. Any smooth function on a closed Y admits a smooth extension to the ambient manifold X. In fact, the smooth version of the Urysohn Lemma (see [Mu1]) is valid: for any two closed disjoint subsets $Y_0, Y_1 \subset X$, there exists a smooth function $f : X \to [0,1]$ such that $f(Y_0) = 0$ and $f(Y_1) = 1$. Moreover, even a stronger version of this claim is available [W2]: we may assume that the infinite jet of f at Y_0

equals to the infinite jet of the identity function $\mathbf{0}$, and the infinite jet of f at Y_1 equals to the infinite jet of the identity function $\mathbf{1}$. This version of the Urysohn Lemma will be useful for the arguments that will follow.

Let v be a traversing vector field on a compact manifold X, and let $A \subset B$ be two closed subsets of $\partial_1 X$. We denote by $X(v, A)$ and $X(v, B)$ the unions of v-trajectories that pass through points of A and B, respectively. Let Θ_A and Θ_B be be the posets of combinatorial types of trajectories from $X(v, A)$ and $X(v, B)$, respectively.

On the way to Theorem 3.3, we need to establish the key technical lemma below.

Lemma 3.3. *Let v be a traversing and boundary generic vector field on a compact manifold X and $A \subset B \subset \partial_1 X$ closed subsets. Consider a smooth function $f : B \cup X(v, A) \to \mathbb{R}$ such that $f(x) < f(x')$ for any two points $x \neq x'$ on the same trajectory, such that x' can be reached from x by moving along the trajectory in the direction of v.*

Then f extends to a smooth function $F : X(v, B) \to \mathbb{R}$ such that $\mathcal{L}_v(F) > 0$ on $X(v, B)$.

Proof. The argument is an induction by the increasing combinatorial types $\omega \in \mathbf{\Omega}^\bullet$ of the v-trajectories that pass through the points of the set $B \setminus A$. With A being fixed, we intend to increase gradually the locus $\tilde{B} \supset A$ for which the desired extension exists, until eventually the growing $\tilde{B}$ will coincide with the given B.

In the proof, we put $X(A) =_{\mathsf{def}} X(v, A)$ and $X(B) =_{\mathsf{def}} X(v, B)$. Since A and B are closed in $\partial_1 X$, bout sets $X(A)$ and $X(B)$ are compact.

Thanks to the property of v to be boundary generic, the set of combinatorial types ω of the v-trajectories in X is finite. So we may assume that, for some even d, all the combinatorial types ω of v-trajectories have the property $|\omega| \leq d$ for some $d \equiv 0 \equiv |\omega| \mod 2$.

Consider all the trajectories through the points of $B \setminus A$ and their combinatorial types residing in a finite set Θ_B. Among these types we pick a *minimal* element ω.

We denote by X_ω the subset of $X(B)$ that is formed by the trajectories of this minimal combinatorial type ω. Let $X_\omega^\partial =_{\mathsf{def}} X_\omega \cap \partial_1 X$. We denote by $\mathcal{T}_\omega$ the Γ-image of X_ω, and by $\mathcal{T}(A)$ the Γ-image of $X(A)$ in $\mathcal{T}(v)$.

By the choice of minimal ω, the trajectories that are the limits of trajectories from X_ω, but are not contained in X_ω, all have combinatorial types residing in the subposet $\omega_\succ \cap \Theta_B$ and thus are contained in $X(A)$.

We aim to show that any given smooth function

$$f : X_\omega^\partial \cup X(A) \to \mathbb{R},$$

with the properties as in the lemma, extends to smooth function

$$F : X_\omega \cup X(A) \to \mathbb{R}$$

so that $\mathcal{L}_v(F) > 0$ on $X_\omega \cup X(A)$. Since replacing f with $f + const$ produces equivalent extension problems, we may assume that $f > 0$ without lost of generality.

First we notice that, for each trajectory $\gamma \subset X_\omega$, there is a smooth strictly monotone function $F_\gamma : \gamma \to \mathbb{R}$ that takes the given increasing values of the discrete function $f|_\gamma : \gamma \cap X_\omega^\partial \to \mathbb{R}$. This interpolating construction is based on a standard monotone block-function $\varphi_{a,b} : [0, a] \to [0, b]$ that smoothly depends on the two non-negative parameters a, b. The infinite jet of $\varphi_{a,b}$ at 0 coincides with the jet of the function x, the infinite jet of $\varphi_{a,b}$ at a coincides with the jet of the function $x + b$, and $\frac{d}{dx}\varphi_{a,b}(x) > 0$ in the interval $[0, a]$. Fig. 3.1, diagram B, shows the four-points interpolation $\phi_{\vec{x},\vec{y}}$ that uses three block-functions of the type $\varphi_{a,b}$.

Since we have chosen $f : X_\omega^\partial \cup X(A) \to \mathbb{R}$ to be positive, we get $F_\gamma > 0$ as well.

Let $V(A)$ be an open regular neighborhood of $\mathcal{T}(A)$ in $\mathcal{T}(v)$. Put $U(A) =_{\mathsf{def}} \Gamma^{-1}(V(A))$.

Let $\tilde{f}_A : U(A) \to \mathbb{R}$ be a smooth extension of $f : X(A) \cup B \to \mathbb{R}$ into a neighborhood $U(A)$ of $X(A)$. We choose $V(A)$ so small that, by continuity, $d\tilde{f}_A(v) > \delta > 0$ and $\tilde{f}_A > 0$ in $U(A)$. Also, by the very construction of the extension $\tilde{f}_A$, its restriction to $U(A) \cap B$ coincides with f.

The two sets $\mathcal{T}_\omega$ and $V(A)$ form an open cover of the space $\mathcal{T}_\omega \cup V(A)$.

Let $W =_{\mathsf{def}} \mathcal{T}_\omega \cap (V(A) \setminus \mathcal{T}(A))$ and $K =_{\mathsf{def}} \Gamma^{-1}(W)$. Then $\Gamma : K \to W$ is a v-oriented fibration with fibers being closed segments or singletons. Thus, $\Gamma : K \to W$ is a trivial fibration. At the same time, $\Gamma : K \cap \partial_1 X \to W$ is a covering map with the fiber of cardinality $\sup(\omega) = |\omega| - |\omega|'$. The triviality of $\Gamma : K \to W$ implies that the holonomy of the covering map $\Gamma : K \cap \partial_1 X \to W$ is trivial, and thus $K \cap \partial_1 X$ is a product. Hence, $K \cap \partial_1 X$ is homeomorphic to a product $L \times C$, where C is a set of cardinality $\sup(\omega)$, $L \subset \partial_1 X \cap X_\omega$, and $\Gamma : L \to W$ is a homeomorphism. By the construction of K, its boundary consists of two disjoint compacts: $\partial' K \subset X(A)$ and $\partial'' K =_{\mathsf{def}} X(\omega) \cap \partial U(A)$.

Let $\hat{X} \supset X$ be an open manifold, $\dim \hat{X} = \dim X$, so that v extends to a smooth vector field $\hat{v} \neq 0$ in $\hat{X}$.

We claim that, for any $\epsilon > 0$, there exists a smooth function $\psi^\bullet_\epsilon$ in the vicinity of K in $\hat{X}$ such that:

1) $jet^\infty(\psi^\bullet_\epsilon)|_{\partial' K} = jet^\infty(\mathbf{1})|_{\partial' K}$,

2) $jet^\infty(\psi^\bullet_\epsilon)|_{\partial'' K} = jet^\infty(\mathbf{0})|_{\partial'' K}$, and

3) $|\mathcal{L}_v(\psi^\bullet_\epsilon)| < \epsilon$ in K.[4]

Indeed, by the Whiney version [W2] of the Urysohn lemma, there exists a smooth $\tilde{\psi}^\bullet : K \to [0,1]$ such that 1) and 2) are satisfied. Since the v-flow gives K a product structure $I \times L \subset \mathbb{R} \times L$, there exists a stretching diffeomorphism $\alpha : K \to \tilde{K} \approx \tilde{I} \times L$ (where the interval $\tilde{I} \supset I$) along the v-directed component I, so that the v-directional derivative of $(\alpha^{-1})^*(\tilde{\psi}^\bullet)$, being restricted to $K \subset \tilde{K}$, is ϵ-small. Thus, $\psi^\bullet_\epsilon =_{\mathsf{def}} (\alpha^{-1})^*(\tilde{\psi}^\bullet)|_K$ satisfies property 3) together with properties 1) and 2) above.

Let $\psi^A_\epsilon : U(A) \to [0,1]$ be the function that equals $\mathbf{1}$ on $X(A)$, equals $\psi^\bullet_\epsilon$ on K, and is $\mathbf{0}$ on $X_\omega \setminus (X_\omega \cap U(A))$. It is smooth thanks to the properties:

$$jet^\infty(\psi^\bullet_\epsilon)|_{\partial' K} = jet^\infty(\mathbf{0})|_{\partial' K} \quad \text{and} \quad jet^\infty(\psi^\bullet_\epsilon)|_{\partial'' K} = jet^\infty(\mathbf{1})|_{\partial'' K}.$$

Let $\psi^\omega_\epsilon =_{\mathsf{def}} \mathbf{1} - \psi^A_\epsilon$. The pair $\{\psi^A_\epsilon, \psi^\omega_\epsilon\}$ is a smooth partition of unity, subordinate to the open cover $\{X_\omega \setminus (\bar{X}_\omega \cap X(A)),\ X(A) \cup K\}$ of the compact space $X_\omega \cup X(A)$.

Next, we form the smooth function $F_\omega : X_\omega \setminus (X_\omega \cap X(A)) \to \mathbb{R}$ whose restriction $F_\gamma : \gamma \to \mathbb{R}$ to each v-trajectory $\gamma \subset X_\omega$ is a monotone function, the canonical interpolation (see Fig. 3.1, B) of the given function $f|_\gamma : \gamma \cap X^\partial_\omega \to \mathbb{R}$.

Consider the smooth function $\tilde{F}_\epsilon : K \to \mathbb{R}$, defined by the formula

$$\tilde{F}_\epsilon =_{\mathsf{def}} \psi^\omega_\epsilon \cdot F_\omega + \psi^A_\epsilon \cdot \tilde{f}_A.$$

It smoothly extends in the obvious way to a function $F_\epsilon : X(A) \cup X_\omega \to \mathbb{R}$ so that:

1) $F_\epsilon = F_\omega$ on $X_\omega \setminus (X_\omega \cap U(A))$,

2) $F_\epsilon = f$ on $X(A) \cup (X_\omega \cap \partial_1 X)$.

By choosing an appropriate ϵ, we aim to insure that $dF_\epsilon(v) > 0$ in $X(A) \cup X_\omega$. Evidently, $dF_\epsilon(v) > 0$ on the complement to K. So we need to concentrate on the restriction $dF_\epsilon(v)|_K$.

Let $\bar{K}$ be the closure of K in X. Consider a positive number

$$m =_{\mathsf{def}} \min\big\{ \min_{\bar{K}} dF_\omega(v),\ \min_{\bar{K}} d\tilde{f}_A(v) \big\}.$$

[4] We suspect that, in fact, there exists a smooth $\psi^\bullet$ that satisfies 1) and 2) and has the property $\mathcal{L}_v(\psi^\bullet) = 0$ in K. Its existence would simplify the following arguments.

By the properties of F_ω and $\tilde{f}_A$, we have $m > 0$. Then, using that $\{\psi_\epsilon^A, \psi_\epsilon^\omega\}$ form a partition of unity and by the product rule, we get

$$\mathcal{L}_v(\tilde{F}_\epsilon) = dF_\epsilon(v) \;\geq\; m + d\psi_\epsilon^\omega(v)\cdot F_\omega + d\psi_\epsilon^A(v)\cdot \tilde{f}_A.$$

Therefore, it suffices to insure that RHS of this inequality is positive in order to guarantee that $\mathcal{L}_v(F_\epsilon) > 0$ on $\bar{K}$. Since $d\psi_\epsilon^A(v) = -d\psi_\epsilon^\omega(v)$ on $\bar{K}$, the inequality

$$m + d\psi_\epsilon^\omega(v)\cdot F_\omega + d\psi_\epsilon^A(v)\cdot \tilde{f}_A > 0$$

may be written as

$$m + d\psi_\epsilon^\omega(v)(F_\omega - \tilde{f}_A) > 0.$$

Recall that $|d\psi_\epsilon^\bullet(v)| < \epsilon$ in K. Hence, the choice $\epsilon < \inf_{\bar{K}}\left\{\frac{m}{2|F_\omega - \tilde{f}_A|}\right\}$ insures that $d\tilde{F}_\epsilon(v) > 0$ on $\bar{K}$. Therefore, this choice of ϵ guarantees that $dF_\epsilon(v) > 0$ in $X(A) \cup X_\omega$. For such $\epsilon > 0$, the smooth function $F = F_\epsilon$ delivers the desired extension.

Finally, we form the closed set $A' =_{\mathsf{def}} A \cup X_\omega^\partial \subset \partial_1 X$ and apply the previous arguments to the new pair $B \supset A'$. This completes the inductive step $A \Rightarrow A \cup (X_\omega \cap \partial_1 X)$ and the proof. □

By letting $A = \emptyset$ and $B = \partial_1 X$ in Lemma 3.3, we get its instant implication:

Corollary 3.2. *Let v be a traversing and boundary generic field on a compact manifold X. Consider a smooth function $f : \partial_1 X \to \mathbb{R}$ such that $f(x) < f(x')$ for any two points $x \neq x'$ on the same trajectory, such that x' can be reached from x by moving in the direction of v.*

Then f extends to a smooth function $F : X \to \mathbb{R}$ such that $\mathcal{L}_v(F) > 0$ on X. ◇

The following conjecture, if true, would strengthen Theorem 3.3 below. It is closely linked to Conjecture 3.1.

Conjecture 3.2. *Let $\partial\mathcal{E}_d \subset \mathbb{R}\times\mathbb{R}^d$ be a smooth hypersurface, given by the polynomial equation $P(u,\vec{x}) =_{\mathsf{def}} u^d + \sum_{i=0}^{d-1} x_i u^i = 0$ of an even degree d, and let $\mathcal{E}_d$ be the domain in $\mathbb{R}\times\mathbb{R}^d$, given by the polynomial inequality $\{P(u,\vec{x}) \leq 0\}$.*

We denote by $\gamma^\bullet(\vec{x}) \subset \mathbb{R}\times\{\vec{x}\}$ a segment/singleton with the following two properties:

- $P(u,\vec{x})|_{\gamma^\bullet(\vec{x})} \leq 0,$

- *no larger segment* $\tilde{\gamma}^\bullet(\vec{x}) \supset \gamma^\bullet(\vec{x})$ *has the property* $P(u, \vec{x})|_{\tilde{\gamma}^\bullet(\vec{x})} \leq 0$.

Consider a smooth diffeomorphism $\phi : \partial\mathcal{E}_d \to \partial\mathcal{E}_d$ *which maps each set* $\gamma^\bullet(\vec{x}) \cap \partial\mathcal{E}_d$ *to a similar set* $\gamma^\bullet(\vec{x}') \cap \partial\mathcal{E}_d$, *while preserving the multiplicity of the* P*-roots and their order in the two sets.*

Let $F : \mathcal{E}_d \to \mathbb{R}$ *be a smooth function such that* $\frac{\partial F}{\partial u} = 0$ *in* $\mathcal{E}_d$. *We denote by* f *the restriction of* F *to* $\partial\mathcal{E}_d$.

Then the function $\phi^*(f)$ *extends to a* smooth *function* $G : \mathcal{E}_d \to \mathbb{R}$ *such that* $\frac{\partial G}{\partial u} = 0$ *in* $\mathcal{E}_d$. ◇

There is a special case of Conjecture 3.2 that we will validate next.

First we treat the case of a boundary generic vector field in the vicinity of the locus $\partial_2^- X(v)$.

Let Q denote the hypersurface $\{u^2 + x_0 = 0\}$ in $\mathbb{R}^1 \times \mathbb{R}^1 \times \mathbb{R}^{n-1}$. The functions $u : Q \to \mathbb{R}$ and $\vec{y} : Q \to \mathbb{R}^{n-1}$ are smooth coordinates on the hypersurface Q. Let $X \subset \mathbb{R}^1 \times \mathbb{R}^1 \times \mathbb{R}^{n-1}$ be the domain defined by $\{u^2 + x_0 \geq 0\}$.

The causality map $\alpha =_{\mathsf{def}} C_{\partial_u}$ takes each point $q = (u, x_0, \vec{y}) \in Q$ to the point $\alpha(q) = (-u, x_0, \vec{y})$.

We denote by $K \subset Q$ the locus $\{u = 0\}$ and by $\pi : \mathbb{R} \times \mathbb{R}^n \to \mathbb{R}^n$ the projection $(u, x_0, \vec{y}) \to (x_0, \vec{y})$. Let $\mathbb{R}^n_+$ be the half-space, defined by $\{x_0 \leq 0\}$.

Lemma 3.4. *Let a function* $f : Q \to \mathbb{R}$ *be of the class* $C^{2k}(Q, \mathbb{R})$ *and invariant under the involution* $\alpha : Q \to Q$. *Then there exists a function* $g : (\mathbb{R}^n)_+ \to \mathbb{R}$ *in the variables* $(x_0, \vec{y})$ *such that:*

- *the restriction of* $\pi^*(g)$ *to* Q *coincides with* f,
- $g \in C^k((\mathbb{R}^n)_+, \mathbb{R})$.

Proof. Put $x = x_0$. We denote by $|\vec{w}|$ the l_1-norm of the vector $\vec{w}$.

Consider the Taylor expansion of $f(u, \vec{y})$ at a point $a = (0, 0, \vec{y}) \in K$. By the Taylor formula, there exists a polynomial $T^{2k}_{f,a}(\Delta u, \Delta\vec{y})$ of degree $\leq 2k$, an open neighborhood $U(f, a) \subset Q$ of point $a \in Q$, and a positive constant $C = C(U(f, a))$ (depending on the estimates of the order $2k + 1$ partial derivatives of f in $U(f, a)$) such that

$$|f(\Delta u, \vec{y} + \Delta\vec{y}) - T^{2k}_{f,a}(\Delta u, \Delta\vec{y})| < C(|\Delta u| + |\Delta\vec{y}|)^{2k+1} \tag{3.11}$$

for all $(\Delta u, \vec{y} + \Delta\vec{y}) \in U(f, a)$.

Since $f(\alpha(u, \vec{y})) = f((u, \vec{y}))$, there exists a function $g : \mathbb{R}^n_+ \to \mathbb{R}$ such that $\pi^*(g)|_Q$ coincides with f: just put $g((\pi(u, \vec{y})) =_{\mathsf{def}} f(u, \vec{y})$.

Using that $f(\alpha(u, \vec{y})) = f((u, \vec{y}))$ identically, $T^{2k}_{f,a}(\Delta u, \Delta\vec{y})$ has terms of even degrees in Δu only. We introduce the polynomial $\tilde{T}^{2k}_{f,a}((\Delta u)^2, \Delta\vec{y})$ in the variables $(\Delta u)^2, \vec{y}$ by the formula $\tilde{T}^{2k}_{f,a}((\Delta u)^2, \Delta\vec{y}) =_{\mathsf{def}} T^{2k}_{f,a}(\Delta u, \Delta\vec{y})$. Then we represent $\tilde{T}^{2k}_{f,a}(\Delta x, \Delta\vec{y})$ as a sum of a polynomial $\tilde{T}^{k}_{f,a}(\Delta x, \Delta\vec{y})$ of degree $\leq k$ in the variables $x, \vec{y}$ and a polynomial $\tilde{R}^{>k}_{f,a}(\Delta x, \Delta\vec{y})$, comprised of monomials whose degrees exceed k.

Then there exists a positive constant C' such that

$$|\tilde{R}^{>k}_{f,a}(\Delta x, \Delta\vec{y})| < C'(|\Delta x| + |\Delta\vec{y}|)^{k+1}$$

for all $(\Delta x, \vec{y} + \Delta\vec{y})$ in a sufficiently small open neighborhood $V(f, a) \subset \mathbb{R}^n$ of $\pi(a)$.

Therefore, in some open neighborhood $W(f, a) \subset \mathbb{R}^n$ of $\pi(a)$, the inequality (3.11) can be rewritten as

$$\begin{aligned} &\left|g(\Delta x, \vec{y} + \Delta\vec{y}) - \tilde{T}^{k}_{f,a}(\Delta x) - \tilde{T}^{>k}_{f,a}(\Delta x)\right| \\ &\leq \left|g(\Delta x, \vec{y} + \Delta\vec{y}) - \tilde{T}^{k}_{f,a}(\Delta x)\right| + \left|\tilde{T}^{>k}_{f,a}(\Delta x)\right| \\ &< C(\sqrt{|\Delta x|} + |\Delta\vec{y}|)^{2k+1} + C'(|\Delta x| + |\Delta\vec{y}|)^{k+1}. \end{aligned} \tag{3.12}$$

Note that the positive function

$$\psi(|\Delta x|, |\Delta\vec{y}|) =_{\mathsf{def}} (\sqrt{|\Delta x|} + |\Delta\vec{y}|)^{2k+1}/(|\Delta x| + |\Delta\vec{y}|)^{k+1}$$

is bounded from above in an open neighborhood $\mathcal{U} = \mathcal{U}(k)$ of $(0, 0)$ in the plane.

Hence, in the vicinity of $\pi(a) = (0, \vec{y})$, the inequality (3.12) transforms into the desired Taylor inequality

$$\left|g(\Delta x, \vec{y} + \Delta\vec{y}) - \tilde{T}^{k}_{f,a}(\Delta x)\right| < \tilde{C}(|\Delta x| + |\Delta\vec{y}|)^{k+1},$$

where the constant $\tilde{C} =_{\mathsf{def}} C \cdot \sup_{\mathcal{U}} \psi(|\Delta x|, |\Delta\vec{y}|) + C' > 0$. Therefore, $g \in C^k(\mathbb{R}^n_+, \mathbb{R})$. □

Definition 3.2. (Property A) Let v be a traversing boundary generic vector field on a compact connected smooth manifold with boundary.

We say that v has property A if each v-trajectory is transversal to $\partial_1 X$ at *some* point (that is, the combinatorial types ω of trajectories start or end with 1), or has the combinatorial type $\omega = (2)$. ♢

For a traversally generic v, the Property A is equivalent to the requirement

$$X\big(v, (33)_{\succeq} \cup (4)_{\succeq}\big) = \emptyset. \tag{3.13}$$

In particular, if each connected component of $\partial_1 X$ is *concave* or *convex* with respect to the v-flow, then the property A is satisfied. Equivalently, if $\partial_3 X(v) = \emptyset$, then all the combinatorial types of v-trajectories are of the form $(11), (1\underbrace{2\dots2}_{k}1), (2)$, so property A is valid.

In particular, Property A is valid for any gradient vector field of a Morse function f on a closed manifold M, being restricted to the compliment X to a disjoint union of sufficiently small convex balls, centered on the f-critical points.

We are now in position to prove the main result of this chapter, dealing with a topological rigidity of boundary value problems for a rather general class of ODEs. This result is a solution of the Holographic Extension Problem, as formulated in (3.4).

Theorem 3.3. (The Smooth Holography Theorem)
Let X_1, X_2 be two compact connected smooth $(n+1)$-manifolds with boundary, equipped with traversing and boundary generic vector fields v_1, v_2, respectively.

- *Then any smooth diffeomorphism $\Phi^\partial : \partial_1 X_1 \to \partial_1 X_2$, such that*

$$\Phi^\partial \circ C_{v_1} = C_{v_2} \circ \Phi^\partial,$$

extends to a homeomorphism $\Phi : X_1 \to X_2$ which maps v_1-trajectories to v_2-trajectories so that the field-induced orientations of trajectories are preserved. The restriction of Φ to each trajectory is a smooth diffeomorphism.

- *If v_1 has Property A from Definition 3.2, then the homeomorphism Φ is a smooth diffeomorphism. In particular, this is the case for any traversing and boundary generic vector field v_1 such that $\partial_3 X_1(v_1) = \emptyset$.*
- *In general, the conjugating homeomorphism $\Phi : X_1 \to X_2$ is a smooth diffeomorphism outside the closed subsets $X_1\big(v_1, (33)_{\succeq} \cup (4)_{\succeq}\big) \subset X_1$ and $X_2\big(v_2, (33)_{\succeq} \cup (4)_{\succeq}\big) \subset X_2$. If the fields are traversally generic, then the set $X_i(v_i, (33)_{\succeq})$ is of codimension 4. The set $X_i(v_i, (4)_{\succeq})$ is of codimension 3.*

Proof. The argument is a modification of the Theorem 3.1 proof, but is somewhat simpler, due to a better control, provided by the *diffeomorphism* Φ^∂. Recall that, in Theorem 3.1, Φ^∂ is just a PD-map; on the other hand, the PD-version of holography has been established only for the *traversally generic* v-flows: only for them we have good local models that led to the conclusion that $\mathcal{T}(v)$ are Whitney stratified trajectory spaces, and thus, triangulable.

Let us stress that, in the following arguments, we assume the knowledge of the spaces X_i, equipped with the oriented 1-foliations $\mathcal{F}(v_i)$ and with the appropriate Lyapunov functions f_i, or with the singular n-dimensional foliations $\mathcal{G}(f_i)$ they produce. Thus, we follow the framework in (3.4).

(1) As in the proof of Theorem 3.1, by Lemma 3.1, the combinatorial type $\omega(\gamma)$ of each v_i-trajectory $\gamma \subset X_i$ can be recovered from the causality map

$$C_{v_i} : \partial_1^+ X_i(v_i) \to \partial_1^- X_i(v_i),$$

where $i = 1, 2$. As a result, the information encoded in C_{v_i} is sufficient for a reconstruction of the $\mathbf{\Omega}^\bullet$-stratified trajectory space $\mathcal{T}(v_i)$—the target space of the finitely ramified quotient map $\Gamma_i : \partial_1^\pm X_i(v_i) \to \mathcal{T}(v_i)$, which is evidently determined by C_{v_i}. Therefore, any homeomorphism $\Phi^\partial : \partial_1 X_1 \to \partial_1 X_2$ that commutes with the causality maps gives rise to a *homeomorphism* $\Phi^\mathcal{T} : \mathcal{T}(v_1) \to \mathcal{T}(v_2)$. It preserves the $\mathbf{\Omega}^\bullet$-stratifications of the two spaces. Recall that the topology in $\mathcal{T}(v_i)$ is defined to be the weakest topology for which the obvious map $\Gamma_i : X_i \to \mathcal{T}(v_i)$ is continuous.

Since the stratifications $\{\partial_j X_i(v_i)\}_j$ can be recovered from the causality maps C_{v_i} and the diffeomorphism Φ^∂ commutes with the causality maps, we get that $\Phi^\partial(\partial_j X_1(v_1)) = \partial_j X_2(v_2)$ for all $j > 0$.

We aim now to lift $\Phi^\mathcal{T}$ to a desired homeomorphism (diffeomorphism) $\Phi : X_1 \to X_2$.

Since v_2 is traversing vector field, by Corollary 3.2 (see also Lemma 5.6 from [K7]), there exists a smooth Lyapunov function $f_2 : X_2 \to \mathbb{R}$ such that $\mathcal{L}_{v_2}(f_2) > 0$ everywhere in X_2. We use f_2 to form an auxiliary function

$$f_1^\partial =_{\mathsf{def}} (\Phi^\partial)^*(f_2) : \partial_1 X_1 \to \mathbb{R}.$$

Since Φ^∂ commutes with the causality maps C_{v_1} and C_{v_2}, we conclude that Φ^∂ maps each v_1-ordered finite set $\gamma \cap \partial_1 X_1$ to v_2-ordered set $\Phi^\mathcal{T}(\gamma) \cap \partial_1 X_2$. Therefore, if $f_2(x) < f_2(y)$ for some $x, y \in \Phi^\mathcal{T}(\gamma) \cap \partial_1 X_2$, then $f_1^\partial((\Phi^\partial)^{-1}(x)) < f_1^\partial((\Phi^\partial)^{-1}(y))$. As a result, $f_1^\partial : \partial_1 X_1 \to \mathbb{R}$ satisfies the hypothesis of Corollary 3.2. Applying this corollary, we produce a smooth function $f_1 : X_1 \to \mathbb{R}$ which extends f_1^∂ and has the property $df_1(v_1) > 0$ everywhere in X_1.

With the Lyapunov function $f_1 : X_1 \to \mathbb{R}$ for v_1 in place, we are ready to define the homeomorphism $\Phi : X_1 \to X_2$ that extends Φ^∂. It takes a typical v_1-trajectory $\gamma \subset X_1$ to the v_2-trajectory $\gamma' \subset X_2$ that projects, with the help of Γ_2, to the point $\Phi^{\mathcal{T}}(\gamma) \in \mathcal{T}(v_2)$. The restriction of Φ to each trajectory γ is given by the formula

$$\phi^{12}_\gamma =_{\mathsf{def}} \big(f_2|_{\Gamma_2^{-1}(\Phi^{\mathcal{T}}(\gamma))}\big)^{-1} \circ (f_1|_\gamma). \tag{3.14}$$

The formula makes sense since, thanks to the property $f_1^\partial = (\Phi^\partial)^*(f_2^\partial)$, the ranges of $f_1 : \gamma \to \mathbb{R}$ and $f_2 : \Gamma_2^{-1}(\Phi^{\mathcal{T}}(\gamma)) \to \mathbb{R}$ coincide and the two functions deliver diffeomorphisms between their domains and ranges.

Here, as usual, we abuse notations: "γ" stands for both a v-trajectory in X and for the corresponding point $[\gamma] = \Gamma(\gamma)$ in the trajectory space $\mathcal{T}(v)$.

(2) Now we introduce the desired 1-to-1 continuous map $\Phi : X_1 \to X_2$ by the formula $\Phi(x) =_{\mathsf{def}} x'$, where x' belongs to the v_2-trajectory over the point $\Phi^{\mathcal{T}}(\gamma_x) \in \mathcal{T}(v_2)$ such that $\phi^{12}_{\gamma_x}(x) = x'$. By the very construction of the function $f_1 : X_1 \to \mathbb{R}$ from the pull-back of f_2^∂, we get $\Phi|_{\partial X_1} = \Phi^\partial$.

Φ is a homeomorphism, since $\Phi^{\mathcal{T}}$ is a homeomorphism (thanks to Φ^∂ being a homeomorphism which commutes with the two causality maps) and the restriction of Φ to each v_1-trajectory γ is the homeomorphism ϕ^{12}_γ from (3.14), which depends in a continuous way on the set $\gamma \cap \partial_1 X_1$. Indeed, as an infinite sequence of v_1-trajectories $\{[\gamma_j] \in \mathcal{T}(v_1)\}_j$ converges to a v_1-trajectory $[\gamma]$, the sequence of finite sets $\{\gamma_j \cap \partial_1 X_1\}_j$ will be contained in an arbitrary small neighborhood in X_1 of some finite subset $A_\gamma \subseteq \gamma \cap \partial_1 X_1$. This implies that $\lim_{j\to\infty} \phi^{12}_{\gamma_{x_j}} = \phi^{12}_{\gamma_x}|_{B_\gamma}$, where B_γ is the minimal segment of γ that contains A_γ. Thus, Φ is 1-to-1 and continuous. Moreover, the restriction of Φ to each trajectory is a smooth orientation-preserving diffeomorphism.

Similarly, by the symmetry of our constructions, Φ^{-1} has these properties as well.

(3) In fact, thanks to the smooth dependence of solutions of a non-singular ODE on their initial values, in the cases described by the property A (perhaps, always), Φ is a smooth diffeomorphism.

To validate this claim, as usual, we embed X_i properly in a larger open manifold $\hat{X}_i$ and extend v_i to vector field $\hat{v}_i$ on $\hat{X}_i$ so that $d\hat{f}_i(\hat{v}_i) > 0$ for an appropriate smooth extension $\hat{f}_i : \hat{X}_i \to \mathbb{R}$ of the function f_i. We denote by $\mathcal{F}(\hat{v}_i)$ the corresponding smooth oriented 1-dimensional foliation on $\hat{X}_i$. It is transversal to the smooth n-dimensional foliation $\mathcal{G}(\hat{f}_i)$, defined by the constant level smooth hypersurfaces $\{\hat{f}_i^{-1}(c)\}_{c\in\mathbb{R}}$. Let $\hat{L}_i^c =_{\mathsf{def}} \hat{f}_i^{-1}(c) \subset \hat{X}_i$ denote a typical smooth leaf of $\mathcal{G}(\hat{f}_i)$. Note that when c is a critical value

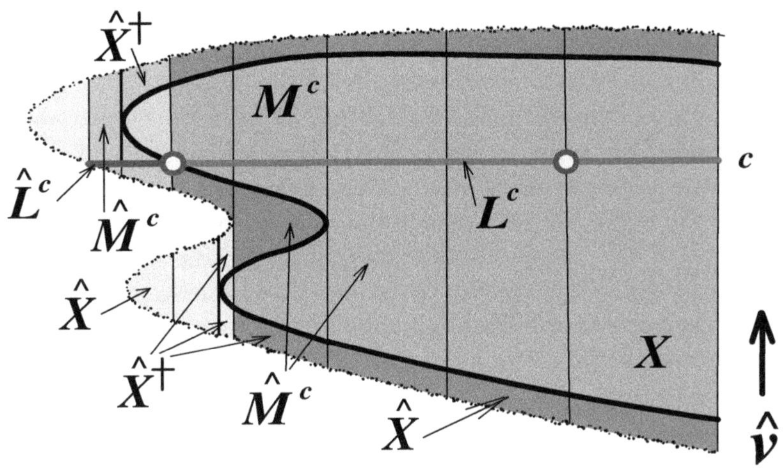

Fig. 3.2 Foliations $\mathcal{F}(\hat{v})$ (represented by the vertical lines) and $\mathcal{G}(\hat{f})$ (represented by the horizontal line/manifold) in $\hat{X}$, singular foliations $\mathcal{F}(v)$ and $\mathcal{G}(f)$—the traces of $\mathcal{F}(\hat{v})$ and $\mathcal{G}(\hat{f})$ in X, and various loci $\hat{L}^c, L^c, \hat{M}^c, M^c$, $\hat{X}^\dagger$, relevant to the arguments below.

of $f_i|_{\partial_1 X_i}$, the locus $L_c = f_i^{-1}(c)$ may not be a smooth hypersurface in X_i, while $\hat{L}_i^c$ is smooth in $\hat{X}_i$.

As we continue the arguments, the reader may find it helpful to keep an eye on Fig. 3.2. The open sets $\{\hat{M}_i^c =_{\mathsf{def}} \bigcup_{x \in \hat{L}^c} \hat{\gamma}_x\}_{c \in \mathbb{R}}$ cover X_i. Thus, $\{\partial \hat{M}_i^c =_{\mathsf{def}} \hat{M}_i^c \cap \partial X_i\}_{c \in \mathbb{R}}$ is an open cover of ∂X_i. Put $M_i^c =_{\mathsf{def}} \bigcup_{x \in L^c} \hat{\gamma}_x$.

Finally, we introduce the set $\hat{X}_i^\dagger$ as the union of all $\hat{v}_i$-trajectories through $\partial_1 X_i$. Hence, $\hat{X}_i^\dagger$ is a closed subset of $\hat{X}_i$ and contains X_i.

By a construction, similar to the one of Φ (see (3.14)), the diffeomorphism Φ^∂ extends to a homeomorphism $\Phi^\dagger : \hat{X}_1^\dagger \to \hat{X}_2^\dagger$. Indeed, each trajectory $\hat{\gamma} \subset \hat{X}_i^\dagger$ is determined by a point $z \in \partial X_i$. Let $\hat{\gamma} =_{\mathsf{def}} \hat{\gamma}_z$. If a leaf $\hat{L}_i^c$ hits $\hat{\gamma}_z$, then the intersection $\hat{L}_i^c \cap \hat{\gamma}_z$ is a singleton. Therefore, we may define the desired extension $\Phi^\dagger$ by the formula $\Phi^\dagger(x) =_{\mathsf{def}} \hat{\gamma}_{\Phi^\partial(z)} \cap \hat{L}_2^c$, where $x \in \hat{X}_1^\dagger$, $c = \hat{f}_1(x)$, and $z \in \hat{\gamma}_x \cap \partial X_1$. Since Φ^∂ conjugates the two causality maps, this definition of $\Phi^\dagger$ does not depend on the choice of $z \in \hat{\gamma}_x \cap \partial X_1$.

If $x \in \hat{X}_1$ is such that there exists $z \in \hat{\gamma}_x \cap \partial X_1$ with the multiplicity $m(z)$ of tangency between $\hat{\gamma}_z$ and ∂X_1 being *odd*, then using the local models of boundary generic fields from Lemma 2.4, we see that any $\hat{v}_1$-trajectory in the vicinity of z hits ∂X_1 (since any real polynomial of an odd degree has a real root). Therefore, in the vicinity of such x, the homeomorphism $\Phi^\dagger$ extends further to a homeomorhism $\hat{\Phi} : \hat{X}_1 \to \hat{X}_2$.

Since each v_1-trajectory, but a singleton, is bounded by two points of odd multiplicity, the only exceptions are the cases when $\hat{\gamma}_x \cap \partial X_1$ is a singleton of an even multiplicity $m(x)$; in the vicinity of such x, $\hat{X}_1$ and $X_1^\dagger$ differ. For these x's, we need an additional reasoning for the existence of an extension of $\Phi^\dagger$ to a germ-homeomorphism $\hat{\Phi} : \hat{X}_1 \to \hat{X}_2$ that maps $\hat{v}_1$-trajectories to $\hat{v}_2$-trajectories. It is also based on the local models of boundary generic vector fields from Lemma 2.4. In fact, Lemma 3.4 provides this reasoning for the points $z \in \partial_2^- X_1(v_1) \setminus \partial_3 X_1(v_1)$, where the field v_1 is strictly convex.

Given two smooth manifolds Y_1 and Y_2, a map $\Psi : Y_1 \to Y_2$ is smooth if and only if its composition with each local coordinate in Y_2 is a smooth function in the local coordinates on Y_1.

The leaves of the smooth foliations $\mathcal{F}(\hat{v}_i)$ and $\mathcal{G}(\hat{f}_i)$ can be locally defined by freezing complementary groups of the appropriate smooth local coordinates in $\hat{X}_i$. Recall that $\hat{\Phi}$ maps the smooth foliation $\mathcal{F}(\hat{v}_1)$ to the smooth foliation $\mathcal{F}(\hat{v}_2)$, the restriction of $\hat{\Phi}$ to the leaves-trajectories being smooth diffeomorphisms. The homeomorphism $\hat{\Phi}$ also maps the smooth foliation $\mathcal{G}(\hat{f}_1)$ to the smooth foliation $\mathcal{G}(\hat{f}_2)$ (remember, $\hat{\Phi}^*(f_2) = f_1$). If the restrictions $\{\hat{\Phi} : \hat{L}_1^c \to \hat{L}_2^c\}_{c \in \mathbb{R}}$ of $\hat{\Phi}$ to the leaves of $\mathcal{G}(\hat{v}_1)$ are smooth maps, we may conclude that the homeomorphism $\Phi : X_1 \to X_2$ is a smooth map.

Let us validate now that $\{\hat{\Phi} : \hat{L}_1^c \to \hat{L}_2^c\}_{c \in \mathbb{R}}$ are indeed smooth maps, *provided* Property A is available. (In the absence of Property A, it would be sufficient to validate Conjecture 3.2.) Since Φ^∂ is a smooth diffeomorphism, the image $\Phi^\partial(z) \in \partial X_2$ depends smoothly on $z \in \partial_1 X_1$. Therefore, the image point $\Phi(x) \in X_2$ depends smoothly on a point $z \in \gamma_x \cap \partial M_1^c$, where $c = f_1(x)$ (as long as $\gamma_z \cap L_1^c \neq \emptyset$).

This does not imply that $\Phi(x)$ depends smoothly on x (in fact, a priori, Φ is only a Hölder map with a Hölder exponent $1/m$, where m is the maximal tangency order of γ's to $\partial_1 X_1$).

However, assuming the validity of the Property A from Definition 3.2, we can overcome this difficulty. When the $\hat{v}_1$-trajectory $\hat{\gamma}$ through a point $x \in X_1$ is transversal to ∂X_1 at *some* point $z \in \partial X_1$, then, in the vicinity of x, the $\hat{v}_1$-induced map $p_1^\partial : \partial \hat{M}_1^c \to \hat{L}_1^c$, $c = f_1(x)$, admits a *smooth* local section $\sigma_1 : \hat{L}_1^c \to \partial \hat{M}_1^c$ which is *transversal* to the fibers of $p_1 : \hat{M}_1^c \to$

$\hat{f}_1^{-1}(c)$. That section is delivered by the boundary ∂X_1 in the vicinity of z. In such a case, Φ is smooth in the vicinity of x, since the composition $p_2^\partial \circ \Phi^\partial \circ \sigma_1 : \hat{L}_1^c \to \hat{L}_2^c$ is a smooth map. This conclusion applies to all v_1-trajectories γ that are bounded by at least one point of multiplicity 1. The exceptions are the trajectories bounded by *two* points of odd multiplicities that both exceed 1, that is, by the trajectories whose combinatorial type belongs to the poset $(33)_\succeq \subset \mathbf{\Omega}^\bullet$. Other exceptions to the transversality case may occur for the trajectories whose combinatorial types belong to the poset $(4)_\succeq \subset \mathbf{\Omega}^\bullet$. They include all the combinatorial types $(2k)$, where $k \geq 4$.

In the special case of trajectories of the combinatorial type $(2) \in \mathbf{\Omega}^\bullet$, the local differentiability of Φ in the vicinity of $z \in \partial_2^- X_1(v_1) \setminus \partial_3 X_1(v_1)$ follows from Lemma 3.4. In the special smooth coordinates $(u, x_0, \vec{y})$, where $\vec{y} = (y_1, \ldots, y_{n-1})$, in the vicinity of such point z, the boundary ∂X_1 is given by an equation $\{u^2 + x_0 = 0\}$, while X_1 by the inequality $\{u^2 + x_0 \geq 0\}$. Each $\hat{v}_1$-trajectory is specified by freezing the coordinates $(x_0, \vec{y})$. The smooth hypersurfaces $\{\hat{L}_1^c\}$ are transversal to the $\hat{v}_1$-trajectories. Since $\hat{\Phi}^\partial$ maps $\partial_2 X_1(v_1) \setminus \partial_3 X_1(v_1)$ to $\partial_2 X_2(v_2) \setminus \partial_3 X_2(v_2)$, a similar system of smooth coordinates is available in the vicinity of $\Phi^\partial(z)$. We use the symbol "$\,'\,$" to denote them.

The following argument, based on Lemma 3.4, uses its notations.

In the appropriate local coordinates $(u, x_0, \vec{y})$ and $(u', x_0', \vec{y}')$, the smooth diffeomorphism $\Phi^\partial : Q \to Q'$ of two quadratic hypersurfaces maps the p_1 folding locus K to the p_2 folding locus K' and commutes with the two causality maps $\alpha : Q \to Q$ and $\alpha' : Q' \to Q'$.

The local coordinate function $x_0' : X_2 \to \mathbb{R}$ pulls back to a smooth α'-invariant function $p_2^*(x_0') : Q' \to \mathbb{R}$. Since Φ^∂ is a smooth diffeomorphism, the further pull-back $\phi = (\Phi^\partial)^*(p_2^*(x_0')) : Q \to \mathbb{R}$ is a smooth function on Q. Because Φ^∂ commutes with α and α', ϕ is α-invariant. Therefore, by Lemma 3.4, ϕ is a restriction to Q of a smooth u-independent function χ in the variables $x_0, \vec{y}$. Similarly, using that Φ^∂ commutes with α and α', we conclude that $\psi = (\Phi^\partial \circ p_2)^*(\vec{y}') : Q \to \mathbb{R}^{n-1}$ is a smooth and α-invariant map. Therefore, ψ is a restriction to Q of a smooth map $\theta : X_1 \to \mathbb{R}^{n-1}$ that depends only on the coordinates $(x_0, \vec{y})$.

At the same time, $\hat{\Phi}^*(\hat{f}_2) = \hat{f}_1$. The functions $(\hat{f}_2, x_0', \vec{y}')$ form a smooth local system of coordinates. By the arguments above, the pull-back under $\hat{\Phi}$ of these coordinates are smooth on $\hat{X}_1$. Therefore, Φ is a smooth homeomorphism in the vicinity of K. By the same token, exchanging the roles of X_1 and X_2, Φ^{-1} is smooth as well.

This concludes the validation of smoothness of Φ in the vicinity of points of the combinatorial type $(2) \in \mathbf{\Omega}^\bullet$, as well as the proof of Theorem 3.3. □

Let v be a boundary generic traversing vector field on a smooth compact connected manifold X with boundary. We denote by $\mathsf{Diff}^\infty(\partial_1 X, C_v, f^\partial)$ the group of smooth diffeomorphisms of $\partial_1 X$ that commute with the causality map C_v and preserve the Lyapunov function $f^\partial : \partial_1 X \to \mathbb{R}$. Equivalently, for $\Psi \in \mathsf{Diff}^\infty(\partial_1 X, C_v, f^\partial)$ we have $\Psi^*(f^\partial) = f^\partial$ and, for any pair $x \succ_v y$ where $x, y \in \partial X$, we have $\Psi(x) \succ_v \Psi(y)$.

Consider the natural restriction maps $\mathcal{R}^\partial : \mathsf{Diff}^\infty(X) \to \mathsf{Diff}^\infty(\partial X)$. The famous Cerf Theorem [Cerf] claims that any orientation-preserving diffeomorphism of S^3 is isotopic to the identity and thus extends to a diffeomorphism of the 4-ball D^4. Thus, the map $\mathcal{R} : \mathsf{Diff}^\infty(D^4) \to \mathsf{Diff}^\infty_+(S^3)$ is onto.

At the same time, by [ChMa], Corollary 1.8, the restriction map $\mathcal{R}^\partial_0 : \mathsf{Diff}^\infty_0(X) \to \mathsf{Diff}^\infty_0(\partial X)$ does not admit a continuous (group-theoretical) section! Here $\mathsf{Diff}^\infty_0(X)$ denotes the path-connected component of the space $\mathsf{Diff}^\infty(X)$ that contains the identity map. Similarly, $\mathsf{Diff}^\infty_0(\partial X)$ denotes the path-connected component of the space $\mathsf{Diff}^\infty(\partial X)$ that contains the identity map.

In contrast with the previous remarkable fact, Theorem 3.3 leads to the following claim.

Corollary 3.3. *Let v be a traversing boundary generic vector field on a compact connected smooth manifold X. Assume that v has* Property A.

Over the subgroup $\mathsf{Diff}^\infty(\partial_1 X, C_v, f^\partial) \subset \mathsf{Diff}^\infty(\partial_1 X)$, there exists a continuous section

$$\sigma : \mathsf{Diff}^\infty(\partial_1 X, C_v, f^\partial) \to \mathsf{Diff}^\infty(X, \mathcal{F}(v), f)$$

of the natural restriction homomorphism $\mathcal{R}^\partial : \mathsf{Diff}^\infty(X) \to \mathsf{Diff}^\infty(\partial_1 X)$. Moreover, σ is a group monomorphism.

Proof. Using Theorem 3.3, the proof mirrors the proof of Corollary 3.1, with the obvious adjustments to the category of smooth maps. The main point is that, if we know and fix the grid of two foliations, $\mathcal{F}(v)$ and $\mathcal{G}(f)$, then by the proof of Theorem 3.3, the extension of any diffeomorphism $\Phi^\partial : \partial_1 X \to \partial_1 X$, that preserves the trace $\mathcal{F}^\partial(v) = \mathcal{F}(v) \cap \partial_1 X$ of $\mathcal{F}(v)$ and the level sets of the function f^∂ invariant, is unique. Therefore, assuming the validity of Property A, the composition $\Phi^\partial_1 \circ \Phi^\partial_2$ of any two such boundary diffeomorphisms, being again a diffeomorphism that preserves the same boundary confined structure, extends to a unique diffeomorphism $\Phi_1 \circ \Phi_2$. Therefore the section σ is a group monomorphism. □

Example 3.1. The Lyapunov function $f : X \to \mathbb{R}$ on the surface X in Fig. 3.3 is the distance from the center of the middle hole, and v is the radial vector field that emanates from the center of that hole. By Corollary 3.3, there is a group-theoretical section $\sigma : \mathsf{Diff}^\infty(\partial_1 X, C_v, f^\partial) \to \mathsf{Diff}^\infty(X, \mathcal{F}(v), f)$. In particular, the obvious action of the dihedral group D_3 on $\partial_1 X$ in Fig. 3.3 lifts by σ to a D_3-action on X. ◇

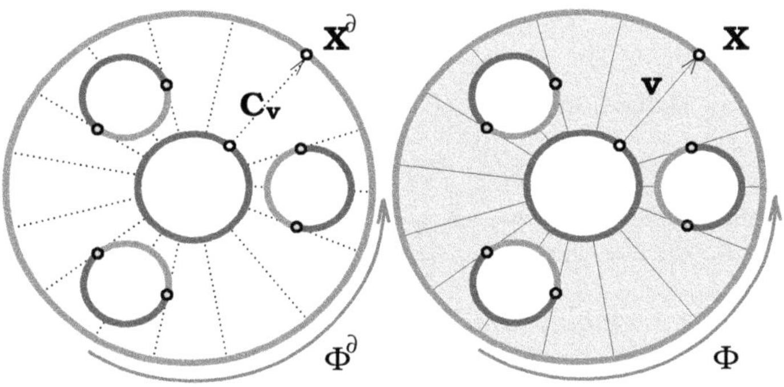

Fig. 3.3 The surface X is a disk with four holes. Any smooth action of the dihedral group D_3 on $\partial_1 X$, which commutes with the causality map C_v and preserves f^∂, lifts to a D_3-action on X that preserves the "radial" foliation $\mathcal{F}(v)$ and the Lyapunov function f.

Corollary 3.4. *Let X_1, X_2 be two smooth compact connected $(n+1)$-manifolds with boundary, equipped with traversing and boundary generic vector fields v_1, v_2, respectively. Then any diffeomorphism $\Phi^\partial : \partial X_1 \to \partial X_2$ such that*

$$\Phi^\partial \circ C_{v_1} = C_{v_2} \circ \Phi^\partial$$

generates a stratification-preserving homeomorphism $\Phi^{\mathcal{T}} : \mathcal{T}(v_1) \to \mathcal{T}(v_2)$ of the two $\mathbf{\Omega}^\bullet$-stratified trajectory spaces.

If v_1 has property A *from Definition 3.2, then $\Phi^{\mathcal{T}}$ induces an isomorphism*

$$(\Phi^{\mathcal{T}})^* : C^\infty(\mathcal{T}(v_2)) \to C^\infty(\mathcal{T}(v_1))$$

of the algebras of smooth functions on the trajectory spaces (see Definition 2.7)—the two trajectory spaces are "diffeomorphic".

Proof. By the proof of Theorem 3.3, there exists a diffeomorphism $\Phi : X_1 \to X_2$ which takes v_1-trajectories to v_2-trajectories and extends Φ^∂, while preserving their combinatorial tangency patterns. Therefore, Φ maps every smooth function $f : X_2 \to \mathbb{R}$ that is constant on each v_2-trajectory to a continuous function $f \circ \Phi : X_1 \to \mathbb{R}$ that is constant on each v_1-trajectory. When v_1 has property A from Definition 3.2, then Φ is a smooth diffeomorphism; so $\Phi^*(f)$ is a smooth function, also constant along the v_1-trajectories.

Similar argument applies to the inverse homeomorphism/diffeomorphism $(\Phi)^{-1} : X_2 \to X_1$. Thus, the two trajectory spaces are diffeomorphic in the sense of Definition 2.8. □

Corollary 3.5. *For a boundary generic and traversing vector field v on X, the topological type of the pair $(X, \mathcal{F}(v))$ can be recovered from each of the following v-induced structures on its boundary $\partial_1 X$:*

(1) the causality map $C_v : \partial_1^+ X(v) \to \partial_1^- X(v)$,
(2) the poset $(\mathcal{C}^\partial(v), \succ)$, whose elements are the points of $\partial_1 X$,
(3) the category $\mathbf{Cat}^\partial(v)$, determined by the poset $(\mathcal{C}^\partial(v), \succ)$.

- *In particular, all the topological invariants of X* [5] *can be recovered from each of the structures* (1) – (3) *on $\partial_1 X$.*
- *When v has Property A from Definition 3.2, the smooth topological type of the pair $(X, \mathcal{F}(v))$ can be recovered from each of the structures* (1) – (3) *on $\partial_1 X$. As a result, all the invariants of the smooth structure on X* [6] *can be reconstructed from each of the structures* (1) – (3) *on $\partial_1 X$.*

Proof. Consider two manifolds X_1 and X_2 which carry traversing boundary generic vector fields v_1 and v_2. Assume that the two manifolds share a common boundary: $\partial_1 X_1 = \partial_1 X_2$. If the two vector fields induce identical causality maps, then, according to Theorem 3.3, the diffeomorphism $\Phi^\partial =_{\mathsf{def}} \mathsf{id}_{\partial_1 X}$, extends to a homeomorphism $\Phi : X_1 \to X_2$ such that the oriented foliation $\mathcal{F}(v_1)$ is mapped by Φ to the oriented foliation $\mathcal{F}(v_2)$, the homeomorphism Φ being a diffeomorphism on each leaf.

[5] such as its rational Pontryagin classes or relative signatures of Wall [Wal]
[6] such as all the characteristic classes of the tangent bundle $\tau(X)$

The equivalence of the three structures in the statement of the corollary is on the level of definitions.

When v_1 has property A, by Theorem 3.3, the homeomorphism Φ may be assumed to be a smooth diffeomorphism. $\square$

Remark 3.5. Unfortunately, Corollary 3.5 and its proof are not very constructive: they are just claims of pure existence of an extension Φ of a given conjugating diffeomorphism Φ^∂. It is not clear how to build the triple $(X, \mathcal{F}(v), f)$ (in the spirit of diagram (3.2) and the recipe depicted in Fig. 2.3, which works well for extending Φ^∂ to a *homeomorphism* Φ) only from the boundary data $(\partial_1 X, \mathcal{F}(v) \cap \partial_1 X, f^\partial)$. In Theorem 3.4 and Corollary 3.6, we will try to overcome this deficiency, which arises only in the smooth category. $\diamondsuit$

This is the time to revisit Example 1.2 that describes a traversally generic vector field v and its causality map C_v on a connected compact surface X with a circular boundary. The statement of Corollary 3.5 is not completely obvious even for such field in $2D$. Even the claim about the reconstruction of the topological type of X from the knowledge of C_v has some subtlety: according to the Morse formula for vector fields [Mo] and using that $v \neq 0$, the Euler number

$$\chi(X) = \chi(\partial_1^+ X(v)) - \chi(\partial_2^+ X(v)) = q - \#(\partial_2^+ X(v)).$$

Thus, to determine the topological type of X, we need to know not only $q = \chi(\partial_1^+ X)$ (which we do), but also the cardinality $\#(\partial_2^+ X)$. The latter can be extracted by examining the discontinuity points of G_v, the graph of C_v. This presumes that the polarity "$\pm$" of each of the $2q$ points from $\partial_2 X$ can be recovered from C_v or G_v. We leave to the reader to discover the recipe.

Example 3.2. Consider a regular neighborhood T of a knot $K \subset S^3$. Let X be the compliment to a solid torus T in S^3. By the Gordon-Luecke Theorem [GL], the topological type of X determines the knot $K \hookrightarrow S^3$, up to a homeomorphism of the ambient sphere. Let v be a traversing and boundary generic vector field on X (since X is a connected manifold with boundary, by [K1], such v exists). Therefore, by Corollary 3.5, the causality map $C_v : \partial_1^+ X(v) \to \partial_1^- X(v)$ (equivalently, the partial order $\succ_v$ on the 2-torus $\partial_1 T$, generated by C_v) determines the knot $K \hookrightarrow S^3$, up to a homeomorphism of the ambient sphere. $\diamondsuit$

Example 3.3. Let us spell out the claims of Corollary 3.5 in the case of a smooth compact connected 4-dimensional manifold X with boundary. For a *boundary generic* vector field v on X, according Corollary 3.5, only the loci $\{X(v,(\omega)_{\succeq})\}_{\omega\in\mathbf{\Omega}^\bullet}$ with ω's, *different from* $(1,\omega_2,\dots,\omega_q)$, $(\omega_1,\dots,\omega_{q-1},1)$, or (2), may be obstacles for the smoothness of the Φ-reconstruction of the pair $(X,\mathcal{F}(v))$.

For a *transversally generic* v, this leaves us with a potential failure of smoothness of Φ only along the locus $X(v,(33)_{\succeq})\cup X(v,(4)_{\succeq}))$. However, $X(v,(33)_{\succeq})=\emptyset$ for the dimensional reasons. Thus, the smoothness of reconstruction of $(X,\mathcal{F}(v))$ may fail only in the vicinity of the points-trajectories, forming the finite set $X(v,(4))\subset\partial_1X$.

Therefore, if $X(v,(4)_{\succeq})=\emptyset$, in accordance with Corollary 3.5, for a traversally generic v, we conclude that all the Gauge invariants of X (for example, see [FM] for their description) like the solution spaces of the Yang-Mills equations $d_A(*F_A)=0$ can be recovered from the causality map $C_v:\partial_1^+X(v)\to\partial_1^-X(v)$. As a practical matter, this recovery could be very challenging... ◇

Now we will describe which boundary data will allow to reconstruct the pair $(X,\mathcal{F}(v)$ in a *canonical way*, thus addressing the problem of Holographic Reconstruction (3.2) in a more constructive way.

Let $f:X\to\mathbb{R}$ be a Lyapunov function for v. The pair $(\mathcal{F}(v),f)$ gives rise to an *embedding* $\alpha:X\hookrightarrow\mathcal{T}(v)\times\mathbb{R}$, defined by the formula $\alpha(x)=([\gamma_x],f(x))$, where $x\in X$ and $[\gamma_x]\in\mathcal{T}(v)$ denotes the point-trajectory through x (see Fig. 2.3). Consider also the restriction α^∂ of the embedding α to the boundary ∂_1X.

The next assertion mirrors the content of Theorem 6.12 from [K7] and Lemma 4.1 from [K9]. It should be compared with Lemma 2.7.

Theorem 3.4. *For a traversing boundary generic vector field v on a compact connected smooth X, the boundary data $(\partial_1X,C_v,f^\partial)$ allow for a canonical (rigid) reconstruction of the triple $(X,\mathcal{F}(v),f)$ via a homeomorphism $\alpha:X\to\alpha(X)\subset\mathbb{R}\times\mathcal{T}(v)$, determined by the homeomorphism $\alpha^\partial:\partial_1X\to\alpha^\partial(\partial_1X)$.*

Proof. As we have seen, the causality map C_v determines the trajectory space $\mathcal{T}(v)$ canonically.

The dependence $x\rightsquigarrow[\gamma_x]$ is continuous by the definition of the quotient topology in $\mathcal{T}(v)$. Note that α maps each v-trajectory γ to the line $[\gamma]\times\mathbb{R}$.

Moreover, for any $c \in \mathbb{R}$, each leaf $\mathcal{G}_c = f^{-1}(c)$ is mapped by α to the slice $\mathcal{T}(v) \times c$ of $\mathcal{T}(v) \times \mathbb{R}$.

Given a finite subset $A \subset \mathbb{R}$, we denote by $[A]$ the minimal closed interval that contains A. With the help of α, each trajectory $\gamma \in \mathcal{F}(v)$ may be identified with the closed interval $[f^\partial(\gamma \cap \partial_1 X)] \subset \mathbb{R}$, and the vector field v with the constant vector field ∂_u on $\mathbb{R}$.

Evidently, the image of $\alpha^\partial : \partial_1 X \hookrightarrow \mathcal{T}(v) \times \mathbb{R}$ bounds the compact image

$$\alpha(X) = \coprod_{[\gamma] \in \mathcal{T}(v)} [f^\partial(\gamma \cap \partial_1 X)]$$

and thus determines it canonically. Therefore, $\alpha(X)$ depends on the causality map C_v and on the function f_1^∂ only! Note that α is a continuous 1-to-1 map of a compact space, and thus, a homeomorphism onto its image. Moreover, the topological type of X depends only on C_v: the apparent dependence of $\alpha(X)$ on f^∂ is not crucial, since the space $\mathsf{Lyap}(v)$ of Lyapunov functions, for a given v, is convex. □

The remaining issue is to make sense of the claim "the embedding α is a diffeomorphism". The next subsection addresses this issue.

3.2.3 *Recovering the algebra $C^\infty(X, \mathbb{R})$ in terms of subalgebras of $C^\infty(\partial_1 X, \mathbb{R})$*

Given a compact smooth manifold X, we denote by $C^\infty(X)$ the algebra of smooth functions on X.

In this subsection, we use the following classical property of these functional algebras: for any compact smooth manifolds X, Y, there is an algebra isomorphism

$$C^\infty(X \times Y) \approx C^\infty(X) \hat{\otimes} C^\infty(Y),$$

where $\hat{\otimes}$ denotes an appropriate completion of the algebraic tensor product $C^\infty(X) \otimes C^\infty(Y)$.

It is known that the (normed) algebra $C^0(X)$ of continuous real-valued functions on a compact space X determines its topological type [GRS]. In this context, X is interpreted as the space of maximal ideals of the algebra $C^0(X)$. In a similar spirit, the algebra $C^\infty(X)$ of smooth functions on a compact smooth manifold X (the algebra $C^\infty(X)$ is considered in the Whitney topology [W3]) determines the *smooth* topological type of X

[KMS], [Na]. Again, X is viewed as the space of maximal ideals of the algebra $C^\infty(X)$.

Recall that a harmonic function h on a compact connected Riemannian manifold X is uniquely determined by its restriction to the smooth boundary $\partial_1 X$ of X. In other words, the Dirichlet boundary value problem has a unique solution in the space of harmonic functions. Therefore, the vector space $\mathcal{H}(X)$ of harmonic functions on X is rigidly determined by its restriction (trace) $\mathcal{H}^\partial(X) =_{\mathsf{def}} \mathcal{H}(X)|_{\partial_1 X}$ to the boundary $\partial_1 X$. Until the end of the section, this fact will serve us as a beacon, as we pose the following question:

Which algebras of smooth functions on the boundary $\partial_1 X$ can be used to reconstruct the algebra $C^\infty(X)$ and thus the smooth topological type of X?

Remembering the flexible nature of smooth functions (in contrast with the rigid harmonic ones), at the first glance, we should anticipate the obvious answer "None!". However, when X carries an additional geometric structure, then the question, surprisingly, may have a positive answer. The geometric structure on X that does the trick is a vector field (i.e., an ordinary differential equation), drawn from a massive class of traversing and boundary generic vector fields.

We will introduce two subalgebras,

$$\mathcal{A}(v) =_{\mathsf{def}} C^\infty(\partial_1 X, v) \text{ and } \mathcal{B}(f) =_{\mathsf{def}} (f^\partial)^*(C^\infty(\mathbb{R})),$$

of the algebra $C^\infty(\partial_1 X)$. They depend only on v and f, respectively.

By Theorem 3.5, $\mathcal{A}(v)$ and $\mathcal{B}(f)$ will allow for a reconstruction of $C^\infty(X)$. In fact, the boundary data, generated by these subalgebras, lead to a unique "rigid" reconstruction

$$C^\infty(X) \approx C^\infty(\partial_1 X, v) \,\hat{\otimes}\, (f^\partial)^*(C^\infty(\mathbb{R})).$$

As a result, the pair $\mathcal{A}(v)$, $\mathcal{B}(f)$, "residing on the boundary", determines the smooth topological type of the bulk X and of the 1-dimensional foliation $\mathcal{F}(v)$, generated by the v-flow.

As we noticed in Chapter 2, the trajectory space $\mathcal{T}(v)$, although a singular space, carries a surrogate smooth structure [K3]. By Definition 2.7, a function $h : \mathcal{T}(v) \to \mathbb{R}$ is smooth if its pull-back $\Gamma^*(h) : X \to \mathbb{R}$ is a smooth function on X. Thus, it makes sense to consider the algebra $C^\infty(\mathcal{T}(v))$ of smooth functions on $\mathcal{T}(v)$. As a subalgebra of $C^\infty(X)$, the algebra $C^\infty(\mathcal{T}(v))$ is formed exactly by the smooth functions $g : X \to \mathbb{R}$, whose directional derivatives $\mathcal{L}_v g = 0$. If we multiply v by a non-vanishing conformal factor $\lambda : X \to \mathbb{R}_+$, then $\mathcal{L}_v g = 0$ if and only if $\mathcal{L}_{\lambda \cdot v}\, g = 0$.

Therefore, the algebra $C^\infty(\mathcal{T}(v))$ depends only on the conformal class of v; in other words, on the foliation $\mathcal{F}(v)$.

In the same spirit, we may talk about diffeomorphisms $\Phi^{\mathcal{T}} : \mathcal{T}(v) \to \mathcal{T}(v)$, as maps that induce isomorphisms of the algebra $C^\infty(\mathcal{T}(v))$ (see Definition 2.8).

If two (v-invariant) functions from $C^\infty(\mathcal{T}(v))$ take different values at a point $[\gamma] \in \mathcal{T}(v)$, then they must take different values on the finite set $\gamma \cap \partial_1 X \subset \partial_1 X$. Therefore, the obvious restriction homomorphism $res^{\partial}_{\mathcal{T}} : C^\infty(\mathcal{T}(v)) \to C^\infty(\partial_1 X)$, induced by the inclusion $\partial_1 X \subset X$, is a *monomorphism*. We denote its image by $C^\infty(\partial_1 X, v)$. Thus, we get an isomorphism

$$res^{\partial}_{\mathcal{T}} : C^\infty(\mathcal{T}(v)) \overset{\approx}{\to} C^\infty(\partial_1 X, v).$$

We think of the subalgebra $C^\infty(\partial_1 X, v) \subset C^\infty(\partial_1 X)$ as an integral part of the boundary data for the holography problems we are tackling.

Let $\pi_k : J^k(X, \mathbb{R}) \to X$ be the vector bundle of k-jets of smooth maps from X to $\mathbb{R}$. We choose a continuous family semi-norms $| \sim |_k$ in the fibers of the jet bundle π_k and use it to define a sup-norm $\| \sim \|_k$ for the sections of π_k. We denote by jet^k the obvious map $C^\infty(X, \mathbb{R}) \to J^k(X, \mathbb{R})$ that takes each function h to the collection of its k-jets $\{jet^k_x(h)\}_{x \in X}$.

The Whitney topology [W3] in the space $C^\infty(X) = \{h : X \to \mathbb{R}\}$ is defined in terms of the countable family of the norms $\{\|jet^k(h)\|_k\}_{k \in \mathbb{N}}$ of such π_k-sections $jet^k(h)$. This topology insures the uniform convergence of functions and their partial derivatives over compact subsets of X.

Note also that

$$\|jet^k(h_1 \cdot h_2)\|_k \leq \|jet^k(h_1)\|_k \cdot \|jet^k(h_2)\|_k$$

for any $h_1, h_2 \in C^\infty(X)$.

Any subalgebra $\mathcal{A} \subset C^\infty(X)$ inherits a topology from the Whitney topology in $C^\infty(X)$. In particular, the subalgebra $C^\infty(\mathcal{T}(v)) \approx C^\infty(X, v)$ does.

As a locally convex vector spaces, $C^\infty(\mathcal{T}(v))$ and $C^\infty(\mathbb{R})$ are nuclear ([DS], [Ga]) so that the topological tensor product $C^\infty(\mathcal{T}(v)) \hat{\otimes} C^\infty(\mathbb{R})$ (over $\mathbb{R}$) is uniquely defined as the completion of the algebraic tensor product $C^\infty(\mathcal{T}(v)) \otimes C^\infty(\mathbb{R})$ [Grot].

We interpret $C^\infty(\mathcal{T}(v)) \hat{\otimes} C^\infty(\mathbb{R})$ as the algebra of "smooth" functions on the product $\mathcal{T}(v) \times \mathbb{R}$ and denote it by $C^\infty(\mathcal{T}(v) \times \mathbb{R})$.

Let us consider two subalgebras, $f^*(C^\infty(\mathbb{R})) \subset C^\infty(X)$ and $(f^\partial)^*(C^\infty(\mathbb{R})) \subset C^\infty(\partial_1 X)$, the second one is assumed to be "known" as a part of the boundary data.

Lemma 3.5. *The intersection $C^\infty(\mathcal{T}(v)) \cap (f)^*(C^\infty(\mathbb{R})) = \underline{\mathbb{R}}$, the space of constant functions on X.*

Proof. If a smooth function $h : X \to \mathbb{R}$ is constant on each v-trajectory γ and belongs to $(f)^*(C^\infty(\mathbb{R}))$, then it must be constant on each connected leaf of $\mathcal{G}(f)$ that intersects γ. Thus, such h is constant on the maximal closed *connected* subset $A_\gamma \subseteq f^{-1}(f(\gamma))$ that contains γ. Each trajectory γ, homeomorphic to a closed interval, has an open neighborhood such that, for any trajectory γ' from that neighborhood, we have $A_\gamma \cap A_{\gamma'} \neq \emptyset$. Since X is connected, any pair γ, γ' of trajectories may be connected by a path $\delta \subset X$. Using the compactness of δ, we conclude that the function h must be a constant along δ. Therefore, h is a constant globally. □

Lemma 3.6. *The restriction operator $H_f^\partial : f^*(C^\infty(\mathbb{R})) \to (f^\partial)^*(C^\infty(\mathbb{R}))$ to the boundary $\partial_1 X$ is an epimorphism of algebras. If the range $f^\partial(\partial_1 X)$ of f^∂ is a connected closed interval in $\mathbb{R}$ (which is the case for a connected $\partial_1 X$), then H_f^∂ is an isomorphism.*

Proof. The restriction operator H_f^∂ is an algebra epimorphism, since any composite function $\phi \circ f^\partial$, where $\phi \in C^\infty(\mathbb{R})$, is the restriction to $\partial_1 X$ of the function $\phi \circ f$.

On the other hand, when $f^\partial(\partial_1 X)$ is a connected subset of $\mathbb{R}$, we claim that H_f^∂ is a monomorphism. Indeed, take a function $\phi \in C^\infty(\mathbb{R})$, such that $\phi \circ f^\partial \equiv 0$, but $\phi \circ f$ is not identically zero on X. Then there is $x \in X$ such that $\phi \circ f(x) \neq 0$. On the other hand, by the hypothesis, $f(x) = f^\partial(y)$ for some $y \in \partial_1 X$. The assumption $f^\partial \circ \phi \equiv 0$ implies that $\phi(f^\partial(y)) = 0$. This contradiction validates the claim that H_f^∂ is a monomorphism. Therefore, when the image $f^\partial(\partial_1 X)$ is a connected interval, H_f^∂ is an isomorphism of algebras. □

Consider the homomorphism of algebras

$$\mathsf{P} : C^\infty(\mathcal{T}(v)) \otimes f^*(C^\infty(\mathbb{R})) \to C^\infty(X)$$

that takes every finite sum $\sum_i h_i \otimes (f \circ g_i)$, where $h_i \in C^\infty(\mathcal{T}(v)) \subset C^\infty(X)$ and $g_i \in C^\infty(\mathbb{R})$, to the finite sum $\sum_i h_i \cdot (g_i \circ f) \in C^\infty(X)$. I.e., P coverts the tensor products to products of functions. Recall that, by Lemma 3.5, $C^\infty(\mathcal{T}(v)) \cap (f)^*(C^\infty(\mathbb{R})) = \underline{\mathbb{R}}$, the constants on X.

Take any connected component $\mathcal{G}_c^\bullet$ of the hypersurface $\mathcal{G}_c = f^{-1}(c)$ and denote by $\mathcal{H}_c^\bullet$ the set of all the trajectories that hit $\mathcal{G}_c^\bullet$.

If $\sum_i h_i \cdot (g_i \circ f) \equiv 0$ on $\mathcal{G}_c^\bullet$, then ether:

(1) all the functions $\{h_i\}_i$ are linearly independent on $\mathcal{G}_c^\bullet$ and all the function $\{(g_i \circ f)\}_i$ vanish on $\mathcal{G}_c^\bullet$, i.e., $\{g_i(c) = 0\}_i$, or

(2) $\{h_i\}_i$ are linearly dependent on $\mathcal{G}_c^\bullet$ and thus on the set $\mathcal{H}_c^\bullet$.

In the first case, $\sum_i h_i \otimes (f \circ g_i) = 0$ on $\mathcal{H}_c^\bullet$.

In the second case, we can express one of these functions, say h_1, as a linear combination of the rest, and thus the sum $\sum_i h_i \otimes (g_i \circ f)$ is reducible on $\mathcal{H}_c^\bullet$ to a similar sum with fewer terms. Recycling this alternative several times, we eventually will reduce the original element $\sum_i h_i \otimes (f \circ g_i) \in \mathsf{ker}(\mathsf{P})$ with the irreducible one (having the minimal number of summunds) and all the functions $\{h_i\}$ being linearly independent on $\mathcal{H}_c^\bullet$. Again if, such irreducible on $\mathcal{H}_c^\bullet$ sum belongs to the kernel $\mathsf{ker}(\mathsf{P})$, it must vanish there.

Since the sets $\{\mathcal{H}_c^\bullet\}_{c \in \mathbb{R}}$ cover X (see Fig. 3.2), we conclude that $\mathsf{ker}(\mathsf{P}) = 0$. Therefore, P is a monomorphism.

Let us compare the, so called, projective crossnorms $\{\| \sim \|_k\}_{k \in \mathbb{Z}_+}$ (see (3.15)) of an element

$$\phi = \sum_i h_i \otimes (f \circ g_i) \in C^\infty(X)$$

and the norms of the element $\mathsf{P}(\phi) = \sum_i h_i \cdot (f \circ g_i)$. By comparing the Taylor polynomial of the product of two smooth functions with the product of their Taylor polynomials, we get that, for all $k \in \mathbb{Z}_+$,

$$\|\phi\|_k =_{\mathsf{def}} \inf\Big\{ \sum_i \|h_i\|_k \cdot \|(f \circ g_i)\|_k \Big\} \geq \|\mathsf{P}(\phi)\|_k, \tag{3.15}$$

where inf is taken over all representations of the element $\phi \in C^\infty(\mathcal{T}(v)) \otimes f^*(C^\infty(\mathbb{R}))$ as a sum $\sum_i h_i \otimes (f \circ g_i)$. Note that the elements of the form $h \otimes (f \circ g)$ span $C^\infty(\mathcal{T}(v)) \otimes f^*(C^\infty(\mathbb{R}))$, and for each k, the map P, being applied to such basic elements, is an "isometry":

$$\|\mathsf{P}(h \otimes (f \circ g))\|_k = \|h \otimes (f \circ g)\|_k.$$

By the inequality in (3.15), P is a bounded (continuous) operator. As a result, by continuity, P extends to an algebra homomorphism

$$\hat{\mathsf{P}} : C^\infty(\mathcal{T}(v)) \,\hat{\otimes}\, f^*(C^\infty(\mathbb{R})) \to C^\infty(X)$$

whose source is the completion of the algebraic tensor product $C^\infty(\mathcal{T}(v)) \otimes f^*(C^\infty(\mathbb{R}))$.

Lemma 3.7. *The embedding $\alpha : X \to \mathcal{T}(v) \times \mathbb{R}$ (introduced prior to Theorem 3.4) induces an algebra epimorphism*

$$\alpha^* : C^\infty(\mathcal{T}(v)) \hat{\otimes} C^\infty(\mathbb{R}) \xrightarrow{\mathsf{id}\,\hat{\otimes}\, f^*} C^\infty(\mathcal{T}(v)) \hat{\otimes} f^*(C^\infty(\mathbb{R})) \xrightarrow{\hat{\mathsf{P}}} C^\infty(X). \tag{3.16}$$

Moreover, the map $\hat{\mathsf{P}}$ is an isomorphism.

Proof. First, we claim that the subalgebra $\mathsf{P}\big(C^\infty(\mathcal{T}(v)) \otimes f^*(C^\infty(\mathbb{R}))\big) \subset C^\infty(X)$ satisfies the three hypotheses of Nachbin's Theorem [Na], listed below. Therefore, by [Na], the P-image of $C^\infty(\mathcal{T}(v)) \otimes f^*(C^\infty(\mathbb{R}))$ is *dense* in $C^\infty(X)$. Let us validate these three hypotheses.

(1) *For each $x \in X$, there is a function $q \in C^\infty(\mathcal{T}(v)) \otimes f^*(C^\infty(\mathbb{R}))$ such that $q(x) \neq 0$.*
Just take $q = f \circ (t + c)$, where $c > \min_X f$ and $t : \mathbb{R} \to \mathbb{R}$ is the identity.

(2) *For each $x, y \in X$, there is a function $q \in C^\infty(\mathcal{T}(v)) \otimes f^*(C^\infty(\mathbb{R}))$ such that $q(x) \neq q(y)$ (i.e., the algebra $C^\infty(\mathcal{T}(v)) \otimes f^*(C^\infty(\mathbb{R}))$) separates the points of X)*

If $f(x) \neq f(y)$, $q = f$ will do. If $f(x) = f(y)$, but $[\gamma_x] \neq [\gamma_y]$, then there is a v-invariant function $h \in C^\infty(\mathcal{T}(v))$ such that $h(x) = 1$ and $h(y) = 0$. To construct such h, we take a transversal section $S_x \subset \hat{X}$ of the $\hat{v}$-flow in the vicinity of x such that all the $\hat{v}$-trajectories through S_x are distinct from the trajectory γ_y. We pick a smooth function $\tilde{h} : S_x \to \mathbb{R}$ such that h is supported in $\mathsf{int}(S_x)$, vanishes with all its derivatives along the boundary ∂S_x, and $\tilde{h}(x) = 1$. Let $\mathcal{S}$ denote the set of $\hat{v}$-trajectories through S_x. Of course, $\tilde{h}$ extends to a smooth function $h^\dagger : \mathcal{S} \to \mathbb{R}$ so that $h^\dagger$ is constant along each trajectory from $\mathcal{S}$. We denote by $h^\ddagger$ the obvious extension of $h^\dagger$ by the zero function. Finally, the restriction h of $h^\ddagger$ to X separates x and y.

(3) *For each point $x \in X$ and vector $w \in T_xX$, there is a function $q \in C^\infty(\mathcal{T}(v)) \otimes f^*(C^\infty(\mathbb{R}))$ such that $dq_x(w) \neq 0$.*
Let us decompose $w = av + bw^\dagger$, where $a, b \in \mathbb{R}$ and the vector $w^\dagger$ is tangent to the hypersurface $S_x = \hat{f}^{-1}(f(x))$. Then, if $a \neq 0$, then $df(w) \neq 0$. If $a = 0$, then there is a function $\tilde{h} : S_x \to \mathbb{R}$ which, with all its derivatives, is compactly supported in the vicinity of x in S_x and such that $d\tilde{h}_x(w^\dagger) \neq 0$. As in the case (2), this function extends to a desired function $h \in C^\infty(\mathcal{T}(v))$. Now put $q = h \otimes 1$.

As a result, the image of $\mathsf{P} : C^\infty(\mathcal{T}(v)) \otimes f^*(C^\infty(\mathbb{R})) \longrightarrow C^\infty(X)$ is dense. Therefore, $\hat{\mathsf{P}}$ and, thus, $(\alpha)^* : C^\infty(\mathcal{T}(v))\, \hat{\otimes}\, C^\infty(\mathbb{R}) \longrightarrow C^\infty(X)$ are epimorphisms.

Let us show that $\hat{\mathsf{P}}$ also is a monomorphism. Take a typical element $\theta = \sum_{i=1}^\infty h_i \otimes (f \circ g_i) \in C^\infty(X, v)\, \hat{\otimes}\, f^*(C^\infty(\mathbb{R}))$, viewed as a sum that converges in all the norms $\| \sim \|_k$ from (3.15). We aim to prove that if $\hat{\mathsf{P}}(\theta) = \sum_{i=1}^\infty h_i \cdot (f \circ g_i)$ vanishes on X, then $\theta = 0$.

For each point $x \in \mathsf{int}(X)$, there is a small closed cylindrical solid $H_x \subset \mathsf{int}(X)$ that contains x and consists of segments of trajectories through a small n-ball $D^n \subset f^{-1}(f(x))$, transversal to the flow. Thus, the product structure $D^1 \times D^n$ of the solid H_x is given by the v-flow and the Lyapunov function $f : X \to \mathbb{R}$. We localize the problem to the cylinder H_x.

Consider the commutative diagram

$$\begin{array}{ccc} C^\infty(X, v)\, \hat{\otimes}\, f^*(C^\infty(\mathbb{R})) & \xrightarrow{\hat{\mathsf{P}}} & C^\infty(X). \\ \downarrow \mathsf{res}'\hat{\otimes}\mathsf{res}'' & & \downarrow \mathsf{res} \\ C^\infty(D^n)\, \hat{\otimes}\, C^\infty(D^1) & \xrightarrow{\approx \hat{\mathsf{Q}}} & C^\infty(H_x), \end{array} \tag{3.17}$$

where $\mathsf{res} : C^\infty(X) \to C^\infty(H_x)$ is the natural homomorphism,

$$(\mathsf{res}'\hat{\otimes}\mathsf{res}'')\Big(\sum_{i=1}^\infty h_i \otimes (f \circ g_i)\Big) =_{\mathsf{def}} \sum_{i=1}^\infty h_i|_{D^n} \otimes (f \circ g_i)|_{D^1},$$

and $\hat{\mathsf{Q}}\big(\sum_{i=1}^\infty \tilde{h}_i \otimes \tilde{g}_i\big) =_{\mathsf{def}} \sum_{i=1}^\infty \tilde{h}_i \cdot \tilde{g}_i$ for $\tilde{h}_i \in C^\infty(D^n)$, $\tilde{g}_i \in C^\infty(D^1)$.

Since $\hat{\mathsf{Q}}$ is an isomorphism for products of manifolds (see [GG]) and $\hat{\mathsf{P}}(\theta) = 0$, it follows from (3.17) that $\theta \in \ker(\mathsf{res}'\hat{\otimes}\mathsf{res}'')$ for any cylinder H_x. After reshuffling terms in the sum, as we did in the proof of Lemma 3.7, one may assume that all the functions $\{h_i|_{D^n}\}_i$ are linearly independent. Using that the functions $h_i|_{D^n}$ and $(f \circ g_i)|_{D^1}$ depend of the complementary groups of coordinates in H_x, we conclude that these functions must vanish for any $H_x \subset \mathsf{int}(X)$. As a result, $\theta = 0$ globally in $\mathsf{int}(X)$ and, by continuity, θ vanishes in X. □

Consider now the "known" homomorphism of algebras

$$(\alpha^\partial)^* : C^\infty(\mathcal{T}(v))\, \hat{\otimes}\, C^\infty(\mathbb{R}) \xrightarrow{\approx \mathsf{res}^\partial_{\mathcal{T}}\, \hat{\otimes}\, (f^\partial)^*}$$
$$\longrightarrow C^\infty(\partial_1 X, v)\, \hat{\otimes}\, (f^\partial)^*(C^\infty(\mathbb{R})) \xrightarrow{\hat{\mathsf{R}}^\partial} C^\infty(\partial_1 X), \tag{3.18}$$

utilizing the boundary data. In (3.18), by the definition of algebra $C^\infty(\partial_1 X, v)$, the map $\mathsf{res}^\partial_{\mathcal{T}} : C^\infty(\mathcal{T}(v)) \to C^\infty(\partial_1 X, v)$ is an isomorphism, and $\hat{\mathsf{R}}^\partial$ denotes the completion of the bounded homomorphism R^∂

that takes each element $\sum_i h_i \otimes (f^\partial \circ g_i)$, where $h_i \in C^\infty(\partial_1 X, v)$ and $g_i \in C^\infty(\mathbb{R})$, to the sum $\sum_i h_i \cdot (g_i \circ f^\partial)$.

The next lemma shows that the hypotheses of Theorem 3.5 below are not restrictive, even when $\partial_1 X$ has many connected components.

Lemma 3.8. *Any traversing vector field v on a connected X admits a Lyapunov function $f : X \to \mathbb{R}$ such that $f(X) = f(\partial_1 X)$.*

Proof. Note that, for any Lyapunov function f, the image $f(\partial_1 X)$ is a disjoint union of finitely many closed intervals $\{I_k = [a_k, b_k]\}_k$, where the index k reflects the natural order of intervals in $\mathbb{R}$ (note that different connected components of $\partial_1 X$ may contribute to the same interval). We will show how to decrease, step by step, the number of these intervals by deforming the original function f. Note that the local extrema of any Lyapunov function on X occur on its boundary $\partial_1 X$ and away from the locus $\partial_2 X(v)$ where v is tangent to $\partial_1 X$. Consider a pair of points $A_{k+1}, B_k \in \partial_1 X \setminus \partial_2 X(v)$ such that $f(A_{k+1}) = a_{k+1}$ and $f(B_k) = b_k$, where $a_{k+1} > b_k$. Then we can increase f in the vicinity of its local maximum B_k so that the B_k-localized deformation $\tilde{f}$ of f has the property $\tilde{f}(B_k) > f(A_{k+1})$ and $\tilde{f}$ is a Lyapunov function for v. This construction decreases the number of intervals in $\tilde{f}(\partial_1 X)$ in comparison to $f(\partial_1 X)$ at least by one. □

We are ready to state the main result of this section ([K15]).

Theorem 3.5. *Assuming that the image $f^\partial(\partial_1 X)$ is a connected interval of $\mathbb{R}$,*[7] *the algebra $C^\infty(X)$ is isomorphic to the subalgebra*

$$C^\infty(\partial_1 X, v)\,\hat{\otimes}\,(f^\partial)^*(C^\infty(\mathbb{R})) \subset C^\infty(\partial_1 X)\hat{\otimes} C^\infty(\partial_1 X).$$

Moreover, by combining (3.16) with (3.18), we get a commutative diagram

$$\begin{array}{ccc} C^\infty(\mathcal{T}(v))\,\hat{\otimes}\, f^*(C^\infty(\mathbb{R})) & \xrightarrow{\hat{\mathsf{R}}} & C^\infty(X) \\ \downarrow \mathsf{id}\,\hat{\otimes}\, H^\partial_f & & \downarrow \mathsf{res} \\ C^\infty(\partial_1 X, v)\,\hat{\otimes}\,(f^\partial)^*(C^\infty(\mathbb{R})) & \xrightarrow{\hat{\mathsf{R}}^\partial} & C^\infty(\partial_1 X), \end{array} \tag{3.19}$$

whose vertical homomorphism $\mathsf{id}\,\hat{\otimes}\, H^\partial_f$ and the horizontal homomorphism $\hat{\mathsf{R}}$ are isomorphisms, and the vertical epimorphism res *is the obvious restriction operator.*

As a result, we get an algebra isomorphism

$$\mathcal{H}(v,f) : C^\infty(\partial_1 X, v)\,\hat{\otimes}\,(f^\partial)^*(C^\infty(\mathbb{R})) \approx C^\infty(X). \tag{3.20}$$

[7] which is the case for a connected $\partial_1 X$

Proof. Consider the commutative diagram (3.19). Its upper-right conner is "unknown", while the lower row is "known" and represents the boundary data. The homomorphism res is obviously an epimorphism. By Lemma 3.7, the left vertical arrow $\mathsf{id}\,\hat{\otimes}\, H^\partial_f$ is an isomorphism. Since, by Lemma 3.7, $\hat{\mathsf{R}}$ is an isomorphism, it follows that $\hat{\mathsf{R}} \circ (\mathsf{id}\,\hat{\otimes}\, H^\partial_f)^{-1}$ must be an isomorphism as well. In particular, $\hat{\mathsf{R}}^\partial$ is an epimorphism, whose kernel consists of the smooth functions on X whose restrictions to $\partial_1 X$ vanish. If $z \in C^\infty(X)$ is a smooth function such that zero is its regular value and $z^{-1}(0) = \partial_1 X$ ($z > 0$ in $\mathsf{int}(X)$), then the kernel of res is the principle ideal $\mathsf{m}(z)$, generated by z. Therefore, by the commutativity of (3.19), the kernel of the homomorphism $\hat{\mathsf{R}}^\partial$ must be also a principle ideal M_∂, generated by the element $(\hat{\mathsf{R}} \circ (\mathsf{id}\,\hat{\otimes}\, H^\partial_f))^{-1}(z)$. □

Theorem 3.5 leads to the following claim.

Corollary 3.6. *If the range $f^\partial(\partial_1 X)$ is a connected interval in $\mathbb{R}$, then the two topological algebras $C^\infty(\partial_1 X, v) \subset C^\infty(\partial_1 X)$ and $(f^\partial)^*(C^\infty(\mathbb{R})) \subset C^\infty(\partial_1 X)$ determine, up to an isomorphism, the algebra $C^\infty(X)$, and thus determine the smooth topological type of the manifold X.*

Proof. We call a maximal ideal of an algebra $\mathcal{A}$ nontrivial if it is different from $\mathcal{A}$.

By Theorem 3.5, the algebra $C^\infty(X)$ is determined by the two algebras on $\partial_1 X$, $C^\infty(\partial_1 X, v)$ and $(f^\partial)^*(C^\infty(\mathbb{R}))$, up to an isomorphism. In turn, the algebra $C^\infty(X)$ determines the smooth topological type of X, viewed as a ringed space. This fact is based on interpreting X as the space $\mathcal{M}(C^\infty(X))$ of nontrivial maximal ideals of the algebra $C^\infty(X)$ [KMS].

Let $\mathsf{m}^\partial_v \lhd C^\infty(\partial_1 X, v)$ and $\mathsf{m}^\partial_f \lhd (f^\partial)^*(C^\infty(\mathbb{R}))$ be a pair of nontrivial maximal ideals. Note that $\mathsf{m}^\partial_v = \mathsf{m}^\partial_v([\gamma])$ consists of functions from $C^\infty(\partial_1 X, v)$ that vanish on the locus $\gamma \cap \partial_1 X$, and $\mathsf{m}^\partial_f = \mathsf{m}^\partial_f(c)$ consists of functions from $(f^\partial)^*(C^\infty(\mathbb{R}))$ that vanish on the locus $\partial_1 X \cap f^{-1}(c)$, where $c \in f(\partial_1 X) \subset \mathbb{R}$. We denote by $\langle \mathsf{m}^\partial_v, \mathsf{m}^\partial_f \rangle$ the maximal ideal of $C^\infty(\partial_1 X, v)\,\hat{\otimes}\,(f^\partial)^*(C^\infty(\mathbb{R}))$ that contains both ideals $\mathsf{m}^\partial_v \hat{\otimes} 1$ and $1 \hat{\otimes} \mathsf{m}^\partial_f$.

If the range $f^\partial(\partial_1 X)$ is a connected interval of $\mathbb{R}$ and $\langle \mathsf{m}^\partial_v, \mathsf{m}^\partial_f \rangle$ is a nontrivial ideal, then $\gamma \cap f^{-1}(c) \neq \emptyset$. Otherwise, $\gamma \cap f^{-1}(c) = \emptyset$. Therefore, with the help of the isomorphism $\mathcal{H}(v, f)$ from (3.20), the nontrivial maximal ideals of $C^\infty(X)$ (which by [KMS] correspond to points $x = \gamma \cap f^{-1}(c) \in X$) are of the form $\mathcal{H}(v, f)\big(\langle \mathsf{m}^\partial_v, \mathsf{m}^\partial_f \rangle\big)$. □

Corollary 3.7. *Let the range $f^\partial(\partial_1 X)$ be a connected interval of $\mathbb{R}$.*

With the isomorphism $\mathcal{H}(v, f)$ from (3.20) being fixed, any algebra isomorphism $\Psi^\partial : C^\infty(\partial_1 X) \to C^\infty(\partial_1 X)$ that preserves the subalgebras $C^\infty(\partial_1 X, v)$ and $(f^\partial)^*(C^\infty(\mathbb{R}))$ extends canonically to the algebra isomorphism $\Psi : C^\infty(X) \to C^\infty(X)$.

Thus, an action of any group G of such isomorphisms Ψ^∂ extends canonically to a G-action on the algebra $C^\infty(X)$ and, via it, to a G-action on X by smooth diffeomorphisms.

Proof. By [Mr], any algebra isomorphism $\Psi : C^\infty(X_1) \to C^\infty(X_2)$ is induced by a unique smooth diffeomorphism $\Phi : X_1 \to X_2$. With this fact in hand, by Theorem 3.5 and Theorem 3.3, the proof is on the level of definitions. □

The crucial question: "How to characterize intrinsically the trace $C^\infty(\partial_1 X, v)$ of the algebra $C^\infty(\mathcal{T}(v)) \approx \mathsf{ker}\{\mathcal{L}_v : C^\infty(X) \to C^\infty(X)\}$ in the algebra $C^\infty(\partial_1 X)$?" remains unanswered.

Keeping in mind the claim of Corollary 3.6, let us revisit Question 3.1 and Conjecture 3.1 (which claims that $C^\infty(\partial_1 X, v) = \mathsf{M}(v)^{C_v}$).

Recall that functions from $C^\infty(\partial_1 X, v)$ are constant along each C_v-"trajectory" $\gamma^\partial = \gamma \cap \partial_1 X$ of the causality map. Furthermore, any smooth function $\psi : \partial_1 X \to \mathbb{R}$ that is constant on each finite set γ^∂ gives rise to a unique *continuous* function ϕ on X that is constant along each v-trajectory γ. However, as we remarked before, such functions ϕ may not be automatically *smooth* on X (a priori, they are just Hölderian with some control of the Hölder exponent that depends on the dimension of X only)!

Remark 3.6. If Conjecture 3.1 is true, then the subalgebra $C^\infty(\partial_1 X, v) \subset C^\infty(\partial_1 X)$ is isomorphic to the algebra $\mathsf{M}(v)^{C_v}$ and thus is determined by the causality map C_v.

On the other hand, if v_1 and v_2 satisfy Property A, then under the hypotheses of Theorem 3.3, there is a smooth diffeomorphism $\Phi : X_1 \to X_2$, which extends a given diffeomorphism Φ^∂ on $\partial_1 X_1$ and takes the oriented foliation $\mathcal{F}(v_1)$ to $\mathcal{F}(v_2)$. Then Φ^* must take $C^\infty(\partial_1 X_2, v_2)$ to $C^\infty(\partial_1 X_1, v_1)$; so the two subalgebras must be isomorphic. By the construction of Φ, we get that $\Phi^* : (f_2^\partial)^*(C^\infty(\mathbb{R})) \to (f_1^\partial)^*(C^\infty(\mathbb{R}))$ is an algebra isomorphism as well. Therefore, under the hypotheses of Theorem 3.3, the hypotheses of Corollary 3.6, are satisfied.

The Holography Theorem 3.3 and Corollary 3.6 have been established assuming Property A from Definition 3.2. If one assumes the validity of

Conjecture 3.1, then, by Corollary 3.6, we may drop Property A from the hypotheses of the Holography Theorem. Indeed, the subalgebras $C^\infty(\partial_1 X, v)$ and $(f^\partial)^*(C^\infty(\mathbb{R}))$ would acquire a description in terms of C_v and f^∂. This would deliver an independent proof of a natural generalization of Corollary 3.5. ◇

3.3 Applications of holography theorems

The next theorem suggests that traversing vector fields and their causality maps give rise to *new representations* of smooth manifolds with boundary. The reader may compare these representations with, say, common representations of smooth manifolds as preimages of the origin $0 \in \mathbb{R}^l$ for smooth maps $f : \mathbb{R}^k \to \mathbb{R}^l$ $(k \geq l)$ that are transversal to 0.

Theorem 3.6. *For $n \geq 2$, any compact connected smooth $(n+1)$-dimensional manifold X with a connected boundary $\partial_1 X$ can be represented, up to a homeomorphism, by a semi-continuous map $C : D^n \to \partial_1 X \setminus \mathsf{int}(D^n)$ of the standard n-ball $D^n \subset \partial_1 X$ to the closure of its compliment in $\partial_1 X$. The C-fibers are finite, and a generic C-fiber is of cardinality 1.*

In other words, the map C captures the topological type of X. Away from a closed subcomplex $K \subset X$ of codimension 3, it captures the smooth topological type of X.

When the boundary $\partial_1 X$ is diffeomorphic to the standard n-sphere, then the topological type of X is encoded via a map C between a pair of standard n-balls.

Proof. Consider any compact connected smooth manifold X with a connected boundary $\partial_1 X$. By Theorem 3.2 and Corollary 3.3 from [K1] (alternatively, by Corollaries 5.6-5.7 from [K7]), for $n > 1$, there is an open set $\mathcal{D}(X)$ of traversally generic vector fields v, such that $\partial_1^+ X(v)$ is diffeomorphic to the standard ball $D^n_+ \subset \partial_1 X$.

Let us outline crudely a few key steps that led to these theorems and corollaries. Starting with a gradient like vector field v of a Morse function $f : X \to \mathbb{R}$, we isotop X to a smooth submanifold $X_1 \subset X$ so that all the critical points of f are outside X_1. Then we restrict f and v to X_1. Next, we perturb f so that it becomes a Morse function on $\partial_1 X_1$. The critical points of $f|_{\partial_1 X_1}$ are of two types: the ones that belong to $\partial_1^+ X_1(v)$ and the ones that belong to $\partial_1^- X_1(v)$. We denote by Σ^+ the finite set of critical points of the first type and by Σ^- the finite set of critical points of the second type. We take a standard ball $D^n \subset \partial_1 X_1$ that contains Σ^+

and no points from Σ^-. Then we argue that v on X_1 can be deformed to another f-gradient-like vector field v_1 so that $\partial_1^+ X_1(v_1) = D^n$. Finally, employing Theorem 2.1, by a perturbation $\tilde{v}_1$ of v_1, we may assume that $\tilde{v}_1$ is transversally generic with respect to $\partial_1 X_1$, while preserving the smooth topological type of $\partial_1^+ X_1(v_1) = D^n \approx \partial_1^+ X_1(\tilde{v}_1)$. Such vector fields $\tilde{v}_1$ form an open set $\mathcal{D}(X_1)$ in the space of traversally generic vector fields on X_1.

When the boundary $\partial_1 X \approx \partial_1 X_1$ is spherical, then $\partial_1^- X(\tilde{v}_1)$ is the complimentary smooth ball D^n_-. According to Corollary 3.5, for any vector field $\tilde{v} \in \mathcal{D}(X_1)$, the topological type of the manifold X_1 is determined by the semi-continuous causality map $C_{\tilde{v}} : D^n_+ \to \partial_1^- X(\tilde{v})$ (equivalently, by its graph $\Gamma(C_{\tilde{v}}) \subset D^n_+ \times \partial_1^- X(\tilde{v})$).

For a traversally generic $\tilde{v}$, the locus $X(\tilde{v}, (33)_\succeq)$ is of codimension 4 in X. Also, the locus $X(\tilde{v}, (4)_\succeq)$ is of codimension 3. Therefore, by Corollary 3.5, this map $C_{\tilde{v}}$ captures, away from the locus $X(\tilde{v}, (33)_\succeq \cup X(\tilde{v}, (4)_\succeq)$ of codimension ≥ 3, the smooth topological type of X. $\square$

Remark 3.7. Compare this description of a $(n+1)$-manifold X with the spherical boundary, via a map $C_v : D^n_+ \to D^n_-$ between two standard balls, with the description of the trajectory space $\mathcal{T}(v)$, given by the Origami Theorem 3.1 and Theorem 4.1 from [K9]. For a specially designed traversally generic v, the Origami Theorem represents the trajectory space $\mathcal{T}(v)$ as the continuous image of a ball D^n, where $n = \dim(\partial_1 X)$. $\diamondsuit$

Example 3.4. Let $f : Y \to \mathbb{R}$ be a smooth Bott-Morse function [B] on a closed smooth manifold Y, and let M be the f-critical locus. Different components of M may have different dimensions; the restriction $f|_M$ is locally constant. Let $\nu(M, Y) = (TY|_M)/TM$ be the normal bundle to M.

Our goal here is to use the gradient flow $v = \nabla f$ of f on the complement $X =_{\mathsf{def}} Y \setminus U_M$ to an appropriate regular neighborhood U_M of M in Y, to set up the causality map $C_v : \partial_1^+ X(v) \to \partial_1^- X(v)$. In such a setting, v is traversing on X. Thanks to the quadratic nature of f in the normal to M direction, we get $\partial_3 X(v) = \emptyset$. Therefore, Property A holds, and the scattering data determine the smooth type of X.

The technicalities of realizing this plan require some explanation. Let us fix a smooth diffeomorphism $\mathcal{N} : \nu(M, Y) \hookrightarrow Y$ that identifies the space of the normal bundle $\nu(M, Y)$ with a regular neighborhood U_M of M in Y so that $\mathcal{N}|_M = \mathsf{id}_M$. We fix also the local exponential diffeomorphisms $\{exp_c : T_c\mathbb{R} \hookrightarrow \mathbb{R}\}_{c \in \mathbb{R}}$, where $T_c\mathbb{R}$ denotes the tangent line to $\mathbb{R}$ at $c \in \mathbb{R}$.

Let $\Psi(x, \xi) =_{\mathsf{def}} (exp^{-1}_{f(x)} \circ f \circ \mathcal{N})(x, \xi)$, where $x \in M$ and a vector

$\xi \in \nu_x(M, Y)$. Then the value of the quadratic differential $Hess(f)|_M$—the Hessian—on a vector $\xi \in \nu(M, Y)$ at a point $x \in M$ is given by the formula

$$Hess_x(f)(\xi) =_{\mathsf{def}} \lim_{t \to 0} \frac{\Psi(x, t\xi)}{t^2}.$$

For each $x \in M$, $Hess_x(f)$ is a quadratic form in $\xi \in \nu_x(M, Y)$. The quadratic differential $Hess(f)|_M$ does not depend on the choice of local diffeomorphism $\mathcal{N}$.

By definition, for a Bott-Morse function f, the Hessian $Hess(f)$ is non-degenerate on the fibers of a normal bundle $\nu(M, Y)$. Moreover, $\nu(M, Y)$ splits into a direct sum of two subbundles, $\nu^+(M, Y)$ and $\nu^-(M, Y)$, so that the restrictions of the quadratic form $Hess(f)$ to $\nu^+(M, Y)$ is positive-definite, and to $\nu^-(M, Y)$ is negative-definite.

At each point $x \in M$, the equation $\{Hess_x(f)(\xi) = 0\}$, where $\xi \in \nu_x(M, Y)$, defines a quadratic cone $Q_x \subset \nu_x(M, Y)$ with the apex at x. If, for a particular connected component M' of M, the Hessian is negative definite, the cone $Q_x = x$, where $x \in M'$.

Let Q be a hypersurface in $\nu(M, Y)$, formed by the union $\bigcup_{x \in M} Q_x$ of these cones.

With the help of $Hess(f)|_M$, we introduce a Riemanian metric g_ν on $\nu(M, Y)$ as follows. By definition,

$$g_\nu \,|_{\nu^+(M,Y)} = Hess(f) \,|_{\nu^+(M,Y)}, \;\; g_\nu \,|_{\nu^-(M,Y)} = -Hess(f) \,|_{\nu^-(M,Y)},$$

and the two subbundles are othogonal: $g_\nu \,|_{\nu^+(M,Y)} \perp g_\nu \,|_{\nu^-(M,Y)}$.

By picking a Riemannian metric g_M on M, we get the Riemannian metric $g^\# =_{\mathsf{def}} g_\nu \oplus g_M$ on the space $\nu(M, Y)$ of the normal fibration. Employing g_ν, we form a closed ϵ-disk bundle $\nu^\epsilon(M, Y) \subset \nu(M, Y)$. With the help of the diffeomorphism $\mathcal{N}$, we map it to $U_\epsilon =_{\mathsf{def}} \mathcal{N}(\nu^\epsilon(M, Y))$, a regular neighborhood of M in Y, and transfer the Riemannian metric $g^\#$ to U_ϵ. Let us denote by $g_\epsilon^\dagger$ this metric on U_ϵ and by g_ϵ an extension of $g_\epsilon^\dagger$ to the entire Y.

Now, we consider the gradient field $\nabla_{g_\epsilon}(f)$ of f. The gradient $\nabla_{g_\epsilon}(f)$, being restricted to $X_\epsilon =_{\mathsf{def}} Y \setminus \mathrm{int}(U_\epsilon)$, has f as its Lyapunov function. Put $v_\epsilon =_{\mathsf{def}} \nabla_{g_\epsilon}(f)|_{X_\epsilon}$. Then v_ϵ is a traversing vector field on X_ϵ.

Let $\tilde{Q} =_{\mathsf{def}} \mathcal{N}(Q)$. Away from M, $\tilde{Q}$ is a smooth hypersurface, which is transversal to the boundary $\partial_1 U_{\epsilon/2}$. It divides $\partial_1 U_{\epsilon/2}$ into two domains: $\partial_1^+ U_{\epsilon/2}$ that contains the $\mathcal{N}$-image of the spherical bundle $S_{\epsilon/2}(\nu^+(M, Y))$ and $\partial_1^- U_\epsilon$ that contains the $\mathcal{N}$-image of the spherical bundle $S_{\epsilon/2}(\nu^-(M, Y))$.

Because of the quadratic nature of f in the vicinity of the singular locus M, the vector field v_ϵ is boundary generic with respect to $\partial_1 U_{\epsilon/2}$. Furthermore,

$$\partial_1^\pm X_{\epsilon/2}(v_\epsilon) = \partial_1^\mp U_{\epsilon/2}(v), \quad \partial_2 X_{\epsilon/2}(v_\epsilon) = \partial_1 U_{\epsilon/2}(v) \cap \tilde{Q},$$

and $\partial_3 X_{\epsilon/2}(v_\epsilon) = \emptyset$.

Therefore, the property A is valid, and by Theorem 3.3, the causality map $C_{v_\epsilon} : \partial_1^+ X_\epsilon(v_\epsilon) \to \partial_1^- X_\epsilon(v_\epsilon)$ allows for a reconstruction of $(X_\epsilon, \mathcal{F}(v_\epsilon))$, up to a smooth diffeomorphism.

It seems that not only the smooth topological type of the pair $(X_\epsilon, \mathcal{F}(v_\epsilon))$ stabilizes for all sufficiently small $\epsilon > 0$, but even the causality maps $\{C_{v_\epsilon}\}$ "stabilize" in the sense that C_{v_ϵ} and $C_{v_{\epsilon'}}$ are conjugated via a diffeomorphism $\Phi^\partial : \partial_1 U_\epsilon(M, Y) \to \partial_1 U_{\epsilon'}(M, Y)$ for all sufficiently small $\epsilon, \epsilon' > 0$. $\diamondsuit$

3.3.1 *Time-dependent vector fields and the holography for space-time capsules*

Let us consider now any *time-dependent* vector field $u(t)$, $t \in \mathbb{R}$, on a n-dimensional manifold Y *without* boundary. Then $u(t)$ gives rise to a non-vanishing vector field $v = v(t) =_{\mathsf{def}} (u(t), 1)$ on the manifold $Y \times \mathbb{R}$. Note that v is traversing with the Lyapunov function $T : Y \times \mathbb{R} \to \mathbb{R}$, given by the formula $T(y, t) = t$. We call a pair (y, t) an event, since we think of T as time, and of Y as space.

Let $X \subset Y \times \mathbb{R}$ be a compact smooth submanifold of codimension zero. We call it space-time capsule. We call the events in the capsule X internal and those in $Y \times \mathbb{R} \setminus \mathsf{int}(X)$ external.

Since $dT(v) > 0$ and X is compact, any v-trajectory $\gamma(t)$ that passes through a point of X is contained in X for a compact set of instances $t \in \mathbb{R}$. In other words, v is *traversing* on X with T playing the role of Lyapunov function.

Assume that $X \subset Y \times \mathbb{R}$ is such that v is boundary generic with respect to $\partial_1 X$. In view of Theorem 6.7 from [K7] (see also [K2]), this assumption can be satisfied by a small perturbation $\tilde{v}$ of v. In fact, such perturbation $\tilde{v}$ can be of the form $(\tilde{u}, 1)$, since the property of a field to be boundary generic/traversally generic depends only on its direction, and not on its magnitude.

We call the boundary $\partial_1 X$ the event horizon of the capsule X. We stress that the event horizon is chosen as an independent set of boundary data, not directly related to the time-dependent dynamic system $u(t)$ on the manifold Y.

Thus $u(t)$ defines the causality map $C_v : \partial_1^+ X(v) \to \partial_1^- X(v)$ which takes each "entrance" point $x_0 = (t_0, y_0)$ of the v-flow on the event horizon $\partial_1 X$ to the closest along the v-trajectory trough x_0 "exit" point $x_1 = (t_1, y_1)$ on $\partial_1 X$.

We can think of the event x_0 as the cause of the event x_1, so that C_v becomes, indeed, the causality map or the causality relation on the event horizon $\partial_1 X$.

The Holography Theorem 3.3 and Corollary 3.4 have the following immediate, but important interpretation, which applies to time-dependent vector fields.

Theorem 3.7. (The Causal Holography Principle)
Let $u(t)$, $t \in \mathbb{R}$, be a time-dependent smooth vector field on a connected n-dimensional smooth manifold Y without boundary.

For any compact $(n+1)$-dimensional smooth event capsule $X \subset Y \times \mathbb{R}$ such that the field $v = \{v(t) =_{\mathsf{def}} (u(t), 1)\}_{t \in \mathbb{R}}$ is boundary generic with respect to $\partial_1 X$, the causality relation on the event horizon $\partial_1 X$ determines the pair $(X, \mathcal{F}(v))$, up to a homeomorphism $\Phi : X \to X$ which is the identity on the event horizon.

When v has Property A from Definition 3.2, then the causality relation on the event horizon $\partial_1 X$ determines the pair $(X, \mathcal{F}(v))$, up to a smooth diffeomorphism $\Phi : X \to X$ which is the identity on the event horizon. $\diamondsuit$

Remark 3.8. We do not claim that the reconstruction of the event capsule X in Theorem 3.7 from the causality map C_v allows also for a reconstruction of its slicing by the fixed-time frames! However, adding the time function $T : \partial_1 X \to \mathbb{R}$ to the boundary data, the reconstruction of the slicing by the fixed-time frames becomes also possible. $\diamondsuit$

In turn, Theorem 3.7 has another interpretation:

Corollary 3.8. (The topological rigidity of the boundary value problem for space-time capsules)
Let Y be a connected smooth n-manifold without boundary, and $X \subset Y \times \mathbb{R}$ a compact smooth event capsule. Let $u_1(t), u_2(t)$, $t \in \mathbb{R}$, be two time-dependent smooth vector fields on Y such that $u_1(y,t) = u_2(y,t)$ for all "external" events $(y,t) \in (Y \times \mathbb{R}) \setminus \mathsf{int}(X)$. Assume that the fields $v_1 = (u_1, 1)$

and $v_2 = (u_2, 1)$ *are boundary generic in* X *and that the two causality maps,* $C_{v_1} : \partial_1^+ X(v_1) \to \partial_1^- X(v_1)$ *and* $C_{v_2} : \partial_1^+ X(v_2) \to \partial_1^- X(v_2)$*, are identical.*

Then the two dynamical systems, generated by v_1 *and* v_2 *on* $Y \times \mathbb{R}$*, are topologically equivalent via a homeomorphism which is an identity on the event horizon.*

When v_1 *has Property* A *(in particular, when the locus* $\partial_3 X(v_1) = \emptyset$*), then the two dynamical systems are equivalent via a smooth diffomorphism that is the identity on the event horizon.* ◇

In search for further applications of the Holography Theorem 3.3, let us pay a brief visit to the classical Lagrangian/Hamiltonian mechanics.

Let $\mathcal{L} : TM \times \mathbb{R} \to \mathbb{R}$ be a smooth function, where TM is the tangent bundle of a n-dimensional smooth manifold M, and the multiplier $\mathbb{R}$—the time line. We assume that $\mathcal{L}(q, v, t)$ is a convex function in $v \in T_qM$ for all $q \in M$ and $t \in \mathbb{R}$, that is,

$$\mathcal{L}(q, \tau v_1 + (1-\tau)v_2, t) \;\leq\; \tau\mathcal{L}(q, v_1, t) + (1-\tau)\mathcal{L}(q, v_2, t)$$

for all $\tau \in [0,1]$ and $v_1, v_2 \in T_qM$. We call such $\mathcal{L}$ a Legandrian function.

For each moment of time $t \in \mathbb{R}$, the function $\mathcal{L}_t = \mathcal{L}(\sim, t)$ gives rise to a bundle map $\mathbf{L}\mathcal{L}_t : TM \to T^*M$, called the Legendrian transformation, whose target is the cotangent bundle of M and whose construction is described below.

For a point $(q, v) \in T_qM$, we define $\mathbf{L}\mathcal{L}_t(q, v)$ as a pair $(q, p) \in T_q^*M$, where the, so called, momenta p are introduced by

$$p(w) = \frac{\partial \mathcal{L}}{\partial v}(q, v, t) = \frac{d}{d\tau}\, \mathcal{L}(q,\, v + \tau \cdot w,\, t)\Big|_{\tau=0}$$

for all vectors $w \in T_qM$.

Consider local coordinates $(q, v) = (q^1, \dots, q^n,\, v_1, \dots, v_n)$ in TM and the corresponding canonical coordinates

$$(q^1, \dots, q^n,\, p_1, \dots, p_n) =_{\mathsf{def}} (q^1, \dots, q^n,\, \mathbf{L}\mathcal{L}_t(q, v_1), \dots, \mathbf{L}\mathcal{L}_t(q, v_n))$$

in T^*M.

The Euler-Lagrange equations

$$\Big\{\frac{d}{dt}\frac{\partial \mathcal{L}}{\partial v_i}\big(q(t), \dot{q}(t), t\big) - \frac{\partial \mathcal{L}}{\partial q_i}\big(q(t), \dot{q}(t), t\big) = 0\Big\}_{i \in [1,n]}$$

describe the curve $\gamma = q(t)$ in M which minimizes the path integral $\int_{t_0}^{t_1} \mathcal{L}\, dt$.

As in the case of Reimannian geometry, we assume that the restriction of $\mathcal{L}_t$ to each fiber T_qM is bounded below by a positive definite quadratic form minus a constant. Then the Legendrian transformation $\mathbf{L}\mathcal{L}_t : TM \to T^*M$ is a diffeomorphism (for example, see [Can]). In such a case, the Hamiltonian function $\mathcal{H} : T^*M \times \mathbb{R} \to \mathbb{R}$ is defined by

$$\mathcal{H}(q, p, t) =_{\mathsf{def}} p(v) - \mathcal{L}(q, v, t),$$

where $v = (\mathbf{L}\mathcal{L}_t)^{-1}(p)$ and $(\mathbf{L}\mathcal{L}_t)^{-1} : T^*M \to TM$ denotes the inverse of the Legendrian transformation $\mathbf{L}\mathcal{L}_t$. Using the natural isomorphism $T^{**}M \approx TM$ and applying the Legendrian transformation $\mathbf{L}$ to $\mathcal{H}$, we get $(\mathbf{L}\mathcal{L})^{-1} = \mathbf{L}\mathcal{H}$.

The space T^*M comes equipped with the tautological Liouville 1-form β, a section of the bundle $T^*(T^*M) \to T^*M$. By its definition, for each point $x = (q, p) \in T^*_qM$ and a tangent vector $(\dot{q}, \dot{p}) \in T_x(T^*M)$,

$$\beta_x(\dot{q}, \dot{p}) =_{\mathsf{def}} p(\dot{q}). \tag{3.21}$$

The 1-form β gives rise to a skew-symmetric symplectic 2-form $\omega = -d\beta$ on T^*M. With the help of ω, any Hamiltonian function $\mathcal{H} : T^*M \times \mathbb{R} \to \mathbb{R}$ produces a unique (time-dependent) Hamiltonian vector field $v^\sharp_{\mathcal{H}} \in T(T^*M)$ via the formula

$$\big\{\omega(v^\sharp_{\mathcal{H}}, \sim)(t) = d\mathcal{H}(\sim)(t)\big\}_{t\in\mathbb{R}}. \tag{3.22}$$

In canonical local coordinates, the Euler-Lagrange equations transform into the Hamilton system of ODEs:

$$\dot{q} = \frac{\partial \mathcal{H}}{\partial p}, \quad \dot{p} = -\frac{\partial \mathcal{H}}{\partial q} = \frac{\partial \mathcal{L}}{\partial q}, \quad \frac{\partial \mathcal{H}}{\partial t} = -\frac{\partial \mathcal{L}}{\partial t}. \tag{3.23}$$

In local coordinates (q, p, t), we consider the vector field

$$v_{\mathcal{H}} =_{\mathsf{def}} (v^\sharp_{\mathcal{H}}, 1) = (\dot{q}, \dot{p}, 1) = \Big(\frac{\partial \mathcal{H}}{\partial p}, -\frac{\partial \mathcal{H}}{\partial q}, 1\Big), \tag{3.24}$$

on $T(T^*M \times \mathbb{R})$, which projects onto a (time-dependent) Hamiltonian vector field $v^\sharp_{\mathcal{H}}$ on T^*M.

Applying Theorem 3.7 to the Hamiltonian vector field $v_{\mathcal{H}}$ from (3.24), we get the following claim.

Corollary 3.9. *For a smooth manifold M without boundary, consider the smooth (time-dependent) Hamiltonian system (3.23) on $T^*M \times \mathbb{R}$. Assume that:*

- *$c \in \mathbb{R}$ is a regular value of a smooth function $F : T^*M \times \mathbb{R} \to \mathbb{R}$,*

- *the space-time capsule*
$$X =_{\mathsf{def}} \{x \in T^*M \times \mathbb{R} |\; F(x) \leq c\}$$
is compact in $T^*M \times \mathbb{R}$,
- *the field* $v_{\mathcal{H}}$ *from (3.24) is boundary generic with respect to the event horizon*
$$\partial_1 X =_{\mathsf{def}} \{x \in T^*M \times \mathbb{R} |\; F(x) = c\}.$$

Then the causality map/relation $C_{v_{\mathcal{H}}}$ *on the event horizon* $\partial_1 X$ *allows for a reconstruction of the pair* $(X, \mathcal{F}(v_{\mathcal{H}}))$*, up to a homeomorphism of* X *which is the identity on* $\partial_1 X$.

If $v_{\mathcal{H}}$ *has Property* A *(for example, if* $\partial_3 X(v_{\mathcal{H}}) = \emptyset$*), then the reconstruction of the pair* $(X, \mathcal{F}(v_{\mathcal{H}}))$ *is possible, up to a smooth diffeomorphism which is the identity on* $\partial_1 X$. ◇

Applying Corollary 3.3 to the setting of Corollary 3.9, we get the following assertion.

Corollary 3.10. (Extending symmetry inside of a space-time capsule)
Under the hypotheses of Corollary 3.9, any action of a group G *of smooth diffeomorphisms on the manifold* $\mathbf{c}X =_{\mathsf{def}} (T^*M \times \mathbb{R}) \setminus \mathsf{int}(X)$*, which preserve the* $v_{\mathcal{H}}$*-trajectories and the time slices* $\{T^{-1}(t)\}_{t \in \mathbb{R}}$*, extends to a* G*-action with similar properties on the space-time capsule* X.

In other words, any G*-symmetry that preserves the* $v_{\mathcal{H}}$*-trajectories and the time slices* $\{T^{-1}(t)\}_{t \in \mathbb{R}}$ *outside of the space-time capsule* X *extends to a* G*-symmetry with similar properties of the space-time capsule* X. ◇

The main unresolved issue here is captured by the following question.

Question 3.2. *How abundant are the Hamiltonian systems* $\{v_{\mathcal{H}}\}$ *that are traversing and boundary generic (alternatively, traversally generic) with respect to a given event horizon* $\partial_1 X \subset T^*M \times \mathbb{R}$? ◇

It follows from [K2] that if a vector field $v_{\mathcal{H}}$ is traversing and boundary generic with respect to $\partial_1 X$, then, for any Hamiltonian function $\tilde{\mathcal{H}}$ that is C^∞-close to $\mathcal{H}$, the vector field $v_{\tilde{\mathcal{H}}}$ also will be traversing and boundary generic with respect to $\partial_1 X$.

We know that any traversing vector field v can be C^∞-approximated by a traversally generic field on X (Theorem 3.5 from [K2]). Thus, the open question is whether such an approximation is possible within the space of Hamiltonian fields.

3.4 Holographic structures on closed manifolds

"That's well and good in practice, but how does it work in theory?"

Shmuel Weinberger

In this section, we formalize the holographic properties of traversally generic flows v by considering more abstract structures on closed smooth manifolds Y, structures that mimic the causality maps C_v, generated by pairs (X, v), where $\partial_1 X = Y$.

The reconstruction procedure $(\partial_1 X, C_v) \Rightarrow (X, \mathcal{F}(v))$, whose existence is claimed by Corollary 3.5, presumes that the manifold X and the traversally generic vector field v on it, giving rise to the boundary-confined data $(\partial_1 X, C_v)$, *do exist.*

Thanks to formula (2.12), for a traversally generic v, we have semi-local models for the causality map C_v in the vicinity of each C_v-"trajectory"

$$\gamma_x^\partial =_{\mathsf{def}} \big(x,\, C_v(x),\, C_v(C_v(x)),\, \dots,\, C_v^{(d)}(x)\big) \in (\partial_1 X)^{d+1},$$

where $d = \#\{\gamma_x \cap \partial_1 X\}$, all the iterative C_v-images of x are distinct, and $x \neq C_v(x')$ for all $x' \neq x$.

Still, the question *"Which pairs $(\partial_1 X, C_v)$ arise from traversing or even traversally generic vector fields on some manifold X?"* deserves attention.

Similarly, for a traversally generic v on a $(n+1)$-dimensional X, we can model the smooth types of the $\mathbf{\Omega}^\bullet_{|\sim|' \leq n}$-stratified trajectory spaces $\mathcal{T}(v)$ *locally* (see Theorem 2.2 and Corollary 2.2).

Again, the question "*Which $\mathbf{\Omega}^\bullet_{|\sim|' \leq n}$-stratified cellular complexes $\mathcal{T}$ can be produced as trajectory spaces $\mathcal{T}(v)$ of traversally generic pairs (X, v)?*" is wide open. Note that the local structure of such $\mathcal{T}$ is known (see Theorem 2.3).

In the definitions below, we will develop a language that is appropriate for tackling these questions. This language is reminiscent of the language of smooth Haefliger structures on manifolds/spaces [Haef]. However, it differs from it in a number of important ways. For example, in our setting, the connecting transformations from one holographic chart to another are special diffeomorphisms that are well-defined on the intersections of the two charts, not just *germs* of diffeomorphisms as in [Haef].

First, we need to reinforce some old notations. For any $\omega \in \mathbf{\Omega}^\bullet_{|\sim|' \leq n}$, consider the product $\mathbb{R} \times \mathbb{R}^{|\omega|'} \times \mathbb{R}^{n-|\omega|'} \approx \mathbb{R}^{n+1}$. We denote by $(u, \vec{x}, \vec{y})$ a typical point in the product. We form a cylinder

$$\Pi_\omega(\epsilon, r) =_{\mathsf{def}} \{(u, \vec{x}, \vec{y}) : \; \|\vec{x}\| < \epsilon, \; \|\vec{y}\| < r\} \subset \mathbb{R} \times \mathbb{R}^{|\omega|'} \times \mathbb{R}^{n-|\omega|'}$$

whose base is an open polydisk $\{\|\vec{x}\| < \epsilon, \; \|\vec{y}\| < r\}$.

We denote by π the obvious projection of $\mathbb{R} \times \mathbb{R}^{|\omega|'} \times \mathbb{R}^{n-|\omega|'}$ onto $\mathbb{R}^{|\omega|'} \times \mathbb{R}^{n-|\omega|'} \approx \mathbb{R}^n$.

We also consider the very familiar polynomial model from (2.11):

$$P_\omega(u, \vec{x}) =_{\mathsf{def}} \prod_{i \in \, sup(\omega)} \Big[(u - \alpha_i)^{\omega(i)} + \sum_{k=0}^{\omega(i)-1} x_{i,k}(u - \alpha_i)^k\Big],$$

where the variables $\vec{x} =_{\mathsf{def}} \{x_{i,k}\}$ are considered in the lexicographic order, and $\alpha_i < \alpha_{i+1}$ for all i.

In what follows, we assume that $\|\vec{x}\| < \epsilon$, where $\epsilon > 0$ has been chosen so small that all the real roots of $P_\omega(u, \vec{x})$ belong to a union of *disjoint* intervals $I_i \subset \mathbb{R}$, centered on the roots $\{\alpha_i\}$. For each ω, we use such ϵ to pick an appropriate cylinder $\Pi_\omega(\epsilon, r)$.

Let us consider the following two types of semi-algebraic sets:

$$\mathbf{Z}_\omega =_{\mathsf{def}} \{P_\omega(u, \vec{x}) \leq 0\} \cap \Pi_\omega(\epsilon, r) \;\; \subset \mathbb{R} \times \mathbb{R}^n \tag{3.25}$$

$$\mathsf{W}_\omega =_{\mathsf{def}} \{P_\omega(u, \vec{x}) = 0\} \cap \Pi_\omega(\epsilon, r) \;\; \subset \mathbb{R} \times \mathbb{R}^n, \tag{3.26}$$

together with the obvious projection $\pi_\omega : \mathbf{Z}_\omega \to \mathbb{R}^n$.

We assume that $\mathbb{R} \times \mathbb{R}^n$ is oriented; from now and on, this orientation is fixed. The nonsingular hypersurface $\mathsf{W}_\omega \subset \mathbb{R} \times \mathbb{R}^n$ *is oriented* with the help of the normal gradient vector field $\nabla(P_\omega(u, \vec{x}))$ and the preferred orientation of $\mathbb{R} \times \mathbb{R}^n$.

For each vector $\mathbf{x} =_{\mathsf{def}} (\vec{x}, \vec{y}) \in \mathbb{R}^n$, we consider the connected components $\{\pi^{-1}_{\omega,\rho}(\mathbf{x})\}_\rho$ of the fiber $\pi^{-1}_\omega(\mathbf{x}) \subset \mathbf{Z}_\omega$ over $\mathbf{x}$ (note that $\pi^{-1}_\omega(\mathbf{x})$ does not depend on the coordinates $\vec{y}$).

We will be dealing with:

(i) a family of elements $\{\omega_\beta \in \mathbf{\Omega}^\bullet_{|\sim|' \leq n}\}_\beta$, indexed by β,

(ii) a family of model polynomials $\{P_\beta(u_\beta, \vec{x}^\beta) =_{\mathsf{def}} P_{\omega_\beta}(u_\beta, \vec{x}^\beta)\}_\beta$ from (2.11), and

(iii) a family of appropriate cylinders $\{\Pi_\beta =_{\mathsf{def}} \Pi_{\omega_\beta}(\epsilon_\beta, r_\beta)\}_\beta$, equipped with the projections $\pi_\beta : \Pi_\beta \to \mathbb{R}^n$. The latter ones are induced by the obvious projection $\pi : \mathbb{R} \times \mathbb{R}^n \to \mathbb{R}^n$.

Definition 3.3. Let Y be a closed smooth n-manifold. A smooth (oriented) proto-holographic atlas on Y is an open cover $\{U_\beta\}_\beta$ (its elements are not necessarily connected sets) of Y, together with the following structures (see Fig. 2.2 and the introduction following it of the poset $\mathbf{\Omega}^\bullet$):

- for each index β, an element $\omega_\beta \in \mathbf{\Omega}^\bullet_{|\sim|' \leq n}$ is chosen,
- for each index β, there exists a homeomorphism h_β from the cover element U_β onto a union of *some* connected components of the nonsingular real n-dimensional semi-algebraic set as in (3.26)

$$\mathsf{W}_\beta =_{\mathsf{def}} \{P_\beta(u_\beta, \vec{x}^\beta) = 0\} \cap \Pi_\beta,$$

which resides in the "solid" $\mathbf{Z}_\beta$ as in (3.25),
- for any pair of indices β, β', there is a (orientation preserving) smooth diffeomorphism

$$h_{\beta,\beta'} : h_\beta(U_\beta \cap U_{\beta'}) \to h_{\beta'}(U_\beta \cap U_{\beta'})$$

of the hypersurfaces in $\mathbb{R} \times \mathbb{R}^n$ such that

$$h_{\beta'}\,|_{U_\beta \cap U_{\beta'}} = h_{\beta,\beta'} \circ h_\beta\,|_{U_\beta \cap U_{\beta'}},$$

- for each $\mathbf{x} \in \pi_\beta(\Pi_\beta) \subset \mathbb{R}^n$ and each connected component $\pi^{-1}_{\beta,\,\rho}(\mathbf{x})$ of the fiber $\pi^{-1}_\beta(\mathbf{x}) \subset \mathbf{Z}_\beta$, the diffeomorphism $h_{\beta,\beta'}$ takes the finite set $h_\beta(U_\beta \cap U_{\beta'}) \cap \pi^{-1}_{\beta,\,\rho}(\mathbf{x})$ to the finite set $h_{\beta'}(U_\beta \cap U_{\beta'}) \cap \pi^{-1}_{\beta',\,\rho'}(\mathbf{x}')$ and preserves the natural orders of points in the two fibers. Here $\mathbf{x}' \in \mathbb{R}^n$ and $\pi^{-1}_{\beta',\,\rho'}(\mathbf{x}')$ denotes some connected component of the fiber of the map $\pi_{\beta'} : \mathbf{Z}_{\beta'} \to \mathbb{R}^n$ over $\mathbf{x}'$.

We say that a proto-holographic atlas is holographic if all the maps $h_\beta : U_\beta \to \mathsf{W}_\beta$ are onto. ◇

Remark 3.9. The last bullet of Definition 3.3 imposes conditions on $h_{\beta,\beta'}$ which are a relaxed version of the following "better looking", but more stringent conditions: "$h_{\beta,\beta'}$ takes each fiber of the obvious projection

$$\pi_\beta : h_\beta(U_\beta \cap U_{\beta'}) \subset \mathbb{R} \times \mathbb{R}^n \to \mathbb{R}^n$$

to a fiber of

$$\pi_{\beta'} : h_{\beta'}(U_\beta \cap U_{\beta'}) \subset \mathbb{R} \times \mathbb{R}^n \to \mathbb{R}^n$$

and preserves the natural orders of points in the two fibers". However, the geometry of traversing flows favors the relaxed constraint from Definition 3.3. ◇

Example 3.5. Here is the most trivial example of a smooth proto-holographic atlas. Pick a smooth structure on a closed manifold Y by specifying an open cover $\{U_\beta\}_\beta$, where each U_β is homeomorphic to an open n-ball D^n, together with some homeomorphisms $\phi_\beta : U_\beta \to D^n \subset \mathbb{R}^n$, and the linking diffeomorphisms $\{\phi_{\beta,\beta'} : \phi_\beta(U_\beta \cap U_{\beta'}) \to \phi_{\beta'}(U_\beta \cap U_{\beta'})\}_{\beta,\beta'}$, which satisfy the usual cocycle conditions. We associate the same element $\omega = (11)$ with each U_β. Then

$$P_{\omega_\beta}(u_\beta, \vec{x}^\beta) = (u-1)(u-2)$$

and $\Pi_\beta = \{(u, \vec{y}) : \; \|\vec{y}\| < r\}$ for all β. Thus, each W_β is a union of two n-balls,

$$D^n_- = \{(u, \vec{y}) : \; u = 1, \; \|\vec{y}\| < r\} \text{ and } D^n_+ = \{(u, \vec{y}) : \; u = 2, \; \|\vec{y}\| < r\},$$

while $\mathbf{Z}_\beta$ is the cylinder $\{(u, \vec{y}) : \; 1 \leq u \leq 2, \; \|\vec{y}\| < r\}$. We identify D^n with D^n_- so that $\phi_\beta = h_\beta$ takes U_β homeomorphically onto D^n_-. The verification of all the properties, required for a proto-holographic atlas, is straightforward.

The next example is a variation on the same theme. This time, each element U_β of a proto-holographic atlas is homeomorphic to a disjoint union of *two* open n-balls D^n. We associate the same element $\omega = (11)$ with each U_β, but now

$$h_\beta : U_\beta \to \mathsf{W}_\beta = D^n_- \coprod D^n_+$$

is surjective—the atlas is holographic. Note that, if there exists a smooth holographic structure on Y which is amenable to such a choice of charts $\{h_\beta : U_\beta \to D^n_- \coprod D^n_+\}$, then Y is a covering space with the fiber of cardinality 2 over some smooth manifold Y_0. ◇

Definition 3.4.

- A proto-holographic atlas $\{V_\sigma\}_\sigma$ on Y is called a refinement of a proto-holographic atlas $\{U_\beta\}_\beta$ if each set V_σ is contained in some set U_β, where $\omega_\sigma \succeq \omega_\beta$ in $\mathbf{\Omega}^\bullet_{|\sim|' \leq n}$, and the map $h_\sigma : V_\sigma \to \mathsf{W}_\sigma$ is a restriction of the map $h_\beta : U_\beta \to \mathsf{W}_\beta$ to V_σ; moreover, the diffeomorphism $h_{\sigma,\sigma'}$ is obtained from the diffeomorphism

$$h_{\beta,\beta'} : h_\beta(U_\beta \cap U_{\beta'}) \to h_{\beta'}(U_\beta \cap U_{\beta'})$$

  by restricting it to the hypersurface $h_\sigma(V_\sigma \cap V_{\sigma'})$.
- Two smooth proto-holographic atlases on Y are called equivalent if they admit a common refinement.
- A proto-holographic structure on Y is the equivalence class of smooth proto-holographic atlases on Y. ◇

Given a proto-holographic/holographic structure $\mathcal{H}$ on Y, any diffeomorphism $\Phi : Y \to Y$ induces the pull-back proto-holographic/holographic structure $\Phi^*\mathcal{H}$ on Y. It is defined by the formulas:

$$\Phi^*(U_\beta) = \Phi^{-1}(U_\beta), \quad \Phi^* h_\beta : \Phi^{-1}(U_\beta) \xrightarrow{\Phi} U_\beta \xrightarrow{h_\beta} \mathsf{W}_\beta, \quad \Phi^* h_{\beta,\beta'} = h_{\beta,\beta'}.$$

Thus the group of smooth diffeomorphisms $\mathsf{Diff}(Y)$ acts on the set of all proto-holographic/holographic structures $\mathcal{H}(Y)$ on Y. Consider $\mathcal{H}^\sharp(Y)$, the orbit set of this action. We have little to say about the modular space $\mathcal{H}^\sharp(Y)$, perhaps, an interesting object to study...

Remark 3.10. The proto-holographic structures do not behave contravariantly under general smooth maps of manifolds. However, a weaker version of the concept does behave contravariantly under general smooth maps of manifolds. We call it appropriately "weak proto-holographic structure".

In order to formulate this weaker version, in the second bullet of Definition 3.3, we relax the requirement that $h_\beta : U_\beta \to \mathsf{W}_\beta$ is a homeomorphism onto a union of some components of the semi-algebraic set W_β; instead, we assume that $h_\beta : U_\beta \to \mathsf{W}_\beta$ is just a *continuous map*, while keeping the rest of the bullets enforced and adding the cocycle condition

$$h_{\beta'',\beta} \circ h_{\beta',\beta''} \circ h_{\beta,\beta'} \big|_{h_\beta(U_\beta \cap U_{\beta'} \cap U_{\beta''})} = \mathsf{id}$$

for each triple of indices β, β', β''. ◇

• Given a map $\pi : A \to B$ of two sets and a subset $C \subset A$, we denote by $\pi^!(C)$ the set $\pi^{-1}(\pi(C)) \subset A$. Evidently, $\pi^!(C) \supset C$. When $\pi : A \to B$ is a map of topological spaces, we denote by $\pi^{!\bullet}(C)$ the union of connected components of the space $\pi^!(C)$ that have nonempty intersections with C.

Definition 3.5. Let Y be a closed smooth n-manifold. Consider the real semi-algebraic set $\mathbf{Z}_\beta =_{\mathsf{def}} \{P_\beta(u_\beta, \vec{x}^\beta) \leq 0\} \cap \Pi_\beta \subset \mathbb{R} \times \mathbb{R}^n$.

We say that a smooth holographic atlas $\{U_\beta, h_\beta\}_\beta$ on Y is (orientably) fillable if:

- for any pair of indices β, β', the (orientation-preserving) C^∞-diffeomorphism

  $$h_{\beta,\beta'} : h_\beta(U_\beta \cap U_{\beta'}) \to h_{\beta'}(U_\beta \cap U_{\beta'})$$

  of the (∇P_β-oriented) hypersurfaces in $\mathbb{R} \times \mathbb{R}^n$ extends to a (orientation-preserving) C^∞-diffeomorphism

  $$H_{\beta,\beta'} : \mathbf{Z}_\beta^{\beta'} \to \mathbf{Z}_{\beta'}^{\beta}$$

  of the domains

  $$\mathbf{Z}_\beta^{\beta'} =_{\mathsf{def}} \pi_\beta^{!\bullet}\big(h_\beta(U_\beta \cap U_{\beta'})\big) \text{ and } \mathbf{Z}_{\beta'}^{\beta} =_{\mathsf{def}} \pi_{\beta'}^{!\bullet}\big(h_{\beta'}(U_\beta \cap U_{\beta'})\big),$$

  so that $H_{\beta,\beta'}$ maps every connected component of each fiber of the projection

  $$\pi_\beta : \mathbf{Z}_\beta^{\beta'} \subset \mathbb{R} \times \mathbb{R}^n \xrightarrow{\pi} \mathbb{R}^n$$

to a connected component of a fiber of the projection

$$\pi_{\beta'} : \mathbf{Z}^{\beta}_{\beta'} \subset \mathbb{R} \times \mathbb{R}^n \stackrel{\pi}{\to} \mathbb{R}^n$$

and preserves their orientations,

- for any triple of indices β, β', β'',

$$H_{\beta'',\, \beta} \circ H_{\beta',\, \beta''} \circ H_{\beta,\, \beta'} \big|_{\mathbf{Z}_\beta \cap \pi^{!}_{\beta}(U_\beta \cap U_{\beta'} \cap U_{\beta''})} = \mathsf{id}.$$

We say that a holographic structure $\mathcal{H}$ on Y is (orientably) fillable, if $\mathcal{H}$ is represented by a holographic atlas which is (orientably) fillable. ♢

Remark 3.11. Note that the quotient $\frac{P_{\beta'} \circ H_{\beta,\beta'}}{P_\beta}$ of the two functions from Definition 3.5 must be positive on the set $\pi^{!\bullet}_{\beta}(h_\beta(U_\beta \cap U_{\beta'}))$. ♢

Remark 3.12. Given two proto-holographic/holographic atlases on closed smooth manifolds Y_1 and Y_2 of equal dimensions, one can form a proto-holographic/holographic atlas on $Y_1 \coprod Y_2$ by considering the obvious disjoint union of atlases. Similarly, given two holographic fillable atlases on Y_1 and Y_2, the obvious union of atlases is a holographic fillable atlas on $Y_1 \coprod Y_2$.

At the first glance, these observations *could* lead to the notion of cobordism between two *proto*-holographic atlases/structures $\mathcal{H}_1$ and $\mathcal{H}_2$ on Y_1 and Y_2, respectively: it is tempting to define $\mathcal{H}_1$ to be cobordant to $\mathcal{H}_2$ if $\mathcal{H}_1 \coprod -\mathcal{H}_2$ is fillable. Alas, such a "cobordism" is *not* an equivalence relation! ♢

Theorem 3.8.

- *Any traversally generic vector field v on a compact (oriented) manifold X with boundary gives rise to a smooth holographic (orientably) fillable structure $\mathcal{H}(v)$ on the boundary $\partial_1 X$.*
- *Conversely, any smooth holographic (orientably) fillable structure $\mathcal{H}$ on a closed manifold Y arises from a traversally generic field v on a (oriented) manifold X whose boundary is diffeomorphic to Y.*

Proof. Consider a traversally generic field v on X. By combining Lemma 2.3 with formula (2.11) (or by Lemma 6.4 from [K7]), X has a v-adjusted cover $\{\mathbf{V}_\beta\}_\beta$ by connected sets $\mathbf{V}_\beta$ which admit special coordinates $(u_\beta, \vec{x}^\beta, \vec{y}^\beta)$. With each set $\mathbf{V}_\beta$ we associate the combinatorial type $\omega_\beta \in \mathbf{\Omega}^\bullet_{|\sim|' \leq n}$ of the core trajectory $\gamma_\beta \subset \mathbf{V}_\beta$, given by the equations

$\{\vec{x}^\beta = 0, \vec{y}^\beta = 0\}$. In these coordinates, $\partial_1 X \cap \mathbf{V}_\beta$ is given by a polynomial equation $\{P_\beta(u_\beta, \vec{x}^\beta) = 0\}$ (as in the second bullet of Definition 3.3) and $\mathbf{V}_\beta$ by the inequality $\{P_\beta(u_\beta, \vec{x}^\beta) \leq 0\}$. These two local descriptions are coupled with the P_β-correlated inequalities $\|\vec{x}^\beta\| < \epsilon_\beta, \|\vec{y}^\beta\| < r_\beta$, designed to partition real roots of u_β-polynomials $\{P_\beta(u_\beta, \vec{x}^\beta)\}_{\vec{x}^\beta}$ into disjoint groups, each group being localized to the vicinity of a root of $P_\beta(u_\beta, \vec{0})$; the groups being labeled by the distinct roots (see formula (2.11)).

Let $H_\beta : \mathbf{V}_\beta \to \mathbf{Z}_\beta \subset \mathbb{R}^{n+1}$ be the diffeomorphism, given by the local coordinates $(u_\beta, \vec{x}^\beta, \vec{y}^\beta)$. We denote by h_β its restriction to the set $U_\beta =_{\mathsf{def}} \mathbf{V}_\beta \cap \partial_1 X$.

By fixing the coordinates $(\vec{x}^\beta, \vec{y}^\beta)$ and a marker $u^\star_\beta \in \mathbb{R}$ such that $P_\beta(u^\star_\beta, \vec{x}^\beta) \leq 0$, we are picking the H_β-images of the v-trajectories $\gamma \subset \mathbf{V}_\beta \subset X$, residing in the set

$$\mathbf{Z}_\beta =_{\mathsf{def}} \{P_\beta(u_\beta, \vec{x}^\beta) \leq 0,\ \|\vec{x}^\beta\| \leq \epsilon_\beta,\ \|\vec{y}^\beta\| \leq r_\beta\}.$$

We define a diffeomorphism

$$H_{\beta,\beta'} : H_\beta(\mathbf{V}_\beta \cap \mathbf{V}_{\beta'}) \to H_{\beta'}(\mathbf{V}_\beta \cap \mathbf{V}_{\beta'})$$

by the formula $H_{\beta'} \circ H_\beta^{-1}|_{H_\beta(\mathbf{V}_\beta \cap \mathbf{V}_{\beta'})}$. Due to the nature of the local coordinates $(u_\beta, \vec{x}^\beta, \vec{y}^\beta)$, the diffeomorphism $H_{\beta,\beta'}$ takes each connected component of every fiber of the u-directed projection

$$\pi_\beta : \mathbf{Z}_\beta = H_b(\mathbf{V}_\beta) \to \mathbb{R}^n$$

to a connected component of a fiber of the u-directed projection

$$\pi_{\beta'} : \mathbf{Z}_{\beta'} = H_{b'}(\mathbf{V}_{\beta'}) \to \mathbb{R}^n.$$

It preserves the u-induced orientation of the π-fibers. Note that $\pi_\beta^{!\bullet}(h_\beta(U_\beta \cap U_{\beta'})) = H_\beta(\mathbf{V}_\beta \cap \mathbf{V}_{\beta'})$ and $\pi_{\beta'}^{!\bullet}(h_{\beta'}(U_\beta \cap U_{\beta'})) = H_{\beta'}(\mathbf{V}_\beta \cap \mathbf{V}_{\beta'})$.

Evidently, for any triple β, β', β'', we have the cocycle identity

$$H_{\beta'',\beta} \circ H_{\beta',\beta''} \circ H_{\beta,\beta'}\, |_{Z_\beta \cap H_\beta(\mathbf{V}_\beta \cap \mathbf{V}_{\beta'} \cap \mathbf{V}_{\beta''})} = \mathsf{id}.$$

Consider the triple

$$U_\beta =_{\mathsf{def}} \mathbf{V}_\beta \cap \partial_1 X,\ \ h_\beta =_{\mathsf{def}} H_\beta|_{U_\beta},\ \text{ and } h_{\beta,\beta'} =_{\mathsf{def}} H_{\beta,\beta'}|_{h_\beta(U_\beta \cap U_{\beta'})}.$$

Thus, $\{U_\beta, h_\beta, h_{\beta,\beta'}\}$ form a holographic atlas on $\partial_1 X$, which is fillable. The validation of this claim is on the level of definitions.

Now we turn to the validation of the second claim in Theorem 3.8. Consider a fillable holographic atlas $\{U_\beta, h_\beta, h_{\beta,\beta'}\}_\beta$ on Y (as in Definition 3.5). In particular, the diffeomorphisms

$$H_{\beta,\beta'} : \mathbf{Z}_\beta^{\beta'} =_{\mathsf{def}} \pi_\beta^{!\bullet}\big(h_\beta(U_\beta \cap U_{\beta'})\big) \longrightarrow \mathbf{Z}_{\beta'}^{\beta} =_{\mathsf{def}} \pi_{\beta'}^{!\bullet}\big(h_{\beta'}(U_\beta \cap U_{\beta'})\big)$$

are available.

Let X denote a quotient of the space $\coprod_\beta \mathbf{Z}_\beta$ by the following equivalence relation: $z \in \mathbf{Z}_\beta$ is equivalent to $z' \in \mathbf{Z}_{\beta'}$ if $z \in \mathbf{Z}_\beta^{\beta'}$ and $z' = H_{\beta,\beta'}(z) \in \mathbf{Z}_{\beta'}^{\beta}$.

The cocycle condition

$$H_{\beta'',\,\beta} \circ H_{\beta',\,\beta''} \circ H_{\beta,\,\beta'} \,|_{\mathbf{Z}_\beta \cap \pi_\beta^{!\bullet}(U_\beta \cap U_{\beta'} \cap U_{\beta''})} = \mathsf{id}$$

implies that if $z \sim z'$ and $z' \sim z''$, then $z'' \sim z$.

We denote by $q : \coprod_\beta \mathbf{Z}_\beta \to X$ the quotient map that takes each point z to its equivalence class $q(z)$. Evidently, for each index β, $q : \mathbf{Z}_\beta \to q(\mathbf{Z}_\beta)$ is a homeomorphism.

It is on the level of definitions to verify that X is a compact $(n+1)$-manifold with the smooth structure being defined by the atlas

$$\{q(\mathbf{Z}_\beta),\ q^{-1} : q(\mathbf{Z}_\beta) \to \mathbb{R} \times \mathbb{R}^n\}_\beta$$

and with the boundary $\partial X = \cup_\beta\, q(\mathsf{W}_\beta)$ being diffeomorphic to Y (recall that $\mathsf{W}_\beta =_{\mathsf{def}} \{P_\beta(u_\beta, \vec{x}^\beta) = 0\} \cap \Pi_\beta \subset \mathbb{R} \times \mathbb{R}^n$).

Moreover, X admits a 1-dimensional foliation $\mathcal{F}$ with oriented leaves. Indeed, each chart $q(\mathbf{Z}_\beta)$ comes equipped with a 1-dimensional foliation $\mathcal{F}_\beta$, generated as the q-image of the connected components of fibers of the projection $\pi_\beta : \mathbf{Z}_\beta \to \mathbb{R}^n$. Since the gluing diffeomorphisms $\{H_{\beta,\beta'}\}$ take components of the fibers of the maps $\pi : \mathbf{Z}_\beta^{\beta'} \to \mathbb{R}^n$ to components of the fibers of the maps $\pi : \mathbf{Z}_{\beta'}^{\beta} \to \mathbb{R}^n$ and preserve their orientation, these local foliations $\{\mathcal{F}_\beta\}_\beta$ produce a globally defined foliation $\mathcal{F}$ on X with oriented leaves.

We can equip X with an auxiliary Riemannian metric g. Let v be the unit vector field tangent to the leaves of $\mathcal{F}$ and coherent with their orientation. By the very construction of $\mathcal{F}$ from the traversally generic building blocks $\{\mathbf{Z}_\beta\}_\beta$, the field v is traversally generic.

Thus, via this construction, we have built the desired pair (X, v) from a given fillable holographic atlas. □

Corollary 3.11. *If a closed smooth manifold Y admits a (orientably) fillable holographic structure, then Y is a boundary of a (orientable) smooth manifold X.* ◇

Any proto-holographic structure $\mathcal{H}$ on Y generates some familiar "proto-objects". Among them is the stratification $\{\partial_j Y\}_{0 \le j \le n}$ of Y by submanifolds of codimension j, an analogue of the Morse stratification

$\{\partial_j X(v)\}_{1 \leq j \leq n+1}$. Note that the indexation of strata is shifted by -1, as compared with the indexation we used routinely for Morse strata.

When $\mathcal{H}$ is holographic, we can also construct the surrogate "trajectory space" $\mathcal{T}(\mathcal{H})$ and the partially-defined "causality map" $C_{\mathcal{H}} : \partial_0^+ Y \to \partial_0^- Y$. Constructions of such $\mathcal{H}$-induced objects are described explicitly in the proof of the following lemma.

Lemma 3.9. *Every smooth proto-holographic structure $\mathcal{H}$ on a closed n-dimensional manifold Y gives rise to a stratification*

$$\{\partial_j^{\pm} Y =_{\mathsf{def}} \partial_j^{\pm} Y(\mathcal{H})\}_{0 \leq j \leq n}$$

of Y by submanifolds $\partial_j^+ Y$, $\partial_j^- Y$ of codimension j such that:

- $\partial_0^+ Y \cup \partial_0^- Y = Y$,
- $\partial(\partial_j^+ Y) = \partial(\partial_j^- Y) = \partial_{j+1}^+ Y \cup \partial_{j+1}^- Y$ *for all j.*

Every smooth holographic structure $\mathcal{H}$ on Y gives rise to:

- *a causality map $C_{\mathcal{H}} : \partial_0^+ Y \to \partial_0^- Y$ (possibly discontinuous) whose fixed-point set is $\partial_1^- Y \cup \dots \cup \partial_n^- Y$,*
- *the $C_{\mathcal{H}}$-trajectory space $\mathcal{T}(\mathcal{H})$ is the target of a finitely ramified map $\Gamma_{\mathcal{H}} : Y \to \mathcal{T}(\mathcal{H})$. Locally, $\mathcal{T}(\mathcal{H})$ is modeled after the standard cellular complexes $\{\mathsf{T}_\omega\}_\omega$ as in Corollary 2.2 (or Theorem 7.4 from [K7]), where $\omega \in \mathbf{\Omega}^\bullet_{|\sim|' \leq n}$.*

Proof. Consider a proto-holographic atlas $\mathcal{H} = \{U_\beta, h_\beta, h_{\beta,\beta'}\}$ on Y. Recall that any point $z_\beta^\star = (u_\beta^\star, \vec{x}_\star^\beta, \vec{y}_\star^\beta) \in \mathsf{W}_\beta \subset \partial \mathbf{Z}_\beta$ has a multiplicity $j(z_\beta^\star)$, defined as the multiplicity of the appropriate root $u_\beta^\star$ of the u_β-polynomial $P_\beta(u_\beta, \vec{x}^\beta)$.

By Lemma 3.1, the multiplicity $j(z_\beta^\star)$ can be detected just in terms of the $z_\beta^\star$-*localized* maximal number of the atomic portions (see Fig. 2.2) of the fibers of the map $\pi : \mathsf{W}_\beta \to \mathbb{R}^n$. Here we localize the fibers of $\pi : \mathsf{W}_\beta \to \mathbb{R}^n$ to the vicinity of $z_\beta^\star$ in the hypersurface W_β. Indeed, if $j(z_\beta^\star)$ is an even number, then any small neighborhood of $z_\beta^\star$ contains ∂_{u_β}-trajectories in $\mathbf{Z}_\beta$ of the combinatorial type $(122\dots21)$, where the number of 2's is $(j(z_\beta^\star) - 2)/2$. If $j(z_\beta^\star)$ is an odd number, then any small neighborhood of $z_\beta^\star$ contains trajectories in $\mathbf{Z}_\beta$ of the combinatorial type $(122\dots2)$, where the number of 2's is $(j(z_\beta^\star) - 1)/2$ (see the arguments in the beginning of the proof of Lemma 3.1). Note that the odd and even multiplicities are distinguished by their local topology.

Thus, for any $y \in U_\beta \subset Y$, the point $z_\beta =_{\mathsf{def}} h_\beta(y)$ acquires multiplicity $j_\beta(z_\beta)$. By the property of $h_{\beta,\beta'}$, described in the fourth bullet of Definition

3.3, the points in the fibers of $\pi_\beta : h_\beta(U_\beta \cap U_{\beta'}) \to \mathbb{R}^n$, which belong to the same connected component of the corresponding fiber $\pi_\beta : \mathbf{Z}_\beta \to \mathbb{R}^n$, are mapped to the points in the fibers of $\pi_{\beta'} : h_{\beta'}(U_\beta \cap U_{\beta'}) \to \mathbb{R}^n$, which also belong to a connected component of the corresponding fiber $\pi_{\beta'} : \mathbf{Z}_{\beta'} \to \mathbb{R}^n$. Therefore, employing this property of diffeomorphism $h_{\beta,\beta'}$ and interpreting the multiplicities in terms of the cardinalities of the π-fibers, we get $j_{\beta'}(h_{\beta'}(y)) = j_\beta(h_\beta(y))$ when $y \in U_\beta \cap U_{\beta'}$.

Consider the stratification $\{\partial_j^\pm \mathsf{W}_\beta =_{\mathsf{def}} \partial_j^\pm \mathbf{Z}_\beta(\partial_\beta)\}_j$ of W_β, generated by the constant vector field $\partial_\beta =_{\mathsf{def}} \frac{\partial}{\partial u_\beta}$.

A proto-holographic atlas $\mathcal{H}$ helps to define a stratification $\{\partial_j Y =_{\mathsf{def}} \partial_j Y(\mathcal{H})\}_{0 \leq j \leq n}$ of Y by closed submanifolds: just put

$$\partial_j Y \cap U_\beta =_{\mathsf{def}} h_\beta^{-1}\big(\partial_j \mathbf{Z}_\beta(\partial_\beta) \cap h_\beta(U_\beta)\big).$$

The stratification $\{\partial_j Y(\mathcal{H})\}_{0 \leq j \leq n}$ can be refined to a stratification

$$\{\partial_j^\pm Y =_{\mathsf{def}} \partial_j Y^\pm(\mathcal{H})\}_{0 \leq j \leq n}$$

by compact submanifolds such that $\partial_j Y = \partial_j^+ Y \cup \partial_j^- Y$ and $\partial_{j+1} Y = \partial(\partial_j^+ Y) = \partial(\partial_j^- Y)$ for all j.

Let us describe its construction. By Theorem 3.9, the stratification $\{\partial_j^\pm \mathsf{W}_\beta =_{\mathsf{def}} \partial_j^\pm \mathbf{Z}_\beta\}_j$ of W_β, generated by the field ∂_β, is characterized in terms of the polynomial P_β as follows: $(u_\beta, \vec{x}^\beta, \vec{y}^\beta) \in \partial_j^+ \mathbf{Z}_\beta$ if $\frac{\partial^k}{\partial u^k} P_\beta(u_\beta, \vec{x}^\beta) = 0$ for all $k < j$, and $\pm \frac{\partial^j}{\partial u^j} P_\beta(u_\beta, \vec{x}^\beta) < 0$.

In fact, the sign of the derivative $\frac{\partial^j}{\partial u^j} P_\beta(u_\beta, \vec{x}^\beta)$ at a root u_β of multiplicity j can be determined by the simple combinatorial rule: consider the real divisor $D_{P_\beta}(\vec{x}^\beta)$ of the u_β-polynomial $P_\beta(\sim, \vec{x}^\beta)$ and decompose its support $\sup(D_{P_\beta}(\vec{x}^\beta))$ into "atoms" and "strings" (as in Fig. 2.2). Then any root of even multiplicity, which is an atom, is assigned the polarity "$-$", any root of even multiplicity which is inside of a string is assigned the polarity "$+$". Any root of odd multiplicity, forming the lower end of a string, is assigned the polarity "$+$", while any root of odd multiplicity, forming the upper end of a string, is assigned the polarity "$-$" (these signs are opposite to the signs of the lowest order non-trivial derivatives $\frac{\partial^{j+1}}{\partial u^{j+1}} P_\beta(u_\beta, \vec{x}^\beta)$).

Therefore, the polarity $\pm$ of a point $z \in \partial_j \mathsf{W}_\beta$ is determined by the ordered sequence of intersection points of the ∂_β-trajectory $\gamma_z \subset \mathbf{Z}_\beta$ trough $z = (u_\beta, \vec{x}^\beta, \vec{y}^\beta)$ with the hypersurface W_β, each intersection point z being considered with its multiplicity $j_\beta(z)$. As the combinatorial rule above implies, the polarities attached to each intersection $z \in \gamma_z \cap \mathsf{W}_\beta$ depend only on the multiplicities and the order of points from $\gamma_z \cap \mathsf{W}_\beta$ along γ_z. Again, by the fourth bullet of Definition 3.3, this kind of data are preserved under the smooth change of coordinates $h_{\beta,\beta'} : h_\beta(U_\beta \cap U_{\beta'}) \to h_{\beta'}(U_\beta \cap U_{\beta'})$. Therefore, the local formulas $\{\partial_j^\pm Y \cap U_\beta =_{\mathsf{def}} h_\beta^{-1}(\partial_j^\pm \mathbf{Z}_\beta \cap h_\beta(U_\beta))\}_\beta$ produce the desired globally-defined stratifications.

Given a holographic atlas $\mathcal{H}$, let us turn to the construction of the causality map $C_\mathcal{H} : \partial_0^+ Y \to \partial_0^- Y$. Take a point $y \in U_\beta \cap \partial_0^+ Y$ to the point $h_\beta(y) \in h_\beta(U_\beta) = \mathsf{W}_\beta$. Then, in the fiber $\pi_\beta^{-1}(h_\beta(y))$ of the map $\pi_\beta : h_\beta(U_\beta) \subset \mathbb{R} \times \mathbb{R}^n \to \mathbb{R}^n$, take the point $h_\beta(y)$ to the point z with the minimal first coordinate $u_\beta(z) > u_\beta(h_\beta(y))$ and such that $P_\beta < 0$ in the open interval $(h_\beta(y), z) \subset \mathbb{R} \times \mathbb{R}^n$; if no such z exists, put $z = h_\beta(y)$. Let $C_\mathcal{H}(y) =_{\mathsf{def}} h_\beta^{-1}(z)$. Since $\mathcal{H}$ is a holographic structure, the map $C_\mathcal{H}$ is well-defined for all points in $\partial_0^+ Y \cap U_\beta$.

Note that the model causality map $C_{\partial_\beta} : \mathsf{W}_\beta^+ \to \mathsf{W}_\beta^-$ has the locus

$$\partial_2^- \mathsf{W}_\beta \cup \dots \cup \partial_{n+1}^- \mathsf{W}_\beta =_{\mathsf{def}} \partial_2^- \mathbf{Z}_\beta \cup \dots \cup \partial_{n+1}^- \mathbf{Z}_\beta$$

for its fixed point set.

This local construction of $C_\mathcal{H}(y)$ does not depend on the choice of a chart U_β which contains y. Indeed, let $y \in U_\beta \cap U_{\beta'}$, $z_\beta = h_\beta(y)$, and $F_\beta(y) = \pi_\beta^{-1}(\pi_\beta(z_\beta)) \subset \mathbf{Z}_\beta$. Similar notations are introduced for the map $h_{\beta'}$. According to the fourth bullet in Definition 3.3, the connecting diffeomorphism $h_{\beta,\beta'}$ maps each portion $F_\beta^\bullet(y)$ of the fiber $F_\beta(y) \cap \mathsf{W}_\beta$ that belongs to the same connected component of the fiber $F_\beta(y) \subset \mathbf{Z}_\beta$ onto the corresponding portion $F_{\beta'}^{'\bullet}(y)$ of the fiber $F'_{\beta'}(y) \cap \mathsf{W}_{\beta'}$ that belongs to the same connected component of the fiber $F'_{\beta'}(y) \subset \mathbf{Z}_{\beta'}$. Moreover, $h_{\beta,\beta'}$ respects the u_β-induced order of points in $F_\beta^\bullet(y)$ and the $u_{\beta'}$-induced order of points in $F_{\beta'}^{'\bullet}(y)$. The two locally-defined causality maps,

$$C_{\mathcal{H},\beta} : \partial_0^+ Y \cap U_\beta \to \partial_0^- Y \cap U_\beta \quad \text{and} \quad C_{\mathcal{H},\beta'} : \partial_0^+ Y \cap U_{\beta'} \to \partial_0^- Y \cap U_{\beta'},$$

operate within the sets $h_\beta^{-1}(F_\beta^\bullet(y))$ and $h_{\beta'}^{-1}(F_{\beta'}^{'\bullet}(y))$, respectively. Thus, by employing the same combinatorial rules (which depend only on the intrinsically-defined polarities of the corresponding points), by the third bullet in Definition 3.3, we get that $C_{\mathcal{H},\beta}(y) = C_{\mathcal{H},\beta'}(y)$.

With the given holographic atlas $\mathcal{H}$ on Y in place, we declare two points $y_1, y_2 \in Y$ to be equivalent if there exists $y_0 \in Y$ such that both y_1 and y_2 can be obtained as images of y_0 under iterations of the causality map $C_\mathcal{H}$. We denote by $\mathcal{T}(\mathcal{H})$ the quotient space of Y by this equivalence relation and call it the trajectory space of $\mathcal{H}$. The obvious finitely-ramified map $\Gamma_\mathcal{H} : Y \to \mathcal{T}(\mathcal{H})$ helps to define the quotient topology on the trajectory space $\mathcal{T}(\mathcal{H})$.

Note that, for each $y \in U_\beta$, the vicinity of the point $\Gamma_{\mathcal{H}}(y)$ in $\mathcal{T}(\mathcal{H})$, is modeled after the standard cell complex T_ω as in Corollary 2.2 (see also Theorems 7.3 and 7.4 from [K7]), where $\omega \in (\omega_\beta)_{\preceq} \subset \mathbf{\Omega}^\bullet_{|\sim|' \leq n}$. Indeed, the obvious projections $\mathsf{W}_\beta \subset \mathbf{Z}_\beta \to \mathsf{T}_{\omega_\beta}$, with the help of h_β^{-1}, provide for the local models of the space $\Gamma_{\mathcal{H}}$. □

Corollary 3.12. *For any traversally generic vector field v on X, the $\mathbf{\Omega}^\bullet_{|\sim|' \leq n}$-stratified topological type of the trajectory space $\mathcal{T}(v)$ depends only on the fillable holographic structure $\mathcal{H}(v)$ that v generates on the boundary $\partial_1 X$.*

Proof. By Theorem 3.8, the fields v_i $(i = 1, 2)$, give rise to fillable holographic structures $\mathcal{H}(v_i)$. By Lemma 3.9, $\mathcal{H}(v_i)$ canonically produces the causality map $C_{\mathcal{H}(v_i)}$ (which happen to coincide with the causality map C_{v_i}). By the hypotheses, $\mathcal{H}(v_1) = \mathcal{H}(v_2)$, which implies the equality $C_{\mathcal{H}(v_1)} = C_{\mathcal{H}(v_2)}$. Since the spaces $\mathcal{T}(v_i)$ depend only on $C_{\mathcal{H}(v_i)} = C_{v_i}$, we conclude that $\mathcal{T}(v_1)$ and $\mathcal{T}(v_2)$ are homeomorphic as $\mathbf{\Omega}^\bullet_{|\sim|' \leq n}$-stratified spaces. □

Remark 3.13. In principle, in accordance with Theorem 3.8, a given fillable holographic structure $\mathcal{H}$ on Y, may have many realizations $\{\mathcal{H}(v)\}$ by traversally generic vector fields v on a variety of manifolds X whose boundary is Y. However, Holography Theorem 3.3 claims that the topological (often the smooth) type of X, together with the v-induced oriented foliation $\mathcal{F}(v)$, is determined by $\mathcal{H}$ via the causality map $C_{\mathcal{H}} = C_v$. ◇

The next definition should be compared with Definition 2.7. We stress that the notions of smooth functions on the trajectory spaces are subtly different in the two definitions: in Definition 2.7, a function on $\mathcal{T}(v)$ is smooth if its pull-back to X is smooth, while in Definition 3.6, a function on $\mathcal{T}(\mathcal{H})$ is defined to be smooth if its pull-back to Y "$= \partial X$" is smooth, a weaker requirement.

Definition 3.6. Let $\mathcal{H}$ be a smooth holographic structure on Y and $\mathcal{T}(\mathcal{H})$ its trajectory space.

A continuous function $h : \mathcal{T}(\mathcal{H}) \to \mathbb{R}$ is called $\mathcal{H}-$smooth, if the composite function $Y \stackrel{\Gamma_{\mathcal{H}}}{\to} \mathcal{T}(\mathcal{H}) \stackrel{h}{\to} \mathbb{R}$ is smooth on Y.

We denote by $C^\infty(\mathcal{T}(\mathcal{H}))$ the algebra of $\mathcal{H}$-smooth functions on $\mathcal{T}(\mathcal{H})$. Note that $C^\infty(\mathcal{T}(\mathcal{H})) \approx C^\infty(Y)^{C_{\mathcal{H}}}$, the $C_{\mathcal{H}}$-invariant subspace of $C^\infty(Y)$. ◇

Since the map $\Gamma_{\mathcal{H}} : Y \to \mathcal{T}(\mathcal{H})$ is onto, the algebra $C^\infty(\mathcal{T}(\mathcal{H}))$ maps isomorphically onto a subalgebra $C^\infty_{\mathcal{H}}(Y)$ of $C^\infty(Y)$ via the induced monomorphism $\Gamma^*_{\mathcal{H}}$.

If a holographic structure $\mathcal{H}$ on $\partial_1 X$ is induced by a traversally generic vector field v on X, then the algebra $C^\infty_{\mathcal{H}}(\partial_1 X) \approx C^\infty(\partial_1 X)^{C_v}$ from Definition 3.6 contains the subalgebra $C^\infty(X, v) \approx \mathsf{ker}(\mathcal{L}_v)$ (see Question 3.1 and Conjecture 3.1 for a discussion of the potential difference between the two algebras).

Corollary 3.13. *Let $\mathcal{H}$ be a holographic structure on a closed smooth manifold Y.*

Then $\mathcal{H}$ determines the algebra $C^\infty(\mathcal{T}(\mathcal{H}))$ of $\mathcal{H}$-smooth functions on the trajectory space $\mathcal{T}(\mathcal{H})$ as a subalgebra $C^\infty_{\mathcal{H}}(Y)$ of $C^\infty(Y)$.

Moreover, any diffeomorphism $\Phi : Y \to Y$ induces an algebra isomorphism

$$\Phi^* : C^\infty_{\mathcal{H}}(Y) \to C^\infty_{\Phi^*\mathcal{H}}(Y),$$

where $\Phi^*\mathcal{H}$ denotes the Φ-induced holographic structure.

Proof. By Lemma 3.9, the obvious maps $\Gamma_{\mathcal{H}} : Y \to \mathcal{T}(\mathcal{H})$ and $\Gamma_{\Phi^*\mathcal{H}} : Y \to \mathcal{T}(\Phi^*\mathcal{H})$ are defined, via the proto-causality maps $C_{\mathcal{H}}$ and $C_{\Phi^*\mathcal{H}}$, in terms of the structures $\mathcal{H}$ and $\Phi^*\mathcal{H}$, respectively.

It is on the level of definitions to check that $\Phi \circ \Gamma_{\Phi^*\mathcal{H}} = \Gamma_{\mathcal{H}}$. Therefore, using that Φ is a diffeomorphism, we get that $\Phi^* : C^\infty_{\mathcal{H}}(Y) \to C^\infty_{\Phi^*\mathcal{H}}(Y)$ is an isomorphism of algebras. □

Question 3.3. *Does the pair of algebras $C^\infty_{\mathcal{H}}(Y) \subset C^\infty(Y)$ determine the holographic structure $\mathcal{H}$ on Y?* ◇

Now, let us combine Holography Theorem 3.3 with Theorem 3.8 into a single proposition.

Theorem 3.9. (The Generalized Holographic Principle)

- *For a traversally generic vector field v on a (oriented) compact connected smooth manifold X, the topological type of the pair $(X, \mathcal{F}(v))$ is determined by the v-induced (orientably) fillable holographic structure $\mathcal{H}(v)$ on $\partial_1 X$.*
- *When, in addition, v has Property* A, *then $\mathcal{H}(v)$ determines the smooth topological type of the pair $(X, \mathcal{F}(v))$.*

- *Conversely, any (orientably) fillable holographic structure $\mathcal{H}$ on a closed manifold Y is of the form $\mathcal{H}(v)$ for a traversally generic vector field v on some (orientable) compact manifold X such that $\partial_1 X = Y$. Moreover, the topological type of the pair $(X, \mathcal{F}(v))$ is determined by $\mathcal{H}$.*
- *If a fillable holographic structure $\mathcal{H}$ on Y is based on the combinatorial types $\omega \notin \{(33)_{\succeq} \cup (4)_{\succeq}\}$, then $\mathcal{H}$ determines the smooth topological type of the pair $(X, \mathcal{F}(v))$, where $\partial_1 X = Y$ and $\mathcal{H}(v) = \mathcal{H}$.*

Proof. By Theorem 3.8, any traversally generic field v on X gives rise to a fillable holographic structure $\mathcal{H}(v)$ on $\partial_1 X$.

On the other hand, by the same Theorem 3.8, any given fillable holographic structure $\mathcal{H}$ on a closed manifold Y produces a traversally generic vector field v on some X such that $\partial_1 X = Y$.

By Lemma 3.9, the causality map C_v can be described intrinsically in terms of $\mathcal{H}(v)$ as the map $C_{\mathcal{H}(v)} : \partial_1^+ X(\mathcal{H}(v)) \to \partial_1^- X(\mathcal{H}(v))$. By Holography Theorem 3.3, C_v allows for a reconstruction of the topological type of the pair $(X, \mathcal{F}(v))$. Therefore, $C_{\mathcal{H}(v)}$ does the same job.

Finally, v is as in the second bullet of Theorem 3.3, if and only if its trajectories have the combinatorial types $\omega \notin \{(33)_{\succeq} \cup (4)_{\succeq}\}$. Thus if a fillable $\mathcal{H}$ is based on the combinatorial types $\{\omega \notin (33)_{\succeq} \cup (4)_{\succeq}\}$, then any $\mathcal{H}$-generated v has Property A and hence its causality map C_v allows for a reconstruction of the smooth type of $(X, \mathcal{F}(v))$. □

Remark 3.14. Theorem 3.6 implies that the universe of fillable holographic structures even on the standard sphere S^n is at least as rich as the universe of smooth types of closed $(n+1)$-manifolds! ◇

Historically, the smooth structures on homotopy spheres have been a favorite subject in topology. To support the tradition, we will make few simple observations about the holographic structures on homotopy spheres and their relations to the underlying smooth structures. Stated loosely, our main remark here is that "the universe" of fillable holographic structures on smooth homotopy spheres maps surjectively on the group Θ_k of the homotopy k-spheres, considered up to an h-cobordism [KM], so that the map respects the connected sum operations in its source and target.

Let us recall one old appealing result of C.T.C. Wall, the classification of smooth highly connected even-dimensional manifolds:

Theorem 3.10. *([Wa]) For $n \geq 3$, the diffeomorphism classes of $(n-1)$-connected $2n$-manifolds X, whose boundaries are homotopy spheres, are in* *natural bijections with the isomorphism classes of $\mathbb{Z}$-valued non-degenerate $(-1)^n$-symmetric bilinear forms that admit quadratic refinements with values in the homotopy group $\pi_n(\mathsf{BSO}(n)) \approx \mathbb{Z}$ for $n \equiv 0 \mod 2$, and in the homotopy group $\pi_n(\mathsf{BSO}(n)) \approx \mathbb{Z}_2$ for $n \equiv 1 \mod 2$.* ♢

Remark 3.15. If a homotopy sphere Σ^{2n-1}, $n \geq 3$, bounds a *framed* smooth $(2n)$-manifold Y, then by surgery, Σ^{2n-1} bounds a framed $(n-1)$-*connected* manifold X (cf. [Kos], Theorem 2.2 on page 201). Therefore, all such homotopy spheres are bounded by manifolds X that satisfy the hypotheses of Wall's Theorem 3.10. ♢

Let $\mathcal{HS}^{\{\mathsf{fill}\}}_{2n-1}$ denote the set of equivalence classes of orientably fillable holographic structures on smooth homotopy $(2n-1)$-spheres, the spheres being considered up to a smooth orientation-preserving diffeomorphism.

Let $\mathcal{HS}^{\{\mathsf{fill}\}}_{2n-1,\, n-1}$ denote the set of equivalence classes of orientably fillable holographic structures on smooth homotopy $(2n-1)$-spheres Σ that bound a $(n-1)$-connected manifold, where $n \geq 3$.

Theorem 3.9, being combined with Theorem 3.10, leads to the following result.

Corollary 3.14. *For $n \geq 3$, there is a map $\Xi_n : \mathcal{HS}^{\{\mathsf{fill}\}}_{2n-1} \to \mathcal{B}_n$, where $\mathcal{B}_n$ denotes the set of isomorphism classes of non-degenerate (over the rationals numbers) $\mathbb{Z}$-valued $(-1)^n$-symmetric bilinear forms on free finite-dimensional $\mathbb{Z}$-modules.*

Moreover, any non-degenerate over $\mathbb{Z}$ bilinear form $\Phi : M \otimes M \to \mathbb{Z}$ on a free finite-dimensional $\mathbb{Z}$-module M that admits a quadratic refinement μ as in (3.27) is in the image $\Xi_n(\mathcal{HS}^{\{\mathsf{fill}\}}_{2n-1,\, n-1})$.

Proof. Let $\mathcal{H}$ be an orientably fillable holographic structure on a smooth homotopy sphere Σ^{2n-1}. By Theorem 3.9, such $\mathcal{H}$ is of the form $\mathcal{H}(v)$ for some traversally generic v on an oriented smooth manifold X whose boundary is Σ^{2n-1}. Moreover, by Theorem 3.9, the topological type of X is determined by $\mathcal{H} = \mathcal{H}(v)$. Let

$$\Phi_X : \big(H_n(X;\mathbb{Z})/\mathrm{Tor}\big) \otimes \big(H_n(X;\mathbb{Z})/\mathrm{Tor}\big) \to \mathbb{Z}$$

be the bilinear $(-1)^n$-symmetric intersection form on X (whose topological type depends on $\mathcal{H}$ only). Since $\partial_1 X$ is a homotopy sphere, by the Poincaré duality, the form Φ_X is non-degenerate over the rationals. Thus, we have mapped any orientably fillable holographic structure $\mathcal{H}$ on a homotopy sphere Σ^{2n-1} to the corresponding bilinear form $\Phi_X =_{\mathsf{def}} \Xi_n(\mathcal{H})$.

Let us show that any bilinear $(-1)^n$-symmetric *non-degenerate over* $\mathbb{Z}$ form $\Phi : M \otimes M \to \mathbb{Z}$ on a finitely-dimensional free $\mathbb{Z}$-module M (we chose $M \approx H_n(X;\mathbb{Z})/\mathrm{Tor}$) can be produced in this way, provided that Φ is enhanced by a quadratic form $\mu : M \to \mathbb{Z}/\langle 1-(-1)^n\rangle$. Recall that the relations between Φ and μ are described by the formulas:

$$\Phi(a,a) = \mu(a) + (-1)^n\mu(a),$$
$$\Phi(a,b) = \mu(a+b) - \mu(a) - \mu(b) \mod \langle 1-(-1)^n\rangle \tag{3.27}$$

for any $a, b \in M$ [Wa]. Note that the RHS of the formula for $\Phi(a,a)$ is well-defined as an element of $\mathbb{Z}$.

By Theorem 3.10, any such pair (Φ, μ) is realized on some smooth orientable $(n-1)$-connected $2n$-manifold X which bounds a homotopy sphere Σ^{2n-1}. For such X, the homology $H_n(X;\mathbb{Z}) \approx M$ (a free $\mathbb{Z}$-module), and Φ becomes the intersection form. The quadratic form $\mu : H_n(X;\mathbb{Z}) \to \mathbb{Z}/\langle 1-(-1)^n\rangle$ is defined by the self-intersections of immersed spheres $S^n \propto X$ [Wa3].

By Lemma 2.1 (see Theorems 3.4 and 3.5 from [K2]), X admits a traversally generic vector field v. By Theorem 3.9, v induces an orientably fillable holographic structure $\mathcal{H}(v)$ on Σ^{2n-1}. Therefore, $\Phi = \Xi_n(\mathcal{H}(v))$. □

Let a manifold X_1 bound a homotopy k-sphere Σ_1, and a manifold X_2 a homotopy k-sphere Σ_2. Consider two traversally generic fields: v_1 on X_1 and v_2 on X_2. By attaching a 1-handle $\mathbf{H} =_{\mathsf{def}} D^1 \times D^k$ to $\Sigma_1 \coprod \Sigma_2 \subset \partial(X_1 \coprod X_2)$, we form a new manifold $X =_{\mathsf{def}} X_1 \#_\partial X_2$ that bounds the homotopy sphere $\Sigma =_{\mathsf{def}} \Sigma_1 \# \Sigma_2$. This 1-surgery on $X_1 \coprod X_2$ can be performed so that the fields v_1 and v_2 extend over the 1-handle $\mathbf{H}$ to form a traversally generic vector field v on X so that the following diffeomorphisms are valid:

- $\partial_1^+ X(v) \approx \partial_1^+ X_1(v_1) \coprod \big(\partial_1^+ X_2(v_2) \setminus \mathsf{int}(D^k)\big)$,
- $\partial_1^- X(v) \approx \big(\partial_1^- X_1(v_1) \setminus \mathsf{int}(D^k)\big) \coprod \partial_1^- X_2(v_2)$,
- $\partial_2^- X(v) \approx \partial_2^- X_1(v_1) \coprod \partial_2^- X_2(v_2)$,
- $\partial_2^+ X(v) \approx \partial_2^+ X_1(v_1) \coprod \partial_2^+ X_2(v_2) \coprod S^{k-1}$,
- $\partial_j^\pm X(v) \approx \partial_j^\pm X_1(v_1) \coprod \partial_j^\pm X_2(v_2)$ for all $j > 2$.

These formulas describe that the handle $\mathbf{H}$ is "slanted" and has a "bottleneck" S^{k-1} with respect to the v-flow. The construction of the field v uses a v_1-adjusted regular neighborhood U_1 of a v_1-trajectory γ_1 of the combinatorial type (11) and a v_2-adjusted regular neighborhood U_2 of a v_2-trajectory γ_2 of the combinatorial type (11). Outside of $U_1 \coprod U_2$, the field v

coincides with $v_1 \coprod v_2$. A priori, the field v on $X =_{\mathsf{def}} X_1 \#_\partial X_2$ depends on the choice of the base trajectories γ_1 and γ_2, or, equivalently, on the choice of base points $s_1^+, s_1^- = C_{v_1}(s_1^+) \in \partial_1 X_1$ and $s_2^+, s_2^- = C_{v_2}(s_2^+) \in \partial_1 X_2$, as well as on the order of the pair $\{X_1, X_2\}$. Note that the trajectory space $\mathcal{T}(v)$ is obtained from $\mathcal{T}(v_1) \coprod \mathcal{T}(v_2)$ by identifying a pair of closed k-balls, centered on $[\gamma_1]$ and $[\gamma_2]$, respectively.

This boundary connected sum construction produces a holographic fillable structure

$$\mathcal{H}(v; s_1^+, s_2^-) =_{\mathsf{def}} \mathcal{H}(v_1; s_1^+, s_1^-) \#_\partial \mathcal{H}(v_2; s_2^+, s_2^-) \tag{3.28}$$

on $\Sigma = \Sigma_1 \# \Sigma_2$.

Lemma 3.10. *Let $\mathcal{H}(v_i; s_i^+, s_i^-)$ be a based orientably fillable holographic structure on a homotopy $(2n-1)$-sphere Σ_i, $i = 1, 2$. Let the orientably fillable holographic structure $\mathcal{H}(v; s_1^+, s_2^-)$ on the homotopy sphere $\Sigma = \Sigma_1 \# \Sigma_2$ be defined by (3.28).*

Then the bilinear $\mathbb{Z}$-form $\Phi_{\mathcal{H}(v)} = \Xi_n(\mathcal{H}(v))$ that, according to Corollary 3.14, corresponds to the orientably fillable holographic structure $\mathcal{H}(v)$ from (3.28) on the homotopy sphere Σ is the direct sum $\Phi_{\mathcal{H}(v_1)} \oplus \Phi_{\mathcal{H}(v_2)}$ of the forms $\Phi_{\mathcal{H}(v_1)}$ and $\Phi_{\mathcal{H}(v_2)}$. □

Proof. Employing the construction from the proof of Corollary 3.14, the validation of the lemma is at the level of definitions. □

We leave to the reader to discover a generalization of the boundary connected sum construction $\#_\partial$ from (3.28) for any pair of based fillable holographic structures $\mathcal{H}_1, \mathcal{H}_2$ on homotopy k-spheres. This generalization will produce a based holographic structure $\mathcal{H} =_{\mathsf{def}} \mathcal{H}_1 \#_\partial \mathcal{H}_2$ on $\Sigma =_{\mathsf{def}} \Sigma_1 \# \Sigma_2$, utilizing two given based holographic structures: $\mathcal{H}_1$ on Σ_1 and $\mathcal{H}_2$ on Σ_2. Each of these two structures $\mathcal{H}_i$ $(i = 1, 2)$ is enhanced by a choice of a pair of base points $s_i^+, s_i^- \in \Sigma_i$. Both points are contained in some element U_i^{base} of the combinatorial type (11) from the atlas $\mathcal{H}_i$; their images, $h_i^{\mathrm{base}}(s_i^+)$ and $h_i^{\mathrm{base}}(s_i^-)$, under the atlas homeomorphism $h_i^{\mathrm{base}}: U_i^{\mathrm{base}} \to \mathbb{R}^n \times \mathbb{R}$, project to the same point in $\mathbb{R}^n$.

Again, the trajectory space $\mathcal{T}(\mathcal{H}_1 \#_\partial \mathcal{H}_2)$ is obtained from the trajectory spaces $\mathcal{T}(\mathcal{H}_1)$ and $\mathcal{T}(\mathcal{H}_2)$ by gluing them along some closed k-balls, D_1^k and D_2^k, where the ball D_i^k resides in the maximal (i.e., indexed by $\omega = (11)$) stratum of $\mathcal{T}(\mathcal{H}_i)$.

The operation

$$(\mathcal{H}_1; s_1^+, s_1^-),\ (\mathcal{H}_2; s_2^+, s_2^-) \Rightarrow (\mathcal{H}_1 \#_\partial \mathcal{H}_2; s_1^+, s_2^-) \tag{3.29}$$

is evidently associative.

The order of two holographic structures $\mathcal{H}_1$ and $\mathcal{H}_2$ in the operation "$\#_\partial$" does matter: we think of Σ_2 as residing "above" Σ_1, at least in the vicinity of points $s_1^- \in \Sigma_1$ and $s_2^+ \in \Sigma_2$ where the "virtual" 1-handle $\mathbf{H}$, responsible for the operation, is attached.

The holographic structure $\mathcal{H}_1 \#_\partial \mathcal{H}_2$ may indeed change when the base points are chosen from different connected components of the pure strata of the combinatorial type (11) in Σ_1 or in Σ_2.

Let $\mathcal{H}^\odot$ denote the fillable based holographic structure on S^k, generated by a constant (convex) vector field $v^\parallel$ on the standard ball $B^{k+1} \subset \mathbb{R}^{k+1}$. Note that $\mathcal{H}^\odot$ is not a neutral element for the operation $\#_\partial$. In fact, $\mathcal{H} \#_\partial \mathcal{H}^\odot \neq \mathcal{H}$ since the trajectory spaces $\mathcal{T}(\mathcal{H})$ and $\mathcal{T}(\mathcal{H} \#_\partial \mathcal{H}^\odot)$ are not homeomorphic (although $\mathcal{T}(\mathcal{H})$ is a deformation retract of $\mathcal{T}(\mathcal{H} \#_\partial \mathcal{H}^\odot)$). Probably, this observation calls for considering (fillable) holographic structures up to the connected sums with the "bubbles" $\mathcal{H}^\odot$.

We denote by $\neg\mathcal{H}$ the (fillable) based holographic structure, obtained from a given (fillable) based holographic structure $\mathcal{H}$ on a homotopy sphere Σ by flipping the orientation of Σ. It consists of a mirror image of Σ together with the mirror image of the (fillable) atlas of $\mathcal{H}$. The partial order of points in Σ, defined by the causality map $C_\mathcal{H}$, is preserved by the reflection.

Let Θ_k be the set of smooth homotopy k-spheres, considered up to an h-cobordism, with the addition induced by the connected sum operation. For $k \neq 4$, thanks to the h-cobordism theorem (see [Mi1], [P1], and [P2]), Θ_k is an abelian group. The role of opposite element $-(\Sigma^k)$ is played by the homotopy sphere Σ^k, taken with the opposite orientation [KM].

The next proposition claims that holographic structures on any homotopy k-sphere are "hiding" inside of the holographic structures on the standard k-sphere.

Proposition 3.2. *Let Σ^k, $k \neq 4$, be a homotopy k-sphere. Then, for any (fillable) based holographic structure $\mathcal{H}$ on Σ^k, there exists a (fillable) based holographic structure $\mathcal{H}^\dagger$ on the standard sphere S^k such that $\mathcal{H}^\dagger = \mathcal{H} \#_\partial \neg\mathcal{H}$, the connected sum.*

Proof. Given a based holographic structure $(\mathcal{H};\, s^+, s^-)$ on Σ^k, represented by an atlas $\mathcal{A} =_{\mathsf{def}} \{h_\beta : U_\beta \to \mathbb{R}^k \times \mathbb{R}\}_\beta$, consider the based holographic structure $(\neg\mathcal{H};\, s^+, s^-)$ represented by the same atlas $\mathcal{A}$ on Σ^k with the orientation of Σ^k being reversed. The holographic structure $\mathcal{H} \#_\partial \neg\mathcal{H}$ is defined on the homotopy sphere $\Sigma^k \#(-\Sigma^k)$. By the h-cobordism theorem, one has a diffeomorphism $\Sigma^k\#(-\Sigma^k) \approx S^k$ for any $k \neq 4$ [KM]. Therefore, the holographic structure $\mathcal{H}^\dagger = \mathcal{H}\#_\partial {}^\neg\mathcal{H}$ is defined on S^k.

The same argument applies to the based *fillable* holographic structures on the homotopy spheres, as spelled out in the proof of Proposition 3.3. □

Proposition 3.3. *Any homotopy sphere Σ^k, $k \neq 4$, admits a fillable holographic structure.*

The forgetful map $\mathcal{K}_k : \mathcal{HS}_k^{\{\mathsf{fill}\}} \to \Theta_k$ that takes each pair
(a smooth homotopy k-sphere Σ, a fillable holographic structure on it)
to Σ is onto and respects the connected sum operations in its source and target.

Proof. Homotopy spheres Σ^k are stably parallelizable manifolds (see [Kos], Corollary 8.6). Thus, their Pontryagin and Stiefel-Whitney characteristic classes vanish. As a result, their Pontryagin and Stiefel-Whitney characteristic numbers vanish as well. This implies that Σ^k represents the zero element in the oriented bordism group Ω_k^{SO} [Wa2]. Therefore, any homotopy sphere Σ^k bounds an oriented connected smooth manifold X.

By Lemma 2.1, X admits a traversally generic vector field v. It generates a fillable holographic structure $\mathcal{H}(v)$ on Σ^k. As a result, any homotopy sphere admits a fillable holographic structure.

By picking a pair of based points, we convert this structure to a based one. Thus, the forgetful map $\mathcal{K}_k$ is surjective and respects the connected sum operations. □

Example 3.6. According to Brieskorn ([Bries]), the intersections $\Sigma^7(l)$ of the complex hypersurfaces

$$V(l) = \{z_1^2 + z_2^2 + z_3^2 + z_4^3 + z_5^{6l-1} = 0\} \subset \mathbb{C}^5, \quad l \in [1, 28],$$

with a sufficiently small sphere $S^9_\epsilon \subset \mathbb{C}^5$, centered at the origin $\mathbf{0} \in \mathbb{C}^5$, realize all 28 Milnor's exotic spheres from the group Θ_7.

By resolving the isolated singularity of $V(l)$ at $\mathbf{0}$, we get a nonsingular (smooth) 8-manifold $\hat{V}(l)$ together with the resolution map $r : \hat{V}(l) \to V(l)$. Let $X(l)$ denote the preimage $r^{-1}(V(l) \cap D^{10}_\epsilon)$, where $D^{10}_\epsilon \subset \mathbb{C}^5$ is the ϵ-ball, centered on $\mathbf{0}$. The compact oriented smooth 8-dimensional manifold $X(l)$ bounds $\Sigma^7(l)$.

By [KM], there exists a hypersurface $K^6(l) \subset S^7$, diffeomorphic to S^6, which cuts S^7 into two complimentary pieces, $\tilde{\Sigma}^7_+(l)$ and $\tilde{\Sigma}^7_-(l)$, each of which is diffeomorphic to $\Sigma^7(l) \setminus \mathsf{int}(D^7)$. The separating sphere $K^6(l)$ resides in the 1-handle that is used to form the connected sum

$S^7 \approx \Sigma^7(l)\#(-\Sigma^7(l))$. Note that S^7 is the boundary of the manifold $X(l)\#_\partial X(l)$.

By Proposition 3.3, for any fillable holographic structure $\mathcal{H}(l)$ on $\Sigma^7(l)$ (one may think of $\mathcal{H}(l)$ as being generated by some traversally generic vector field $v(l)$ on $X(l)$), there exists a fillable holographic structure $\mathcal{H}^\dagger(l)$ on S^7 such that $\mathcal{H}^\dagger(l) = \mathcal{H}(l)\,\#_\partial\, ^\neg\mathcal{H}(l)$. Thus, any "exotic" holographic structure $\mathcal{H}(l)$ "hides in the plain view" in a holographic structure on the standard sphere S^7. ◇

By Proposition 3.3, for $k \neq 4$, the set $\mathcal{HS}_k^{\{\mathsf{fill}\}}$ of fillable holographic structures on homotopy k-spheres is at least as rich as their image Θ_k.

For $k \neq 3$, the group Θ_k is finite with occasionally a quite complex torsion [KM]. For example, the order of Θ_{15} is $16,256 = 2^7(2^7 - 1)$. Thus, $\mathcal{HS}_{15}^{\{\mathsf{fill}\}}$ maps onto a group of order $16,256$.

Thanks to the pioneering works of Kervaire & Milnor [KM], Browder [Br], and more recently, of Hill, Ravenel, & Hopkins [HHR], the groups Θ_k, $k \neq 4$, are well-understood in terms of the stable homotopy groups of spheres π_k^{st} and of the J_k-homomorphism

$$J_k : \pi_k(\mathsf{SO}(N)) \to \pi_{k+N}(S^N),$$

where $N \gg k$. Presently, the only remaining "problematic" dimensions are $k = 4$ and $k = 126$.

Against the background of our investigations of traversing flows, the fibers of the forgetful map $\mathcal{K}_k : \mathcal{HS}_k^{\{\mathsf{fill}\}} \to \Theta_k$ are interesting objects to study.

3.5 Causality maps as measure preserving transformations

In this section, we show how special closed differential n-forms Θ on a compact connected smooth $(n+1)$-dimensional manifold X generate a measure μ_Θ on the boundary $\partial_1 X$ such that the causality map C_v becomes a measure-preserving transformation (Theorem 3.4). For a traversing vector field v, the differential form Θ helps also to define, in the spirit of [S], a measure μ_Θ on the space of trajectories $\mathcal{T}(v)$ (see Definition 3.16).

As the lemma below testifies, the causality maps $C_v : \partial_1^+ X(v) \to \partial_1^- X(v)$, although discontinuous, have some "positive features": they preserve certain v-amenable n-dimensional measures on the boundary $\partial_1 X$.

Lemma 3.11. *Let X be a compact smooth oriented $(n+1)$-dimensional manifold with boundary, carrying a boundary generic traversing vector field*

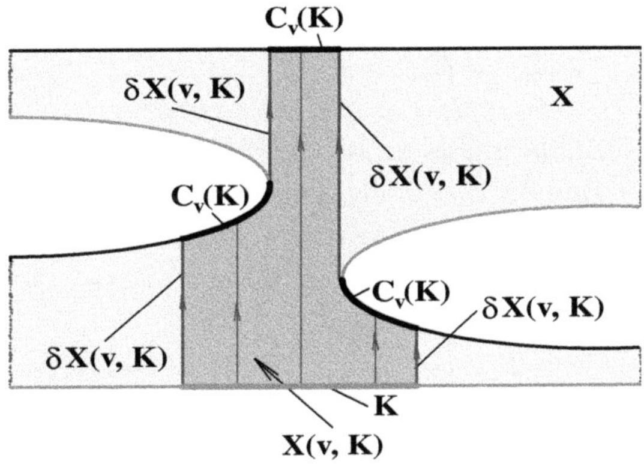

Fig. 3.4 The shaded set $X(v, K)$ and its boundary $\partial X(v, K) = K \cup C_v(K) \cup \delta X(v, K)$ for a codimension zero submanifold $K \subset \partial_1^+ X$.

v. We denote by Ω^∂ a volume n-form on $\partial_1 X$, consistent with its orientation [Mos]. Let $\Theta \in \bigwedge^n(T^*X)$ be a differential n-form on X, subject to the constraints:

(1) $\Theta(v \wedge w) = 0$ for any poly-vector $w \in \bigwedge^{n-1}(TX)$ (we call such form Θ is "v-horizontal"),
(2) $d\Theta = 0$,
(3) the function $(\Theta/\Omega^\partial) : \partial_1 X \to \mathbb{R}$ is nonnegative on $\partial_1^+ X(v)$ and nonpositive on $\partial_1^- X(v)$.

Let μ^∂ be a Lebesgue measure on $\partial_1 X$.

- *Then the Lie derivative $\mathcal{L}_v(\Theta) = 0$.*
- *By restricting the form Θ to the boundary $\partial_1 X$, for any μ^∂-measurable set $K \subset \partial_1^+ X(v)$, we get*

$$\int_K \Theta = \Big| \int_{C_v(K)} \Theta \Big|. \tag{3.30}$$

Proof. Using the Cartan identity $\mathcal{L}_v(\Theta) = d(v \rfloor \Theta) + v \rfloor d\Theta$ for the Lie derivative and properties (1) and (2) of Θ, we conclude that $\mathcal{L}_v(\Theta) = 0$; that is, the form Θ is invariant under the v-flow. Thanks to property (1), Θ is a horizontal form.

Let v be a boundary generic traversing vector field and Θ an n-form on X as in the hypotheses of Lemma 3.9. For any Lebesgue-measurable

$K \subset \partial_1 X$, we *define* its measure $\mu_\Theta(K)$ by the formula

$$\mu_\Theta(K) =_{\mathsf{def}} \int_{K \cap \partial_1^+ X(v)} \Theta - \int_{K \cap \partial_1^- X(v)} \Theta. \tag{3.31}$$

Note that this formula makes sense since, for a boundary generic v, the sets $\partial_1^\pm X(v)$ are smooth manifolds, and the intersection $K \cap \partial_1^\pm X(v)$ of two Lebesgue measurable sets is again a Lebesgue measurable set.

For any set $A \subset \partial_1^+ X(v)$, we denote by $X(v, A)$ the set, formed by the v-trajectories through the points of A.

Consider the locus $\partial_2^+ X(v) \subset \partial(\partial_1^+ X(v))$, the closure of points of the boundary $\partial_1 X$, where the v-flow is simply tangent to the boundary and the boundary is *concave* with respect to the flow in the sense of [K1] (see Definition 2.3). For a boundary generic traversing vector field v, let $\mathcal{X}$ denote the set $X(v, \partial_2^+ X(v))$, the union of v-trajectories that contain points from the locus $\partial_2^+ X(v)$. Then $\mathcal{X} \cap \mathrm{int}(\partial_1^+ X(v))$ is the discontinuity locus of the causality map C_v.

For a boundary generic v, Lemmas 2.3 and 2.4 (see also Lemmas 3.1 and 3.4 from [K2]) provide us with semi-local real algebraic models of the domain and range of the causality map C_v, as well as with semi-local models of the causality map itself. The expression "semi-local" here means "in the vicinity of each v-trajectory". In particular, these models imply that the Lebesgue measure $\mu^\partial(C_v(A)) = 0$ for any set $A \subset \partial_1^+ X$ such that $\mu^\partial(A) = 0$. They imply also that the C_v-image of a Lebesgue measurable set $K \subset \partial_1^+ X$ is Lebesgue measurable in $\partial_1^- X$.

In order to prove that $\int_K \Theta + \int_{C_v(K)} \Theta = 0$ for any Lebesgue measurable $K \subset \partial_1^+ X(v)$, it will suffice to show that $\int_N \Theta + \int_{C_v(N)} \Theta = 0$ for all n-dimensional compact piecewise differentiable ("PD" for short) manifolds $N \subset \partial_1^+ X(v)$.

For any PD-submanifold $N \subset \partial_1^+ X$, we form the set $X(v, N)$. Again, thanks to the semi-local models of Lemmas 2.3 and 2.4, the locus $X(v, N)$ is a piecewise differentiable manifold with boundary. Its oriented boundary $\partial X(v, N)$ is formed by the three PD-manifolds: N, $C_v(N) \subset \partial_1 X$, and the rest, which we denote $\delta X(v, N)$. The latter locus is built out of segments of v-trajectories (see Fig. 3.4). Thus, by the Stokes Theorem,

$$\int_{X(v,N)} d\Theta = \int_N \Theta + \int_{C_v(N)} \Theta + \int_{\delta X(v,N)} \Theta.$$

Since $d\Theta = 0$, we get $\int_{X(v,N)} d\Theta = 0$. Using that Θ is vertical and $\delta X(v, N)$ consists of v-trajectories, we get $\int_{\delta X(v,N)} \Theta = 0$, which implies formula (3.30) and the measure-defining formula (3.31). $\square$

Definition 3.7. Let v be a smooth non-vanishing vector field on a compact orientable $(n+1)$-dimensional manifold X and ξ a n-dimensional distribution, transversal to v. Consider a differential n-form Θ on X.

(1) We say that Θ is integrally dual to v, if:

- $d\Theta = 0$
- the kernel $\mathsf{ker}(\Theta)$ of Θ is a 1-dimensional distribution on X,
- $v \in \mathsf{ker}(\Theta)$,
- $\pm\Theta|_{\partial_1^\pm X(v)} \geq 0$.

(2) We call Θ integrally adjusted to the pair (v, ξ), if Θ integrally dual to v and $\Theta|_\xi > 0$ with respect to the orientation of ξ, induced by v and the orientation of X. (It follows from the property $\Theta|_\xi > 0$ that the kernel $\mathsf{ker}(\Theta)$ is a 1-dimensional distribution.) ◊

With Definition 3.7 in place, we can rephrase Lemma 3.9 as follows.

Proposition 3.4. *For a traversing vector field v on a compact smooth $(n+1)$-manifold X, any integrally dual to v n-form Θ is v-invariant.*

Moreover, Θ defines a measure μ_Θ on the boundary $\partial_1 X$ (see (3.31)) such that the causality map C_v is the measure preserving transformation. ◊

Lemma 3.12. *Let X be a smooth compact oriented $(n+1)$-manifold.*

For a given non-vanishing vector field v on X, an integrally dual form Θ is unique, up to the multiplication by a smooth positive function $h : X \to \mathbb{R}_+$ such that $dh \wedge \Theta = 0$ in X.

Proof. Consider the n-th exterior power $\bigwedge^n T^*X \to X$ of the cotangent bundle $T^*X \to X$. The linear constraints on a differential form $\theta \in \bigwedge^n T^*X$, imposed by the property $\{v \in \mathsf{ker}(\theta)\}$, define a 1-dimensional subbundle Λ_v of the $(n+1)$-dimensional bundle $\bigwedge^n T^*X$. Let us denote by $\Gamma(\Lambda_v)$ the linear space of smooth sections-forms Θ of the line bundle $\Lambda_v \to X$. Since $\dim(\Lambda_v) = 1$, any two sections $\Theta, \Theta' \in \Gamma(\Lambda_v)$ differ by a functional multiple, i.e., $\Theta' = h \cdot \Theta$.

Consider the kernel $\mathcal{K}_v$ of the differential

$$d :\ \Gamma(\Lambda_v) \longrightarrow \Gamma(\bigwedge^{n+1} T^*X) \approx C^\infty(X, \mathbb{R}).$$

Let $\Gamma^\star(\Lambda_v)$ denote the subspace of nowhere vanishing sections from $\Gamma(\Lambda_v)$. Then any two integrally dual to v n-forms $\Theta, \Theta' \in \Gamma^\star(\Lambda_v) \cap \mathcal{K}_v$, and

$\Theta' = h\Theta$, where the functional coefficient $h : X \to \mathbb{R}$ is positive/negative in X.

Using that $d\Theta = 0$, we get $0 = d(\Theta') = d(h\Theta) = dh \wedge \Theta$, where $h \neq 0$. Thus, h is such that $dh \wedge \Theta = 0$ identically on X. The requirement $\pm h \cdot \Theta|_{\partial_1^\pm X(v)} \geq 0$ implies that $h : \partial_1 X \to \mathbb{R}$ must be nonnegative on $\partial_1 X$ and thus $h > 0$ on X. □

Remark 3.16. We stress that Lemma 3.12 does not claim the *existence* of an integrably dual or integrably adjusted form Θ for a given $v \neq 0$ or (v, ξ), respectively. It follows from Theorem 3.14 below that, for any non-vanishing vector field v, there is no *local* obstruction to the existence of integrally dual form Θ. In fact, at least any invariant Calabi vector field (as in Definition 3.14, the second bullet) admits such a form.

On the other hand, by Lemma 3.14, the existence of an integrably dual form Θ is equivalent to an existence of a v-invariant volume form Ω on X. ◇

Lemma 3.13. *For a given non-vanishing vector field v, the space $\mathcal{D}(v)$ of n-forms Θ, integrably dual to v, is convex in the space of all n-forms, provided $\mathcal{D}(v) \neq \emptyset$.*

Similarly, for a given non-vanishing vector field v and a n-distribution ξ, transversal to v, the space $\mathcal{D}(v, \xi)$ of n-forms Θ, integrably adjusted to (v, ξ), is convex in the space of all n-forms, provided $\mathcal{D}(v, \xi) \neq \emptyset$.

Proof. Consider two integrably dual to v forms Θ_1, Θ_2. Evidently, if Θ_1 and Θ_2 are closed, so is their linear combination $\Theta = t\Theta_1 + (1-t)\Theta_2$, where $t \in [0, 1]$. Also, if $v \in \mathsf{ker}(\Theta_1)$ and $v \in \mathsf{ker}(\Theta_2)$, then $v \in \mathsf{ker}(\Theta)$. The positivity condition $\pm\Theta|_{\partial_1^\pm X(v)} \geq 0$ follows since $t \geq 0$ and $1 - t \geq 0$.

If Θ_1 and Θ_2 are integrably dual to (v, ξ), then, in addition, $\Theta|_\xi > 0$ since $\Theta_1|_\xi > 0$ and $\Theta_2|_\xi > 0$. □

Definition 3.8. Let v be a non-vanishing vector field on a smooth oriented $(n+1)$-manifold X. We say that v is intrinsically nildivergent if there exists a volume $(n+1)$-form Ω on X such that $d(v \rfloor \Omega) = 0$. ◇

Example 3.7. Let $v = \sum_{i=1}^{n+1} a_i\, \partial_{x_i}$ in coordinates $(x_1, \dots, x_{n+1})$ on $\mathbb{R}^{n+1}$. A volume form $\Omega = f\, dx_1 \wedge \cdots \wedge dx_{n+1}$ produces nildivergent form $\Theta = v \rfloor \Omega$, if the function f satisfies the equation

$$\langle \nabla f, v \rangle = -\mathrm{div}(v)\, f.$$

Here $\langle\ ,\ \rangle$ denotes the Euclidean scalar product, ∇f the gradient of f, and $\mathrm{div}(v) = \sum_{i=1}^{n+1} \frac{\partial a_i}{\partial x_i}$. ◇

Let us describe one standard mechanism that produces n-forms Θ, integrally dual to a given vector field $v \neq 0$.

Lemma 3.14. *A non-vanishing vector field v on a $(n+1)$-dimensional X is intrinsically nildivergent with the help of $(n+1)$-form Ω if and only if Ω is v-invariant. Then the form $\Theta = v \rfloor \Omega$ is integrally dual to v (see Definition 3.7), and Θ is v-invariant as well. Moreover, if X is oriented, then there is a n-distribution ξ on X, is transversal to v, and such that $\Theta|_\xi > 0$.*

Proof. Since $d\Omega = 0$, by the Cartan formula, we get $\mathcal{L}_v(\Omega) = d(v \rfloor \Omega)$. Thus $d(v \rfloor \Omega) = 0$ if and only if Ω is v-invariant. By Definition 3.8, an intrinsically nildivergent form $\Theta = v \rfloor \Omega$ is closed.

Since $v \rfloor \Theta = v \rfloor (v \rfloor \Omega) = 0$, we get $v \in \mathsf{ker}(\Theta)$. Moreover, since Ω is a volume form, $\dim(\mathsf{ker}(\Theta)) = 1$. As in Lemma 3.9, it follows that Θ is a v-invariant form.

Let us pick a metric g on X so that Ω is its volume form. Let ν be the unit vector field, inward normal in the metric g to $\partial_1 X$ in X. Then the quotient $(v \rfloor \Omega)|_{\partial_1 X}/(\nu \rfloor \Omega)|_{\partial_1 X}$ of the two n-forms, being restricted to $\partial_1 X$, equals to $\cos(\angle_g(v, \nu))$, a non-negative function on $\partial_1^+ X(v)$ and non-positive on $\partial_1^- X(v)$ by the very definition of these two loci. The g-orthogonal to v subbundle $v_g^\perp \subset T_*X$ plays the role of the desired distribution ξ. □

Lemma 3.15. *Any diffeomorphism $\Phi : X \to X$ transforms a form Θ, intergrally dual to a given vector field v, into the form $\Phi^*(\Theta)$, intergrally dual to the vector field $\Phi_*^{-1}(v)$.*

In particular, if v is nildivergent with the help of a $(n+1)$-volume form Ω and a diffeomorphism Φ is such that $\Phi^*(\Omega) = \pm\Omega$, then the vector field $\Phi_*(v)$ is nildivergent. Thus, the group of volume-preserving diffeomorphisms $\mathsf{Diff}(X, \Omega)$ of X acts naturally on the space of nildivergent vector fields.

Proof. If $v \rfloor \Theta = 0$, then by naturality, $\Phi_*^{-1}(v) \rfloor \Phi^*(\Theta) = 0$. Also by naturality, $d(\Phi^*(\Theta)) = \Phi^*(d\Theta) = 0$.

Evidently, any diffeomorphism Φ maps $\partial_1^\pm X(v)$ to $\partial_1^\pm X(\Phi_*(v))$. On the other hand, if $\pm\Theta(w) \geq 0$ for a polyvector $w \in \bigwedge^n T_*(\partial_1^\pm X(v))$, then, by naturality, $\pm\Phi^*(\Theta)(\Phi_*^{-1}(w)) \geq 0$. Therefore, if $\pm\Theta|_{\partial_1^\pm X(v)} \geq 0$, then $\pm\Phi^*(\Theta)|_{\partial_1^\pm X(\Phi_*^{-1}(v))} \geq 0$.

If $d(v \rfloor \Omega) = 0$ and $\Phi^*(\Omega) = \pm\Omega$, then

$$d(\Phi_*^{-1}(v) \rfloor \Phi^*(\Omega)) = d(v \rfloor \pm \Omega) = 0.$$

Thus, $\Phi_*^{-1}(v)$ is nildivergent, provided that v is and Φ preserves, up to a sign, the volume form. Hence, the group of volume-preserving diffeomorphisms $\mathsf{Diff}(X, \Omega)$ of X acts naturally on the space of nildivergent vector fields. □

3.6 On proto-billiard maps and the Poincaré return maps

Next, we combine the causality maps $C_v : \partial_1^+ X(v) \to \partial_1^- X(v)$ with measure-preserving involutions τ on the boundary $\partial_1 X$ to introduce proto-billiard maps $B_{v,\tau} : \partial_1^+ X(v) \to \partial_1^+ X(v)$. They give rise to dynamical measure-preserving systems (Theorem 3.11).

In order to introduce some *dynamics* in our considerations of the causality maps of traversing flows, we need to assume the validity of the following property. It mimics an elastic reflection from the boundary of a billiard.

• **The Involution Hypotheses.** Let X be a compact connected smooth $(n+1)$-manifold X with boundary. For a traversing boundary generic vector field v and a n-form Θ, integrably dual to v on X (as in Definition 3.7), we assume that there exists a diffeomorphism $\tau : \partial_1^- X(v) \to \partial_1^+ X(v)$ such that

$$\tau^*\big(\Theta|_{\partial_1^+ X(v)}\big) = \Theta|_{\partial_1^- X(v)}. \tag{3.32}$$

Occasionally, we will assume that τ is the restriction to $\partial_1^+ X(v)$ of a smooth involution $\hat{\tau} : \partial_1 X \to \partial_1 X$, whose fixed point set is the locus $\partial_2 X(v) = \partial(\partial_1^+ X(v)) = \partial(\partial_1^- X(v))$.

In general, such τ may be unavailable (say, when $\partial_1^+ X(v)$ and $\partial_1^- X(v)$ are not diffeomorphic). However, assuming its existence, by Proposition 3.4, the proto-billiard map—the composition

$$B_{v,\tau} =_{\mathsf{def}} \tau \circ C_v : \partial_1^+ X(v) \to \partial_1^+ X(v), \tag{3.33}$$

of the causality map C_v with τ—preserves the measure μ_Θ. As a result, for such τ and Θ, it is possible to talk about the dynamics of μ_Θ-preserving iterations $\{(B_{v,\tau})^{\circ k}\}_{k\in\mathbb{N}}$ of $B_{v,\tau}$.

In the case of billiard maps on a Riemannian manifold M, τ is a smooth involution $\partial_1(SM) \to \partial_1(SM)$, induced by the orthogonal reflection of unit tangent vectors from $TM|_{\partial M}$ with respect to the boundary ∂M.

• In what follows, we say that a property is valid almost everywhere, if it may be violated only for a set of points of zero measure.

Theorem 3.11. *Let X be a compact connected smooth $(n+1)$-manifold, equipped with a traversing and boundary generic vector field v and an integrally dual to it n-form Θ. Let $\tau : \partial_1^- X(v) \to \partial_1^+ X(v)$ be a diffeomorphism which satisfies (3.32).*

Then the proto-billiard map $B_{v,\tau} : \partial_1^+ X(v) \to \partial_1^+ X(v)$ from (3.33) has the following infinite return property*: any point $x \in \partial_1^+ X(v)$ has an open neighborhood $U \subset \partial_1^+ X(v)$ such that, for almost every $x' \in U$, the sets $\{(B_{v,\tau})^{\circ k}(x')\}_k$ intersect U for infinitely many k's.*

Let N be an open neighborhood of $\partial_1 X$ in X and let Ω be a volume form on X. In particular, the return property holds for any intrinsically nildivergent traversing and boundary generic v and for τ, induced by a diffeomorphism $\tilde{\tau} : N \to N$ such that $\tilde{\tau}^*(\Omega) = -\Omega$ and $\tilde{\tau}_*(v) = -v$.

Proof. Combining Theorem 3.4 with the property $\tau^*(\Theta|_{\partial_1^+ X(v)}) = \Theta|_{\partial_1^- X(v)}$ from (3.32), we conclude that the proto-billiard map $B_{v,\tau}$ preserves the measure μ_Θ on $\partial_1^+ X(v)$. Since the volume $\mu_\Theta(\partial_1^+ X(v))$ is finite, the standard argument (see [Walt]) about measure-preserving transformations leads to the infinite return property.

In the case of nildivergent v, we have $\Theta = v \rfloor \Omega$, where Ω is a v-invariant volume form. Assume that a diffeomorphism $\tau : \partial_1^+ X(v) \to \partial_1^- X(v)$ admits a lifting $\tilde{\tau} : N \to N$ so that $\tilde{\tau}^*(\Omega) = -\Omega$ and $\tilde{\tau}_*(v) = -v$. Then, employing Lemma 3.15, we get

$$\tilde{\tau}^*(\Theta) = \tilde{\tau}^*(v \rfloor \Omega) = (-v) \rfloor (-\Omega) = v \rfloor \Omega = \Theta$$

in N. Thus, $B_{v,\tau}$ preserves the measure μ_Θ on $\partial_1^+ X(v)$, and hence the return property follows. □

In Chapter 4, we will strive to construct an involution τ, subject to (3.32), for the geodesic vector field v^g on the tangent spherical fibration $SM \to M$ whose base is a compact Riemannian manifold (M, g).

Definition 3.9. Given a non-vanishing vector field v on a smooth manifold X, a closed 1-form α is called Lyapunov, if $\alpha(v) > 0$. ◇

Definition 3.10. We say that a smooth vector field v is gradient-like, if there are a smooth function $f : X \to \mathbb{R}$ and a Riemannian metric g in the vicinity of the zero locus $Z(v)$ of v so that:

(1) $df(v) > 0$ in $X \setminus Z(v)$, and

(2) v is the gradient $\nabla_g(f)$ in the vicinity of $Z(v)$. ◇

Definition 3.11. (Lyapunov genus) Given a smooth vector field v on a manifold X, we consider an open cover $\mathcal{U} = \{U_i\}_i$ of X such that, in each U_i, v admits a smooth Lypunov function $f_i : U_i \to \mathbb{R}$, i.e., $df_i(v) \geq 0$ in U_i and $df_i(v) > 0$ in $U_i \setminus (U_i \cap Z(v))$.

We call the minimal cardinality of such covers $\mathcal{U}$ the Lyapunov genus of v and denote it by $\mathsf{Lyap}(v)$. Insisting that the elements of open covers $\mathcal{U} = \{U_i\}_i$ are connected sets, leads to a similar notion of the connected Lyapunov genus $\mathsf{Lyap}^\bullet(v) \geq \mathsf{Lyap}(v)$. ◇

By definition, for any *traversing* vector field v, $\mathsf{Lyap}(v) = 1$.

For example, for the vector field on the annulus, depicted in Fig. 3.6, (a), we get $\mathsf{Lyap}^\bullet(v) = \mathsf{Lyap}(v) = 2$, while for the vector field on the annulus, depicted in Fig. 3.6, (c), we get $\mathsf{Lyap}^\bullet(v) = \mathsf{Lyap}(v) = 1$.

Fig. 3.5 Lorenz strange attractor

Question 3.4. *Given the Lorenz equations*

$$\frac{dx}{dt} = \sigma(y - x)$$

$$\frac{dy}{dt} = x(\rho - z) - y$$

$$\frac{dz}{dt} = xy - \beta z,$$

for some values of parameters σ, ρ, β that lead to a chaotic behavior of this dynamical system (say, for $\rho = 28$, $\sigma = 10$, $\beta = \frac{8}{3}$), what is the Lyapunov genus of such Lorenz flow?

For these special values of parameters, does the yz coordinate plane (or any other smooth surface in $\mathbb{R}^3$) delivers the chopping of the flow into closed segments and singletons, as in Lemma 3.16 or Theorem 3.12? $\diamondsuit$

Lemma 3.16. *Let X be a compact manifold and v a smooth non-vanishing vector field on X. If a smooth hypersurface $H \subset X$ bounds a domain $X_1 \subset X$ so that, for any v-trajectory γ, the connected components of $\gamma \cap X_1$ and of $\gamma \cap (X \setminus \mathsf{int}(X_1))$ are singletons or closed intervals, then* $\mathsf{Lyap}(v) \leq 2$.

Proof. The connected components of $\gamma \cap X_1$ and $\gamma \cap (X \setminus \mathsf{int}(X_1))$ are singletons or closed intervals. Thus, the vector fields $v|_{X_1}$ and $v|_{X_2}$ are traversing. By Lemma 2.1 (or Lemma 4.1 from [K1]), each traversing vector field admits a Lyapunov function. Therefore, in the presence of a chopping hypersurface H, we get $\mathsf{Lyap}(v) \leq 2$. $\square$

Definition 3.12. Let v be a smooth non-vanishing vector field on a compact $(n+1)$-manifold X. We say that v has the chopping property if there exists finally many disjoint and transversal the v-flow sections $\{S_\alpha\}_\alpha$, each of which is diffeomorphic to a closed n-ball, so that, for every positive or negative time trajectory γ that is contained in $\mathsf{int}(X)$, the intersection $\gamma \cap (\cup_\alpha S_\alpha)$ is a discrete subset of γ which divides it in closed segments. $\diamondsuit$

Theorem 3.12. *Let v be a smooth non-vanishing boundary generic vector field on a compact connected $(n+1)$-manifold X. Assume that v possesses the chopping property.*

Then there exists a smooth hypersurface $H \subset X$ that bounds a submanifold $X_1 \subset X$ so that, for each v-trajectory γ, the connected components of the sets $\gamma \cap X_1$ and $\gamma \cap (X \setminus \mathsf{int}(X_1))$ are singletons or closed intervals. As a result, $\mathsf{Lyap}(v) \leq 2$.

If $\partial_1 X \neq \emptyset$, then this hypersurface H may be constructed as the final result of a smooth isotopy of the boundary $\partial_1 X$ inside of X.

If $\partial_1 X = \emptyset$, then one can choose H to be a sphere, bounding the standard ball.

Proof. First, we consider the case of $\partial_1 X \neq \emptyset$. The v-trajectories of a non-vanishing vector field v may be of several types (see Fig. 3.6, a):

(i) the trajectories that originate and terminate at the boundary $\partial_1 X$ (they are homeomorphic to the interval $[0, 1]$ or to a singleton),

(ii) the trajectories that originate/terminate at $\partial_1 X$, but never reach it again in positive/negative time (they are homeomorphic to the semi-open intervals $[0, 1)$ or $(0, 1]$),

(iii) the trajectories that belong to $\mathsf{int}(X)$ (they are homeomorphic to the open interval $(0, 1)$ or to the circle).

Our next goal is, by choosing the appropriate chooping hypersurface, to eliminate the types (ii) and (iii).

Consider the locus $\partial_2^- X(v) \subset \partial_1 X$ for which the v-trajectory through $x \in \partial_2^- X(v)$ is the singleton x. Let V be an open small neighborhood of $\partial_2^- X(v)$ in X which consists of v-trajectories that are singletons or closed segments only.

Consider the transversal ball-shaped sections $\{S_\alpha \subset \mathsf{int}(X)\}_\alpha$ of the $\hat{v}$-flow that deliver its chopping property from Definition 3.12. By the definition, for each positive and negative time trajectory γ of the types (ii) or (iii), $\gamma \cap (\coprod_\alpha S_\alpha)$ is a discrete set in γ, so that γ is chopped by $\coprod_\alpha S_\alpha$ into infinitely many closed segments.

We may choose V so small, that the v-trajectories through $\coprod_\alpha S_\alpha$ miss V. Thus we can concentrate on the trajectories that belong to $X \setminus V$.

We engulf each S_α into a sufficiently flat (as a pancake) in the v-direction $(n+1)$-dimensional closed ball $B_\alpha^{n+1} \subset \mathsf{int}(X)$ so that all the balls $\{B_\alpha^{n+1}\}_\alpha$ are disjoint. Moreover, each positive and negative time trajectory γ of the types (ii) and (iii) is such that $\gamma \cap (\coprod_\alpha B_\alpha^{n+1})$ and $\gamma \cap (X \setminus (\coprod_\alpha \mathsf{int}(B_\alpha^{n+1})))$ both are disjoint unions of closed segments due to chopping property of the section $\coprod_\alpha S_\alpha$.

Next, we delete from the interior of X the open balls $\{B_\alpha^{n+1}\}$. We denote by $X' \subset X$ the resulting smooth manifold.

By the construction above, each v-trajectory in X', is chopped by the boundary $\partial_1 X \coprod (\coprod_\alpha \partial B_\alpha^{n+1})$ into closed segments. Thus, each v-trajectory in X' is either a closed segment with its ends in $\partial_1 X'$, or a singleton, residing in $\partial_1 X$.

By Lemma 3.16, such a v-flow on X' admits a Lyapunov function $f : X' \to \mathbb{R}$.

For each ball B_α^{n+1}, we drill a tunnel H_α in X'. The tunnel $H_\alpha \subset X' \setminus \mathsf{int}(B_\alpha^{n+1})$ originates at a point of $\partial_1 X$ and reaches B_α^{n+1} (H_α is a solid 1-handle attached to $\partial_1 X$ and ∂B_α^{n+1}). We choose the tunnels so that they

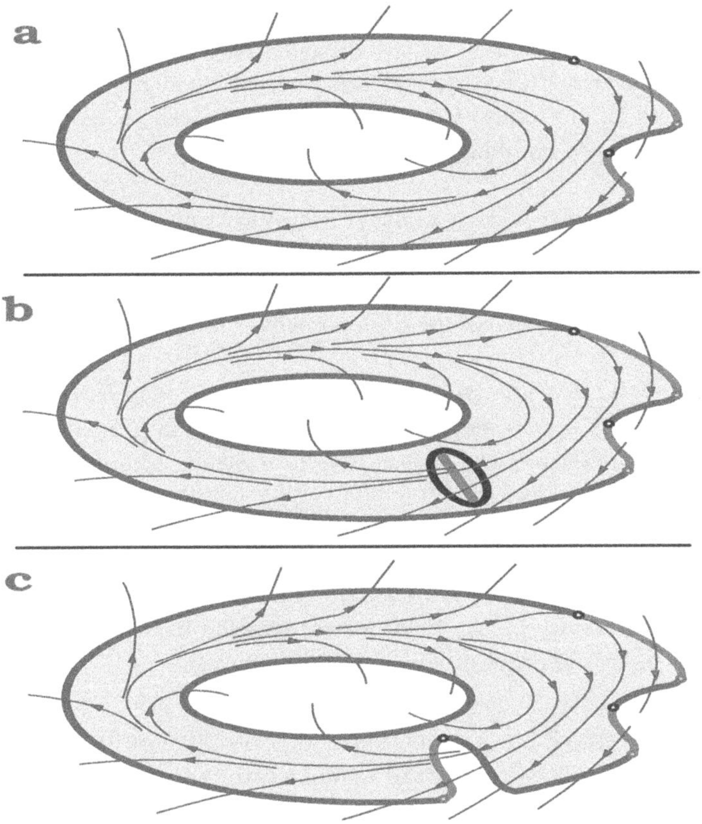

Fig. 3.6 (a) A vector flow with trapped trajectories on a manifold X with boundary. (b) Finding disrupting ball-shaped hypersection(s) $\{S\}$ of the flow. (c) Expelling the disrupting section(s) by an isotopy based on attaching a 1-handle to $\partial_1 X$ and to the boundary of the ball that engulfs the section S.

do not intersect each other. We delete the union $\coprod_\alpha H_\alpha$ from X', round the corners, and denote the resulting smooth manifold by X''. Evidently, X'' is diffeomorphic to X via a diffeotopy in X (see Fig. 3.6, c). Let us restrict the vector field v to X''. We denote by $\tilde{v}$ the restriction. Clearly, $f|_{X''}$ is still a Lyapunov function for $\tilde{v}$.

Therefore, X'' is diffeotopic to X, and the vector field $\tilde{v}$ is traversing with the help of the Lyapunov function $f|_{X''}$. For the same reason, the restriction of v to $X \setminus \mathsf{int}(X'')$ is also traversing with the help of a Lyapunov function $g : \coprod_\alpha (B^{n+1}_\alpha \cup H_\alpha) \to \mathbb{R}$. Hence, $\mathsf{Lyap}(v) \leq 2$.

The case of a closed X is similar and actually simpler. We construct the $(n+1)$-balls $\{B_\alpha^{n+1}\}$ as in the previous case. Their boundaries form a chopping surface H. By appropriate 1-surgeries on H, performed in the complements to the balls, we can incapsulate all these balls in a single $(n+1)$-ball, whose boundary will do the chopping job. □

Corollary 3.15. *Let X be a compact connected smooth manifold. Let v be a boundary generic non-vanishing vector field on X. Assume that each positive and negative time non-cyclic trajectory (of type* (iii)*) that is contained in the interior of X asymptotically approaches one of finitely many cyclic trajectories $\{c_\alpha\}$.*

If $\partial_1 X \neq \emptyset$, then the chopping hypersurface H may be constructed as the final result of a smooth isotopy of the boundary $\partial_1 X$ inside of X.

If $\partial_1 X = \emptyset$, then one can choose H to be a sphere, bounding the standard ball. In both cases, $\mathsf{Lyap}(v) \leq 2$.

Proof. For each v-cycle c_α, we pick a small transversal to v n-ball $S_\alpha \subset \mathsf{int}(X)$ so that all the balls are disjoint. Then each trajectory of types (ii) or (iii), is chopped by $\coprod_\alpha S_\alpha$ into closed segments (see Fig. 3.6). Thus the arguments as in the proof of Theorem 3.12 apply. □

When the v-flow on X is periodic, one has a more constructive way (as compared to Theorem 3.12) to manufacture the chopping hypersurface $H \subset X$. Proposition 3.7 and Theorem 3.10 from [R] imply directly the following claim.

Lemma 3.17. *Let X be a closed manifold, and v a smooth non-vanishing vector field on X which generates a $T_\star$-*periodic *flow $\{\Phi_t\}_{t\in\mathbb{R}}$. Assume that there exists a smooth involution $\tau : X \to X$ such that $\Phi_{-t} = \tau \circ \Phi_t \circ \tau$ for all $t \in \mathbb{R}$.*

Then the τ-fixed set $X^\tau = A \coprod \Phi_{T_\star/2}(A)$, where A is a closed codimension 1 submanifold of X and a global section of the flow $\{\Phi_t\}_{t\in\mathbb{R}}$.[8]

Thus, X^τ is the "chopping" hypersurface, and $\mathsf{Lyap}(v) = 2$. ◇

For a non-vanishing v on a closed manifold, let us consider the following construction which, in the spirit of Lemma 3.17, produces the desired proto-billiard "reflection map" $\tau : \partial_1^- X(v) \to \partial_1^+ X(v)$.

[8] As a result, X fibers over the circle S^1 with the fiber A so that each v-trajectory is mapped by a submersion onto S^1.

Let Y be a *closed* smooth $(n+1)$-dimensional manifold, equipped with a volume form Ω, and a non-vanishing vector field v. All these structures on Y are presumed to be smooth. In addition, assume that the following three properties hold:

$$(3.34)$$

- A smooth chopping hypersurface H, as in Theorem 3.12, separates Y into two compact smooth manifolds, X_1 and X_2, so that the vector fields $v|_{X_1}$ and $v|_{X_2}$ are traversing,
- v is boundary generic with respect to the hypersurface $H \subset Y$,
- $d(v \rfloor \Omega) = 0$. (Thus, v is nildivergent on Y; if we assume that Ω is v-invariant, then the property $d(v \rfloor \Omega) = 0$ follows.) ♢

Thanks to Theorem 3.12, there exist Lyapunov functions $f_i : X_i \to \mathbb{R}$ for the vector fields $v_i =_{\mathsf{def}} v|_{X_i}$. Thus, v_i is traversing and boundary generic on X_i.

Note that $\partial_1^{\pm} X_1(v_1) = \partial_1^{\mp} X_2(v_2)$. Therefore, under hypotheses (3.34), the two causality maps,

$$\begin{aligned} C_{v_1} &: \partial_1^{+} X_1(v_1) \to \partial_1^{-} X_1(v_1) = \partial_1^{+} X_2(v_2) \\ C_{v_2} &: \partial_1^{+} X_2(v_2) \to \partial_1^{-} X_2(v_2) = \partial_1^{+} X_1(v_1), \end{aligned} \tag{3.35}$$

are available. Their composition produces the Poincaré return map

$$P_v =_{\mathsf{def}} C_{v_2} \circ C_{v_1} : \partial_1^{+} X_1(v_1) \to \partial_1^{+} X_1(v_1). \tag{3.36}$$

Note that the locus $\partial_2^{-} X_1(v_1) = \partial_2^{+} X_2(v_2)$ is the fixed point set of the map C_{v_1}, but not of C_{v_2}; similarly, the locus $\partial_2^{+} X_1(v_1) = \partial_2^{-} X_2(v_2)$ is the fixed point set of the map C_{v_2}, but not of C_{v_1}.

Corollary 3.16. *Under the hypotheses (3.34), the Poincaré return map* $P_v : \partial_1^{+} X_1(v_1) \to \partial_1^{+} X_1(v_1)$ *preserves the measure* μ_Θ*, induced as in (3.31) by the closed* v*-invariant* n*-form* $\Theta = v \rfloor \Omega$ *on the hypersurface* $\partial_1^{+} X_1(v_1) = \partial_1^{-} X_1(v_2)$.

As a result, any point $x \in \partial_1^{+} X(v_1)$ *has an open neighborhood* $U \subset \partial_1^{+} X(v_1)$ *such that, for almost every* $x' \in U$*, the sets* $\{(P_v)^k(x')\}_k$ *intersect* U *for infinitely many* k*'s.*

Proof. Under the hypotheses (3.34), the closed n-form $\Theta = v \rfloor \Omega$ has all the properties, listed in Lemma 3.9, on both manifolds, X_1 and X_2. By this lemma, both maps, C_{v_1} and C_{v_2}, preserve the measure μ_Θ, induced by the restriction of the form Θ to $\partial_1^{+} X(v_1)$, and so does their composition, the

Poincaré return map P_v. By standard arguments about measure-preserving transformations (see [Walt]), we get the return property of P_v, as stated in the corollary. $\square$

The dynamics of P_v-iterations that preserve the measure μ_Θ on the locus $\partial_1^+ X_1(v_1)$ will preoccupy us for some time...

Remark 3.17. Let $\mathcal{F}(v), \mathcal{F}(v_1), \mathcal{F}(v_2)$ be the 1-dimensional oriented foliations, produced by the vector fields v, $v_1 =_{\mathsf{def}} v|_{X_1}, v_2 =_{\mathsf{def}} v|_{X_2}$, respectively. Although, according to Holography Theorem 3.4, each map C_{v_i} $(i = 1, 2)$ allows for a reconstruction of the topological type of the pair $(X_i, \mathcal{F}(v_i))$, the Poincaré return map P_v *alone* seems to be insufficient for a reconstruction of the pair $(Y, \mathcal{F}(v))$.

Also, while each trajectory space $\mathcal{T}(v_i) = \partial_1^+ X_1(v_i)/\{x \sim C_{v_i}(x)\}$ is a separable compact space (a CW-complex for a traversally generic v_i), the quotient $\mathcal{T}(v) =_{\mathsf{def}} \partial_1^+ X_1(v_i)/\{x \sim P_v(x)\}$ typically is a pathological (non-separable) space. $\diamondsuit$

Theorem 3.13. (Double Holography) *Under the first two hypotheses from (3.34), the two causality maps C_{v_1} and C_{v_2} from (3.35) are sufficient for a reconstruction of the topological type of the pair $(Y, \mathcal{F}(v))$.*

Proof. Recall that C_{v_i} allows for a reconstruction of the trajectory space $\mathcal{T}(v_i)$ as the quotient space

$$\partial_1 X_i/\{x \sim C_{v_i}(x) \mid x \in \partial_1^+ X_i(v_i)\}.$$

This construction produces a continuous map $\Gamma_i^\partial : \partial_1 X_i \to \mathcal{T}(v_i)$, the restriction of the obvious map $\Gamma_i : X_i \to \mathcal{T}(v_i)$ to the boundary $\partial_1 X_i$. In particular, each point $x \in \partial_1 X_1 = \partial_1 X_2$ generates a pair of unique trajectories $[\gamma_x^{\{1\}}] \in \mathcal{T}(v_1)$ and $[\gamma_x^{\{2\}}] \in \mathcal{T}(v_2)$ that contain x.

Let $\mathcal{H}_i$ denote the 1-dimensional "foliation" of $\mathcal{T}(v_i) \times \mathbb{R}$ by the fibers of the obvious trivial fibration $\pi_i : \mathcal{T}(v_i) \times \mathbb{R} \to \mathcal{T}(v_i)$ (see Fig. 2.3).

As in Theorem 3.4, the two Lyapunov functions $\{f_i : X_i \to \mathbb{R}\}_{i=1,2}$ help to realize $(X_i, \mathcal{F}(v_i))$ as the pull-back of $(\mathcal{T}(v_i) \times \mathbb{R}, \mathcal{H}_i)$ under the embedding $\alpha_i : X_i \to \mathcal{T}(v_i) \times \mathbb{R}$, given by the formula $\alpha_i(x) = (\Gamma_i(x), f_i(x))$. By Theorem 3.4, the knowledge of the imbedding α_i^∂ (which is produced using C_{v_i} and $f_i|_{\partial_1 X_i}$ only) is sufficient for a reconstruction of the topological type of the pair $(X_i, \mathcal{F}(v_i))$.

Now, using the identity map, we glue $\alpha_1(X_1)$ and $\alpha_2(X_2)$ together along their boundaries $\alpha_1(\partial_1 X_1)$ and $\alpha_2(\partial_1 X_2)$ by identifying the pairs

$([\gamma_x^{\{1\}}], f_1(x))$ with $([\gamma_x^{\{2\}}], f_2(x))$ for any $x \in \partial_1 X_1 = \partial_1 X_2$. The result of the gluing is a manifold homeomorphic to Y. The two v-oriented foliations $\mathcal{F}(v_1)$ and $\mathcal{F}(v_2)$ match continuously (in fact, piecewise differentiably) along $\partial_1 X_1$, thus forming a topological 1-dimensional foliation on Y, which is homeomorphic to $\mathcal{F}(v)$. □

3.7 Calabi's vector fields, v-harmonizing metrics, and the associated minimal foliations

In this section, we are preoccupied with intrinsically harmonic Lyapunov functions $f : X \to \mathbb{R}$ and closed Lyapunov 1-forms α on X, specially adjusted to the given vector field v (Theorem 3.14). They go hand in hand with so called v-harmonizing Riemannian metrics g on X (see Definition 3.15). The v-harmonizing pairs (g, α) or (g, df) each produces a pair of mutually orthogonal minimal (i.e., taut) foliations $\mathcal{F}(v)$, $\mathcal{G}(\alpha)$ of dimensions 1 and n, respectively (Corollary 3.17). By Theorem 3.15, for a traversing boundary generic v, there exists a v-harmonizing pair (g, df), such that the n-form $\Theta = *_g(df)$ defines a measure on $\partial_1 X$ with respect to which the causality map C_v is measure-preserving. Here "$*_g$" denotes the Hodge star operator.

Assuming that the v-harmonizing metric g and differential df are v-invariant, Theorem 3.17 describes the residual structures on the boundary $\partial_1 X$ that allow for a reconstruction of X, v and g, up to a diffeomorphism of X.

Let us fix a closed 1-form α on X. Recall the notion of Schwartzman's asymptotic cycles $A(\mu)$ (see [Sch], or [F], page 199), as it applies to the case of a non-vanishing v-flow on a *closed* smooth manifold X.

Consider the v-invariant Borel measure μ on X. For any de Rham cohomology class $[\eta] \in H^1(X; \mathbb{R})$, we define a homology class $A(\mu, v) \in H_1(X; \mathbb{R})$ by the formula

$$\langle A(\mu, v), [\eta] \rangle =_{\mathsf{def}} \int_X \eta(v)\, \mu. \tag{3.37}$$

In fact, this definition depends only on the cohomology class $[\eta]$ of the closed form η since the measure μ is v-invariant [Sch].

As a special case of Proposition 10.44 from [F], we instantly get the following claim.

Proposition 3.5. *Let X be a closed smooth manifold equipped with a v-invariant Borel measure μ. If a closed 1-form α is Lyapunov for the flow*

v, then it satisfies the homological property: $\langle A(v,\mu), [\alpha]\rangle > 0$, for any Schwartzman's asymptotic cycle $A(v,\mu) \in H_1(X;\mathbb{R})$ defined by (3.37). $\diamondsuit$

A closed and co-closed nonsingular 1-form α on a compact Riemannian $(n+1)$-manifold X produces a beautiful geometric structure: a pair of mutually orthogonal (singular) foliations $\mathcal{F}_\alpha$ and $\mathcal{G}_\alpha$ of dimensions 1 and n, respectively. These foliations are minimal [K6], [Su], [Su1] provided that α has the following global property:

• **The Calabi Condition ([Ca]):** Given closed 1-form α, through each point $x \in X$, there exists a smooth path γ—either a loop or a segment with its ends residing in $\partial_1 X$—such that

$$\alpha(\dot{\gamma}) > 0. \tag{3.38}$$

Calabi calls such 1-forms α transitive. In fact, in [Ca] he studied closed 1-forms α that may have Morse type singularities (different from extrema) and are transitive (now the path γ avoids the singularities of α). He proves that such transitive α is intrinsically harmonic, i.e., there exists a Riemannian metric g on X so that α is also co-closed. In fact, the transitivity of α is also necessary for its intrinsic harmonicity.

Let θ be a closed 1 form on the solid torus $S^1 \times D^n$, the pull-back of the canonical 1-form on the circle S^1 under the obvious projection $S^1 \times D^n \to S^1$, defined by the product structure. Similarly, consider the obvious function $D^1 \times D^n \to D^1 \subset \mathbb{R}$. Let θ denote the differential of this function.

Definition 3.13. Assume that a smooth compact manifold X is oriented and equipped with a non-vanishing vector field v.

- We call an orientation-preserving diffeomorphism $\kappa : S^1 \times D^n \to \hat{X}$ on its image a Calabi tube, if $\theta(\kappa_*^{-1}(\hat{v})) > 0$. Similarly, an orientation-preserving diffeomorphism $\kappa : D^1 \times D^n \to \hat{X}$ on its image is called a Calabi tube, if $\theta(\kappa_*^{-1}(\hat{v})) > 0$.
- We call a Calabi tube v-invariant if the orientation-preserving diffeomorphisms $\kappa : S^1 \times D^n \to \hat{X}$ or $\kappa : D^1 \times D^n \to \hat{X}$ are such that $\hat{v}$ is tangent to every curve $\{\kappa(S^1 \times u)\}_{u \in D^n}$ or to every curve $\{\kappa(D^1 \times u)\}_{u \in D^n}$, respectively.
- We call a v-invariant Calabi tube balanced if the function $(\kappa^{-1})^*(\theta)(v)$ is constant.

$\diamondsuit$

Definition 3.14. (Calabi vector fields)

- We say that a non-vanishing vector field v on X is a Calabi field, if X admits a cover by Calabi tubes (whose images reside in $\hat{X}$).
- We say that a non-vanishing vector field v on X is an invariant Calabi field, if X admits a cover by v-invariant Calabi tubes.
- We say that an invariant Calabi vector field v on X is an balanced, if X admits a cover by balanced v-invariant Calabi tubes. ◇

Example 3.8. Let Y be a manifold that admits a smooth S^1-action whose main orbit-type is S^1. Let v be a vector field, generated by this action. Consider compact codimension zero smooth submanifold $X \subset Y$ which is contained in the locus $Y^\circ \subset Y$, formed by orbits of the main orbit-type, and restrict v to X. Then $v|_X$ is an invariant Calabi vector field. ◇

Lemma 3.18. *If v is an invariant Calabi vector field with respect to a cover of X by Calabi tubes, then v is a balanced invariant Calabi vector field (with respect to a differently parametrized Calabi cover). As a result, the second and third bullets in Definition 3.14 produce equivalent notions.*

Proof. Let $x : D^1 \times D^n \to D^1$ and $y : D^1 \times D^n \to D^n$ be the obvious coordinates on the product. Let $\kappa^{-1} : U_{\hat{\gamma}^\star} \to D^1 \times D^n$ be as in Definition 3.13, second bullet, and put $\theta = dx$.

Our goal is to construct a new diffeomorphism $\tilde{\kappa}^{-1} : U_{\hat{\gamma}^\star} \to D^1 \times D^n$ so that the image of each v-trajectory $\gamma \subset U_{\hat{\gamma}^\star}$ is still the fiber of the projection $y : D^1 \times D^n \to D^n$ and the function $[(\tilde{\kappa}^{-1})^*\theta](v)$ is constant on the trajectories γ.

Let $\kappa_*^{-1}(v) =_{\mathsf{def}} f(x, y)\, \partial_x$. Thus, $\theta(\kappa_*^{-1}(v)) = dx(\kappa_*^{-1}(v))$ is a function $f(x, y) > 0$. Consider the auxiliary function

$$g(x, y) = \int_0^x \frac{dt}{f(t, y)} \Big/ \int_0^1 \frac{dt}{f(t, y)},$$

which is strictly increasing in $x \in [0, 1]$ and has the property $g(0, y) = 0$, $g(1, y) = 1$. We define the diffeomorphism $\phi : D^1 \times D^n \to D^1 \times D^n$ by the formula $(x', y') = \phi(x, y) =_{\mathsf{def}} (g(x, y), y)$. Then ϕ transforms $f(x, y)\partial_x$ into $\partial_{x'}$. Thus the diffeomorphism ϕ has the property $[(\tilde{\kappa}^{-1})^*\theta](v) = 1$.

The case of a toroidal Calabi tube $U_{\hat{\gamma}^\star}$ is similar: $x : S^1 \times D^n \to S^1$ is a circular-valued map, viewed as a function with the period 1, and $f(x + 1, y) = f(x, y)$ for all x, y. Under these assumptions, the same formulas deliver the desired diffeomorphism $\phi : S^1 \times D^n \to S^1 \times D^n$ with the property $[(\tilde{\kappa}^{-1})^*\theta](v) = 1$. □

Lemma 3.19. *Any traversing vector field v is an invariant balanced Calabi field.*

Proof. By the definition of a traversing vector field (see Definition 2.1 and [K1]), each v-trajectory γ is a closed segment. We consider a closed segment $\hat{\gamma} \subset \hat{X}$ of the $\hat{v}$-trajectory that contains γ in its interior and such that the boundary $\partial\hat{\gamma} \subset \hat{X} \setminus X$. We take a disk-shaped smooth transversal section D^n of the $\hat{v}$-flow at a point $o \in \gamma$ and form a small $\hat{v}$-invariant tubular neighborhood $U_{\hat{\gamma}}$ of $\hat{\gamma}$ in $\hat{X}$ by taking the union of $\hat{v}$-trajectories through D^n and "trimming" this sheaf at the two ends, as described below. We denote by $V_{\hat{\gamma}}$ the trimmed set of $\hat{v}$-trajectories that pass through the sphere ∂D^n. Put $\delta U_{\hat{\gamma}} =_{\mathsf{def}} \partial(U_{\hat{\gamma}}) \setminus V_{\hat{\gamma}}$.

Since $\partial\hat{\gamma} \cap X = \emptyset$, we may pick the tube $U_{\hat{\gamma}} \supset \hat{\gamma}$ so narrow (equivalently, the section D^n so small) that $X \cap (\delta U_{\hat{\gamma}}) = \emptyset$. This choice helps us to introduce a product structure $\kappa_{\hat{\gamma}} : I \times D^n \approx U_{\hat{\gamma}}$ in $U_{\hat{\gamma}}$, where I denotes a closed interval, so that:

(1) $\kappa_{\hat{\gamma}}(I) \times o = \hat{\gamma}$, where the $\kappa_{\hat{\gamma}}|_I$ is an orientation-preserving diffeomorphism,

(2) $U_{\hat{\gamma}}$ consists of segments of $\hat{v}$-trajectories,

(3) each slice $\kappa_{\hat{\gamma}}(t \times D^n)$, where $t \in I$, is transversal to the $\hat{v}$-flow, and

(4) $\kappa_{\hat{\gamma}}$ is an orientation-preserving diffeomorphism with respect to the orientation of $U_{\hat{\gamma}}$, induced by the preferred orientation of $\hat{X}$.

This choice of the tube $U_{\hat{\gamma}}$ satisfies all the properties, listed in Definition 3.13, the second bullet. Thus, $U_{\hat{\gamma}}$ is a v-invariant Calabi tube, which contains γ. By Lemma 3.18, there is a reparametrization of $U_{\hat{\gamma}}$ so that $U_{\hat{\gamma}}$ becomes balanced.

By the compactness of X, it admits a finite subcover by v-invariant balanced Calabi's tubes $\{U_{\hat{\gamma}}\}_{\hat{\gamma}}$. Therefore, v is an invariant balanced Calabi field. □

Given a Riemmanian metric g on a $(n+1)$-dimensional manifold X, we consider the Hodge star operator $*_g : T^*X \to \bigwedge^n T^*X$. The Hodge $*$-operator is a bundle isomorphism. In local coordinates and with respect to a local basis $\{e_j^\star\}_j$ in T^*X and some associated dual local basis $\{\eta_j^\star\}_j$ in $\bigwedge^n T^*X$, the operator $*_g$ is given by the $(n+1) \times (n+1)$-matrix $\mathsf{G} = \sqrt{\det(\mathsf{g})} \cdot (\mathsf{g})^{-1}$, where $\mathsf{g} = (g_{jk})$ is the scalar product matrix. Remarkably, for $n \geq 2$, the operator $*_g$ determines the metric g since $\det(\mathsf{G}) = \det(\mathsf{g})^{(n-1)/2}$ ([Ca]).

Recall that the co-derivative operator δ acts on differential p-forms on X

by the formula

$$\delta =_{\mathsf{def}} (-1)^{(n+1)(p+1)+1} (*_g) \circ d \circ (*_g).$$

We say that a p-form α is co-closed if $\delta\alpha = 0$. The closed and co-closed forms α are harmonic, i.e., they satisfy the Laplace equation $(d+\delta)^2\alpha = 0$; however, on manifolds with boundary, not any harmonic form is closed and co-closed!

Given a 1-form α on (X, g), we denote by α_{tan} the 1-form, a section of the bundle $TX|_{\partial_1 X}$, that coincides with α on $T(\partial_1 X)$ and vanishes on the normal vector field $\nu_g(\partial_1 X, X)$. By definition, $\alpha_{\mathsf{norm}} =_{\mathsf{def}} \alpha - \alpha_{\mathsf{tan}}$, as sections of $TX|_{\partial_1 X}$.

The basic relation between closed and co-closed forms and the de Rham cohomology is a bit subtle [CTGM]. For example, in terms of 1-forms α, the relation is given by

$$\begin{aligned} H^1(X;\mathbb{R}) &\approx \{\alpha|\ d\alpha = 0,\ \delta\alpha = 0,\ \alpha_{\mathsf{norm}} = 0\}, \\ H^1(X;\partial X;\mathbb{R}) &\approx \{\alpha|\ d\alpha = 0,\ \delta\alpha = 0,\ \alpha_{\mathsf{tan}} = 0\}. \end{aligned} \tag{3.39}$$

The main ideas for proving the next theorem can be found in [K6], as a special case of Theorem C. However, in Theorem 3.14 below, the given ingredient is the vector field v, not a closed 1-form α as in [K6].

Theorem 3.14. *Let v be a Calabi vector field (see Definition 3.14, the first bullet) on a compact smooth $(n+1)$-manifold X, $n \geq 2$.*

(1) *then there exists a smooth 1-form α and a metric g on X such that:*

- $\alpha(v) > 0$
- $d(*_g\alpha) = 0$,
- $\dim(\mathsf{ker}(*_g\alpha)) = 1$,
- $\alpha \wedge (*_g\alpha) = *_g(1)$ *is a volume form on X.*

(2) *Let v be a v-*invariant *Calabi vector field on X, $\dim X \geq 3$. Then, in addition to the properties in* (1), *one may choose α to be a v-invariant 1-form. The n-form $\Theta =_{\mathsf{def}} *_g\alpha$ is integrally dual to v; i.e., in addition to the properties in* (1), *we have $v \in \mathsf{ker}(\Theta)$. Furthermore, one may choose the metric g on X to be v-invariant.*

(3) *If v is a* traversing *vector field, then v is a balanced v-invariant Calabi field. Moreover, $\alpha = df$, an exact 1-form. For an appropriate g, the harmonic n-form $\Theta = *_g\alpha$ is integrally dual of v, and $f : X \to \mathbb{R}$ is a harmonic Lyapunov function for v. Also, one may choose the metric g on X to be v-invariant.*

Proof. Here is a general plan for proving the theorem:

(i) starting with the Calabi vector field v, we construct v-amenable 1-form α (in the third bullet case, $\alpha = df$, where f is a Lyapunov function for v) and a closed n-form Θ so that $\alpha \wedge \Theta > 0$;

(ii) then we construct a metric g for which $*_g(\alpha) = \Theta$.

We consider a cover $\hat{\mathcal{U}}$ of X by Calabi's tubes $U_{\hat{\gamma}}$, where $\hat{\gamma}$ runs over the set of all $\hat{v}$-trajectories that have a nonempty intersection with X. Using compactness of X, we pick a finite subcover $\{U_{\hat{\gamma}_i}\}_i$ of $\hat{\mathcal{U}}$ so that $X = \bigcup_i (U_{\hat{\gamma}_i} \cap X)$.

As before, we divide Calabi's tubes into two types: for the first type, the core $\hat{\gamma}$ of $U_{\hat{\gamma}}$ is a closed segment, for the second type, the core is a simple loop. For each of the tubes, we fix a product structure, given by a diffeomorphism $\kappa_{\hat{\gamma}}$ with the properties as in Definition 3.13, the first bullet.

For each tube of the first type, we consider a function $\tilde{f}_{\hat{\gamma}} : U_{\hat{\gamma}} \to I$, the pull-back by $\kappa_{\hat{\gamma}}^{-1}$ of the obvious function $I \times D^n \to I$. By the definition of Calabi tube, $d\tilde{f}_{\hat{\gamma}}(v) > 0$ in $U_{\hat{\gamma}}$. Similarly, for each tube $U_{\hat{\gamma}}$ of the second type, with the help of $\kappa_{\hat{\gamma}}^{-1}$, we produce a 1-form $\tilde{\alpha}_{\hat{\gamma}}$ on $U_{\hat{\gamma}}$ such that $\tilde{\alpha}_{\hat{\gamma}}(v) > 0$.

Let $\tilde{\psi} : D^n \to \mathbb{R}_+$ be a smooth non-negative bell-shaped function with the support in the interior of D^n and such that all its partial derivatives of any order vanish at the boundary ∂D^n. Using the $(\kappa_{\hat{\gamma}}^{-1})$-induced projection $p : U_{\hat{\gamma}} \to D^n$, we form the pull-back function $\psi = \tilde{\psi} \circ p$ and multiply $\tilde{f}_{\hat{\gamma}}$ by ψ to get a smooth function $f_{\hat{\gamma}} : U_{\hat{\gamma}} \to \mathbb{R}$ with the support in the interior of $U_{\hat{\gamma}}$. Thus $df_{\hat{\gamma}}(\hat{v}) > 0$ in the interior of $U_{\hat{\gamma}}$ and $df_{\hat{\gamma}}(\hat{v}) \geq 0$ globally.

Similarly, for Calabi tubes of the toroidal kind, we put $\alpha_{\hat{\gamma}} = \psi \cdot \tilde{\alpha}_{\hat{\gamma}}$, where $\psi : U_{\hat{\gamma}} \to \mathbb{R}_+$ is the pull-back of the bell function $\tilde{\psi} : D^n \to \mathbb{R}_+$ under the $(\kappa_{\hat{\gamma}}^{-1})$-induced projection $U_{\hat{\gamma}} \to D^n$. Thanks to the choice of $\tilde{\psi}$, this 1-form $\alpha_{\hat{\gamma}}$ is well-defined globally. Again, $\alpha_{\hat{\gamma}}(\hat{v}) > 0$ in the interior of $U_{\hat{\gamma}}$ and $\alpha_{\hat{\gamma}}(\hat{v}) \geq 0$ globally. Note that the 1-form $\alpha_{\hat{\gamma}}$ is not closed!

For Calabi tubes of both kinds, we introduce the n-form $\Theta_{\hat{\gamma}}$ on $U_{\hat{\gamma}}$ as the pull-back, under the map $(\kappa_\gamma)^{-1}$, of the standard volume form vol_{D^n} on D^n, being multiplied by the bell function $\tilde{\psi} : D^n \to \mathbb{R}_+$. Evidently, $\Theta_{\hat{\gamma}}$ extends trivially on X. Since $\Theta_{\hat{\gamma}}$ depends only on the coordinates in D^n, for the dimensional reason, we get $d\Theta_{\hat{\gamma}} = 0$.

Let Ω^∂ be the volume form on the boundary $\partial_1 X$. The restriction of the function $\pm\Theta_{\hat{\gamma}}/\Omega^\partial$ to $\partial_1^\pm X(v) \setminus \partial_2 X(v)$ is positive. Indeed, in the case of $\partial_1^- X(v) \setminus \partial_2 X(v)$ (or of $\partial_1^+ X(v) \setminus \partial_2 X(v)$), $\hat{v}$ points outside (inside) of both domains, X and $U_{\hat{\gamma}}^\dagger$. Here $U_{\hat{\gamma}}^\dagger \subset U_{\hat{\gamma}}$ denotes the subtube that is the preimage of $\gamma \subset \hat{\gamma}$ under the projection $U_{\hat{\gamma}} \to \hat{\gamma}$, delivered by the product structure of the Calabi tube $U_{\hat{\gamma}}$.

We define the global 1-form α on X by the formula $\sum_i \alpha_{\hat{\gamma}_i}$, and the global n-form Θ by the formula $\sum_i \Theta_{\hat{\gamma}_i}$. Since $d\Theta_{\hat{\gamma}_i} = 0$ by its construction, Θ is a closed form. Again, since $\alpha_{\hat{\gamma}_i}(\hat{v}) > 0$ in the open set $U_{\hat{\gamma}_i}$ for all i, we conclude that $\alpha(v) > 0$ in X.

Finally, with the candidates α and Θ in place, we consider the Hodge star bundle isomorphism $*_g : T^*X \to \bigwedge^n T^*X$, where the Riemmanian metric g on X to be determined. By [Ca], for $n \geq 2$, the operator $*_g$ determines the metric g. Therefore, it suffices to pick any g such that $*_g(\alpha) = \Theta$. By the construction of α and Θ, we get $\alpha \wedge *_g\alpha > 0$. By Lemma 1 from [Ca], such a metric g exists. Moreover, Calabi's argument (see [Ca], pages 110-112) insures that $\alpha \wedge (*_g\alpha) = vol_g =_{\mathsf{def}} *_g(1)$, the g-induced volume $(n+1)$-form on X. For the reader's convenience, we will sketch his argument below.

With respect to such a choice of g, the form Θ is closed, its kernel is 1-dimensional, $\alpha(v) > 0$, and $\alpha \wedge (*_g\alpha) = vol_g$. Thus, claim (1) from of the theorem is valid.

Note that, for each *v-invariant* Calabi tube, $v \in \mathsf{ker}(\Theta_{\hat{\gamma}_i})$ and $\mathcal{L}_v(\Theta_{\hat{\gamma}_i}) = 0$ for all i. Thus, for an invariant Calabi vector field v, in addition, we get $v \in \mathsf{ker}(\Theta)$ and $\mathcal{L}_v(\Theta) = 0$.

According to Lemma 3.18, by choosing an appropriate parametrization of the invariant Calabi tubes from the cover $\hat{\mathcal{U}}$, we may assume that each $\alpha_{\hat{\gamma}_i}(v)$ is constant. Therefore, α is a v-invariant form, i.e., $\mathcal{L}_v\alpha = \sum_i \mathcal{L}_v(\alpha_{\hat{\gamma}_i}) = 0$.

Next, we are going to show that we may choose the metric g in (2) to be v-invariant. To achieve this, we need to revisit the argument in [Ca].

We notice that, for a given v and α and Θ as above, the choice of g is far from being unique. Using the product structure in an invariant balanced Calabi tube $U_{\hat{\gamma}}$, we introduce there local coordinates $\{x_i\}_{i\in[0,n]}$ such that:

(i) $dx_0 = \alpha$, where $x_0 : U_{\hat{\gamma}} \to S^1$ or $x_0 : U_{\hat{\gamma}} \to \mathbb{R}^1$, depending on the type of the tube,

(ii) $v = \partial_{x_0}$ (using that $U_{\hat{\gamma}}$ is balanced), and

(iii) $dx_1 \wedge \dots \wedge dx_n = \Theta$.

Then we define the dual basis in $\bigwedge^n T^*U_{\hat{\gamma}}$ as

$$\eta_0 =_{\mathsf{def}} dx_1 \wedge \dots \wedge dx_n,\; \eta_1 =_{\mathsf{def}} -dx_0 \wedge dx_2 \wedge \dots \wedge dx_n,$$

$$\eta_2 =_{\mathsf{def}} dx_0 \wedge dx_1 \wedge dx_3 \dots \wedge dx_n, \quad \dots, \text{ etc.}$$

Again, following [Ca], for $n \geq 2$, this choice of bases $\{dx_i\} \in T^*U_{\hat{\gamma}}$ and $\{\eta_i\} \in \bigwedge^n T^*U_{\hat{\gamma}}$ defines a unique candidate for the local star operator $*_{g_{\hat{\gamma}}} : T^*U_{\hat{\gamma}} \to \bigwedge^n T^*U_{\hat{\gamma}}$ that takes each dx_i to $\phi_{\hat{\gamma}} \cdot \eta_i$, where the smooth functions $\{\phi_{\hat{\gamma}} : U_{\hat{\gamma}} \to \mathbb{R}_+\}_{\hat{\gamma}}$ form a finite partition of unity, subordinate to the cover $\hat{\mathcal{U}}$ and such that $\mathcal{L}_v(\phi_{\hat{\gamma}}) = 0$. We think of $*_{g_{\hat{\gamma}}}$ as a section of the vector bundle $\mathsf{Homeo}(T^*X,\ \bigwedge^n T^*X) \to X$ with the support in $U_{\hat{\gamma}}$.

We notice that all the forms $\{dx_i\}$ and $\{\eta_i\}$ are v-invariant. Moreover, since each $\phi_{\hat{\gamma}}$ does not depend on x_0, $\{\phi_{\hat{\gamma}} \cdot \eta_i\}_i$ are v-invariant as well. Therefore, the $U_{\hat{\gamma}}$-localized star operators $*_{g_{\hat{\gamma}}} : T^*X \to \bigwedge^n T^*X$ must be also v-invariant.

Finally, we pick the v-invariant operator-section

$$*_g =_{\mathsf{def}} \sum_{\hat{\gamma}} *_{g_{\hat{\gamma}}} \in \mathsf{Homeo}(T^*X,\ \bigwedge^n T^*X)$$

which has the desired properties: $*_g(\alpha) = \Theta$, and $\alpha \wedge \Theta = vol_g$. As a result, the corresponding metric g must be v-invariant as well.

Therefore, claim (2) is valid.

When v is traversing, by Lemma 3.19, X admits a cover by v-invariant Calabi tubes, which are cylinders only. As a result, $\alpha = df$, where $f = \sum_i f_i$. Hence, α is exact! Moreover, since $d\alpha = 0$ and $d\Theta = 0$, both α and Θ are harmonic in g. By the constructions of α and Θ above, they satisfy all the properties from claim (3) of Theorem 3.14, including the property $\mathcal{L}_v g = 0$. □

The properties of the forms Θ and α, listed in Theorem 3.14, motivate the following definition.

Definition 3.15. For a non-vanishing vector field v on a smooth compact $(n+1)$-manifold X, consider the space $\mathcal{H}ar(v)$ of smooth Riemannian metrics g on X, paired with smooth 1-forms α, such that:

(1) $\alpha(v) > 0$,
(2) $d\alpha = 0$,
(3) the n-form $\Theta =_{\mathsf{def}} *_g(\alpha)$ is closed (thus α is *harmonic* in the metric g),
(4) $v \in K(\Theta)$, the kernel of Θ,
(5) $\alpha \wedge \Theta$ is the g-induced $(n+1)$-volume form on X,
(6) the function $\Theta/\Omega^\partial \geq 0$ on $\partial_1^+ X(v)$, and $\Theta/\Omega^\partial \leq 0$ on $\partial_1^- X(v)$, where Ω^∂ denotes a volume n-form on $\partial_1 X$, consistent with its orientation.

For a given v, we say that (g, α) is a **harmonizing pair**, if all the six properties above are valid. Note that if (g, α) is a harmonizing pair for some v, then (g, α) is also a harmonizing pair for $\lambda \cdot v$, where $\lambda : X \to \mathbb{R}_+$ is a positive smooth function. $\diamondsuit$

Remark 3.18. Note the main difference between Definition 3.15 and the list of properties in claim (1) of Theorem 3.14: the form α must be closed in Definition 3.15, while the form $\Theta = *_g \alpha$ is required to be closed in both Definition 3.15 and Theorem 3.14. $\diamondsuit$

By Theorem 3.14, any traversing boundary generic vector field v admits a v-harmonizing pair (g, df).

The proof of the corollary below can be found in [K6], as a special case of Theorem C. See also [Su] and the proof of Corollary 3.18 below for a sketch of the arguments that validate Theorem C from [K6] and the second bullet in Corollary 3.17.

Corollary 3.17. *For any non-vanishing vector field v and a v-harmonizing pair (g, α) on X (as in Definition 3.15), the following properties hold:*

- *the 1-foliation $\mathcal{F}(v)$, determined by v, is formed by the geodesic curves in the metric g,*
- *the n-foliation $\mathcal{G}(\alpha)$, defined by the closed 1-form α, consists of leaves $\mathcal{L}$ that minimize the g-induced n-volume among all sufficiently small perturbations of the foliation $\mathcal{L}$ that are fixed on $\partial_1 X$.*
- *the leaves of $\mathcal{F}(v)$ and of $\mathcal{G}(\alpha)$ are mutually orthogonal in g.* $\diamondsuit$

Corollary 3.18. *Let v be a traversing vector field on compact smooth $(n+1)$-manifold X with boundary, $f : X \to \mathbb{R}$ a Lyapunov function for v, and (g, df) a v-harmonizing pair.*

Then the Plateau problem for each of the $(n-1)$-dimensional boundaries $f^{-1}(c) \cap \partial_1 X$ has a smooth solution $f^{-1}(c)$ in (X, g) for any regular value c of $f|_{\partial_1 X}$. If $H_n(X; \mathbb{R}) = 0$, then this n-volume minimizing solution is unique.

In fact, for all sufficiently close (in the C^∞-topology) to v traversing vector fields $\tilde{v}$ and the corresponding $\tilde{v}$-harmonizing pairs $(\tilde{g}, df)$, the Plateau problem for the boundary $f^{-1}(c) \cap \partial_1 X$ still has a smooth solution.

Proof. The proof may be extracted from [K6], Theorem C and D (which deal with a more general setting than the one required here). Let us sketch its main trust.

Since f is a Lyapunov function for v and v is traversing, f attends is extrema on the boundary $\partial_1 X$. For any non-critical for $f|_{\partial_1 X}$ value c, the leaves $\mathcal{G}_c = f^{-1}(c)$, where $c \in f(X)$, of the foliation $\mathcal{G}(df)$ are nonsingular. By the construction of a v-harmonizing metric g, we get $vol_g(\mathcal{G}_c) = \int_{\mathcal{G}_c} \Theta$, where $\Theta = *_g(df)$. Consider any small smooth perturbation $\mathcal{H}$ of the hypersurface $\mathcal{G}_c$ such that $\mathcal{H} \cap \partial_1 X = \mathcal{G}_c \cap \partial_1 X$. Like $\mathcal{G}_c$, $\mathcal{H}$ is transversal to v. By the Stokes' theorem, $\int_{\mathcal{H}} \Theta = \int_{\mathcal{G}_c} \Theta$. On the other hand, using the g-orthogonality of v to $\mathcal{G}_c$, we conclude that at each point $x \in \mathcal{H}$, where $T_x\mathcal{H}$ is not tangent to $\mathcal{G}(df)$, the n-volume form $dg|_{T_x\mathcal{H}} > \Theta|_{T_x\mathcal{H}}$. Therefore, $vol_{g|}(\mathcal{H}) > vol_{g|}(\mathcal{G}_c)$, unless $\mathcal{H}$ is tangent to $\mathcal{G}(df)$ almost everywhere, in which case $\mathcal{H} = \mathcal{G}_c$. Hence, $\mathcal{G}_c$ minimizes the g-induced n-volume locally, provided that the boundary $\mathcal{G}_c \cap \partial_1 X = \mathcal{H} \cap \partial_1 X$ is fixed.

If $H_n(X; \mathbb{R}) = 0$, then $\mathcal{H} \cup -\mathcal{G}_c$ is a trivial cycle for any $\mathcal{H}$ that shares with $\mathcal{G}_c$ its boundary $\mathcal{G}_c \cap \partial_1 X$. Thus, $\int_{\mathcal{H}} \Theta = \int_{\mathcal{G}_c} \Theta$. Therefore, the previous argument works for such $\mathcal{H} \neq \mathcal{G}_c$: hence $vol_{g|}(\mathcal{H}) > vol_{g|}(\mathcal{G}_c)$. This implies that the volume minimizing solution $\mathcal{H}$ of the Plateau problem for the contour $f^{-1}(c) \cap \partial_1 X$ is unique, provided $H_n(X; \mathbb{R}) = 0$. □

Theorem 3.14, being combined with Lemma 3.9, directly implies the following theorem.

Theorem 3.15. *For any traversing boundary generic vector field v on a smooth compact $(n+1)$-manifold X, there exists a v-invariant metric g and a smooth Lyapunov function $f : X \to \mathbb{R}$ such that (g, df) is a v-harmonising pair.*

For any such pair $(g, df) \in \mathcal{H}ar(v)$, the measure μ_Θ on $\partial_1 X$, induced by the closed n-form $\Theta =_{\mathsf{def}} *_g(df)$ on X (see (3.31)), is preserved under the causality map C_v. The form Θ is v-invariant. ◇

In combination with Theorem 3.15, Corollary 3.17 leads to the following claim.

Corollary 3.19. *Let v be a traversing boundary generic vector field on X, and (g, df) a v-harmonizing pair. Assume that no pair of distinct points $a, b \in \partial_1 X$ admits two distinct geodesics in g that connect a and b.*[9]

Then the knowledge of the causality map $C_v : \partial_1^+ X(v) \to \partial_1^- X(v)$ and of the v-harmonizing metric g on X allows for the unique reconstruction of the foliation $\mathcal{F}(v)$.

Proof. Consider a geodesic curve $[\gamma(g)]$ that connects two points, $x \in \partial_1 X$ and $C_v(x) \in \partial_1 X$. By Corollary 3.17, the segment $[\gamma]$ of the v-trajectory

[9]For example, this is the case when (X, g) is contained in a larger Riemannian manifold $(\hat{X}, \hat{g})$ such that any two distinct points in $\hat{X}$ belong to a single geodesic in $\hat{g}$.

that connects x and $C_v(x)$ is a geodesic curve. By the uniqueness hypotheses, we get $[\gamma(g)] = [\gamma]$. Thus the foliation $\mathcal{F}(v)$ can be reconstructed from C_v and the v-harmonizing metric g. □

Lemma 3.20. *Let v be a traversing boundary generic vector field on a compact connected smooth Riemannian $(n+1)$-manifold (X, g) with boundary. Then the measure μ_Θ on $\partial_1 X$ can be recovered from the g-induced volume n-form Ω^∂ on the boundary $\partial_1 X$ and the function*

$$\phi^\partial =_{\mathsf{def}} (\Theta|_{\partial_1 X})/(\Omega^\partial) : \partial_1 X \to \mathbb{R},$$

the cos *of the angle formed by v and the inner normal ν_g to $\partial_1 X$.*

Proof. Let $K(df)$ denotes the kernel of the form df. By the definition of the auxiliary function $\phi^\partial : \partial_1 X \to [-1, 1]$ and using that $\Theta|_{K(df)}$ and Ω^∂ on $\partial_1 X$ both are the g-induced volume n-forms, we have $\Theta(w) = \phi^\partial(x) \cdot \Omega^\partial(w)$ for any $x \in \partial_1 X$ and $w \in \Lambda^n T_x(\partial_1 X)$. □

For a traversing boundary generic vector field v on X, let $\mathcal{Y}_2(v)$ denotes the set of v-trajectories that pass through the tangency locus $\partial_2 X(v)$. Let $\mathcal{K}(v) =_{\mathsf{def}} \mathcal{Y}_2(v) \cap \partial_1^+ X(v)$. For such a v, the set $\mathcal{K}(v)$ is a compact $(n-1)$-dimensional CW-complex. Therefore, $\mu_\Theta(\mathcal{K}(v)) = 0$ and the obvious map $\Gamma^\partial : \partial_1^+ X(v) \to \mathcal{T}(v)$, whose target is the trajectory space $\mathcal{T}(v)$, is a homeomorphism on the complement to the zero-measure set $\mathcal{K}(v)$.

This observation motivates the following definition.

Definition 3.16. Let v be a boundary generic traversing vector field on a compact connected smooth $(n+1)$-manifold X with boundary, and let Θ be a differential n-form that is integrally dual of v (as in Definition 3.7).

- We introduce a measure $\tilde{\mu}_\Theta$ on the trajectory space $\mathcal{T}(v)$ by the formula

  $$\tilde{\mu}_\Theta(A) =_{\mathsf{def}} \mu_\Theta((\Gamma^\partial)^{-1}(A)),$$

  where $A \subset \mathcal{T}(v)$ is such that its Γ^∂-preimage is Lebesgue-measurable in the compact manifold $\partial_1^+ X(v)$.
- Then the integral $\int_{\partial_1^+ X(v)} \Theta$ can be interpreted as the volume of the trajectory space $\mathcal{T}(v)$ with respect to the measure $\tilde{\mu}_\Theta$ on $\mathcal{T}(v)$, induced by the n-form Θ. ◇

The next assertion has the flavor of the classical Santaló formula [S].

Theorem 3.16. *Let v be a smooth traversing vector field on a smooth compact connected manifold with boundary.*

• *For any v-harmonizing pair (g, df), the g-induced volume form Ω^∂ on $\partial_1 X$, and the* cos*-function $\phi^\partial : \partial_1 X \to [-1,1]$ (as in Lemma 3.20) allow for a computation of the Θ-induced volume of the trajectory space $\mathcal{T}(v)$ via each of the two formulas:*

$$vol_\Theta(\mathcal{T}(v)) = \pm \int_{\partial_1^\pm X(v)} \phi^\partial \cdot \Omega^\partial.$$

Letting $g^\partial = g|_{\partial_1 X}$, we get $vol_\Theta(\mathcal{T}(v)) \leq vol_{g^\partial}(\partial_1^\pm X(v))$, which implies

$$vol_\Theta(\mathcal{T}(v)) \;\leq\; \frac{1}{2} vol_{g^\partial}(\partial_1 X). \tag{3.40}$$

• *Assuming that $f(\partial_1 X) \subset [0,1]$,*[10] *for any such v-harmonizing pair (g, df), we get the* "holographic" isoperimetric inequality*:*

$$vol_g(X) \;\leq\; vol_{g^\partial}(\partial_1 X). \tag{3.41}$$

Proof. Put $\Theta = *_g(df)$. Examining Definition 3.16, for each v-harmonizing pair (g, df), the measure $\tilde{\mu}_\Theta$ on $\mathcal{T}(v)$ can be reconstructed from the following data:

- the locus $\partial_1^+ X(v)$,
- the map $\Gamma^\partial : \partial_1^+ X(v) \to \mathcal{T}(v)$ (whose generic fiber is a singleton),
- the volume n-form Ω^∂ on $\partial_1^+ X(v)$, induced by the metric g^∂,
- the "cos" function ϕ^∂.

Therefore, we conclude that $vol_\Theta(\mathcal{T}(v)) = \int_{\partial_1^+ X(v)} \phi^\partial \cdot \Omega^\partial$, the volume of the trajectory space, can be reconstructed from the data in the first, third, and fourth bullet.

Since $df(v) > 0$ on X, the function f attends its extrema on $\partial_1 X$. Thus, if $f : \partial X \to [0,1]$, then $f : X \to [0,1]$. Therefore, by Stokes' Theorem, we get

$$vol_g(X) = \int_X df \wedge \Theta = \int_{\partial_1 X} f \cdot \Theta \leq \int_{\partial_1 X} 1 \cdot |\phi^\partial| \cdot \Omega^\partial \leq vol_{g^\partial}(\partial_1 X).$$

In other words, the volume of the bulk X does not exceed the surface area of its boundary[11]. □

Combining Theorem 3.16 with Theorem 3.7, we get the following claim, which superficially is remenicent of Bekenstein-Hawken formula for the volumes of event horizons [Bek], [Hawk]...

[10]which is always possible to achieve by an affine transformation of the target $\mathbb{R}$

[11]This fact may please our fellow physicists, concerned with the areas of the event horizons.

For a function $T : X \to \mathbb{R}$, we denote by $\mathsf{var}(T) = \sup_X T - \inf_X T$ and call it the variation of T on X.

Corollary 3.20. *Let $u(t)$, $t \in \mathbb{R}$, be a time-dependent smooth vector field on a connected n-dimensional smooth manifold Y without boundary. Let $T : Y \times \mathbb{R} \to \mathbb{R}$ denote the time coordinate. Consider a compact smooth space-time capsule $X \subset Y \times \mathbb{R}$ as in Theorem 3.7.*

Then, for any v-harmonizing pair (g, dT) on X such that the variation $\mathsf{var}(T|_X) \leq 1$, we get the "holographic" inequality

$$vol_g(X) \;\leq\; vol_{g^\partial}(\partial_1 X)$$

between the volumes of the capsule and its boundary. ◇

Definition 3.17. Let X be a compact connected smooth manifold with boundary and v a traversing vector field on it. We say that Lyapunov function $f : X \to \mathbb{R}$ is balanced if $df(v) = 1$. ◇

By Theorem 3.14, any traversing vector field admits a balanced Lyapunov function.

Let $\mathsf{Diff}^\dagger(X, \partial_1 X)$ denote the group of the smooth diffeomorphisms of X that are identities on the boundary $\partial_1 X$ and whose differentials are the identities on the bundle $TX|_{\partial_1 X}$. The group $\mathsf{Diff}^\dagger(X, \partial_1 X)$ acts naturally on the space $\mathcal{R}(X)$ of smooth Riemannian metrics on X.

Theorem 3.17. *Let v be a traversing boundary generic vector field on a compact connected smooth manifold X with boundary.*

*There exists a v-harmonizing pair (g, df) (as in Definition 3.15), where the metric g is v-*invariant *and the Lyapunov function f is balanced.*

We assume that v has Property A *from Definition 3.2. Then the following boundary-confined data:*

- *the causality map $C_v : \partial_1^+ X(v) \to \partial_1^- X(v)$,*
- *the restriction f^∂ of the Lyapunov function f to the boundary,*
- *the restriction g^∂ of the Riemannian metric g to the boundary $\partial_1 X$,*
- *the angle-function $\theta_g : \partial_1 X \to S^1$, generated by v and the inner g-normal vector field ν to $\partial_1 X$ in X,*

allow for a reconstruction of the smooth topological type of X and of the harmonizing metric g on it, up to the natural $\mathsf{Diff}^\dagger(X, \partial_1 X)$*-action on the space $\mathcal{R}(X)$ of Riemannian metrics on X.*

Proof. By Theorem 3.14, there is a v-harmonizing pair (g, df).

Next, we recreate our usual two-foliation background, a pair of smooth foliations $(\mathcal{G}(\hat{f}),\, \mathcal{F}(\hat{v}))$ which delivers a "coordinate grid" for $X \subset \hat{X}$ so that the intersections $\mathcal{G}(\hat{f}) \cap \partial_1 X$ and $\mathcal{F}(\hat{v}) \cap \partial_1 X$ provide the holographic data from which $(X, \mathcal{F}(v))$ may be recovered.

By Definition 3.15, the integrable distribution $K(df) \subset TX$ of the kernels of $d\hat{f}$ is the g-orthogonal compliment $K_v^\perp$ to the field v. At the same time, the integrable distribution $K(\Theta) \subset TX$ by the kernels of the n-form $\Theta = *_g(d\hat{f})$ contains the field v. The leaves of $\mathcal{G}(\hat{f})$ and $\mathcal{F}(\hat{v})$ are g-orthogonal in X. We denote by $g^\perp$ the restriction of the metric g to the n-dimensional distribution $K_v^\perp$, and by $g^\uparrow$ the restriction of g to the 1-dimensional distribution $K_v =_{\mathsf{def}} K(\Theta)$. Therefore, knowing the restrictions $g^\perp$ and $g^\uparrow$ is sufficient for determining the metric g on X.

By Theorem 3.15, the pair (df, g) is v-invariant, and so is the form Θ. Thus the pairs $(K_v^\perp, g^\perp)$ and $(K_v, g^\uparrow)$ are v-invariant. On the other hand, by the v-invariant property of g, if we know $g|_{f^{-1}(c)}$ in the vicinity of a v-trajectory γ for one particular f^∂-regular value $c \in \mathbb{R}$, then we know all the restrictions $\{g|_{f^{-1}(c')}\}_{c' \in \mathbb{R}}$ in vicinity of γ, provided $f^{-1}(c') \cap \gamma \neq \emptyset$ and $f^{-1}(c) \cap \gamma \neq \emptyset$.

Similarly, by the v-invariance of (df, g), if we know the restriction of g to the γ-tangent line at one particular point, we know the restriction of g to the γ-tangent line at any other point along γ.

By the Holography Theorem 3.3 or Corollary 3.5, the map C_v determines the pair $(X, \mathcal{F}(v))$, up to a diffeomorphism $\Phi : X \to X$ that is the identity on $\partial_1 X$. Recall that the property of Φ being a diffeomorphism (and not just a homeomorphism) depends on Property A. By the proof of Corollary 3.5, the map C_v determines the triple $(X, \mathcal{F}(v), \mathcal{G}(f))$, up to a diffeomorphism $\Phi : X \to X$ that is the identity on $\partial_1 X$.

Let $(g_\tau)^\partial$ denote the restriction of the metric g to $T_*(\partial_1 X)$, and $(g_\nu)^\partial$ to the normal (in the metric g) bundle $\nu(\partial_1 X, X)$.

Since, by the hypotheses, the diffeomorphism Φ is assumed to fix the boundary $\partial_1 X$ and respect the map $\theta : \partial_1 X \to S^1$, where $\theta =_{\mathsf{def}} \angle_g(v, \nu)$, the Φ-action on the bundle $TX|_{\partial_1 X}$ must be trivial.

Let $(g^\uparrow)^\partial$ denotes the restriction of the metric $g^\uparrow$ to the foliation $\mathcal{F}(\hat{v})|_{\partial_1 X}$, and let $(g^\perp)^\partial$ denotes the restriction of $g^\perp$ to the foliation $\mathcal{G}(\hat{f})|_{\partial_1 X}$. The knowledge of g^∂ and θ makes it possible to determine the orthogonal decomposition $g|_{\partial X} = (g^\uparrow)^\partial \oplus (g^\perp)^\partial$ along $\partial_1 X$. Note that the plane, spanned by the vectors $\nu(x)$ and $v(x)$ at $x \in \partial_1 X$, is g-

orthogonal to the subspace $T_x(\partial_1 X) \cap K^{\perp}_{v(x)}$. The orthogonal (2×2)-matrix $A(\theta)$, representing the rotation on the angle θ, connects the decomposition $g|_{\partial_1 X} = (g^{\uparrow})^{\partial} \oplus (g^{\perp})^{\partial}$ to the decomposition $g|_{\partial_1 X} = (g_{\nu})^{\partial} \oplus (g_{\tau})^{\partial}$. Therefore, knowing $(g_{\nu})^{\partial} \oplus (g_{\tau})^{\partial}$ and θ determines $(g^{\uparrow})^{\partial} \oplus (g^{\perp})^{\partial}$.

By the v-invariant property of (g, df), the decomposition $g|_{\partial_1 X} = (g^{\uparrow})^{\partial} \oplus (g^{\perp})^{\partial}$ spreads uniquely to an orthogonal decomposition $g = g^{\uparrow} \oplus g^{\perp}$ in X.

Therefore, the quadruple $(f^{\partial},\, g^{\partial},\, \theta_g,\, C_v)$ determines g, up to the natural Φ-action, where $\Phi \in \mathsf{Diff}^{\dagger}(X, \partial_1 X)$. □

3.8 Contact holography: recovering contact geometry from boundary data

3.8.1 *In this section...*

In this section, we study contact 1-forms β and contact structures ξ_β they generate on compact connected smooth $(2n+1)$-manifolds X with boundary. We focus on some residual structures along the boundary $\partial_1 X$ that allow to reconstruct X and the contact form β or the contact structure ξ_β on X, up to a smooth diffeomorphism. When such a reconstruction is possible, it deserves the adjective "holographic".

We understand better the reconstruction of contact forms β than the reconstruction of associated contact structures ξ_β. Thus, our results mainly focus on contact forms.

First, let us describe the general plan of this long section and some of its results informally. All these results can be found in [K16].

The section consists of the following four subsections:

(1) Contact Holography: recovering contact geometry from the boundary data. We are exhibiting differences between generic non-vanishing vector fields and Reeb vector fields, produced by contact forms (see Propositions 3.6, 3.7).

The main results of this section are the holography Theorem 3.18, together with its Corollary 3.29. They claim that certain list $\{C_{v_\beta}, \beta^{\partial}, (f^{\bullet})^{\partial}\}$ of boundary-confined data is sufficient for a reconstruction of the bulk X and a contact form β on it, up to a diffeomorphism $\Phi : X \to X$ which is the identity on the boundary $\partial_1 X$. These results lead to Theorem 3.19 asserting the existence of a group-theoretical section

$$\sigma : \mathsf{Diff}^{\infty}_{+}(\partial_1 X;\, C_{v_\beta}, \beta^{\partial}, (f^\bullet)^{\partial}) \to \mathsf{Diff}^{\infty}_{+}(X; \beta, f^\bullet)$$

with values in the group of β-contactomorphisms $\mathsf{Diff}^{\infty}_{+}(X; \beta, f^\bullet)$ (see Section 4 for the definitions of all these groups and their decorations). The map σ is the right inverse of the obvious restriction map which acts in the direction opposite to σ.

(2) On contact forms and contact vector fields on manifolds with boundary.
The main results of this section is Corollary 3.31 dealing with special 1-parameter families of contact forms in the spirit of Gray Stability Theorem [Gray]. In this section, we analyze the restrictions imposed on Reeb fields and their flows by the existence of flow-invariant boundary $\partial_1 X$ (see Corollary 3.30).

(3) Non-squeezing of $2n$-symplectic volumes, Morse wrinkles of Reeb fields, and holography.
This section contains the non-squeezing Theorem 3.21. It imposes constraints on contact embeddings Ψ of $(2n+1)$-compact manifolds with boundary into a given contact $(2n+1)$-manifold, which carries a traversing Reeb vector field. It is an analogue of Gromov's non-squeezing Theorem [Gr2] in Symplectic Geometry. The new phenomenon is that the $(2n-1)$-volume of the equatorial locus in the target restricts the $(2n-1)$-volume of the equatorial locus in the source (thus preventing "a camel caring a contact load from passing through the eye of a needle"), provided that we a priori restrict the number of times the Reeb trajectories in the target can hit the boundary of the Ψ-image.

Then we introduce numerical invariants of the Reeb flows on contact manifolds with boundary. They are based on volumes of the Morse strata of the boundary, which measure how "wrinkled" the boundary is with respect to the Reeb flow. We study the holographic properties of these invariants under contact embeddings of equi-dimensional manifolds (Proposition 3.9, Theorem 3.22).

(4) Shadows of Legandrian submanifolds of (X, β) on the screen $\partial_1^+ X(v_\beta)$ and on the trajectory space $\mathcal{T}(v_\beta)$.
In this section, we consider Legandrian submanifolds L of a given contact compact manifold (X, β) whose Reeb vector field v_β is traversing. Proposition 3.10 describes the structures on the boundary $\partial_1 X$, sufficient for reconstructing L. These structures include the $(-v_\beta)$-flow generated Lagrangian "shadow" $L^\dagger \subset \partial_1 X$ of L. For sub-Legandrian

links, in Corollary 3.35, we give a boundary-localized criterion when the v_β-flow through L interacts with a v_β-concave portion of the boundary $\partial_1 X$.

3.8.2 *Recovering contact forms from boundary data*

Let Y be a smooth submanifold of a smooth manifold X and α a differential k-form on X, a section of the bundle $\bigwedge^k T^*X \to X$. Then we denote by $\alpha|_Y$ the restriction of α to Y, a section of the bundle $\bigwedge^k T^*Y \to Y$.

• We use the unconventional notation "$\alpha||_Y$" to denote the restriction of the section $\alpha : X \to \bigwedge^k T^*X$ to the section $\alpha|| : Y \to \bigwedge^k T^*X$.

We start with one lemma, not related to Contact Geometry.

Lemma 3.21. *Let v be a smooth traversing vector field and β a 1-form on a compact smooth manifold X such that $\beta(v) = 1$. Then v admits a Lyapunov function $f : X \to \mathbb{R}$ such that $df(v) = 1$. Moreover, for any v-trajectory γ and any closed interval $[a,b] \subset \gamma$, we have $\int_{[a,b]} df = \int_{[a,b]} \beta$.*

Proof. Let $\dim X = d$. By definition of a traversing vector field, there exists a Lyapunov function $h : X \to \mathbb{R}$ such that $dh(v) > 0$. We embed properly X in a d-dimensional open manifold $\hat{X}$ so that v extends to $\hat{v}$ and h to $\hat{h}$, subject to the constraint $d\hat{h}(\hat{v}) > 0$ in $\hat{X}$. By shrinking $\hat{X} \supset X$ if necessary, we may also extend β to $\hat{\beta}$ in $\hat{X}$ so that $\hat{\beta}(\hat{v}) = 1$ in the vicinity of $X \subset \hat{X}$.

Using that each v-trajectory γ extends to a closed segment of the unique $\hat{v}$-trajectory $\hat{\gamma} \subset \hat{X}$, we construct a tubular cover $\{U_\alpha \subset \hat{X}\}_\alpha$ of X such that there are diffeomorphisms $\{\phi_\alpha : D^{d-1} \times D^1 \to U_\alpha\}_\alpha$ that map each segment $a \times D^1$, $a \in D^{d-1}$, to a segment of a $\hat{v}$-trajectory. Moreover, $\phi_\alpha(D^{d-1} \times \partial D^1) \subset \hat{X} \setminus X$. Using compactness of X, we choose a finite subcover of the cover $\{U_\alpha\}_\alpha$. Abusing notations, we denote it $\{U_\alpha\}_\alpha$.

Let $g_\alpha : U_\alpha \to \mathbb{R}$ be defined by the formula

$$g_\alpha(\phi_\alpha(a \times t)) = \int_{\phi_\alpha([a\times 0),\,(a\times t])} \hat{\beta} =_{\mathsf{def}} \int_{\phi_\alpha(a\times 0)}^{\phi_\alpha(a\times t)} \hat{\beta}, \tag{3.42}$$

where $a \in D^{d-1}$. Note that $dg_\alpha|_\gamma = \hat{\beta}|_\gamma$, where $\gamma = \{\phi_\alpha(a \times t)\}_t$. Since $\hat{\beta}(\hat{v}) = 1$, we get $dg_\alpha(\hat{v}) = 1$. The finite cover $\{U_\alpha\}_\alpha$ gives rise to a finite cover $\{V_\alpha\}_\alpha$ of the compact trajectory space $\mathcal{T}(v)$.

Recall that a function $\psi : \mathcal{T}(v) \to \mathbb{R}$ is defined to be smooth, if its pull-back $\Gamma^*(\psi) : X \to \mathbb{R}$ is smooth. We use a smooth partition of unity

$\{\psi_\alpha : V_\alpha \to [0,1]\}_\alpha$ to introduce a smooth function

$$f =_{\mathsf{def}} \sum_\alpha \psi_\alpha \cdot g_\alpha.$$

By its construction, each function ψ_α has the property $d\psi_\alpha(v) = 0$. Thus,

$$df(v) = \sum_\alpha \big(\psi_\alpha \cdot dg_\alpha(v) + g_\alpha \cdot d\psi_\alpha(v)\big) = \sum_\alpha \psi_\alpha = 1.$$

By a similar formula, any closed interval $[a,b] \subset \gamma$, we get

$$\int_{[a,b]} df = \sum_\alpha \Big(\int_{[a,b]} \psi_\alpha \cdot dg_\alpha + \int_{[a,b]} g_\alpha \cdot d\psi_\alpha\Big) = \sum_\alpha \Big(\psi_\alpha \int_{[a,b]} dg_\alpha\Big)$$
$$= \sum_\alpha \psi_\alpha \cdot \int_{[a,b]} \beta = \int_{[a,b]} \beta. \tag{3.43}$$

□

Corollary 3.21. *Any traversing vector field v admits a Lyapunov function g such that $dg(v) = 1$.*

Proof. If $f : X \to \mathbb{R}$ is a Lyapunov function for v, then the 1-form $\beta = \frac{df}{df(v)}$ satisfies the hypotheses of Lemma 3.21. By the lemma, there exists a Lypunov function $g : X \to \mathbb{R}$ such that $dg(v) = 1$. □

Let us recall now a few basic definitions from the field of Contact Geometry.

Definition 3.18. A coorientable contact structure on a $(2n+1)$-dimensional smooth manifold X is a $2n$-dimensional distribution ξ of cooriented hyperplanes in the tangent bundle TX. The distribution ξ is generated by the kernels of a differential 1-form β on X such that $\beta \wedge (d\beta)^n > 0$ everywhere. Equivalently, the restriction $(d\beta)^n|_\xi > 0$ everywhere.

An coorientable almost contact structure ξ on a $(2n+1)$-dimensional smooth manifold X consists of:

• a $2n$-dimensional distribution η of cooriented hyperplanes in the tangent bundle TX, generated by the kernels of a differential 1-form β on X,

• a non-vanishing vector field v such that $\beta(v) = 1$ (thus, v is transversal to the distribution η), and

• a non-degenerate skew-symmetric 2-form ω on η (equivalently, a complex structure $J : \eta \to \eta$ such that $J^2 = -\mathsf{id}$). ◇

If $\lambda : X \to \mathbb{R}_+$ is a positive smooth function, then by a direct computation,

$$(\lambda\beta) \wedge (d(\lambda\beta))^n = \lambda^{n+1}\beta \wedge \Big(d\beta + \frac{d\lambda}{\lambda} \wedge \beta\Big)^n = \lambda^{n+1}[\beta \wedge (d\beta)^n] > 0.$$

We denote by ξ^∂ the restriction of the distribution ξ to the boundary $\partial_1 X$ (not to be confused with the intersection $\xi \cap \partial_1 X$). Since the normal (quotient) 1-bundle $\nu = TX|_{\partial_1 X} / T(\partial_1 X)$ is trivial, the distribution ξ^∂ is a $(2n)$-dimensional orientable subbundle of the bundle $\nu \oplus T(\partial_1 X) \to \partial_1 X$.

Definition 3.19. A vector field v_β on X is called the Reeb field for a contact 1-form β, if $\beta(v_\beta) = 1$ and $v_\beta \in \mathsf{ker}(d\beta)$. ♢

The property $\beta \wedge (d\beta)^n \neq 0$ implies that the kernel $\mathsf{ker}(d\beta)$ is 1-dimensional. Thus, any contact form β generates a *unique* Reeb vector field v_β.

Example 3.9. For the standard contact form $\beta = dz + x\, dy$ on $\mathbb{R}^3$, the Reeb vector field $v_\beta = \partial_z$. ♢

Example 3.10. For each integer n, consider the contact differential form

$$\beta_n = cos(n\theta_1)\, d\theta_2 + sin(n\theta_1)\, d\theta_3$$

on the torus T^3 with the angular coordinates $\theta_1, \theta_2, \theta_3$. The contact distribution ξ_n is given by the kernel $\mathsf{ker}(\beta_n)$ and the Reeb vector field by

$$v_{\beta_n} = cos(n\theta_1)\, \partial_{\theta_2} + sin(n\theta_1)\, \partial_{\theta_3}.$$

♢

The following claim is well-known.

Lemma 3.22. *For a contact structure ξ, generated by a contact 1-form β, there is a 1-to-1 correspondence between the space $C^\infty(X)$ of smooth functions and the space $\mathcal{R}eeb(X, \xi)$ of Reeb vector fields amenable to ξ. The correspondence is given by the formula $f \longrightarrow v_{\{e^f \cdot \beta\}}$.*

Proof. Let us fix a contact form β such that $\mathsf{ker}(\beta) = \xi$. Then any other contact form, producing ξ can be written as $e^f \cdot \beta$ for some smooth function f. Let us compare $e^f \cdot \beta$ and $e^g \cdot \beta$, where g is another smooth function. By the definition of Reeb fields, we get $e^f \cdot \beta(v_{\{e^f \beta\}}) = 1 = e^g \cdot \beta(v_{\{e^g \beta\}})$. If $v_{\{e^f \cdot \beta\}} = v_{\{e^g \cdot \beta\}}$, this is equivalent to the equation $e^f = e^g$, or to the equation $f = g$. Therefore, the correspondence $f \longrightarrow v_{\{e^f \cdot \beta\}}$ is a continuous bijection. □

The next few lemmas are dealing with the way a contact form β and its differential $d\beta$ interact with the boundary $\partial_1 X$.

Lemma 3.23. *Let β be a contact form on a $(2n+1)$-dimensional manifold X and v_β its boundary generic Reeb vector field.*

Then $(d\beta)^n|_{\mathsf{int}(\partial_1^+ X(v_\beta))} > 0$, while $(d\beta)^n|_{\mathsf{int}(\partial_1^- X(v_\beta))} < 0$. Thus, $d\beta|_{\mathsf{int}(\partial_1^+ X(v_\beta))}$ is a symplectic 2-form on $\mathsf{int}(\partial_1^+ X(v_\beta))$.

Proof. Since $\beta \wedge (d\beta)^n$ is the volume form on X, by the definition of $\partial_1^+ X(v_\beta)$, we conclude that $v_\beta \rfloor (\beta \wedge (d\beta)^n)|_{\partial^+ X(v_\beta)} > 0$ in $\mathsf{int}(\partial_1^+ X(v_\beta))$.

On the other hand, since $v_\beta \rfloor (d\beta)^n = 0$, we get

$$v_\beta \rfloor (\beta \wedge (d\beta)^n) = (v_\beta \rfloor \beta)(d\beta)^n - \beta \wedge (v_\beta \rfloor (d\beta)^n) = (d\beta)^n.$$

Therefore, $(d\beta)^n|_{\mathsf{int}(\partial_1^+ X(v_\beta))} > 0$. As a result, $d\beta|_{\mathsf{int}(\partial_1^+ X(v_\beta))}$ is a symplectic form. □

Corollary 3.22. *Let β be a contact form on $(2n+1)$-dimensional X, whose Reeb vector field v_β is boundary generic. Then $\int_{\partial_2 X(v_\beta)} \beta \wedge d\beta^{n-1} > 0$.*

Proof. Since $(d\beta|_{\mathsf{int}(\partial_1^+ X(v_\beta))})^n > 0$, by the definition of the Morse stratum $\partial_2 X(v_\beta)$ and Stokes' Theorem,

$$\int_{\partial_2 X(v_\beta)} \beta \wedge (d\beta)^{n-1} = \int_{\partial_1^+ X(v_\beta)} (d\beta)^n > 0.$$

□

Corollary 3.23. *Let v be a traversing vector field on a $(2n+1)$-dimensional compact connected smooth manifold X. Consider a $(2n)$-dimensional domain $Y \subset \partial_1^+ X(v)$ and a neighborhood U of Y in X. Let δ^U be a non-vanishing 1-form on U.*

If $(d\delta^U)^n|_Y \leq 0$, then no contact form β with the Reeb vector field v is an extension of the given form δ^U. ◇

Lemma 3.24. *Let v be a traversing vector field on an $(2n+1)$-dimensional compact connected oriented smooth manifold X with boundary. Then X admits a 1-form β such that $\beta(v) = 1$ and $d\beta(v) = 0$.*

Proof. By definition of a traversing vector field, there exists a Lyapunov function $h : X \to \mathbb{R}$ such that $dh(v) > 0$.

As before, we embed properly X in a $(2n+1)$-dimensional open manifold $\hat{X}$ so that v extends to $\hat{v}$ and h to $\hat{h}$, subject to the constraint $d\hat{h}(\hat{v}) > 0$

in $\hat{X}$. By Corollary 3.21, we may assume that there is another Lyapunov function $\hat{f} : \hat{X} \to \mathbb{R}$ such that $d\hat{f}(\hat{v}) = 1$ in the vicinity of X in $\hat{X}$.

Using that each v-trajectory γ extends to a closed segment of the unique $\hat{v}$-trajectory $\hat{\gamma} \subset \hat{X}$, we construct a tubular cover $\{U_\alpha \subset \hat{X}\}_\alpha$ of X as in the proof of Lemma 3.21. The finite cover $\{U_\alpha\}_\alpha$ gives rise to a finite cover $\{V_\alpha\}_\alpha$ of the compact trajectory space $\mathcal{T}(v)$.

In the coordinates $\{x_1, x_2, \dots x_{2n}\}$ on the standard ball $D^{2n} \subset \mathbb{R}^{2n}$, we consider the closed 1-form

$$\rho =_{\mathsf{def}} \frac{1}{2} \cdot \sum_{i=1}^{n} (x_{i+1} dx_i - x_i dx_{i+1}).$$

Using the diffeomorphism $\phi_\alpha : D^{2n} \times D^1 \to U_\alpha$ as in the proof of Lemma 3.21, we transfer ρ to the 1-form $\rho_\alpha =_{\mathsf{def}} \phi_\alpha^\dagger(\rho)$ and the differential dx_{2n+1} to the differential $\phi_\alpha^\dagger(dx_{2n+1}) = df$ on U_α. Put

$$\beta_\alpha =_{\mathsf{def}} df + \rho_\alpha.$$

We use a smooth partition of unity $\{\psi_\alpha : V_\alpha \to [0,1]\}_\alpha$ by smooth functions on $\mathcal{T}(v)$ (i.e., by v-invariant smooth functions ψ_α on X) to introduce a smooth 1-form $\beta =_{\mathsf{def}} \sum_\alpha \psi_\alpha \beta_\alpha$. By its construction, ψ_α has the property $d\psi_\alpha(v) = 0$. Therefore,

$$\beta(v) = \sum_\alpha \psi_\alpha \cdot \beta_\alpha(v) = df(v) \cdot \sum_\alpha \psi_\alpha = df(v) = 1.$$

Thus, β is a non-vanishing 1-form.

At the same time, since $d(df) = 0$, $\psi_\alpha(v) = 0$, $v \rfloor d\psi_\alpha = 0$, and $\rho_\alpha(v) = 0$, $v \rfloor d\rho_\alpha = 0$, we get:

$$v \rfloor d\beta = v \rfloor d\big(\sum_\alpha \psi_\alpha \cdot \beta_\alpha\big) = v \rfloor d\big(\sum_\alpha \psi_\alpha \cdot \rho_\alpha\big) =$$

$$= v \rfloor \big(\sum_\alpha d\psi_\alpha \wedge \rho_\alpha\big) + v \rfloor \big(\sum_\alpha \psi_\alpha \, d\rho_\alpha\big) = 0.$$

If we would know that $\beta \wedge (d\beta)^n > 0$ in X, then we would conclude that β is a contact form and v is its Reeb vector field. However, the property $\beta \wedge (d\beta)^n > 0$ does not hold automatically! The next Proposition 3.6 spells out some hypotheses that insure the desired property. □

Proposition 3.6. *Let v be a traversing boundary generic vector field on a connected compact smooth $(2n+1)$-manifold with boundary. Let $\xi(v)$ be a $(2n)$-subbundle of the tangent bundle TX, transversal to v. Consider any 1-form β such that $\beta(v) = 1$ and $v \rfloor d\beta = 0$. (By Lemma 3.24, such a form β exists.)*

If $\partial_2^- X(v) = \emptyset$ (i.e., if the vector field v is concave*), then:*

- *β is a contact form,*
- *v is its Reeb vector field,*
- *$\xi(v)$ admits a complex structure amenable to the symplectic form $d\beta$,*

if and only if, $(d\beta)^n\big|_{\mathsf{int}(\partial_1^+X(v))} > 0$.[12]

Proof. We notice that $v \rfloor (d\beta) = 0$ and $v \rfloor \beta = 1$ imply that $\mathcal{L}_v\beta = 0$. Thus, β is v-invariant. By the same two properties of β, we get

$$v \rfloor (\beta \wedge (d\beta)^n) = (v \rfloor \beta)(d\beta)^n - \beta \wedge (v \rfloor (d\beta)^n) = (d\beta)^n.$$

Hence, if $(d\beta)^n|_{\mathsf{int}(\partial_1^+X(v))} > 0$, then $\big(v \rfloor (\beta \wedge (d\beta)^n)\big)|_{\mathsf{int}(\partial_1^+X(v))} > 0$.

Let $w_1, w_2, \dots w_{2n}$ be a frame in the tangent space $T_x\partial_1^+X(v)$ whose orientation agrees with the orientation of $\partial_1 X(v)$, induced by the preferred orientation of X. Then $v \wedge w_1 \wedge w_2 \wedge \dots \wedge w_{2n}$ is a positive volume. Thus, $(\beta \wedge (d\beta)^n)(v \wedge w_1 \wedge w_2 \wedge \dots \wedge w_{2n}) > 0$ on $\mathsf{int}(\partial_1^+X(v))$.

Using that β is v-invariant, we conclude that $\beta \wedge (d\beta)^n$ is a positive volume form along each v-trajectory γ such that $\gamma \cap \mathsf{int}(\partial_1^+X(v)) \neq \emptyset$.

If $\partial_2^-X(v) = \emptyset$ (this implies that $\partial_j X(v)) = \emptyset$ for all $j \geq 3$), any trajectory γ has this property. Therefore, in such a case, we conclude that $\beta \wedge (d\beta)^n > 0$ everywhere in X. As a result, β is a contact form with the Reeb vector field $v_\beta = v$ and the bundle $\xi_\beta \approx \xi(v)$ admits a symplectic structure, induced by $d\beta$ (equivalently, a complex structure $J : \xi_\beta \to \xi_\beta$, consistent with symplectic form $d\beta$).

Conversely, if β, subject to the properties $\beta(v) = 1$ and $v \rfloor d\beta = 0$, is such that $\beta \wedge (d\beta)^n > 0$ in X, then v is the Reeb field for β, and, by Lemma 3.23, $(d\beta)^n\big|_{\mathsf{int}(\partial_1^+X(v))} > 0$. □

Remark 3.19. Proposition 3.6 suggests that should be a geometric connection between the cohomology classes, with the values in the groups $H^{*+1}(X; \pi_*(\mathsf{SO}(2n)/\mathsf{U}(n))$ that obstruct the existence of complex (symplectic) structure on the distribution $\xi(v)$ and the geometry of the loci in $\partial_1^+X(v)$, where $(d\beta)^n|_{\partial_1^+X(v)}$ changes sign. Note that this flip in sign can be expressed as the global property $\int_{\partial_1^+X(v)} |(d\beta)^n| > \int_{\partial_1^+X(v)} (d\beta)^n$. ◇

As Corollary 3.26 below testifies, not any traversing vector field v is the Reeb field for some contact form β! Some obstructions for finding such β are of the homotopy-theoretical nature (like in Remark 3.19), others (see Proposition 3.7) are of more obscure geometrical nature.

[12]Thus, if the bundle $\xi(v)$ does not admit a complex structure, then the restriction of the $2n$-form $(d\beta)^n$ to $\mathsf{int}(\partial_1^+X(v))$ cannot be positive.

At the first glance, it seems probable that any traversing vector field v on a $(2n+1)$-dimensional manifold X is a Reeb vector field for a contact form β, provided that a transversal to v $(2n)$-dimensional distribution ξ admits a complex/symplectic structure. However, this conjecture is false, as the next simple example testifies.

Example 3.11. Consider the 3-fold $X = S \times [0,1]$ where S is an *oriented closed* surface. Let v be a constant vector on X, tangent to the fibers of the projection $S \times [0,1] \to S$. Then Proposition 3.6 tells us that such a v is not a Reeb vector field of any contact form β on X. Indeed, any transversal to v plane field ξ is orientable and thus admits a complex/symplectic structure. However, $d\beta$ cannot be positive on $\partial_1^+ X(v) = S \times \{0\}$ since $\int_S d\beta = 0$. $\diamondsuit$

Let us generalize Example 3.11.

Definition 3.20. A $2k$-shell is the product $S \times [0,1]$, where S is a connected oriented *closed* smooth manifold of dimension $2k$, $k \leq n$.

Let v be a smooth non-vanishing vector field on a $(2n+1)$-dimensional compact manifold X and let ξ be a $(2n)$-distribution, transversal to v and admitting a complex/symplectic structure.

We say that (X, v, ξ) contains a contact $(2k+1)$-shell if there is a regular embedding $\Psi : S \times [0,1] \to X$ that maps each oriented segment $s \times [0,1]$, $s \in S$, to an oriented segment of a v-trajectory and such that a normal $(2n-2k)$-bundle $\nu(\Psi)$ of $\Psi(S \times [0,1])$ in X is a complex/symplectic subbundle of the complex/symplectic $2n$-bundle $\xi|_{\Psi(S \times [0,1])}$. $\diamondsuit$

Equivalently, (X, v, ξ) contains a contact $(2k+1)$-shell $\Psi : S \times [0,1] \to X$ if and only if there is a regular embedding $\psi : S \to X$ such that there exists a symplectic $(2n-2k)$-subbundle μ of $\xi|_{\psi(S)}$ whose fibers are transversal to $\psi(S)$; i.e., $\mathbb{R}_v \oplus \mu$ is a normal bundle of $\psi(S)$ in X, where $\mathbb{R}_v$ stands for the line bundle spanned by v.

Proposition 3.7. *If v_β is the Reeb vector field of a contact form β on a compact $(2n+1)$-dimensional manifold X, then $(X, v_\beta, \mathsf{ker}\beta)$ does not contain any contact $(2k+1)$-shells for all $k \in [1,n]$.*

In particular, the sets $\partial_1^\pm X(v_\beta)$ have no closed connected components.

Proof. Let β be a contact form whose Reeb field is v_β. Let $\xi_\beta = \mathsf{ker}(\beta)$. Assume that (X, v_β, ξ_β) contains a contact $(2k)$-shell $\Psi(S \times [0,1])$. Thanks to Definition 3.20, the normal bundle $\nu(\Psi)$ is a complex/symplectic subbundle of $\xi_\beta|_{\Psi(S \times [0,1])}$.

Consider the form $\Theta^\bullet =_{\mathsf{def}} \Psi^*(\beta \wedge (d\beta)^k)$ on $S \times [0,1]$. Since $\beta \wedge (d\beta)^n$ is a volume form and $(d\beta)^{n-k}|_{\nu(\Psi)}$ is the volume form by Definition 3.20, the form $\Theta^\bullet$ must be a volume form on $S \times [0,1]$. Thus, $\beta^\bullet =_{\mathsf{def}} \Psi^*(\beta)$ is a contact form on $S \times [0,1]$ and the transfer $v^\dagger$ of v_β under the map Ψ is the Reeb vector field for $\beta^\bullet$. Moreover, $\partial_1^+(S \times [0,1])(v^\dagger) = S \times \{0\}$.

On the other hand, Proposition 3.6 tells us that such $v^\dagger$ cannot be the Reeb vector field of the contact form $\beta^\bullet$. Indeed, the form $(d\beta^\bullet)^k$, being exact, cannot be positive on the closed manifold $\partial_1^+(S \times [0,1])(v^\dagger) = S \times \{0\}$ since $\int_S (d\beta^\bullet)^k = \int_S d\big(\beta^\bullet \wedge (d\beta^\bullet)^{k-1}\big) = 0$. This contradiction proves the proposition. □

Corollary 3.24. *Let β be a contact form on an oriented smooth $(2n+1)$-manifold X. Let the Reeb vector field v_β of β be transversal to a smooth connected oriented hypersurface $Y \subset X$, an open $(2n)$-manifold.*

Then Y does not contain any closed *smooth oriented $(2k)$-dimensional submanifold S, $k \in [1,n]$, whose normal bundle $\nu(S)$ in Y admits a complex/symplectic structure, amenable to the symplectic form $d\beta|_Y$.*

In particular, this claim is valid for $Y = \mathsf{int}(\partial_1^+ X(v_\beta))$.

Proof. Let U be a smooth regular neighborhood of Y in X. Consider the $(-v_\beta)$ induced projection $Q^\beta : U \to Y$. We denote its differential by Q^β_*. Since v_β is transversal to Y, the bundle map $Q^\beta_*| : \xi_\beta|_Y \to TY$ is a symplectomorphism with respect to forms $d\beta|_{\xi_\beta}$ and $d\beta|_Y$. If the normal bundle $\nu(S)$ of S in Y is a symplectic subbundle of TY, then $(Q^\beta_*)^{-1}\big(\nu(S)\big) \cap \xi_\beta$ is a symplectic subbundle of $\xi_\beta|_{(Q^\beta)^{-1}(S)}$. It is on the level of definitions to check that $(Q^\beta)^{-1}(S)$ is a contact shell in $U \subset X$. Now the corollary follows from Proposition 3.7. □

Motivated by these results and feeling a bit adventurous, let us state the following conjecture.

Conjecture 3.3. *Let v be a non-vanishing (perhaps, traversing?) vector field on a compact $(2n+1)$-dimensional manifold X, such that a normal to v $(2n)$-dimensional distribution ξ admits a complex/symplectic structure.*

Then v is the Reeb vector field of a contact form on X if and only if (X, v, ξ) does not contain any contact $(2k+1)$-shells for all $k \in [1,n]$. ◇

We denote by $\mathcal{V}(X)$ the space of all vector fields on X in the C^∞-topology.

Proposition 3.8. *Let $\mathcal{V}_{\{\neq 0,\, \xi \text{ is symp.}\}}(X)$ be the space of smooth non-vanishing vector fields v on X that admit a transversal to them symplectic $(2n)$-distribution $\xi \subset TX$, where $\dim(X) = 2n+1$. Let $k \in [1, n]$.*

Then the subset $\mathcal{W}_k(X)$ of $\mathcal{V}_{\{\neq 0,\, \xi \text{ is symp.}\}}(X)$, consisting of pairs (v, ξ) that admit contact $(2k+1)$-shells, is open in the C^∞-topology.

Therefore, if $\mathcal{W}_k(X) \neq \emptyset$, then $\mathcal{V}_{\{\neq 0,\, \xi \text{ is symp.}\}}(X)$ contains an open nonempty set of pairs (v, ξ), where v is not the Reeb vector field for any contact form β on X, subject to to $(d\beta)^n|_\xi > 0$.

Proof. Let $(v_\star, \xi_\star) \in \mathcal{V}_{\{\neq 0,\, \xi \text{ is symp.}\}}(X)$. Let $\Psi_\star : S \times [0,1] \to X$ be a contact $(2k+1)$-shell for the triple $(X, v_\star, \xi_\star)$. We denote by $\tilde{\nu}(\Psi_\star)$ the symplectic subbundle of $\xi_\star$ that extends the normal to $\Psi_\star(S \times [0,1])$ bundle $\nu(\Psi_\star)$ into a neighborhood $U_\star \subset X$ of $\Psi_\star(S \times [0,1])$. Then the set of $v \in \mathcal{V}(X)$, transversal to $\xi_\star$, is open. With the manifold $\Psi_\star(S \times \{0\})$ and the neighborhood $U_\star$ being fixed, each v sufficiently close to $v_\star$ defines a new shell $\Psi(v) : S \times [0,1] \to X$ by integrating the field v, subject to the initial conditions in $\Psi_\star(S \times \{0\})$. Indeed, for any v sufficiently close to $v_\star$, the symplectic distribution $\tilde{\nu}(\Psi_\star)$ is still transversal to $\Psi(v)(S \times [0,1])$. □

Let Z be a compact smooth $(2n+1)$-dimensional manifold and $f : Z \to \mathbb{R}$ a Morse function. Let $\tilde{v}$ be a f-gradient-like vector field on Z. We denote by X the complement in Z to a disjoint union of small open $(2n+1)$-balls $\{B_z\}$ that are centered on the critical points $\{z\}$ of f and are convex in the local Euclidean metric g_z, adjusted to the local Morse coordinates at z.

A similar construction makes sense for any closed 1-form α on Z with Morse type singularities. In their vicinity, $\alpha = df$, where f has the standard Morse form in the Morse coordinates at each singularity. Recall that a vector field v is of the gradient type on $X = Z \setminus \bigcup_{z \in \mathsf{sing}(\alpha)} B_z$, if $\alpha(v) > 0$ on X.

Corollary 3.25. *If Z is a closed manifold, then the gradient-like vector field v on $X = Z \setminus \bigcup_{z \in \mathsf{crit}(f)} B_z$ is not a Reeb field of any contact form.*

Proof. Since any Morse function $f : Z \to \mathbb{R}$ on a closed manifold Z has maxima and minima $\{z\}$, the spheres $S_z = \partial B_z$ surrounding such critical points give rise to contact $(2n+1)$-shells. By Proposition 3.7, v cannot be a Reeb vector field. □

Corollary 3.26. *Let Z be a* compact *smooth manifold of an odd dimension that exceeds 7. Consider a Morse function $f : Z \to \mathbb{R}$ such that* $\mathsf{crit}(f) \neq \emptyset$ *and its gradient like vector field v.*

Then v is not a Reeb vector field of any contact form on the manifold

$$X = Z \setminus \bigcup_{z \in \mathsf{crit}(f)} B_z.$$

Moreover, for any 1-form β on X such that $\beta(v) = 1$ and $v \rfloor d\beta = 0$, the form $(d\beta)^n|_{\mathsf{int}(\partial_1^+ X(v))}$ must flip sign.

Proof. For any f-critical point z, consider the sphere $S_z^{2n} = \partial B_z$. Let $a \in H^{2n}(S_z^{2n}; \mathbb{Z})$ be a generator. Let $\tilde{K}(S^{2n})$ denote the reduced complex K-theory of the sphere.

For each complex vector bundle η over S^{2n}, the n-th Chern class $c_n(\eta)$ is a multiple of $(n-1)!\, a$. Moreover, for each m such that $m \equiv 0 \mod (n-1)!$, there is a unique $[\eta] \in \tilde{K}(S^{2n})$ such that $c_n(\eta) = ma$ (see [Huse], 9.8 Corollary).

On the other hand, for any Morse singularity, the Euler class $\chi(\xi(v)) = (-1)^{\mathsf{ind}_z(f)}\, 2$, where $\mathsf{ind}_z(f)$ is the Morse index of z. This can be seen by applying the Gaussian map to v along S_z^{2n}, while considering the orthogonal to v hyperplane field $v^\perp$. Therefore, if the real $2n$-bundle $\xi(v)$ has a complex structure, then for each f-critical point z, we must have $(-1)^{\mathsf{ind}_z(f)}\, 2 \equiv 0 \mod (n-1)!$. As a result, a necessary condition for this construction to lead to a contact structure on the $(2n+1)$-dimensional X is

$$(-1)^{\mathsf{ind}_z(f)}\, 2 \equiv 0 \mod (n-1)! \text{ for each } z \in \mathsf{crit}(f).$$

Unfortunately, this arithmetic condition can be satisfied only for $n \leq 3$. Therefore, no gradient field v of a Morse function $f : Z \to \mathbb{R}$ can produce a contact form β on $X = Z \setminus (\bigcup_{z \in \mathsf{crit}(f)} D_z)$ such that $v_\beta = v|_X$ in all dimensions $\dim X \geq 9$.

If we identify a small vector $u \in T_z Z$ with the point $exp_{g(z)}(u) \in B_z$ in the Euclidean metric $g(z)$ in the vicinity of z, in terms of the Hessian of f at z, we get

$$\partial_1^+ X(v) = \coprod_{z \in \mathsf{crit}(f),\ u \in T_z Z} \{Hess(f)_z(u,u) \geq 0\} \cap \partial B_z,$$

and

$$\partial_2 X(v) = \partial_2^+ X(v) = \coprod_{z \in \mathsf{crit}(f),\ u \in T_z Z} \{Hess(f)_z(u,u) = 0\} \cap \partial B_z.$$

Therefore, v is concave in relation to $\partial_1 X$ and Proposition 3.6 is applicable. Hence, by the argument above, for any 1-form β such that $\beta(v) = 1$ and $v \rfloor d\beta = 0$, the form $(d\beta)^n|_{\mathsf{int}(\partial_1^+ X(v))}$ must flip sign, provided $\dim X \geq 9$. $\square$

Corollary 3.26 admits the following strait-forward generalization.

Corollary 3.27. *Let $\tilde{v}$ be a vector field on an odd-dimensional manifold Y such that the zeros of $\tilde{v}$ form a finite set $\mathcal{Z}_{\tilde{v}}$. Let $U(\mathcal{Z}_{\tilde{v}})$ be an open regular neighborhood of $\mathcal{Z}_{\tilde{v}}$ in Y. Put $X = Z \setminus U(\mathcal{Z}_{\tilde{v}})$ and let $v = \tilde{v}|_X$.*

Then a necessary condition for v being a Reeb vector field of a contact form β on X is

$$(-1)^{\mathsf{ind}_z(v)}\, 2 \equiv 0 \mod (n-1)! \text{ for each } z \in \mathcal{Z}_{\tilde{v}},$$

where $\mathsf{ind}_z(v)$ denotes the local index of the vector field v at z. ♢

The next assertion would be useful on many occasions.

Conjecture 3.4. *Let X be a compact connected oriented smooth manifold with boundary, $\dim(X) = 2n+1$. Let β be a contact form on X such that its Reeb vector field v_β is transversal to some $\hat{v}_\beta$-invariant codimension one foliation $\mathcal{G}$ on an open $(2n+1)$-manifold $\hat{X}$ properly containing X.*[13]

Then there is a smooth isotopy of $\{\phi_t : X \to X\}_{t\in[0,1]}$ such that the Reeb vector field v_{β_1} of the contact form $\beta_1 =_{\mathsf{def}} \phi_1^*(\beta)$ is traversing on $\phi_1(X) \subset X$. ♢

Since a connected manifold with boundary admits a handle decomposition without handles of the top dimension $2n+1$, there is no homotopy-theoretical obstruction for extending any almost contact structure (see Definition 3.18) from the base of each handle to its interior (so the preconditions for applying the h-principle are valid). Therefore, by Gromov's Theorem [Gr1] (see also its enforcement, the fundamental Theorem 1.1 of Borman, Eliashberg, and Murphy [BEM]), the contact form β and its cooriented contact structure ξ on X exist. However, we do not know whether the appropriate transversal to v_β and $\hat{v}_\beta$-invariant foliation $\mathcal{G}$ exists as well. The challenge is to construct v_β possessing the chopping property as in Definition 3.12. This would allow to use Theorem 3.12 to validate the conjecture.

Question 3.5. *A C^∞-small perturbation β' of a given contact form β is a contact form. Is a C^∞-small perturbation v' of a Reeb vector field v_β a Reeb vector field for some contact form? For some contact form C^∞-close to β?* ♢

For a traversing vector field v on X, every v-trajectory $\gamma \subset X$ has a neighborhood $\hat{U}(\gamma) \subset \hat{X}$ and a t-parametric family of diffeomorphisms

[13] By [Su1], there is a metric on X in which the v_β-trajectories are geodesics.

$\{\phi^t : \hat{U}(\gamma) \to \hat{U}(\gamma)\}_{t\in[-\epsilon,\epsilon]}$, where $\epsilon > 0$ depends on γ, such that $\phi^0 = \mathsf{id}$, and $\frac{d}{dt}\phi^t(a)\big|_{t=0} = v(a)$ for all $a \in \hat{U}(\gamma) \cap X$.

Definition 3.21. Let v be a vector field on X. We say that a contact form β is v-invariant if $(\phi^t)^*(\beta) = \beta$ in $\hat{U}(\gamma) \cap X \cap (\phi^t)(X)$ for all sufficiently small t. Equivalently, $\mathcal{L}_v \beta = 0$.

Similarly, the contact structure ξ is v-invariant, if $(\phi^t)^*(\beta) = \lambda_t \cdot \beta$, where $\lambda_t : X \to \mathbb{R}_+$ is a smooth positive function (i.e., $(\phi^t)^*(\xi) = \xi$ as distributions). ♢

Definition 3.22.

- We call two contact forms β_1 and β_2 on X Reeb-equivalent ($\mathcal{R}$-equivalent for short), if they share the same Reeb vector field v. Similarly, we call contact structures ξ_1 and ξ_2 $\mathcal{R}$-equivalent if they are generated by $\mathcal{R}$-equivalent contact forms.
- We call contact forms β_1 and β_2 Reeb-conformally-equivalent ($\mathcal{RC}$-equivalent for short), if their Reeb vector fields v_1 and v_2 are proportional: i.e., $v_2 = \lambda \cdot v_1$, where $\lambda : X \to \mathbb{R}_+$ is a smooth positive function. Similarly, we call contact structures ξ_1 and ξ_2 $\mathcal{RC}$-equivalent if they are generated by $\mathcal{RC}$-equivalent contact forms. ♢

Remark 3.20. Given a contact form β, if a smooth positive function $\lambda : X \to \mathbb{R}_+$ is such that $\mathcal{L}_{v_\beta}\lambda = 0$, then the form $\beta_1 =_{\mathsf{def}} \lambda^{-1}\beta$ is $\mathcal{RC}$-equivalent to the form β. Moreover, the Reeb vector field $v_{\beta_1} = \lambda v_\beta$. As a result, positive functions from the algebra $C^\infty(\mathcal{T}(v_\beta))$ of smooth functions on the trajectory space $\mathcal{T}(v_\beta)$ act on a given β by multiplication and transform it into contact forms $\mathcal{RC}$-equivalent to β. ♢

Remark 3.21. Evidently, $\mathcal{RC}$-equivalent contact structures/forms *share the dynamics* of their Reeb flows. ♢

Lemma 3.25. *Let $\eta : X \to \mathbb{R}$ be a smooth function such that $\mathcal{L}_{v_\beta}\eta \equiv d\eta(v_\beta) = 0$. Consider the 1-form $\beta_1 = \beta + d\eta$.*

Then β_1 is $\mathcal{R}$-equivalent to β.

Proof. By definition, $d\eta(v_\beta) = 0, v_\beta \rfloor d\beta = 0$, and for dimensional reasons $(d\eta \wedge (d\beta)^n)|_{\mathsf{ker}\beta} = 0$. Thus, $d\eta \wedge (d\beta)^n = 0$. Therefore,

$$\beta_1 \wedge (d\beta_1)^n = (\beta + d\eta) \wedge (d\beta)^n = \beta \wedge (d\beta)^n > 0.$$

Also $\beta_1(v_\beta) = \beta(v_\beta) + d\eta(v_\beta) = 1$ and, since $d\beta_1 = d\beta$, we get $v_\beta \rfloor d\beta_1 = 0$. Hence $v_{\beta_1} = v_\beta$. □

Lemma 3.26. *For a contact form β on X, any smooth diffeomorphism $\Phi : X \to X$, that preserves the $\mathcal{RC}$-equivalence class of the contact form β, also preserves the Morse stratifications $\{\partial_j X(v_\beta)\}_j, \{\partial_j^\pm X(v_\beta)\}_j$ of the boundary (see (2.1)).*

Proof. Recall that β uniquely defines the Reeb field v_β, and the conformal class of v_β determines the v_β-generated Morse stratification on the boundary.

If $\Phi^*(\beta) = \beta_1$ and $v_{\beta_1} = \lambda \cdot v_\beta$ for some smooth function $\lambda : X \to \mathbb{R}_{>0}$ (i.e., β_1, β are $\mathcal{RC}$-equivalent), then the two Morse stratifications coincide, since they are defined in terms of applying iteratively the Lie derivatives $\mathcal{L}_{v_\beta}$ and $\mathcal{L}_{v_{\beta_1}} = \lambda \cdot \mathcal{L}_{v_\beta}$ to an auxiliary function $h : \hat{X} \to \mathbb{R}$ that determines the boundary $\partial_1 X$ as in (2.8) [K1]. Such a function h has the three properties:

- 0 is a regular value of h,
- $h^{-1}(0) = \partial_1 X$
- $h^{-1}((-\infty, 0]) = X.$ (3.44)

□

We adopt the notation $Op(Y)$ for the vicinity of a topological subspace Y in the ambient space X.

The following relative version of the Gray's theorem [Gray] may be found in [EM], page 95. Lemma 3.32 will generalize a bit the claim of Lemma 3.27 below (for this reason, we skip the proof of Lemma 3.27).

Lemma 3.27. *Let $\{\xi_t\}_{t\in[0,1]}$ be a smooth family of contact structures on a compact connected manifold X with boundary such that $\xi_t = \xi_0$ in the vicinity $Op(\partial_1 X)$ of $\partial_1 X$ in X.*

Then there is an isotopy $\phi_t : X \to X$, fixed in $Op(\partial_1 X)$, such that $(\phi^t)^*(\xi_0) = \xi_t$ for all $t \in [0, 1]$. ◊

Let X be a compact connected smooth $(2n+1)$-dimensional manifold X with boundary.

We denote by $\mathsf{Cont.forms}(X)$ the space of smooth contact forms β on X, and by $\mathsf{Cont.forms}(X, \partial_1 X; Op(\beta^\partial))$ the space of smooth contact forms β on X, such that the section $\beta^\partial = \beta| : Op(\partial_1 X) \to T^*(X)|_{Op(\partial_1 X)}$ is fixed in the vicinity of $\partial_1 X$.

Similarly, we denote by $\mathsf{Cont}(X)$ the space of smooth contact structures ξ on X, and by $\mathsf{Cont}(X, \partial_1 X; Op(\xi^\partial))$ the space of smooth contact structures ξ on X, such that the distribution $\xi \subset T_*(X)|_{Op(\partial_1 X)}$ is fixed.

Corollary 3.28. *Let contact forms β_1 and β_2 on X generate contact structures ξ_1 and ξ_2 which belong to the same path-connected component of the space $\mathsf{Cont}(X, Op(\xi^\partial))$. Assume that the Reeb vector field v_1 of the contact form β_1, is traversing on X.*

Then the Reeb vector field v_2 of a contact form β_2, is traversing as well. As a result, any v_2-trajectory is homeomorphic to a closed interval or to a singleton (in particular, v_2 has no closed or trapped trajectories).

Proof. Consider a smooth path in $\mathsf{Cont}(X, Op(\xi^\partial))$ that connects ξ_1 to ξ_2. By Lemma 3.27, there exists a smooth orientation-preserving diffeomorphism Φ, identical on $Op(\partial_1 X)$, such that $\lambda \cdot \beta_1 = \Phi^*(\beta_2)$ for a smooth positive function $\lambda : X \to \mathbb{R}_+$. Let us check that $v_2 =_{\mathsf{def}} \Phi_*(\lambda^{-1} v_1)$ is the Reeb vector field for β_2:

$$v_2 \rfloor d\beta_2 = \Phi_*(\lambda^{-1} v_1) \rfloor d\beta_2 = \lambda^{-1} v_1 \rfloor \Phi^*(d\beta_2) =$$

$$= v_1 \rfloor \Big(\frac{d\lambda}{\lambda} \wedge \beta_1 + d\beta_1\Big) = v_1 \rfloor \Big(\frac{d\lambda}{\lambda} \wedge \beta_1\Big).$$

Since $\left(\frac{d\lambda}{\lambda} \wedge \beta_1\right)\big|_{\mathsf{ker}(\beta_1)} = 0$, the 1-form $(v_2 \rfloor d\beta_2)|_{\mathsf{ker}\beta_1} = 0$. Also, $(v_2 \rfloor d\beta_2)$ vanishes on v_1:

$$v_1 \rfloor (v_2 \rfloor d\beta_2) = v_1 \rfloor \big(v_1 \rfloor (\frac{d\lambda}{\lambda} \wedge \beta_1)\big) = 0.$$

Therefore, $v_2 \rfloor d\beta_2 = 0$. Moreover,

$$\beta_2(v_2) = \beta_2(\Phi_*(\lambda^{-1} \cdot v_1)) = \Phi^*(\beta_2)(\lambda^{-1} \cdot v_1) = \lambda \cdot \beta_1(\lambda^{-1} \cdot v_1) = 1.$$

Hence, $v_2 = \Phi_*(\lambda \cdot v_1)$ is the Reeb vector field for the form β_2. Therefore, the Reeb vector field v_2 is traversing as well. □

We are ready to formulate the main result of this section. It reflects the scheme in (3.6).

Theorem 3.18. (Conjugate holography for contact forms and their Reeb fields)

Let X_1, X_2 be compact connected oriented smooth $(2n+1)$-dimensional manifolds with boundaries, equipped with contact forms β_1, β_2. We assume that their Reeb vector fields v_1, v_2 are traversing and boundary generic.

We denote by $f_1^\bullet, f_2^\bullet$ the Lyapunov functions such that $df_1^\bullet(v_1) = 1 = df_2^\bullet(v_2)$ (whose existence is guarantied by Lemma 3.21). In addition, we assume that the v_2 possesses Property A from Definition 3.2.[14]

[14]Recall that the property $\partial_3 X_2(v_2) = \emptyset$ implies Property A.

Assume that a smooth orientation-preserving diffeomorphism $\Phi^\partial : \partial_1 X_1 \to \partial_1 X_2$ *commutes with the two causality maps:*

$$C_{v_2} \circ \Phi^\partial = \Phi^\partial \circ C_{v_1}.$$

- *Then* Φ^∂ *extends to a smooth orientation-preserving diffeomorphism* $\Phi : X_1 \to X_2$ *such that* Φ *maps the oriented foliation* $\mathcal{F}(v_1)$ *to the oriented foliation* $\mathcal{F}(v_2)$*, and the pull-back contact form* $\beta_1^\dagger = \Phi^*(\beta_2)$ *is* $\mathcal{RC}$*-equivalent to* β_1 *(in the sense of Definition 3.22).*
- *If, in addition,* $(\Phi^\partial)^*(f_2^\bullet)^\partial = (f_1^\bullet)^\partial$ *and* $(\Phi^\partial)^*(\beta_2|_{\partial_1 X_2}) = \beta_1|_{\partial_1 X_1}$*, then there exists a smooth diffeomorphism* $\Phi : X_1 \to X_2$ *which extends* Φ^∂ *and such that* $(\Phi)^*(\beta_2) = \beta_1$*.*

Proof. Our arguments rely heavily on the Holography Theorem 3.3. Assuming Property A, it claims the existence of an orientation-preserving diffeomorphism $\Phi : X_1 \to X_2$ that extends Φ^∂ and maps the v_1-oriented foliation $\mathcal{F}(v_1)$ to the v_2-oriented foliation $\mathcal{F}(v_2)$.

In the process of extending the diffeomorphism Φ^∂ to a homeomorphism $\Phi : X_1 \to X_2$, following the scheme in (3.6), *we assume implicitly the existence of the singular foliations* $\mathcal{F}(v_i)$ *and* $\mathcal{G}(f_i)$ *and of the Lyapunov functions* $f_i : X_i \to \mathbb{R}$, $i = 1, 2$.

It remains to sort out what happens to the contact forms β_1, β_2 under this diffeomorphism Φ or its modifications.

Since $\mathcal{L}_{v_i}\beta_i = 0$, the contact form β_i is v_i-invariant, and thus is determined by its boundary confined section $\beta_i^\partial : \partial_1 X_i \to T^*X_i|_{\partial X_i}$, *provided that the* v_i*-flow is known* (as in the scheme (3.6)).

We observe that $f_1 = \Phi^*(f_2)$ by the very construction of Φ. Let $v_1^\dagger =_{\mathsf{def}} \Phi_*^{-1}(v_2)$. Since Φ maps $\mathcal{F}(v_1)$ to $\mathcal{F}(v_2)$, we conclude that $v_1^\dagger$ must be positively-proportional to v_1. Therefore, for the vector field $v_1^\dagger$, we get $df_1(v_1^\dagger) = \Phi^*(df_2)(v_1^\dagger) = df_2(v_2) = 1$. At the same time, $\beta_1(v_1) = 1$.

Consider the 1-form $\beta_1^\dagger =_{\mathsf{def}} \Phi^*(\beta_2)$ and denote by $\xi_1^\dagger$ its kernel. Using that Φ preserves the orientations, we see that

$$\Phi^*(\beta_2) \wedge \Phi^*(d\beta_2)^n = \Phi^*(\beta_2 \wedge (d\beta_2)^n) > 0$$

in X_1. Therefore, the hyperplane distribution $\xi_1^\dagger$ is a cooriented contact structure with the contact form $\beta_1^\dagger$.

By the arguments as in the proof of Corollary 3.28, $v_1^\dagger = \Phi_*^{-1}(v_2)$ is the Reeb vector field for $\beta_1^\dagger$. Thus, $\beta_1^\dagger$ is $v_1^\dagger$-invariant. Since $v_1^\dagger = \lambda \cdot v_1$, where $\lambda : X_1 \to \mathbb{R}_+$ is a positive smooth function, β_1 and $\beta_1^\dagger$ are $\mathcal{RC}$-equivalent contact forms, and ξ_1 and $\xi_1^\dagger$ are $\mathcal{RC}$-equivalent contact structures on X_1

(see Definition 3.22). Equivalently, $\xi_2^\dagger =_{\mathsf{def}} \Phi_*(\xi_1)$ is $\mathcal{RC}$-equivalent to ξ_2. This proves the first assertion of the theorem.

Now we turn to the second assertion. The special Lypunov functions $f_1^\bullet, f_2^\bullet$ are essential for constructing the desired extension Φ of Φ^∂. We can build the extension Φ of Φ^∂ as in Theorem 3.3, using the map $\Phi^{\mathcal{T}} : \mathcal{T}(v_1) \to \mathcal{T}(v_2)$ (induced by Φ^∂) and the restrictions $(f_1^\bullet)^\partial = (\Phi^\partial)^*((f_2^\bullet)^\partial)$ and $(f_2^\bullet)^\partial$ of the Lyapunov functions $f_1^\bullet$ and $f_2^\bullet$. By the very construction of Φ, we get $\Phi^*(f_2^\bullet) = f_1^\bullet$. Since $df_i^\bullet(v_i) = 1$ and Φ maps v_1-trajectories to v_2-trajectories, it follows that $\Phi_*(v_1) = v_2$. By the hypotheses, $(\Phi^\partial)^*(\beta_2|_{\partial_1 X_2}) = \beta_1|_{\partial_1 X_1}$ and since $\beta_1(v_1) = 1 = \beta_2(v_2)$, we conclude that $(\Phi^\partial)^*(\beta_2||_{\partial_1 X_2}) = \beta_1||_{\partial_1 X_1}$ (see the very beginning of this section for the notation "$\beta_i||$"), except, perhaps, along the locus $\partial_2 X_2$, where v_2 is tangent to $\partial_1 X_2$. However, since Φ^∂ is a smooth diffeomorphism, by continuity, we get that $(\Phi^\partial)^*(\beta_2||_{\partial_1 X_2}) = \beta_1||_{\partial_1 X_1}$ without exception. Thus, using that β_i is v_i-invariant, we get that $(\Phi)^*(\beta_2) = \beta_1$. □

Recall that this construction of the extension Φ in the first assertion of Theorem 3.18 is implicit and far from being canonical. For example, it depends on the choice of extension of $f_1^\partial =_{\mathsf{def}} (\Phi^\partial)^*(f_2^\partial)$ to a smooth function $f_1 : X_1 \to \mathbb{R}$, which is strictly monotone along the v_1-trajectories. The uniqueness (rigidity) of the extension Φ may be achieved, if we assume the *full knowleadge* of the manifolds X_i, equipped with the foliation grids $\mathcal{F}(v_i), \mathcal{G}(f_i)$ and the Lyapunov functions f_i or $f_i^\bullet$ (see Corollary 3.29).

The next theorem, an instant corollary of Theorem 3.18, claims that the *smooth topological type* of the triple $\{X,\ \mathcal{F}(v_\beta),\ \mathcal{G}(f^\bullet)\}$ and of the contact form β may be reconstructed from the appropriate boundary-confined data.

Corollary 3.29. (Reeb flow holography)

Let X be a compact connected smooth $(2n+1)$-dimensional manifold with boundary, equipped with a contact form β. We assume that the Reeb vector field $v = v_\beta$ is traversing and boundary generic and possesses Property A. *By Lemma 3.21, using β, we construct a Lyapunov function $f^\bullet : X \to \mathbb{R}$ such that $df^\bullet(v) = 1$.*

Then the following boundary-confined data:

- *the causality map $C_v : \partial_1^+ X(v) \to \partial_1^- X(v)$,*
- *the restriction $(f^\bullet)^\partial : \partial_1 X \to \mathbb{R}$ of the Lyapunov function $f^\bullet : X \to \mathbb{R}$,*
- *the restriction $\beta|_{\partial_1 X}$ of the contact form β,*

are sufficient for a reconstruction of the triple $(X, f^\bullet, \beta)$ (and thus, of the Reeb vector field v_β as well), up to diffeomorphisms $\Phi : X \to X$ which are the identity map on $\partial_1 X$.

Proof. Assume that we have two sets of structures, $\{X, \mathcal{F}(v_{\beta_1}), \mathcal{G}(f_1^\bullet), \beta_1\}$ and $\{X, \mathcal{F}(v_{\beta_2}), \mathcal{G}(f_2^\bullet), \beta_2\}$, which agree along $\partial_1 X$. By Theorem 3.18, being applied to $\Phi^\partial = \mathsf{id}_{\partial_1 \mathsf{X}}$, the following data

$$\begin{aligned} &\{\partial_1 X,\ \mathcal{F}(v_{\beta_1}) \cap \partial_1 X,\ (f_1^\bullet)^\partial,\ \beta_1^\partial\} = \\ &= \{\partial_1 X,\ \mathcal{F}(v_{\beta_2}) \cap \partial_1 X,\ (f_2^\bullet)^\partial,\ \beta_2^\partial\} \end{aligned} \tag{3.45}$$

on the boundary $\partial_1 X$ allow an extension of Φ^∂ to a diffeomorphism $\Phi : X \to X$ which maps $\{X, \mathcal{F}(v_{\beta_1}), \mathcal{G}(f_1^\bullet), \beta_1\}$ to $\{X, \mathcal{F}(v_{\beta_2}), \mathcal{G}(f_2^\bullet), \beta_2\}$. Therefore, we have reconstructed the triple $(X,\ f^\bullet,\ \beta)$, up to a diffeomorphism Φ, which is the identity on $\partial_1 X$. Note that the reconstructions of the contact form β implies the reconstruction of the Reeb field v_β and of the contact structure ξ_β. □

3.8.3 *Boundary data of scattering maps C_{v_β} and the group-theoretical sections of β-contactomorphisms*

Although the construction of an extension $\Phi : X \to X$ of a diffeomorphism $\Phi^\partial : \partial_1 X \to \partial_1 X$, subject to the hypotheses of Corollary 3.29, is not canonical, we will argue that by fixing one such extension $\Phi_\star$, Corollary 3.29 makes it possible to extend any other diffeomorphism $\tilde{\Phi}^\partial : \partial_1 X \to \partial_1 X$, subject to the hypotheses of Corollary 3.29, in a *rigid way*, so that the reconstruction works well for *families* of such diffeomorphisms $\tilde{\Phi}^\partial$'s.

Following the scheme in (3.6), we assume now that the quadruple $(X, \mathcal{F}(v_\beta), f^\bullet, \beta)$ is *known and fixed*. Then, given *any* diffeomorphism $\Phi^\partial : \partial_1 X \to \partial_1 X$ that respects the boundary data (3.45), we can extend it *canonically* to a diffeomorphism $\Phi : X \to X$ that keeps $\{\mathcal{F}(v), f^\bullet, \beta\}$ invariant. Indeed, any point $x \in X$ is determined by the unique pair $\big(\gamma_x \cap \partial_1 X,\ f^\bullet(x)\big)$, subject to the requirement $f^\bullet(x) \in [f^\bullet(\gamma_x \cap \partial_1 X)]$. Now put $\Phi(x) = x'$, where the point $x' \in X$ is uniquely defined by the pair

$$\Big(\Phi^\partial(\gamma_x \cap \partial_1 X),\ \Phi^\partial\big((f^\bullet)^{-1}(f^\bullet(x)) \cap \partial_1 X\big)\Big).$$

By the properties of Φ^∂, the latter pair is of the form

$$\big(\gamma_{x'} \cap \partial_1 X,\ \big((f^\bullet)^{-1}(f^\bullet(x')) \cap \partial_1 X\big)\big),$$

where $f^\bullet(x') \in f^\bullet([\gamma_{x'} \cap \partial_1 X])$.

Assuming the validity of Property A, as in Theorem 3.18, we get that Φ is a smooth diffeomorphism.

Since Φ^∂ preserves the boundary data $\{(f^\bullet)^\partial, \beta^\partial\}$ and the contact form β is v_β-invariant, we get that Φ must preserve both β and its Reeb vector field v_β. Thus, the extension Φ is a contactomorphism.

Consider the topological groups $\mathsf{Diff}^\infty_+(X)$ and $\mathsf{Diff}^\infty_+(\partial_1 X)$ of smooth orientation-preserving diffeomorphisms and the natural restriction homomorphism $\mathcal{R} : \mathsf{Diff}^\infty_+(X) \to \mathsf{Diff}^\infty_+(\partial_1 X)$. Similarly, we may consider the groups ${}^\circ\mathsf{Diff}^\infty_+(X)$ and ${}^\circ\mathsf{Diff}^\infty_+(\partial_1 X)$ that are the path connected components that contain the identity of the groups $\mathsf{Diff}^\infty_+(X)$ and $\mathsf{Diff}^\infty_+(\partial_1 X)$.

Recall that the famous Cerf Theorem [Cerf] claims that any orientation-preserving diffeomorphism of S^3 is isotopic to the identity, and thus, extends to a diffeomorphism of the 4-ball D^4. Therefore, the homomorphism $\mathcal{R} : \mathsf{Diff}^\infty_+(D^4) \to \mathsf{Diff}^\infty_+(S^3)$ is onto.

However, remarkably, for any smooth compact connected X with boundary, the restriction map $\mathcal{R} :\ {}^\circ\mathsf{Diff}^\infty_+(X) \to {}^\circ\mathsf{Diff}^\infty_+(\partial_1 X)$ does not admit a *continuous* section ([ChMa], Corollary 1.8)!

Let $\mathsf{Diff}^\infty_+(X; \beta, f^\bullet)$ denote the subgroup of the group $\mathsf{Diff}^\infty_+(X)$, whose elements preserve the contact form β and the Lyapunov function $f^\bullet : X \to \mathbb{R}$ such that $df^\bullet(v_\beta) = 1$.

Let $\mathsf{Diff}^\infty_+(\partial_1 X;\, C_{v_\beta}, \beta^\partial, (f^\bullet)^\partial)$ denote the subgroup of $\mathsf{Diff}^\infty_+(\partial_1 X)$ whose elements commute with the causality map $C_{v_\beta} : \partial_1^+ X(v_\beta) \to \partial_1^- X(v_\beta)$, preserve the restriction to the boundary $(f^\bullet)^\partial : \partial_1 X \to \mathbb{R}$ of the Lyapunov function $f^\bullet$ and the section $\beta^\partial : \partial_1 X \to T^*(X)|_{\partial_1 X}$.

In contrast with the fact that $\mathcal{R} : \mathsf{Diff}^\infty_+(X) \to \mathsf{Diff}^\infty_+(\partial_1 X)$ does not admit a group-theoretical section [ChMa], we get the following claim.

Theorem 3.19. *Under the hypotheses of Corollary 3.29, the restriction homomorphism* $\mathcal{R} : \mathsf{Diff}^\infty_+(X) \to \mathsf{Diff}^\infty_+(\partial_1 X)$ *admits a group-theoretical continuous section* σ *over the subgroup*

$$\mathsf{Diff}^\infty_+\big(\partial_1 X;\, C_{v_\beta}, \beta^\partial, (f^\bullet)^\partial\big) \subset \mathsf{Diff}^\infty_+\big(\partial_1 X\big).$$

Moreover, the image $\sigma\big(\mathsf{Diff}^\infty_+\big(\partial_1 X;\, C_{v_\beta}, \beta^\partial, (f^\bullet)^\partial\big)\big)$ *belongs to the subgroup* $\mathsf{Diff}^\infty_+(X; \beta, f^\bullet)$ *of* $\mathsf{Diff}^\infty_+(X)$.

Proof. Follows instantly from the previous arguments and the proof of Theorem 3.18. □

3.8.4 *On contact forms and contact vector fields on manifolds with boundary*

We present a few basic properties of contact forms and contact vector fields we will rely upon.

Let α be a differential k-form on a smooth manifold X and let $Z \subset X$ be a smooth submanifold.

• We use somewhat unconventional notations: we write "$\alpha|_Z$" for the restriction of the form $\alpha : X \to \bigwedge^k T^*X$ to the form $\alpha|_Z : Z \to \bigwedge^k T^*Z$, and we write "$\alpha||_Z$" for the restriction of α to the section of the bundle $\bigwedge^k T^*X|_Z \to Z$.

Let $\alpha^\partial =_{\mathsf{def}} \alpha||_{\partial_1 X}$. In particular, $\beta^\partial =_{\mathsf{def}} \beta||_{\partial_1 X}$ is the restriction of the section-form β to the section of the cotangent bundle $T^*X \to \partial_1 X$ (not to be confused with the restriction of β to the submanifold $\partial_1 X \subset X$).

Definition 3.23. A smooth vector field u on X is called a contact field for a given contact 1-form β, if $\mathcal{L}_u\beta = \lambda \cdot \beta$ for some smooth function $\lambda : X \to \mathbb{R}$. ◇

Since $\mathcal{L}_u(\mu\beta) = (d\mu(u) + \mu\lambda)\beta$, where λ, μ are smooth functions and $\mathcal{L}_u(\beta) = \lambda\beta$, the notion of contact vector field depends only on the contact structure ξ_β, generated by a contact form β.

To derive Corollary 3.31 below of Theorem 3.18, we will rely on a couple of lemmas.

Lemma 3.28. *Let v be a traversing vector field on a compact connected oriented smooth $(2n+1)$-dimensional manifold X with boundary. Assume that two cooriented contact structures ξ_1 and ξ_2 on X share the same Reeb vector field v of their contact forms β_1 and β_2.*[15]

If the distributions-subbundles $\xi_1^\partial \subset T_*(X)\,|_{\partial_1 X}$ and $\xi_2^\partial \subset T_*(X)\,|_{\partial_1 X}$ coincide, then $\xi_1 = \xi_2$.

Proof. Let β_i $(i = 1, 2)$ be a contact 1-form whose kernel is the hyperplane distribution ξ_i. By the definition of the Reeb vector field $v = v_i$, we have $v\rfloor d\beta_i = 0$ and $v\rfloor\beta_i = 1$. The Cartan's identity $\mathcal{L}_v\beta_i = v\rfloor d\beta_i + d(v\rfloor\beta_i)$ implies that the directional derivative $\mathcal{L}_v\beta_i = 0$, i.e., β_i is v-invariant. Thus, the 1-form β_i^∂, viewed as a section of the bundle $T^*X|_{\partial_1^+ X(v_i)}$, spreads by the v-flow uniquely along each v-trajectory to produce the invariant 1-form β_i. Using that v is traversing, we conclude that β_i^∂ determines β_i

[15] Hence, ξ_1 and ξ_2 are $\mathcal{R}$-equivalent.

everywhere in X. As a result, the distributions ξ_1, ξ_2 are v-invariant. Thus, ξ_1 is determined by ξ_1^∂ and ξ_2 is determined by ξ_2^∂. Therefore, if $\xi_1^\partial = \xi_2^\partial$, then $\xi_1 = \xi_2$. □

Given a contact form β, any vector field on X can be written as

$$u = h \cdot v_\beta + w,$$

where v_β is the Reeb vector field, $h : X \to \mathbb{R}$ is a smooth function, and the vector field $w \in \xi_\beta$.

Then $\mathcal{L}_u\beta = \lambda \cdot \beta$ for some smooth function $\lambda : X \to \mathbb{R}$ if and only if

$$\mathcal{L}_u\beta \equiv \mathcal{L}_{h \cdot v_\beta + w}\beta = h \cdot \mathcal{L}_{v_\beta}\beta + (v_\beta \rfloor \beta)\, dh + \mathcal{L}_w\beta = \lambda \cdot \beta.$$

Using that $\mathcal{L}_{v_\beta}\beta \equiv 0$ and $v_\beta \rfloor \beta = 1$, this is equivalent to the condition $1 \cdot dh + \mathcal{L}_w\beta = \lambda \cdot \beta$, or, by Cartan's formula, to the condition

$$w \rfloor d\beta = -dh + \lambda \cdot \beta. \tag{3.46}$$

Since $v_\beta \rfloor d\beta = \mathbf{0}$, (3.46) implies that

$$0 = -w \rfloor (v_\beta \rfloor d\beta) = v_\beta \rfloor (w \rfloor d\beta) = v_\beta \rfloor (-dh + \lambda \cdot \beta) = -dh(v_\beta) + \lambda.$$

Thus, $\lambda = dh(v_\beta)$, and the equation (3.46) transforms into the equation

$$w \rfloor d\beta = -dh + dh(v_\beta) \cdot \beta \tag{3.47}$$

with respect to a vector field $w \in \xi_\beta$.

Restricting (3.47) to the hypersurface $\{h = c\}$ gives the equations

$$\begin{aligned} (w \rfloor d\beta)\,|_{h^{-1}(c)} &= -dh(v_\beta) \cdot \beta\,|_{h^{-1}(c)}, \\ (w \rfloor \beta)\,|_{h^{-1}(c)} &= 0. \end{aligned} \tag{3.48}$$

Lemma 3.29. *Let β be a contact form. For any choice of the smooth function h, there exists a unique vector field $w \in \mathsf{ker}(\beta)$ that satisfies equation (3.47).*

Proof. Since $d\beta|_{\mathsf{ker}(\beta)}$ is a non-degenerated skew-symmetric bilinear form, for any 1-form α, viewed as a section of the dual bundle $(\mathsf{ker}(\beta))^* \to X$, there exists a section $w(\alpha)$ of the bundle $\mathsf{ker}(\beta) \to X$ such that $w(\alpha) \rfloor d\beta = \alpha$. Therefore, the restriction of equation (3.47) to $\mathsf{ker}(\beta)$ gives the *solvable* equation

$$(w \rfloor d\beta)|_{\mathsf{ker}(\beta)} = -(dh)|_{\mathsf{ker}(\beta)} \tag{3.49}$$

with respect to the vector field $w \in \mathsf{ker}(\beta)$.

By the same token, the solution w of (3.49) is unique.

If w is the solution of (3.49), then it is automatically a solution of (3.47), since $v_\beta \rfloor (w \rfloor d\beta) = v_\beta \rfloor (-dh + dh(v_\beta)) \cdot \beta$ produces the identity $0 = 0$. □

Lemma 3.30. *Let β be a contact form. For any choice of smooth function h, the vector field $w \in \mathsf{ker}(\beta)$ that satisfies (3.47) is tangent to any hypersurface $\{h = c\}_{c\in\mathbb{R}}$.*

Proof. It follows from (3.47) that $0 = w \rfloor (w \rfloor d\beta) = w \rfloor (-dh + dh(v_\beta)\cdot\beta)$. Since $w \rfloor \beta = 0$, we get $w \rfloor dh = 0$. Thus, w is tangent to any hypersurface $h^{-1}(c) \subset X$, where $c \in \mathbb{R}$. □

By Lemma 3.30, the vector field $u = h \cdot v_\beta + w$ is also tangent to $\partial_1 X$, since $h|_{\partial_1 X} = 0$. Thus, we getting the following claim.

Corollary 3.30. *Let β be a contact form on X. For any smooth function $h : \hat{X} \to \mathbb{R}$ that has the properties listed in (3.44) and for any vector field $w \in \mathsf{ker}(\beta)$ that satisfies the equation*

$$w \rfloor d\beta = -dh + dh(v_\beta) \cdot \beta,$$

the one-parameter family diffeomorphisms $\{\phi_t(w) : X \to X\}_t$ that integrates the vector field w is well-defined for all $t \in \mathbb{R}$.

Also, the one-parameter family diffeomorphisms $\{\psi_t(u) : X \to X\}_t$ that integrates the contact vector field $u = h \cdot v_\beta + w$, satisfying the equation

$$\mathcal{L}_u\beta = dh(v_\beta) \cdot \beta,$$

is well-defined for all $t \in \mathbb{R}$. ◇

Lemma 3.31. *For a vector field $w \in \mathsf{ker}(\beta)$, $\mathcal{L}_w\beta = 0$ if and only if $w = 0$.*

Proof. By the formula $\mathcal{L}_w\beta = w\rfloor d\beta + d(w\rfloor\beta) = w\rfloor d\beta$, we get $\mathcal{L}_w\beta = 0$ if and only if $w\rfloor d\beta \equiv 0$. Since the symplectic form $d\beta|_{\mathsf{ker}\beta}$ is non-degenerated, it follows that $w = 0$. □

The next example exhibits the simplest case of these contact forms and fields.

Example 3.12.

• Let $X = \{x^2 + y^2 + z^2 \leq 1\} \subset \mathbb{R}^3$ and $\beta = dz + x\,dy$. The Reeb vector field v_β is ∂_z. It is tangent to the 2-sphere $\partial_1 X$ along its equator $E = \{z = 0\} \cap X$. Thus, the restriction of β to E is equal to the restriction of the form $x\,dy$ to E. If θ is the natural parameter on the circle E, then $\beta|_E = cos^2(\theta)\,d\theta$. It is non-negative everywhere on the equator E and vanishes for $\theta = \pm\pi/2$, or at the points $(x, y, z) = (0, \pm 1, 0)$.

If in (3.47) we pick $h = (x^2 + y^2 + z^2 - 1)/2$, then the equation (3.47) reduces to

$$w \rfloor (dx \wedge dy) = -(x\,dx + y\,dy + z\,dz) + z(dz + x\,dy).$$

Coupling this equation with the equation $w \rfloor (dz + x\, dy) = 0$, we get

$$w = (xz - y)\, \partial_x + x\, \partial_y - x^2\, \partial_z$$

for the vector field $w \in \xi_\beta$ which is tangent to the concentric spheres. Note that the fixed point set of the w-flow is the z-axis. Put

$$u = h \cdot v_\beta + w = (xz - y)\partial_x + x\, \partial_y + \frac{1}{2}\big(- x^2 + y^2 + z^2 - 1\big)\partial_z.$$

The u-flow preserves the distribution ξ_β and the sphere $\partial_1 X$. It has exactly two fixed points: $(0, 0, \pm 1)$.

• We pick the shell $\{1 \leq x^2 + y^2 + z^2 \leq 4\}$ for the role of X and the same contact form $\beta = dz + x\, dy$. Then $\partial_2^+ X(v_\beta)$ is the equator

$$E_1 = \{x^2 + y^2 + z^2 = 1\} \cap \{z = 0\}$$

and $\partial_2^- X(v_\beta)$ is the equator

$$E_2 = \{x^2 + y^2 + z^2 = 4\} \cap \{z = 0\}.$$

The same computation leads to $\beta|_{E_1} \geq 0$ and $\beta|_{E_2} \leq 0$. In other words, $\pm\beta|_{\partial_2^\pm X(v_\beta)} \geq 0$.

Now the function h is $(x^2 + y^2 + z^2 - 4)(x^2 + y^2 + z^2 - 1)$. We leave to the reader to solve the equation $w \rfloor d\beta = -dh + dh(v_\beta) \cdot \beta$ for $w \in \xi_\beta$.

• The next example is also given by the contact form $\beta = dz + x\, dy$, but the function $h = z$. The choice of $X \subset \mathbb{R}^3$ is unimportant. In this case, $v_\beta = \partial_z$, and (3.47) reduces to

$$w \rfloor (dx \wedge dy) = -dz + (dz + x\, dy) = x\, dy.$$

Thus, $w = x\, \partial_x \in \xi_\beta$ and $u = x\, \partial_x + z\partial_z$. The vector field w is tangent to the planes $\{z = c\}_c$, and the vector field u is tangent to the plane $\{z = 0\}$. The u-flow keeps the form β invariant. The plane $\{x = 0\}$ is the fixed point set of the w-flow, and the line $\{x = 0, z = 0\}$ is the fixed point set of the u-flow. ◇

Example 3.13. • Let $X = \{x_1^2 + y_1^2 + x_2^2 + y_2^2 + z^2 \leq 1\} \subset \mathbb{R}^5$ and $\beta = dz + x_1 dy_1 + x_2 dy_2$. We pick $h = (x_1^2 + y_1^2 + x_2^2 + y_2^2 + z^2 - 1)/2$. The Reeb vector field $v_\beta = \partial_z$ and the 3-form

$$\beta \wedge d\beta = dz \wedge dx_1 \wedge dy_1 + dz \wedge dx_2 \wedge dy_2 + x_1\, dy_1 \wedge dx_2 \wedge dy_2 + x_2\, dy_2 \wedge dx_1 \wedge dy_1.$$

Let E be the equator $\{z = 0\} \cap \{x_1^2 + y_1^2 + x_2^2 + y_2^2 = 1\}$, a 3-sphere $\partial_2^- X(v_\beta) \subset \partial_1 X = S^4$. Thus

$$\beta \wedge d\beta|_E = (x_1\, dy_1 \wedge dx_2 \wedge dy_2 + x_2\, dy_2 \wedge dx_1 \wedge dy_1)|_E.$$

A somewhat lengthy computation in the spherical coordinates $\psi, \theta \in [0, \pi], \phi \in [0, 2\pi]$ on the 3-sphere E leads to the formula

$$\beta \wedge d\beta|_E = \big[(sin^2\psi\, cos^2\psi + sin^4\psi\, sin^2\theta\, cos^2\phi)\, sin\theta\big]\; d\psi \wedge d\theta \wedge d\phi\; \geq 0.$$

Therefore, $(\beta \wedge d\beta)|_{\partial_2^- X(v_\beta)} \geq 0$. Note that $(\beta \wedge d\beta)||_{\partial_2^- X(v_\beta)} = 0$ along the circle $\{x_1 = 0\} \cap \{x_2 = 0\} \cap E$ in the (y_1, y_2)-plane. $\diamondsuit$

These examples motivate the following conjecture.

Conjecture 3.5. *Let β be a smooth contact form on a compact connected manifold X and v_β its boundary generic Reeb vector field v_β. Then:*

- *For an even j, $\beta|_{\mathsf{int}(\partial_j^+ X(v_\beta))}$ is a contact form away from a zero measure set. Moreover,*

$$\mp\big[\beta \wedge (d\beta)^{\frac{2n-j}{2}}\big]|_{\partial_j^\pm X(v_\beta)} \geq 0, \;\; and\; \beta \wedge (d\beta)^{\frac{2n-j}{2}}||_{\partial_{j+1} X(v_\beta)} \equiv 0\,.$$

- *For an odd j, $d\beta|_{\mathsf{int}(\partial_j^+ X(v_\beta))}$ is a symplectic form away from a zero measure set. Moreover,*

$$\pm(d\beta)^{\frac{2n+1-j}{2}}|_{\partial_j^\pm X(v_\beta)} \geq 0, \;\; and\; (d\beta)^{\frac{2n+1-j}{2}}||_{\partial_{j+1} X(v_\beta)} \equiv 0\,.$$

$\diamondsuit$

Recall that the Gray Stability Theorem [Gray] claims that, for any smooth family $\{\xi_t\}_{t\in[0,1]}$ of contact structures on a *closed* smooth manifold X, there exists a smooth isotopy $\{\phi_t\}_{t\in[0,1]}$ such that $(\phi_t)^*(\xi_t) = \xi_0$ for all $t \in [0,1]$. In fact, we may assume that the manifold X is *compact* in this claim, provided that the t-derivative $\dot\beta_t =_{\mathsf{def}} \frac{d}{dt}\beta_t$ of the contact forms β_t has the property $\dot\beta_t||_{\partial_1 X} = dh_t||_{\partial_1 X}$ (as sections along $\partial_1 X$).

Lemma 3.32. *Let X be a compact connected smooth $(2n+1)$-dimensional manifold X with boundary. Let $\{h_t : \hat X \to \mathbb{R}\}_{t\in[0,1]}$ be a family of smooth functions, each of which has 0 as its regular value, $h_t^{-1}(0) = \partial_1 X$, and $h_t^{-1}((-\infty, 0]) = X$.*

Consider a family $\{\xi_t\}_{t\in[0,1]}$ of cooriented contact structures on X, generated by a smooth family of contact forms $\{\beta_t\}_{t\in[0,1]}$ such that the t-derivative $\dot\beta_t = dh_t$ along $\partial_1 X$.[16]

Then there is a smooth t-family of diffeomorphism $\{\phi_t : X \to X\}_{t\in[0,1]}$, such that $(\phi_t)^*(\xi_t) = \xi_0$; in particular, ϕ_1 is a contactomorphism that maps ξ_0 to ξ_1.

[16] The property $\dot\beta_t||_{\partial_1 X} = dh_t||_{\partial_1 X}$ implies that $\beta_t|_{\partial_1 X}$ is constant.

Proof. We apply the Moser trick [Mo], slightly adjusted to the category of manifolds with boundary. We are looking for a t-family $\{\phi_t : X \to X\}$ of diffeomorphisms such that $(\phi_t)^*(\beta_t) = \lambda_t \cdot \beta_0$, where $\lambda_t : X \to \mathbb{R}_+$ is a smooth t-family of positive functions. Let $\{w_t\}$ be a t-family of vector fields on X *to be constructed*, such that the w_t-flow gives rise to the desired family of diffeomorphisms $\phi_t : X \to X$.

We use [Ge], Lemma 2.19, which claims that

$$\frac{d}{dt}(\phi_t^*(\beta_t)) = \phi_t^*(\dot{\beta}_t + \mathcal{L}_{w_t}\beta_t).$$

By differentiating the identity $\phi_t^*(\beta_t) = \lambda_t \cdot \beta_0$ with respect to t, the desired property of the vector fields w_t can be stated as

$$\phi_t^*(\dot{\beta}_t + \mathcal{L}_{w_t}\beta_t) = \dot{\lambda}_t \cdot \beta_0 = \frac{\dot{\lambda}_t}{\lambda_t} \cdot \phi_t^*(\beta_t).$$

By introducing the auxiliary function

$$\mu_t =_{\mathsf{def}} \frac{d}{dt}(log(\lambda_t)) \circ (\phi_t)^{-1},$$

with the help of Cartan's formula, we get

$$\phi_t^*\big(\dot{\beta}_t + d(\beta_t(w_t)) + w_t \rfloor d\beta_t\big) = \phi_t^*(\mu_t \cdot \beta_t). \tag{3.50}$$

If we choose $w_t \in \xi_t = \mathsf{ker}(\beta_t)$, this equation will be satisfied if $\dot{\beta}_t + w_t \rfloor d\beta_t = \mu_t \cdot \beta_t$, or equivalently, if

$$w_t \rfloor d\beta_t = \mu_t \cdot \beta_t - \dot{\beta}_t. \tag{3.51}$$

Plugging the Reeb vector field v_{β_t} of β_t in (3.51), we get

$$\dot{\beta}_t(v_{\beta_t}) = \mu_t. \tag{3.52}$$

Hence, we can use (3.52) to determine the function μ_t.

With μ_t in place, the non-degeneracy of the 2-form $d\beta_t|_{\xi_t}$ and the fact that (thanks to (3.52)) $v_{\beta_t} \in \mathsf{ker}(\mu_t \cdot \beta_t - \dot{\beta}_t)$ allows us to find a unique solution $w_t \in \xi_t$ of (3.51), written as

$$w_t \rfloor d\beta_t = \dot{\beta}_t(v_{\beta_t}) \cdot \beta_t - \dot{\beta}_t. \tag{3.53}$$

It remains to verify that this w_t is tangent to the boundary $\partial_1 X$, provided that $\dot{\beta}_t = dh_t$ along $\partial_1 X$. Applying the operation $w_t \rfloor$ to both sides of (3.53), we get $0 = \dot{\beta}_t(w_t)$. By the hypothesis, $\dot{\beta}_t||_{\partial_1 X} = dh_t||_{\partial_1 X}$. Therefore, $0 = \dot{\beta}_t(w_t)||_{\partial_1 X} = dh_t(w_t)||_{\partial_1 X}$. It follows that $dh_t(w_t)|_{\partial_1 X} = dh_t(w_t)|_{h_t^{-1}(0)} = 0$, which implies that w_t is tangent to $h_t^{-1}(0) = \partial_1 X$. Therefore, the w_t-flow preserves the manifold X and the diffeomorphisms $\phi_t : X \to X$ are well-defined for all $t \in [0,1]$. Thus, $\phi_1^*(\beta_1) = \lambda_1 \cdot \beta_0$ and $\phi_1^*(\xi_1) = \xi_0$. □

Corollary 3.31. *We adopt the notations and hypotheses of Lemma 3.32: let a t-family of contact forms* $\{\beta_t\}_{t\in[0,1]}$ *on* X *be such that*

$$\dot{\beta}_t||_{\partial_1 X} = dh_t||_{\partial_1 X}.$$

Let $\{\xi_t\}_{t\in[0,1]}$ *be the corresponding t-family of contact structures.*

Assume that the Reeb vector field v_0 *of the contact form* β_0 *is traversing.*

Then the Reeb vector field v_1 *of a contact form* β_1*, is traversing as well. As a result, any* v_1*-trajectory is homeomorphic to a closed interval or to a singleton (in particular,* v_1 *has no closed or trapped trajectories).*

If, in addition, v_0 *is boundary generic, then the causality maps* C_{v_1} *and* C_{v_0} *are conjugate with the help of the diffeomorphism* $\phi_1^\partial = \phi_1| : \partial_1 X \to \partial_1 X$ *as in Lemma 3.32.*

Moreover, the knowledge of the boundary data $\{C_{v_{\beta_0}}, (f_0^\bullet)^\partial, \beta_0|_{\partial_1 X}\}$ *is sufficient for reconstructing the smooth topological type of* $\{X, \xi_1\}$*, provided that* v_{β_0} *satisfies Property* A.

Proof. By Lemma 3.32, for an appropriate diffeomorphism $\phi_1 : X \to X$, we get $\phi_1^*(\beta_1) = \lambda_1 \cdot \beta_0$. By arguments as in the proof of Corollary 3.28, we get that $v_1 = (\phi_1)_*(\lambda_1^{-1} v_0)$ is the Reeb vector field for β_1. This implies that, with the help of $\phi_1 : X \to X$, the smooth dynamics of the v_0-flow and the v_1-flow are similar. In particular, if v_0 is traversing, then so is v_1. If v_0 is boundary generic, then so is v_1.

For the same reason, the causality maps C_{v_1} and C_{v_0}, which depend only on the conformal classes of vector fields v_0 and v_1, are conjugate with the help of ϕ_1^∂.

The last assertion follows by combining Theorem 3.18 (or Corollary 3.29) with Lemma 3.32, provided that v_{β_0} satisfies Property A. □

3.8.5 *Non-squeezing of symplectic volumes, Morse wrinkles of Reeb fields, and holography*

Let us introduce now some new numerical invariants of contact forms on manifolds with boundary.

Definition 3.24. Let v_β be a traversing Reeb vector field for a contact 1-form β on X. The Reeb diameter of (X, β) is defined by

$$\mathsf{diam}_{\mathcal{R}}(\beta) =_{\mathsf{def}} \sup_{\gamma\subset X} \Big\{ \int_\gamma \beta \Big\},$$

where γ runs over the v_β-trajectories. ◊

Lemma 3.33. *Consider an orientation-preserving submersion $\Psi : X \to Y$ of two compact equidimentional manifolds. Let β_Y be a contact form on Y whose Reeb vector field v_Y is traversing. We assume that $\Psi(\partial_1 X)$ is boundary generic relative to v_Y.*

Then the form $\beta_X =_{\mathsf{def}} \Psi^*(\beta_Y)$ is a contact form on X. Its Reeb vector field v_X is traversing with the help of the Lyapunov function $f_X = \Psi^*(f_Y)$, where $df_Y(v_Y) > 0$. Moreover, the Reeb diameters satisfy the inequality

$$\mathsf{diam}_{\mathcal{R}}(\beta_X) \leq \mathsf{diam}_{\mathcal{R}}(\beta_Y).$$

If Ψ is a covering map (i.e., an immersion that is also surjective), then

$$\mathsf{diam}_{\mathcal{R}}(\beta_X) = \mathsf{diam}_{\mathcal{R}}(\beta_Y).$$

Proof. Since Ψ is submersion and $\Psi(\partial_1 X)$ is boundary generic relative to v_Y, for any v_Y-trajectory γ (which is a closed interval or a singleton) the set $\Psi^{-1}(\gamma)$ is either a disjoint union of finitely many closed intervals $\tilde{\gamma}$, or a finite set. Moreover, $\Psi : \tilde{\gamma} \hookrightarrow \gamma$ is an immersion. Thus, $\int_{\tilde{\gamma}} \beta_X = \int_{\tilde{\gamma}} \Psi^*(\beta_Y) = \int_{\Psi(\tilde{\gamma})} \beta_Y \leq \int_\gamma \beta_Y$. In the case of a covering map Ψ, we get $\Psi : \tilde{\gamma} \to \gamma$ is a diffeomorphism. In such a case, $\int_{\tilde{\gamma}} \beta_X = \int_\gamma \beta_Y$. Since any v_X-trajectory $\tilde{\gamma}$ is obtained from some v_Y-trajectory γ as described above, the claims about the Reeb diameters follow. □

Definition 3.25. For a traversing Reeb vector field, we introduce the volume of the space of Reeb trajectories $\mathcal{T}(v_\beta)$ by the formula

$$vol_{(d\beta)^n}(\mathcal{T}(v_\beta)) =_{\mathsf{def}} \int_{\partial_1^+ X(v)} (d\beta)^n. \tag{3.54}$$

◇

Among all the contact forms β that represent a given contact structure ξ, we may consider only β's such that their Reeb vector fields v_β are traversing and $\mathsf{diam}_{\mathcal{R}}(\beta) = 1$. Of course, if $\mathsf{diam}_{\mathcal{R}}(\beta) = \lambda > 0$ for some contact form β and its traversing Reeb vector field v_β, then $\mathsf{diam}_{\mathcal{R}}(\lambda^{-1}\beta) = 1$ for the traversing Reeb vector field λv of the contact form $\lambda^{-1}\beta$. Thus, without lost of generality, we may assume that $\mathsf{diam}_{\mathcal{R}}(\beta) = 1$ in Theorem 3.20.

Theorem 3.20. *Let X be a compact connected smooth $(2n+1)$-dimensional manifold with boundary, equipped with a contact 1-form β. Let v_β be a traversing boundary generic Reeb vector field for β.*

Then we have an "isoperimetric" inequality of volumes

$$\int_X \beta \wedge (d\beta)^n \leq \mathsf{diam}_{\mathcal{R}}(\beta) \cdot \int_{\partial_1^+ X(v_\beta)} (d\beta)^n =$$
$$=_{\mathsf{def}} \mathsf{diam}_{\mathcal{R}}(\beta) \cdot vol_{(d\beta)^n}(\mathcal{T}(v_\beta)). \tag{3.55}$$

This enequality can be also written in the "equatorial" form:

$$\int_X \beta \wedge (d\beta)^n \;\leq\; \mathsf{diam}_{\mathcal{R}}(\beta) \cdot \int_{\partial_2 X(v_\beta)} \beta \wedge (d\beta)^{n-1}. \tag{3.56}$$

Proof. The argument is a variation on the theme of Santalo formula [S]. For any $y \in \partial_1^+ X(v_\beta)$ we consider the v-trajectory γ_y through y. The $(-v)$-directed projection $p : X \to \partial_1^+ X(v_\beta)$ is a submersion, away from the set $p^{-1}(\partial_2 X(v_\beta))$ of zero measure in $\partial_1^+ X(v_\beta)$. Thus, we can use the Dieudonne generalization of the Fubini formula [D]:

$$\int_X \beta \wedge (d\beta)^n = \int_{y \in \partial_1^+ X(v_\beta)} \Big(\int_{\gamma_y} \beta\Big)(d\beta)^n \leq \mathsf{diam}_{\mathcal{R}}(\beta) \cdot \int_{\partial_1^+ X(v_\beta)} (d\beta)^n,$$

together with Definition 3.24, to validate (3.55). By Stokes' theorem, applied to $\partial_1^+ X(v_\beta)$, (3.56) follows.

Although this argument is simple, it is somewhat surprising to realize that the $(d\beta)^n$-volume of the trajectory space $\mathcal{T}(v_\beta)$ depends only on the form $\beta \wedge (\partial\beta)^{n-1}|_{\partial_2 X(v_\beta)}$ on the "equator" $\partial_2 X(v_\beta)$. □

Definition 3.26. If the Reeb vector field v_β of a contact form β on X is traversing, then we introduce the $\mathsf{average\ length\ of\ the\ Reeb\ trajectories}$ by

$$\mathsf{av}_{\mathcal{R}}(\beta) \;=_{\mathsf{def}}\; \frac{\int_X \beta \wedge (d\beta)^n}{\int_{\partial_1^+ X(v_\beta)} (d\beta)^n}. \tag{3.57}$$

◇

By (3.56), $\mathsf{av}_{\mathcal{R}}(\beta) \;\leq\; \mathsf{diam}_{\mathcal{R}}(\beta)$.

Multiplying β by a scalar $\lambda \in \mathbb{R}_+$ results in multiplication of $\mathsf{av}_{\mathcal{R}}(\beta)$ and $\mathsf{diam}_{\mathcal{R}}(\beta)$ by λ. Thus, within a given contact structure ξ_β, $\mathsf{av}_{\mathcal{R}}(\beta)$ and $\mathsf{diam}_{\mathcal{R}}(\beta)$ may take any positive value.

Example 3.14. Let $\beta = dz + x\,dy$ and

$$X = \Big\{\frac{x^2}{(A/2)^2} + \frac{y^2}{(B/2)^2} + \frac{z^2}{(C/2)^2} \leq 1\Big\},$$

a solid ellipsoid. Then the inequality (3.55) can be written as

$$\frac{\pi}{6} ABC \leq C \cdot \big(\frac{\pi}{4} AB\big),$$

or as $\frac{\pi}{6} \leq \frac{\pi}{4}$. ◇

Corollary 3.32. *For any $\eta \in C^\infty(\mathcal{T}(v_\beta))$, the contact form $\beta_1 = \beta + d\eta$ shares with β the Reeb diameter $\mathsf{diam}_{\mathcal{R}}(\beta)$, the volume $\int_X \beta \wedge (d\beta)^n$, and the volume $vol_{(d\beta)^n}(\mathcal{T}(v_\beta))$ of the trajectory space.* ◇

Proof. The "exact slantings" $\beta_1 = \beta + d\eta$ of a contact form β do not change the β-lengths of Reeb trajectories, provided the function $\eta \in C^\infty(\mathcal{T}(v_\beta))$.

By Lemma 3.25, β and $\beta + d\eta$ share the same Reeb vector field: $v_\beta = v_{\beta+d\eta}$. Thus, for a trajectory γ, bounded by points a and b, we get

$$\int_\gamma (\beta + d\eta) = \int_\gamma \beta + (\eta(b) - \eta(a)) = \int_\gamma \beta$$

since $\eta(b) = \eta(a)$. As a result, $\mathsf{diam}_\mathcal{R}(\beta + d\eta) = \mathsf{diam}_\mathcal{R}(\beta)$.

Using the Fubini formula, we verify that the $(\beta + d\eta) \wedge (d(\beta + d\eta))^n$-induced and the $(\beta) \wedge (d\beta)^n$-induced volumes of X are equal:

$$\int_X (\beta + d\eta) \wedge (d(\beta + d\eta))^n = \int_{y \in \partial_1^+ X(v_\beta)} \Big(\int_{\gamma_y} (\beta + d\eta)\Big)(d\beta)^n$$

$$= \int_{y \in \partial_1^+ X(v_\beta)} \Big(\int_{\gamma_y} \beta\Big)(d\beta)^n + \int_{y \in \partial_1^+ X(v_\beta)} \Big(\int_{\gamma_y} d\eta\Big)(d\beta)^n.$$

Thanks to the property $\mathcal{L}_{v_\beta}\eta = 0$, $\int_{\gamma_y} d\eta = 0$ for all y, and the latter formula reduces to

$$= \int_{y \in \partial_1^+ X(v_\beta)} \Big(\int_{\gamma_y} \beta\Big)(d\beta)^n = \int_X (\beta) \wedge (d\beta)^n.$$

Also,

$$vol_{(d\beta)^n}(\mathcal{T}(v_\beta)) = \int_{\partial_1^+ X(v_\beta)} (d\beta)^n = \int_{\partial_1^+ X(v_{\beta+d\eta})} (d\beta)^n$$

$$= \int_{\partial_1^+ X(v_{\beta+d\eta})} (d(\beta + d\eta))^n = vol_{(d(\beta+d\eta))^n}(\mathcal{T}(v_{\beta+d\eta}))$$

□

Definition 3.27. Let Y be a smooth manifold and v a non-vanishing vector field on it that admits a Lyapunov function. Consider a compact smooth connected manifold X with boundary, $\dim X = \dim Y$, and a regular imbedding $\Psi : X \to \mathsf{int}(Y)$ such that v is boundary generic relative the hypersurface $\Psi(\partial_1 X)$. We introduce a natural number $c^\bullet(\Psi, v)$ as the half of the maximum number of intersections of v-trajectories with the hypersurface $\Psi(\partial_1 X)$. ◇

In general, $c^\bullet(\Psi, v)$ resists an intrinsic characterization in terms of the pull-back $\Psi^\dagger(v)$ of the vector field v to X. We know only one *lower* bound of $c^\bullet(\Psi, v)$ in terms of the vector field $\Psi^\dagger(v)$ and the combinatorial types $\omega(\gamma^\dagger)$ of its trajectories $\gamma^\dagger$:

$$c^\bullet(\Psi^\dagger(v)) =_{\mathsf{def}} \max_{\gamma^\dagger \subset X} \Big\{\frac{1}{2}|\omega(\gamma^\dagger)|\Big\} \leq c^\bullet(\Psi, v).$$

This estimate follows from the local models of the traversing boundary generic flows in the vicinity of their trajectories (see [K2], [K3] or Chapter 2).

Remark 3.22. Even if the connected source and the target of Ψ both carry *convex* vector fields, the number $c^\bullet(\Psi, v)$ can be arbitrary big. Indeed, just rap a very narrow ($\epsilon \ll 1$) band $Z = I \times [-\epsilon, \epsilon]$ (with rounded corners) k times around the cylinder $C = S^1 \times [-1, 1]$, where I denotes the interval $[0, 1]$ and the vector field v is aligned with the generators of the cylinder.

To get 3-dimensional manifolds, consider the embedding

$$\Psi : X =_{\mathsf{def}} S^1 \times (I \times [-\epsilon, \epsilon]) \hookrightarrow S^1 \times (S^1 \times [-1, 1]) =_{\mathsf{def}} Y,$$

obtained by multiplying the wrapping map $\psi : Z \subset C$ with the identity map on the first multiplayer S^1. Then $c^\bullet(\Psi, v_Y) = k$. $\diamondsuit$

Consider a traversing v-flow on a compact manifold Y and a regular embedding $\Psi : X \to \mathsf{int}(Y)$, where Y is a compact manifold, and $\dim(X) = \dim(Y)$. Assuming that v is boundary generic with respect to the hypersurface $\Psi(\partial_1 X)$. With the help of $(-v)$-flow, we can generate a shadow $\mathcal{X}^\partial(\Psi)$ of $\Psi(X)$ on the boundary of Y. It is produced by forming the space $\mathcal{X}(\Psi)$ of v-trajectories through the points of $\Psi(X)$ (equivalently, through the locus $\partial_1^+\Psi(X)(v)$) and then taking the intersection $\mathcal{X}(\Psi) \cap \partial_1^+ Y(v)$. This constructions presumes that $\Psi(X)$ is "semi-transparent" for the flow (see Fig. 3.7, the left diagram).

The next theorem resembles Gromov's non-squeezing theorem [Gr2] in symplectic geometry, being transferred to the contact geometry environment (see Fig. 3.7). It deals with the embeddings $\Psi : X_1 \to X_2$ (where $\dim X_1 = 2n + 1 = \dim X_2$) that respect the contact forms: i.e., $\beta_1 = \Psi^*(\beta_2)$. It employs the number $c^\bullet(\Psi, v_{\beta_2})$ that measures how "folded" is the image $\Psi(\partial_1 X_1)$ with respest to the v_{β_2}-flow. The "equatorial" inequality of the theorem may be also viewed as delivering a lower bound of the number $c^\bullet(\Psi, v_{\beta_2})$ in terms of the appropriate β_1- and β_2-generated symplectic $2n$-volumes.

Theorem 3.21. (Non-squeezing equatorial property of the Reeb fields)

Consider two compact smooth manifolds X_1, X_2 equipped with contact forms β_1, β_2, respectively. We assume that the Reeb vector fields v_{β_1}, v_{β_2} are boundary generic and v_{β_2} is traversing. Let $\Psi : X_1 \hookrightarrow X_2$ be a smooth regular embedding such that $\beta_1 = \Psi^*(\beta_2)$.

Then

$$\int_{X_1} \beta_1 \wedge (d\beta_1)^n \;\leq\; \int_{X_2} \beta_2 \wedge (d\beta_2)^n,$$

$$\mathsf{diam}_{\mathcal{R}}(\beta_1) \;\leq\; \mathsf{diam}_{\mathcal{R}}(\beta_2).$$

We also get the inequality between the symplectic $2n$-volumes of the two trajectory spaces (or, equivalently, between the contact-like volumes of the tangency locii, the "equators"):

$$0 \leq \int_{\partial_2 X_1(v_{\beta_1})} \beta_1 \wedge (d\beta_1)^{n-1} = vol_{(d\beta_1)^n}(\mathcal{T}(v_{\beta_1})) \;\leq$$

$$\leq\; c^\bullet(\Psi, v_{\beta_2}) \cdot vol_{(d\beta_2)^n}(\mathcal{T}(v_{\beta_2})) = c^\bullet(\Psi, v_{\beta_2}) \cdot \int_{\partial_2 X_1(v_{\beta_2})} \beta_2 \wedge (d\beta_2)^{n-1}.$$

Proof. To simplify the notations, put $v_1 = v_{\beta_1}$, $v_2 = v_{\beta_2}$.

Using that $\beta_1 = \Psi^*(\beta_2)$, we see that $\Psi_*(v_1) = v_2|_{\Psi(X_1)}$. Moreover, the combinatorial tangency pattern to $\partial_1 X_1$ of every v_1-trajectory γ_1 coincides with the tangency to $\Psi(\partial_1 X)$ pattern of the portion $\Psi(\gamma_1)$ of the unique v_2-trajectory $\gamma_2 \supset \Psi(\gamma_1)$. Note that the tangency pattern of γ_2 to $\Psi(\partial_1 X)$ could be different.

The validation of the first two inequalities is straightforward. Since $\Psi^*(\beta_2) = \beta_1$, by naturality, we get $\int_{X_1} \beta_1 \wedge (d\beta_1)^n = \int_{\Psi(X_1)} \beta_2 \wedge (d\beta_2)^n$. Evidently, $\int_{\Psi(X_1)} \beta_2 \wedge (d\beta_2)^n \leq \int_{X_2} \beta_2 \wedge (d\beta_2)^n$. Thus,

$$\int_{X_1} \beta_1 \wedge (d\beta_1)^n \leq \int_{X_2} \beta_2 \wedge (d\beta_2)^n.$$

It follows instantly that $\mathsf{diam}_{\mathcal{R}}(\beta_1) \leq \mathsf{diam}_{\mathcal{R}}(\beta_2)$, since the Ψ-image of any v_{β_1}-trajectory γ is a segment of a v_{β_2} trajectory γ' that contains $\Psi(\gamma)$.

Now we are going to validate the equatorial inequality (see Fig. 3.7). Let $\mathcal{W}_2(\Psi)$ denote the $2n$-dimensional set comprised of v_2-trajectories through the locus $\Psi(\partial_2 X_1(v_1)) \coprod \partial_2 X_2(v_2)$. Consider the set.

$$\mathcal{V}_1^\circ(\Psi) =_{\mathsf{def}} \Psi\big(\partial_1^+ X_1(v_1)\big) \setminus \big(\Psi(\partial_1^+ X_1(v_1)) \cap \mathcal{W}_2(\Psi)\big)$$

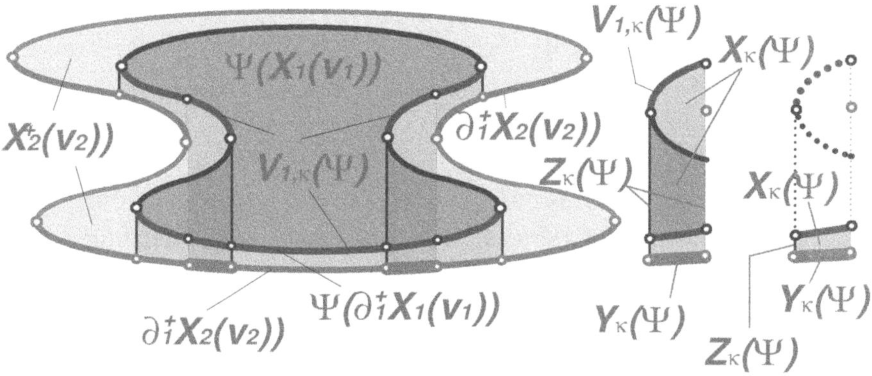

Fig. 3.7 The figure helps with the notations in the proof of Theorem 3.21. In the figure, v_{β_1} and v_{β_2} are the constant vertical vector fields. The two right diagrams represent two different "slices" of the main diagram on the left; the second slice on the right is contained in the first slice. Here, in the last claim of the theorem, the number $c^\bullet(\Psi, v_{\beta_2}) = 2$.

along which the continuous dependence of the v_2-flow on the initial conditions in $\Psi(\partial_1^+ X_1(v_1))$ holds. Let $\mathcal{V}_{1,\kappa}^\circ(\Psi)$ be a connected component of the set $\mathcal{V}_1^\circ(\Psi)$, and let $\mathcal{V}_{1,\kappa}(\Psi)$ be the closure of $\mathcal{V}_{1,\kappa}^\circ(\Psi)$. We denote by $\partial_1^+ X_{1,\kappa}^\circ(v_1) \subset \partial_1^+ X(v_1)$ the Ψ-preimage of $\mathcal{V}_{1,\kappa}^\circ(\Psi)$.

Keeping in mind that $\Psi_*(v_1) = v_2|_{\Psi(X_1)}$, consider the set $\mathcal{X}_\kappa^\circ(\Psi)$, formed by the union of portions of the $(-v_2)$-trajectories that originate at the set $\mathcal{V}_{1,\kappa}^\circ(\Psi)$ and terminate at the points of the set $\partial_1^+ X_2(v_2)$. Let $\mathcal{X}_\kappa(\Psi)$ be the closure of $\mathcal{X}_\kappa^\circ(\Psi)$.

Let $\mathcal{Y}_\kappa(\Psi) =_{\mathsf{def}} \mathcal{X}_\kappa(\Psi) \cap \partial_1^+ X_2(v_2)$ be the set where the v_2-trajectories through $\mathcal{V}_{1,\kappa}(\Psi)$ terminate. Then the boundary of the set $\mathcal{X}_\kappa(\Psi)$ consists of three parts: $\mathcal{V}_{1,\kappa}(\Psi)$, $\mathcal{Y}_\kappa(\Psi)$, and the rest, denoted by $\mathcal{Z}_\kappa(\Psi)$. The $2n$-dimensional Lebesgue measures of mutual intersections of these three parts is zero.

We notice that $\mathcal{Z}_\kappa(\Psi)$ consists of portions of $(-v_2)$-trajectories. Since $v_2 \rfloor d\beta_2 = 0$, we conclude that $\int_{\mathcal{Z}_\kappa(\Psi)} (d\beta_2)^n = 0$.

Now we apply Stokes' theorem to the zero form $d\big((d\beta_2)^n\big)$ on $\mathcal{X}_\kappa(\Psi)$ to get

$$0 = \int_{\mathcal{X}_\kappa(\Psi)} d((d\beta_2)^n) = \int_{\mathcal{Y}_\kappa(\Psi)} (d\beta_2)^n - \int_{\mathcal{V}_{1,\kappa}(\Psi)} (d\beta_2)^n.$$

Therefore, using that $\Psi(\partial_1^+ X_{1,\kappa}^\circ(v_1)) = \mathcal{V}_{1,\kappa}(\Psi)$, by the definitions of the relevant sets, we get

$$\int_{\mathcal{Y}_\kappa(\Psi)} (d\beta_2)^n = \int_{\mathcal{V}_{1,\kappa}(\Psi)} (d\beta_2)^n = \int_{\Psi(\partial_1^+ X_{1,\kappa}^\circ(v_1))} (d\beta_2)^n.$$

Thus, by the latter equality,

$$\int_{\partial_1^+ X_{1,\kappa}^\circ(v_1)} (d\beta_1)^n = \int_{\Psi(\partial_1^+ X_{1,\kappa}^\circ(v_1))} (d\beta_2)^n = \int_{\mathcal{Y}_\kappa(\Psi)} (d\beta_2)^n \leq \int_{\partial_1^+ X_2(v_2))} (d\beta_2)^n,$$

since $\mathcal{Y}_\kappa(\Psi)$ is a part of the locus $\partial_1^+ X_2(v_2))$ on which, by Lemma 3.23, $(d\beta_2)^n \geq 0$.

Next, we notice that the domains $\{\mathcal{Y}_\kappa(\Psi)\}_\kappa$ may overlap massively in $\partial_1^+ X_2(v_2))$ and the multiplicity of the overlap is bounded from above by the maximal number of times a v_2-trajectory, passing through the set $\Psi(X_1)$, hits the set $\Psi(\partial_1^+ X_1(v_1))$. By Definition 3.27, this natural number is $c^\bullet(\Psi, v_{\beta_2})$.

Adding the inequalities for all κ and using that $\{\partial_1^+ X_{1,\kappa}^\circ(v_1)\}_\kappa$ form a partition of $\partial_1^+ X_1(v_1)$, modulo the set of $2n$-dimensional measure zero, we get

$$\int_{\partial_1^+ X_1(v_1)} (d\beta_1)^n = \sum_\kappa \int_{\partial_1^+ X_{1,\kappa}(v_1)} (d\beta_1)^n \leq c^\bullet(\Psi, v_{\beta_2}) \cdot \int_{\partial_1^+ X_2(v_2))} (d\beta_2)^n.$$

Finally, in view of Definition 3.25 and by Stokes' Theorem, we get

$$\int_{\partial_2 X_1(v_1)} \beta_1 \wedge (d\beta_1)^{n-1} = vol_{(d\beta_1)^n}(\mathcal{T}(v_1))$$

and

$$\int_{\partial_2 X_1(v_2)} \beta_2 \wedge (d\beta_2)^{n-1} = vol_{(d\beta_2)^n}(\mathcal{T}(v_2)).$$

□

Example 3.15. Let Y be a compact connected smooth $(2n+1)$-manifold with boundary, equipped with a contact form β_Y such that its Reeb vector field v_Y is traversing and boundary generic. Consider the solid ellipsoid

$$X = \Big\{ \sum_{i=1}^{2n+1} \frac{x_i^2}{a_i^2} \leq 1 \Big\}$$

in the Euclidean space E^{2n+1}, equipped with the standard contact form

$$\beta = dx_{2n+1} + \frac{1}{2} \sum_{i=1}^{n} (x_{i+1}\, dx_i - x_i\, dx_{i+1}).$$

Then Theorem 3.21 claims that, for any embedding $\Psi : X \to Y$ such that $\beta = \Psi^*(\beta_Y)$, the following three inequalities are valid:

- $$\frac{\pi^{n+\frac{1}{2}}}{\Gamma(n+\frac{3}{2})}\, a_1 a_2 \dots a_{2n+1} \leq \int_Y \beta_Y \wedge (d\beta_Y)^n,$$

- $$a_{2n+1} \leq \mathsf{diam}_{\mathcal{R}}(\beta_Y),$$

- $$\frac{\pi^{n}}{\Gamma(n+1)}\, a_1 a_2 \dots a_{2n} \leq c^{\bullet}(\Psi, v_{\beta_2}) \cdot \int_{\partial_1^+ Y(v_Y)} (d\beta_Y)^n$$
$$= c^{\bullet}(\Psi, v_{\beta_2}) \cdot \int_{\partial_2 Y(v_Y)} d\beta_Y \wedge (d\beta_Y)^{n-1},$$

where $\Gamma(\sim)$ is the Gamma function.

In particular, for X being a ball of radius r, we get the following constraints on the injectivity radius r of a contact embedding:

$$\frac{\pi^{n+\frac{1}{2}}}{\Gamma(n+\frac{3}{2})}\, r^{2n+1} \leq \int_Y \beta_Y \wedge (d\beta_Y)^n,$$

$$r \leq \mathsf{diam}_{\mathcal{R}}(\beta_Y),$$

$$\frac{\pi^{n}}{\Gamma(n+1)}\, r^{2n} \leq c^{\bullet}(\Psi, v_{\beta_2}) \cdot \int_{\partial_2 Y(v_Y)} d\beta_Y \wedge (d\beta_Y)^{n-1}.$$

$\diamondsuit$

Example 3.16. Consider a solid X shaped as a sand clock (see Fig. 3.8) which is vertically alined with the z-axis in the xyz-space $\mathbb{R}^3$. Let $\beta = dz + x\,dy$. We consider a 2-parameter family of transformations

$$\{A_{\lambda\mu} : x \to \lambda x,\ y \to \lambda y,\ z \to \mu z\}_{\lambda,\mu \in \mathbb{R}_+}.$$

We compare two contact 3-folds (X, β) and $(A_{\lambda\mu}(X), \beta)$. Then $c^{\bullet}(\Psi, v_2) \geq 2$. If $c^{\bullet}(\Psi, v_2) = 2$, unless $\{\lambda < 2,\ \mu < 1,\ \lambda^2\mu < 1\}$, no contact embedding Ψ of X into $A_{\lambda\mu}(X)$ exists. $\diamondsuit$

Let v_β be a boundary generic Reeb vector field of a contact form β on a compact $(2n+1)$-manifold X. For any boundary generic v_β, we have $\partial_j X(v_\beta) = \partial(\partial^+_{j-1} X(v_\beta))$ for all $j \geq 1$. Thus, for an odd j, by Stokes' theorem,

$$\int_{\partial_j X(v_\beta)} (d\beta)^{\frac{2n+1-j}{2}} = \int_{\partial^+_{j-1} X(v_\beta)} d\big((d\beta)^{\frac{2n+1-j}{2}}\big) = \int_{\partial^+_{j-1} X(v_\beta)} \mathbf{0} = 0.$$

In contrast, for an even j, we get

$$\int_{\partial_j X(v_\beta)} \beta \wedge (d\beta)^{\frac{2n-j}{2}} = \int_{\partial^+_{j-1} X(v_\beta)} d\big(\beta \wedge (d\beta)^{\frac{2n-j}{2}}\big) =$$

$$= \int_{\partial^+_{j-1} X(v_\beta)} (d\beta)^{\frac{2n+2-j}{2}}.$$

Conjecturally, the last integral is positive (see Conjecture 3.5 and Examples 3.12-3.13).

Motivated by these observations, we introduce some numerical measures of the stratified concavity/convexity of the Reeb flows v_β on X relative to its boundary $\partial_1 X$:

$$\kappa^+_j(\beta) =_{\mathsf{def}} \int_{\partial^+_j X(v_\beta)} (d\beta)^{\frac{2n+1-j}{2}} \quad (= \kappa_{j+1}(\beta))$$
$$\kappa_j(\beta) = 0$$

for an odd $j \in [1, 2n-1]$; and (3.58)

$$\kappa^+_j(\beta) =_{\mathsf{def}} \int_{\partial^+_j X(v_\beta)} \beta \wedge (d\beta)^{\frac{2n-j}{2}}$$
$$\kappa_j(\beta) =_{\mathsf{def}} \int_{\partial_j X(v)} \beta \wedge (d\beta)^{\frac{2n-j}{2}} \quad (= \kappa^+_{j-1}(\beta))$$

for an even $j \in [2, 2n]$. (3.59)

Of course, these quantities depend on the choice of the contact form β that represents a given contact structure ξ and whose Reeb vector field v_β is boundary generic. They measure how "wrinkled" $\partial_1 X$ is in relation to the Reeb flow v_β. By the very definition of $\kappa^+_j(\beta), \kappa_j(\beta)$, they belong to the category of "boundary data".

Note that the inequality (3.55) can be written in terms of these quantities as

$$\kappa^+_0(\beta) \leq \mathsf{diam}_{\mathcal{R}}(\beta) \cdot \kappa^+_1(\beta)$$

and the inequality (3.56) as

$$\kappa^+_0(\beta) \leq \mathsf{diam}_{\mathcal{R}}(\beta) \cdot \kappa_2(\beta),$$

where $\kappa^+_0(\beta) =_{\mathsf{def}} \int_X \beta \wedge (d\beta)^n$.

3.8.6 *Invariance of the volumes $\{\kappa_j(\beta)\}$ under special deformations of contact forms β*

Definition 3.28. Let $\mathcal{B} = \{\beta_t\}_{t\in[0,1]}$ be a family of contact forms on a compact connected smooth $(2n+1)$-manifold X with boundary.

- We say the family $\mathcal{B}$ is variationally boundary exact if

$$\frac{\partial}{\partial t}\beta_t\Big|_{\partial_1 X} = d\eta_t$$

for a family of smooth functions $\{\eta_t : \partial_1 X \to \mathbb{R}\}_{t\in[0,1]}$.

- We say the family $\mathcal{B}$ is variationally boundary $(2n+1-j)$-exact if, for some smooth family of $(2n-j)$-forms $\{\zeta_t\}_{t\in[0,1]}$ on $\partial_1 X$,

$$\Big\{\frac{\partial}{\partial t}\Big[\beta_t \wedge (d\beta_t)^{\frac{2n-j}{2}}\Big]\Big\}\Big|_{\partial_1 X} = d\zeta_t, \text{ where } j \in [2, 2n] \text{ is even}, \quad (3.60)$$

$$\Big\{\frac{\partial}{\partial t}\Big[(d\beta_t)^{\frac{2n+1-j}{2}}\Big]\Big\}\Big|_{\partial_1 X} = d\zeta_t, \text{ where } j \in [1, 2n-1] \text{ is odd}. \quad (3.61)$$

Occasionally, we compliment these requirements with the additional one:

$$\zeta_t\big|_{\partial_{j+1}X(v_{\beta_t})} = 0. \quad (3.62)$$

$\diamondsuit$

Lemma 3.34. *Any variationally boundary exact family $\mathcal{B}$ of contact forms is variationally boundary $(2n+1-j)$-exact.*

If the functions η_t are such that $\eta_t|_{\partial_{j+1}X(v_t)} = 0$, then $\zeta_t|_{\partial_{j+1}X(v_t)} = 0$.

Proof. Note that, for any t-family differential forms, their exterior differential $d =_{\mathsf{def}} d_t$ commutes with the partial derivative operator $\frac{\partial}{\partial t}$ and with the restriction of forms to the boundary $\partial_1 X$. Consider the derivative

$$\frac{\partial}{\partial t}\Big[(d\beta_t)^k\Big] = k\,\frac{\partial}{\partial t}\Big(d\beta_t\Big) \wedge \big[(d\beta_t)^{k-1}\big] = k\, d\Big(\frac{\partial}{\partial t}(\beta_t)\Big) \wedge \big[(d\beta_t)^{k-1}\big].$$

Let $\eta_t : \partial_1 X \to \mathbb{R}$ be a smooth family of functions from Definition 3.28. Replacing $\frac{\partial}{\partial t}(\beta_t)\big|_{\partial_1 X}$ with $d\eta_t$ in these formulas and using that $d(d\eta_t) = 0$, we see that $\frac{\partial}{\partial t}\Big[(d\beta_t)^k\Big]\Big|_{\partial_1 X} = 0$. Similarly,

$$\frac{\partial}{\partial t}\Big[\beta_t \wedge (d\beta_t)^k\Big] = \frac{\partial}{\partial t}\Big[\beta_t\Big] \wedge (d\beta_t)^k + \beta_t \wedge \frac{\partial}{\partial t}\Big[(d\beta_t)^k\Big]$$

Since $\frac{\partial}{\partial t}\Big[(d\beta_t)^k\Big]\Big|_{\partial_1 X} = 0$, this formula collapses to $\frac{\partial}{\partial t}\Big[\beta_t\Big] \wedge (d\beta_t)^k\Big|_{\partial_1 X}$.

Replacing $\frac{\partial}{\partial t}\Big[\beta_t\Big]\Big|_{\partial_1 X}$ with $d\eta_t$ once more, we get

$$\frac{\partial}{\partial t}\Big[\beta_t\Big] \wedge (d\beta_t)^k\Big|_{\partial_1 X} = d\eta_t \wedge (d\beta_t)^k\Big|_{\partial_1 X} = d\Big[\eta_t \cdot (d\beta_t)^k\Big]\Big|_{\partial_1 X} = d\zeta_t.$$

Thus, (3.60) is valid.

If the functions η_t are such that $\eta_t|_{\partial_{j+1}X(v_t)} = 0$, then $\zeta_t|_{\partial_{j+1}X(v_t)} = 0$; so the property (3.62) holds as well. □

Proposition 3.9. *Let $\mathcal{B} = \{\beta_t\}_{t\in[0,1]}$ be a variationally boundary $(2n+1-j)$-exact family of contact forms on a compact connected smooth $(2n+1)$-manifold X. Assume that their Reeb vector fields v_{β_t} are boundary generic for $t = 0, 1$, and for almost all $t \in (0, 1)$.*

- *Then the integral $\kappa_j(\beta_t)$ ($j \equiv 0 \mod 2$) from (3.59) and $\kappa^+_{j-1}(\beta_t)$ ($j \equiv 0 \mod 2$) from (3.58) are t-independent. If, in addition, the property (3.62) holds, then $\kappa^+_j(\beta_t)$ ($j \equiv 0 \mod 2$) is also invariant within the family $\mathcal{B}$.*
- *If $\mathcal{B} = \{\beta_t\}_{t\in[0,1]}$ is variationally boundary exact family, then $\{\kappa_j(\beta_t)\}$ are $\mathcal{B}$-invariant for all even $j \in [2, 2n]$. If in addition, (3.62) holds for the functions η_t, then $\kappa^+_j(\beta_t)$ ($j \equiv 0 \mod 2$) is $\mathcal{B}$-invariant.*

Proof. Let d be the exterior derivative operator on $X \times [0, 1]$ and d_t the exterior derivative operator on a typical slice $X \times \{t\}$. Let $v_t =_{\mathsf{def}} v_{\beta_t}$.

Consider the following sets:

$$\mathcal{Y}_j =_{\mathsf{def}} \bigcup_{t\in[0,1]} \partial_j X(v_t) \times \{t\} \subset X \times [0, 1],$$

$$\mathcal{Y}^+_j =_{\mathsf{def}} \bigcup_{t\in[0,1]} \partial^+_j X(v_t) \times \{t\} \subset X \times [0, 1],$$

Although $\mathcal{Y}_j$ may have singularities, located in some t-slices, by the hypothesis, such special t's have measure zero in $[0, 1]$; so one can integrate a smooth differential form against the rectifiable chain $\mathcal{Y}_j$.

We start with an even j. Put $s =_{\mathsf{def}} \frac{2n-j}{2}$. Then

$$\kappa_j(\beta_1) - \kappa_j(\beta_0) =_{\mathsf{def}} \int_{\partial_j X(v_1)} \beta_1 \wedge (d_1\beta_1)^s - \int_{\partial_j X(v_0)} \beta_0 \wedge (d_0\beta_0)^s$$

$$\overset{Stokes}{=} \int_{\mathcal{Y}_j} \mathsf{d}\big(\beta_t \wedge (d_t\beta_t)^s\big) = \int_{\mathcal{Y}_j} (d_t\beta_t)^{s+1} + \int_{\mathcal{Y}_j} \frac{\partial}{\partial t}\Big[\beta_t \wedge (d_t\beta_t)^s\Big] \wedge dt$$

Each of these two integrals vanishes. Indeed, dt is not present in the $(2s+2)$-form $(d_t\beta_t)^{s+1}$. Thus its restriction to the $(2s+2)$-chain $\mathcal{Y}_j \subset$

$\partial_1 X \times [0,1]$ vanishes. The second integral is zero as well by the variationally boundary $(2n+1-j)$-exact hypothesis:

$$\int_{\mathcal{Y}_j} \frac{\partial}{\partial t}\big(\beta_t \wedge (d_t\beta_t)^s\big) \wedge dt \stackrel{Fubini}{=} \int_{[0,1]} \Big(\int_{\partial_j X(v_{\beta_t})} \frac{\partial}{\partial t}\Big[\beta_t \wedge (d_t\beta_t)^s\Big]\Big) dt$$

$$= \int_{[0,1]} \Big(\int_{\partial_j X(v_{\beta_t})} d_t\zeta_t \Big) dt = 0,$$

since integrating the exact form $d_t\zeta_t$ over the cycle $\partial_j X(v_{\beta_t}) \subset \partial_1 X$ results in zero. Therefore, $\kappa_j(\beta_1) = \kappa_j(\beta_0)$ for this particular even $j \in [2, 2n]$.

By Lemma 3.34, the second bullet's claim follows for all *even* $j \in [2, 2n]$.

Using Stokes' theorem identity $\kappa_j(\beta) = \kappa^+_{j-1}(\beta)$ $(j \equiv 0 \mod 2)$, we conclude that, for *odd* $j \in [1, 2n-1]$, the quantities $\{\kappa^+_j(\beta)\}_j$ are $\mathcal{B}$-deformation invariants as well.

If we add property (3.62) to property (3.60), then a similar computation shows that $\kappa^+_j(\beta_1) = \kappa^+_j(\beta_0)$ for an even j. This time, Stokes' theorem is applied to $\mathcal{Y}^+_j$. Its boundary consists of three parts: $\partial^+_j X(v_0), \partial^+_j X(v_1)$, and their complement $\delta\mathcal{Y}^+_j = \mathcal{Y}_{j+1}$. Therefore,

$$\kappa^+_j(\beta_1) - \kappa^+_j(\beta_0) = \int_{\mathcal{Y}^+_j} \mathsf{d}\big(\beta_t \wedge (d_t\beta_t)^s\big) \pm \int_{\delta\mathcal{Y}^+_j} \beta_t \wedge (d_t\beta_t)^s$$

$$= \int_{\mathcal{Y}^+_j} (d_t\beta_t)^{s+1} + \int_{\mathcal{Y}^+_j} \frac{\partial}{\partial t}\Big[\beta_t \wedge (d_t\beta_t)^s\Big] \wedge dt \pm \int_{\delta\mathcal{Y}^+_j} \beta_t \wedge (d_t\beta_t)^s.$$

Since the form dt is not present in the first and third integrals, they vanish. As a result,

$$\kappa^+_j(\beta_1) - \kappa^+_j(\beta_0) = \int_{[0,1]} \Big(\int_{\partial^+_j X(v_{\beta_t})} \frac{\partial}{\partial t}\Big[\beta_t \wedge (d_t\beta_t)^s\Big]\Big) dt$$

$$= \int_{[0,1]} \Big(\int_{\partial^+_j X(v_{\beta_t})} d_t\zeta_t \Big) dt \stackrel{Stokes}{=} \int_{[0,1]} \Big(\int_{\partial_{j+1} X(v_{\beta_t})} \zeta_t \Big) dt = 0$$

by (3.62). □

Consider now the following quantities (see Definition 3.26):

$$\begin{aligned}
\mathcal{K}^+_j(\xi) &=_{\mathsf{def}} \inf_{\{\beta \rightsquigarrow \xi \mid \mathsf{av}_{\mathcal{R}}(\beta)=1\}} \big\{|\kappa^+_j(\beta)|\big\} \quad \text{for an odd } j \in [1, 2n-1],\\
\mathcal{K}^+_j(\xi) &=_{\mathsf{def}} \inf_{\{\beta \rightsquigarrow \xi \mid \mathsf{av}_{\mathcal{R}}(\beta)=1\}} \big\{|\kappa^+_j(\beta)|\big\} \quad \text{for an even } j \in [2, 2n],\\
\mathcal{K}_j(\xi) &=_{\mathsf{def}} \inf_{\{\beta \rightsquigarrow \xi \mid \mathsf{av}_{\mathcal{R}}(\beta)=1\}} \big\{|\kappa_j(\beta)|\big\} \quad \text{for an even } j \in [2, 2n]. \quad (3.63)
\end{aligned}$$

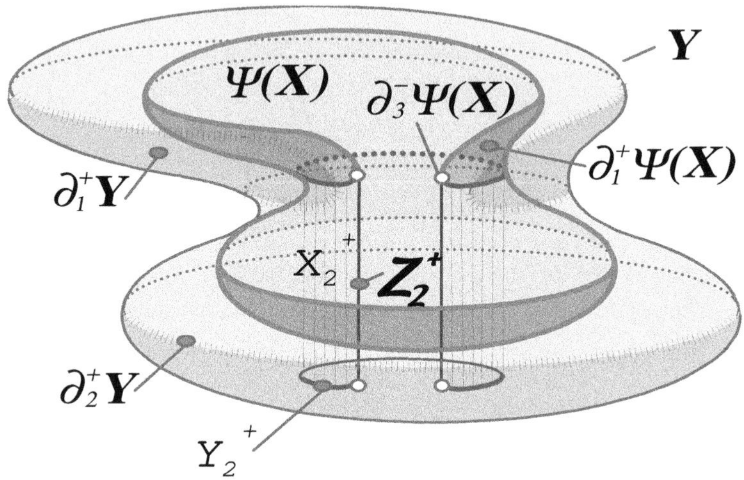

Fig. 3.8 By Theorem 3.22, $\kappa_1^+(\beta_X) = \kappa_2(\beta_X) = -\int_{\mathcal{Y}_1^+} d\beta_Y$. The figure illustrates the geometry of 3-folds X and Y and of an embedding $\Psi : X \to Y$ that leads to this formula. Although the relevant for the formula loci are: the surface $\Psi(\partial_1^+ X)$, the solid $\mathcal{X}_1^+ \subset Y$, and the surface $\mathcal{Y}_1^+ \subset \partial_1^+ Y$, they are not shown in the figure due to "artistic limitations"; instead, we have chosen to show similar loci of one dimension lower: the curve $\Psi(\partial_2^+ X)$, the surface $\mathcal{X}_2^+$, and the curve $\mathcal{Y}_2^+$.

They measure how "wrinkled" the Morse stratification $\{\partial_j^{\pm} X(v_\beta)\}_j$ could be with respect to a given contact structure $\xi = \mathsf{ker}\beta$ on X. As a practical matter, the computation of $\{\mathcal{K}_j^+(\xi)\}_j, \{\mathcal{K}_j(\xi)\}_j$ could be quite challenging...

Theorem 3.22. (Holography for the κ_j^+-volumes under the contact embeddings)

Let X and Y be compact smooth $(2n+1)$-manifolds, equipped with contact forms β_X, β_Y and their Reeb vector fields v_X, v_Y, respectively. We assume that v_X and v_Y are boundary generic and v_Y is traversing. Let $\Psi : X \hookrightarrow \mathsf{int}(Y)$ be a smooth embedding such that $\beta_X = \Psi^*(\beta_Y)$. For $j \in [1, 2n+1]$, consider the $(-v_Y)$-guided projections

$$\Pi_Y : \Psi\big(\partial_j^+ X(v_X)\big) \to \partial_1^+ Y(v_Y)$$

and denote their images by $\mathcal{Y}_j^+$.[17]

Then the integrals $\{\kappa_j^+(\beta_X)\}_j$ from (3.58) and $\{\kappa_j(\beta_X)\}_j$ from (3.59) can be recovered from the Π_Y-projections $\{\mathcal{Y}_j^+\}_j$ and from the knowledge

[17]Here, as before, we assume that $\Psi(X)$ is "semi-transparent" for the $(-v_Y)$-flow.

of the contact form $\beta_Y^{\partial_+} =_{\mathsf{def}} \beta_Y||_{\partial_1^+ Y(v_Y)}$. *Specifically, for an odd* $j \in [1, 2n-1]$,

$$\kappa_j^+(\beta_X) = -\int_{\mathcal{Y}_j^+} (d\beta_Y)^{\frac{2n+1-j}{2}}$$

and, for an even $j \in [2, 2n]$,

$$\kappa_j(\beta_X) = -\int_{\mathcal{Y}_{j-1}^+} (d\beta_Y)^{\frac{2n+2-j}{2}}.$$

Proof. We consider the union $\mathcal{X}_j^+$ of downward $(-v_Y)$-directed trajectories through the points of $\Psi(\partial_j^+(X)(v_X))$. Note that $\dim(\mathcal{X}_j^+) = 2n+2-j$.

Let

$$\mathcal{Y}_j^+ =_{\mathsf{def}} \mathcal{X}_j^+ \cap \partial_1^+ Y(v_Y)$$

—the Π_Y-image of $\partial_j^+ X(v_X)$—, and let (see Fig. 3.8)

$$\mathcal{Z}_j^+ =_{\mathsf{def}} \partial(\mathcal{X}_j^+) \setminus \big(\Psi\big(\partial_j^+ X(v_X)\big) \cup \mathcal{Y}_j^+\big).$$

We claim that, for an odd j, $\int_{\mathcal{Z}_j^+} (d\beta_Y)^{\frac{2n+1-j}{2}} = 0$ since $v_Y \in \mathsf{ker}(d\beta_Y)$ and $\mathcal{Z}_j^+$ of dimension $2n+1-j$ consists of portions of v_Y-trajectories.

(a) For an odd $j \in [1, 2n-1]$, we get

$$\kappa_j^+(\beta_X) = \int_{\partial_j^+ X(v_X)} (d\beta_X)^{\frac{2n+1-j}{2}} = \int_{\partial_j^+ X(v_X)} \Psi^*(d\beta_Y)^{\frac{2n+1-j}{2}} =$$

$$= \int_{\Psi\left(\partial_j^+(X)(v_X)\right)} (d\beta_Y)^{\frac{2n+1-j}{2}}.$$

By Stokes' theorem, applied to $\mathcal{X}_j^+$, and using that $d\big((d\beta_Y)^{\frac{2n+1-j}{2}}\big) = 0$ and that $\int_{\mathcal{Z}_j^+} (d\beta_Y)^{\frac{2n+1-j}{2}} = 0$, we get

$$\kappa_j^+(\beta_X) = \int_{\Psi\left(\partial_j^+(X)(v_X)\right)} (d\beta_Y)^{\frac{2n+1-j}{2}} = -\int_{\mathcal{Y}_j^+} (d\beta_Y)^{\frac{2n+1-j}{2}}.$$

This validates the claim of Theorem 3.22 for odd j's.

(b) For an even $j \in [2, 2n]$, by Stokes' theorem, we get

$$\kappa_j(\beta_X) = \int_{\partial_j X(v_X)} \beta_X \wedge (d\beta_X)^{\frac{2n-j}{2}} = \int_{\partial_{j-1}^+ X(v_X)} (d\beta_X)^{\frac{2n+2-j}{2}}$$

$$= \int_{\partial_{j-1}^+ X(v_X)} \Psi^*\big((d\beta_Y)^{\frac{2n+2-j}{2}}\big) = \int_{\Psi\left(\partial_{j-1}^+(X)(v_X)\right)} (d\beta_Y)^{\frac{2n+2-j}{2}}.$$

Now we are facing the situation, described in the case (a). Therefore,

$$\kappa_j(\beta_X) = -\int_{\mathcal{Y}^+_{j-1}} (d\beta_Y)^{\frac{2n+2-j}{2}},$$

and thus is determined by $\mathcal{Y}^+_{j-1}$—the Π_Y-projection of the locus $\Psi(\partial^+_{j-1}(X)(v_X))$. □

Example 3.17. Let $\dim X = \dim Y = 5$. We adopt the notations and hypotheses of Theorem 3.22. Then $\kappa_1(\beta_X) = 0 = \kappa_3(\beta_X)$ and

$$\kappa_1^+(\beta_X) = -\int_{\mathcal{Y}_1^+} (d\beta_Y)^2 = \kappa_2(\beta_X), \quad \kappa_3^+(\beta_X) = -\int_{\mathcal{Y}_3^+} d\beta_Y = \kappa_4(\beta_X).$$

In contrast, we do not know whether $\kappa_2^+(\beta_X)$ and $\kappa_4^+(\beta_X)$ can be recovered in terms of the Π_Y-generated boundary data residing in $\partial_1 Y$. ◇

Combining Proposition 3.9 with Theorem 3.22 leads to the following claim.

Corollary 3.33. *Let X and (Y, β_Y) be as in Theorem 3.22. Consider a t-family $\{\Psi_t : X \hookrightarrow \mathsf{int}(Y)\}_{t\in[0,1]}$ of orientation-preserving contact embeddings such that, for an even $j \in [2, 2n]$, the t-family of forms $\{\Psi_t^*(\beta_Y \wedge (d\beta_Y)^{\frac{2n-j}{2}})\}$ is variationally boundary $(2n+1-j)$-exact in the sense of Definition 3.28.*

Then

$$\int_{\mathcal{Y}^+_{j-1}(t)} (d\beta_Y)^{\frac{2n+2-j}{2}} = -\kappa_j(\Psi_t^*(\beta_Y)) = -\kappa^+_{j-1}(\Psi_t^*(\beta_Y))$$

is t-independent. ◇

3.8.7 *Shadows of Legandrian submanifolds of (X, β) on the screen $\partial_1^+ X(v_\beta)$ and on the trajectory space $\mathcal{T}(v_\beta)$*

Let $\dim X = 2n+1$. We consider a cooriented contact structure (X, ξ), its contact form β, and its traversing (boundary generic) Reeb vector field v_β.

Recall that $L \subset X$ is sub-Legendrian manifold, if $\beta|_L \equiv 0$; that is, if the tangent space T_xL is contained in the distribution ξ_β.

Remark 3.23. Certain properties of a contact pair (X, β_X) are inherited by any contact pair (Y, β_Y) that incapsulates (X, β_X) so that $\beta_Y|_X = \beta_X$. For example, if (X, β_X), $\dim X = 2n+1$, contains a Legandrian sphere S^n or a Legandrian torus T^n, then so does (Y, β_Y). If (X, β_X) contains a

Legandrian sphere S^n that bounds an embedded ball D^{n+1}, then so does (Y, β_Y). In particular, if a 3-fold (X, β_X) contains a Legandrian loop L that bounds a 2-ball $B^2 \subset X$ (such β_X is called overtwisted [E]), then any contact 3-fold (Y, β_Y) that incapsulates (X, β_X) is also overtwisted. $\diamondsuit$

Our next goal is to describe the images $L^\dagger$ of sub-Legendrian compact submanifolds $L \subset X$ ($\dim(L) \leq n$) both in $\partial_1^+ X(v_\beta)$ under the $(-v_\beta)$-flow and in the trajectory space $\mathcal{T}(v_\beta)$ under the obvious quotient map $\Gamma : X \to \mathcal{T}(v_\beta)$. Note that these shadows of L do not change when we apply a t-family (t being sufficiently small) of local diffeomorphisms ϕ^t (as in Definition 3.21), generated by the Reeb vector field v_β. Since β is invariant under ϕ^t, we conclude that $\phi^t(L)$ is also sub-Legendrian.

We denote by $\mathcal{I}_x(d\beta)$, $x \in X$, an isotropic space of $d\beta|_{\xi_x}$, that is, a maximal vector subspace of $\mathsf{ker}_x(\beta)$ such that $d\beta(w_1, w_2) = 0$ for any $w_1, w_2 \in \mathcal{I}_x(d\beta) \subset \mathsf{ker}_x(\beta)$. For a contact form β, $\dim(\mathcal{I}_x(d\beta)) = n$. The tangent space of a Legandrian manifold L is isotropic in relation to $d\beta$ since $\beta|_L = 0$, which implies that $d\beta|_L = 0$.

By Lemma 3.23, $\mathsf{int}(\partial_1^+ X(v_\beta))$ is an open symplectic manifold with the help of symplectic form $\omega^\dagger =_{\mathsf{def}} d\beta|_{\partial_1^+ X(v_\beta)}$. Recall that the form $\omega^\dagger$ degenerates along the boundary $\partial_2 X(v_\beta) = \partial(\partial_1^+ X(v_\beta))$, since there $v_\beta \in \mathsf{ker}(d\beta) \cap T(\partial_1 X)$.

Let $\mathcal{X}_2(v_\beta)$ denote the union of the downward $(-v_\beta)$-trajectories through the locus $\partial_2 X(v_\beta)$, and $\mathsf{L}(v_\beta)$ the union of the downward $(-v_\beta)$-trajectories through the points of a sub-Legandrian submanifold $L \subset X$.

Put $L^\dagger =_{\mathsf{def}} \mathsf{L}(v_\beta) \cap \partial_1^+ X(v_\beta)$ (see Fig. 3.9).

Definition 3.29. We say that a sub-Legandrian manifold $L \subset X$ is in a weakly general position to the v_β-flow, if $L \setminus (L \cap \mathcal{X}_2(v_\beta))$ is dense in L. $\diamondsuit$

Lemma 3.35. *The locus* $\mathsf{int}(L^\dagger) = L^\dagger \setminus (L^\dagger \cap \mathcal{X}_2(v_\beta))$ *is an open (possibly disconnected)* Lagrangian *submanifold of the open* $(2n)$*-manifold*

$$\partial_1^+ X(v_\beta) \setminus (\mathcal{X}_2(v_\beta) \cap \partial_1^+ X(v_\beta)).$$

That is, for any $x \in \mathsf{int}(L^\dagger)$, *the tangent space* $T_x(\mathsf{int}(L^\dagger)) \subset \mathcal{I}_x(\omega^\dagger)$, *an isotropic subspace of the symplectic form* $\omega^\dagger = d\beta|_{\mathsf{int}(\partial_1^+ X(v_\beta))}$ *at* x.

Proof. By Lemma 3.23, the 2-form $\omega^\dagger = d\beta|_{\partial_1^+ X(v_\beta)}$ is symplectic on $\mathsf{int}(\partial_1^+ X(v_\beta))$. Since $d\beta$ is v_β-invariant, for any $x \notin \partial_2 X(v_\beta)$ and for any

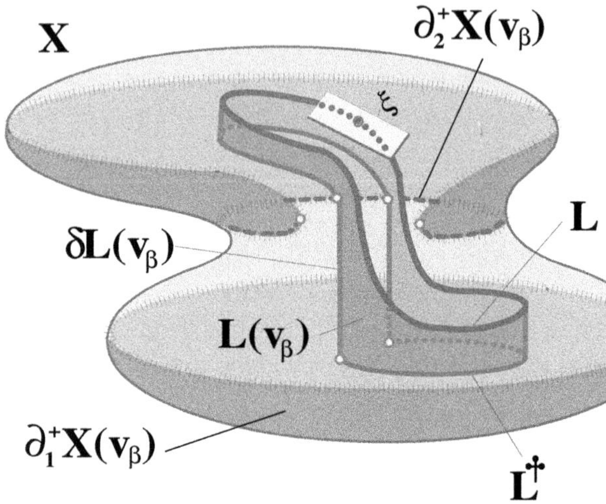

Fig. 3.9 The $(-v_\beta)$-shadow $L^\dagger \subset \partial_1^+ X(v_\beta)$ of a closed Legandrian manifold $L \subset X$, shown as a bold loop in the interior of the bulk X. In the figure, $L^\dagger$ consists of two arcs. The surface $\mathsf{L}(v_\beta)$ consists of $(-v_\beta)$-trajectories falling down from L. Note the locus $\delta\mathsf{L}(v_\beta) \subset X$ that consists of $(-v_\beta)$-trajectories falling down from $L^\dagger \cap \partial_2^+ X(v_\beta)$.

isotropic space $\mathcal{I}_x(d\beta) \subset \mathsf{ker}_x(\beta)$, its isomorphic image under the differential of the $(-v_\beta)$ projection on $\mathsf{int}(\partial_1^+ X(v_\beta))$, is an isotropic space for the 2-form $\omega^\dagger$. Therefore, $\mathsf{int}(L^\dagger)$ is a Lagrangian submanifold of the open manifold $\partial_1^+ X(v_\beta) \setminus (\mathcal{X}_2(v_\beta) \cap \partial_1^+ X(v_\beta))$. □

Remark 3.24. Since L is embedded in X and v_β is transversal to ξ, $\mathsf{int}(L^\dagger)$ is *immersed* in the open $(2n)$-manifold $\partial_1^+ X(v_\beta) \setminus (\mathcal{X}_2(v_\beta) \cap \partial_1^+ X(v_\beta))$. ♢

Lemma 3.36. *Let γ be a Reeb trajectory through a point $a_\star \in X$, which hits $\partial_1^+ X(v_\beta)$ transversally at a point $b_\star$. Let $L^\dagger$ be a given Lagrangian submanifold (with respect to the symplectic form $\omega^\dagger =_{\mathsf{def}} d\beta|_{\mathsf{int}(\partial_1^+ X(v_\beta))}$) of $\partial_1^+ X(v_\beta)$, which contains $b_\star$. Then, in the vicinity of $a_\star$, there exists a unique germ of Legendrian submanilold $L \subset X$ that contains $a_\star$ and whose $(-v_\beta)$-governed projection is the germ of $L^\dagger$ at $b_\star$.*

Proof. In the vicinity of $a_\star$, by the Darboux theorem, we can pick coordinates $(z, x_1, y_1, \ldots, x_n, y_n)$ such that: they vanish at $a_\star$, $\beta = dz + \sum_{i=1}^n x_i\, dy_i$, $d\beta = \sum_{i=1}^n dx_i \wedge dy_i$, and $v_\beta = \partial_z$. In these coordinates, the germ at $a_\star$ of the Legendrian submanilold L in question is given, in a parametric form, by some smooth functions $z(\vec{t})$, $x_1(\vec{t})$, $y_1(\vec{t})$, $\dots$, $x_n(\vec{t})$, $y_n(\vec{t})$, where $\vec{t} = (t_1, \dots, t_n) \in \mathbb{R}^n$. We assume that the tangent to L vectors

$$\tau_i = \Big(\frac{\partial z}{\partial t_i}, \frac{\partial x_1}{\partial t_i}, \frac{\partial y_1}{\partial t_i}, \dots, \frac{\partial x_n}{\partial t_i}, \frac{\partial y_n}{\partial t_i}\Big),$$

$1 \leq i \leq n$, are linearly independent in the vicinity of $a_\star$. Put

$$\tilde{\tau}_i = \Big(\frac{\partial x_1}{\partial t_i}, \frac{\partial y_1}{\partial t_i}, \dots, \frac{\partial x_n}{\partial t_i}, \frac{\partial y_n}{\partial t_i}\Big).$$

We first will prove the assertion of the lemma for the case when $\partial_1^+ X(v_\beta)$ is equal to the hyperplane $H = \{z = const\}$, where the point $\gamma \cap H$ resides in the interval of γ, bounded by $a_\star$ and $b_\star$.

The property of L being Legendrian can be expressed by the set of equations:

$$\{\beta(\tau_i) = 0\}_{1 \leq i \leq n} = \Big\{-\frac{\partial z}{\partial t_i} = \sum_{q=1}^{n} x_q \frac{\partial y_q}{\partial t_i}\Big\}_{1 \leq i \leq n}, \tag{3.64}$$

If these equations are solvable, then the system

$$\{d\beta(\tau_i, \tau_j) = d\beta(\tilde{\tau}_i, \tilde{\tau}_j) = 0\}_{1 \leq i < j \leq n}, \tag{3.65}$$

must be solvable as well.

We can solve the equations in (3.64) for $z(\vec{t})$, provided that the RHS of equation (3.64) is a gradient vector field whose components are $\Big\{\sum_{q=1}^{n} x_q \frac{\partial y_q}{\partial t_i}\Big\}_{1 \leq i \leq n}$, or that

$$\Big\{\frac{\partial \tilde{\tau}_i}{\partial t_j} = \frac{\partial \tilde{\tau}_j}{\partial t_i}\Big\}_{1 \leq i < j \leq n}. \tag{3.66}$$

Moreover, the solution $z(\vec{t})$ of (3.64) with the property $z(\vec{0}) = a_\star$ will be unique.

By a direct computation, the group of equations (3.65) coincides with the group of equations in (3.66). In turn, the equations in (3.66) claim that $\{d\beta(\tilde{\tau}_i, \tilde{\tau}_j) = 0\}_{1 \leq i \neq j \leq n}$. In other words, they claim that the vectors $\{\tilde{\tau}_i\}_{1 \leq i \leq n}$ belong to some isotropic n-space $\mathcal{I}so(d\beta|_H) \subset H$, so that the n-manifold $L_H^\dagger$, given by the functions $\{x_1(\vec{t}), y_1(\vec{t}), \dots, x_n(\vec{t}), y_n(\vec{t})\}$ is Lagrangian in the hypersurfuce H. The same argument shows that the germ of any Lagrangian submanifold $L_H^\dagger \subset H$, passing through the point $\vec{0} \in H$, is the shadow, under the z-projection, of the germ of unique Legandrian L, containing the origin of $\mathbb{R}^{2n+1}$.

It remains to prove that any germ of a Lagrangian $L^\dagger_H \subset H$ (with respect to $d\beta|_H$) at the point $\gamma \cap H$, with the help of $(-v_\beta)$-flow, projects onto a germ of a Lagrangian $L^\dagger \subset \partial_1^+ X(v_\beta)$ (with respect to $d\beta|_{\partial_1^+ X(v_\beta) \setminus \partial_2 X(v_\beta)})$ at the point $b_\star = \gamma \cap \partial_1^+ X(v_\beta)$. The latter observation follows from the transversality of γ to $\partial_1^+ X(v_\beta)$ at $b_\star$ and the v_β-invariance of the form $d\beta$. □

For a given Lagrangian submanifold $L^\dagger \subset \mathsf{int}(\partial_1^+ X(v_\beta))$, consider the (discontinuous) function $\Psi^\beta_{L^\dagger} : L^\dagger \to \mathbb{R}_+$, defined by

$$\Psi^\beta_{L^\dagger}(x) = \int_{[x,\, C_{v_\beta}(x)]} \beta,$$

where C_{v_β} is the causality map. Note that, for a connected component $L^\ddagger$ of $L^\dagger \subset \partial_1^+ X(v_\beta) \setminus (\mathcal{X}_2(v_\beta) \cap \partial_1^+ X(v_\beta))$, the function $\Psi^\beta_{L^\ddagger} = \Psi^\beta_{L^\dagger}\big|_{L^\ddagger}$ is continuous and smooth.

Put $\mathcal{D}^\beta_{L^\ddagger} =_{\mathsf{def}} \sup_{x \in L^\ddagger} \{\Psi^\beta_{L^\ddagger}(x)\}$. One may think of this number as the "Reeb elevation" of the component $L^\ddagger$. Evidently, $\mathsf{diam}_{\mathcal{R}}(\beta) \geq \mathcal{D}^\beta_{L^\ddagger}$.

Let us revisit a given Legandrian manifold $L \subset X$ whose $(-v_\beta)$-projection is $L^\dagger$. For any $x \in \mathsf{int}(L^\dagger)$, take the first point $y(x) \in \gamma_x$ that belongs to L and consider the function

$$\psi^\beta_L(x) =_{\mathsf{def}} \int_{[x, y(x)]} \beta,$$

where $[x, y(x)] \subset \gamma_x$ is the segment of the trajectory γ_x, bounded by x and $y(x)$.

Evidently, $\psi^\beta_L(x) \leq \Psi^\beta_{L^\ddagger}(x)$ for any $x \in L^\ddagger \subset L^\dagger$.

Corollary 3.34. *Let β be a contact form whose Reeb vector field v_β is traversing and boundary generic. Let $L \subset X$ be a Legandrian submanifold in weakly general position with "the waterfall" $\mathcal{X}_2(v_\beta)$, the set of $(-v_\beta)$-trajectories through the locus $\partial_2 X(v_\beta)$. Let $L^\dagger$ be its $(-v_\beta)$-guided projection on $\partial_1^+ X(v_\beta)$.*

Consider the connected components $\{L^\ddagger_\kappa\}_\kappa$ of the Lagrangian manifold $\mathsf{int}(L^\dagger)$, immersed in $\partial_1^+ X(v_\beta) \setminus (\mathcal{X}_2(v_\beta) \cap \partial_1^+ X(v_\beta))$. In each component $L^\ddagger_\kappa$ we pick a base point b_κ, distinct from the points of self-intersection of $L^\dagger$.

Then the knowledge of the shadow $L^\dagger$, together with the base points $\{b_\kappa\}_\kappa$ and set of values $\{\psi^\beta_L(b_\kappa)\}_\kappa$, makes it possible to reconstruct the Legandrian embedding $L \subset X$.

Proof. The reconstruction of L resembles the process of analytic continuation.

Let $\pi_{v_\beta} : L \to L^\dagger$ be the $(-v_\beta)$-guided projection. Let $L^\dagger_\times \subset L^\dagger$ denotes the subset consisting of points that have multiple π_{v_β}-preimages in L, and let $\mathcal{L}^\dagger_2 =_{\mathsf{def}} L^\dagger \cap \mathcal{X}_2(v_\beta)$. By the hypothesis, the set $L^\dagger \setminus (L^\dagger_\times \cup \mathcal{L}^\dagger_2)$ is open and dense in $L^\dagger$.

Knowing the point $b_\kappa \in L^\ddagger_\kappa$ makes it possible, with the help of the value $\psi^\beta_L(b_\kappa)$, to locate a unique point $a_\kappa \in L \cap \gamma_{b_\kappa}$. There is a unique germ L_κ of L in the vicinity of a_κ that projects onto the germ of $L^\ddagger_\kappa$ at b_κ. Let $U^\dagger_\kappa$ be an open neighborhood of b_κ in $L^\dagger$ over which the reconstruction of L is achieved.

Take a point $x \in L^\dagger$ that belongs to the boundary of $U^\dagger_\kappa$ and such that $x \notin L^\dagger_\times \cup \mathcal{L}^\dagger_2$. Then by Lemma 3.36, there exists a unique lift $L^\bullet$ to X of $L^\dagger$ over some open neighborhood $U_x \subset L^\dagger$ of x, such that the point $y = \gamma_x \cap L$ belongs to $L^\bullet$. Since, over $U_x \cap U^\dagger_\kappa$, both $L^\bullet$ and L are the lifts of $L^\dagger$ that share the point y, again by Lemma 3.36, we conclude that $L^\bullet = L$ over U_x. As a result, the reconstruction of L over $U_x \cup U^\dagger_\kappa$ is available. Therefore, the maximal open set in $L^\dagger$ over which the lift of $L^\dagger$, subject to the constraint "a_κ belongs to the lift", is possible is a connected component $L^\ddagger$ of the open set $L^\dagger \setminus (L^\dagger_\times \cup \mathcal{L}^\dagger_2)$. Since L is in general position with $\mathcal{X}_2(v_\beta)$, the union of such lifts, taken over all κ's, is a dense subset $L \setminus (L \cap \pi^{-1}_{v_\beta}(\mathcal{L}^\dagger_2 \cup L^\dagger_\times))$ of L. Therefore, taking the closure of this union, produces the given $L \subset X$. □

Remark 3.25. Note that the very existence of Legandrian $L \subset X$ insured that the assembly instructions in the proof of Corollary 3.34 were consistent.

Even when the Reeb vector field v_β is convex (so the waterfall $\mathcal{X}_2(v_\beta) = \partial_2^- X(v_\beta)$) and the v_β-trajectories γ_x depend continuously on $x \in \partial_1^+ X(v_\beta)$, we do not know whether any closed Lagrangian submanifold $L^\dagger \subset \partial_1^+ X(v_\beta)$ is the π_{v_β}-shadow of a Legandrian manifold $L \subset X$. Corollary 3.34 claims only that, if such L exists, then choosing a point $a_\star \in L$ makes it unique.

It seems that to tackle this question we need to be able to estimate from above the function $\Psi^\beta_{L^\dagger}(x)$ (or at least the Reeb elevation $\mathsf{elev}_{\mathcal{R}}(L^\dagger)$ of $L^\dagger$) in terms of the form β in the vicinity of $L^\dagger \subset \partial_1^+ X(v_\beta)$. ♢

Proposition 3.10. *Let β be a contact form on X and v_β its traversing boundary generic Reeb vector field. Assume that Property* A *from Definition 3.2 is valid. Let $\phi : (L, \partial L) \hookrightarrow (X, \partial_1 X)$ be a Legandrian embedding of a compact manifold L.*

Then the following boundary-confined data:

- *the causality map* $C_{v_\beta} : \partial_1^+ X(v_\beta) \to \partial_1^- X(v_\beta)$,
- *the restriction* $(f^\bullet)^\partial : \partial_1 X \to \mathbb{R}$ *of the Lyapunov function* $f^\bullet : X \to \mathbb{R}$ *such that* $df^\bullet(v_\beta) = 1$,
- *the restriction* $\beta^\partial = \beta||_{\partial_1 X}$ *of the contact form* β,
- *the Lagrangian (with respect to the symplectic form* $d\beta|_{\mathsf{int}(\partial_1^+ X(v_\beta))}$*) immersion*

$$\mathsf{int}(L^\dagger) \subset \partial_1^+ X(v_\beta) \setminus (\mathcal{X}_2(v_\beta) \cap \partial_1^+ X(v_\beta)),$$

- *the set of base points* $\{b_\kappa\}_\kappa$ *and the values* $\{\psi_L^\beta(b_\kappa)\}_\kappa$ *as in Corollary 3.34,*

allow to reconstruct the Legandrian embedding $\phi : L \hookrightarrow X$, *up to a diffeomorphism of* X *that is the identity on* $\partial_1 X$.

Proof. By Corollary 3.29, the hypotheses in the first three bullets allow for a reconstruction of $(X, v_\beta, f^\bullet, \beta)$ up to a diffeomorphism that is the identity on $\partial_1 X$. With this quadruple in place, the base points $\{b_\kappa \in \mathsf{int}(L^\dagger)\}_\kappa$ and the values $\{\psi_L^\beta(b_\kappa)\}_\kappa$, determine uniquely a point $y(b_\kappa) \in L$ on each trajectory γ_{b_κ}. By Corollary 3.34, the assertion follows. □

Lemma 3.37. *Let* X *be a compact connected smooth* $2n+1$*-dimensional manifold,* β *a contact form, and* v_β *its traversing boundary generic Reeb vector field.*

Let $L \subset X$ *be a sub-Legandrian* $(2k+1)$*-dimensional closed submanifold of* X, *where* $2k+1 \leq n$. *We assume that* L *is in weakly general position with respect to the waterfall* $\mathcal{X}_2(v_\beta)$. *We denote by* $\mathsf{L}(v_\beta)$ *the set of downward* $(-v_\beta)$*-trajectories originating in* L, *and by* $\delta\mathsf{L}(v_\beta)$ *the set of downward trajectories, originating in* L *and hitting the locus* $\partial_2^+ X(v_\beta)$ *(see Fig. 3.9).*

Then, for any $k > 0$, *we get*

$$\int_{\mathsf{L}(v_\beta)} (d\beta)^{k+1} = 0, \qquad \text{and} \qquad \int_{L^\dagger} \beta \wedge (d\beta)^k = 0.$$

In contrast, for $k = 0$, *we get*

$$\int_{\mathsf{L}(v_\beta)} d\beta = 0, \qquad \text{and} \qquad \int_{L^\dagger} \beta = -\int_{\delta\mathsf{L}(v_\beta)} \beta \leq 0.$$

Proof. Since $T_x L \subset \mathsf{ker}\beta$ we see that $\beta \wedge (d\beta)^k|_L = 0$. Thus, $\int_L \beta \wedge (d\beta)^k = 0$.

Note that, at a generic point $x \in \mathsf{L}(v_\beta)$, the tangent space $T_x(\mathsf{L}(v_\beta))$ is spanned by $v_\beta(x)$ and the image of a tangent space $T_y L$ under $(-v_\beta)$-flow that takes $y \in L$ to x. Then, $(d\beta)^{k+1}|_{T_x(\mathsf{L}(v_\beta))} = 0$, since $v_\beta \in \mathsf{ker}(d\beta)$. Therefore, $\int_{\mathsf{L}(v_\beta)} (d\beta)^{k+1} = 0$.

Let $\delta\mathsf{L}(v_\beta)$ be the set of downward trajectories originating in L and intersecting $\partial_2^+ X(v_\beta)$ (see Fig. 3.9). Again, at a generic point $x \in \delta\mathsf{L}(v_\beta)$, the tangent space $T_x(\delta\mathsf{L}(v_\beta))$ is spanned by $v_\beta(x)$ and a $2k$-dimensional subspace $\mathcal{K}_x$ of $T_x L$. Thus, for $k > 0$, we get $\beta \wedge (d\beta)^k|_{T_x(\delta\mathsf{L}(v_\beta))} = 0$ since $v_\beta \in \mathsf{ker}\beta$ and $d\beta|_{\mathcal{K}_x} = 0$. Therefore, $\int_{\delta\mathsf{L}(v_\beta)} \beta \wedge (d\beta)^k = 0$.

The boundary $\partial(\mathsf{L}(v_\beta)) = \delta\mathsf{L}(v_\beta) \cup L^\dagger \cup -L$. In fact, $L^\dagger$, away from the set $\delta\mathsf{L}(v_\beta) \cap \partial_1^+ X(v_\beta)$ of measure zero, is an immersed submanifold of $\partial_1^+ X(v_\beta)$.

Recall that $d(\beta \wedge (d\beta)^k) = (d\beta)^{k+1}$. Therefore, for $k > 0$, by the Stokes' theorem and using that $\int_L \beta \wedge (d\beta)^k = 0$, $\int_{\delta\mathsf{L}(v_\beta)} \beta \wedge (d\beta)^k = 0$, and $\int_{\mathsf{L}(v_\beta)} (d\beta)^{k+1} = 0$, we conclude that $\int_{\mathsf{L}^\dagger} \beta \wedge (d\beta)^k = 0$ as well.

In the special case $k = 0$, using similar analysis and Stokes' theorem, we get

$$\int_{L^\dagger} \beta = - \int_{\delta\mathsf{L}(v_\beta)} \beta \leq 0$$

since $\beta(v_\beta) = 1$. □

Corollary 3.35. (A holographic property of sub-Legandrian links in relation to a concave boundary)

Let β be a contact form whose Reeb vector field v_β is traversing on X and boundary generic. Let $L \hookrightarrow X$ a sub-Legandrian link[18] *be in weakly general position with respect to the waterfall $\mathcal{X}_2(v_\beta)$. Let $L^\dagger \subset \partial_1^+ X(v_\beta)$ be the image of L under the $(-v_\beta)$-flow guided projection $X \to \partial_1^+ X(v_\beta)$.*

Then $\int_{L^\dagger} \beta < 0$, if and only if, there exists a down-flow trajectory $\gamma_\star$ that originates at L and hits the concavity locus $\partial_2^+ X(v_\beta)$.

On the other hand, if $\int_{L^\dagger} \beta \geq 0$, then $\int_{L^\dagger} \beta = 0$, and there is no such trajectory $\gamma_\star$. In such a case, the $L^\dagger$ is a collection of closed immersed curves.

Proof. Since $\beta(v_\beta) = 1$, $\int_{\delta\mathsf{L}(v_\beta)} \beta > 0$ when $\delta\mathsf{L}(v_\beta)$ has a positive 1-dimensional measure. This is equivalent to the existence of downward trajectory that connects L to $\partial_2^+ X(v_\beta)$.

In the case $k = 0$, by Lemma 3.37, we get that $\int_{\delta\mathsf{L}(v_\beta)} \beta > 0$ if and only if $\int_{L^\dagger} \beta < 0$. □

[18] a collection of embedded disjoint loops

We conclude this section with a few natural questions in hope that our readers may know the answers...

Question 3.6. *Is the Reeb map $\beta \to v_\beta$ from the space of contact forms to the space of non-vanishing vector fields open?* ◇

Question 3.7. *Assuming that a Reeb vector field v_β is traversing, how to differentiate between tight and overtwisted contact structures (see [E], [BEM] for the definition) in terms of boundary data?* ◇

Question 3.8. *Let β be a contact form on a compact manifold X with boundary. Assuming the Reeb vector field v_β is traversing, does any Lagrangian*[19] *submanifold $L^\dagger$ of the open manifold*

$$\partial_1^+ X(v_\beta) \setminus (\mathcal{X}_2(v_\beta) \cap \partial_1^+ X(v_\beta))$$

is the image, under the $(-v_\beta)$-directed projection, of a Legandrian manifold $L \subset X \setminus \mathcal{X}_2(v_\beta)$? The same question, assuming that v_β-flow is convex. ◇

3.9 Contactomorphisms of tangent spherical fibrations

This section may serve as a bridge to Chapter 4.

Consider a smooth manifold M and a Riemmanian metric g on it. The tangent spherical fibration $\pi : S_gM \to M$ generates a contact 1-form β_g on S_gM as described in formula (4.3) and the paragraph that precedes it.

We are going to investigate one particular subgroup of the contactomorphism group of the pair (S_gM, β_g), the subgroup that respects the π-fiber structure.

We start with the following lemma.

Lemma 3.38. *Let X be a smooth manifold. For any smooth integrable k-dimensional distribution ξ on X, there exists a unique k-dimensional foliation $\mathcal{F}$ on X that integrates ξ.*

Proof. The claim of the lemma is a local property. Thus we may assume that ξ is spent by linearly independent smooth vector fields $w_1, \dots, w_k$ with the property $[w_i, w_j] \in \xi$ for all i, j. Our aim is to show that through every point $x \in X$ there exist a unique germ of a k-manifold F that is tangent to ξ.

Let $\gamma_i(x)$ be a unique w_i-trajectory containing a given point x. The uniqueness of $\gamma_i(x)$ is classically based on the C^2-smoothness of w_i.

[19]with respect to the symplectic form $\omega^\dagger =_{\mathsf{def}} d\beta|_{\mathsf{int}(\partial_1^+ X(v_\beta))}$

Put $S_1(x) = \gamma_1(x)$. Let $S_2(x)$ be the germ at x of the smooth surface $\bigcup_{y \in S_1(x)} \gamma_2(y)$. Let $S_3(x) = \bigcup_{y \in S_2(x)} \gamma_3(y)$, and so on, until we form $S_k(x) = \bigcup_{y \in S_{k-1}(x)} \gamma_k(y)$.

If the ξ-integral manifold $F(x)$ through x exists (which it does since ξ is integrable), using that $\{w_i\}$ are tangent to $F(x)$, we conclude that $F(x)$ must contain the uniquely manufactured k-dimensional manifold $S_k(x)$. Therefore, the germ of $F(x)$ must coincide with the germ of $S_k(x)$. □

Definition 3.30. Given a diffeomorphism $\psi : M \to M$ of a Riemannian manifold (M, g), we call the map $D_g^\circ \psi : S_g M \to S_g M$, defined by

$$D_g^\circ \psi : \psi(m, v) \to \big(\psi(m),\ D\psi(v)/\|D\psi(v)\|_g\big),$$

where $\|v\|_g = 1$ and $D\psi$ is the differential of ψ, the g-normalized differential of ψ. ◇

In fact, if ψ is an isometry, then $D_g^\circ \psi$ is a contactomorphism: $(D_g^\circ \psi)^*(\beta_g) = \beta_g$.

Proposition 3.11. *Let (M, g) be a compact smooth Riemannian n-dimensional manifold. Let η_π be the $(n-1)$-distribution on $S_g M$, tangent to the fibers of $\pi : S_g M \to M$.*

- *Any diffeomorphism $\Psi : S_g M \to S_g M$ that preserves the contact form β_g and the distribution η_π must preserve the fibration structure of $\pi : S_g M \to M$ as well. As a result, Ψ produces in a canonical way a diffeomorphism $\psi(\Psi) : M \to M$ so that $\psi(\Psi) \circ \pi = \pi \circ \Psi$.*
- *On the other hand, any g-isometry $\psi : M \to M$ produces a diffeomorphism $D_g^\circ \psi : S_g M \to S_g M$—its normalized differential, which also preserves the contact form β_g and the π-fibration structure, and has the property $\psi \circ \pi = \pi \circ (D_g^\circ \psi)$.*

Proof. Since each spherical π-fiber S_m, $m \in M$, is a Legandrian submanifold, β_g vanishes identically on the subdistribution η_π.

Since $\Psi^*(\beta_g) = \beta_g$ and $\Psi^*(\eta_\pi) = \eta_\pi$ by the hypotheses, the differential $D\Psi$ of Ψ must preserve the $(n-1)$-distribution η_π and $2(n-1)$-distribution ξ_g, the kernel of β_g. Therefore, each image $\Psi(S_m)$ must be tangent to η_π.

Pick a point $x \in S_m$. By Lemma 3.38, $\Psi(S_m)$ must coincide with the unique Legandrian sphere $S_{m'}$ that contains the point $\Psi(x)$ and is tangent to η_π at $\Psi(x)$. Moreover, also by Lemma 3.38, the sphere $S_{m'}$ does not depend on the choice of $x \in S_m$. Thus, we define $\psi(m) = m'$. By its definition, $\pi \circ \psi = \Psi \circ \pi$.

Consider $C^\infty(M)$, the algebra of smooth functions on M, and $C^\infty(S_gM)$, the algebra of smooth functions on S_gM. The pull-back $\pi^*(C^\infty(M)) \approx C^\infty(M)$ is the subalgebra of $C^\infty(S_gM)$. It consists of smooth functions on S_gM that are constant on each π-fiber. On the other hand, using the local product structure of the fibration π, any smooth function on S_gM that is constant on the π-fibers is the pull-back of a smooth function on M. Since Ψ maps fibers to fibers, Ψ defines an isomorphism $\Psi^* : C^\infty(S_gM) \to C^\infty(S_gM)$ which preserves the subalgebra $\pi^*(C^\infty(M))$ invariant. Therefore, $\Psi^*| : \pi^*(C^\infty(M)) \to \pi^*(C^\infty(M))$ defines an isomorphism $\psi^* : C^\infty(M) \to C^\infty(M)$, which can be incorporated in a commutative square diagram whose vertical arrows are the homomorphisms $\pi^* : C^\infty(M) \to C^\infty(S_gM)$.

By a theorem from [Mr], the isomorphism ψ^* is induces by the unique diffeomorphism $\psi : M \to M$. Similarly, the commutative square diagram of algebras implies that $\psi \circ \pi = \pi \circ \Psi$.

On the other hand, any g-isometry $\psi : M \to M$ produces via its differential the diffeomorphism $D_g^\circ\psi : S_gM \to S_gM$ which preserves the contact form β_g and such that $\psi \circ \pi = \pi \circ (D_g^\circ\psi)$. □

Given a compact Riemannian manifold (M, g) and its tangent spherical fibration $\pi : S_gM \to M$, we denote by $\mathsf{Diff}(S_gM, \beta_g, \eta_\pi)$ the group of diffeomorphisms $\Psi : S_gM \to S_gM$ that preserve the contact form β_g and the distribution η_π.

We denote by $\mathsf{Diff}_{\mathsf{S-inv}}(S_gM, \beta_g, \eta_\pi)$ its subgroup that consists of diffeomorphisms that also preserve each spherical π-fiber invariant. We consider also the groups $\mathsf{Diff}^+(S_gM, \beta_g, \eta_\pi)$ and $\mathsf{Diff}^+_{\mathsf{S-inv}}(S_gM, \beta_g, \eta_\pi)$ based on the orientation-preserving diffeomorphisms Ψ.

Proposition 3.12. *The group* $\mathsf{Diff}_{\mathsf{S-inv}}(S_gM, \beta_g, \eta_\pi)$ *consists of a single element, the identity map.*

Proof. Let $a, b \in \mathsf{int}(M)$ be any pair of points that can be connected in M by a single simple geodesic segment γ of some length s such that the lengths of all other geodesic segments that connect a and b exceeds $2s$. We call such geodesic segment γ "short". Let $\tilde\gamma \subset S_gM$ be the π-lift of γ, connecting the point $A = (a, \dot\gamma(a))$ with the point $B = (b, \dot\gamma(b))$. (The length of $\tilde\gamma$ in the Sasaki metric gg is also s). Let us also call such geodesic flow segments $\tilde\gamma$ "short".

Consider the spheres $S_a = \pi^{-1}(a)$ and $S_b = \pi^{-1}(b)$. Let C_a denote the "cylinder" consisting of the v^g-trajectories of length $\leq s$ starting at points of the sphere S_a. Let C_b be the "cylinder" consisting of the $(-v^g)$-trajectories of length $\leq s$ starting at points of the sphere S_b. We claim that $C_a \cap C_b = \tilde{\gamma}$. Indeed, if a point x lies on a v^g-trajectory $\tilde{\gamma}'$ through a point $a' \in S_a$ and on a $(-v^g)$-trajectory $\tilde{\gamma}''$ through a point $b' \in S_b$, then x must belong to a v^g-trajectory $\hat{\gamma} \subset S_g M$ that connects a' to b' and whose length is $\leq 2s$. Using the uniqueness of γ among the trajectories of the length is $\leq 2s$ and the fact that the image of $\hat{\gamma}$ under π connects a to b, we conclude that $\hat{\gamma} = \tilde{\gamma}$.

If a diffeomorphism $\Psi \in \mathsf{Diff}_{\mathsf{S-inv}}(S_g M, \beta_g, \eta_\pi)$, then $\Psi(S_a) = S_a$, $\Psi(S_b) = S_b$, and, since $\Psi(\beta_g) = \beta_g$, it follows that $\Psi_*(v^g) = v^g$ (not point-wise). Therefore, the cylinders C_a and C_b are Ψ-*invariant* and so is their intersection $\tilde{\gamma}$, i.e, $\Psi(\tilde{\gamma}) = \tilde{\gamma}$, as sets. If for a point $x \in \tilde{\gamma}$, $\Psi(x) \neq x$, by the basic property of Ψ, we have that both $\Psi(x)$ and x belong to S_x. However, $\tilde{\gamma} \cap S_x$ may consist of two points at most: any geodesic in M does not have self-intersections of multiplicity ≥ 3. Since γ was assumed not to have self-intersections (a simple curve), we conclude that $\Psi(x) = x$ for all *short* $\tilde{\gamma}$ and all $x \in \tilde{\gamma}$.

Finally, we notice that the union of short $\tilde{\gamma}$'s is dense in $S_g M$ since any point in $\mathsf{int}(S_g M)$ belongs to some short $\tilde{\gamma}$. Therefore, Ψ, being the identity map on a dense subset of $S_g M$, is the identity map of $S_g M$. □

Theorem 3.23. *Let (M, g) be a compact smooth Riemannian n-dimensional manifold, let β_g be the contact form on the tangent spherical fibration $\pi : S_g M \to M$, and let η_π be the distribution, tangent to the π-fibers.*

Then there is a group monomorphism

$$\Pi : \mathsf{Diff}(S_g M, \beta_g, \eta_\pi) \hookrightarrow \mathsf{Diff}(M)$$

whose image $\Pi(\mathsf{Diff}(S_g M, \beta_g, \eta_\pi))$ is the isometry group $\mathsf{Iso}(M, g) \subset \mathsf{Diff}(M)$.

Thus, $\mathsf{Diff}(S_g M, \beta_g, \eta_\pi)$ is a Lie group, compact when M is compact.

Proof. By the very definitions of all three groups, we get an exact sequence:

$$1 \to \mathsf{Diff}_{\mathsf{S-inv}}(S_g M, \beta_g, \eta_\pi) \to \mathsf{Diff}(S_g M, \beta_g, \eta_\pi) \stackrel{\Pi}{\to} \mathsf{Diff}(M) \to .$$

By Proposition 3.11, the homomorphism Π is well-defined. By Proposition 3.12, the Π-kernel $\mathsf{Diff}_{\mathsf{S-inv}}(S_g M, \beta_g, \eta_\pi) = 1$. Thus, Π is a monomorphism.

The image of Π contains the isometry group $\mathsf{Iso}(M, g)$ since any g-isometry $\psi : M \to M$ produces the normalized differential diffeomorphism

$D_g^\circ\psi : S_gM \to S_gM$, which preserves the contact form β_g and such that $\psi \circ \pi = \pi \circ (D_g^\circ\psi)$.

For any short v_g-trajectory γ, $\pi(\gamma)$ is a geodesic arc; thus, we have

$$\delta_g(\pi(\gamma)) =_{\mathsf{def}} g\text{-length}(\pi(\gamma)) = \pm \int_\gamma \beta_g.$$

Therefore, the pair (π, β_γ) determines the distance function δ_g on M.

The fundamental theorem of Richard Palais [Pal] claims that the distance function δ_g, generated by a Riemannian metric g on a C^r-smooth ($r \in [2, \infty]$) manifold M, allows for a reconstruction of the metric g! Therefore, the pair (π, β_g) *determines* g. As a result, any automorphism of the pair (π, β_g) produces a g-isometry. In other words, $\Pi(\mathsf{Diff}(S_gM, \beta_g, \eta_\pi)) \subset \mathsf{Iso}(M, g)$.

Therefore, we proved that $\Pi(\mathsf{Diff}(S_gM, \beta_g, \eta_\pi)) = \mathsf{Iso}(M, g)$.

By a theorem of Myers and Steenrod [MySt], the group $\mathsf{Iso}(M, g)$ is a Lie group, compact when M is compact. Therefore, $\mathsf{Diff}(S_gM, \beta_g, \eta_\pi)$ is a Lie group, compact when M is compact. □

Theorem 3.23 leads to the following corollaries.

Corollary 3.36.

- *If the isometry group* $\mathsf{Iso}(M, g)$ *is finite/trivial, then so is the group* $\mathsf{Diff}(S_gM, \beta_g, \eta_\pi)$. *In particular, if the metric* g *on* M *is sufficiently "bumpy", then the rigidity claim "*$\mathsf{Diff}(S_gM, \beta_g, \eta_\pi) = 1$*" follows.*
- *If the isometry group* $\mathsf{Iso}(M, g)$ *is infinite, then so is the group* $\mathsf{Diff}(S_gM, \beta_g, \eta_\pi)$.
- $\mathsf{Diff}(S_gM, \beta_g, \eta_\pi)/\Pi^{-1}(\mathsf{Iso}_0(M, g))$ *is a discrete group, where* $\mathsf{Iso}_0(M, g)$ *denotes the connected component of* $\mathsf{Iso}(M, g)$ *that contains the identity.* ◇

Corollary 3.37. *The monomorphism* Π *gives rise to a representation*

$$\Pi_\star : \mathsf{Diff}(S_gM, \beta_g, \eta_\pi) \to \mathsf{GL}_{\mathbb{R}}(H_*(M; \mathbb{R})),$$

which acts on the homology $H_*(M; \mathbb{R})$.

- *If the Lefschetz number*

$$L(\Pi_\star(\Phi); \mathbb{R}) =_{\mathsf{def}} \sum_{i=0}^{n} (-1)^i \, \mathsf{trace}\big(\Pi_\star(\Phi)_{*,i} : H_i(M; \mathbb{R}) \to H_i(M; \mathbb{R})\big) \neq 0,$$

then the contactomorphism $\Phi \in \mathsf{Diff}(S_gM, \beta_g, \eta_\pi)$ *has an invariant* π*-fiber.*

- *If* M *is compact, then the image of* $\Pi_\star$ *in* $\mathsf{GL}_{\mathbb{R}}(H_*(M; \mathbb{R}))$ *is a finite group.*

Proof. The first claim is clear since any homeomorphism of M induces a $\mathbb{Z}$-linear transformation of $H_*(M;\mathbb{Z})$ and a $\mathbb{R}$-linear transformation of $H_*(M;\mathbb{R})$.

The second claim follows instantly from the Lefschetz fixed point theorem, being applied to the transformation $\Pi_\star(\Phi) : M \to M$.

By Theorem 3.23, the Π-image belongs to the g-isometry group of M. Thus, interpreting $H_*(M;\mathbb{R})$ as a space of harmonic forms on M in the metric g, we get that $H_*(M;\mathbb{R})$ is an inner-product space and that the $\Pi_\star$-representation acts on it by orthogonal transformations. Since $\Pi_\star$ preserves the lattice $H_*(M;\mathbb{Z})$ in the inner-product space $H_*(M;\mathbb{R})$, such automorphisms of a finitely generated lattice form a finite group. We conclude that the $\Pi_\star$-image of $\mathsf{Diff}(S_gM, \beta_g, \eta_\pi)$ in $\mathsf{GL}_\mathbb{R}(H_*(M;\mathbb{R}))$ is a finite group. □

Thus, when the isometry group $\mathsf{Iso}(M, g)$ is known, we get a computation of the group $\mathsf{Diff}(S_g, \beta_g, \eta_\pi)$. Here are few examples of such cases for 3-folds.

Corollary 3.38. *For the standard Riemannian metric g on the sphere S^3, we get a group isomorphism*

$$\Pi_\star : \mathsf{Diff}(S_g(S^3), \beta_g, \eta_\pi) \approx \mathsf{O}(4).$$

◇

Corollary 3.39. *For the standard Riemannian metric g on the ball D^3, we get a group isomorphism*

$$\Pi_\star : \mathsf{Diff}(S_g(D^3), \beta_g, \eta_\pi) \approx \mathsf{O}(3).$$

◇

Corollary 3.40. *For the standard flat Riemannian metric g on the torus T^3, we get a group isomorphism*

$$\Pi_\star : \mathsf{Diff}(S_g(T^3), \beta_g, \eta_\pi) \approx S^1 \times S^1 \times S^1 \times \mathsf{GL}_3(\mathbb{Z}).$$

◇

Let us list more corollaries of Theorem 3.23, based on few cases of 3-manifolds M for which the weak homotopy type of the group $\mathsf{Diff}(M)$ is known [Hat1].

Waldhausen proved [Wald] that the mapping class group of a closed aspherical oriented Haken 3-manifold M is identified with the automorphism group $\mathsf{Out}(\pi_1(M))$ of the fundamental group $\pi_1(M)$.

In particular, one gets a weak homotopy equivalence $\Pi_\star : \mathsf{Diff}(T^3) \sim S^1 \times S^1 \times S^1 \times \mathsf{GL}_3(\mathbb{Z})$ by Waldhausen's theorem in Hempel's book on 3-manifolds.

Gabai's fundamental theorem [Gab] claims that, for a hyperbolic 3-manifold M, $\mathsf{Diff}(M) \sim \mathsf{Iso}(M, g)$.

Employing Mostov's Rigidity Theorem, we get:

Corollary 3.41. *For the hyperbolic Riemannian metric g on a closed 3-fold M^3, there is a simple homotopy equivalence:*

$$\Pi_\star : \mathsf{Diff}(S_g(M^3), \beta_g, \eta_\pi) \sim \mathsf{Aut}(\pi_1(M^3)).$$

All these examples motivate the following assertion.

Theorem 3.24. *Let (M, g) be a smooth Riemannian manifold such that the natural embedding of spaces $\mathsf{Iso}(M, g) \subset \mathsf{Diff}(M)$ induces the homotopy group isomorphisms $\pi_i(\mathsf{Iso}(M, g)) \approx \pi_i(\mathsf{Diff}(M))$ for all $i \leq k$. Let Z be a finite CW-complex of a dimension $\leq k$.*

Then any Z-parametric continuous family of diffeomorphisms $\{\phi_z : M \to M\}_{z \in Z}$ can be coherently homotoped in the space $\mathsf{Diff}(M)$ to a Z-family of diffeomorphisms $\{\phi'_z : M \to M\}_{z \in Z}$ which lifts to a continuous Z-family of diffeomorphisms $\{\Phi'_z : S_g M \to S_g M\}_{z \in Z}$ that preserve the contact form β_g.

Proof. By applying the standard obstruction theory to the pair $\mathsf{Iso}(M, g) \subset \mathsf{Diff}(M)$ and using that $\pi_i(\mathsf{Iso}(M, g)) \approx \pi_i(\mathsf{Diff}(M))$ for all $i \leq k$, we conclude that any continuous map $\Psi : Z \to \mathsf{Diff}(M)$ can be homotoped to a continuous map $\Psi' : Z \to \mathsf{Iso}(M, g)$, provided $\dim(Z) \leq k$.

By Theorem 3.23, $\Pi : \mathsf{Diff}(S_g M, \beta_g, \eta_\pi) \to \mathsf{Iso}(M, g)$ is a Lie group isomorphism, and thus a homeomorphism. Therefore, $\Psi' : Z \to \mathsf{Iso}(M, g)$ lifts to a continuous map $\Xi : Z \to \mathsf{Diff}(S_g M, \beta_g, \eta_\pi)$ so that $\Pi \circ \Xi = \Psi'$. □

Unfortunately, the groups of diffeomorphisms of compact manifolds are well understood only in few cases. Corollaries 3.38 - 3.41 provide precious few examples where the weak homotopy type of the group $\mathsf{Diff}(M)$ is known.

Theorem 3.25. *For each 3-fold (M, g) from Corollaries 3.38 - 3.41, any diffeomorphism $\phi : M \to M$ can be homotoped in the space of diffeomorphisms $\mathsf{Diff}(M)$ to a diffeomorphism $\phi' : M \to M$ that lifts to a β_g-preserving contactomorphism of $\Phi' : S_g M \to S_g M$.*

Proof. In fact, for all 3-folds in Corollaries 3.38 - 3.41, the Lie group of isometries $\mathsf{Iso}(M, g)$ is weakly homotopy equivalent to the group of diffeomorphisms $\mathsf{Diff}(M)$ [Hat1]!

Therefore, the claim follows from Theorem 3.24. □

In particular, we get the following assertion.

Corollary 3.42. *Let $\pi : S_g(S^3) \to S^3$ be the tangent spherical fibration for the standard metric g on S^3.*

Any orientation preserving diffeomorphism $\phi : S^3 \to S^3$ can be homotoped in $\mathsf{Diff}_+(S^3)$ to a diffeomorpfism ϕ' which lifts to a diffeomorphism $\Phi' : S_g(S^3) \to S_g(S^3)$ preserving the contact form β_g on $S_g(S^3)$. ◇

Definition 3.31. Let $\pi : X \to M$ be a smooth fibration, where X and M are smooth manifolds, $\dim X = 2n - 1$, $\dim M = n$, and the π-fibers are $(n-1)$-manifolds. We denote by η_π the $(n-1)$-distribution tangent to the π-fibers. Let β be a contact form on X.

We call π a Legandrian fibration if all the π-fibers $\{L_m = \pi^{-1}(m)\}_{m \in M}$ are Legandrian submanifols relative to β, that is, if each fiber is tangent to the distribution $\mathsf{ker}(\beta)$.[20] ◇

The model case of Legandrian fibration is provided by the left-invariant contact structure on the Lie group $\mathsf{SU}(2) \approx S^3$. For an appropriate choice of a left-invariant contact form β on $\mathsf{SU}(2)$, the fibers of the Hopf fibration $\pi : S^3 \to S^2$ are Legandrian. Also, for any Riemannian metric g on M, the tangent spherical fibration $\pi : S_g M \to M$ is Legandrian with respect to the contact form β_g on $S_g M$.

Remark 3.26. Note that multiplying the contact form β by a positive smooth function $\mu : X \to \mathbb{R}_+$, leads to the same distribution $\mathsf{ker}(\mu \cdot \beta) = \mathsf{ker}(\beta)$. Thus, a fibration $\pi : X \to M$ remains Legandrian for the contact form $\mu \cdot \beta$. In other words, the notion of Legandrian fibration depends on the contact structure ξ_β generated by the contact form β. ◇

The following bold conjecture is inspired by the results of Pang [Pan].

Conjecture 3.6. *Let $\pi : X \to M$ be a smooth Legandrian fibration with closed connected fibers L. Then L is diffeomorphic to the standard sphere.* ◇

[20] so that the Reeb vector field v_β is transversal to the fibers of π

3.10 Contact de Rham cohomology and Hodge structures transversal to the Reeb foliations

In this section, following [He], we apply general results from [EKA], [EKH] about the transversal to foliations Hodge structures to the case of 1-dimensional foliations generated by the Reeb flows.

Recall that when we consider smooth structures/objects (like functions, differential forms, vector fields, etc.) on a given smooth compact manifold X with boundary this means, that X is embedded into a larger open equidimensional manifold $\hat{X}$ so that a structure in question is a restriction of a similar structure/object on $\hat{X}$ to X. As we change the structure/object on X, we may shrink the ambient $\hat{X}$. Thus, as in the past, one should think of $\hat{X} \supset X$ as a germ.

Let X be a compact smooth m-dimensional manifold and let $\Omega^0(X) = C^\infty(X, \mathbb{R})$ be the algebra of smooth functions on X. Consider the de Rham differential complex of exterior differential forms

$$\Omega^*_{d\mathcal{R}}(X) =_{\mathsf{def}} \tag{3.67}$$
$$\Big\{0 \stackrel{d}{\to} \Omega^0(X) \stackrel{d}{\to} \Omega^1(X) \stackrel{d}{\to} \Omega^2(X) \stackrel{d}{\to} \dots \stackrel{d}{\to} \Omega^m(X) \stackrel{d}{\to} 0\Big\},$$

considered as $\Omega^0(X)$-modules.

For a smooth non-vanishing vector field v on a m-manifold X, consider the vector space $\Omega^k_{\mathsf{b}}(X, v)$ of smooth v-invariant differential k-forms α that are v-horizontal, that is, $v \rfloor \alpha = 0$. The differential $d\alpha$ is evidently v-invariant when α is. Hence, for a v-invariant and horizontal α, we get

$$0 = \mathcal{L}_v \alpha = v \rfloor d\alpha + d(v \rfloor \alpha) = v \rfloor d\alpha,$$

which implies that $d\alpha$ is horizontal as well. When $k = 0$, then $\Omega^0_{\mathsf{b}}(X, v) \approx C^\infty(X, v)$, the space of v-invariant smooth functions on X.

Alternatively, we could define $\Omega^k_{\mathsf{b}}(X, v)$ as the vector space of k-forms α that are v-horizontal and whose differential $d\alpha$ is v-horizontal as well. Some authors call such differential forms "basic".

Therefore, for $m = \dim(X)$ and any vector field $v \neq 0$ on X, we get the horizontal and v-invariant (alternatively, basic) de Rham differential complex

$$\Omega^*_{\mathsf{basic}\, d\mathcal{R}}(X, v) =_{\mathsf{def}} \tag{3.68}$$
$$\Big\{0 \stackrel{d}{\to} \Omega^0_{\mathsf{b}}(X, v) \stackrel{d}{\to} \Omega^1_{\mathsf{b}}(X, v) \stackrel{d}{\to} \Omega^2_{\mathsf{b}}(X, v) \stackrel{d}{\to} \dots \stackrel{d}{\to} \Omega^{m-1}_{\mathsf{b}}(X, v) \stackrel{d}{\to} 0\Big\}$$

of $\Omega^0_{\mathsf{b}}(X, v)$-modules. Here the algebra $\Omega^0_{\mathsf{b}}(X, v)$ consists of smooth functions $f : X \to \mathbb{R}$ such that $\mathcal{L}_v f = 0$. Note that $\Omega^m_{\mathsf{b}}(X, v) = 0$ since no non-zero m-form on an open set can be v-horizontal.

Evidently, the de Rham differential complex of basic forms $\Omega^*_{\mathsf{basic}\, d\mathcal{R}}(X, v)$ maps canonically to the de Rham differential complex $\Omega^*_{d\mathcal{R}}(X)$ of all forms.

Definition 3.32. We denote by $H^*_{\mathsf{basic}\, d\mathcal{R}}(X, v)$ the cohomology groups of the v-basic de Rham differential complex (3.68). $\diamondsuit$

Proposition 3.13. *Let $\{\Psi_t : X \to X\}_{t\in[0,1]}$ be a smooth isotopy, and $\{v_t\}_{t\in[0,1]}$ a 1-parameter family of smooth non-vanishing vector fields on X such that $(\Psi_t)_*(v_t) = \lambda_t \cdot v_0$ for a smooth t-family $\lambda_t : X \to \mathbb{R}_+$ of positive functions such that $\lambda_0 = 1$.*

Then the basic de Rham differential complexes $\Omega^*_{\mathsf{basic}\, d\mathcal{R}}(X, v_0)$ and $\Omega^*_{\mathsf{basic}\, d\mathcal{R}}(X, v_t)$ are isomorphic with the help of Ψ^*_t. As a result, the de Rham basic cohomology are isomorphic:

$$\Psi^*_t : H^*_{\mathsf{basic}\, d\mathcal{R}}(X, v_0) \approx H^*_{\mathsf{basic}\, d\mathcal{R}}(X, v_t).$$

Proof. For a given X, the notion $\{v\rfloor\alpha = 0,\ v\rfloor d\alpha = 0\}$ of a basic form α depends only on the conformal class of the vector field v. Therefore, using that $d(\Psi^*_t(\alpha)) = \Psi^*_t(d\alpha)$ and that $(\Psi_t)_*(v_t) = \lambda_t v_0$, we conclude that Ψ^*_t is an isomorphism that maps basic forms to basic forms and commutes with their differentials. Thus, the basic differential complex $\Omega^*_{\mathsf{basic}\, d\mathcal{R}}(X, v_0)$ is isomorphic to the basic differential complex $\Omega^*_{\mathsf{basic}\, d\mathcal{R}}(X, v_t)$ with the help of Ψ^*_t. $\square$

Theorem 3.26. *Let $\{\beta_t\}_{t\in[0,1]}$ be a smooth family of contact forms on a compact manifold X such that $\beta_t = \beta_0$ in the vicinity of $\partial_1 X$.*

Then the de Rham basic differential complexes $\Omega^*_{\mathsf{basic}\, d\mathcal{R}}(X, v_{\beta_t})$ and $\Omega^*_{\mathsf{basic}\, d\mathcal{R}}(X, v_{\beta_0})$ are isomorphic with the help of a diffeotopy $\{\Phi_t : X \to X\}_{t\in[0,1]}$ which is the identity in the vicinity of $\partial_1 X$.

As a result, the de Rham basic cohomology spaces are isomorphic as well: $\Phi^*_t : H^*_{\mathsf{basic}\, d\mathcal{R}}(X, v_{\beta_0}) \approx H^*_{\mathsf{basic}\, d\mathcal{R}}(X, v_{\beta_t})$.

Proof. Let $\{\xi_t\}_{t\in[0,1]}$ be a smooth family of contact structures, generated by the contact family $\{\beta_t\}_{t\in[0,1]}$ on a compact manifold X. By the hypotheses, $\xi_t = \xi_0$ in the vicinity of $\partial_1 X$. The relative version of the Gray Stability Theorem [Gray] (see Lemma 3.27) claims that, for any smooth family $\{\xi_t\}_{t\in[0,1]}$ of contact structures on X, there exists a smooth isotopy $\{\Phi_t\}_{t\in[0,1]}$ such that $(\Phi_t)^*(\xi_t) = \xi_0$ for all $t \in [0,1]$ and $\Phi_t = \mathsf{id}$ in the vicinity of ∂X. Moreover, $(\Phi_t)^*(\beta_t) = \lambda_t \cdot \beta_0$ for an appropriate smooth family $\{\lambda_t\}_{t\in[0,1]}$ of positive functions on X such that $\lambda_0 = 1$.

A computation as in the proof of Corollary 3.28, implies that $v_{\beta_t} =_{\mathsf{def}} (\Phi_t)_*(\lambda_t^{-1} \cdot v_0) = \lambda_t^{-1} \cdot (\Phi_t)_*(v_0)$ is the Reeb vector field for β_t.

By Proposition 3.13, we conclude that the differential complexes $\Omega^*_{\mathsf{basic}\, d\mathcal{R}}(X, v_{\beta_t})$ and $\Omega^*_{\mathsf{basic}\, d\mathcal{R}}(X, v_{\beta_0})$ are isomorphic with the help of $\Phi_t^* = (\Psi_t^{-1})^*$, where the isotopy Ψ_t is as in Proposition 3.13.

In fact, the same argument works if we assume that $\dot{\beta}_t||_{\partial_1 X} = dh_t||_{\partial_1 X}$ (a relaxed constraint in comparison with the hypothesis "$\beta_t = \beta_0$ in the vicinity of $\partial_1 X$"), where $\dot{\beta}_t = \frac{d}{dt}\beta_t$ (see Lemma 3.32 for the notations and properties of these objects). □

Corollary 3.43. *Let* $\mathsf{Cont}(X; \partial_1 X)$ *be the space of smooth contact forms on* X *that are fixed in the vicinity of* $\partial_1 X$*. Let* $\boldsymbol{\Omega}_{\mathsf{basic}\, d\mathcal{R}}(X, \partial_1 X; \mathsf{Reeb})$ *be the set of isomorphism classes of basic de Rham complexes, produced by various Reeb vector fields on* X*, fixed in the vicinity of* $\partial_1 X$*.*

There exists a surjective map

$$\Upsilon : \pi_0(\mathsf{Cont}(X; \partial_1 X)) \to \boldsymbol{\Omega}_{\mathsf{basic}\, d\mathcal{R}}(X, \partial_1 X; \mathsf{Reeb})$$

whose source is the set of path-connected components of the space $\mathsf{Cont}(X; \partial_1 X)$*.* ◇

Lemma 3.39. *Let* v *be a non-vanishing vector field on a smooth manifold* X*, possibly with boundary. Let* γ *be a positive/negative time* v*-trajectory, originating at a point in the interior of* X*, and such that its closure* $\bar{\gamma} \subset X$ *is a compact set, contained in the interior of* X*.*

Then the field v *does not admit any Lyapunov function* $f : X \to \mathbb{R}$*.*

Proof. Assume, to the contrary, that v admits a Lyapunov function $f : X \to \mathbb{R}$. Since $\bar{\gamma}$ is compact, f must attend its maximum at some point $x_\star \in \bar{\gamma}$. Since $\bar{\gamma}$ is a positively v-invariant set, it contains the positive v-trajectory $\gamma_\star$ through $x_\star$. However, since $df(v)(x_\star) > 0$, $x_\star$ cannot be the maximum of f on $\bar{\gamma}$. □

Proposition 3.14. *Let* β *be a contact form on a smooth compact connected manifold* X*, and let* v_β *be its Reeb vector field.*

- *The homology class* $[d\beta] \in H^2_{\mathsf{basic}\, d\mathcal{R}}(X, v_\beta)$ *is zero if and only if* v_β *admits a closed* 1*-form* α *such that* $\alpha(v_\beta) = 1$*.*

In particular, if $[d\beta] = 0$ *in* $H^2_{\mathsf{basic}\, d\mathcal{R}}(X, v_\beta)$*, then the* v_β*-flow is transversal to a codimension one foliation* $\mathcal{F}_\alpha$*, defined in the vicinity of* X *in* $\hat{X}$*.*

- *If* v_β *admits a Lyapunov function, then* $\partial_1 X \neq \emptyset$ *and the homology class* $[d\beta] \in H^2_{\mathsf{basic}\, d\mathcal{R}}(X, v_\beta)$ *is zero.*

Conversely, assuming that $H^1(X;\mathbb{R}) = 0$, *if* $[d\beta] = 0$ *in* $H^2_{\mathsf{basic}\, d\mathcal{R}}(X, v_\beta)$, *then* v_β *admits a Lyapunov function, and* $\partial_1 X \neq \emptyset$.

Proof. If X is compact and a closed Lyapunov form α is such that $\alpha(v_\beta) = 1$, then the 2-form $d\beta = d(\beta - \alpha)$ is a boundary of a v_β-*horizontal* 1-form $\beta - \alpha$. Indeed, $(\beta - \alpha)(v_\beta) = 1 - 1 = 0$. The boundary $d\beta$ is also horizontal by the basic properties of Reeb vector fields. So, $\beta - \alpha$ is a basic form. As a result, $[d\beta] = 0$ in $H^2_{\mathsf{basic}\, d\mathcal{R}}(X, v)$.

Conversely, if $[d\beta] = 0$ in $H^2_{\mathsf{basic}\, d\mathcal{R}}(X, v)$, then for a basic 1-form α', we get $d\beta = d\alpha'$. Thus, $(\beta - \alpha')(v_\beta) = \beta(v_\beta) = 1$. Therefore, the 1-form $\alpha = \beta - \alpha'$ is closed and $\alpha(v_\beta) = 1$.

Let $\mathcal{F}_\alpha$ be codimension one foliation, defined in the vicinity of X in $\hat{X}$ by the integrable distribution $\{u \in TX|\ \alpha(u) = 0\}$. If $[d\beta] = 0$ in $H^2_{\mathsf{basic}\, d\mathcal{R}}(X, v_\beta)$, then the v_β-flow evidently is transversal to $\mathcal{F}_\alpha$.

This proves the claims in the first bullet.

If X is compact and $df(v_\beta) > 0$ for a smooth Lyapunov function f, then by Corollary 3.21, for another Lyapunov function h, we have $dh(v_\beta) = 1$. Therefore, $d\beta = d(\beta - dh)$ is a boundary of a v_β-*horizontal* 1-form $\alpha = \beta - dh$. Its boundary $d\beta$ is also horizontal by the basic properties of Reeb vector fields; so $\beta - dh$ is a basic form. As a result, $[d\beta] = 0$ in $H^2_{\mathsf{basic}\, d\mathcal{R}}(X, v)$.

Assume now that $H^1(X;\mathbb{R}) = 0$. Since $d(d\beta) = 0$ and $v_\beta \rfloor d\beta = 0$, the form $d\beta$ is a cocycle in $\Omega^2_{\mathsf{basic}\, d\mathcal{R}}(X, v)$. If $[d\beta] = 0$ in $H^2_{\mathsf{basic}\, d\mathcal{R}}(X, v)$, then there is a *basic* 1-form α such that $d\alpha = d\beta$. Using that $H^1(X;\mathbb{R}) = 0$, we get $\beta - \alpha = df$ for a smooth function $f : X \to \mathbb{R}$. Since α is a v_β-basic form, then $df(v_\beta) = \beta(v_\beta) - \alpha(v_\beta) = 1 - 0 = 1$. Thus, df is a Lyapunov function for v_β.

If X is a closed manifold and $v_\beta \neq 0$ admits a Lyapunov function f, then there exists a point $x_\star \in X$ where f attends its extremum. Thus, $df(v_\beta(x_\star)) = 0$, which contradicts to f being a Lyapunov function. Therefore, $\partial_1 X \neq \emptyset$. □

Corollary 3.44. *Let* β *be a smooth contact form on a* closed *manifold* X. *Assume that* $H^1(X;\mathbb{R}) = 0$. *Then the class* $[d\beta] \in H^2_{\mathsf{basic}\, d\mathcal{R}}(X, v)$ *is nontrivial, while the class* $[d\beta] \in H^2_{d\mathcal{R}}(X)$ *is always trivial.*

Proof. Since X is closed, the closure of any v_β-trajectory is compact in $X = \mathsf{int}(X)$. By Lemma 3.39, v_β does not admit a Lyapunov function.

By Proposition 3.14 (see the last paragraph in its proof), the class $[d\beta] \in H^2_{\mathsf{basic}\, d\mathcal{R}}(X, v)$ is nontrivial, provided $H^1(X; \mathbb{R}) = 0$. At the same time, the class $[d\beta] \in H^2_{d\mathcal{R}}(X)$ is always trivial. □

Recall that the vector field v_β, tangent to the fibers of the Hoph fibration, is not transversal to any 2-dimensional foliation on S^3. Thus, $[d\beta] \neq 0$ in $H^2_{\mathsf{basic}\, d\mathcal{R}}(S^3, v_\beta)$.

Definition 3.33. Let X be an oriented closed 3-fold, and ϕ a nonsingular vector flow on X. Following [Good], we say that ϕ has the linking property: if a periodic ϕ-orbit γ bounds an embedded disk $D^2 \subset X$, then the interior of D^2 intersects a periodic ϕ-orbit. ◇

Evidently, if a flow ϕ has a closed trajectory, bounded by an embedded 2-disk and possesses the linking property, then ϕ must have at least two closed trajectories.

Corollary 3.45. *Let X be an oriented closed 3-fold equipped with a contact form β. If $[d\beta] = 0$ in $H^2_{\mathsf{basic}\, d\mathcal{R}}(X, v_\beta)$, then the v_β-flow has the linking property.*

Proof. By [Good], Theorem 1.2, if a nonsingular vector flow ϕ on an oriented closed 3-fold is transversal to a 2-dimensional foliation, then ϕ has the linking property. Therefore, by Proposition 3.14, if $[d\beta] = 0$ in $H^2_{\mathsf{basic}\, d\mathcal{R}}(X, v_\beta)$, then the v_β-flow has the linking property.

In particular, if $[d\beta] = 0$ and the Reeb flow has a closed trajectory, bounded by an embedded disk, then it has another closed trajectory, linked with the first one. □

By [Good], Theorem 2.1, in any homotopy class of nonsingular vector fields on any oriented closed 3-fold, there is an open set of nonsingular vector fields without the linking property. In fact, a similar claim is valid in the higher dimensions. These results motivate the following conjecture.

Conjecture 3.7. *Let X be a closed (compact ?) connected oriented smooth manifold. The set of contact forms β on X for which $[d\beta] \neq 0$ in $H^2_{\mathsf{basic}\, d\mathcal{R}}(X, v_\beta)$ is open in the space of contact forms.* ◇

Recall that, for a non-vanishing vector field v, a closed 1-form α is called Lyapunov, if $\alpha(v) > 0$.

In formula (3.37), we introduced the notion of Schwartzman's asymptotic cycles $A(v_\beta, \mu)$ (see also [F], page 199). Let us see how it applies to the case of Reeb flow v_β on a *closed* manifold X.

Consider the v_β-invariant Borel measure μ_β on X, defined by the volume form $\beta \wedge (d\beta)^n$. We define an element $A(v_\beta, \mu_\beta) \in H_1(X; \mathbb{R})$ by the formula

$$\langle A(v_\beta, \mu_\beta), [\alpha] \rangle =_{\mathsf{def}} \int_X \alpha(v_\beta)\ \beta \wedge (d\beta)^n,$$

where $[\alpha] \in H^1(X; \mathbb{R})$ runs over de Rham cohomology classes and α is a closed 1-form of the class $[\alpha]$. In fact, this definition of $A(v_\beta, \mu_\beta)$ depends only on the contact form β.

As a special case of Proposition 10.44 from [F], we get the following claim.

Proposition 3.15. *Let X be a closed smooth manifold equipped with a contact form β. If a closed 1-form α is Lyapunov for the Reeb flow v_β, then it satisfies the homological property: $\langle A(v_\beta, \mu_\beta),\ [\alpha] \rangle < 0$, where the Schwartzman's asymptotic cycle $A(v_\beta, \mu_\beta) \in H_1(X; \mathbb{R})$.*

Moreover, a similar inequality is valid for any v_β-invariant positive Borel measure μ on X. ◇

Let X be a compact connected smooth manifold equipped with a non-vanishing vector field v. Given a non-vanishing closed k-form η such that $v \rfloor \eta = 0$ (i.e., η is v-horizontal), we may ask whether there exist a k-form τ such that $v \rfloor \tau = 0$ (i.e., τ is v-horizontal) and $\mathcal{L}_v(\tau) = \eta$, and what are the implications of its existence? Informally speaking, we are asking for the implications of existence of the "global integral" or "global potential" $\mathcal{L}_v^{-1}(\eta)$ for a given basic form η.

Definition 3.34. Let X be a smooth manifold equipped with a non-vanishing vector field v.

Let $Fr_\star$ be an ordered collection of k tangent to X vectors at a point $x_\star \in \mathsf{int}(X)$, the vectors being transversal to the positive time trajectory γ through $x_\star$ and linearly independent in the quotient space $T_{x_\star}X/T_{x_\star}\gamma$.

We consider the k-frame field $Fr^\bullet_\gamma$, spread from $Fr_\star$ along γ by the differential ϕ^t_* of the positive time v-flow $\{\phi^t\}_{t \in \mathbb{R}_+}$. We may view $Fr^\bullet_\gamma$ as a section of the pull-back of the bundle $\{\Lambda^k(TX/T\gamma)|_\gamma \to \gamma\}$ under the parametrization map $\phi^t : \mathbb{R} \to \gamma$, where $\phi^t(0) = x_\star$.

Let τ be a v-horizontal differential k-form on X.

• We say that the k-form τ is coarsely bounded along γ in the direction of the frame $Fr_\star$, if the values $\{|\tau(\phi^t_*(Fr_\star))|\}_{t \in \mathbb{R}_+}$ are bounded. ◇

Lemma 3.40. *Let X be a compact connected smooth manifold, equipped with a non-vanishing vector field v. Let η be a v-basic non-vanishing k-form and τ a v-horizontal k-form on X such that $\mathcal{L}_v\tau = \eta$.*

Then τ is not coarsely bounded along any trapped *v-trajectory $\gamma \subset \mathsf{int}(X)$ in the direction of any k-frame $Fr_\star$ such that $\eta(Fr_\star) \neq 0$. In fact, the function $\{t \to \tau(\phi_*^t(Fr_\star))|_\gamma\}$ is a t-polynomial of degree 1.*

On the other hand, for any v-trajectory $\gamma \subset \mathsf{int}(X)$ and any k-frame $Fr_\star$ such that $\eta(Fr_\star) = 0$, the values $\{\tau(\phi_*^t(Fr_\star))|_\gamma\}$ are t-independent.

Proof. Contrary to the claim, assume that τ is coarsely bounded along a trapped v-trajectory $\gamma \subset \mathsf{int}(X)$ in the direction of a frame $Fr_\star$ such that $\eta(Fr_\star) > 0$. We denote by γ_T the portion of γ that corresponds to the time interval $[0, T]$. Then, for a v-invariant frame Fr along γ_T, using that $(\mathcal{L}_v\tau)(Fr) = \eta(Fr)$, we get

$$\int_{[0,T]} \mathcal{L}_v\tau\big(\phi_*^t(Fr_\star)\big)\, dt = \int_{[0,T]} \eta\big(\phi_*^t(Fr_\star)\big)\, dt$$

Using that the form η and the frame $Fr_\gamma^\bullet$ are ϕ^t-invariant,

$$\int_{[0,T]} \mathcal{L}_v\tau\big(\phi_*^t(Fr_\star)\big)\, dt = \eta\big(Fr_\star\big) \cdot T.$$

Since

$$\mathcal{L}_v\tau\big(\phi_*^t(Fr_\star)\big) = \lim_{\epsilon \to 0} \frac{\tau\big(\phi_*^{t+\epsilon}(Fr_\star)\big) - \tau\big(\phi_*^t(Fr_\star)\big)}{\epsilon} = \frac{d}{dt}\tau\big(\phi_*^t(Fr_\star)\big),$$

we get

$$\tau\big(\phi_*^T(Fr_\star)\big) - \tau\big(Fr_\star\big) = \eta\big(Fr_\star\big) \cdot T.$$

Thus,

$$\tau\big(\phi_*^T(Fr_\star)\big)) = \tau\big(Fr_\star\big) + \eta\big(Fr_\star\big) \cdot T$$

is a polynomial of degree 1 in T, unless $\eta\big(Fr_\star\big) = 0$. We see that $\lim_{T\to\infty} \tau\big(\phi_*^T(Fr_\star)\big) = \infty$, provided $\eta(Fr_\star) \neq 0$; so τ is not coarsely bounded along any trapped v-trajectory γ in the direction of any frame $Fr_\star$ such that $\eta(Fr_\star) > 0$.

The same argument shows that for any v-trajectory $\gamma \subset \mathsf{int}(X)$ and any k-frame $Fr_\star$ such that $\eta(Fr_\star) = 0$, the values $\{\tau\big(\phi_*^t(Fr_\star)\big)|_\gamma\}$ are t-independent. □

Corollary 3.46. *Let X be a compact connected smooth manifold equipped with a non-vanishing vector field v. Let η be a basic non-vanishing k-form, and τ a horizontal k-form on X such that $v \rfloor d\tau = \eta$.*

Then τ is not coarsely bounded along any trapped *v-trajectory $\gamma \subset \mathsf{int}(X)$ in the direction of any k-frame $Fr_\star$ such that $\eta(Fr_\star) > 0$.*

On the other hand, if a horizontal form τ is such that the form $v \rfloor d\tau$ is basic and τ is coarsely bounded along any v-trajectory $\gamma \subset \mathsf{int}(X)$ in the direction of any frame $Fr_\star$, then v does not have trapped trajectories, and $\partial_1 X \neq \emptyset$. In such a case, v admits a Lyapunov function.

Proof. Since $v \rfloor d\tau = \eta$ and $v \rfloor \tau = 0$, we get $\mathcal{L}_v(\tau) = d(v \rfloor \tau) + v \rfloor d\tau = \eta$. Thus Lemma 3.40 applies.

In particular, if a v-horizontal form τ is such that the form $v \rfloor d\tau$ is basic and τ is coarsely bounded along any v-trajectory $\gamma \subset \mathsf{int}(X)$ in the direction of any frame $Fr_\star$, then v does not have trapped trajectories.

If a non-vanishing v-flow has no trapped trajectories, then its trajectories are closed segments (whose boundaries reside in $\partial_1 X$) or singletons (residing in $\partial_1 X$), which implies that $\partial_1 X \neq \emptyset$. By Lemma 2.1 (or Lemma 5.6 from [K7]), such a flow admits a Lyapunov function. □

As an instant application of Corollary 3.46, we get the following claim.

Corollary 3.47. *Let β be a contact form and τ a $2k$-form on a compact connected smooth manifold X of dimension $2n+1$ such that*[21]

$$v_\beta \rfloor \tau = 0 \;\; \text{and} \;\; v_\beta \rfloor d\tau = (d\beta)^k.$$

Then τ is not coarsely bounded along each trapped v_β-trajectory $\gamma \subset \mathsf{int}(X)$ in the direction of any frame $Fr_\star$ such that $(d\beta)^k(Fr_\star) > 0$.

On the other hand, if τ is coarsely bounded along any v_β-trajectory $\gamma \subset \mathsf{int}(X)$ in the direction of any frame $Fr_\star$, then v does not have trapped trajectories, and $\partial_1 X \neq \emptyset$. In such a case, v_β admits a Lyapunov function.

Proposition 3.16. *Let β be a contact form on a compact connected smooth manifold X. Assume that $H^{2k-1}_{\mathsf{basic}\, d\mathcal{R}}(X, v_\beta) = 0$ and $[(d\beta)^k] = 0$ in $H^{2k}_{\mathsf{basic}\, d\mathcal{R}}(X, v_\beta)$.*[22]

Then there exists a $(2k-2)$-form τ on X such that $v_\beta \rfloor \tau = 0$ and $v_\beta \rfloor d\tau = (d\beta)^{k-1}$.

If this form τ is coarsely bounded along each v_β-trajectory $\gamma \subset \mathsf{int}(X)$ in the direction of any $(2k-2)$-frame $Fr_\star$, then v_β does not have trapped Reeb trajectories, and $\partial_1 X \neq \emptyset$. In such a case, v_β admits a Lyapunov function.

[21]This τ is v-horizontal, but not basic!
[22]Note that $[(d\beta)^k] = 0$ in $H^{2k}_{d\mathcal{R}}(X)$.

Proof. Since $d[(d\beta)^k] = 0$ and $v_\beta \rfloor (d\beta)^k = 0$, we get that $(d\beta)^k \in \Omega^{2k}_{\mathsf{basic}\, d\mathcal{R}}(X, v_\beta)$ is a basic cocycle.

If $[(d\beta)^k] = 0$ in $H^{2k}_{\mathsf{basic}\, d\mathcal{R}}(X, v)$, then there is a basic $(2k-1)$-form α such that $d\alpha = (d\beta)^k = d(\beta \wedge (d\beta)^{k-1})$. Thus, $d(\alpha - \beta \wedge (d\beta)^{k-1}) = 0$.

Using that $H^{2k-1}_{\mathsf{basic}\, d\mathcal{R}}(X, v) = 0$, we conclude that there is a horizontal $(2k-2)$-form τ such that $d\tau = \alpha - \beta \wedge (d\beta)^{k-1}$. Since α is a v_β-basic form,

$$v_\beta \rfloor d\tau = v_\beta \rfloor \alpha - v_\beta \rfloor (\beta \wedge (d\beta)^{k-1}) = -v_\beta \rfloor (\beta \wedge (d\beta)^{k-1}) = (d\beta)^{k-1}.$$

By the Cartan formula, $\mathcal{L}_{v_\beta} \tau = d(v_\beta \rfloor \tau) + v_\beta \rfloor d\tau = (d\beta)^{k-1}$. Thus, Corollary 3.46 is applicable (replacing k with $k-1$). We conclude that if τ is coarsely bounded along each Reeb trajectory $\gamma \subset \mathsf{int}(X)$ in the direction of any frame $Fr_\star$, then v_β does not have trapped trajectories and $\partial_1 X \neq \emptyset$. In such a case, by Corollary 3.46, v_β admits a Lyapunov function. □

Switching gears, let us briefly describe some main features of the Hodge theory, based on structures *transversal* to a given 1-dimensional foliation. We follow closely the presentations in [Lin] and [Raz].

Consider the 1-dimensional oriented foliation $\mathcal{F}(v_\beta)$, generated by the Reeb vector field v_β of a contact form β on a compact manifold X. We denote by ξ_β the distribution $\mathsf{ker}(\beta)$, transversal to the foliation $\mathcal{F}(v_\beta)$. The 2-form $d\beta$ defines a symplectic structure on the bundle ξ_β. As before, all these structures are defined in the vicinity of a given compact manifold X contained in an equidimentional open manifold $\hat{X}$.

We consider also a v_β-invariant Riemannian metric g on TX and a v_β-invariant almost complex structure $J : \xi_\beta \to \xi_\beta$, $J^2 = -\mathsf{Id}$, that satisfy the following compatibility properties (i.e., a basic almost-Kähler structure):

$$\begin{aligned} v_\beta &\perp_g \xi_\beta, \\ \|v_\beta\|_g &= 1, \\ d\beta(\sim, \sim) = g(J \sim, \sim) &= d\beta(J \sim, J \sim) \;\text{ on }\; \xi_\beta. \end{aligned} \tag{3.69}$$

For $\dim(X) = 2n+1$, the metric g defines the familiar Hodge operator

$$*_g : \Omega^p(X) \to \Omega^{2n+1-p}(X).$$

The non-degenerated 2-form $d\beta : \xi_\beta \times \xi_\beta \to \mathbb{R}$ produces a bundle isomorphism $Q_{d\beta} : \xi_\beta \to \xi^*_\beta$ which leads to an isomorphism

$$\Lambda^p Q_{d\beta} : \Lambda^p \xi_\beta \to \Lambda^p(\xi^*_\beta)$$

for each p. Thus, we get a non-degenerated pairing

$$K^{p*}_{d\beta} : \Lambda^p \xi^*_\beta \times \Lambda^p \xi^*_\beta \to \mathbb{R},$$

produced by $(\Lambda^p Q_{d\beta})^{-1}$. Any basic p-form α can be viewed as a section of the bundle $\Lambda^p \xi^*_\beta$, since v_β belongs to the kernel of α. Therefore, we get a non-degenerated pairing

$$K^{p*}_{d\beta} : \Omega^p_{\mathsf{b}}(X, v_\beta) \times \Omega^p_{\mathsf{b}}(X, v_\beta) \to \mathbb{R}$$

of basic p-forms. We define the basic Hodge operator

$$*_{\mathsf{b}} : \Omega^p_{\mathsf{b}}(X, v_\beta) \to \Omega^{2n-p}_{\mathsf{b}}(X, v_\beta)$$

by the formula

$$\kappa \wedge (*_{\mathsf{b}}\rho) = K_{d\beta}(\kappa, \rho) \cdot \frac{(d\beta)^n}{n!},$$

for any pair $\kappa, \rho \in \Omega^p_{\mathsf{b}}(X, v_\beta)$ of basic p-forms.

The (basic) co-derivative operators are introduced by the formulas

$$\delta_g =_{\mathsf{def}} (-1)^p(*_g) \circ d \circ (*_g), \tag{3.70}$$

$$\delta_{\mathsf{b}} =_{\mathsf{def}} (*_{\mathsf{b}}) \circ d \circ (*_{\mathsf{b}}), \tag{3.71}$$

and the (basic) Laplace operators by the formulas

$$\Delta_g =_{\mathsf{def}} d \circ \delta_g + \delta_g \circ d, \tag{3.72}$$

$$\Delta_{\mathsf{b}} =_{\mathsf{def}} d \circ \delta_{\mathsf{b}} + \delta_{\mathsf{b}} \circ d. \tag{3.73}$$

We call a basic form α basically harmonic if $\Delta_{\mathsf{b}}(\alpha) = 0$. If α is closed, i.e., $d\alpha = 0$, and co-closed, i.e., $\delta_{\mathsf{b}}\alpha = 0$, then the form α is evidently basically harmonic. If $\partial_1 X \neq \emptyset$, the converse may be not true due to boundary effects.

Since d and Δ_{b} commute, the basic harmonic forms form a subcomplex $(\mathsf{Harm}^*_{\mathsf{b}}(X, v), d)$ of the complex $(\Omega^*_{\mathsf{basic}\, d\mathcal{R}}(X, v), d)$. Therefore, we may define $H^k(\mathsf{Harm}^*_{\mathsf{b}}(X, v), d)$, the basic harmonic k-homology of (X, v).

Based on the main theorem from [CTGM], we propose the following conjecture.

Conjecture 3.8. *For a traversing and boundary generic vector field v on a connected compact smooth manifold X with a boundary,*

$$H^k(\mathsf{Harm}^*_{\mathsf{b}}(X, v), d) \approx H^k(X; \mathbb{R}) \oplus H^{k-1}(X; \mathbb{R})$$
$$\approx H^k(\mathcal{T}(v); \mathbb{R}) \oplus H^{k-1}(\mathcal{T}(v); \mathbb{R}).$$

◇

Next, we apply the general results from [KA], [EKA], [EKH], and [He], to the case of 1-dimensional Reeb foliations on *closed* manifolds. Unfortunately, no similar theories have been developed for (singular) foliations on compact manifolds *with boundary*, although the contours of such theories (in the spirit of [CTGM] and Conjecture 3.8) are clear...

For contact structures on compact manifolds with no boundary, this application leads directly to the following assertions [KA]:

Theorem 3.27. *Let β be a contact form on a* closed *orientable $(2n+1)$-manifold X. For any v_β-invariant metric g on X that satisfies properties (3.69), the following claims are valid:*

- *There is an isomorphism*

$$H^*_{\mathsf{basic}\, d\mathcal{R}}(X, v_\beta) \approx \ker(\Delta_{\mathsf{b}}(v_\beta)),$$

  implying that the basic de Rham cohomology is a finite-dimensional space.
- *The basic Hodge star operator induces a duality isomorphism*

$$*_{\mathsf{b}} : H^k_{\mathsf{basic}\, d\mathcal{R}}(X, v_\beta) \approx H^{2n-k}_{\mathsf{basic}\, d\mathcal{R}}(X, v_\beta)$$

  by taking the homology class to the image of its harmonic representative.

Combining Theorem 3.27 with Theorem 3.26 instantly leads to the following assertion.

Theorem 3.28. *Let $\{g_t, \beta_t, J_t\}_{t\in[0,1]}$ be a smooth family of Riemannian metrics g_t, contact forms β_t, and complex operators $J_t : \xi_t \to \xi_t$ on a closed manifold X that satisfy properties (3.69) of basic almost Kähler structures.*

Then the spaces $\ker(\Delta_{\mathsf{b}}(g_t, \beta_t, J_t))$ and $\ker(\Delta_{\mathsf{b}}(g_0, \beta_0, J_0))$ of basically harmonic k-forms for the Reeb vector fields v_{β_t} and v_{β_0} are isomorphic with the help of a diffeotopy $\Phi_t : X \to X$. ◇

Applying Theorem 3.27 to the basic form $(d\beta)^k$ leads to the following assertion.

Corollary 3.48. *Let g be a v_β-invariant metric on a closed orientable $(2n+1)$-dimensional manifold X, which satisfies properties (3.69) of basic almost Kähler structures.*

Then the v_β-basic form $(d\beta)^k$, $k \leq n$, decomposes canonically as the sum of two basic $2k$-forms:

$$(d\beta)^k = h_{(d\beta)^k} + d\theta, \tag{3.74}$$

where the basic form $h_{(d\beta)^k}$ is basically harmonic, that is, $\Delta_{\mathsf{b}}(h_{(d\beta)^k}) = 0$, *and θ is a basic $(2k-1)$-form.*

If the elements

$$[h_{(d\beta)^k}] \in H^{2k}_{\mathsf{basic}\, d\mathcal{R}}(X, v_\beta) \;\; or \;\; [*_{\mathsf{b}}(h_{(d\beta)^k})] \in H^{2n-2k}_{\mathsf{basic}\, d\mathcal{R}}(X, v_\beta)$$

are nontrivial, then so are the elements $[(d\beta)^k]$ *and* $[*_{\mathsf{b}}(d\beta)^k]$. ◇

Recall that, in the classical Kähler geometry on even-dimensional X, the volume of a complex k-dimensional submanifold $Y \subset X$ is given by the formula $vol(Y) = \frac{1}{k!}\int_Y \omega^k$, where ω is the symplectic 2-form on X. In particular, $\int_Y \omega^k > 0$. In sharp contrast with such a phenomenon, in the contact geometry on an odd-dimensional closed X, for Hodge structures transversal to the Reeb foliation $\mathcal{F}(v_\beta)$ and for the symplectic form $\omega = d\beta$, we get the following properties.

Corollary 3.49. *Let X be as in Corollary 3.48, and let $\Sigma \subset X$ be a compact oriented submanifold of dimension $2k$.*

- *Assuming that Σ is closed, the* "harmonic volume" *and the* "symplectic volume" *of Σ both vanish:*

$$\int_\Sigma h_{(d\beta)^k} = \int_\Sigma (d\beta)^k = 0$$

- *If the boundary $\partial\Sigma$ of Σ is tangent to the distribution ξ_β, then for $2k-1 \leq n$ we get:*

$$\int_\Sigma h_{(d\beta)^k} = -\int_{\partial\Sigma} \theta.$$

Proof. The first claim follows by applying Stokes' theorem

$$\int_\Sigma (d\beta)^k = \int_\Sigma d(\beta \wedge (d\beta)^{k-1}) = \int_{\partial\Sigma} \beta \wedge (d\beta)^{k-1} = \int_\emptyset \beta \wedge (d\beta)^{k-1} = 0$$

to formula (3.74) and using that $\int_\Sigma d\theta = 0$ for a closed Σ.

If the boundary $\partial\Sigma$ of an oriented compact submanifold $\Sigma \subset X$ of dimension $\dim(\Sigma) = 2k \leq n$ is tangent to the distribution ξ_β (i.e., if $\partial\Sigma$ a Legandrian submanifold), then

$$0 = \int_{\partial\Sigma} \beta \wedge (d\beta)^{k-1} = \int_\Sigma (d\beta)^k = \int_\Sigma (h_{(d\beta)^k} + d\theta) = \int_\Sigma h_{(d\beta)^k} + \int_{\partial\Sigma} \theta.$$

Therefore,

$$\int_\Sigma h_{(d\beta)^k} = -\int_{\partial\Sigma} \theta = -\int_\Sigma d\theta.$$

□

By the formula above, the integral $\int_\Sigma d\theta$ does not depend on the choice of Σ, whose Legandrian boundary $\partial\Sigma$ is fixed. Also, $\int_{\partial\Sigma} \theta$ does not change if one replaces θ with $\theta + d\eta$. Thus, we get the following claim.

Corollary 3.50. *Let g be a v_β-invariant metric on a closed orientable $(2n+1)$-manifold X, which satisfies properties (3.69) of basic almost Kähler structures.*

For $2k - 1 \leq n$, we get a map

$$\int \theta : \ \{\text{bounded Legandrian } (2k-1)\text{-submanifolds of } X\} \longrightarrow \mathbb{R},$$

which depends on the contact form β and the metric g only. ◇

Note that, when the $(\int \theta)$-image of a Legandrian submanifold $\partial\Sigma \subset X$ is negative, then the harmonic volume of an oriented manifold Σ is positive.

The exterior multiplication of basic forms generates a multiplicative structure in vector space $\Omega^*_{\mathsf{basic}\, d\mathcal{R}}(X, v)$.

Following [Raz], we introduce the symplectic basic Hodge and symplectic basic co-boundary operators:

$$*_{d\beta} =_{\mathsf{def}} -J_{\mathsf{b}} \circ (*_{\mathsf{b}}) \tag{3.75}$$

$$\delta_{d\beta} =_{\mathsf{def}} (-1)^{k+1} (*_{d\beta}) \circ d \circ (*_{d\beta}). \tag{3.76}$$

The basic Lefschetz-Weyl operators are defines by:

$$L =_{\mathsf{def}} d\beta \wedge \sim, \qquad \Lambda =_{\mathsf{def}} (*_{d\beta}) \circ L \circ (*_{d\beta}). \tag{3.77}$$

The symplectic basic Hodge operator $*_{d\beta}$ maps basic forms to basic forms.

Definition 3.35. A basic form α is said to be symplectically harmonic if $d\alpha = 0$ and $\delta_{d\beta}\alpha = 0$. ◇

For the Reeb foliation $\mathcal{F}(v_\beta)$, Raźny's Theorem 2.17 [Raz] reduces to the following claim:

Theorem 3.29. *For a closed oriented $(2n+1)$-dimensional manifold X, equipped with a contact 1-form β, the following two claims are equivalent:*

- (Basic Hard Lefschetz property) *The multiplication by $(d\beta)^k$ map*

  $$L^k : H^{n-k}_{\mathsf{basic}\, d\mathcal{R}}(X, v_\beta) \to H^{n+k}_{\mathsf{basic}\, d\mathcal{R}}(X, v_\beta)$$

  is an isomorphism for all $k \leq n$,

- *Every basic cohomology class has a $\delta_{d\beta}$-closed representative.* ◇

By Proposition 3.14, the homology class $[d\beta] \in H^2_{\mathsf{basic}\, d\mathcal{R}}(X, v_\beta)$ is zero if and only if v_β admits a closed 1-form α such that $\alpha(v_\beta) = 1$. Therefore, for such (X, β), the Basic Hard Lefschetz property fails spectacularly.

Definition 3.36. Let X be an open oriented $(2n+1)$-dimensional manifold, equipped with a contact 1-form β.

- A basic $(n-k)$-form α is called primitive if

$$L^{k+1}\alpha =_{\mathsf{def}} \alpha \wedge (d\beta)^{k+1} = 0.$$

Equivalently, a basic form α is primitive if $\Lambda\alpha = 0$.

- Similarly, a basic cohomology class $[\alpha]$ of degree $(n-k)$ is primitive if $L^{k+1}[\alpha] = 0$. ◇

Note that, on a closed X, $d\beta$ is not primitive since $(d\beta)^n \neq 0$.

Applying Proposition 2.19 [Raz] to our setting, we get the Lefschetz decomposition of basic forms:

Proposition 3.17. *Given a closed oriented smooth $(2n+1)$-dimensional manifold X, equipped with a contact 1-form β, then any basic k-form α can be uniquely decomposed as*

$$\alpha = \sum_{i=0}^{\lceil k/2 \rceil} \rho_{k-2i} \wedge (d\beta)^i,$$

where $\{\rho_{k-2i}\}$ are basic primitive *$(k-2i)$-forms, given by the formula*

$$\rho_{k-2i} = \Big(\sum_r \frac{a_{i,r}}{r!}\, L^r \circ \Lambda^{i+r}\Big)\alpha.$$

Here the constants $a_{i,r}$ depend only on (n, i, r). ◇

As Theorem 2.20 from [Raz], Proposition 3.17 leads to the following claim:

Theorem 3.30. *Assume that a closed oriented smooth $(2n+1)$-dimensional manifold X, equipped with a contact 1-form β, is such that the Reeb 1-foliation $\mathcal{F}(v_\beta)$ satisfies the Basic Hard Lefschetz property. Let $[\alpha]$ be a basic cohomology class of degree k.*

Then the class $[\alpha]$ can be uniquely decomposed as

$$[\alpha] = \sum_{i=0}^{\lceil k/2 \rceil} [\rho_{k-2i}] \wedge [d\beta]^i,$$

where $\{[\rho_{k-2i}]\}$ are basic primitive *$(k-2i)$-cohomology classes.* ◇

In turn, Raźny's Theorem 2.20, [Raz], leads to another important conceptual result, Corollary 4.4 [Raz]. It claims (crudely) that homeomorphic manifolds, both satisfying the Basic Hard Lefschetz property for the appropriate foliations of the same dimension, have isomorphic basic de Rham cohomologies! Applying this result to Reeb foliations, we get the following assertion:

Theorem 3.31. *Let β_0, β_1 be two contact forms on a closed manifold X such that both Reeb foliations, $\mathcal{F}(v_{\beta_1})$ and $\mathcal{F}(v_{\beta_1})$ satisfy the Basic Hard Lefschetz property.*

Then, for each k, their basic de Rham k-cohomologies are isomorphic

$$H^k_{\mathsf{basic}\, d\mathcal{R}}(X, v_{\beta_1}) \approx H^k_{\mathsf{basic}\, d\mathcal{R}}(X, v_{\beta_0}),$$

and so are the spaces of basic harmonic k-forms:

$$\mathsf{ker}(\Delta^k_{\mathsf{b}}(v_{\beta_1})) \approx \mathsf{ker}(\Delta^k_{\mathsf{b}}(v_{\beta_0})).$$

◇

Corollary 3.51. *For a closed manifold X, if the Reeb foliation $\mathcal{F}(v_{\beta_1})$ satisfies the Basic Hard Lefschetz property and the Reeb foliation $\mathcal{F}(v_{\beta_0})$ does not, then the contact forms β_0 and β_1 belong to different path connected components of the space of contact forms* $\mathsf{Cont}(X)$.

Proof. Combining Theorem 3.26 with Theorem 3.31 validates the claim.

□

Proposition 3.18. *Let X, Y be equidimensional compact smooth manifolds, and $\Psi : X \to Y$ a smooth immersion. Let $v_Y \neq 0$ be a vector field on Y. Put $v_X = \Psi^\dagger(v_Y)$, the transfer of v_Y to X.*

Then Ψ induces homomorphisms

$$\Psi^*_{\mathsf{b}} : \Omega^*_{\mathsf{basic}\, d\mathcal{R}}(Y, v_Y) \to \Omega^*_{\mathsf{basic}\, d\mathcal{R}}(X, v_X)$$

of the basic differential complexes. Their basic de Rham and regular de Rham cohomologies form a commutative diagram:

$$\begin{array}{cccc} \Psi^*_{\mathsf{b},H} : & H^*_{\mathsf{basic}\, d\mathcal{R}}(Y, v_Y) & \to & H^*_{\mathsf{basic}\, d\mathcal{R}}(X, v_X). \\ & \downarrow & & \downarrow \\ \Psi^*_H : & H^*_{d\mathcal{R}}(Y) & \to & H^*_{d\mathcal{R}}(X). \end{array}$$

If v_Y is traversing, then so is v_X, and the lower row of the diagram can be replaced by the Ψ-induced homomorphism in singular homology

$$\Psi^*_{\mathcal{T},H} : H^*(\mathcal{T}(v_Y); \mathbb{R}) \to H^*(\mathcal{T}(v_X); \mathbb{R}).$$

Proof. The first claim follows from the naturality of the boundary operator d and the operator $v\rfloor$ under smooth transformations.

The second claim is implied by [K3], Theorem 5.1, or by [K7], Theorem 6.8, which assert that, for a traversing vector field v, the map $X \to \mathcal{T}(v)$ is a homology equivalence. □

For a contact form β and its Reeb vector field v_β, the form $(d\beta)^\ell \in \Omega^{2\ell}(X, v_\beta)$ is closed, v_β-invariant, and horizontal. Therefore, it is basic and determines an element $[(d\beta)^\ell] \in H^{2\ell}_{\mathsf{basic}\, d\mathcal{R}}(X, v_\beta)$. Note that if $[(d\beta)^\ell] = 0$, then $[(d\beta)^k] = 0$ for all $k > \ell$.

Lemma 3.41. *Let X, Y be equidimensional compact smooth manifolds that carry contact forms β_X, β_Y. Let $\Psi : X \to Y$ be a smooth immersion such that $\Psi^*(\beta_Y) = \beta_X$.*

Then, for any $\ell \in [1, n]$, we have the equality

$$\Psi^*([(d\beta_Y)^\ell]) = [(d\beta_X)^\ell]$$

in $H^{2\ell}_{\mathsf{basic}\, d\mathcal{R}}(X, v_X)$.

Proof. The property $\Psi^*(\beta_Y) = \beta_X$ implies that Ψ_* maps the Reeb vector field v_{β_X} the to the Reeb vector field v_{β_Y}. By naturality, the pull-back of a v_{β_Y}-basic differential form on Y is a v_{β_X}-basic differential form on X and the differential operators d_X and d_Y on forms commute with Ψ^*. Therefore, $\Psi^*([(d\beta_Y)^\ell]) = [(d\beta_X)^\ell]$ in $H^{2\ell}_{\mathsf{basic}\, d\mathcal{R}}(X, v_X)$. □

Corollary 3.52. *Under the hypotheses of Lemma 3.41, if $[(d\beta_X)^\ell] \neq 0$ in $H^{2\ell}_{\mathsf{basic}\, d\mathcal{R}}(X, v_{\beta_X})$ and $[(d\beta_Y)^k] = 0$ in $H^{2k}_{\mathsf{basic}\, d\mathcal{R}}(Y, v_{\beta_Y})$ for $k \leq \ell$, then no map Ψ, subject to $\Psi^*(\beta_Y) = \beta_X$, exists.* ◇

Let us conclude with a conjecture, weakly motivated by [Raz], Corollary 4.4. Recall that the corollary claims that *homeomorphic* closed manifolds, both satisfying the Basic Hard Lefschetz property for the appropriate foliations of the same dimension, have isomorphic basic de Rham cohomology. In other words, for closed smooth manifolds that satisfy the Basic Hard Lefschetz property, the basic de Rham cohomology is a topological invariant.

Conjecture 3.9. *Let X be a compact connected smooth manifold and v a (non-vanishing) vector field on X. Assume that v admits a Lyapunov function and is transversally generic.*

Then the v-basic de Rham cohomology $H^*_{\mathsf{basic}\, d\mathcal{R}}(X, v)$ depends only on the $(\mathbf{\Omega}^\bullet)$-stratified topological type of the trajectory space $\mathcal{T}(v)$. ◇

Chapter 4

Causal Holography in Application to the Inverse Scattering Problems and Billiards

"A mathematician is a blind man in a dark room looking for a black cat which isn't there."

Charles Darwin

"If we knew what it was we are doing, it would not be called research, would it?"

Albert Einstein

4.1 In this chapter...

Let M be a compact connected and smooth Riemannian n-manifold with boundary and g a Riemannian metric on M. In this chapter, we apply the holographic causality principles from Chapter 3 to the geodesic flow v^g on the space $SM =_{\mathsf{def}} S_gM$ of unit tangent vectors on M. Our main observation is that the holographic causality is intimately linked to the classical inverse scattering problems, as well as to geodesic billiards. The geodesic scattering is the focus of our investigations in the chapter.

In general, our approach to the inverse scattering problem relies more on the methods of Differential Topology and Singularity Theory, and less on the more analytical methods of Differential Geometry and Operator Theory. Of course, this topological approach has its limitations: by itself it allows only for a reconstruction of the geodesic flow on the tangent spherical fibration from the scattering data.

First, let us explain informally what we mean by the scattering data on a given compact connected smooth Riemannian manifold M with boundary

∂M. For each geodesic curve $\gamma \subset M$ which "enters" M at a point $m \in \partial M$ in the direction of a unitary tangent vector $u \in T_m(M)$ (that points inside of M), we register the first along γ "exit point" $m' \in \partial M$ and the corresponding exit direction, given by an unitary tangent vector $u' \in T_{m'}(M)$ at m' (that points outside of M). Of course, not for any geodesic curve γ on M, starting at $m \in \partial M$, this construction makes sense: for example, $\gamma \setminus m$ may belong to the interior of M. In such a case, the metric g is called trapping.

For a non-trapping g, we call the correspondence

$$C_{v^g} =_{\mathsf{def}} \{(m, v) \Rightarrow (m', v')\}_{m \in \partial M,\, v \in T_m M}$$

"the scattering data".

We strive to restore the metric g on M (say, up to the action of diffeomorphisms of M that are the identity maps on ∂M) from the scattering data. This resembles the problem of reconstructing the mass distribution from the gravitational lensing [TuTo]. Attempts to reconstruct the pair (M, g) from the scattering map C_{v^g} are intimately related to the fundamental problems in tomography, gravitational lensing, geophysical survey, etc.

This restoration is harder when (M, g) has *closed geodesics* or *geodesics of infinite length*, i.e., when the metric g is trapping. Fig. 4.1 shows an example of a metric on the sphere S^2 with many trapped geodesics.

However, for special (M, g), the restoration of g from enhanced scattering data is possible. Such restorations are in the focus of deep research efforts: see [Cr], [Cr1], [We], as well as [SU] - [SU4] and [SUV] - [SUV3]. Perhaps, the paper [SUV3] reflects the relatively modern state of the art.

Moving towards the goal of reconstructing a metric g from the scattering data, we introduce a class of Riemannian metrics which are called non-trapping metrics (see Definition 4.1). Such metrics g have no closed geodesics and geodesics of infinite length. For them, the geodesic flow v^g on the spherical tangent bundle $SM \to M$ is *traversing*.

If one thinks of the spherical tangent fibration $SM \to M$ as the bundle of rays in the fibers of the tangent bundle $TM \to M$, then the construction of SM is g-independent. Alternatively, one may think of SM as subspace of TM, defined as the bundle of unit spheres in the metric g. In such a case, we could use the notation "$S_g M$"; however in this chapter, to simplify notations, we will use the metric neutral notation "SM" unless we need to deal with different metrics at the same time.

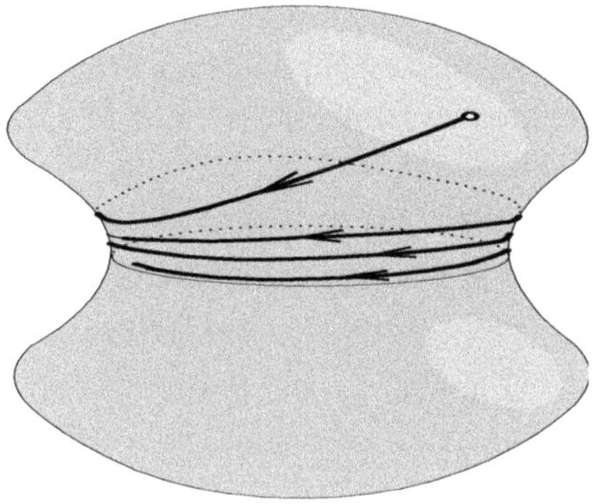

Fig. 4.1 Riemannian metric of a sphere with a massive set of trapped geodesics. In the diagram, a single geodesic raps infinitely many times around the neck of the sphere, while approaching a closed geodesic, the waist of the neck.

We denote by $\mathcal{NT}(M)$ the space of the non-trapping Riemannian metrics on M. They form an open set in the space of all Riemannian metrics (see Theorem 4.1).

In fact, there are examples of two real analytic Riemannian manifolds with isometric boundaries and identical scattering (even lens) data, but with different C^∞-jets of the metric tensors along the boundaries ([Zh], Theorem 4.3)! However, these examples have trapped geodesics; so they employ metrics that are fundamentally different from the ones we study here.

By Lemma 4.2, the metrics g on M, for which the geodesic vector field v^g on SM admit a Lyapunov function, are exactly the non-trapping metrics.

Next, we introduce another class of Riemannian metrics g on M such that the boundary ∂M is "generically curved" in g (see Definition 4.3). For them, the geodesic field v^g is boundary generic in the sense of Definition 2.2. We call such metrics g geodesically boundary generic, or boundary generic for short.

We denote by $\mathcal{NT}^\dagger(M)$ the space of non-trapping and geodesically boundary generic metrics on M. We speculate (see Conjecture 4.2) that, for any M, the space $\mathcal{NT}^\dagger(M)$ is open and dense in $\mathcal{NT}(M)$ and prove that it is, indeed, open (see Theorem 4.1).

We also consider a subspace $\mathcal{NT}^\ddagger(M) \subset \mathcal{NT}^\dagger(M)$, formed by metrics g for which the geodesic vector field v^g on SM is traversally generic in the sense of Definition 2.6.

In fact, for all non-trapping metrics g, the geodesic field v^g on SM allows for an arbitrary accurate C^∞-*approximations* by boundary generic or even traversally generic vector fields w on SM (however, such fields w are not necessarily Hamiltonian).

For every metric $g \in \mathcal{NT}^\dagger(M)$, the geodesic scattering along the boundary ∂M can be expressed in terms of the scattering map

$$C_{v^g} : \partial_1^+(SM) \to \partial_1^-(SM).$$

It acts from a domain $\partial_1^+(SM)$ in the boundary $\partial_1(SM)$ to the complementary domain $\partial_1^-(SM)$, both domains being *diffeomorphic* and actually g-*independent*!

Theorem 4.2, the main result of Section 4.3, claims that, for a non-trapping boundary generic metric g and for a fixed scattering map C_{v^g}, the geodesic flow v^g on SM is topologically rigid. This means that when two scattering maps, $C_{v^{g_1}}$ and $C_{v^{g_2}}$, are conjugated with the help of a smooth diffeomorphism $\Phi^\partial : \partial_1(SM_1) \to \partial_1(SM_2)$, then the *un*parametrized geodesic flows on SM_1 and SM_2 are conjugated with the help of a homeomorphism (often a diffeomorphism) $\Phi : SM_1 \to SM_2$ which extends Φ^∂.

Let $\mathcal{T}(v^g)$ denote the space of geodesics for (M, g). In Theorem 4.3, we prove that, for any metric $g \in \mathcal{NT}^\dagger(M)$, the scattering data are sufficient for a reconstruction of the $\mathbf{\Omega}^\bullet$-stratified topological type of the space of geodesics $\mathcal{T}(v^g)$. In general, $\mathcal{T}(v^g)$ is not a smooth manifold, but for $g \in \mathcal{NT}^\ddagger(M)$, it is a compact CW-complex (Corollary 2.2).

For $g \in \mathcal{NT}^\dagger(M)$, the space $\mathcal{T}(v^g)$ carries some "surrogate smooth structure" (see Definition 2.8). It is also captured by the scattering data.

Theorem 4.4 claims that, for any $g \in \mathcal{NT}^\dagger(M)$, the geodesic scattering map C_{v^g} allows for a reconstruction of the homology spaces $H_*(M;\mathbb{R})$ and $H_*(M, \partial M;\mathbb{R})$, equipped with the Gromov simplicial semi-norms (see Definition 4.8 and [Gr]). In particular, the simplicial volume of the relative fundamental cycle $[M, \partial M]$ can be recovered from the scattering map. Furthermore, if $\dim M \geq 3$, the geodesic scattering map C_{v^g} allows for a reconstruction of the fundamental group $\pi_1(M)$, together with all the homotopy groups $\{\pi_i(M)\}_{i<\dim M}$.

Let (N, g) be a closed smooth locally symmetric Riemannian n-manifold, $n \geq 3$, of negative sectional curvature. In Theorem 4.11, we prove that, if and M is obtained from N by removing the interior of a smooth n-ball so that $\tilde{g} = g|_M$ is non-trapping and boundary generic, then the knowledge of the scattering map $C_{v^{\tilde{g}}} : S^{n-1} \times D^{n-1} \to S^{n-1} \times D^{n-1}$ makes it possible to reconstruct N and the metric g, up to a positive scalar factor.

In Section 4.4, we study the inverse scattering problem in the presence of additional information about the *lengths of geodesic curves* that connect each point $(m, v) \in \partial_1^+(SM)$ to the "scattered" point $C_{v^g}(m, v) \in \partial_1^-(SM)$. This information is commonly called the "lens data".

Our main result here is Theorem 4.12, which claims the strong topological rigidity (see Definition 4.4) of the geodesic flow for given lens data. The proof of the theorem requires additional hypotheses about the metric, which we call "ballanced" (see Definition 4.12).

We apply Theorem 4.12 to the case of manifolds M, obtained from closed manifolds (N, g) by removing special domains $U \subset N$. Then the scattering problem on $(M, g|_M)$ is intimately linked to the geodesic flow on SN (see Theorem 4.13). By combining Theorem 4.13 with some classical results from [CK], [BCG], we are able to reconstruct (N, g) from the scattering and lens data on M, provided that either N is locally symmetric and of negative sectional curvature, or admits a non-vanishing parallel vector field (see Corollary 4.13 and Corollary 4.14).

For a metric $g \in \mathcal{NT}^\dagger(M)$, an involution $\tau : \partial_1^+(SM) \to \partial_1^-(SM)$, induced by the g-orthogonal (ellastic) reflection of tangent vectors with respect to the boundary ∂M, helps to convert the scattering map C_{v^g} to the billiard map

$$B_{v^g} : \partial_1^+(SM) \to \partial_1^+(SM).$$

The dynamics of iterations of the map B_{v^g} is the focus of the theory of billiards, a very active and flourishing field of research.

4.2 Boundary generic metrics of the non-trapping type

Let M be a compact n-dimensional smooth Riemannian manifold with boundary, and g a smooth Riemannian metric on M. Let $SM \to M$ be the tangent spherical bundle. With the help of g, we may interpret the bundle SM is a subbundle of the tangent bundle TM.

The metric g on M induces a locally-defined one-parameter family of diffeomorphisms $\{\psi_g^t : TM \to TM\}$, the geodesic flow. If M has a boundary, then the time parameter t may be restricted. The submanifold $S_gM \subset TM$ of unit spheres is invariant under $\{\psi_g^t\}$, so that the restrictions $\{\psi_g^t : S_gM \to S_gM\}$ make sense. When the choice of g is clear, we abbreviate the notation S_gM to the notation SM.

Let us review briefly few basic constructions and facts about the differential geometry of geodesic flows.

In canonical local coordinates $(q, p) = (q^1, \dots, q^n, p_1, \dots p_n)$ on the tangent space TM, the differential equations of the geodesic flow v^g are:

$$\dot{q}^i = \sum_{j=1}^{n} g_{ij}(q)\, p_j$$

$$\dot{p}_i = -\frac{1}{2} \sum_{j,k=1}^{n} \frac{\partial g_{jk}}{\partial q^i}(q)\, p_j p_k, \tag{4.1}$$

where $g_{ij}(q)$ is the metric tensor of the metric g.

This system can be rewritten in terms of the Hamiltonian function

$$H^g(q, p) =_{\mathsf{def}} \frac{1}{2} \sum_{i,j} g_{ij}(q)\, p_i p_j$$

—the kinetic energy—in the already familiar (c.f. (3.23)) Hamiltonian form:

$$\dot{q}^i = \frac{\partial H^g}{\partial p_i}$$

$$\dot{p}_i = -\frac{\partial H^g}{\partial q^i} \tag{4.2}$$

The projections of the trajectories of (4.1) or (4.2) on M are the geodesic curves.

Let v^g—a section of the secondary tangent bundle $T(TM) \to TM$—be the vector field on the manifold TM, tangent to the trajectories of the (partially defined) geodesic flow $\{\psi^t_g : TM \to TM\}$ that integrates (4.2). Since v^g-flow preserves the function H, v^g is tangent to $SM \subset TM$; thus, $v^g|_{SM} \neq 0$.

Each g-unitary tangent vector u at a point $m \in M \setminus \partial M$ determines a unique geodesic curve $\gamma_{(m,u)} \subset M$ trough m in the direction of u. When $m \in \partial M$, the germ of the geodesic curve $\gamma_{(m,u)}$ is uniquely defined for unit vectors $u \in T_m M$ that point inside of M.

By definition, $\psi^t_g(m, u)$ is the point $(m', u') \in SM$ such that the distance along $\gamma_{(m,u)} \subset M$ from $m' \in \gamma_{(m,u)}$ to m is t, and u' is the tangent vector to $\gamma_{(m,u)}$ at m'. Once more, we stress that $\psi^t_g(m, u)$ may not be well-defined for all $t \in \mathbb{R}$ and all $u \in TM$: some geodesic curves γ may originate at the boundary ∂M and never reach it again, and some tangent vectors $u \in TM|_{\partial M}$ may point outside of M. However, such constraints are familiar to us: our entire enterprise deals with such boundary-induced complications.

Let us revisit now few basic constructions from Symplectic and Contact Geometry (already familiar from Subsection 3.8), produced by Riemannian metrics.

Let β be the tautological Liouville 1-form (locally, "$-p^* dq$") on the cotangent bundle T^*M. By definition,

$$\beta(q, p^*; \dot{q}, \dot{p}^*) = p^*(\dot{q}).$$

The Liouville 1-form β is a smooth section of the bundle $T^*(T^*M) \to T^*M$.

The closed 2-form $\omega = d\beta$ is called the canonical symplectic form on T^*M (locally, "$dq \wedge dp^*$").

The metric g on M gives rise to a bundle isomorphism $\Phi_g : TM \to T^*M$. Consider the pull-backs

$$\beta_g =_{\mathsf{def}} \Phi_g^*(\beta), \quad \omega_g =_{\mathsf{def}} \Phi_g^*(\omega)$$

of the forms β and ω under the diffeomorphism Φ.

Let μ be the unitary (in g) radial vector field, normal to S_gM in TM and tangent to the fibers of $TM \to M$. Then $\beta_g = \pm\, \mu \rfloor \omega_g$.

In local coordinates $(q^1, \ldots, q^n, p_1, \ldots, p_n)$ on TM, these forms may be written as

$$\beta_g = -\sum_{i,j} g_{ij}\, p_i\, dq^j, \tag{4.3}$$

and

$$\omega_g = \sum_{i,j} g_{ij}\, dq^i \wedge dp_j + \sum_{i,j,k} \frac{\partial g_{ij}}{\partial q^k}\, p_i\, dq^j \wedge dq^k. \tag{4.4}$$

• Note that here $\dim(SM) = 2n - 1$, while in Subsection 3.8 the convention was $\dim(X) = 2n + 1$!

By the Liouville theorem, both β_g and $\omega_g = d\beta_g$ are v^g-invariant forms, that is, $\mathcal{L}_{v^g}(\beta_g) = 0$ and $\mathcal{L}_{v^g}(\omega_g) = 0$. Also, we have $\beta_g(v^g) = 1$.

In terms of the canonical symplectic 2-form ω_g and the Hamiltonian function H^g on TM, the geodesic vector field v^g is characterized by the property

$$\omega_g(v^g, \sim) = dH^g(\sim).$$

We denote by $(\omega_g)^n$, the nth exterior power of the 2-form ω_g. It is a $(2n)$-dimensional volume form on TM.

Let

$$\Omega_g =_{\mathsf{def}} \{\beta_g \wedge (\omega_g)^{n-1}\}\big|_{SM} \tag{4.5}$$

be a $(2n-1)$-dimensional volume form Ω_g on SM. In fact, $\Omega_g = \mu \rfloor (\omega_g)^n$.

Next, we introduce a $(2n-2)$-form Θ_g on SM by

$$\Theta_g =_{\mathsf{def}} v^g \rfloor \Omega_g = (v^g \wedge \mu) \rfloor \omega_g^n. \tag{4.6}$$

Both forms, Ω_g and Θ_g, are v^g-invariant.

A given Riemannian metric g on M induces the Sasaki Riemannian metric gg on the manifold TM ([Sa]). For any pair of tangent vectors, $v, u \in T_m M$, we consider the germ γ_u of the geodesic curve through m in the direction of u. Using the g-induced symmetric connection ∇^g on M, we consider the Jacobi vector field $\tilde{v}$ along γ_u, produced by the parallel transport of v along the geodesic γ_u. Consider the natural parameter s along the curve γ_u.

We get a germ at (m, v) of a curve $\delta_{\gamma_u, v} =_{\mathsf{def}} \{s \to \tilde{v}(s)\}$ in TM, which projects on the geodesic γ_u under the map $\pi : TM \to M$. We denote by $W(v, u)$ the velocity vector $\frac{d}{ds}\delta_{\gamma_u, v}(0)$ of $\delta_{\gamma_u, v}$ at the point (m, v). Thus, $W(v, u) \in T_{(m,v)}(TM)$.

The correspondence $H_{(m,v)} : u \to W(v, u)$ produces a linear injection $H_{(m,v)} : T_m(M) \to T_{(m,v)}(TM)$. We denote by $H_{(m,v)}(TM)$ its image and consider the horizontal subbundle $H(TM) \hookrightarrow TM$ of the bundle $T(TM) \to TM$ that this construction delivers. The subbundle, formed by vectors that are tangent to the fibers of $\pi : TM \to M$, is called the vertical subbundle of $T(TM)$ and is denoted by $V(TM)$. With its help, one splits the tangent bundle $T(TM) \to TM$ into a direct sum $H(TM) \oplus V(TM)$.

Then, by the definition of the Sasaki metric gg, this bundle isomorphism

$$H(TM) \oplus V(TM) \approx TM \oplus TM$$

is an isometry with respect to the metric gg on the source and the metric $g \oplus g$ on the target.

Since under the parallel transport the norm of the vectors $\tilde{v}$ is preserved, the curves $\{\delta_{\gamma_u, v}\}_u$ reside in SM, provided $(m, v) \in SM$. Therefore, we get a bundle decomposition $T(SM) \approx H^S(TM) \oplus V(SM)$, where the n-bundle $H^S(TM) \to SM$ is a restriction of $H(TM) \to TM$ to $SM \subset TM$, and $V(SM) \to SM$ is the $(n-1)$-bundle, tangent to the fibers of $SM \to M$.

The Sasaki metric gg has a number of nice properties. For example, in the metric gg, the fibers of the bundle $T(TM) \to TM$ are the geodesic submanifods ([Be], Prop. 1.102) and the integral trajectories of v^g are *geodesic curves* in the gg-metric on $T(TM)$ ([Be], Prop. 1.106).

The Sasaki metric gg induces a metric on the compact submanifold $SM \subset TM$. Abusing notations, we also denote it by "gg".

We notice that this construction $(M, g) \Rightarrow (SM, gg)$ can be iterated so that one gets a sequence of compact Riemannian manifolds:

$$(M, g) \rightsquigarrow (SM, gg) \rightsquigarrow (S(SM), gggg) \rightsquigarrow \dots,$$

and a tower of spherical fibrations:

$$(M, g) \leftarrow (SM, gg) \leftarrow (S(SM), gggg) \leftarrow \dots,$$

where $gggg =_{\mathsf{def}} (gg)(gg)$, etc., and each manifold $S(\dots(SM)\dots)$ being equipped with its own geodesic flow.

As any non-vanishing vector field, the geodesic field $v^g \in T(SM)$ divides the boundary $\partial_1(SM)$ into two loci: $\partial_1^+(SM)$, where v^g points inside of SM or is tangent to its boundary, and $\partial_1^-(SM)$, where v^g points outside of SM or is tangent to its boundary. In fact, $\partial_1^+(SM)$ and $\partial_1^-(SM)$ *do not depend on* g: the first locus is formed by the pairs (m, v), where $m \in \partial M$ and $v \in T_m M$ points inside of M or is tangent to ∂M. Therefore, both $\partial_1^+(SM)$ and $\partial_1^-(SM)$ are homeomorphic to the tangent $(n-1)$-disk bundle of the manifold ∂M. The locus $\partial_2(SM) = \partial(\partial_1^+(SM)) = \partial(\partial_1^-(SM))$ is also g-independent; it is the space of the sphere bundle, associated with the tangent $(n-1)$-bundle $T(\partial M)$.

Definition 4.1. Let (M, g) be a compact connected smooth Riemannian manifold with boundary.

We say that a metric g on M is of the Lyapunov type if there exists a smooth Lyapunov function $F : SM \to \mathbb{R}$ for the geodesic vector field v^g such that $dF(v^g) > 0$.

We denote by $\mathcal{L}yap(M)$ the space of the Lyapunov type Riemannian metrics on M. ◇

The condition $dF(v^g) > 0$ is equivalent to the property $\{\tilde{F}, H^g\} > 0$, where $\tilde{F} : TM \setminus M \to \mathbb{R}$ is a smooth extension of F, the Hamiltonian H^g is defined by equations (4.1) and (4.2), and $\{,\}$ stands for the Poisson bracket of functions on TM.

Lemma 4.1. *The set $\mathcal{L}yap(M)$ of the Lyapunov type Riemannian metrics is* open *in the space $\mathcal{R}(M)$ of all Riemannian metrics on M, considered in the C^∞-topology.*

Proof. Let $\sigma : M \to TM$ denote the zero section. If $dF(v^g) > 0$ on SM, then $F : SM \to \mathbb{R}$ extends in a compact neighborhood U of SM in $TM \setminus \sigma(M)$ to a smooth function $\tilde{F} : U \to \mathbb{R}$ so that $d\tilde{F}(v^g) > 0$ in U. In this neighborhood, $d\tilde{F}(v^{g'})|_U > 0$ for all metrics g', sufficiently close to g. For such metrics g', the space of unit spheres $S_{g'}M \subset TM$ is *fiberwise* close to $S_g M \subset TM$; in particular, we may assume that $S_{g'}M \subset U$. Recall that the geodesic field $v^{g'}$ is tangent to $S_{g'}M$, thus $d\tilde{F}(v^{g'}) > 0$ on $S_{g'}M$. □

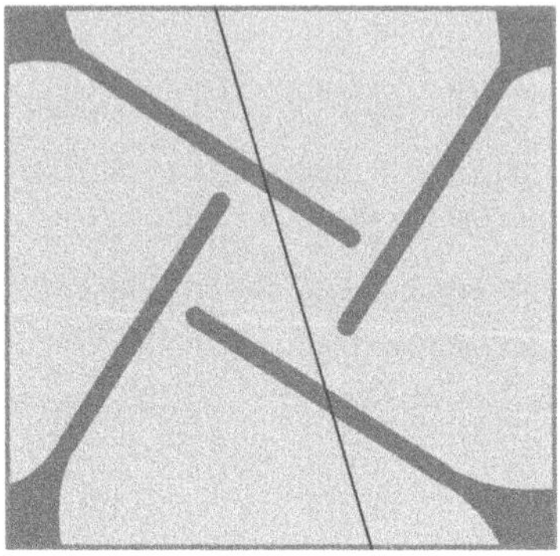

Fig. 4.2 The flat metric on the 2-torus becomes non-trapping on the complement to a curvy dark disk (whose "center" is at the corners of the square fundamental domain). Note that any line hits the curvy disk.

Definition 4.2. A Riemannian metric g on a connected compact manifold M with boundary is called non-trapping if g has no closed geodesics and no geodesics of infinite length (the later are homeomorphic to an open or a semi-open interval). $\diamondsuit$

Lemma 4.2. *Let M be a compact connected smooth manifold with boundary. A metric g on M is non-trapping, if and only if, it is of the Lyapunov type.*

Proof. If a smooth function $F : SM \to \mathbb{R}$ is such that $dF(v^g) > 0$, then each v^g-trajectory is either a singleton residing in $\partial_1(SM)$ or a closed segment with its both ends residing in $\partial_1(SM)$. Thus, such vector fields v^g are *traversing*. Evidently, F prevents v^g from having a closed trajectory. Let $\tilde{\gamma}$ be a trajectory that starts at a point $w \in SM$ and is homeomorphic to a semi-open interval. Thus, $\tilde{\gamma}$ extends beyond any point on $\tilde{\gamma}$ that can be reached from w (i.e., $\tilde{\gamma}$ is trapped in $\mathsf{int}(SM)$). Consider the closure K of $\tilde{\gamma}$. It is a compact and v^g-invariant set. Therefore, F attends its maximum at a point $w_\star \in K$. However, $dF(v^g) > 0$ at $w_\star$ and a germ of a v^g-trajectory through $w_\star$ belongs to K, a contradiction to the assumption that F attends its maximum at $w_\star$. A similar argument rules out the $(-v^g)$-trajectories that are homeomorphic to a semi-open intervals.

Conversely, by Lemma 2.1 (or Lemma 5.6 from [K7]), any traversing vector field (by Definition 2.1, such field has only trajectories that are homeomorphic to closed intervals or singletons) admits a Lyapunov function. In particular, v^g admits a Lyapunov function F.

Thus, g is of the Lyapunov type if and only if the image of any v^g-trajectory $\tilde{\gamma}$ under the map $SM \to M$ is either a singleton in ∂M or a compact geodesic curve γ whose ends reside in ∂M (by our convention, γ does not extends beyond its two ends). In particular, any $g \in \mathcal{L}yap(M)$ has no closed geodesics in M and no geodesics that originate at the boundary ∂M and are trapped in $\mathsf{int}(M)$ for all positive times. □

Example 4.1. Consider a flat metric g on the torus $T^2 = \mathbb{R}^2/\mathbb{Z}^2$. We form a punctured torus M by removing an open disk D^2 from T^2. Let $Q^2 \subset \mathbb{R}^2$ be the fundamental square domain. It is possible to position $D^2 \subset T^2$ so that its lift $\tilde{D}^2$ to $\mathbb{R}^2$ will have intersections with *any* line that passes through Q^2 (see Fig. 4.2). We restrict the flat metric g to $M = T^2 \setminus \mathsf{int}(D^2)$. Evidently, the intersection of any geodesic in T^2 with M is a disjoint union of closed intervals, whose boundaries reside in ∂M, and singletons, also residing in ∂M. For such a choice of $(M, g|_M)$, thanks to Lemma 4.2, the metric $g|_M$ is non-trapping and of the Lyapunov type. Moreover, by Lemma 4.1, any metric g' on M, sufficiently close to this flat metric $g|_M$, is also non-trapping. ♢

Corollary 4.1. *Let (N, g) be an open Riemannian manifold such that no geodesic curve is closed or has an end (homeomorphic to the interval $[0, 1)$) that is contained in a compact set. Let $M \subset N$ be a smooth compact codimension zero submanifold.*

Then the restriction $g|_M$ is a non-trapping metric, and so are all the metrics $\tilde{g}$ on M that are sufficiently close to $g|_M$.

In particular, for any compact domain $M \subset \mathbb{R}^n$, the flat metric g_{E} on M is of the non-trapping type, and so are all the metrics g that are sufficiently close to g_{E}.

Proof. Thanks to the hypothesis about (N, g) and the compactness of M, any geodesic $\gamma \subset N$ through any point in $\mathsf{int}(M)$ hits the boundary ∂M in both directions. By Lemma 4.2, the pair $(M, g|_M)$ admits a Lyapunov function for the geodesic flow $v^{g|_M}$ and is non-trapping.

By Lemma 4.1, any metric $\tilde{g}$ on M, which is sufficiently close to $g|_M$, is also non-trapping. □

We confess to waste some time trying to prove the following assertion.

Conjecture 4.1. *For any closed smooth Riemannian manifold (N, g), there is a finite disjoint union of smooth (g-*convex*) balls whose removal from N delivers a non-trapping metric g_M on their complement M.* ♢

Definition 4.3. Let (M, g) be a compact connected Riemannian manifold with boundary.

- We say that a metric g on M is geodesically boundary generic if the geodesic vector field $v^g \in T(SM)$ is boundary generic (in the sense of Definition 2.2) with respect to the boundary $\partial_1(SM) = SM|_{\partial M}$.
- We say that a metric g on M is geodesically traversally generic if the vector field $v^g \in T(SM)$ is traversally generic (in the sense of Definition 2.6) with respect to the boundary $\partial_1(SM) = SM|_{\partial M}$.

We denote the space of all non-trapping metrics on M by the symbol $\mathcal{NT}(M)$, the space of all non-trapping geodesically boundary generic metrics on M by the symbol $\mathcal{NT}^\dagger(M)$, and the space of all non-trapping geodesically traversally generic metrics on M by the symbol $\mathcal{NT}^\ddagger(M)$. $\diamondsuit$

By the definition, we get the inclusions

$$\mathcal{NT}^\ddagger(M) \subset \mathcal{NT}^\dagger(M) \subset \mathcal{NT}(M).$$

Remark 4.1. If (M, g) is such that there exists a geodesic curve $\gamma \subset M$ whose arc is contained in ∂M, then the metric g is not geodesically boundary generic in the sense of Definition 4.3. For example, the Euclidean metric on $\mathbb{R}^3$ is not geodesically boundary generic with respect to the ruled surface $S = \{z = x^2 - y^2\}$: indeed, S is comprised of lines (geodesics). $\diamondsuit$

In the following sections, we will try to address the following question.

Question 4.1. *How to formulate the property of the geodesic vector field v^g being boundary generic/traversally generic with respect to $\partial_1(SM)$ in terms of the geodesic curves and Jacobi vector fields on M and their interactions with the boundary ∂M?* $\diamondsuit$

Example 4.2. Let M be a domain in the Euclidean plane $\mathbb{R}^2$, bounded by simple smooth closed curves. Then the flat metric g_E is boundary generic on M if and only if ∂M is comprised of strictly concave and convex loops or arcs that are separated by the cubic inflection points. In particular, no line, tangent to the boundary, has the order of tangency that exceed 3 (see Example 4.8 for the details). $\diamondsuit$

Remark 4.2. Of course, not any metric g on M is non-trapping. At the same time, thanks to Lemma 4.1, the non-trapping metrics form an open set.

Examples of geodesically boundary generic metrics are also not so hard to exhibit. They require only a *localized* control of the geometry of ∂M in terms of g (see Lemmas 4.5 and 4.6). For instance, if each connected component of ∂M is either *strictly convex* or *strictly concave* in g, then g is geodesically boundary generic.

In contrast, to manufacture a geodesically traversally generic metric is a more delicate task. In fact, we know only few examples, where non-trapping metrics are proven to be geodesically traversally generic; all these cases feature a convex boundary ∂M. However, we suspect that geodesically traversally generic metrics are abundant (see Conjecture 4.2). In any case, by Theorem 4.1 below, the property of a metric g to be traversally generic is *stable* under its small smooth perturbations. $\diamondsuit$

We have only a weak evidence for the validity of following conjecture; however, the world in which it is valid seems to be a pleasing place...

Conjecture 4.2. *The sets $\mathcal{NT}^\dagger(M)$ and $\mathcal{NT}^\ddagger(M)$ are open and* dense *in the space $\mathcal{NT}(M)$.* $\diamondsuit$

The openness of $\mathcal{NT}^\dagger(M)$ and $\mathcal{NT}^\ddagger(M)$ in $\mathcal{NT}(M)$ follows from the theorem below.

Theorem 4.1. *Let M be a compact smooth connected manifold with boundary.*

In the space $\mathcal{R}(M)$ of all Riemannian metrics on M, the subspaces $\mathcal{NT}^\dagger(M)$ and $\mathcal{NT}^\ddagger(M)$ are open in the C^∞-topology. Each of these spaces is invariant under the natural action of the smooth diffeomorphism group $\mathsf{Diff}(M)$ *on $\mathcal{R}(M)$.*

If a metric g on M is non-trapping, then the geodesic field v^g on SM can be approximated arbitrary well in the C^∞-topology by a traversally generic vector field $\tilde{v} \in \mathcal{V}^\ddagger(SM)$ on SM.

Proof. The construction $\{g \Rightarrow v^g\}$ of the geodesic flow defines a continuous map $\mathcal{GF} : \mathcal{R}(M) \to \mathcal{V}_{\neq 0}(SM)$, where $\mathcal{V}_{\neq 0}(SM)$ denotes the space of all smooth non-vanishing vector fields on SM. Applying Theorem 2.1 (or Theorems 6.6 and 6.7 from [K7]), which deal with traversing or boundary generic vector fields of general nature, to the context of geodesic vector fields, we conclude that the subspace $\mathcal{V}^\ddagger(SM)$, formed by traversally generic (and thus traversing) vector fields, is open in $\mathcal{V}_{\neq 0}(SM)$. Similarly, the boundary generic and traversing vector fields form an open set

$\mathcal{V}^\dagger(SM)$ in $\mathcal{V}_{\neq 0}(SM)$. Moreover, $\mathcal{V}^\ddagger(SM)$ and $\mathcal{V}^\dagger(SM)$ are dense in the space $\mathcal{V}_{\mathsf{trav}}(SM)$ of all traversing vector fields on SM.

Since the germ of a geodesic γ through a point $m \in M$ in the direction of a given tangent vector $u \neq 0$ depends smoothly on metric g, we conclude that $\mathcal{GF}$ is a continuous map. Therefore, the two spaces

$$\mathcal{NT}^\ddagger(M) =_{\mathsf{def}} (\mathcal{GF})^{-1}(\mathcal{V}^\ddagger(SM)) \quad \text{and} \quad \mathcal{NT}^\dagger(M) =_{\mathsf{def}} (\mathcal{GF})^{-1}(\mathcal{V}^\dagger(SM))$$

are open in the space $\mathcal{R}(M)$ of smooth Riemannian metrics.

By definition, for any $g \in \mathcal{NT}(M)$, the geodesic field v^g on SM is traversing. Again, by Theorem 6.7 from [K7], v^g can be approximated by a traversally generic field $\tilde{v} \in \mathcal{V}^\ddagger(SM)$. Note that the projections of $\tilde{v}$-trajectories under the map $SM \to M$ may not stay C^∞-close to the geodesic lines in the original metric g due to the concave boundary effects.

Evidently, by the "naturality" of the geodesic flow, the spaces $\mathcal{NT}^\dagger(M)$, $\mathcal{NT}^\ddagger(M)$ are invariant under the natural action of the diffeomorphism group $\mathsf{Diff}(M)$ on the space $\mathcal{R}(M)$ of Riemannian metrics. □

Remark 4.3. Let M be a codimension zero compact submanifold of a compact Riemannian manifold N such that $M \subset \mathsf{int}(N)$. If a metric g on the ambient N is non-trapping, then its restriction $g|_M$ is non-trapping on M.

We conjecture that $\mathcal{NT}^\dagger(M) \neq \emptyset$ for any compact connected M with boundary. If M admits a *non-trapping* metric g such that each component of ∂M is simply concave or convex (in particular, nowhere in ∂M the normal curvature flips sign), then $\mathcal{NT}^\dagger(M) \neq \emptyset$. For example, a (convex) n-ball from which a number of (convex) n-balls are deleted represents such a manifold M.

At the same time, the question whether $\mathcal{NT}^\ddagger(M) \neq \emptyset$ for a given connected M with boundary remains open!

Example 4.4 describes one simple case of $M \approx S^{n-1} \times [0, 1]$ for which $\mathcal{NT}^\ddagger(M) \neq \emptyset$. ◇

In many cases the construction of a balanced Lyapunov function $F : SM \to \mathbb{R}$ for v^g which satisfies (4.8) is quite explicit in terms of (M, g), as the next example shows.

Example 4.3. Consider the Poincaré model of the hyperbolic space $\mathbb{H}^n$ with its virtual spherical boundary $\partial\mathbb{H}^n$ and hyperbolic metric g. The space $\mathbb{H}^n$ is modeled by the open unit ball in the Euclidean space $\mathbb{R}^n$. Each geodesic curve is an arc of a circle in $\mathbb{R}^n$ that hits the sphere $\partial\mathbb{H}^n$ at a pair of points, where it is orthogonal (in the Euclidean metric) to $\partial\mathbb{H}^n$ (see Fig. 4.4, in which the blob M should be ignored for now).

For each oriented geodesic γ trough a given point x in the direction of a given vector w, consider the distance $d_{\mathbb{H}}(x, w)$ between x and the unique point $y \in \bar{\gamma} \cap \partial\mathbb{H}^n$ that can be reached from x by moving along γ in the direction of $-w$; that is, $d_{\mathbb{H}}(x, w)$ is the length of the circular arc $(y, x) \subset \bar{\gamma}$ in $\mathbb{R}^n$. Evidently, $d_{\mathbb{H}}(x, w)$ is strictly increasing, as one moves along the oriented γ.

Let $M \subset \mathbb{H}^n$ be a compact codimension 0 smooth submanifold, equipped with the induced hyperbolic metric. Then the geodesic field v^g on the space SM admits a Lyapunov function since $d_{\mathbb{H}}(x, w)$ is strictly increasing along the oriented trajectories of v^g. As a result, the hyperbolic metric on M is non-trapping.

Again, by Theorem 4.1, any metric g' on M, sufficiently close to the hyperbolic metric $g|_M$, is also non-trapping. $\diamondsuit$

The construction from Example 4.3 can be generalized. Let M be a compact smooth codimension zero submanifold of a compact connected Riemannian manifold (L, g) with boundary. Assume that every geodesic γ in L such that $\gamma \cap M \neq \emptyset$ intersects ∂L transversally at a pair of points. This condition means that ∂L looks "convex", as seen from M. Then, according to Lemma 4.3 below, by a construction similar to the previous one, the metric $g|_M$ is non-trapping and a balanced Lyapunov function for the geodesic vector field v^g on SM is available.

Lemma 4.3. *Let M be a codimention zero smooth compact submanifold of a compact connected Riemannian manifold (L, g) so that $M \cap \partial L = \emptyset$. Assume that each geodesic in L that intersects M hits ∂L transversally at a pair of distinct points.*

Then the metric $g|_M$ is non-trapping and the geodesic flow on SM admits a balanced Lyapunov function.

Proof. Each geodesics γ in M extends to a unique geodesic γ_L in L, which intersects with ∂L at a pair of points $a(\gamma_L)$ and $b(\gamma_L)$. The orientation of γ picks one of these two points, say $a(\gamma_L)$, as the starting point of γ_L. Consider the function $f(x) =_{\mathsf{def}} \ell_g(x, a(\gamma_L))$, where $x \in \gamma_L$ and $\ell_g(x, a(\gamma_L))$ is the length of the geodesic arc $[x, a(\gamma_L)] \subset \gamma_L$. Let $\tilde{\gamma}_L$ be a lift of γ_L to the spherical bundle SL. Topologically, $\tilde{\gamma}_L$ is a closed interval.

For any point $(m, v) \in SM$, we take the trajectory $\tilde{\gamma}_L((m, v))$ of the geodesic flow on SL through (m, v). We define the function F :

$\tilde{\gamma}_L((m,v)) \to \mathbb{R}$ as the pull-back of $f : \gamma_L \to \mathbb{R}$. In fact, $F((n,v))$ is the length (in the Sasaki metric gg) of the segment $[(a(\gamma_L), w(\gamma_L)), (n,v)]$ of $\tilde{\gamma}_L$, where $w(\gamma_L)$ is the tangent to γ_L vector at the point $a(\gamma_L)$ and v is tangent to γ_L at $n \in L$. Using that $\tilde{\gamma}_L((m,v))$ is transversal to $\partial_1(SL)$ at the point $(a(\gamma_L), w(\gamma_L))$, we get that F is a smooth function of (n,v). Indeed, the solutions ODEs depend smoothly on the initial data, provided that such data vary along a transversal section of the flow. Evidently, $dF(v^g) > 0$ since f is increasing along γ_L. The variation of F along any segment Δ of $\tilde{\gamma}_L$ equals to its length in the metric gg, which, in turn, equals the length of $\pi(\Delta)$ in the metric g. In particular, for any $\Delta \subset \tilde{\gamma}_L \cap SM$, the variation of F along Δ is the length of $\pi(\Delta)$ in g. Thus, F is balanced in M. □

4.3 The geodesic scattering and holography

In this section, we will apply the Holographic Causality Principle (see Theorems 3.3, Corollary 3.5) to geodesic flows on the spaces SM of unit tangent vectors on compact connected Riemannian manifolds M with boundary.

In the rest of this chapter, we will be guided by a single important observation: if a metric g on M is non-trapping, then the causality map

$$C_{v^g} : \partial_1^+(SM) \to \partial_1^-(SM),$$

(introduced in Chapter 3 for smooth traversing vector fields v on compact manifolds X with boundary) is available. This map represents a metric-induced scattering. With the help of C_{v^g}, each unit tangent vector $u \in T^+M|_{\partial M}$ is mapped ("scattered") to a unit tangent vector $u' \in T^-M|_{\partial M}$. Here $T^{\pm}M|_{\partial M} \subset TM|_{\partial M}$ denote the bundle of half-spaces, formed by vectors along ∂M that are tangent to M and point inside/outside of M.

We will employ boundary generic or even traversally generic metrics g in order to control well the local structure of the causality map C_{v^g}.

For each tangent vector $w \in T_xM|_{\partial M}$, consider its orthogonal decomposition $w = n \oplus u$ with respect to the metric g, where n is the exterior normal to ∂M and $u \in T_x(\partial M)$. We denote by $\tau_g(x,w)$ the point $(x, -n \oplus u) \in T_xM|_{\partial M}$.

Lemma 4.4. *The g-independent manifolds $\partial_1^+(SM)(v^g)$ and $\partial_1^-(SM)(v^g)$ are diffeomorphic via the orientation-reversing involution*

$$\tau_g : \partial_1(SM) \to \partial_1(SM).$$

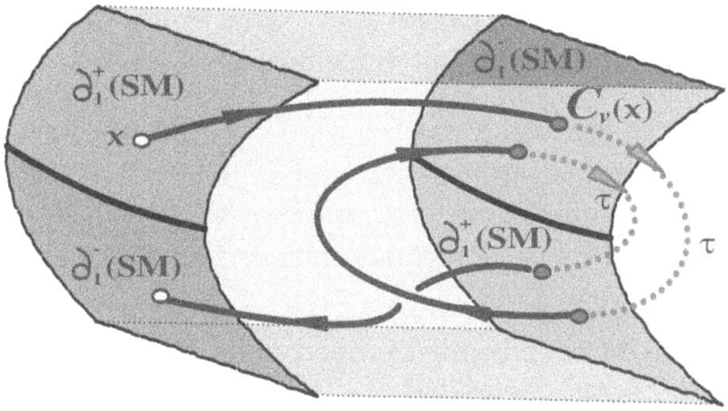

Fig. 4.3 A scattering (causality) map $C_{v^g} : \partial_1^+(SM) \to \partial_1^-(SM)$ and an involution $\tau : \partial_1^-(SM) \to \partial_1^+(SM)$, mimicking the elastic reflection from ∂M. The two bold arcs that separate $\partial_1^+(SM)$ and $\partial_1^-(SM)$ represent the tangent locus $\partial_2(SM)$.

Proof. The strata $\partial_1^{\pm}(SM)(v^g)$ consist of pairs (m, v) where $m \in \partial M$ and the vector $v \in T_m M$ points inside/outside of M. Thus, by the definition of τ_g, we see that it maps $\partial_1^+(SM)(v^g)$ to $\partial_1^-(SM)(v^g)$ by an orientation-reversing diffeomorphism. □

Given a compact Riemannian manifold (M, g), we denote by $\mathcal{F}(v^g)$ the oriented 1-dimensional foliation on SM, produced by the geodesic field v^g.

Let $\psi^t : SM \to SM$ be the v^g-flow generated family of local diffeomorphisms; if M has a boundary then, for each $x \in SM$, the image $\psi^t(x)$ may be well-defined only for some values of t, a time interval in $\mathbb{R}$.

Definition 4.4. Given two compact Riemannian n-manifolds, (M_1, g_1) and (M_2, g_2), consider the geodesic vector fields v^{g_1} on SM_1 and v^{g_2} on SM_2.

- We say that two metrics g_1 and g_2 have topologically conjugate geodesic flows if there is a homeomorphism $\Phi : SM_1 \to SM_2$ such that $(\Phi \circ \psi_1^t)(x) = (\psi_2^t \circ \Phi)(x)$ for all $x \in SM_1$ and all moments $t \in \mathbb{R}$ for which $\psi_1^t(x)$ is well-defined[1]. The restriction of Φ to each v^{g_1}-trajectory is required to be an orientation-preserving diffeomorphism.
- We say that the metrics g_1 and g_2 have topologically weakly conjugate geodesic flows if there is a homeomorphism $\Phi : SM_1 \to SM_2$ such that

[1]This implies that Φ preserves the lengths of the corresponding trajectories in the Sasaki metrics, and thus the lengths of the corresponding geodesic curves in M_1 and M_2 are equal.

it maps each leaf of the foliation $\mathcal{F}(v^{g_1})$ to a leaf of the foliation $\mathcal{F}(v^{g_2})$; Φ is an orientation-preserving diffeomorphism on each leaf. $\diamondsuit$

Both notions of conjugacy come in *different flavors* by requiring that g_i, Φ^∂, and Φ belong to various classes of C^k-smooth objects, where $k = 0, 1, 2, \dots, \infty$. For example, we may consider $\Phi^\partial, (\Phi^\partial)^{-1}$ of the class C^∞, while $\Phi, (\Phi)^{-1}$ may be just homeomorphisms (i.e., of the class C^0).

In the definition below, we assume that a given smooth compact Riemannian manifold M is embedded properly into a larger open manifold $\hat{M}$ and the metric g on M is extended smoothly to a metric $\hat{g}$ on $\hat{M}$. However, the properties, described in the definition, do not depend on a particular germ of extension $(\hat{M}, \hat{g}) \supset (M, g)$.

Definition 4.5. We say that a boundary generic non-trapping metric g on a compact manifold M has Property AA, if any geodesic curve $\gamma \subset M$ either has at least one point of transversal intersection with the boundary ∂M or is quadratically tangent to ∂M at some point where ∂M strictly convex. $\diamondsuit$

Remark 4.4. We will see soon that Property AA implies that the geodesic flow v^g on SM has Property A from Definition 3.2. In other words, the combinatorial tangency types of such v^g-trajectories do not belong to the closed subposet $(33)_{\succeq} \cup (4)_{\succeq} \subset \mathbf{\Omega}^\bullet$.

Recall that Property A reflects the shortcomings of our proof of the *smooth* version of the Holography Theorem 3.3. We suspect that it is a superfluous assumption, and the conjugating homeomorphism Φ in Theorem 3.3 is actually a diffeomorphism. Therefore, Property AA is likely a superfluous constraint as well. $\diamondsuit$

Remark 4.5. Even if we know the smooth type of a manifold M, the inverse scattering problems of the type "$C_{v^g} \overset{??}{\rightsquigarrow} g$" on Riemannian manifolds (M, g) have an unavoidable two-fold *intrinsic ambiguity*.

The first one arises from the action of the M-diffeomorphisms on the space $\mathcal{R}(M)$ that are the identity on ∂M and whose differentials are the identity on $TM|_{\partial M}$. Let us denote by $\mathsf{Diff}(M, \partial M)$ the group of such diffeomorphisms.

Each diffeomorphism $\phi \in \mathsf{Diff}(M, \partial M)$ acts naturally on g, producing a new metric $\phi^*(g)$. Since such ϕ maps geodesics in g to geodesics in $\phi^*(g)$, for any $\phi \in \mathsf{Diff}(M, \partial M)$, the geodesic flows v^g and $v^{\phi^*(g)}$ on $TM \setminus M$ are conjugated with the help of the diffeomorphism $\phi_* = D\phi$. Thus, (M, g) and $(M, \phi^*(g))$ share the same the scattering map $C_{v^g} : \partial_1^+(SM) \to \partial_1^-(SM)$.

Therefore, for a given "random" pair (M, g) and its scattering map C_{v^g}, the best we can hope for is to reconstruct g up to the $\mathsf{Diff}(M, \partial M)$-action.

The second source of ambiguity is more profound; it arises from the fact that distinct metrics g on a given manifold M may share the same set $\mathcal{T}(v^g)$ of un-parametrized geodesics (i.e., the metrics are equigeodesic in the sense of Definition 4.9) and, for a traversing geodesic flow v^g, the scattering C_{v^g} depends only on the oriented 1-dimensional foliation $\mathcal{F}(v^g)$. $\diamondsuit$

The next theorem, the main result of this section, can be found in [K5].

Theorem 4.2. (The weak topological rigidity of the geodesic flow for the inverse scattering problem)

Let (M_1, g_1) and (M_2, g_2) be two compact connected smooth Riemannian n-manifolds with boundaries. Let the metrics g_1, g_2 be geodesically boundary generic, and let g_2 be non-trapping. Assume that the scattering maps

$$C_{v^{g_1}} : \partial_1^+(SM_1) \to \partial_1^-(SM_1) \text{ and } C_{v^{g_2}} : \partial_1^+(SM_2) \to \partial_1^-(SM_2)$$

are conjugated by a smooth diffeomorphism $\Phi^\partial : \partial_1(SM_1) \to \partial_1(SM_2)$.

Then the metric g_1 is also of the non-trapping type. Moreover, the metrics g_1 and g_2 have topologically *weakly conjugate geodesic flows.*

If the metric g_2 possess Property AA *from Definition 4.5, then so does g_1, and the conjugating homeomorphism $\Phi : SM_1 \to SM_2$ is a smooth diffeomorphism.*

Proof. By the hypotheses, g_2 is non-trapping and geodesically boundary generic. Equivalently, by Lemma 2.1, the geodesic vector field v^{g_2} is traversing and boundary generic.

If the causality map $C_{v^{g_1}}$ is not well-defined for some $w \in \partial_1^+(SM_1)$, then, by the conjugating property of Φ^∂, the map $C_{v^{g_2}}$ is not well-defined at $\Phi^\partial(w)$, a contradiction to the property of v^{g_2} being traversing. Therefore, the v^{g_1}-trajectory that connects w with $C_{v^{g_1}}(w)$ is a closed segment or a singleton for any $w \in \partial_1^+(SM_1)$. In other words, v^{g_1} is a traversing field. By Lemma 2.1, any traversing field admits a Lyapunov function. Thus, v^{g_1} admits a Lyapunov function. Therefore, both causality maps $C_{v^{g_1}}$ and $C_{v^{g_2}}$ are well-defined and g_1 is non-trapping.

By applying the Holography Theorem 3.3, we conclude that the two geodesic flows are topologically conjugate (in the sense of Definition 4.4) by a homeomorphism $\Phi : SM_1 \to SM_2$ that extends Φ^∂ and maps the v^{g_1}-trajectories to the v^{g_2}-trajectories, while preserving their orientations.

In other words, Φ maps the oriented 1-foliation $\mathcal{F}(v^{g_1})$ to the oriented 1-foliation $\mathcal{F}(v^{g_2})$. As a result, g_1 and g_2 have topologically weakly conjugate geodesic flows.

When Property A is valid for one of the two manifolds, by the proof of Holography Theorem 3.3, it is valid for the other one; indeed, the combinatorial types of the tangency are respected by Φ^∂. In such a case, by the Holography Theorem 3.3, Φ is a diffeomorphism when v^{g_2} has Property A. By Lemma 4.5 below, Property A for v^{g_1} or v^{g_2} is equivalent to Property AA for g_1 or g_2. $\square$

Similarly to the way Theorem 3.3 directly implied Corollary 3.5, Theorem 4.2 directly implies the next Corollary 4.2.

Corollary 4.2. *Assume that a compact connected smooth manifold M with boundary admits a boundary generic non-trapping Riemannian metric g.*

Then the scattering map $C_{v^g} : \partial_1^+(SM) \to \partial_1^-(SM)$ allows for a reconstruction of the pair $(SM, \mathcal{F}(v^g))$, up to a homeomorphism of SM which is the identity on $\partial_1(SM)$ and an orientation-preserving diffeomorphism on each v^g-trajectory.

If g possess property AA *from Definition 4.5, then it is possible to reconstruct the pair $(SM, \mathcal{F}(v^g))$, up to a smooth diffeomorphism of SM which is the identity on $\partial_1(SM)$.* $\diamondsuit$

Example 4.4. Consider a shell M, produced by removing the interior of a compact strictly convex domain from a compact strictly convex domain in the Euclidean space E^n (so topologically M is the shell $S^{n-1} \times [0,1]$). Any geodesic line γ in M that is tangent to the boundary of the interior convex domain is transversal to the boundary of the exterior convex domain. The intersection $\gamma \cap \partial M$ of any geodesic line $\gamma \subset \mathsf{E}^n$ that is tangent to the boundary of the exterior convex domain is a singleton. For such a pair (M, g_E), the geodesic field v^g on SM is traversally generic; in particular it is boundary generic and non-trapping. Thus Theorem 4.2 is applicable to (M, g_E), as well as to any metric g in M, sufficiently close to g_E. We conclude that the scattering map $C_{v^g} : \partial_1^+(SM) \to \partial_1^-(SM)$ (where $\partial_1^\pm(SM)$ each is diffeomorphic to $S^{n-1} \times S^0 \times D^{n-1}$) allows for a reconstruction of the pair $(SM, \mathcal{F}(v^g))$, up to a diffeomorphism of SM which is the identity on $\partial_1(SM)$. $\diamondsuit$

Example 4.5. The idea of this example is inspired by its special case in [We], Figure 3. The example shows that the requirement for the metrics to be non-trapping is essential for the validity of Theorem 4.2.

We are going to construct a pair of Riemannian manifolds, (M, g) and (M', g'), that have isometric neighborhoods of their boundaries, share the same scattering data, although M and M' even have distinct topological types. We stress that the metric g in this example is *trapping*!

Let (M, g) be a compact Riemannian manifold with boundary. Let $N \subset \text{int}(M)$ be a codimension 1 submanifold which separates M. We denote by $M^\vee$ the domain in M such that $\partial M^\vee = N$ and $M^\vee \cap \partial M = \emptyset$.

Consider a regular neighborhood $U(N, M)$ of N in M. We view $U(N, M)$ as the space of the g-normal 1-bundle $\nu : U(N, M) \to N$. Let $\sigma : U(N, M) \to U(N, M)$ be a fiber-preserving involution such that the fixed set $U(N, M)^\sigma = N$. We assume that the involution σ is an isometry and its restriction to each normal fiber of ν is a flip.

We assume that the metric g has a very special property: there is a smooth *free* involution $\tau : N \to N$ such that any geodesic in M, passing through a point in the interior of $M^\vee$, hits the boundary $N = \partial(M^\vee)$ exactly at a pair of points $m \neq \tau(m)$. Moreover, we require that the involution $\tau : N \to N$ extends to an isometry $\hat{\tau} : U(N, M) \to U(N, M)$ so that $\hat{\tau}^2 = \sigma$.

Under the assumptions above, we may remove the interior of $M^\vee$ from M and identify the pair of points on each τ-orbit in N. The result of this identification is a new Riemannian manifold

$$M' =_{\mathsf{def}} M \setminus \mathsf{int}(M^\vee)/\{m \sim \tau(m),\ m \in N\}$$

which shares with M its boundary: $\partial M' = \partial M$. Thanks to the properties of τ and σ, g produces a well-defined metric g' on M'.

By this construction, for any point $x = (m, w) \in \partial_1^+(SM)$, the corresponding geodesics, γ_x in M and γ'_x in M', are in a natural 1-to-1 correspondence: indeed, to produce γ'_x we delete the portion $\gamma_x^\vee =_{\mathsf{def}} \gamma_x \cap \mathsf{int}(M^\vee)$ from γ_x and identify the pair of points $\gamma_x^\vee \cap N$, a τ-orbit. This recipe applies only to γ_x's that have a nonempty intersection with $\text{int}(M^\vee)$; the rest of γ_x's is contained in $M \setminus \mathsf{int}(M^\vee)$ and is shared by M and M'.

Evidently, the scattering data in M and M' are the same, while M and M' are even topologically different! This fact does not contradict Theorem 4.2, since M has closed geodesics residing in N and thus g is trapping.

Here is one basic example of the situation described above: M is obtained from the unit sphere $S^n \subset \mathbb{R}^{n+1}$ by removing a small polar cup B^n from the Northern hemisphere $S^n_+ \subset S^n$, and g is the restriction to

$M = S^n_+ \setminus \mathsf{int}(B^n)$ of the standard spherical metric on S^n. Let $N \subset S^n$ be the equator such that $N \cap B^n = \emptyset$. The pair $(M^\vee, g^\vee)$ is the Southern hemisphere S^n_- with the standard spherical metric. The free involution $\tau : N \to N$ is the antipodal map on S^{n-1}, $\hat{\tau}$ is the antipodal map on S^n, and the involution $\sigma : S^n \to S^n$ is the reflection with respect to the hyperplane of the equator. ◇

It follows instantly that, under the hypotheses of Theorem 4.2, the scattering maps faithfully distinguish between manifolds whose spherical tangent bundles have distinct topological types.

Corollary 4.3. *Let (M_1, g_1) and (M_2, g_2) be two smooth compact connected Riemannian n-manifolds with boundaries. Let the metrics g_1, g_2 be non-trapping and geodesically boundary generic. Assume that the boundaries ∂M_1 and ∂M_2 are diffeomorphic*[2]*, but the spaces SM_1 and SM_2 are not homeomorphic.*

Then no smooth diffeomorphism $\Phi^\partial : \partial_1(SM_1) \to \partial_1(SM_2)$ conjugates the two scattering maps $C_{v^{g_1}}$ and $C_{v^{g_2}}$. ◇

Example 4.6. We extending here the setting of Example 3.2. Recall that the classical knots are determined by the topological types of their complements in S^3 or $\mathbb{R}^3$ (see Gordon-Luecke Theorem [GL]), while the links are not. In fact, there are infinitely many distinct links whose complements are all homeomorphic to the complement of the Whitehead link. Let us apply Corollary 4.3, while keeping these facts in mind.

Let $L_1 \subset D^3_1 \subset N_1$ and $L_2 \subset D^3_2 \subset N_2$ be two links in closed Riemannian 3-folds (N_1, g_1) and (N_2, g_2), respectively. We denote by U_1 and U_2 the regular tubular neighborhoods of $L_1 \subset D^3_1$ and $L_2 \subset D^3_2$. Put $M_1 = N_1 \setminus \mathsf{int}(U_1)$, $M_2 = N_2 \setminus \mathsf{int}(U_2)$. We assume that L_1 and L_2 have the same number of components; thus there exists a diffeomorphism $\psi : U_1 \to U_2$. We assume also that SM_1 and SM_2 are *not* homeomorphic.

For any choice of boundary generic and non-trapping Riemannian metrics g_1 on M_1 and g_2 on M_2, the scattering maps $C_{v^{g_1}}$ and $C_{v^{g_2}}$ are not conjugated via any smooth diffeomorphism $\Phi^\partial : \partial_1(SU_1) \to \partial_1(SU_2)$.

Therefore, these scattering maps *distinguish* between the links L_1 and L_2 with non-homeomorphic "stabilized" complements SM_1 and SM_2 (when $N_i = S^3$, then SM_i is homeomorphic to $M_i \times S^2$). ◇

[2]This implies that $\partial_1(SM_1)$ and $\partial_1(SM_2)$ are diffeomorphic.

If a pair (M, g) is such that the geodesic field v^g is traversing on SM, then the trajectory space $\mathcal{T}(v^g)$—the space of un-parametrized geodesics on M—is a separable compact space. It is given the quotient topology via the obvious map $\Gamma : SM \to \mathcal{T}(v^g)$. Recall that, for any boundary generic traversing vector field v on X, the obvious map $\Gamma^\partial : \partial_1 X \to \mathcal{T}(v)$ has finite fibers. In particular, for a geodesically boundary generic non-trapping metric g, the obvious map $\Gamma^\partial : \partial_1(SM) \to \mathcal{T}(v^g)$ has finite fibers.

The space $\mathcal{T}(v^g)$, although not a manifold in general, for traversing geodesic flows, inherits a surrogate smooth structure from SM (as in Definitions 2.7 and 2.8). This structure manifests itself in the form of the algebra $C^\infty(\mathcal{T}(v^g))$ of smooth functions $f \in C^\infty(SM)$ such that $\mathcal{L}_{v^g}(f) = 0$.

A Riemannian manifold (M, g) of dimension n produces a $(2n-2)$-bundle $\xi^g = \mathsf{ker}(\beta_g)$ over SM. It is a distribution, transversal to the 1-foliation $\mathcal{F}(v^g)$. The isomorphism class of the bundle ξ^g is well-defined by homotopy class of v^g. As a distribution, ξ^g is not integrable. However, for a non-trapping g, there is a normal to v^g integrable distribution $\tilde{\xi} \subset T(SM)$ generated by the tangent bundle to the $(2n-2)$-foliation $\{\hat{F}^{-1}(c)\}_{c \in \mathbb{R}}$, where $d\hat{F}(v^g) > 0$ on SM. As bundles, $\tilde{\xi}$ and ξ^g are isomorphic.

Let g be non-trapping and $\tilde{\gamma}$ be a v^g-trajectory. Consider two points a and b on $\tilde{\gamma}$, where b can be reached from a by moving in the direction of v^g. We denote by $\phi^t : SM \to SM$ be the partially-defined v^g-flow in the vicinity of the segment $[a, b] \subset \tilde{\gamma}$. Thus, $\phi^{t_\star}(a) = b$ for an appropriate $t_\star > 0$. For any tangent vector $w \in T_a(SM)$, consider its image $\phi^{t_\star}_*(w) \in T_b(SM)$. Since the distribution $\xi^g = \xi_{\beta_g}$ is ϕ^t-*invariant*, we may define an equivalence relation "$\simeq$" among the points of its total space $E(\xi^g)$ of the bundle ξ^g by the formula $w_a \simeq w_b$ if $\phi^{t_\star}(a) = b$ and $\phi^{t_\star}_*(w_a) = w_b$ (one needs to symmetrize this formula with respect to a and b). Then the obvious quotient map $(E(\xi^g)/\simeq) \to \mathcal{T}(v^g)$ defines a vector $(2n-2)$-bundle τ^g over the space of geodesics $\mathcal{T}(v^g)$. Alternatively, if a Lyapunov function $F : SM \to \mathbb{R}$ is such that its differential dF is v^g-invariant (a balanced F will do), the tangent bundle $\mathsf{ker}(dF)$ to the foliation $\mathcal{G}(\hat{F})$ may be used to manufacture an isomorphic copy of the bundle τ^g.

As in the case of general traversing flows, the bundle τ^g may be viewed as a surrogate tangent bundle of the stratified singular space $\mathcal{T}(v^g)$, since its restriction to the open and dense set $\mathcal{T}(v^g, \omega_0)$ of trajectories of the combinatorial tangency type $\omega_0 = (1,1)$ may be identified with the "honest" tangent bundle on the open manifold $\mathcal{T}(v^g, \omega_0)$.

Thanks to the symplectic 2-form $d\beta_g$ on ξ^g, the bundle τ^g admits a complex structure.

Corollary 4.4. *We adopt the notations and hypotheses of Theorem 4.2. Let $F_1 : SM_1 \to \mathbb{R}$ and $F_2 : SM_2 \to \mathbb{R}$ be balanced Lyapunov functions for the geodesic vector fields v^{g_1} and v^{g_2}, respectively.*

Then the geodesic flows weakly conjugating homeomorphism $\Phi : SM_1 \to SM_2$ (which extends Φ^∂) induces a bundle isomorphism $\Phi^* : \xi^{g_2} \to \xi^{g_1}$.

Similar, the Φ^∂-generated quotient map $\Phi^{\mathcal{T}} : \mathcal{T}(v^{g_1}) \to \mathcal{T}(v^{g_2})$ of the geodesic spaces induces the "tangent" bundle isomorphism

$$(\Phi^{\mathcal{T}})^* : \tau^{g_2} \to \tau^{g_1}.$$

Proof. By the proof of Holography Theorem 3.3, the homeomorphism Φ has the property $\Phi^*(F_2) = F_1$, where the balanced Lyapunov function $F_i : SM_i \to \mathbb{R}$ is such that $dF_i(v^{g_i}) = 1$ $(i = 1, 2)$. This property may require to change v_1^g by a conformal factor $\lambda : SM_1 \to \mathbb{R}_+$, which does not affect the causality map: i.e., $C_{\lambda v^{g_1}} = C_{v^{g_1}}$ since $\mathcal{F}(\lambda v^{g_1}) = \mathcal{F}(v^{g_1})$.

Since both geodesic flows are traversing, the quotient space $\mathcal{T}(v^{g_i}) = \partial_1(SM_i)/C_{v^{g_i}}$ is determined by the scattering map $C_{v^{g_i}} : \partial_1^+ X_i \to \partial_1^- X_i$. Because $\Phi^\partial \circ C_{v^{g_1}} = C_{v^{g_2}} \circ \Phi^\partial$, the diffeomorphism $\Phi^\partial : \partial_1(SM_1) \to \partial_1(SM_2)$ generates the homeomorphism $\Phi^{\mathcal{T}} : \mathcal{T}(v^{g_1}) \to \mathcal{T}(v^{g_2})$. Moreover, by the proof of Theorem 4.2, the homeomorphism $\Phi^{\mathcal{T}}$ has the property $\Gamma_2 \circ \Phi = \Phi^{\mathcal{T}} \circ \Gamma_1$, where $\Gamma_i : SM_i \to \mathcal{T}(v^{g_i})$ is the map that takes a point to the v^{g_i}-trajectory through it. Using that dF_i is v^{g_i}-invariant, we get that the distribution ξ^{g_i} is v^{g_i}-invariant, so that $\Gamma_i^*(\tau_i) = \xi^{g_i}$. Therefore, we conclude that $(\Phi^{\mathcal{T}})^* : \tau^{g_2} \to \tau^{g_1}$ is a bundle isomorphism. □

Recall that for any traversing boundary generic field v on a compact smooth manifold X, each v-trajectory γ produces an ordered finite sequence $\omega = (\omega_1, \omega_2, \dots)$ of positive integral multiplicities, associated with the finite ordered set $\gamma \cap \partial_1 X$. In Chapter 2, these sequences ω have been organized in an universal poset $\mathbf{\Omega}^\bullet$.

In view of Corollary 4.4, Theorem 4.2 has an instant, but philosophically interesting implication:

Theorem 4.3. *Assume that M admits a non-trapping geodesically boundary generic Riemannian metric g.*

Then the scattering map $C_{v^g} : \partial_1^+(SM) \to \partial_1^-(SM)$ allows for a reconstruction of:

- *the $\mathbf{\Omega}^\bullet$-stratified topological type of the space $\mathcal{T}(v^g)$ of un-parametrized geodesics on M.*

- *the isomorphism class of the "tangent"* $(2n-2)$*-bundle* τ^g *of the space* $\mathcal{T}(v^g)$.
- *the smooth topological type (see Definition 2.8) of the space* $\mathcal{T}(v^g)$*, provided that the property* AA *from Definition 4.5 is valid.*

Proof. By definition, $\mathcal{T}(v^g)$ is the quotient space of SM under the obvious map Γ that takes each v^g-trajectory to a singleton. Now just apply Theorem 4.2 and Corollary 4.2 to reconstruct the map Γ. By Corollary 4.4, C_{v^g} allows for a reconstruction of the $\mathbf{\Omega}^\bullet$-stratified topological type of the space $\mathcal{T}(v^g)$.

By combining Corollary 4.4 with Theorem 4.5, we get the second claim.

When property AA is valid, Theorem 4.2 leads to a reconstruction, up to an algebra isomorphism, of the algebra $C^\infty(\mathcal{T}(v^g), \mathbb{R})$ of smooth functions on $\mathcal{T}(v^g)$—the kernel of the directional derivative operator $\mathcal{L}_{v^g} : C^\infty(SM, \mathbb{R}) \to C^\infty(SM, \mathbb{R})$. By Definition 2.8, the last claim follows. □

Question 4.2. *Given a boundary generic metric* g *on* M*, how to describe the strata* $\{\partial_j^\pm SM(v^g)\}_j$ *of* $\partial(SM)$ *in terms of the local Riemannian geometry of the pair* $(M, \partial M)$*?* ◇

The next two lemmas represent a step towards answering Question 4.2. The language, in which the answer is given, is reminiscent of the language in [Zh].

As before, we assume that a given smooth compact Riemannian manifold M is embedded properly into a larger open manifold $\hat{M}$, and the metric g on M is extended smoothly to a metric $\hat{g}$ on $\hat{M}$.

Lemma 4.5. *For a boundary generic Riemannian metric* g *on a compact connected* M*, the stratum* $\partial_j(SM)(v^g)$ *consists of pairs* $(x, w) \in SM|_{\partial M}$ *such that the germ of the geodesic curve* $\gamma \subset M$ *through* x *in the direction of* w *is tangent to* ∂M *with multiplicity* $m(x) = j - 1$*. Moreover, the stratum* $\partial_j^\pm(SM)(v^g)$ *also has a similar description*[3] *in terms of the germ of* $\gamma \subset M$ *at* x.

Proof. Take a smooth function $z : \hat{M} \to \mathbb{R}_+$ with the properties:

(1) 0 is a regular value of z,

(2) $z^{-1}(0) = \partial M$,

(3) $z^{-1}((-\infty, 0]) = M$.

We may use the distance functions $-d_g(\sim, \partial M)$ in M and $d_g(\sim, \partial M)$ in $\hat{M} \setminus M$ for the role of z.

[3] Its exact formulation is described in the proof.

Let $\tilde{z} : S\hat{M} \to \mathbb{R}_+$ be the composition of the projection $\pi : S\hat{M} \to \hat{M}$ with z. Evidently, $\tilde{z} : S\hat{M} \to \mathbb{R}$ has the same three properties with respect to $\partial_1(SM)$ as $z : \hat{M} \to \mathbb{R}$ has with respect to ∂M.

For any pair $(x, w) \in \partial_1(SM)$, consider the germ of the geodesic curve $\gamma \subset \hat{M}$ through $x \in \partial M$ in the direction of $w \in T_xM$ and its lift $\tilde{\gamma} \subset S\hat{M}$, the geodesic $v^{\hat{g}}$-flow curve through $(x, w) \in \partial_1(SM)$. Then $\pi : \tilde{\gamma} \to \gamma$ locally is an orientation preserving smooth diffeomorphism of curves. Therefore, the k-jet $jet^k(\tilde{z}|_{\tilde{\gamma}}) = 0$ if and only if k-jet $jet^k(z|_\gamma) = 0$.

By formula (2.8), $(x, w) \in \partial_j(SM)(v^g)$ if and only if $jet^{j-1}(\tilde{z}|_{\tilde{\gamma}}) = 0$; similarly, $(x, v) \in \partial_j^+(SM)(v^g)$ if, in addition, $\frac{\partial^j}{(\partial \tilde{t})^j}(\tilde{z}|_{\tilde{\gamma}})(x, v) \geq 0$ (here $\tilde{t}$ is the natural parameter along $\tilde{\gamma}$).

Thanks to the orientation-preserving smooth diffeomorphism $\pi : \tilde{\gamma} \to \gamma$, these properties of $\tilde{z}|_{\tilde{\gamma}}$ are equivalent to the similar properties $jet^{j-1}(z|_\gamma) = 0$ and $\frac{\partial^j}{(\partial t)^j}(z|_\gamma)(x) \geq 0$ of $z|_\gamma$. The latter ones describe the multiplicity $m(x) =_{\mathsf{def}} j - 1$ of tangency of γ to ∂M at x. □

Corollary 4.5. *Let M be a smooth compact connected Riemannian manifold of dimension n. For a boundary generic (see Definition 4.3) metric g on M, the order of tangency of any geodesic curve to ∂M does not exceed $2n - 1$.*

Proof. For any boundary generic vector field w on SM, the locus $\partial_{\dim(SM)+1}(SM)(w) = \emptyset$. In particular, since $\dim(SM) = 2n - 1$, we get $\partial_{2n}(SM)(v^g) = \emptyset$. Applying Lemma 4.5, the claim follows. □

Corollary 4.6. *Let M be a compact connected orientable smooth n-dimensional manifold with boundary, and g a boundary generic metric g on M. Then the following formula holds*

$$\sum_{j=2}^{2n-1} (-1)^j \chi\big(\partial_j^+(SM)(v^g)\big) = 0,$$

where $\chi(\sim)$ stands for the Euler characteristic of the appropriate space.

Proof. We apply the Morse Formula [Mo] for the index of a vector field to the geodesic field v^g on SM (see [K1]). Since $v^g \neq 0$, we get that the index $\mathsf{ind}(v^g) = 0$. For an orientable M, $\chi((SM)(v^g))$, the 0-term of the Morse Formula, equals $\chi(S^{n-1}) \cdot \chi(M)$, while the 1-term is $\chi(\partial_1^+(SM)(v^g)) = \chi(D^{n-1}) \cdot \chi(\partial M)$. Thus, for $n \equiv 0 \mod 2$, we get $\chi(S^{n-1}) = 0$ and $\chi(\partial M) = 0$, hence $\chi\big((SM)(v^g)\big) - \chi\big(\partial_1^+(SM)(v^g)\big) = 0$. For $n \equiv 1 \mod 2$, we get $\chi(S^{n-1}) = 2$, and $\chi(\partial M) = 2 \cdot \chi(M)$. Hence again $\chi\big((SM)(v^g)\big) -$

$\chi\big(\partial_1^+(SM)(v^g)\big) = 0$. Therefore, the index formula gets contributions only from the terms $\{\partial_j^+(SM)(v^g)\}_{j\geq 2}$. □

Example 4.7.

• Let $n = 2$. Then, for an orientable Riemannian surface (M, g), by Corollary 4.6, we get $\chi\big(\partial_2^+(SM)(v^g)\big) = \chi\big(\partial_3^+(SM)(v^g)\big)$. By Lemma 4.5, this relation reflects the fact that the number of positively curved (with respect to g) arcs in ∂M is equal to the number of positive inflection points in ∂M. By definition, as one crosses a positive inflection point in the direction consistent with the orientation of ∂M, the negative curvature of ∂M switches to the positive one.

• For an orientable Riemannian 3-fold (M, g), we get the relation

$$\chi\big(\partial_2^+(SM)(v^g)\big) + \chi\big(\partial_4^+(SM)(v^g)\big) = \chi\big(\partial_3^+(SM)(v^g)\big) + \chi\big(\partial_5^+(SM)(v^g)\big).$$

In general, the number $\sum_{k=2}^{n-1} \chi\big(\partial_{2k}^+(SM)(v^g)\big)$ can serve as an integral invariant of a boundary generic metric g, being considered up to actions of smooth diffeomorphisms on the space $\mathcal{R}(M)$ of Riemannian metrics. ◇

Definition 4.6. Let M be a compact connected smooth n-manifold M with boundary. Pick $j \in [2, 2n-1]$. We say that a boundary generic metric g on M is:

- boundary j-concave if $\partial_j^-(SM)(v^g) = \emptyset$,
- boundary j-convex if $\partial_j^+(SM)(v^g) = \emptyset$. ◇

For a boundary generic g, if $\partial_j^\pm(SM)(v^g) = \emptyset$, then evidently $\partial_{j+1}(SM)(v^g) = \emptyset$. As we have done for arbitrary traversing boundary generic vector fields, we think of boundary $(j+1)$-concave metrics as *more complex* in comparison with the boundary j-concave ones.

Note that there is no *topological obstruction* to the existence of boundary 2-concave/2-convex metric on a given smooth compact manifold M. Indeed, fix a smooth product structure $\partial M \times [0, 1]$ in a compact regular neighborhood U of ∂M. Pick any Riemannian metric g^∂ on ∂M and a strictly monotone smooth function $\phi : [0, 1] \to [1, 1 \pm 0.5]$, where $\phi(0) = 1$. Using g^∂ and ϕ, we define the warped product metric g_U in U. If $\phi'(0) > 0$, ∂M is convex in g_U; if $\phi'(0) < 0$, ∂M is concave in g_U. Next, by a smooth partition of unity argument, we extend g_U to a Riemannian metric g on M. This implies that $\partial_2^-(SM)(v^g) = \emptyset$ when $\phi'(0) > 0$, and $\partial_2^+(SM)(v^g) = \emptyset$ when $\phi'(0) < 0$.

In contrast, by Proposition 4.1 below, the existence of a 2-convex *non-trapping* metric on M restricts severely its topology [K5].

Question 4.3. *What are the topological obstructions to the existence of a* non-trapping *boundary generic and boundary* j*-covex/*j*-concave smooth metric* g *on a given manifold* M *with boundary?* ◇

In one special case of (M, g), described in Proposition 4.1, the traversal genericity of the geodesic flow on SM comes "for free" at the expense of a very restricted topology of M. In fact, by Proposition 4.1, a random manifold M does not admit a non-trapping metric g such that ∂M is convex! In this proposition, we exhibit a topological obstruction to the boundary 2-convexity of non-trapping metrics g on M.

Proposition 4.1. *Let* M *be a connected smooth manifold with boundary.*

M *admits a* non-trapping *boundary* 2*-convex metric, if and only if, the space* SM *is diffeomorphic to the product* $D(\partial M) \times [0,1]$*, where* $D(\partial M)$ *denotes the space of the tangent disk bundle over the boundary* ∂M*, and the corners of the product* $D(\partial M) \times [0,1]$ *are rounded.*

As a result, SM *fibers over* ∂M *with the fiber* D^n *and over* M *with the fiber* S^{n-1}.

In particular, no connected smooth n*-manifold* M *with boundary, such that* $\pi_{n-1}(\partial M, pt) = 0$[4] *for any choice of the base point* $pt \in \partial M$*, admits a boundary* 2*-convex non-trapping metric.*

Proof. The geodesic vector field v^g is traversing and boundary convex if and only if $\partial_1^+(SM)$ is a smooth deformation retract of SM [K1]. Therefore, SM is diffeomorphic to the product $\partial_1^+(SM) \times [0,1]$ with the conners of the product being rounded. In turn, $\partial_1^+(SM)$ is canonically diffeomorphic to the space of the tangent $(n-1)$-disk bundle $D(\partial M) \to \partial M$.

As a result, SM fibers over ∂M with the fiber $D^n \approx D^{n-1} \times [0,1]$ and over M with the fiber S^{n-1}. This property of SM puts severe restrictions on the topology of M: in particular, the space SM must be homotopy equivalent to ∂M.

Note that $M = D^n$ has the desired property: $M \times S^{n-1} \approx \partial M \times D^n$.

To validate the second claim, consider the homology class $[S] \in H_{n-1}(SM; \mathbb{Z})$ of a fiber S of the fibration $\pi : SM \to M$. Since M is connected and has a boundary, $H_n(M; \mathbb{Z}) = 0$, so π admits a section. As a result, $[S] \neq 0$ in $H_{n-1}(SM; \mathbb{Z})$.

[4]This is the case when ∂M is a union of tori of dimension ≥ 2.

On the other hand, if $\partial_2^+(SM)(v^g) = \emptyset$, there is a continuous retraction $\{r_t : SM \to SM\}_{t\in[0,1]}$ along the $(-v^g)$-trajectories of SM onto $\partial_1^+(SM)(v^g)$ (so that $r_1 : SM \to \partial_1^+(SM)(v^g)$). The later locus is homeomorphic to the space η of the $(n-1)$-disk bundle, associated with the tangent bundle $T(\partial M)$. Hence, $\pi_{n-1}(\eta) = \pi_{n-1}(\partial M) = 0$ by the hypothesis. Therefore, $r_1 : S \to \partial_1^+(SM)(v^g)$ must be contractible in $\partial_1^+(SM)(v^g)$ and thus in SM. This conclusion contradicts with the fact that $[S] \neq 0$. □

The next two propositions deal with non-trapping metrics on M in which the boundary ∂M is *concave*. With some new tools in hand, we will return to the concave setting in Section 4.5. The cases of boundary concave non-trapping metrics are much less restrictive than the cases of non-trapping boundary convex metrics.

Proposition 4.2. *Let M be a connected compact smooth n-manifold with boundary, such that $H_{n-1}(M;\mathbb{Z}) = 0$ and the natural homomorphisms $H_k(\partial M;\mathbb{Z}) \to H_k(M,\mathbb{Z})$ are trivial for all $k \in [1, n-1]$. Let g be a non-trapping metric on M such that ∂M is strictly* concave *in g (in particular, $\partial_2^-(SM)(v^g) = \emptyset$ and $\partial_3(SM)(v^g) = \emptyset$).*

Then there is a geodesic γ in M that is simply tangent to ∂M at two distinct points at least. If the geodesic flow is traversally generic, such double-tangent geodesics form a $(2n-4)$-dimensional family.

Proof. If ∂M is concave in a metric g, then, by Lemma 4.5, $\partial_1(SM)$ is concave with respect to the geodesic flow v^g. If g is non-trapping, then v^g is traversing. Then the combinatorial types of v^g-trajectories in SM may only be of the form (11), (121), or (12...21), where the number of 2's is ≥ 2.

By the spectral sequence argument, applied to the fibration $SM \to M$, $H_{n-1}(M;\mathbb{Z}) = 0$ implies that $H_{2n-2}(SM;\mathbb{Z}) = 0$. Similarly, the triviality of the natural homomorphisms $H_k(\partial M;\mathbb{Z}) \to H_k(M,\mathbb{Z})$ for $k \in [1, n-1]$ implies the triviality of the natural homomorphisms $H_k(\partial_1(SM);\mathbb{Z}) \to H_k(SM,\mathbb{Z})$ for $k \in [1, 2n-2]$. Thus, the hypotheses of Corollary 5.18 from [K7] are satisfied. By that corollary (we apply it to the case $X = SM$, $v = v^g$), the geodesic field v^g cannot have trajectories *only* of the combinatorial types (11) and (121). Thus, we conclude that the v^g-flow must have a trajectory $\tilde{\gamma}$ that is simply tangent to $\partial_1(SM)$ at two points at least. Therefore, the existence of the v^g-trajectory $\tilde{\gamma}$ implies that the set $SM(v^g, (1221)) \neq \emptyset$. The projection of $\tilde{\gamma}$ in M delivers a geodesic γ that is at least twice tangent to ∂M.

Note that, if the set $SM(v^g, (12\dots 21))$ of v^g-trajectories of the combinatorial type $(12\dots 21)$ is non-empty, then so is the set of v^g-trajectories of the combinatorial type (1221).

Since the reduced norm $|(1221)|' = 2$, for a *traversally generic* v^g, we get that $\dim(SM(v^g, (1221))) = (2n-2) - 2 = 2n-4$. □

Proposition 4.3. *Let $n \geq 3$. Let M be a connected compact smooth n-manifold with boundary. We denote by $\beta_1(M)$ its first Betti number. Assume that:*

(1) $\pi_1(M)$ does not admit an epimorphism whose target is a free group in $\beta_1(M)$ generators,

(2) the natural homomorphism $H_1(\partial M; \mathbb{Z}) \to H_1(M; \mathbb{Z})$ is trivial.

Let g be a non-trapping metric on M such that ∂M is strictly concave *in g.*

Then there is a geodesic γ in M that is simply tangent to ∂M at a pair of distinct points. If the geodesic flow v^g is traversally generic, such double tangent geodesics form a $(2n-4)$-dimensional family.

Proof. For $n \geq 3$, by the long exact homotopy sequence of the fibration $SM \to M$, we get $\pi_1(SM) \approx \pi_1(M)$. Also, for $n \geq 3$, by a spectral sequence argument, $\beta_1(SM) = \beta_1(M)$. Similarly, the triviality of the natural homomorphism $H_1(\partial M; \mathbb{Z}) \to H_1(M; \mathbb{Z})$ implies the triviality of the natural homomorphism $H_1(\partial_1(SM); \mathbb{Z}) \to H_1(SM; \mathbb{Z})$.

Thus, Corollary 5.17 from [K7] applies to the geodesic field v^g. Under the concavity of ∂M hypotheses (equivalently, under the concavity of $\partial_1(SM)$ with respect to the geodesic flow), we conclude that the v^g-flow must have a trajectory $\tilde{\gamma}$ that is simply tangent to $\partial_1(SM)$ at two points at least. The projection of $\tilde{\gamma}$ in M delivers a geodesic γ that is at least twice tangent to ∂M.

As in Corollary 4.5, for a traversally generic geodesic flow, the existence of $\tilde{\gamma}$ leads to $\dim(SM(v^g, (1221))) = 2n-4$. □

Note that if $\partial_{j+1}(SM)(v^g) = \emptyset$, then both loci $\partial_j^{\pm}(SM)(v^g)$ are closed submanifolds of $\partial_1(SM)$. By Theorem 2.2, they are equipped with the normal $(j-1)$-framing. Therefore, combining Lemma 4.5 with this observation, leads directly to the following claim.

Corollary 4.7. *Let $j \in [2, 2n-1]$. Given a compact connected smooth n-manifold M with boundary, any boundary generic metric g on M such* *that no geodesic curve is tangent to* ∂M *with multiplicity* $j + 1$[5] *defines an element* $\theta_j^+(g)$ *in the cohomotopy group* $\pi^{j-1}(\partial_1(SM))$. *The element* $\theta_j^+(g)$ *is produced via the Pontryagin-Thom construction, being applied to the closed normally framed submanifold* $\partial_j^+(SM)(v^g) \subset \partial_1(SM)$.

In fact, $\theta_j^+(g)$ *depends only on the connected component of the space of boundary generic metrics on* M *with no* $(j+1)$*-tangent to* ∂M *geodesics.*

◇

Now, let us rephrase the arguments in the proof of Lemma 4.5 in terms of the exponential maps and the appropriate symmetric tensors.

We embed isometrically (M, g) in an open Riemannian pair $(\hat{M}, \hat{g})$.

For each point $x \in \partial M$, consider the $\hat{g}$-induced exponential map $\exp_x^{\hat{g}} : T_xM \to \hat{M}$ and its locally-defined inverse $\ln_x^{\hat{g}} : \hat{M} \to T_xM$. We may assume that $\exp_x^{\hat{g}}$ is well-defined for all sufficiently small $w \in T_xM$ and $\ln_x^{\hat{g}}$ for all $y \in \hat{M}$ that are sufficiently close to x. The local maps $\{\exp_x^{\hat{g}}\}_{x \in \partial M}$ can be organized into a single map $\exp^{\hat{g}} : TM|_{\partial M} \to \hat{M}$, well-defined in a neighborhood of the zero section $\sigma : \partial M \to TM|_{\partial M}$.

Let a function $z : \hat{M} \to \mathbb{R}$ be as in the proof of Lemma 4.5 (see properties (1)-(3)). We denote by $H_x : T_xM \to \mathbb{R}$ the pull-back of z under the exponential map $\exp_x^{\hat{g}}$. Similarly, we denote by $H : TM|_{\partial M} \to \mathbb{R}$ the pull-back of the function $z : \hat{M} \to \mathbb{R}$ under the map $\exp^{\hat{g}}$.

We pick a vector $w \in T_xM$, where $x \in \partial M$. By Lemma 4.5, the order of tangency of the hypersuface $\partial M = \{z = 0\} \subset \hat{M}$ with the geodesic curve $\exp_x^{\hat{g}}(tw) \subset \hat{M}$, $t \in \mathbb{R}$, is equal to the order of tangency of the hypersurface $\{H_x(w) = 0\} \subset T_xM$ with the line $\{tw\}_{t \in \mathbb{R}} \subset T_xM$. Therefore, the property $(x, w) \in \partial_j(SM)(v^g)$ is equivalent to the property $jet_{t=0}^{j-1}(H_x(tw)) = 0$, where $jet_{t=0}^{j-1}$ denotes the $(j-1)$-jet of the t-function $H_x(tw)$ at $t = 0$, and w is a unitary tangent vector at x.

Let $\sigma : M \to TM$ denote the zero section of tangent bundle $\pi : TM \to M$. Let $\pi^* : T^*M \to M$ be the cotangent bundle.

Definition 4.7. Let $\phi : TM \to \mathbb{R}$ be a smooth function. Let $jet_0^k(\phi|_{\pi^{-1}(x)})$ denotes for the k-jet at the point $\sigma(x) = 0 \in T_xM$ of the function ϕ, being restricted to the π-fiber T_xM.

We say that two smooth functions $f, h \in C^\infty(TM, \mathbb{R})$ share the same vertical k-jet field along the zero section σ and write "$f \equiv_\pi^k h$", if

$$jet_0^k(f|_{\pi^{-1}(x)}) = jet_0^k(h|_{\pi^{-1}(x)})$$

for all $x \in M$.

[5]These metrics are both $(j+1)$-concave and $(j+1)$-convex.

We denote by $\mathcal{J}^k(\pi)$[6] the quotient space of the functional space $C^\infty(TX, \mathbb{R})$ by the equivalence relation " $\equiv^k_\pi$ ". $\diamondsuit$

Consider the k-jet space $Jet^k(\mathbb{R}^n, \mathbb{R})$, jets being taken at the origin $0 \in \mathbb{R}^n$. We interpret $Jet^k(\mathbb{R}^n, \mathbb{R})$ as the space of real polynomials in n variables of degree k at most. Let $\mathcal{S}^j((\mathbb{R}^n)^*)$ denotes the space of homogeneous polynomials of degree j in n variables.

Any jet $\tau \in Jet^k(\mathbb{R}^n, \mathbb{R})$ has a unique representation as a sum $\tau = \tau_0 + \ldots + \tau_k$, where the homogeneous summands $\tau_j \in \mathcal{S}^j((\mathbb{R}^n)^*)$. Therefore, the k-jets admit a direct sum decomposition:

$$Jet^k(\mathbb{R}^n, \mathbb{R}) \approx \bigoplus_{j=0}^{k} \mathcal{S}^j((\mathbb{R}^n)^*).$$

The general linear group $\mathsf{GL}(n)$ acts on $\mathbb{R}^n$, and thus on the space $Jet^k(\mathbb{R}^n, \mathbb{R})$. Each subspace $\mathcal{S}^j((\mathbb{R}^n)^*)$ in the decomposition above is invariant under this action.

Let $\Pi : P^*M \to M$ be the principle $\mathsf{GL}(n)$-bundle, associated with the cotangent bundle $\pi^* : T^*M \to M$. We can form the bundle $Jet^k_{\mathsf{vert}}(\pi) \to M$, associated with the bundle $\Pi : P^*M \to M$. Its total space is $P^*M \times_{\mathsf{GL}(n)} Jet^k(\mathbb{R}^n, \mathbb{R})$ and its fiber is the space $Jet^k(\mathbb{R}^n, \mathbb{R})$.

Similarly, we may construct the bundle

$$\mathcal{S}^j(\pi^*) = P^*M \times_{\mathsf{GL}(n)} \mathcal{S}^j((\mathbb{R}^n)^*)$$

over M, associated with the cotangent bundle $\pi^* : T^*M \to M$. Employing the $\mathsf{GL}(n)$-equivariant isomorphism

$$\beta : Jet^k(\mathbb{R}^n, \mathbb{R}) \approx \bigoplus_{j=0}^{k} \mathcal{S}^j((\mathbb{R}^n)^*),$$

we get a bundle isomorphism $B : Jet^k_{\mathsf{vert}}(\pi) \approx \bigoplus_{j=0}^k \mathcal{S}^j(\pi^*)$.

Given a smooth bundle $\eta : E(\eta) \to M$, we denote by $\Gamma^\infty(\eta)$ the space of its smooth sections over M.

By the definition of the space $\mathcal{J}^k(\pi)$, there is the obvious bijection $\mathcal{E} : \mathcal{J}^k(\pi) \longrightarrow \Gamma^\infty(Jet^k_{\mathsf{vert}}(\pi))$. We use it to form the composite map

$$\mathsf{T}^k : C^\infty(TM, \mathbb{R}) \xrightarrow{\equiv^k_\pi} \mathcal{J}^k(\pi) \xrightarrow{\mathcal{E}} \Gamma^\infty(Jet^k_{\mathsf{vert}}(\pi)) \approx \bigoplus_{j=0}^{k} \Gamma^\infty(\mathcal{S}^j(\pi^*)).$$

[6] not to be confused with the standard notation "$J^k(\pi)$" that refers to the equivalence classes of local sections of the bundle π, a "horizontal" construction!

Therefore, for a fixed k, each function $f \in C^\infty(TM, \mathbb{R})$ gives rise to a sequence of sections $\{T_j^k(f) \in \Gamma^\infty(\mathcal{S}^j(\pi^*))\}_{j\in[0,k]}$.

In particular, $T_0^k(f) = f|_{\sigma(M)}$, $T_1^k(f) = d_\pi f|_{\sigma(M)}$, where d_π stands for the differential along the π-fibers.

We call the formal sum $\mathsf{T}^k(f) = \sum_{j=0}^k T_j^k(f)$ the vertical Taylor polynomial of the function $f : TM \to \mathbb{R}$ of degree k.

Each of the sections $T_j^k(f)$ can be evaluated at any point $(x, w) \in TM$ by forming the symmetric contravariant tensor $T_j^k(f)(x) \in \mathcal{S}^j(T_x^*M)$ and evaluating it at a generic vector $w \in T_xM$. We denote by $\tau_j^k(f)(x, w)$ the result of this evaluation.

Now we choose $k = 2n - 1 = \dim(SM)$ and consider the vertical Taylor expansion $T^k(H) = \sum_{j=0}^k T_j^k(H)$ of the function

$$H(x, w) =_{\mathsf{def}} z(\exp_x^{\hat{g}}(w)), \tag{4.7}$$

on TM, where $x \in \hat{M}$ resides in the vicinity of ∂M and $T_j^k(H)(x, w)$ is a homogeneous polynomial in the variables $w^* \in T_x^*\hat{M}$ of degree j. As before, we denote by $\tau_j^k(x, w) =_{\mathsf{def}} \tau_j^k(H)(x, w)$ the result of evaluation of the symmetric differential j-form $T_j^k(H)$ at the vector $w \in T_x\hat{M}$. Note that $\tau_0^k(x, w) = z(x)$.

In light of the arguments that precede Definition 4.7 and using the homogeneity of $\tau_j^k(x, w)$ in w, we conclude that the property $(x, w) \in \partial_j(SM)(v^g)$ is equivalent to the requirement $\{\tau_l^k(x, w) = 0\}_{0<l<j}$, where w is a unit tangent vector and $k \geq j$, where $k = 2n - 1$. Moreover, (x, w) belongs to the pure stratum $\partial_j^\circ(SM)(v^g)$ if and only if $\{\tau_l^k(x, w) = 0\}_{0<l<j}$ and $\tau_j^k(x, w) \neq 0$.

Although the function H depends on the choice of the auxiliary function z, the solution set $\{\tau_l^k(x, w) = 0\}_{0<l<j}$ does not. Indeed, replacing the function z with the new function $z \cdot q$, where $q|_{\partial M} > 0$, and using the multiplicative properties of the local Taylor expansions, leads to a new system of constraints $\{\tilde{\tau}_l^k(x, w) = 0\}_{0<l<j}$ which exhibit a "solvable triangular pattern", as compared with the original system $\{\tau_l^k(x, w) = 0\}_{0<l<j}$.

Remark 4.6. We fix $k = \dim(SM)$ and let $j \leq k$. For each $x \in \partial M$, the homogeneous in w equations $\{\tau_l^k(x, w) = 0\}_{l<j}$ of degree l define a (z-independent) real algebraic subvariety $\mathcal{V}_{<j}(x)$ of the unit sphere $S_x \subset T_xM$. The pure stratum $\partial_j^\circ(SM)(v^g)$ gives rise to the semialgebraic set

$$\mathcal{V}_j^\circ(x) =_{\mathsf{def}} \mathcal{V}_{<j}(x) \setminus \mathcal{V}_{<j+1}(x).$$

The subvariety $\mathcal{V}_{<j+1}(x)$ separates $\mathcal{V}_{<j}(x)$ into two semialgebraic sets: $\mathcal{V}^+_{<j}(x)$, where $\tau^k_j(x,w) \geq 0$, and $\mathcal{V}^-_{<j}(x)$, where $\tau^k_j(x,w) \leq 0$. These loci reflect the generalized concavity and convexity of ∂M (in the metric g) at $x \in \partial M$ in the direction of a given unit vector $w \in S(T_xM)$.

Note that, for given x and j, the sets $\mathcal{V}_{<j}(x)$, $\mathcal{V}^+_{<j}(x)$, $\mathcal{V}^-_{<j}(x)$ may be empty. Thus, with ∂M and g being fixed, we regard the requirements

$$\mathcal{V}_{<j}(x) \neq \emptyset,\ \mathcal{V}^+_{<j}(x) \neq \emptyset,\ \mathcal{V}^-_{<j}(x) \neq \emptyset$$

as constraints, imposed by the metric g on $x \in \partial M$. $\diamondsuit$

Let the function H be given by (4.7). We will focus now on the restrictions of the smooth functions $\tau^{2n-1}_j(H) : T\hat{M} \to \mathbb{R}$ to the spherical subspace $S\hat{M}$ in the vicinity of the boundary $\partial_1(SM)$. Together, $\{\tau^{2n-1}_j(H)\}_{0 \leq j \leq 2n-1}$ will generate a smooth map $\vec{\tau}(H) : S\hat{M} \to \mathbb{R}^{2n}$.

Let $y_0, \dots, y_{2n-1}$ denote the standard coordinates in $\mathbb{R}^{2n}$. Let L_j be the subspace of $\mathbb{R}^{2n}$ defined by the equations $\{y_0 = 0, \dots, y_j = 0\}$. Consider the complete flag

$$\mathsf{F}(2n) = \{\mathbb{R}^{2n} \supset L_0 \supset \ldots \supset L_{2n-2} \supset L_{2n-1} = \{0\}\}.$$

Let us focus on the interaction of the map $\vec{\tau}^{\,2n-1}(H) : S\hat{M} \to \mathbb{R}^{2n}$ with the flag $\mathsf{F}(2n)$. The considerations above imply that

$$\partial_j(SM)(v^g) = (\vec{\tau}^{\,2n-1}(H))^{-1}(L_{j-1}).$$

As in Lemma 2.2, the definition of boundary generic geodesic field $v^{\hat{g}} \in T(S\hat{M})$ with respect $\partial_1(SM) = \{\tau^{2n-1}_0(H) = 0\}$ implies the linear independence of the differential 1-forms $d\tau^{2n-1}_0(H), \dots, d\tau^{2n-1}_{j-1}(H)$ along the locus

$$\partial_j(SM)(v^g) = \{\tau^{2n-1}_0(H) = 0,\ \dots\ ,\ \tau^{2n-1}_{j-1}(H) = 0\}.$$

In turn, the linear independence of these forms is equivalent to the non-vanishing of the j-form

$$d\tau^{2n-1}_0(H) \wedge \ \ldots\ \wedge d\tau^{2n-1}_{j-1}(H) \in \bigwedge^j T^*(SM)$$

along the locus $\partial_j(SM)(v^g)$ (c.f. (2.9)).

In light of the considerations above, we get a "more constructive" version of Lemma 4.5.

Lemma 4.6. *Let* $n = \dim M$. *Consider an auxiliary function* $z : \hat{M} \to \mathbb{R}$ *with properties (1)-(3) from the proof of Lemma 4.5 and the associated function*

$$H : T\hat{M}|_{\partial M} \stackrel{\exp^{\hat{g}}}{\longrightarrow} \hat{M} \stackrel{\hat{z}}{\longrightarrow} \mathbb{R}.$$

The metric $\hat{g}$ *on* $\hat{M}$ *is boundary generic with respect to* $\partial M = z^{-1}(0)$ *if and only if the map* $\vec{\tau}^{\,2n-1}(H) : S\hat{M} \to \mathbb{R}^{2n}$ *is transversal to each space* L_j *from the flag* $\mathsf{F}(2n)$.[7] *Moreover,* $\partial_j(SM)(v^g) = \vec{\tau}(H))^{-1}(L_{j-1})$.

In other words, $\hat{g}$ *is boundary generic if and only if, for each* $j \in [1, 2n]$, *the differential* j*-form*

$$d\tau_0^{2n-1}(H) \wedge d\tau_1^{2n-1}(H) \wedge \ldots \wedge d\tau_{j-1}^{2n-1}(H) \in \bigwedge^{j} T^*(SM)$$

does not vanish along the locus

$$\partial_j(SM)(v^g) = \{\tau_0^{2n-1}(H) = 0,\ \tau_1^{2n-1}(H) = 0,\ \ldots,\ \tau_{j-1}^{2n-1}(H) = 0\}.$$

$\diamondsuit$

Example 4.8. Let M be a domain in $\mathbb{R}^2$, bounded by a smooth simple curve ∂M, given by an equation $\{z(x_1, x_2) = 0\}$. We assume that 0 is a regular value of the smooth function $z : \mathbb{R}^2 \to \mathbb{R}$ and that $z \leq 0$ defines M.

Since $\mathbb{R}^2$ carries the Eucledian metric $g = g_{\mathsf{E}}$, the exponential map is given by the formula $\exp_{\vec{x}}(\vec{w}) = \vec{x} + \vec{w}$, where $\vec{x} \in \mathbb{R}^2$ and $\vec{w} \in T_x(\mathbb{R}^2) \approx \mathbb{R}^2$.

For each point $\vec{x} = (x_1, x_2)$, we consider the Taylor expansion $T(\vec{x}, \vec{w})$ of the function $H(\vec{x}, \vec{w}) = z(\vec{x} + \vec{w})$ as a sum:

$$T(\vec{x}, \vec{w}) = \tau_0(\vec{x}) + \tau_1(\vec{x}, \vec{w}) + \tau_2(\vec{x}, \vec{w}) + \tau_3(\vec{x}, \vec{w}) + \ldots,$$

where $w = (w_1, w_2) \in T_x(\mathbb{R}^2)$, and the summands are of the form:

$$\begin{aligned}
\tau_0(\vec{x}) &= z(\vec{x}),\\
\tau_1(\vec{x}, \vec{w}) &= \tau_1^1(\vec{x})w_1 + \tau_2^1(\vec{x})w_2,\\
\tau_2(\vec{x}, \vec{w}) &= \tau_{11}^2(\vec{x})w_1^2 + \tau_{12}^2(\vec{x})w_1w_2 + \tau_{22}^2(\vec{x})w_2^2,\\
\tau_3(\vec{x}, \vec{w}) &= \tau_{111}^3(\vec{x})w_1^3 + \tau_{112}^3(\vec{x})w_1^2w_2 + \tau_{122}^3(\vec{x})w_1w_2^2 + \tau_{222}^3(\vec{x})w_2^3.
\end{aligned}$$

The manifold SM is diffeomorphic to the product $M \times S^1$. The boundary $\partial_1(SM)$ is a 2-torus, given by the equation $\{\tau_0(\vec{x}) = 0\}$.

The locus $\partial_2 SM(v^g) = S(\partial M) \subset \partial_1(SM)$ is given by the pair of homogeneous in w_1, w_2 equations:

$$\begin{aligned}
\tau_0(\vec{x}) &= 0\\
\tau_1^1(\vec{x})w_1 + \tau_2^1(\vec{x})w_2 &= 0,
\end{aligned}$$

where (w_1, w_2) is a unitary vector. For the locus $\partial_1^+(SM)(v^g)$, we get

$$\tau_1^1(\vec{x})w_1 + \tau_2^1(\vec{x})w_2 \geq 0.$$

[7] By the properties of z, $\vec{\tau}(H)$ is transversal to L_0 and $\vec{\tau}(H)^{-1}(L_0) = \partial_1(SM)$.

For a *boundary generic* relative to g_{E} curve ∂M, the locus $\partial_2 SM(v^g)$ is represented by two smooth "parallel" loops in the 2-torus $\partial_1(SM)$. They separate $\partial_1(SM)$ into two annuli, $\partial_1^+(SM)(v^g)$ and $\partial_1^-(SM)(v^g)$.

The locus $\partial_3(SM)(v^g) \subset \partial_2(SM)(v^g)$ is given by the homogeneous in w_1, w_2 equations

$$\begin{aligned} \tau_0(\vec{x}) &= 0, \\ \tau_1^1(\vec{x})w_1 + \tau_2^1(\vec{x})w_2 &= 0, \\ \tau_{11}^2(\vec{x})w_1^2 + \tau_{12}^2(\vec{x})w_1w_2 + \tau_{22}^2(\vec{x})w_2^2 &= 0, \end{aligned}$$

where (w_1, w_2) is a unitary vector. Depending on the geometry of ∂M in the vicinity of x, they may have no nontrivial in $\vec{w}$ solutions. For example, this is the case when the quadratic form in the LHS of the third equation is non-trivial and not divisible by the linear form in LHS of the second equation. On the other hand, for a given x, the system has a pair of nontrivial solutions when the quadratic form is indefinite and the LHS of the second equation divides the LHS of the third one. Alternatively, a pair of non-trivial solutions is available when the quadratic form in the LHS of the third equation is trivial.

For a boundary generic relative to g_{E} curve ∂M, the locus $\partial_3(SM)(v^g)$ is a collection of an even number of points.

The locus $\partial_4(SM)(v^g)$ is given by adding the cubic in w_1, w_2 equation $\tau^3(\vec{x}, \vec{w}) = 0$ to the previous homogeneous system of three equations. For a boundary generic metric g_{E} in relation to ∂M, the resulting system must have only the trivial solution $w_1 = 0, w_2 = 0$ for all $\vec{x} \in \partial M$.

For a boundary generic relative to g_{E} curve ∂M, the differential 1-form $d\tau_0 \in T^*(SM)$ does not vanish at the points of the surface $\{\tau_0 = 0\}$, the differential 2-form $d\tau_0 \wedge d\tau_1 \in \bigwedge^2 T^*(SM)$ does not vanish at the points of the curve $\{\tau_0 = 0, \tau_1 = 0\}$, and the differential 3-form $d\tau_0 \wedge d\tau_1 \wedge d\tau_2 \in \bigwedge^3 T^*(SM)$ does not vanish at the points of the finite locus $\{\tau_0 = 0, \tau_1 = 0, \tau_2 = 0\}$.

Therefore, a boundary generic relative to g_{E} curve ∂M may be either strictly convex, or the union of several arcs, along which the strict concavity and convexity of ∂M alternate. These arcs are bounded by finitely many points of cubic inflection. No tangent to ∂M line has tangency of an order that exceed 3. $\diamondsuit$

The validity of the next conjecture would imply "one half" of Conjecture 4.2, concerned with the space of boundary generic metrics being dense in the space of all metrics.

Conjecture 4.3. *Let $\hat{M}$ be an open smooth n-manifold with a Riemannian metric $\hat{g}$. Consider a smooth compact submanifold $\alpha : N \subset M$ of codimension one. Then there is a small smooth isotopy $A : N \times [0,1] \to M$ of the imbedding α such that the metric $\hat{g}$ is boundary generic[8] with respect to the submanifold $A(N \times \{1\})$.* ♢

The next proposition claims that these refined stratified convexity/concavity properties of the boundary ∂M with respect to a given non-trapping metric g on M can be recovered from the scattering data.

Corollary 4.8. *Assume that a compact connected smooth n-manifold M admits a geodesically boundary generic non-trapping Riemannian metric g.*

Then the scattering map

$$C_{v^g} : \partial_1^+(SM) \to \partial_1^-(SM)$$

allows for a reconstruction of the loci $\{\partial_j^{\pm} SM(v^g)\}_{j\in[1,2n-1]}$.

As a result, for each $j > 0$, the locus $\{(x,v) \in SM|_{\partial M}\}$, such that the germ of the geodesic curve $\gamma \subset M$ through x in the direction of v is tangent to ∂M with the multiplicity $j-1$, can be reconstructed from the scattering map.

Proof. By the proof of Holography Theorem 3.1 (see also Theorem 3.3), the causality map $C_v : \partial_1^+ X(v) \to \partial_1^- X(v)$ of any boundary generic traversing field v on a smooth compact manifold X with boundary allows for a reconstruction of all the strata $\{\partial_j^{\pm} X(v)\}$. With this fact in hand, the claims of the corollary are the immediate implications of Lemma 4.5 and Theorem 4.2. □

Assume that a metric g on M is geodesically boundary generic (as in Definition 4.3). For an oriented geodesic curve $\gamma \subset M$, consider its intersections $\{x_i\}_i$ with ∂M. We denote by m_i the multiplicity of the intersection x_i (see the proof of Lemma 4.5).

Let $m(\gamma) =_{\mathsf{def}} \sum_i m_i$, and $m'(\gamma) =_{\mathsf{def}} \sum_i (m_i - 1)$ (cf. (2.5)).

The next Corollary 4.9 describes the ways in which geodesic arcs can be inscribed in M, provided that the metric on M is *traversally generic* as in Definition 4.3.

For example, if a metric on a surface M is traversally generic, then any geodesic curve may have two simple tangencies to ∂M at most. Similarly,

[8]That is, the geodesic flow $v^{\hat{g}}$ on $S\hat{M}$ is boundary generic with respect to the submanifold ΣN that is produced by restricting the spherical bundle $S\hat{M} \to \hat{M}$ to $A(N \times \{1\})$.

any geodesic curve may have one cubic tangency at most. No tangencies of multiplicity ≥ 4 are permitted.

Corollary 4.9. *Assume that a n-dimensional compact smooth manifold M admits a traversally generic Riemannian metric g. Then any geodesic curve in M has at most $2n-2$ simple points of tangency to the boundary ∂M.*

In general, any geodesic curve $\gamma \subset M$ interacts with the boundary ∂M so that the multiplicity and the reduced multiplicity of γ satisfy the inequalities:

$$m(\gamma) \leq 4n-2 \quad \text{and} \quad m'(\gamma) \leq 2n-2.$$

Moreover, these inequalities hold for any metric g' sufficiently close in the C^∞-topology to g.

Proof. Since, by Lemma 4.5, the multiplicity of tangency of a geodesic curve γ to ∂M at a point x and the multiplicity of tangency of its lift $\tilde{\gamma} \subset SM$ to $\partial_1(SM)$ at the point $(x, \dot{\gamma}(x))$ are equal, the claim follows from the basic properties of any traversally generic vector field, being applied to the geodesic field v^g (see Theorem 3.5 from [K2], or Theorem 6.7 from [K7], and Remark 2.2).

The last claim follows from two facts: (1) the geodesic field v^g depends smoothly on g, and (2) that the space $\mathcal{V}^\ddagger(SM)$ of traversally generic vector fields is open in the space $\mathcal{V}(SM)$ of all vector fields (see Theorem 2.1 or [K2], Theorem 3.5). □

The next corollary should be compared with Theorem D from [Cr2], in a way, a stronger than Corollary 4.10 result, but which uses more boundary data (like the distance "across M" between points of ∂M).

Corollary 4.10. *Let $M \subset \mathbb{H}^n$ be a codimension zero compact smooth submanifold of the hyperbolic space, such that the metric $g = (g_{\mathbb{H}})|_M$ is geodesically boundary generic.*

Then the metric g is non-trapping, and the scattering map C_{v^g} : $\partial_1^+(SM) \to \partial_1^-(SM)$ allows for a reconstruction of the pair $(SM, \mathcal{F}(v^g))$, up to a homeomorphism (a diffeomorphism when Property AA *is valid) of SM which is the identity on $\partial_1(SM)$.*

Proof. In view of Corollary 4.2 and Example 4.2, the hyperbolic metric on M is non-trapping. By Theorem 4.2, the corollary follows. □

Definition 4.8. Let X be a topological space. We consider singular k-dimensional simplexes $\{f_i : \Delta^k \to X\}$ and singular k-chains $\{c = \sum a_i f_i\}$, where the coefficients $a_i \in \mathbb{R}$. Put $\|c\|_{\mathbf{\Delta}} =_{\mathsf{def}} \sum_i |a_i|$. Following Gromov [Gr], for any homology class $h \in H_k(X;\mathbb{R})$, we define its simplicial semi-norm $\|h\|_{\mathbf{\Delta}}$ as $inf_{\{c\}}\{\|c\|_{\mathbf{\Delta}}\}$, the infimum is taken over all cycles c that represent h.

In fact, $\| \sim \|_{\mathbf{\Delta}}$ defines a semi-norm on the vector space $H_k(X;\mathbb{R})$.

A similar definition of the simplicial semi-norm is available for for the relative homology $H_k(X,Y;\mathbb{R})$, where Y is a closed subspace of X. $\diamondsuit$

Recall a few fundamental facts from [Gr] about the simplicial volume $\|[X]\|_{\mathbf{\Delta}}$. If an oriented closed n-manifold X has the Ricci curvature bounded from *below* by the negative number $(-1)/(n-1)$, then

$$\|[X]\|_{\mathbf{\Delta}} < n! \cdot vol(X).$$

On the other hand, for every $n \in \mathbb{N}$, there is a constant $C_n \in \mathbb{R}_+$ such that the following holds: if an oriented closed n-manifold X has the Ricci curvature bounded from *above* by $\delta \in \mathbb{R}_{<0}$, then

$$\|[X]\|_{\mathbf{\Delta}} > C_n \cdot |\delta|^{n/2} \cdot vol(X).$$

We will use the simplicial semi-norms $\| \sim \|_{\mathbf{\Delta}}$ in Section 4.5, where we will employ them to estimate from below the size of the space of geodesics of given tangency patterns.

In the next theorem, we say that manifolds M and M' share the same stable topological (smooth) type, if $M \times S^{n-1}$ and $M' \times S^{n-1}$ are homeomorphic (diffeomorphic).

Theorem 4.4. *Assume that a compact connected smooth n-manifold M with boundary admits a boundary generic non-trapping Riemannian metric g. Then the following statements are valid:*

- *The geodesic scattering map $C_{v^g} : \partial_1^+(SM) \to \partial_1^-(SM)$ allows for a reconstruction of the homology groups $H_*(M;\mathbb{Z})$ and $H_*(M,\partial M;\mathbb{Z})$, as well as for the reconstruction of the homotopy groups $\{\pi_i(M)\}_{i<n}$.*
- *Also, the Gromov simplicial semi-norms $\| \sim \|_{\mathbf{\Delta}}$ on $H_*(M;\mathbb{R})$ and on $H_*(M,\partial M;\mathbb{R})$ can be reconstructed form C_{v^g}.*
  In particular, the simplicial volume $\|[M,\partial M]\|_{\mathbf{\Delta}}$ can be recovered form the scattering map C_{v^g}.
- *If, in addition, M has a trivial tangent bundle, then the stable topological (smooth when Property* AA *from Definition 4.5 holds) type of M is also reconstructable from the scattering map C_{v^g}.*

Proof. By Theorem 4.2, the topological type of SM can be reconstructed from the scattering map C_{v^g} and, therefore, the cohomology/homology of SM with arbitrary coefficients. Moreover, the Gromov simplicial semi-norms $\| \sim \|_\Delta$ on $H_*(SM;\mathbb{R})$ and $H_*(SM, \partial_1(SM);\mathbb{R})$, being invariants of the homotopy type of the pair $(SM, \partial_1(SM))$, can be recovered form C_{v^g} because the topological types of SM and $\partial_1(SM)$ can be recovered.

Since M has the homotopy type of a $(n-1)$-dimensional CW-complex $(\partial M \neq \emptyset)$, we get $H^n(M;\mathbb{Z}) = 0$. Thus, the spherical fibration $\pi : SM \to M$ admits a continuous section σ.

The cohomology/homology spectral sequence of the spherical fibration $\pi : SM \to M$ is trivial, since $H^n(M;\mathbb{Z}) = 0$ and $H_n(M;\mathbb{Z}) = 0$. Thus, with the help of a section $\sigma : M \to SM$, we get a decomposition

$$H_*(SM;\mathbb{Z}) \approx H_*(M;\mathbb{Z}) \oplus H_{*-n+1}(M;\mathbb{Z}).$$

In particular, $H_*(SM;\mathbb{Z}) \approx H_*(M;\mathbb{Z})$ for all $* < n-1$ and $H_{n-1}(SM;\mathbb{Z}) \approx H_{n-1}(M;\mathbb{Z}) \oplus \mathbb{Z}$, where a generator of $\mathbb{Z}$ is produced by the fundamental class $[S]$ of the spherical fiber. As a result, $H_*(M;\mathbb{Z})$, a direct summand of $H_*(SM;\mathbb{Z})$, can be recovered from $H_*(SM;\mathbb{Z})$ and the element $[S] \in H_{n-1}(SM;\mathbb{Z})$, and thus from the scattering map C_{v^g}.

The simplicial semi-norm of a homology class does not increase under continuous maps of spaces [Gr]. Note also that $\|[S]\|_\Delta = 0$. Therefore, with the help of the section σ and the map π, the natural homomorphisms

$$\sigma_* : H_*(M;\mathbb{R}) \to H_*(SM;\mathbb{R}), \quad \sigma_* : H_*(M, \partial M;\mathbb{R}) \to H_*(SM, \partial(SM);\mathbb{R}),$$

induced by the section σ, are *isometries* with respect to the semi-norm $\| \sim \|_\Delta$.

Similar considerations of spectral sequences, using that $H_0(M, \partial M;\mathbb{Z}) = 0$, are valid for $H_*(M, \partial M;\mathbb{Z})$, a direct summand of $H_*(SM, \partial_1(SM);\mathbb{Z})$.

Therefore, the Gromov simplicial semi-norms on $H_*(M, \partial M;\mathbb{Z})$ and $H_*(M;\mathbb{Z})$ can be recovered form C_{v^g} as well. In particular, the simplicial volume $\|[M, \partial M]\|_\Delta$ is determined by C_{v^g}.

The long exact homotopy sequence of the fibration $SM \to M$ identifies $\pi_i(M)$ with $\pi_i(SM)$ for all $i < n$. As a result, the homotopy groups $\{\pi_i(M)\}_{i<n}$ can be recovered from the scattering map C_{v^g} as well.

For a trivial tangent bundle TM, the "C_{v^g}-reconstructible" space SM is diffeomorphic to the product $M \times S^{n-1}$. Thus, the stable smooth type of M (the stabilization being understood as the multiplication with the sphere S^{n-1}) is "reconstructible" as well. □

Definition 4.9. Given two compact Riemannian n-manifolds, (M_1, g_1) and (M_2, g_2), we say that g_1 and g_2 are equigeodesic, if there is a diffeomorphism $\phi : M_1 \to M_2$ such that for any geodesic γ in the metric g_1, considered as an *un-parametrized curve*, the image $\phi(\gamma)$, considered as an un-parametrized curve, is a geodesic in the metric g_2. Moreover, a similar property holds for inverse diffeomorphism ϕ^{-1}. ◇

In fact, *locally* any two metrics on $\mathsf{int}(M)$ are equigeodesic. However, in the vicinity of points $m \in \partial M$, this is not the case: for example, if ∂M is concave in g_1 at m, but convex in g_2 at m, then g_1 and g_2 are not equigeodesic in the vicinity of m.

If g_1 and g_2 on M are equigeodesic, then their projectivized Weyl tensors [We1] coincide.

Equigeodesic metrics g_1 and g_2 have topologically weakly conjugate geodesic flows, were the conjugation diffeomorphism $\Phi : S_{g_1}M_1 \to S_{g_2}M_2$ from Definition 4.4 is induced by the diffeomorphism $\phi : M_1 \to M_2$ from Definition 4.9 and its normalized differential $D^{\odot}\phi$. Recall that the normalized differential is defined as a composition of the differential $D\phi : TM_1 \setminus \sigma(M_1) \to TM_2 \setminus \sigma(M_2)$ with the obvious radial projection $p : TM_2 \setminus \sigma(M_2) \to S_{g_2}M_2$, where $\sigma : M_i \to TM_i$ denotes the zero section.

Definition 4.10. A Riemannian metric g on a manifold M is called geodesically rigid if any metric g' on M which shares with g the same set of un-parametrized geodesic curves is *proportional* to g.

A manifold M is called geodesically rigid if any two equigeodesic metrics on M are proportional. ◇

For example, a closed connected surface is geodesically rigid if and only if its Euler characteristic is negative [MatT]. Note that a particular metric on some M could be geodesically rigid, while M itself is not geodesically rigid.

In general, the existence of *non*-proportional equigeodesic metrics on a *closed* M seems to be an exception: it puts substantial restrictions on the topology of M (see [Mat1], Theorem 2, Theorem 7, and Corollary 2).

One might hope that, for a random (i.e., "bumpy") metric $g \in \mathcal{NT}^{\dagger}(M)$ on a connected manifold M with boundary, the reconstruction of (M, g) from the appropriate boundary data is possible, up to an action of diffeomorphisms $\phi : M \to M$ that are the identity on ∂M (for special interesting cases of such reconstructions see [Cr], [Cr1], [We], and especially [SU4] and

[SUV3]). This wishful thinking is also supported by [Mat], in particular, by his Theorem 4.7 (below).

By Theorem 4.2, if two scattering maps are smoothly conjugate, then the two metrics are geodesic flow weakly topologically conjugate (see Definition 4.4), provided they are boundary generic and non-trapping.

In general, we do not know when two metrics on a given M, that have (weakly) conjugate geodesic flows, are isometric (up to a positive scalar). For example, C. Croke proved in [Cr3] that, for $n \geq 2$, the flat product metric g on $D^n \times S^1$ can be recovered from its scattering map C_{v^g} (even in the class of *trapping* metrics).

In Theorem 4.11, for an interesting case of non-trapping metrics of negative sectional curvature, we will get a pleasing positive answer. This theorem should be compared with more general results of P. Stefanov, G. Uhlmann, and A. Vasy [SUV1], [SUV2], [SUV3], obtained by powerful analytic techniques.

For complete (in particular, for closed) Riemannian manifolds, the investigation of metrics that have topologically conjugate geodesic flows or are equigeodesic in the sense of Definition 4.9 led to a variety of strong results [BCG], [Cr], [Cr1], [CEK], [Mat], [Mat1] [SUV], [SU4]. Let us describe their spirit: under certain conditions, imposed on metrics a priori, the (weak) topological conjugacy of geodesic flows implies an *isometry* of the underlying metrics! In some cases, to get the isometry, one needs to know also the lengths of geodesic curves (the, so called, "lens data").

In particular, in the work of C. Croke, P. Eberlein, and B. Kleiner, the following result was established.

Theorem 4.5. [*CEK*] *Let* (N_1, g_1) *and* (N_2, g_2) *be two* closed *Riemannian manifolds,* $\dim N_1 = \dim N_2 \geq 3$. *Assume that both manifolds have non-positive sectional curvatures and that one of them has rank* $k \geq 2$.

We denote by $\{\psi_i^t\}_{t\in\mathbb{R}}$, $i = 1, 2$, *the* g_i*-induced geodesic flow on* SN_i.

If there is a homeomorphism $\Phi : SN_1 \to SN_2$ *such that* $\Phi \circ \psi_1^t = \psi_2^t \circ \Phi$ *for all* $t \in \mathbb{R}$, *then* g_1 *and* g_2 *are isometric.* ◇

Here the rank of a Riemannian manifold (N, g) is the maximal dimension of a subspace of the tangent space $T_x N$ (to any point x) on which the curvature is identically zero (equivalently, the dimension of the maximal subalgebra of commuting Killing fields on N).

Definition 4.11. Given a point p of Riemannian manifold M, we say that (M, p) admit a local geodesic inversion symmetry, if there exists a neighborhood U_p and a diffeomorphism $\Psi_p : U_p \to U_p$ such that Ψ_p fixes p and reverses the geodesics γ through p, i.e., $\Psi(\gamma(t)) = \gamma(-t)$.

By definition, a locally symmetric Riemannian n-manifold M admits a geodesic inversion symmetry which is an isometry about each point $p \in M$. ♢

G. Besson, G. Courtois, and S. Gallot proved an equally striking theorem:

Theorem 4.6. [*BCG*] *Let (N_1, g_1) be a closed locally symmetric (see Definition 4.11) manifold of a negative sectional curvature and of dimension ≥ 3.*

Then any Riemannian manifold (N_2, g_2) whose geodesic flow is C^1-conjugate[9] *to that of (N_1, g_1) is isometric to (N_1, g_1).* ♢

More important results of this flavor are contained the following theorems of V.S. Matveev (see [KrMa] for generalizations of Theorem 4.7).

Theorem 4.7. [*Mat1*], [*Mat2*] *For all $n \geq 4$, in the space of all Riemannian metrics on a closed n-manifold M, taken in the C^2-topology, the geodesically rigid metrics (see Definition 4.10) form an open and dense set.*

For $n = 3$, this claim is valid if the C^2-topology is replaced by the C^∞-topology. ♢

Also, it seems likely that in the space of all Levi-Civita connections (taken in the C^∞-topology) on a closed manifold M, the geodesically rigid connections form an open and dense set.

Theorem 4.8. [*Mat*] *Every closed connected manifold that admits a Riemannian metric of negative sectional curvature, is geodesically rigid.* ♢

Motivated by the spirit of these theorems (as far as we understand, they do not apply directly to the geodesic flows on manifolds with boundary), we move towards linking the Riemannian manifolds that have topologically (weakly) conjugate traversing geodesic flows with the problem of inverse geodesic scattering. In its more daring formulation, the problem of inverse geodesic scattering aims to reconstruct a metric g on M from the scattering map $C_{v^g} : \partial_1^+(SM) \to \partial_1^-(SM)$ and, perhaps, some additional boundary

[9] that is, the geodesic flows conjugating map $\Phi : SN_2 \to SN_1$ is a C^1-diffeomorphism.

confined data. The reconstruction is thought up to the natural action of $\mathsf{Diff}(M, \partial M)$—the group of diffeomorphisms that are the identity on ∂M—on the space of Riemannian metrics $\mathcal{R}(M)$ (see Remark 4.5).

As we move towards this goal, let us revisit Theorem 3.18 which relies on basic Contact Geometry structures, as described in Section 3.8.

Let β_g be the Liouville contact 1-form on SM, produced by a metric g as in formula (4.3). We denote by

$$(\beta_g)^{\partial^+} =_{\mathsf{def}} \beta_g||_{\partial_1^+(SM)}$$

the restriction of the section $\beta_g : SM \to T^*(SM)$ of the cotangent bundle to the section of the bundle $T^*(SM)|_{\partial_1^+(SM)} \to \partial_1^+(SM)$ (not to be confused with the restriction of the form β_g to the submanifold $\partial_1^+(SM) \subset SM$).

The next theorem is an instant corollary of Theorem 3.18.

Theorem 4.9. *Let g be a boundary generic and non-trapping Riemannian metric on a compact connected manifold M with boundary. Assume that any geodesic curve in M is either transversal to ∂M at* some *point or is a singleton quadratically tangent to ∂M.*

We choose a balanced Lyapunov function $F : SM \to \mathbb{R}$ for the geodesic vector field v^g so that $dF(v^g) = 1$. Let $F^\partial =_{\mathsf{def}} F|_{\partial_1(SM)}$.

Then the scattering map $C_{v^g} : \partial_1^+(SM) \to \partial_1^-(SM)$, together with the function $F^\partial : \partial_1(SM) \to \mathbb{R}$ and the 1-form $(\beta_g)^{\partial^+}$, allow for a reconstruction of SM, the geodesic field v^g, and the contact 1-form β_g, up to a diffeomorphism $\Phi : SM \to SM$ which is the identity on $\partial_1(SM)$ and preserves the 1-form $(\beta_g)^{\partial^+}$. ◇

Changing the trust, let us recall a generalization of the famous Mostov Rigidity Theorem.

Theorem 4.10. [*Most*] *Let (N_1, g_1) and (N_2, g_2) be two closed locally symmetric Riemannian manifolds with negative sectional curvatures and of dimension ≥ 3.*

Given an isomorphism $\psi : \pi_1(N_1) \to \pi_1(N_2)$ of their fundamental groups, there exists a unique isometry *$\Psi : (N_1, c \cdot g_1) \to (N_2, g_2)$, where $c > 0$ is a constant, so that Ψ induces the isomorphism ψ.* ◇

In the spirit of Theorem 1.3 from [BCG] and [CEK], by combining Mostov Theorem 4.10 with Theorem 4.2, we get Theorem 4.11 below. It and its Corollary 4.11 should be compared with Theorem D from [Cr2], which deals with domains in the hyperbolic space $\mathbb{H}^n$.

Theorem 4.11. *Let $n \geq 3$. Consider two closed locally symmetric Riemannian n-manifolds (N_1, g_1) and (N_2, g_2) with negative sectional curvatures. Let a connected manifold M_i $(i = 1, 2)$ be obtained from N_i by removing the interior of a smooth codimension zero submanifold $U_i \subset N_i$ such that the induced homomorphism $\pi_1(M_i) \to \pi_1(N_i)$ of the fundamental groups is an isomorphism*[10]*.*

Assume that the restriction of the metric g_i to M_i is boundary generic and non-trapping. Assume also that the two geodesic scattering maps

$$C_{v^{g_1}} : \partial_1^+(SM_1) \to \partial_1^-(SM_1), \quad C_{v^{g_2}} : \partial_1^+(SM_2) \to \partial_1^-(SM_2)$$

are conjugated via a smooth diffeomorphism $\Phi^\partial : \partial_1(SM_1) \to \partial_1(SM_2)$.

Then Φ^∂ determines a unique diffeomorphism $\phi : N_1 \to N_2$ such that $\phi^*(g_2) = c \cdot g_1$ for a constant $c > 0$.[11]

Proof. By Theorem 4.2, the spaces SM_1 and SM_2 are homeomorphic via an extension Φ of Φ^∂. In particular, Φ induces an isomorphism $\Phi_* : \pi_1(SM_1) \to \pi_1(SM_2)$ of the fundamental groups. Each space SM_i fibers over M_i with the spherical fiber S^{n-1}. Thus, when $n \geq 3$, the fundamental groups $\pi_1(M_1)$ and $\pi_1(M_2)$ are isomorphic via an isomorphism $\phi_* : \pi_1(M_1) \to \pi_1(M_2)$, which is determined by Φ_*. In fact, ϕ_* is induced by taking a section $\sigma_1 : M_1 \to SM_1$ (the bundle $TM_1 \to M_1$ admits a non-vanishing section since $\partial M_1 \neq \emptyset$), composing σ_1 with Φ, and then applying the projection $SM_2 \to M_2$.

By the hypotheses, the inclusion homomorphisms $\alpha_i : \pi_1(M_i) \to \pi_1(N_i)$ are isomorphisms. Therefore, $\pi_1(N_1)$ and $\pi_1(N_2)$ are isomorphic via the isomorphism $\hat{\phi}_* = \alpha_2 \circ \phi_* \circ \alpha_1^{-1}$. By Mostow Rigidity Theorem 4.2, for a given $\hat{\phi}_*$, there exists a unique diffeomorphism $\hat{\phi} : N_1 \to N_2$ such that $\hat{\phi}^*(g_2) = c \cdot g_1$, where c is a positive constant.

We claim that the diffeomorphism $\hat{\phi} : N_1 \to N_2$ is determined by the conjugating diffeomorphism $\Phi^\partial : \partial_1(SM_1) \to \partial_1(SM_2)$. Indeed, consider the obvious map $\Gamma_i : SM_i \to \mathcal{T}(v^{g_i})$, where $\mathcal{T}(v^{g_i})$ denotes the trajectory space of the geodesic v^{g_i}-flow. Since v^{g_i} is traversing, Γ_i is a quasifibration (see [May] for the notion of a quasifibration) with contractible fibers. Therefore, $(\Gamma_i)_* : \pi_1(SM_i) \to \pi_1(T(v^{g_i}))$ is an isomorphism. By Corollary 4.3 and Theorem 4.3, we get a commutative diagram $\Phi^{\mathcal{T}} \circ \Gamma_1 = \Gamma_2 \circ \Phi$, where $\Phi^{\mathcal{T}} : \mathcal{T}(v^{g_1}) \to \mathcal{T}(v^{g_2})$ is a homeomorphism which is determined by Φ^∂.

[10]For example, by a general position argument, this the case when U_i has a spine of codimension 3 at least. In particular, U_i may be a disjoint union of n-balls.

[11]Here we able to overcome the curse of Property AA!

In fact, $\Phi^{\mathcal{T}} \circ \Gamma_1^\partial = \Gamma_2^\partial \circ \Phi^\partial$, where $\Gamma_i^\partial =_{\mathsf{def}} \Gamma_i|_{\partial_1(SM_i)}$. Thus, the isomorphism $\Phi_* : \pi_1(SM_1) \to \pi_1(SM_2)$ is determined by the diffeomorphism Φ^∂. As a result, $\phi_* : \pi_1(M_1) \to \pi_1(M_2)$ and hence $\hat{\phi}_* : \pi_1(N_1) \to \pi_1(N_2)$ are determined by Φ^∂. Therefore, by Mostow Rigidity Theorem 4.10, the corresponding unique diffeomorphism $\hat{\phi} : N_1 \to N_2$ that delivers the isometry between g_2 and $c \cdot g_1$ is determined by Φ^∂. □

Example 4.9. Let $n \geq 3$. We adopt the settings of Theorem 4.11: consider two closed locally symmetric Riemannian n-manifolds, (N_1, g_1) and (N_2, g_2), with negative sectional curvatures. Let $U_1 \subset N_1, U_2 \subset N_2$ be smooth n-balls, and let $M_i =_{\mathsf{def}} N_i \setminus \mathsf{int}(U_i)$. Under the hypotheses of Theorem 4.11, we consider the diffeomorphism $\Phi^\partial : \partial_1(SM_1) \to \partial_1(SM_2)$ that conjugates the two scattering maps $C_{v^{g_1}}, C_{v^{g_2}}$. Then we fix diffeomorphisms $\partial_1(SM_i) \approx S^{n-1} \times S^{n-1}$. With their help, the scattering maps $C_{v^{g_i}} : \partial_1^+(SM_i) \to \partial_1^-(SM_i)$, $i = 1, 2$, become maps whose sources and targets are both diffeomorphic to the handle $S^{n-1} \times D^{n-1}$.

Using these identifications, by Theorem 4.11, the scattering maps $C_{v^{g_i}} : S^{n-1} \times D^{n-1} \to S^{n-1} \times D^{n-1}$, with the help of the conjugating diffeomorphism $\Phi^\partial : S^{n-1} \times S^{n-1} \to S^{n-1} \times S^{n-1}$, allow to identify the metric g_2 on $N_2 = M_2 \cup_{\partial U_2} U_2$, with the metric $c \cdot g_1$, where c is a positive scalar. In other words, the non-conjugated scattering maps $C_{v^{g_i}}$ distinguish (up to a constant c) between different locally symmetric spaces (N_1, g_1) and (N_2, g_2) of negative sectional curvatures. ♢

The next immediate corollary of Theorem 4.11 is inspired by the image of geodesic motion of a bouncing particle in the complement M to a number of disjoint balls, placed in a closed hyperbolic manifold N of a dimension greater than two. We assume that the balls are placed so "dense" in N that every geodesic curve hits some ball in both directions. Under these assumptions, the collisions of the probe particle with the boundary ∂M "feel the shape of N".

Corollary 4.11. *Let (N_1, g_1) and (N_2, g_2) be two closed hyperbolic manifolds of dimension $n \geq 3$. Let M_i be produced from N_i by removing the interiors of an equal number of disjoint smooth n-balls, so that the geodesic flow on each M_i is boundary generic*[12] *and non-trapping.*

If the scattering maps $C_{v^{g_1}}$ and $C_{v^{g_2}}$ on $\partial_1(SM_1)$ and $\partial_1(SM_2)$ (note that $\partial_1(SM_i)$ is diffeomorphic to $\coprod_{l=1}^k S^{n-1} \times S^{n-1}$) are conjugated with

[12]This is the case when the balls are g_i-convex in N_i.

the help of a diffeomorphism $\Phi^\partial : \partial_1(SM_1) \to \partial_1(SM_2)$, *then* (N_1, g_1) *is isometric to* $(N_2,\, c \cdot g_2)$, *where* c *is a positive constant.* ◇

Let (N, g) be a closed locally symmetric Riemannian manifold whose sectional curvature is negative, and let $U \subset N$ be a smooth codimension zero submanifold. Take $M = N \setminus \mathsf{int}(U)$ and assume that $g|_M$ is boundary generic and non-trapping.

By the Mostow Theorem 4.10, the isometry group $\mathsf{Iso}(N, g)$ is isomorphic to the group $\mathsf{Aut}(\pi_1(N))$ of automorphisms of the fundamental group $\pi_1(N)$.

For an isometry $\psi : N \to N$, put $M_\psi =_{\mathsf{def}} N \setminus \mathsf{int}(\psi(U))$. Evidently any isometry $\psi : N \to N$ induces a diffeomorphism $\Psi : SN \to SN$ which commutes with the geodesic flow on SN. Moreover,

$$\Phi =_{\mathsf{def}} \Psi| : SM \to SM_\psi$$

conjugates the *parametrized* geodesic flows, delivered by the metrics $g^M =_{\mathsf{def}} g|_M$ and $g^{M_\psi} =_{\mathsf{def}} g|_{M_\psi}$. As a result, the scattering maps $C_{v^{g^M}}$ and $C_{v^{g^{M_\psi}}}$ are conjugated with the help of the diffeomorphism

$$\Phi^\partial =_{\mathsf{def}} \Phi| : \partial_1(SM) \to \partial_1(SM_\psi).$$

Is this construction provides *the only way* in which the scattering maps on the complements of isometric domains in N can be conjugated? More accurately, we pose the following question.

Question 4.4. *Let* (N, g) *be a closed locally symmetric Riemannian manifold of a negative sectional curvature. Consider two smooth codimension zero submanifolds* $U, V \subset N$. *Let* $\psi : U \to V$ *be an isometry, and denote by* $\Psi : SU \to SV$ *the diffeomorphism, induced by* ψ *and its differential.*

We form the submanifolds $M =_{\mathsf{def}} N \setminus \mathsf{int}(U)$ *and* $L =_{\mathsf{def}} N \setminus \mathsf{int}(V)$.

Assume that the metrics $g_M =_{\mathsf{def}} g|_M$ *and* $g_L =_{\mathsf{def}} g|_L$ *are boundary generic and non-trapping. Also assume that the scattering maps* $C_{v^{\{g_M\}}}$ *and* $C_{v^{\{g_L\}}}$ *are conjugated with the help of the diffeomorphism* $\Psi^\partial =_{\mathsf{def}} \Psi|_{\partial_1(SU)}$.

Can we conclude that there exists an isometry $\tilde{\psi} : N \to N$ *such that* $\tilde{\psi}|_U = \psi$? ◇

4.4 The inverse geodesic scattering problem in presence of length data

Consider a non-trapping metric g on a compact smooth manifold M with boundary. Now, we will enhance the scattering data $C_{v^g} : \partial_1^+(SM) \to$

$\partial_1^-(SM)$ by adding information about the length $\ell(\tilde{\gamma})$ (equivalently, travel time) along each v^g-trajectory $\tilde{\gamma}$, or more accurately, about the length of the segment of v^g-trajectory, bounded by $x \in \partial_1^+(SM)$ and $C_{v^g}(x) \in \partial_1^-(SM)$. This new data is commonly called "lens data".

Any balanced Lyapunov function $F : SM \to \mathbb{R}$ (i.e., $dF(v^g) = 1$) captures the lens data:

$$\Big\{\ell(\tilde{\gamma}_x) = F(C_{v^g}(x)) - F(x) = \int_x^{C_{v^g}(x)} \beta_g\Big\}_{x \in \partial_1^+(SM)}, \tag{4.8}$$

where β_g denotes the contact form on SM, generated by g.

On the other hand, the knowledge of the lens data determines a smooth balanced Lyapunov function $F : SM \to \mathbb{R}$ only up to adding a smooth function $H \in C^\infty(\mathcal{T}(v^g))$ (i.e., such that $\mathcal{L}_{v^g} H = 0$).

Definition 4.12. We say that the Lyapunov function $F : SM \to \mathbb{R}$ of v^g is weakly balanced if, for each point $x \in \partial_1^+(SM)$, the variation $F(C_{v^g}(x)) - F(x)$ is equal (up to an universal constant) to the arc length $\ell_g(x)$. $\diamondsuit$

The next lemma is a variation of Lemma 3.18 and Lemma 3.21.

Lemma 4.7. *Let (X, g) be a compact connected smooth Riemannian manifold with boundary. Let v be a smooth traversing and boundary generic vector field, and α a smooth 1-form on X such that $\alpha(v) > 0$. Assume that for each $x \in \partial_1^+ X(v)$, we have*

$$\int_x^{C_v(x)} \|v\|_{\mathsf{g}}\, d\mathsf{g} = \int_x^{C_v(x)} \alpha.$$

Here $d\mathsf{g}$ denotes the measure on the trajectory γ_x, induced by the metric $\mathsf{g}|_{\gamma_x}$.

Then there exists a homeomorphism $\Psi : X \to X$ with the following properties:

(4.9)

- $\Psi|_{\partial_1 X}$ *is the identity,*
- *each v-trajectory γ is invariant under Ψ,*
- *for each γ, $\Psi| : \gamma \to \gamma$ is a smooth diffeomorphism,*
- $(\Psi|_\gamma)^*(\alpha)(v) = \|v\|_{\mathsf{g}}$.

If v is such that any v-trajectory γ is either transversal to $\partial_1 X$ at some *point, or $\gamma \cap \partial_1 X$ is a singleton of multiplicity 2, then the homeomorphism Ψ is a smooth diffeomorphism.*

Proof. Consider the segment $I_x^{C_v(x)}$ of the v-trajectory through $x \in \partial_1^+ X(v)$, bounded by points x and $C_v(x)$, and two strictly monotone smooth functions on $I_x^{C_v(x)}$:

$$K(y) = \int_x^y \alpha, \quad L(y) = \int_x^y \|v\|_{\mathsf{g}}\, d\mathsf{g}.$$

We form the monotone function $\Psi(y) =_{\mathsf{def}} (K^{-1} \circ L)(y)$. It maps the arc $I_x^{C_v(x)}$ to itself. Indeed, for each $x \in \partial_1^+ X(v)$ and $y \in I_x^{C_v(x)}$, $\Psi(y)$ is well-defined by the identity

$$\int_x^y \alpha = \int_x^{\Psi(y)} \|v\|_{\mathsf{g}}\, d\mathsf{g}.$$

By lemma's hypotheses, $\Psi(x) = x$ and $\Psi(C_v(x)) = C_v(x)$ for all $x \in \partial_1^+ X(v)$. Thus, $\Psi(I_x^{C_v(x)}) = I_x^{C_v(x)}$. In particular, Ψ is the identity on the boundary $\partial_1 X$. Let us verify that the map Ψ is continuous along γ_x, despite the possible discontinuity of C_v.

For any pair $y \in I_x^{C_v(x)}$, $y' \in I_{C_v(x)}^{C_v(C_v(x))}$, put

$$\tilde{K}(y') =_{\mathsf{def}} \int_x^{y'} \alpha = L + \int_{C_v(x)}^{y'} \alpha,$$

$$\tilde{L}(y') =_{\mathsf{def}} \int_x^{y'} \|v\|_{\mathsf{g}}\, d\mathsf{g} = L + \int_{C_v(x)}^{y'} \|v\|_{\mathsf{g}}\, d\mathsf{g},$$

where L denotes the integrals $\int_x^{C_v(x)} \|v\|_{\mathsf{g}}\, d\mathsf{g} = \int_x^{C_v(x)} \alpha$, the latter equality is valid by the lemma hypotheses. Since $\tilde{K}$ and K, as well as $\tilde{L}$ and L, differ by the same shift L, we have $(K^{-1} \circ L)(y') = (\tilde{K}^{-1} \circ \tilde{L})(y')$. By the continuity of g and α, we get that $\tilde{K}^{-1} \circ \tilde{L}$ is continuous in the vicinity of $C_v(x) \in \gamma_x$, along the interval $I_x^{C_v(x)} \cup I_{C_v(x)}^{C_v(C_v(x))}$. Thus $\Psi(y)$ and $\Psi(y')$ are close for any two points $y, y' \in \gamma_x$ that are sufficiently close to the point $C_v(x)$. Therefore, $\Psi| : \gamma_x \to \gamma_x$ is a homeomorphism.

Note that the derivative $\Psi' > 0$ in $I_x^{C_v(x)}$ since, by the definition of Ψ, $\Psi'(y) = \|v(y)\|_{\mathsf{g}} / \alpha(v(y))$ for all $y \in \gamma_x$. Therefore, the restriction of Ψ to any v-trajectory γ is a smooth diffeomorphism.

In order to show that $\Psi : X \to X$ is a homeomorphism, we embed X into an open manifold $\hat{X}$ and extend smoothly the vector field v, the form α, and the metric g into a open neighborhood U of X in $\hat{X}$ so that the extensions $\hat{v}$, $\hat{\alpha}$, $\hat{\mathsf{g}}$ have the property $\hat{\alpha}(\hat{v}) > 0$ in U. In what follows, we may adjust the size of U according to the needs of our arguments. As usually, we will treat $\hat{X} = U$, $\hat{v}$, $\hat{\alpha}$, $\hat{\mathsf{g}}$ as a germs. Any v-trajectory γ is contained in a unique $\hat{v}$-trajectory $\hat{\gamma}$.

In the next paragraph, we will rely on few basic facts about boundary generic traversing flows, as described in Section 2. Let γ_0 be a v-trajectory. Since v is boundary generic, the intersection $\gamma_0 \cap \partial_1 X$ is a finite ordered collection of points $\{x_1, x_2, \dots, x_k\}$. Thus $C_v(x_i) = x_{i+1}$ for all $i \in [1, k-1]$, unless $\gamma_0 \cap \partial_1 X$ is a singleton ($k = 1$), in which case $C_v(x_1) = x_1$. Let V be a cylindrical neighborhood of $\hat{\gamma}_0$ in $\hat{X}$ which consists of $\hat{v}$-trajectories. By choosing an appropriate U, we may assume that $\hat{\gamma}_0 \cap X = \gamma_0$. We may also assume that V admits a system of coordinates $(u, \vec{w})$ such that the $\hat{v}$-trajectories are given by the equations $\{\vec{w} = \vec{const}\}$. For any $x \in \hat{\gamma}$, we denote by S_x the hypersurface $\{u = u(x)\}$. For any $\hat{\gamma} \subset V$, the intersection $\hat{\gamma} \cap \partial_1 X$ consists of finitely many points, while $\hat{\gamma} \cap X$ consists of finitely many closed intervals or singletons. Picking V sufficiently narrow, the set $\hat{\gamma} \cap \partial_1 X$ may be divided into at most k disjoint (possibly empty) subsets $A_{\hat{\gamma},i}$ that correspond to the elements $x_i \in \gamma_0 \cap \partial_1 X$. The cardinality of each set $A_{\hat{\gamma},i}$ does not exceed $m(x_i)$, the multiplicity of tangency of γ_0 to $\partial_1 X$ at x_i, and $\#(A_{\hat{\gamma},i}) \equiv m(x_i) \mod 2$.

Let $\{S_i =_{\mathsf{def}} S_{x_i}\}_{1 \leq i \leq k}$ be disjoint transversal sections (where $x_i \in S_i$) of the $\hat{v}$-flow in the vicinity of $\gamma_0 \subset V$. By the choice of V and by the construction of the non-empty sets $A_{\hat{\gamma},i} \subset \hat{\gamma}$, the points of $A_{\hat{\gamma},i}$ are located in the vicinity of the point $y_i =_{\mathsf{def}} S_i \cap \hat{\gamma}$.

Let us compare $\Psi(x)$ and $\Psi(y)$ for a pair of points $x \in \gamma_0$ and $y \in \hat{\gamma} \cap X$ that are close in X. We may assume that, for some index i, $x \in [x_i, x_{i+1}) \subset \hat{\gamma}_0$ and that $z =_{\mathsf{def}} S_x \cap \hat{\gamma}$ and y are close in $\hat{\gamma}$. Since $y, z \in \hat{\gamma}$, by the previous argument, $\Psi(z)$ and $\Psi(y)$ are close. So, it suffices to compare $\Psi(z) = (K^{-1} \circ L)(z)$ and $\Psi(x) = (K^{-1} \circ L)(x)$. Here, by the definition, $K(x) = \int_{x_i}^{x} \alpha$, $L(x) = \int_{x_i}^{x} \|v\|_{\mathsf{g}}\, d\mathsf{g}$, and $K(z) = \int_{z^\star}^{z} \alpha$, $L(z) = \int_{z^\star}^{z} \|v\|_{\mathsf{g}}\, d\mathsf{g}$, where $z^\star$ is the lowest point in the connected component of the set $\hat{\gamma} \cap X$ that contains z. By the choice of V, for some $j \leq i$, $z^\star$ is a point of the set $A_{\hat{\gamma},j}$, and thus is close to both $x_j \in \gamma_0$ and to the point $z_j^\star =_{\mathsf{def}} S_j \cap \hat{\gamma}$. By the continuity of α and g and by the choice of $V \supset \gamma_0$, the value $K_\dagger(x) =_{\mathsf{def}} \int_{x_j}^{x} \alpha$ is close to the value

$$K_\ddagger(z) =_{\mathsf{def}} \int_{z_j^\star}^{z} \alpha = \int_{z_j^\star}^{z^\star} \alpha + K(z),$$

and the value $L_\dagger(x) =_{\mathsf{def}} \int_{x_j}^{x} \|v\|_{\mathsf{g}}\, d\mathsf{g}$ is close to

$$L_\ddagger(z) =_{\mathsf{def}} \int_{z_j^\star}^{z} \|v\|_{\mathsf{g}}\, d\mathsf{g} = \pm \int_{z_j^\star}^{z^\star} \|v\|_{\mathsf{g}}\, d\mathsf{g} + L(z).$$

Therefore, $(K_\dagger^{-1} \circ L_\dagger)(x)$ is close to $(K_\ddagger^{-1} \circ L_\ddagger)(z)$. Since, by the lemma hypotheses, $\int_{x_j}^{x_i} \alpha = \int_{x_j}^{x_i} \|v\|_{\mathsf{g}}\, d\mathsf{g}$, we get that $(K_\dagger^{-1} \circ L_\dagger)(x)$ is close to $(K^{-1} \circ L)(x)$. Since $\int_{z_j^\star}^{z^\star} \alpha$ and $\int_{z_j^\star}^{z^\star} \|v\|_{\mathsf{g}}\, d\mathsf{g}$ are small, we conclude that $(K_\ddagger^{-1} \circ L_\ddagger)(z)$ is close to $(K^{-1} \circ L)(z)$. As a result, $\Psi(x)$ and $\Psi(z)$ are close in V. Thus $\Psi : X \to X$ is continuous. By the same token, $\Psi^{-1} : X \to X$ is continuous as well.

The argument validating that Ψ is a smooth diffeomorphism (under the lemma hypotheses in its last paragraph), is similar to the one in the proof of Theorem 3.3. It employs that the boundary $\partial_1 X$, at a point of transversal intersection $\gamma \cap \partial_1 X$, is a smooth section σ of the v-flow, together with the hypotheses that the metric g and the form α are smooth. This implies that the transformation Ψ, defined by the formula

$$\int_{\gamma_x \cap \sigma}^{x} \alpha = \int_{\gamma_x \cap \sigma}^{\Psi(x)} \|v\|_{\mathsf{g}}\, d\mathsf{g},$$

and the similarly defined transformation Ψ^{-1} are smooth on $\hat{X}$. □

For $i = 1, 2$, let $\{\psi_i^t : SM_i \to SM_i\}_t$ denote the geodesic flow transformations. Recall that, for each $x \in SM_i$, $\{\psi_i^t(x)\}$ are partially-defined for an appropriate interval $\mathcal{I}_i(x)$ of moments $t \in \mathbb{R}$.

The next theorem strengthens Theorem 4.2 at the cost of adding lens data to the set of boundary-confined data.

Theorem 4.12. (The strong topological rigidity of the geodesic flow for the inverse scattering problem in the presence of lengths data)

Let (M_1, g_1) and (M_2, g_2) be two smooth compact connected Riemannian n-manifolds with boundaries. Let the metric g_1 be boundary generic, and g_2 be non-trapping and boundary generic. Assume that the scattering maps

$$C_{v^{g_1}} : \partial_1^+(SM_1) \to \partial_1^-(SM_1) \;\; \textit{and} \;\; C_{v^{g_2}} : \partial_1^+(SM_2) \to \partial_1^-(SM_2)$$

are conjugated by a smooth diffeomorphism $\Phi^\partial : \partial_1(SM_1) \to \partial_1(SM_2)$ that preserves the lens data (lengths) as in formula (4.8):

$$\left\{\ell_{g_2}(\Phi^\partial(w)) = \ell_{g_1}(w)\right\}_{w \in \partial_1^+(SM_1)}.$$

Then:

- *g_1 is a non-trapping metric.*

- Φ^∂ *extends to a homeomorphism* $\Phi : SM_1 \to SM_2$ *such that*

$$\Phi \circ \psi_1^t(x) = \psi_2^t \circ \Phi(x)$$

for each $x \in SM_1$ *and all* $t \in \mathbb{R}$ *for wich* $\psi_1^t(x)$ *is well-defined.*

- *the restriction of* Φ *to each* v^{g_1}*-trajectory is a smooth diffeomorphism.*
- *If* g_2 *is such that no geodesic curve* $\gamma \subset M_2$ *is cubically tangent to* ∂M_2 *at a pair of distinct points, and no geodesic curve is a singleton of multiplicity* ≥ 4*, then the conjugating homeomorphism* $\Phi : SM_1 \to SM_2$ *is a smooth diffeomorphism.*

Proof. The proof is a modification of the arguments from the Holography Theorem 3.3. As in that theorem, we start with a balanced Lyapunov function $F_2 : SM_2 \to \mathbb{R}$ such that $dF_2(v^{g_2}) = 1$. We consider the pull-back

$$F_1^\partial =_{\mathsf{def}} (\Phi^\partial)^*(F_2|_{\partial X_2}) : \partial_1(SM_1) \to \mathbb{R},$$

and using Lemma 3.3, extend F_1^∂ to a smooth function $F_1 : SM_1 \to \mathbb{R}$ so that $dF_1(v^{g_1}) > 0$. Thus, g_1 is a non-trapping metric.

By Lemma 3.21, we can modify $F_1 \rightsquigarrow \tilde{F}_1$ so that $d\tilde{F}_1(v^{g_1}) = 1$. Abusing notations, we denote this modified $\tilde{F}_1$ by F_1.

Since $dF_1(v^{g_1}) = 1 = dF_2(v^{g_2})$ and by the hypothesis $\big\{\ell_{g_2}(\Phi^\partial(w)) = \ell_{g_1}(w)\big\}_{w \in \partial_1^+(SM_1)}$, for any $w \in \partial_1^+(SM_1)$, we get the equality

$$\int_w^{C_{v^{g_1}}(w)} dF_1 = \int_{\Phi^\partial(w)}^{C_{v^{g_2}}(\Phi^\partial(w))} dF_2. \tag{4.10}$$

Let $\Gamma_i : SM_i \to \mathcal{T}(v^{g_i})$ and $\Gamma_i^\partial : \partial_1(SM_i) \to \mathcal{T}(v^{g_i})$, $i = 1, 2$, be the obvious projections. Then, as in the proof of the Holography Theorem 3.3, we construct the conjugating homeomorphism $\Phi : SM_1 \to SM_2$ for the scattering maps so that: Φ maps each v^{g_1}-trajectory $\tilde{\gamma}$ to the v^{g_2}-trajectory that is suspended over the point

$$\Gamma_2^\partial \circ \Phi^\partial(\tilde{\gamma} \cap \partial_1(SM_1)) = \Phi^{\mathcal{T}} \circ \Gamma_1(\tilde{\gamma}) \in \mathcal{T}(v^{g_2}).$$

The restriction $\Phi|_{\tilde{\gamma}}$ is a diffeomorphism defined by

$$\Phi|_{\tilde{\gamma}} : x \to F_2^{-1} \circ S_w \circ F_1(x),$$

where a point $w \in \tilde{\gamma}_x \cap \partial_1^+(SM_1)$, and $S_w : \mathbb{R} \to \mathbb{R}$ is the parallel shift that takes $F_1(w)$ to $F_2(\Phi^\partial(w))$. Thanks to property (4.10), for each x, this map does not depend on the choice of $w \in \tilde{\gamma}_x \cap \partial_1^+(SM_1)$. For the same reason, $\Phi|_{\tilde{\gamma}}(x)$ depends continuously on x.

By the construction of Φ, the lengths of the geodesic flows segments are preserved by Φ: i.e., for each $x \in SM_1$ and an appropriate t,

$$t = \int_x^{\psi_1^t(x)} dF_1 = \int_x^{\psi_1^t(x)} \beta_{g_1} = \int_{\Phi(x)}^{\psi_2^t(\Phi(x))} \beta_{g_2} = \int_{\Phi(x)}^{\psi_2^t(\Phi(x))} dF_2.$$

Here β_{g_1}, β_{g_2} are the Liouville 1-forms produced by the metrics g_1, g_2.

Thus, $\Phi \circ \psi_1^t(x) = \psi_2^t \circ \Phi(x)$ for each $x \in SM_1$ and all $t \in \mathbb{R}$ for which $\psi_1^t(x)$ is well-defined.

If g_2 is such that no geodesic curve $\gamma \subset M_2$ is cubically tangent to ∂M_2 at a pair of distinct points and no geodesic curve is a singleton of multiplicity 4 (we named these features of g_2 "Property AA"), then by Lemma 4.5, v^{g_2}, and thus v^{g_1} both possess Property A from Definition 3.2. Therefore, by Lemma 4.7, the conjugating homeomorphism $\Phi : SM_1 \to SM_2$ is a smooth diffeomorphism. □

Theorem 4.12 leads to the next Theorem 4.13. Crudely speaking, the latter theorem claims that if two closed Riemannian manifolds N_1 and N_2 are isometric in the vicinity of two regions U_1 and U_2 and the scattering and lens data for the complimentary non-trapping regions $M_1 = N_1 \setminus U_1$ and $M_2 = N_2 \setminus U_2$ are smoothly conjugated, then the geodesic flows of N_1 and N_2 are (strongly) conjugated as well.

Theorem 4.13. *Let (N_1, g_1) and (N_2, g_2) be two closed smooth Riemannian n-manifolds. For $i = 1, 2$, let U_i be a codimension zero compact submanifold of N_i with a smooth boundary ∂U_i. Put $M_i =_{\mathsf{def}} N_i \setminus \mathsf{int}(U_i)$. Consider an open neighborhood V_i of U_i in N_i.*

Assume that the metric $g_2|_{M_2}$ is non-trapping, boundary generic, and satisfies Property AA[13].

Let $\psi : V_1 \to V_2$ be an isometry *with respect to $g_1|_{V_1}$ and $g_2|_{V_2}$. Consider the diffeomorphism $\Psi : SV_1 \to SV_2$, induced by ψ, and its restriction*

$$\Psi^\partial : \partial_1(SU_1) \to \partial_1(SU_2).$$

Assume that the scattering maps

$$C_{v^{g_1|M_1}} : \partial_1^+(SM_1) \to \partial_1^-(SM_1) \text{ and } C_{v^{g_2|M_2}} : \partial_1^+(SM_2) \to \partial_1^-(SM_2)$$

are conjugated with the help of the diffeomorphism Ψ^∂ and, for each $w \in \partial_1^+(SM_1)$, the length data agree: $\ell_{g_2|_{M_2}}(\Psi^\partial(w)) = \ell_{g_1|_{M_1}}(w)$.

Then the geodesic flows on SN_1 and SN_2 are smoothly topologically conjugated in the sense of Definition 4.4.

[13]If U_2 is strictly concave in N_2, then these hypotheses reduce to the requirement that $g_2|_{M_2}$ is non-trapping and weakly balanced.

Proof. Let $i = 1, 2$ and let $\tilde{\gamma}$ be a v^{g_i}-trajectory in SN_i. Put

$$\tilde{\gamma}_{M_i} =_{\mathsf{def}} \tilde{\gamma} \cap SM_i,\ \tilde{\gamma}_{U_i} =_{\mathsf{def}} \tilde{\gamma} \cap SU_i, \quad \text{and } \tilde{\gamma}_{V_i} =_{\mathsf{def}} \tilde{\gamma} \cap SV_i.$$

For each point $x \in \partial_1(SU_1)$, we denote by $\tilde{\gamma}(x)$ the v^{g_1}-trajectory through x in SN_1.

Since $\psi : V_1 \to V_2$ is an isometry, the diffeomorphism Ψ maps the v^{g_1}-trajectory $\tilde{\gamma}_{V_1}(x)$ to the v^{g_2}-trajectory $\tilde{\gamma}_{V_2}(\Psi(x))$, while preserving the natural parameterizations on the two curves in the Sasaki metric $\mathsf{g}_i|_{SV_i}$. On the other hand, by Theorem 4.12, the diffeomorphism $\Phi : SM_1 \to SM_2$, which extends $\Phi^\partial =_{\mathsf{def}} \Psi^\partial$, maps each trajectory $\tilde{\gamma}_{M_1}(x)$ to the trajectory $\tilde{\gamma}_{M_2}(\Psi(x)) = \Phi(\tilde{\gamma}_{M_1}(x))$, while preserving the $(\mathsf{g}_i|_{SM_i})$-induced natural parameterizations on the two curves.

Let us denote by Σ_i the union of all *segments* of the v^{g_i}-trajectories in $SM_i \cap SV_i = S(M_i \cap V_i)$ that have a nonempty intersection with $\partial_1(SM_i)$. Thus, any point of Σ_i is connected to a point of $\partial_1(SM_i)$ by an arc $\delta \subset \Sigma_i$ of a v^{g_i}-trajectory. Note that $\bar{\Sigma}_i$, the closure of Σ_i in SN_i, is a compact set, containing $\partial_1(SM_i)$. We denote by $\Sigma_i^\bullet$ the set $SU_i \cup \bar{\Sigma}_i$.

Since $\Psi|_{\partial(SM_1)} = \Phi|_{\partial(SM_1)}$ and both diffeomorphisms, Φ and Ψ, preserve the natural parameterizations along the trajectories in their source and the target spaces, we conclude that $\Phi|_{\Sigma_1} = \Psi|_{\Sigma_1}$ as maps. It is possible that Ψ and Φ may differ on $SV_1 \setminus \Sigma_1^\bullet$.

Consider the homeomorphism $\Xi : SN_1 \to SN_2$, defined by the formula

$$\Xi =_{\mathsf{def}} \Psi \bigcup_{\Psi^\partial} \Phi : \ SU_1 \bigcup_{\partial_1(SU_1)} SM_1 \to SU_2 \bigcup_{\partial_1(SU_2)} SM_2.$$

By the properties of Φ an Ψ, the homeomorphism Ξ maps the v^{g_1}-trajectories in SN_1 to the v^{g_2}-trajectories in SN_2.

By the definitions of the sets $\Sigma_i, \Sigma_i^\bullet$ and using that $\Phi|_{\Sigma_1} = \Psi|_{\Sigma_1}$, we may also interpret Ξ as the homeomorphism

$$\Psi \bigcup_{\Psi|_{\partial(\Sigma_1^\bullet)}} \Phi : \ \Sigma_1^\bullet \bigcup_{\partial(\Sigma_1^\bullet)} (SM_1 \setminus \mathsf{int}(\Sigma_1)) \to \Sigma_2^\bullet \bigcup_{\partial(\Sigma_2^\bullet)} (SM_2 \setminus \mathsf{int}(\Sigma_2)).$$

Since $\Phi|_{\Sigma_1} = \Psi|_{\Sigma_1}$, the maps Φ and Ψ are smooth diffeomorphisms; since $\Sigma_1 \supset \partial_1(SU_1)$, we conclude that Ξ is a smooth diffeomorphism in the vicinity of $\partial_1(SU_1)$.

By Theorem 4.12, the diffeomorphism $\Phi = \Xi|_{SM_1}$ preserves the lens data, i.e., it has the property $\Xi^*(d\mathsf{g}_2|_{\Xi(\gamma^\#)}) = d\mathsf{g}_1|_{\gamma^\#}$ for every $(v^{g_1|_{M_1}})$-trajectory $\gamma^\# \subset SM_1$. By the hypotheses, $\psi : U_1 \to U_2$ is an isometry. Thus, for every $(v^{g_1|_{U_1}})$-trajectory $\gamma^\dagger \subset SU_1$, we have $\Xi^*(d\mathsf{g}_2|_{\Xi(\gamma^\dagger)}) = d\mathsf{g}_1|_{\gamma^\dagger}$.

Let $\tilde{\gamma}$ be an integral curve of the geodesic field v^{g_1} on SN_1. The set $\tilde{\gamma}_{U_1} = \tilde{\gamma} \cap SU_1$ is the union of trajectories $\{\gamma^\dagger\}$ of the geodesic field $v^{g_1|_{U_1}}$, while the set $\tilde{\gamma}_{M_1} = \tilde{\gamma} \cap SM_1$ is the union of $(v^{g_1|_{M_1}})$-trajectories $\{\gamma^\#\}$. Therefore, by the arguments above, $\Xi^*(dg_2|_{\Xi(\tilde{\gamma})}) = dg_1|_{\tilde{\gamma}}$ for all v^{g_1}-trajectories $\tilde{\gamma}$ in SN_1. This implies that $\Xi \circ \phi_1^t = \phi_2^t \circ \Xi$ for all moments $t \in \mathbb{R}$, where ϕ_i^t denotes the geodesic flow on SN_i.

Thus, the metrics g_1 and g_2 are geodesic flow strongly conjugate by the smooth diffeomorphism $\Xi : SN_1 \to SN_2$. □

In the next three corollaries, we combine Theorem 4.13 with a number of classical results claiming the possibility to reconstruct the metric on special closed manifolds from the corresponding geodesic flows.

Corollary 4.12. *Let the Riemannian n-manifolds $(N_1, g_1) \supset (M_1, g_1|)$, $(N_2, g_2) \supset (M_2, g_2|)$ and the isometry $\psi : V_1 \to V_2$ be as in Theorem 4.13.*

If the scattering maps $C_{v^{(g_1|_{M_1})}}$ and $C_{v^{(g_2|_{M_2})}}$ are conjugated with the help of the diffeomorphism $\Psi^\partial =_{\mathsf{def}} \Psi|_{\partial_1(SU_1)}$ (where Ψ is induced by the isometry ψ) and if, for each $w \in \partial_1^+(SM_1)$, the length data agree, i.e., if $\ell_{g_2|_{M_2}}(\Psi^\partial(w)) = \ell_{g_1|_{M_1}}(w)$, then the manifolds $(M_1, g_1|_{M_1})$ and $(M_2, g_2|_{M_2})$ share the same volume.

Proof. By Theorem 4.13, the metrics g_1 and g_2 are geodesic flow strongly conjugate by a diffeomorphism $\Xi : SN_1 \to SN_2$ from the class $C^\infty(SN_1, SN_2)$. By [CK], Proposition 1.2, the manifolds (N_1, g_1) and (N_2, g_2) share the same volume. Since $\psi : U_1 \to U_2$ is an isometry, the manifolds $(M_1, g_1|_{M_1})$ and $(M_2, g_2|_{M_2})$ share the same volume as well. □

A vector field w on a manifold with affine connection is called parallel if the covariant derivative $\nabla_v w = 0$ for any vector field v. Recall that the existence of a non-vanishing parallel vector field on a Riemannian manifold (N, g) is equivalent to N being isometric to the quotient of the Riemannian product $L \times \mathbb{R}$ by an isometry subgroup $\mathsf{G} \subset \mathsf{Iso}(L) \times \mathsf{Iso}_+(\mathbb{R})$, where L is a simply connected complete Riemannian manifold ([CK]).

Corollary 4.13. *Let the pairs of Riemannian n-manifolds, $(N_1, g_1) \supset (M_1, g_1|)$ and $(N_2, g_2) \supset (M_2, g_2|)$ be as in Theorem 4.13. In addition, assume that N_2 admits a non-vanishing* parallel *vector field.*

If the scattering maps $C_{v^{(g_1|_{M_1})}}$ and $C_{v^{(g_2|_{M_2})}}$ are conjugated with the help of the diffeomorphism $\Psi^\partial =_{\mathsf{def}} \Psi|_{\partial_1(SU_1)}$ and if, for each $w \in \partial_1^+(SM_1)$, the lens data agree: $\ell_{g_2|_{M_2}}(\Psi^\partial(w)) = \ell_{g_1|_{M_1}}(w)$, then the manifolds (N_1, g_1) and (N_2, g_2) are isometric.

Proof. By Theorem 4.13, the metrics g_1 and g_2 are geodesic flow conjugate by a smooth diffeomorphism $\Xi : SN_1 \to SN_2$. By [CK], Theorem 1.1, the manifolds (N_1, g_1) and (N_2, g_2) are isometric. □

Corollary 4.14. *Let $n \geq 3$. Let the pairs of Riemannian n-manifolds, $(N_1, g_1) \supset (M_1, g_1|)$ and $(N_2, g_2) \supset (M_2, g_2|)$, be as in Theorem 4.13. Assume, in addition, that (N_2, g_2) is locally symmetric and of a negative sectional curvature.*

If the scattering maps $C_{v^{g_1|M_1}}$ and $C_{v^{g_2|M_2}}$ are conjugated with the help of the diffeomorphism $\Psi^\partial =_{\mathsf{def}} \Psi|_{\partial(SU_1)}$ and if, for each $w \in \partial_1^+(SM_1)$, the length data agree: $\ell_{g_2|_{M_2}}(\Psi^\partial(w)) = \ell_{g_1|_{M_1}}(w)$, then the spaces (N_1, g_1) and (N_2, g_2) are isometric.

Proof. By Theorem 4.13, the C^1-diffeomorphism $\Xi : SN_1 \to SN_2$ conjugates the geodesic flows on SN_1 and SN_2. Since (N_2, g_2) is a locally symmetric space of negative sectional curvature, the Besson-Courtois-Gallaot Theorem 4.1 applies. So, for an appropriate constant $c > 0$, we get that there exists an isometry $\chi : (N_1, c \cdot g_1) \to (N_2, g_2)$. Using that $\psi : U_1 \to U_2$ is an isometry, we conclude that $c = 1$. □

4.5 Holography of geodesic flows, the v^g-harmonizing metrics, and billiards' dynamics

The Sections 4.6 - 4.9 partially mirror Section 3.7, with the general traversing vector field v being replaced by the geodesic vector field v^g produced by a non-trapping metric g. Most of the results of these sections are contained in [K11].

First, we describe the nature of the main assertions in the next few subsections.

As before, let (M, g) be a compact connected smooth Riemannian manifold with boundary, where g is a non-trapping metric. Let SM be the space of the unit spherical tangent to M bundle, and v^g the geodesic vector field on SM. We continue to study the scattering maps $C_{v^g} : \partial_1^+ SM \to \partial_1^- SM$, generated by the v^g-flow, and the dynamics of the billiard maps $B_{v^g,\tau} : \partial_1^+ SM \to \partial_1^+ SM$. Here $\tau : \partial_1^- SM \to \partial_1^+ SM$ denotes an involution, mimicking the elastic reflection from the boundary ∂M.

Mirroring [K11], we will get a variety of holography theorems that address the inverse scattering problems for C_{v^g} and theorems that describe the dynamics of $B_{v^g,\tau}$. Our main tools here are Lyapunov functions

$F : SM \to \mathbb{R}$ for v^g and special harmonizing Riemannian metrics $g^\bullet$ on SM, the metrics in which dF is a harmonic 1-form. For such metrics $g^\bullet$, we will get a family of isoperimetric inequalities of the type

$$vol_{g^\bullet}(SM) \leq vol_{g^\bullet|}(\partial_1(SM))$$

and formulas for the average volume of the minimal hypesufaces $\{F^{-1}(c)\}_{c \in F(SM)}$. We investigate the interplay between the harmonizing metrics $g^\bullet$ and the classical Sasaki metric gg on SM [Sa].

Assuming ergodicity of the billiard map $B_{v^g,\tau}$, we also get Santaló-Chernov type formulas for the average length of free geodesic segments in M and for the average variation of the Lyapunov function F along the v^g-trajectories. Our effort in the core of this section is in the spirit of "Integral Geometry and Geometric Probability" by Santaló ([S], Chapter 19) and works by Vidal Abascal [V]-[V2].

- *Let us now outline some results in the order they appear in the next few sections.*

In Section 4.7, we apply the established properties of general traversing vector flows from Chapter 3 (Sections 3.5 - 3.7) to non-trapping boundary generic geodesic flows v^g on the space of unit tangent spherical bundle $SM \to M$, where M is a compact connected manifold with boundary.

Corollary 4.15 provides numerous examples of non-trapping metrics on codimension zero compact submanifolds of the hyperbolic and the Euclidean spaces.

In Theorem 4.14, for a boundary generic non-trapping metric g on M, we construct a v^g-harmonizing and v^g-invariant metric $g^\bullet$ on the space SM and a balanced (see Definition 4.12) Lyapunov function $F : SM \to \mathbb{R}$, so that 1-dimensional foliation $\mathcal{F}(v^g)$ and orthogonal to it $(2n-2)$-dimensional foliation $\mathcal{G}(F) =_{\mathsf{def}} \{F^{-1}(c)\}_{c \in \mathbb{R}}$ are minimal in $g^\bullet$. Moreover, we prove that the scattering map $C_{v^g} : \partial_1^+(SM) \to \partial_1^-(SM)$ preserves the measure, defined by the harmonic form $\Theta =_{\mathsf{def}} *_{g^\bullet}(dF)$.

Let g be as above. In Theorem 4.15, for a given v^g-invariant volume form Ω on SM and a Lyapunov function F, we use the form $\Theta =_{\mathsf{def}} v^g \rfloor \Omega$ to construct a v^g-harmonizing metric $g^\bullet$ such that $*_{g^\bullet}(dF) = \Theta$.

Let $vol_\Theta(\mathcal{T}(v^g))$ denote the Θ-induced volume of the space of geodesics $\mathcal{T}(v^g)$. Then we prove the inequality

$$vol_\Theta(\mathcal{T}(v^g)) \;\leq\; \frac{1}{2} vol_{g^\bullet|}(\partial_1(SM)).$$

Normalizing the Lyapunov function F so that $F(SM) \subset [0,1]$, we get the isoperimetric inequality

$$vol_{g^\bullet}(SM) \leq vol_{g^\bullet|}(\partial_1(SM)).$$

In holography Theorem 4.16, we reconstruct the space SM, the vector field v^g, and the v^g-invariant harmonizing metric $g^\bullet$, up to a diffeomorphism of SM, in terms of some enhanced scattering data.

In Section 4.8, we study generalized billiard maps $B_{v^g,\tau} : \partial_1^+(SM) \to \partial_1^+(SM)$ on billiard tables (M, g), were g boundary generic and non-trapping (see Theorem 4.17 and Corollary 4.16). We employ the restrictions to SM of the contact Liouville 1-form β_g and the symplectic 2-form $\omega_g = d\beta_g$.

In holography Theorem 4.3, we show how to reconstruct the form β_g from its restriction $\beta_g^{\partial^+}$ to the boundary $\partial_1(SM)$, the restriction $F|_{\partial(SM)}$ of the Lyapunov function $F : SM \to \mathbb{R}$, and the scattering map C_{v^g}.

In Section 4.9, we study the dynamics of ergodic (see Definition 4.15) billiard maps $B_{v^g,\tau}$ on the billiards (M, g), where g is boundary generic and non-trapping. We apply the classical Birkhoff Theorem [Bi] to the measure μ_Θ on $\partial_1(SM)$, generated by an appropriately constructed closed and v^g-invariant $(2n-2)$-form Θ (see Theorem 4.18).

In Theorem 4.19, we compute the spacial and time averages of the variation Δ_F of the Lyapunov function $F : SM \to \mathbb{R}$ along the v^g-trajectories; for the ergodic billiards, both computations produce the same result.

Theorem 4.20 is a version of these computations for a given Lyapunov function F and the form $\Theta = \omega_g^{n-1}$, the $(n-1)^{st}$ exterior power of the symplectic 2-form ω_g on SM. It employs a v^g-harmonizing metric $g^\bullet$ on the space SM. In contrast, Theorem 4.21 is a result of a similar calculation of the average length of the v^g-trajectories in the Sasaki metric gg ([Sa]) on SM. It utilizes the same form $\Theta = \omega_g^{n-1}$. In this case, the results of the computation can be expressed directly in terms of the volumes $vol_g(M)$ and $vol_{g|_{\partial M}}(\partial M)$ (see formulas (4.24) and (4.25)). For any non-trapping g, this leads to the following inequality (Corollary 4.19):

$$vol_g(M) \leq c(n) \cdot \mathsf{gd}(M, g) \cdot vol_{g|}(\partial M),$$

where $c(n) > 0$ is a well-known universal constant, and $\mathsf{gd}(M, g)$, the geodesic diameter, denotes the maximal length of free geodesic arcs in M (see Fig. 4.5). The later inequality resembles one classical inequality from [Cr], where $\mathsf{gd}(M, g)$ is replaced by the diameter of M.

In Theorem 4.23, we derive formulas for computing the volume $A_{g^\bullet}(c)$ of a taut slice $F^{-1}(c)$ ($c \in \mathbb{R}$) in the v^g-harmonizing metric $g^\bullet$, as well as the average value of $A_{g^\bullet}(c)$.

4.6 The v^g-harmonizing metrics and their isoperimetric inequalities

The lemma below describes conditions under which the geodesic scattering map C_{v^g} is a μ_Θ-measure-preserving transformation for an appropriate choice of a closed $(2n-2)$-form Θ on SM. In a way, this lemma it is a special case of Lemma 3.11 (see Fig. 3.4).

Lemma 4.8. *Assume that a metric g on a compact connected Riemannian n-manifold M with boundary is non-trapping and boundary generic.*

Pick a $(2n-2)$-volume form Ω^∂ on $\partial_1(SM)$, consistent with its orientation. Let Θ be a differential $(2n-2)$-form on SM such that:

- $\dim(\mathsf{ker}(\Theta)) = 1$,
- *the geodesic field $v^g \in \mathsf{ker}(\Theta)$,*
- $d\Theta = 0$,
- *the function $\Theta/\Omega^\partial : \partial_1(SM) \to \mathbb{R}$ is positive (negative) only on the interior of $\partial_1^+(SM)$ (of $\partial_1^-(SM)$).*

Then the scattering map $C_{v^g} : \partial_1^+(SM) \to \partial_1^-(SM)$ preserves the Θ-induced measure μ_Θ on $\partial_1(SM)$, i.e.,

$$\Big|\int_{C_{v^g}(K)} \Theta\Big| = \int_K \Theta$$

for any Lebesgue-measurable set $K \subset \partial_1^+(SM)$.

Proof. Since the metric g on M is non-trapping, by Lemma 4.2, the geodesic field v^g is traversing on SM (see also [K5], Lemma 2.2). Since the metric g is boundary generic relative to ∂M, by Definition 4.3, the field v^g is boundary generic relative to $\partial_1(SM)$. Thus, Lemma 3.11 is applicable to v^g. By that lemma, we get that C_{v^g} preserves the measure μ_Θ. □

Corollary 4.15. *Let $(N, g^\dagger)$ be a closed hyperbolic or flat n-dimensional Riemannian manifold, and let M be a compact smooth n-manifold, equipped with a submersion $p : M \to N$ such that the natural homomorphism $p_* : \pi_1(M) \to \pi_1(N)$ of the fundamental groups is trivial. Denote by g the p-induced pull-back to M of the metric $g^\dagger$ on N.*

Then g is a non-trapping metric, and (M, g) admits a balanced Lyapunov function $F : SM \to \mathbb{R}$.

In particular, any compact smooth domain M in the hyperbolic space $\mathbb{H}^n$ or in the Euclidean space $\mathbb{E}^n$ admits a non-trapping hyperbolic or Euclidean metric with a balanced Lyapunov function $F : SM \to \mathbb{R}$.

Proof. Since the homomorphism $p_* : \pi_1(M) \to \pi_1(N)$ is trivial, M admits a lifting $\tilde{p} : M \to \tilde{N}$ to the universal cover $\tilde{N} = \mathbb{H}^n$ (or $\tilde{N} = \mathbb{E}^n$) of N. This lifting is also a submersion. Hence, it suffices to treat the case of a submersion $\tilde{p} : M \to \mathbb{H}^n$ or of a submersion $\tilde{p} : M \to \mathbb{E}^n$ which pulls back the hyperbolic or Euclidean metric to form the metric g on M.

Let us consider the hyperbolic case first. We recycle the constructions familiar from Example 4.3. In the Poincaré model, $\mathbb{H}^n$ is identified with the interior of the unit Euclidean ball B^n, and the geodesics in $\mathbb{H}^n$ with the arcs of circles in B^n (or diameters through the origin) that are orthogonal to its boundary ∂B^n.

Let $B^n(r) \subset B^n$ be a concentric Euclidean ball of a radius $r \in (0, 1)$. Since M is compact, we may assume that $\tilde{p}(M) \subset B^n(r_0)$ for some $r_0 < 1$. Then there is $r_\star \in (r_0, 1)$ such that any geodesic arc $\hat{\gamma}$ that intersects with $B^n(r_0)$ hits $\partial B^n(r_\star)$ at a pair of points and is *transversal* to $\partial B^n(r_\star)$ at the intersections. Let us explain informally the last claim (we leave the details to the reader): in the hyperbolic metric, as $r_\star \to 1$, $B^n(r_0)$ becomes infinitesimally small relative to $B^n(r_\star)$, and the geodesic arcs through $B^n(r_0)$ approach the diameters of B^n; at the same time, in the Euclidean metric, $\partial B^n(r_\star)$ approaches ∂B^n.

Now, picking $r_\star$ very close to 1 and the ball $B^n(r_\star)$ for the role of L in Lemma 4.3, we validate the corollary (in the hyperbolic case) by pulling back, with the help of the bundle map $S\tilde{p} : SM \to S\mathbb{H}^n$, the Lyapunov function on $SB^n(r_0)$, built in the lemma (see Fig. 4.4, the left diagram).

The Euclidean case is similar: we pick an Euclidean ball $B^n(r_0) \supset \tilde{p}(M)$, where the radius $r_0 \in (0, +\infty)$. Then there is $r_\star \in (r_0, +\infty)$ such that any line $\hat{\gamma}$ that intersects with $B^n(r_0)$ hits the sphere $\partial B^n(r_\star)$ at a pair of points and is transversal there to $\partial B^n(r_\star)$. Specifically, for a pair $(\vec{a}, \vec{v})$, where $\vec{a} \in \mathbb{E}^n$ is such that $\|\vec{a}\| \leq r_0 < r_\star$ and $\vec{v} \in T_{\vec{a}}(\mathbb{E}^n)$ has the norm 1, we may pick the composition of the appropriately normalized submersion $S\tilde{p} : SM \to S\mathbb{E}^n$ with the following Lyapunov function (see the right diagram in Fig. 4.4)

$$F(\vec{a}, \vec{v}) =_{\mathsf{def}} -\big(r_\star^2 - \|\vec{a}\|^2 + \langle \vec{a}, \vec{v} \rangle^2\big)^{\frac{1}{2}} + \langle \vec{a}, \vec{v} \rangle. \tag{4.11}$$

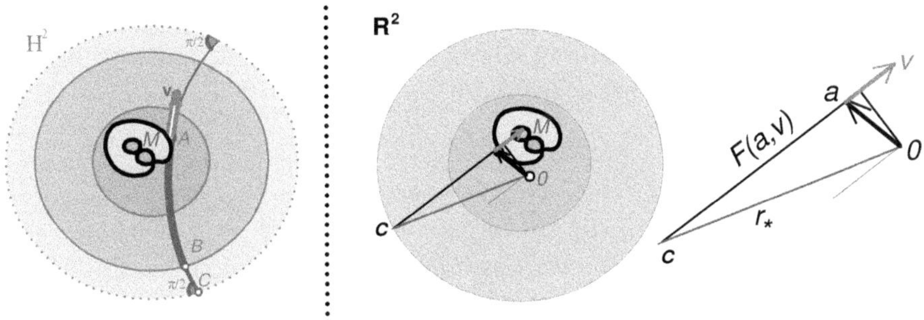

Fig. 4.4 Constructing Lyapunov functions $F(a, v)$ for geodesic flows on $S_g\mathbb{H}^2$ (on the left) and on $S_g\mathbb{R}^2$ (on the right). Here the geodesic trajectory $\gamma_{(a,v)}$ in $\mathbb{H}^2$ or in $\mathbb{R}^2$ is determined by an unit vector $v \in T_a(\mathbb{H}^2)$ or $v \in T_a(\mathbb{R}^2)$, where $\|v\|_g = 1$ and a point $a \in \mathbb{H}^2$ or $a \in \mathbb{R}^2$, respectively. In the case of $\mathbb{H}^2$, the value $F(a, v)$ equals to the hyperbolic length of the circular arc AB (or the Euclidean length of the arc AC). In the case of $\mathbb{R}^2$, the value $F(a, v)$ equals to the Euclidean length of the segment AC. The (overlapping) blobs inside of the smallest disks of the Euclidean radius r_0 show the $\tilde{p}$-image of a manifold M under its submersions in $\mathbb{H}^2$ or in $\mathbb{R}^2$.

This composition serves as a Lyapunov function on SM for any submersion p such that $\tilde{p}(M) \subset B^n(r_0)$, where $r_0 < r_\star$. □

The next theorem is a direct application of Theorem 3.14, claim (3), and Theorem 3.15, together with Corollary 3.17, to the non-trapping boundary generic geodesic flows.

Theorem 4.14. *Let g be a boundary generic non-trapping Riemannian metric on a smooth compact connected n-manifold M with boundary. Then the following claims hold:*

- *The geodesic vector field v^g admits a balanced Lyapunov function $F : SM \to \mathbb{R}$ and a v^g-harmonizing pair $(g^\bullet, dF)$, where $g^\bullet$ is a v^g-invariant Riemmanian metric on SM.*
- *In the metric $g^\bullet$, the leaves of the 1-foliation $\mathcal{F}(v^g)$ are geodesic curves*[14]*, and the leaves $\{F = c\}_{c\in\mathbb{R}}$ of the orthogonal $(2n-2)$-foliation $\mathcal{G}(F)$ are the $(2n-2)$-volume-minimizing relative hypersurfaces in $\big(SM,\, \partial_1(SM)\big)$.*

[14] As they are also in the Sasaki metric gg.

- *The* $(2n-2)$*-form* $\Theta =_{\text{def}} *_{g^\bullet}(dF)$ *has all the properties from Lemma 4.8. As a result, such a form* Θ *defines a measure* μ_Θ *on* $\partial_1(SM)$*, which is preserved by the scattering map* $C_{v^g} : \partial_1^+(SM) \to \partial_1^-(SM)$*.* ◇

Theorem 4.15. *Let* g *be a boundary generic non-trapping Riemannian metric on a smooth compact connected* n*-manifold* M *with boundary, and let* $F : SM \to \mathbb{R}$ *be a Lyapunov function for the geodesic vector field* v^g*.*

Let Θ *be a* $(2n-2)$*-form on* SM *with the properties as in Lemma 4.8. In particular, we may choose a nil-divergent form* $\Theta = v^g \rfloor \Omega$*, where* Ω *is any* v^g*-invariant volume form on* SM*.*

Then the following claims hold:

- *There exists a Riemannian metric* $g^\bullet$ *on* SM *such that* $*_{g^\bullet}(dF) = \Theta$*.*
- *The function* F *may be chosen to be balanced and the metric* $g^\bullet$ *to be* v^g*-invariant.*
- *In the* Θ*-induced measure, the volume of the space* $\mathcal{T}(v^g)$ *of geodesics on* M *satisfies the inequality:*

$$vol_\Theta(\mathcal{T}(v^g)) =_{\text{def}} \int_{\partial_1^\pm(SM)} \Theta \;\leq\; vol_{(g^\bullet)^\partial}(\partial_1^\pm(SM)),$$

where $(g^\bullet)^\partial$ *denotes the restriction of the metric* $g^\bullet$ *to the boundary* $\partial_1(SM)$*. As a result,*

$$vol_\Theta(\mathcal{T}(v^g)) \leq \frac{1}{2} vol_{(g^\bullet)^\partial}(\partial_1(SM)). \tag{4.12}$$

- *Assuming that the Lyapunov function* $F : SM \to \mathbb{R}$ *is chosen so that* $F(\partial_1(SM)) \subset [0,1]$*, the following isoperimetric inequality holds:*

$$vol_{g^\bullet}(SM) \leq vol_{(g^\bullet)^\partial}(\partial_1(SM)). \tag{4.13}$$

Proof. By the properties of the non-trapping metric g on M, the geodesic field v^g on SM is traversing ([K5], Lemma 2.2). Thus there exists a smooth function $F : SM \to \mathbb{R}$ with the property $dF(v^g) > 0$ ([K1], Lemma 4.1). By Theorem 4.14 (a corollary of Theorem 3.14), F may be chosen to be balanced. Also by Theorem 4.14 (c.f. Theorem 3.17), there exists a v^g-harmonizing and v^g-invariant metric $g^\bullet$ on SM in which the form dF is *co-closed.* In particular, along the locus $\partial_1^+(SM)$, the field v^g points inside of SM, so that the closed form $\Theta = *_{g^\bullet}(dF)|_{\partial_1^+(SM)}$ is positively proportional to the $g^\bullet$-induced volume form in the interior of $\partial_1^+(SM)$. The coefficient of proportionality is $\cos\big(\angle(\nu, v^g)\big)$, where ν is the inward normal (in the metric $g^\bullet$) to $\partial_1(SM)$ in SM.

By Theorem 4.14 (or Corollary 3.17), the leaves of the 1-foliation $\mathcal{F}(v^g)$ are geodesic curves in the v^g-harmonizing metric $g^\bullet$, and the leaves of the orthogonal $(2n-2)$-foliation $\mathcal{G}(F) =_{\mathsf{def}} \{F^{-1}(c)\}_{c\in\mathbb{R}}$ are the volume-minimizing hypersurfaces H (among all hypersurfaces Σ with the property $\Sigma\cap\partial(SM) = H\cap\partial(SM)$). This validates the claims in the first two bullets of the theorem.

The last two bullets follow directly from Theorem 3.14 (or Theorem 3.16). □

Remark 4.7. Note that $vol_\Theta(\mathcal{T}(v^g)) = vol_\Theta(\mathcal{T}(-v^g))$, so that the volume of $\mathcal{T}(v^g)$ can be also expressed in terms of integration over the locus $\partial_1^+(SM)(-v^g) = \partial_1^-(SM)(v^g)$. ◇

Theorem 4.16. *Let g be a boundary generic non-trapping Riemannian metric on a smooth compact connected n-manifold M with boundary. Assume that any geodesic curve in M is either transversal to ∂M at* some *point or is simply tangent to ∂M.*

We choose a v^g-invariant and v^g-harmonising pair $(g^\bullet, dF)$ on SM with a balanced Lyapunov function F.[15]

Then the scattering map $C_{v^g} : \partial_1^+(SM) \to \partial_1^-(SM)$, the restriction $F^\partial : \partial_1(SM) \to \mathbb{R}$ of F and of the metric $g^\bullet$ to the boundary $\partial_1(SM)$, and the $g^\bullet$-induced angle map $\theta : \partial_1(SM) \to S^1$ allow for a reconstruction of:

- *the space SM,*
- *the geodesic vector field v^g,*
- *and of the v^g-harmonizing metric $g^\bullet$ on SM,*

up to the natural action of diffeomorphisms $\Phi : SM \to SM$ that are the identity map on $\partial_1(SM)$ and whose differentials are the identity map on the bundle $T(SM)|_{\partial_1(SM)}$.

Proof. By Theorem 3.15 and Theorem 3.17, the knowledge of $F^\partial : \partial_1(SM) \to \mathbb{R}$, the metric $(g^\bullet)^\partial$ on SM, and angle map $\theta : \partial_1(SM) \to S^1$, allow for a reconstruction of the smooth topological type of SM, the foliation $\mathcal{F}(v^g)$, and the metric $g^\bullet$, provided that $g^\bullet$ is v^g-invariant. Using that $dF(v^g) = 1$, we can reconstruct v^g from $\mathcal{F}(v^g)$. In general, by Theorem

[15]By Theorem 4.14, such a pair $(g^\bullet, dF)$ exists.

3.17, the reconstruction is possible up to a homeomorphism $\Phi : SM \to SM$ that is fixed on $\partial_1(SM)$ and whose restriction to each v^g-trajectory is an orientation-preserving diffeomorphism. Under the hypotheses of Theorem 4.16 and by Theorem 3.3, Φ is a smooth diffeomorphism. With the help of the Φ-invariant map $\theta : \partial_1(SM) \to S^1$, we conclude that the differential $D\Phi$ must act trivially on $T(SM)|_{\partial_1(SM)}$.

Assuming that the smooth topological type of SM is known, we may drop the assumption that any geodesic curve in M is either transversal to ∂M at some point or is simply tangent to ∂M ("Property AA"): indeed, the knowledge of the metric $g^\bullet$ on the complement to the locus $SM(v^g, (33)_{\succeq} \cup (4)_{\succeq})$ of codimension 3 at least, by continuity, is sufficient for the reconstruction of $g^\bullet$ everywhere. □

Moving away from the non-trapping metrics, we propose the following few notions.

Definition 4.13. Let N be a smooth compact manifold and let $g^\bullet$ be a Riemmanian metric on SN such that $\|v^g\|_{g^\bullet} = 1$ everywhere. Consider a 1-form $\alpha = \alpha(g^\bullet)$ on SN, characterized by the two properties:

$$(1)\ \mathsf{ker}(\alpha) \perp_{g^\bullet} v^g, \text{ and } (2)\ \alpha(v^g) = 1.$$

We say that a Riemmanian metric g on N is geodesically harmonic, if there exists a Riemmanian metric $g^\bullet$ on SN as above and such that the 1-form α is closed and co-closed.

A Riemmanian metric g on N is invariantly geodesically harmonic, if it is geodesically harmonic and the metric $g^\bullet$ is v^g-invariant.

• We denote by $\mathcal{H}ar(N)$ ($\mathcal{H}ar_{inv}(N)$) the space of (invariantly) geodesically harmonic metrics on N. ◇

By Theorem 4.14, any non-trapping and boundary generic metric g on a connected compact M is *invariantly geodesically harmonic.* Indeed, just follow the well-traveled path (see the proof of Theorem 3.14): take a well-balanced Lyapunov function F, put $\alpha = dF$, construct a closed v^g-invariant n-form Θ, whose kernel is spanned by v^g and such that $\alpha \wedge \Theta > 0$, and finally, construct a v^g-invariant metric $g^\bullet$ so that $\Theta = *_{g^\bullet}(\alpha)$. Then $\mathsf{ker}(\Theta) \perp_{g^\bullet} \mathsf{ker}(dF)$. Thus, g is invariantly geodesically harmonic in the sense of Definition 4.13.

Remark 4.8. The geodesically harmonic metrics on *closed* manifolds M seem to be a *rare phenomenon.* For instance, if $H^1(SM; \mathbb{R}) = 0$ (for

$\dim(M) \geq 3$, this is equivalent to $H^1(M;\mathbb{R}) = 0$) no such metric $g^\bullet$ is available, since the cohomology class of the closed form α must be nontrivial.

Recall also that the v^g-*invariant* Sasaki metrics gg on SM are extremely rare. According to [Be], Proposition 1.104, gg is v^g-invariant if and only if the sectional curvature of (M, g) is identically 1. ♢

Remark 4.9. Note that if (N, g) is (invariantly) geodesically harmonic, then any smooth codimension zero proper submanifold $(M, g|_M) \subset (N, g)$ (with boundary generic metric $g|_M$) is automatically (invariantly) geodesically harmonic. ♢

Let us describe numerically how far a given metric $g^\bullet$ on SM is from being "harmonic". Speaking informally, we would like to measure the "distance" between the standard contact structure β_g on SM (see (4.3)) and taught foliations $\mathcal{G}$ on SM that are transversal to v^g.

Let $\| \sim \|_2^{g^\bullet}$ denote the L_2-norm of differential forms on SM in a given metric $g^\bullet$. Let the 1-form $\alpha_\bullet$ be as in Definition 4.13. Then the quantity

$$\delta(g^\bullet) =_{\mathsf{def}} \sqrt{\left(\|d\alpha_\bullet\|_2^{g^\bullet}\right)^2 + \left(\|d(*_{g^\bullet}(\alpha_\bullet))\|_2^{g^\bullet}\right)^2} \tag{4.14}$$

measures the failure of a candidate metric $g^\bullet$ to deliver the geodesic harmonicity of g.

Evidently, if g is geodesically harmonic, then $\delta(g^\bullet) = 0$ for an appropriate metrics $g^\bullet$.

Definition 4.14. Let (M, g) be a connected compact Riemannian manifold. Consider the number

$$\mathcal{D}^\bullet(g) =_{\mathsf{def}} \inf_{\{g^\bullet\}} \{\delta(g^\bullet)\}, \tag{4.15}$$

where $\delta(g^\bullet)$ is defined by (4.14) and $g^\bullet$ runs over all Riemannian metrics on SM such that $vol_{g^\bullet}(SM) = 1$.

If $g^\bullet$ runs over all v^g-*invariant* Riemannian metrics on SM such that $vol_{g^\bullet}(SM) = 1$, we get a similar to (4.15) quantity $\mathcal{D}^\bullet_{inv}(g) \geq \mathcal{D}^\bullet(g)$. ♢

Recall that $\mathcal{D}^\bullet_{inv}(g) = \mathcal{D}^\bullet(g) = 0$ for any non-trapping g on a compact connected manifold with boundary.

Conjecture 4.4. *If $\mathcal{D}^\bullet(g) = 0$, then g is geodesically harmonic.*

If $\mathcal{D}^\bullet_{inv}(g) = 0$, then g is invariantly geodesically harmonic. ♢

In the spirit of Conjecture 4.4, $\mathcal{D}^\bullet(g)$ should measure the failure of geodesic harmonicity for a given metric g on M.

Problem 4.1.

- *Estimate* $\delta(gg)$ *(see Definition 4.14) for the Sasaki metric* gg *on* SM.
- *Compute* $\delta(gg)$ *for the* v^g*-invariant Sasaki metric* gg *on any compact symmetric space* (N, g) *of rank one. By [Be], these are exactly the Riemmanian manifolds* S^n, $\mathbb{RP}^n$, $\mathbb{CP}^n$, $\mathbb{HP}^n$, *and* $\mathbb{C}a\mathbb{P}^2$ *with constant sectional curvature* 1. ◇

4.7 The holography of billiard maps for non-trapping metrics

In this section, we will present some direct applications of the results from Chapter 3 to the geodesic flows of non-trapping metrics on manifolds with boundary.

Let (M, g) be a compact connected smooth Riemannian n-manifold with boundary and v^g the geodesic vector field on the tangent unitary bundle SM.

Any tangent vector $\dot{q} \in T_qM$, where $q \in \partial M$, is a sum of $a\vec{n} + b\vec{t}$, where $\vec{n}$ is the inner normal to ∂M in M (with respect to the metric g), $\vec{t} \in T_q(\partial M)$, and $a, b \in \mathbb{R}$.

Consider a smooth involution $\tau_g : \partial_1(SM) \to \partial_1(SM)$ that takes any tangent to M vector $\dot{q} = a\vec{n} + b\vec{t}$ at $q \in \partial M$, to the vector $\tau_g(\dot{q}) = -a\vec{n} + b\vec{t}$, the orthogonal reflection of $\dot{q}$ with respect to the hyperplane $T_q(\partial M)$. The involution τ_g induces a g-isometry of each sphere $\{S_qM \subset T_qM\}_{q \in \partial M}$ with respect to its equator. Evidently, τ_g maps $\partial_1^\pm(SM)$ to $\partial_1^\mp(SM)$ (see Fig. 4.3).

One can generalize the construction of the mirror reflection τ_g : $\partial_1(SM) \to \partial_1(SM)$ in the spirit of Finsler structures (see [Car] for their definitions and properties) as follows. Consider any smooth involution $\tau : \partial_1(SM) \to \partial_1(SM)$ which is a map of the spherical fibration $\{\eta : \partial_1(SM) = SM|_{\partial M} \to \partial M\}$ such that it is the identity on its base ∂M and the τ-fixed locus $\partial_1(SM)^\tau$ is $S(T(\partial M))$, the unit spherical tangent fibration of ∂M.

For example, we may consider a new smooth Riemannian metric $\tilde{g}$ on the fibers of the tangent vector bundle $TM|_{\partial M} \to \partial M$ and the $\tilde{g}$-generated spherical fibration $\tilde{S}M \to \partial M$. For each point $x \in \partial_1(SM)$, we take the ray ℓ_x in T_xM through x and the origin, then produce the unique point $y \in \tilde{S}M$ that belongs to ℓ_x, then apply the reflection $\tau_{\tilde{g}}$ to y and, finally, produce the point $\tau(x) = \ell_{\tau_{\tilde{g}}(y)} \cap SM$.

The results of this section rely on the existence of a differential $(2n-2)$-form Θ on SM, subject to the following properties (which mirror the list in Definition 3.7):

(4.16)

- $d\Theta = 0$,
- $\dim(\mathsf{ker}(\Theta)) = 1$,
- $v^g \in \mathsf{ker}(\Theta)$,
- $\pm\Theta|_{\partial_1^\pm(SM)} \geq 0$,
- $\tau^*(\Theta|_{\partial_1(SM)}) = \Theta|_{\partial_1(SM)}$ with respect to a given smooth involution $\tau : \partial_1(SM) \to \partial_1(SM)$ as above.

The form $(\omega_g)^n$, the nth exterior power of the symplectic 2-form $\omega_g = d\beta_g$, is a $(2n)$-dimensional volume form on TM, $\dim M = n$. Let

$$\Omega_g =_{\mathsf{def}} \{\beta_g \wedge (\omega_g)^{n-1}|_{SM}\}$$

be a $(2n-1)$-dimensional volume form on SM. In fact, $\Omega_g = \mu \rfloor (\omega_g)^n$, where μ is the unitary radial vector field normal to SM in TM.

We introduce a $(2n-2)$-form Θ_g on SM by the formula

$$\Theta_g =_{\mathsf{def}} v^g \rfloor \Omega_g = (v^g \wedge \mu) \rfloor \omega_g^n. \tag{4.17}$$

Both forms, Ω_g and Θ_g, are v^g-invariant.

These constructions lead to the following lemma.

Lemma 4.9. *Let (M, g) be a compact connected n-manifold with boundary. The $(2n-2)$-form $\Theta_g =_{\mathsf{def}} v^g \rfloor \Omega_g$ is equal to the form $\pm(\omega_g)^{n-1}|_{SM}$ and has all the properties, with respect to the involution $\tau_g : \partial_1(SM) \to \partial_1(SM)$, listed in (4.16).*

Proof. We start the validation of the properties of Θ_g in the order they are listed in (4.16). First, we notice that $v^g \in \mathsf{ker}((\omega_g)^{n-1}|_{SM})$, since $dH(w) = \omega_g(v^g, w)$ and the Hamiltonian $H \equiv 1$ on SM. Therefore,

$$v^g \rfloor (\beta_g \wedge (\omega_g)^{n-1}) = (v^g \rfloor \beta_g) \wedge (\omega_g)^{n-1} = 1 \cdot (\omega_g)^{n-1}$$

on SM. Since $d\omega_g = 0$, we get that $(\omega_g)^{n-1}$ is a closed form.

Using that

$$\Omega_g = \beta_g \wedge (\omega_g)^{n-1}$$

is a volume form on SM, we conclude that the form $(\omega_g)^{n-1}$ does not vanish on SM. Thus, we get $\dim(\mathsf{ker}((\omega_g)^{n-1}|_{SM})) = 1$.

In the vicinity of each point $x \in \partial M$, we may pick a local coordinate system $(q^1, \dots, q^n)$ on M so that ∂M is given by the equation $\{q^1 = 0\}$, the gradient field ∂_{q^1} is g-orthogonal to ∂M and has length 1. In other words, we may choose q^1 on M to be the g-induced distance function from ∂M. The "vertical" coordinates $(p_1, \dots, p_n)$ on TM are correlated in the standard way with the coordinates $(q^1, \dots, q^n)$ on M. Then the reflection involution is given by the formula

$$\tau_g((q^1, q^2 \dots, q^n,\ p_1, p_2, \dots, p_n)) = (-q^1, q^2 \dots, q^n,\ -p_1, p_2, \dots, p_n).$$

Applying τ_g to formulas (4.3) and (4.4) and letting $q^1 = 0$, implies that

$$\tau_g^*(\beta_g|_{\partial_1(SM)}) = \beta_g|_{\partial_1(SM)} \quad \text{and} \quad \tau_g^*(\omega_g|_{\partial_1(SM)}) = \omega_g|_{\partial_1(SM)}.$$

As before, we denote the Sasaki metric on TM by gg. Let ν be the unitary vector field, normal in $gg|_{SM}$ to $\partial_1(SM)$ in SM, and let n be the unitary vector field, g-normal to ∂M in M.

We introduce the $(2n-2)$-volume form

$$\Omega_g^\partial =_{\mathsf{def}} \nu \rfloor \Omega_g = \big(\nu \rfloor (\mu \rfloor \omega_g^n)\big)$$

on $\partial_1(SM)$. Because the involution τ_g is an orientation reversing isometry on each spherical fiber of $\{SM|_{\partial M} \to \partial M\}$ and is the identity on the base ∂M, we conclude that $\tau_g^*(\Omega_g^\partial) = -\,\Omega_g^\partial$.

We notice that the normal vector $\nu = \nu(q, p)$ must be orthogonal in the Sasaki metric gg to any vector θ, tangent to the fiber of $\{SM \to M\}$ over a point $q \in \partial M$. In fact, ν is tangent to the v^g-trajectory through the point (q, n), where $n \in T_qM$ is a g-normal unit vector to ∂M in M at $q \in \partial M$. Therefore, representing v^g in the local coordinates $(q, p, \dot q, \dot p)$ on TM as $(\dot q, \dot p)$, we get $\langle \nu, v^g \rangle_{gg} = \langle n, \dot q \rangle_g$. As a result, $v^g = (\dot q, \dot p)$ points inside of SM if and only if $\dot q$ points inside of M. Thus, the restriction of the $(2n-2)$-form $\Theta_g = v^g \rfloor \Omega_g$ to $\partial_1^\pm(SM)$ is nonnegative/nonpositive. Indeed, Ω_g^∂ is the volume form on $\partial_1(SM)$ and

$$\Theta_g / \Omega_g^\partial = \cos(\underset{gg}{\angle}(v^g, \nu)) = \cos(\underset{g}{\angle}(\dot q, n))$$

on $\partial_1(SM)$. At the same time, $\tau_g^*(\Theta|_{\partial_1(SM)}) = \Theta|_{\partial_1(SM)}$ since

$$\Theta|_{\partial_1(SM)} / \Omega^\partial = \cos(\underset{gg}{\angle}(v^g, \nu)) = \cos(\underset{g}{\angle}(\dot q, n)), \text{ and}$$

$$\tau_g^*(\Theta_g\,|_{\partial_1(SM)})/\tau_g^*(\Omega_g^\partial) = \cos(\underset{gg}{\angle}((\tau_g)_*(v^g), (\tau_g)_*(\nu)))$$
$$= \cos(\underset{g}{\angle}(\tau_g(\dot q), -n)) = \cos(\underset{g}{\angle}(\dot q, n)).$$

□

The next lemma is an abstract of Lemma 4.9; unlike the previous lemma, its validation is on the level of definitions.

Lemma 4.10. *Let (M, g) be a n-dimensional compact Riemannian manifold. Assume that a volume form Ω on SM is v^g-invariant. Then the $(2n-2)$-form $\Theta =_{\mathsf{def}} v^g \rfloor \Omega$, is integrally dual (see Definition 3.7) to the geodesic vector field v^g on SM. Thus, with the help of Ω, the vector field v^g is nil-divergent.*

Let N be a regular neighborhood of $\partial_1(SM)$ in SM and $\tilde\tau : N \to N$ a smooth involution with the properties:

- *$\tilde\tau(\partial_1(SM)) = \partial_1(SM)$,*
- *the $\tilde\tau$-fixed set $\partial_1(SM)^{\tilde\tau} = S(\partial M)$, where $S(\partial M)$ denotes the spherical tangent bundle of ∂M,*
- *$\tau^*(\Omega) = -\,\Omega$ along $\partial_1(SM)$.*

Then $\tilde\tau^*(\Theta|_{\partial_1(SM)}) = \Theta|_{\partial_1(SM)}$, and $\pm\Theta|_{\partial_1^\pm(SM)} \geq 0$. ◇

Combining Lemma 4.10 with Lemma 4.8 leads instantly to the following result.

Theorem 4.17. *Let g be a smooth non-trapping Riemannian metric on a compact connected manifold M with boundary. Let N be a regular neighborhood of $\partial_1(SM)$ in SM. Choose a volume form Ω on SM such that $\mathcal{L}_{v^g}\Omega = 0$ and put $\Theta = v^g \rfloor \Omega$. Assume that a smooth involution $\tilde\tau : N \to N$ satisfies the three bullets in Lemma 4.10 with respect to the form Θ.*

- *Then the form Θ is integrally dual (see Definition 3.7) to v^g.*
- *The billiard map B_{v^g}—the composition of the scattering map $C_{v^g} : \partial_1^+(SM) \to \partial_1^-(SM)$ with the diffeomorphism $\tau =_{\mathsf{def}} \tilde\tau| : \partial_1^-(SM) \to \partial_1^+(SM)$—preserves the Θ-induced measure μ_Θ on $\partial_1^+(SM)$.*
- *There exists a v^g-harmonizing pair $(g^\bullet, dF)$ on SM such that $\Theta = v^g \rfloor \Omega$ coincides with $*_{g^\bullet}(dF)$.* ◇

Corollary 4.16. *Under the assumptions of Theorem 4.17, the Billiard Map*

$$B_{v^g,\tau} =_{\mathsf{def}} \tau \circ C_{v^g} : \partial_1^+(SM) \to \partial_1^+(SM)$$

has the infinite return property (as in Theorem 3.11). ◇

4.8 The averages of ergodic billiards for non-trapping metrics

The chaotic dynamics of iterations of billiard maps has been a subject of a well-established and flourishing research industry (see [Si], [Si1], [BFK], and [CM]). Since the Euclidean billiards with strictly concave boundary are dispersing, in what follows, the reader may replace any assumption about the ergodicity of a billiard (M, g) by the assumption that billiard is Euclidean and its boundary is strictly concave [CM], [BFK].

Let us recall few standard definitions, related to the dynamics of measure-preserving maps, as they apply to the billiard maps.

Definition 4.15. Let (M, g) be a connected compact Riemannian n-manifold with boundary. The billiard map

$$B_{v^g,\tau} : \partial_1^+(SM)(v^g) \to \partial_1^+(SM)(v^g)$$

is ergodic with respect to a given $(2n-2)$-form Θ on SM as in Theorem 4.17, if the invariance of any Lebesgue-measurable set $K \subset \partial_1^+(SM)$ under the billiard map B_{v^g} implies that ether the measure $\mu_\Theta(K)$ or the measure $\mu_\Theta(\partial_1^+(SM) \setminus K)$ is zero. $\diamondsuit$

A natural inclination is to anticipate that, for a "random" pair (M, g), the metric g being non-trapping, boundary generic, and "bumpy", the billiard map B_{v^g} is ergodic. This thinking is supported by the theory of Sinai billiards [Si], [Si1]. For them, the smooth boundary ∂M is strictly concave, so that the billiard is dispersing, which implies not only ergodicity of the billiard map, but also its mixing property and positive entropy [Si], [Si1].

Let us recall now the content of the famous Birkhoff Theorem [Bi], as it applies to the billiard maps, with the measure μ_Θ on $\partial_1^+(SM)$ being induced by the appropriate $(2n-2)$-form Θ on SM. We assume that $\Theta|_{\partial_1(SM)}$ is invariant under an involution $\tau : \partial_1(SM) \to \partial_1(SM)$, a *generalized* billiard reflection; so the billiard map $B_{v^g,\tau} : \partial_1^+(SM) \to \partial_1^+(SM)$ preserves the measure μ_Θ.

Let $f \in L^1(\partial_1^+(SM), \mu_\Theta)$ be an integrable (with respect to the measure μ_Θ) function $f : \partial_1^+(SM) \to \mathbb{R}$. Its time average is defined by the formula

$$\hat{f}(z) =_{\mathsf{def}} \lim_{m\to\infty} \frac{1}{m} \sum_{k=0}^{m-1} f\big((B_{v^g,\tau})^{\circ k}(z)\big),$$

where $(B_{v^g,\tau})^{\circ k}$ denotes the k-th iteration of the billiard map.

In fact, the limit $\hat{f}(z)$ exists almost[16] for all $z \in \partial_1^+(SM)$, and $\hat{f} \in L^1(\partial_1^+(SM), \mu_\Theta)$. Moreover, $\hat{f}$ is an invariant function (that is, if $B_{v^g} \circ \hat{f} = \hat{f}$ almost everywhere) and

$$\int_{\partial_1^+(SM)} \hat{f}\, d\mu_\Theta = \int_{\partial_1^+(SM)} f\, d\mu_\Theta.$$

The space average of f is defined as

$$\bar{f} =_{\mathsf{def}} \frac{1}{\mu_\Theta(\partial_1^+(SM))} \int_{\partial_1^+(SM)} f\, d\mu_\Theta.$$

In particular, if the billiard map $B_{v^g,\mathcal{T}}$ is ergodic, then $\hat{f}$ must be *constant* almost everywhere: indeed, any level set $\hat{f}^{-1}((-\infty, c))$ is $(B_{v^g,\mathcal{T}})$-invariant for any $c \in \mathbb{R}$. As a result, $\hat{f} = \bar{f}$ almost everywhere (for example, see [Bi], [Tab]).

Therefore, using that $\mu_\Theta(\mathcal{T}(v^g)) =_{\mathsf{def}} \mu_\Theta(\partial_1^+(SM))$ and that, by Theorem 4.17, $B_{v^g,\mathcal{T}}$ is a μ_Θ-measure-preserving transformation, we get the following version of the Birkhoff Theorem [Bi]:

Theorem 4.18. *Let (M, g) be a compact connected smooth Riemannian n-manifold with boundary, the metric g being non-trapping and boundary generic. Let a $(2n-2)$-form Θ on SM be as in Theorem 4.17 or in list (4.16).*

If the billiard map $B_{v^g,\mathcal{T}}$ is ergodic in the measure μ_Θ, then

$$\lim_{m\to\infty} \frac{1}{m} \sum_{k=0}^{m-1} f\big((B_{v^g,\mathcal{T}})^{\circ k}(z)\big) = \frac{1}{\mu_\Theta(\mathcal{T}(v^g))} \int_{\partial_1^+(SM)} f\, d\mu_\Theta \tag{4.18}$$

for any given function $f \in L^1(\partial_1^+(SM), \mu_\Theta)$ and almost all points $z \in \partial_1^+(SM)$. ♢

For ergodic billiard maps $B_{v^g,\mathcal{T}}$, some "*metric-flavored*" holographic properties hold: as Theorem 4.20 and Corollary 4.21 below claim, for boundary generic non-trapping metrics g on M, the g-induced volume of the space M can be recovered from the volume of the space of geodesics and the average length $\mathsf{av}(\ell)$ of a free geodesic.

A free geodesic in M is a segment $[\gamma]$ of a geodesic curve γ in M such that the boundary $\partial[\gamma] \subset \partial M$ and $\mathsf{int}([\gamma]) \cap \partial M = \emptyset$.

In our treatment, we are guided by the Santalo formula [S], and the results in [Ch], [CM], [Tab].

[16]the exceptional z's forming a set of μ_Θ-measure 0.

For flat compact billiards M in the Euclidean space E^n or in the flat torus T^n, the average length $\mathsf{av}(\ell)$ of a free trajectory is given by a beautiful formula (see [S], [Ch]):

$$\mathsf{av}(\ell) = \frac{vol_{\mathsf{E}}(S^{n-1})}{vol_{\mathsf{E}}(B^{n-1})} \cdot \frac{vol_{\mathsf{E}}(M)}{vol_{\mathsf{E}}(\partial M)} = 2\sqrt{\pi} \cdot \frac{\Gamma(\frac{n+1}{2})}{\Gamma(\frac{n}{2})} \cdot \frac{vol_{\mathsf{E}}(M)}{vol_{\mathsf{E}}(\partial M)}, \quad (4.19)$$

in terms of the Euclidean volume $vol_{\mathsf{E}}(B^{n-1})$ of the unit ball B^{n-1} and of the volume $vol_{\mathsf{E}}(S^{n-1})$ of the unit sphere $S^{n-1} \subset \mathsf{E}^n$.

These results admit the following generalizations.

Theorem 4.19. *Let (M, g) be a compact connected Riemannian n-manifold with boundary, such that the metric g is boundary generic and non-trapping. Let $F : SM \to \mathbb{R}$ be a Lyapunov function for the geodesic field v^g. For each $z \in \partial_1^+(SM)$, consider the variation*

$$\Delta_F(z) =_{\mathsf{def}} F(C_{v^g}(z)) - F(z)$$

of F along the segment $[z, C_{v^g}(z)]$ of the v^g-trajectory $\tilde{\gamma}_z$ through z. This construction gives rise to a well-defined measurable function $\Delta_F : \partial_1^+(SM) \to \mathbb{R}$.

Consider a differential $(2n-2)$-form Θ with the properties as in (4.16). With these ingredients in place, the following statements hold:

- *The average of the variation of F along the v^g-trajectories can be calculated via the Santalo-type formula*

$$\mathsf{av}(\Delta_F) =_{\mathsf{def}} \frac{\int_{\partial_1^+(SM)} \Delta_F \cdot \Theta}{\int_{\partial_1^+(SM)} \Theta} = \frac{\int_{SM} dF \wedge \Theta}{\int_{\partial_1^+(SM)} \Theta}$$
$$=_{\mathsf{def}} \frac{vol_{dF\wedge\Theta}(SM)}{vol_\Theta(\mathcal{T}(v^g))}. \quad (4.20)$$

- *If the billiard map B_{v^g} is ergodic with respect to the measure μ_Θ (for example, if g is Euclidean and ∂M is strictly concave in g), then the average of the variation Δ_F can be also calculated via the formula*

$$\mathsf{av}(\Delta_F) = \lim_{m\to\infty} \frac{1}{m} \sum_{k=0}^{m-1} \Delta_F\big((B_{v^g,\tau})^{\circ k}(z)\big) \quad (4.21)$$

for almost all $z \in \partial_1^+(SM)$.

Proof. If the metric g is non-trapping, the geodesic vector field v^g is traversing. Thus there exists a Lyapunov function $F : SM \to \mathbb{R}$ for v^g.

By the properties listed in (4.16), the billiard map $B_{v^g,\tau}$ preserves the measure μ_Θ on $\partial_1^+(SM)$, and $dF \wedge \Theta$ may serve as the volume form on SM (not to be confused with the v^g-invariant volume form Ω that gave rise to Θ via the formula $\Theta =_{\mathsf{def}} v^g \rfloor \Omega!$).

Since the scattering maps C_{v^g} and C_{-v^g} are continuous and smooth away from a set of vanishing Lebesgue measure, the function Δ_F is a Lebesgue-measurable, and thus μ_Θ-measurable.

Thus, using that $d\Theta = 0$ and by the Stokes' formula, we get

$$\int_{\partial_1^+(SM)} \Delta_F \cdot \Theta = \int_{\partial_1^+(SM)}^{\partial_1^-(SM)} F \cdot \Theta \stackrel{Stokes}{=} \int_{SM} dF \wedge \Theta =_{\mathsf{def}} vol_{dF\wedge\Theta}(SM)$$

By definition, $vol_\Theta(\mathcal{T}(v^g))$—the volume of the space of geodesics—is the integral $\int_{\partial_1^+(SM)} \Theta$. Therefore,

$$\mathsf{av}(\Delta_F) = \frac{\int_{\partial_1^+(SM)} \Delta_F \cdot \Theta}{\int_{\partial_1^+(SM)} \Theta} = \frac{\int_{SM} dF \wedge \Theta}{\int_{\partial_1^+(SM)} \Theta} =_{\mathsf{def}} \frac{vol_{dF\wedge\Theta}(SM)}{vol_\Theta(\mathcal{T}(v^g))},$$

which validates formula (4.20).

When $B_{v^g,\tau}$ is ergodic in μ_Θ, by applying Theorem 4.18 to the function $f =_{\mathsf{def}} \Delta_F$, we validate formula (4.21). □

For traversing flows v^g, their Lyapunov functions F attend extrema on the boundary $\partial(SM)$. Hence, the global variation $\mathsf{var}(F)$ of F on SM is equal to its variation $\mathsf{var}(F^\partial)$ on $\partial_1(SM)$. On the other hand, evidently, $\Delta_F(z) \leq \mathsf{var}(F)$ for all z. This leads to the following direct corollary of Theorem 4.19.

Corollary 4.17. *Let (M, g) be a compact connected Riemannian n-manifold with boundary, such that the metric g is boundary generic and non-trapping. Consider the variation $\mathsf{var}(F^\partial) =_{\mathsf{def}} \mathsf{var}(F|_{\partial_1(SM)})$ of the Lyapunov function $F : SM \to \mathbb{R}$ on the boundary $\partial_1(SM)$.*

Then, for any Θ as in Theorem 4.19,

$$\mathsf{var}(F^\partial) \cdot vol_\Theta(\mathcal{T}(v^g)) \geq vol_{dF\wedge\Theta}(SM).$$

In particular, if $\mathsf{var}(F^\partial) = 1$, then $vol_\Theta(\mathcal{T}(v^g)) \geq vol_{dF\wedge\Theta}(SM)$. ◇

Corollary 4.18. *Under the ergodicity hypotheses from the second bullet of Theorem 4.19, knowing the Θ-induced volume of the space of geodesics*

$\mathcal{T}(v^g)$, together with the limit

$$\lim_{m\to\infty} \frac{1}{m} \sum_{k=0}^{m-1} \Delta_F\big((B_{v^g,\tau})^{\circ k}(z)\big)$$

for almost any $z \in \partial_1^+(SM)$, allows to determine the $(dF \wedge \Theta)$-induced volume of the space SM.

Both $vol_\Theta(\mathcal{T}(v^g))$ and the limit above may be recovered in terms of the data, confined to the boundary $\partial_1(SM)$. ◇

Now we will derive two theorems, Theorem 4.20 and Theorem 4.21, that are very much in the spirit of Theorem 4.19, but require some additional analysis.

We denote by $\ell_{gg}(z)$ the length (in the Sasaki metric gg) of the segment $[z, C_{v^g}(z)]$ of the v^g-trajectory $\tilde{\gamma}_z$ through $z \in \partial_1^+(SM)$. In fact, $\ell_{gg}(z)$ is the length of the free geodesic segment $\pi([z, C_{v^g}(z)])$ in the metric g on M, where $\pi : SM \to M$ is the obvious projection.

In general, for *any* Riemannian metric $g^\bullet$ on SM, we denote by $\ell_{g^\bullet}(z)$ the length, in the metric $g^\bullet$, of the segment $[z, C_{v^g}(z)]$ of the v^g-trajectory $\tilde{\gamma}_z$ through $z \in \partial_1^+(SM)$. We denote by $\ell_{g^\bullet} : \partial_1^+(SM) \to \mathbb{R}_+$ the measurable function that this construction generates.

With these notations fixed, for the classical billiard maps based on the elastic reflections τ, we get a stronger than Theorem 4.19 claim.

Theorem 4.20. *Let (M, g) be a compact connected Riemannian n-manifold with boundary, such that the metric g is boundary generic and non-trapping.*

Let $F : SM \to \mathbb{R}$ be a Lyapunov function for the geodesic vector field v^g. Consider the differential $(2n-2)$-form $\Theta_g =_{\mathsf{def}} \omega_g^{n-1}$, where ω_g is the restriction of the symplectic form on TM to SM.

With these ingredients in place, the following statements hold:

- *The measure μ_{Θ_g} on $\partial_1^+(SM)$, defined by the closed $(2n-2)$-form Θ_g with the help of formula (3.31), is invariant under the billiard map B_{v^g}.*
- *There exists a v^g-harmonizing metric $g^\bullet$ on SM such that $*_{g^\bullet}(dF) = \Theta_g$. Moreover, in $g^\bullet$, the v^g-trajectories are geodesic curves, and the leaves of the foliation $\mathcal{G} = \{F^{-1}(c)\}_{c\in\mathbb{R}}$ are the $(2n-2)$-volume minimizing (in their relative to $\partial_1(SM) \cap F^{-1}(c)$ class) hypersurfaces.*

- *The average value of the length function $\ell_{g^\bullet}$ on the space of v^g-trajectories can be calculated via the formula*

$$\mathsf{av}(\ell_{g^\bullet}) =_{\mathsf{def}} \frac{\int_{\partial_1^+(SM)} \ell_{g^\bullet} \cdot \omega_g^{n-1}}{\int_{\partial_1^+(SM)} \omega_g^{n-1}}$$

$$= \frac{\int_{SM} dF \wedge \omega_g^{n-1}}{\int_{\partial_1^+(SM)} \omega_g^{n-1}} = \frac{vol_{g^\bullet}(SM)}{vol_{\omega_g^{n-1}}(\mathcal{T}(v^g))}. \tag{4.22}$$

- *If the billiard map B_{v^g} is ergodic with respect to the measure $\mu_{\omega^{n-1}}$ (say, if ∂M is strictly concave in g), then the average value of the function $\ell_{g^\bullet}$ can be also calculated via the formula*

$$\mathsf{av}(\ell_{g^\bullet}) = \lim_{m\to\infty} \frac{1}{m} \sum_{k=0}^{m-1} \ell_{g^\bullet}\big((B_{v^g})^{\circ k}(z)\big) \tag{4.23}$$

for almost all $z \in \partial_1^+(SM)$.

Proof. Our argument is based essentially on Lemma 4.9. Thanks to that lemma, the form $\beta_g \wedge \omega_g^{n-1}|_{SM}$ is the volume form Ω_{gg} on SM, and the form $\Theta_g = v^g \rfloor \Omega_{gg} = \omega_g^{n-1}|_{SM}$ on SM has all the desired properties from the list (4.16). In particular, the form $\Theta_g = \omega_g^{n-1}$ produces a measure $\mu_{\omega_g^{n-1}}$ on $\partial_1^+(SM)$. By Theorem 4.17, the billiard map $B_{v^g} : \partial_1^+(SM) \to \partial_1^+(SM)$ preserves this measure.

The second bullet is validated by Theorem 4.14, which claims, in particular, that there exists a v^g-harmonizing metric $g^\bullet$ on SM such that $*_{g^\bullet}(dF) = \omega_g^{n-1}$ and $\mathsf{ker}(dF) \perp_{g^\bullet} \mathsf{ker}(\omega_g^{n-1})$. Thus $dF \wedge \omega_g^{n-1}$ is the volume form $dg^\bullet$, produced by $g^\bullet$. In the metric $g^\bullet$, by its construction, we have $\|dF\|^*_{g^\bullet} = 1$; hence $dF(v^g) = \|v^g\|_{g^\bullet}$.

Now we will validate formulas (4.22) from the third bullet. Since $\|dF\|^*_{g^\bullet} = 1$, the variation $\Delta_F(z) = F(C_{v^g}(z)) - F(z) = \ell_{g^\bullet}(z)$. Therefore, we get

$$\int_{\partial_1^+(SM)} \ell_{g^\bullet} \cdot \omega_g^{n-1} = \int_{\partial_1^+(SM)} \Delta_F \cdot \omega_g^{n-1} =$$

$$= \int_{\partial_1^+(SM)}^{\partial_1^-(SM)} F \cdot \omega_g^{n-1} \overset{Stokes}{=} \int_{SM} dF \wedge \omega_g^{n-1} = vol_{g^\bullet}(SM).$$

Finally, formula (4.23) in the last bullet follows by applying (4.18) to $f = \ell_{g^\bullet}$ and the billiard map. □

Remark 4.10. There is a tension between Theorem 4.20, which deals with taught foliations $\mathcal{G}$ on SM, and Theorem 4.19, which deals with fillable contact structures induced by β_g on SM (see [ET] for the relevant definitions). $\diamondsuit$

Let us revisit the classical case of the Sasaki metric gg on SM (cf. [S], [Ch]).

Theorem 4.21. *Let (M, g) be a compact connected Riemannian n-manifold with boundary, such that the metric g is boundary generic and non-trapping.*

Consider the measurable function $\ell_{gg} : \partial_1^+(SM) \to \mathbb{R}_+$, defined as the length in Sasaki metric gg of the segment $[z, C_{v^g}(z)]$ of the v^g-trajectory $\tilde{\gamma}_z$ through $z \in \partial_1^+(SM)$.[17]

Let $\Theta_g =_{\mathsf{def}} \omega_g^{n-1}$, where ω_g is the restriction of the symplectic form on TM to SM.

With these ingredients in place, the following statements hold:

- *The measure μ_{Θ_g} on $\partial_1^+(SM)$, defined by the form Θ_g with the help of formula (3.31), is invariant under the billiard map B_{v^g}.*
- *The average value of the function ℓ_{gg}, which is equal to the average length of a free geodesic segment in M, can be calculated via the formula*

$$\begin{aligned}\mathsf{av}(\ell_{gg}) &=_{\mathsf{def}} \frac{\int_{\partial_1^+(SM)} \ell_{gg} \cdot \omega_g^{n-1}}{\int_{\partial_1^+(SM)} \omega_g^{n-1}} = \frac{\int_{SM} \beta_g \wedge \omega_g^{n-1}}{\int_{\partial_1^+(SM)} \omega_g^{n-1}} \\ &= \frac{vol_{gg}(SM)}{vol_{\omega_g^{n-1}}(\mathcal{T}(v^g))} = \frac{vol_{\mathsf{E}}(S^{n-1})}{vol_{\mathsf{E}}(B^{n-1})} \cdot \frac{vol_g(M)}{vol_{g|}(\partial M)} \\ &= 2\sqrt{\pi} \cdot \frac{\Gamma(\frac{n+1}{2})}{\Gamma(\frac{n}{2})} \cdot \frac{vol_g(M)}{vol_{g|}(\partial M)}. \end{aligned} \tag{4.24}$$

- *If the billiard map B_{v^g} is ergodic with respect to the measure $\mu_{\omega^{n-1}}$, then $\mathsf{av}(\ell_{gg})$ can be also calculated, for almost all $z \in \partial_1^+(SM)$, via the formula*

$$\mathsf{av}(\ell_{gg}) = \lim_{m \to \infty} \frac{1}{m} \sum_{k=0}^{m-1} \ell_{gg}\big((B_{v^g})^{\circ k}(z)\big). \tag{4.25}$$

Proof. As in Theorem 4.20, the argument is based on Lemma 4.9. Thanks to the lemma, $\beta_g \wedge \omega_g^{n-1}|_{SM}$ is the Sasaki volume form Ω_{gg} on SM, the form

[17] Recall that $\ell_{gg}(z)$ equals the length $\ell_g(z)$ of the free geodesic segment $\pi([z, C_{v^g}(z)])$ in the metric g on M, where $\pi : SM \to M$ is the obvious projection.

$\Theta_g =_{\mathsf{def}} v^g \rfloor \Omega_{gg} = \omega_g^{n-1}|_{SM}$ on SM has all the desired properties from the list (4.16). In particular, the form $\Theta_g = \omega_g^{n-1}$ produces the measure $\mu_{\omega_g^{n-1}}$ on $\partial_1^+(SM)$. By Theorem 4.17, the billiard map $B_{v^g} : \partial_1^+(SM) \to \partial_1^+(SM)$ preserves this measure.

Let ν be the inner unitary normal to $\partial(SM)$ in SM (with respect to gg) vector field. Let $\Omega_{gg}^{\partial} =_{\mathsf{def}} \nu \rfloor \Omega_{gg}$ be the Sasaki volume form on $\partial_1(SM)$.

Recall that any v^g-trajectory $\tilde{\gamma} \subset SM$ is a geodesic curve in the metric gg ([Be], Proposition 1.106). By the definition of the Sasaki metric, $\tilde{\gamma}$ is horizontal (i.e., tangent to the horizontal distribution $H(TM)$ on TM, defined by the g-induced connection) and orthogonal to each spherical fiber of the fibration $\pi : SM \to M$. Therefore, π projects $\tilde{\gamma}$ onto a geodesic curve $\gamma \subset M$ in the metric g and the lengths of the two geodesics are equal: $\ell_{gg}(\tilde{\gamma}) = \ell_g(\gamma)$.

Consider the discontinuous map $\Pi : SM \to \partial_1^+(SM)$ that takes any point $x \in SM$ to the maximal point $\Pi(x)$ in the finite set $\tilde{\gamma}_x \cap \partial_1^+(SM)$, so that $\Pi(x)$ lies below x on the trajectory $\tilde{\gamma}_x$ (the order in $\tilde{\gamma}_x$ is defined by v^g). Away from the zero measure set $\Pi^{-1}\big(\partial_2(SM)(v^g)\big)$, the map Π is a smooth fibration with the fiber being a closed segment. Thus we may integrate the 1-form β_g along the Π-fibers. This integration leads to the following equations:

$$\int_{\partial_1^+(SM)} \ell_{gg} \cdot \omega_g^{n-1} = \int_{\partial_1^+(SM)} \Big(\int_z^{C_{v^g}(z)} \beta_g \Big) \cdot \omega_g^{n-1}$$
$$\overset{Fubini}{=} \int_{SM} \beta_g \wedge \omega_g^{n-1} = \pm\, vol_{gg}(SM). \tag{4.26}$$

The equality marked "Fubini" is a special case of the generalized Fubini formula, proven by Dieudonné [D]. Under the name of "projection formula" it can be found in [BT], Proposition 6.15.[18] In our context, the projection formula uses that $v^g \in \mathsf{ker}(\omega_g^{n-1})$, $\beta_g(v^g) = 1$, and $\Theta|_{\mathsf{ker}(\beta_g)} > 0$.

The metric gg induces the Euclidean metric g_{E} on the fibers of the fibration $\pi : TM \to M$. Let $\mathcal{H}_g =_{\mathsf{def}} H(SM)$ denote the horizontal n-distribution on SM. It is the intersection with SM of the horizontal n-distribution $H(TM)$ on TM that is defined by the g-induced connection on M (as described in paragraphs that preceded Lemma 4.9). Recall that $v^g \in \mathcal{H}_g$.

Let $\rho_{\mathsf{E}}(S^{n-1})$ denote the $(n-1)$-form on SM that coincides with the Euclidean volume form on the π-fibers $\{S^{n-1}\}$ and vanishes on the polyvectors that contain vectors from the horizontal distribution $\mathcal{H}_g$.

[18] See [Her] or [S], page 349, formulas (19.64a) and (19.64b), for its generalization.

Therefore, by another instance of the Fubini formula, being applied to the forms $\rho_{\mathsf{E}}(S^{n-1})$ and $\pi^*(dg)$, we get

$$vol_{gg}(SM) = vol_{\mathsf{E}}(S^{n-1}) \cdot vol_g(M).$$

The manifold $\partial_1^+(SM)$ fibers over the boundary ∂M with the fiber being a hemisphere $S_+^{n-1} \subset S^{n-1} \subset \mathsf{E}^n$.

As in the proof of Lemma 4.9, the restriction of the form $\Theta_g = \omega_g^{n-1}$ to the boundary $\partial_1(SM)$, is proportional to the gg-induced volume $(2n-2)$-form $\Omega_{gg}^\partial = -\nu \rfloor (\beta_g \wedge \omega_g^{n-1})$ on $\partial_1(SM)$ with the coefficient of proportionality $\cos(\angle_{gg}(v^g, \nu)) = \langle v^g, \nu\rangle_{gg}$. This function $\langle v^g, \nu\rangle_{gg} \geq 0$ exactly on the locus $\partial_1^+(SM)$.

By definition, $vol_{\Theta_g}(\mathcal{T}(v^g)) = \mu_{\Theta_g}(\partial_1^+(SM))$. Therefore, the gg-induced volume of the space of geodesics is given by

$$vol_{\Theta_g}(\mathcal{T}(v^g)) =_{\mathsf{def}} \int_{\partial_1^+(SM)} \Theta_g = \int_{\partial_1^+(SM)} \langle v^g, \nu\rangle_{gg} \cdot \Omega_{gg}^\partial.$$

Consider the gg-orthogonal decomposition $T(SM) \approx H(SM) \oplus V(SM)$ of the tangent bundle into horizontal and vertical distributions on SM. Then both vectors, v^g and the normal to $\partial_1(SM)$ vector $\nu \in T_{(m,v)}(SM)$ at the point $(m, v) \in \partial_1(SM)$, are horizontal. We can interpret the vector ν as a inner normal vector to the hemisphere $S_+^{n-1} \subset B^n \subset T_{(m,v)}(SM)$, pointing towards the center of the unit ball B^n. Then $\langle v^g, \nu\rangle_{gg} = \langle v, \nu\rangle_g$.

Put $dg^\partial =_{\mathsf{def}} dg|_{\partial M}$. Examining the fibration $\pi : \partial_1^+(SM) \to \partial M$ with the fiber S_+^{n-1}, we get that the volume form $\Omega_{gg}^\partial = \pi^*(dg^\partial) \wedge dg_{\mathsf{E}}(S_+^{n-1})$. Note that

$$\langle v, \nu\rangle_g \cdot dg_{\mathsf{E}}(S_+^{n-1}) = dg_{\mathsf{E}}(B^{n-1}),$$

the Euclidean volume form on the equatorial ball $B^{n-1} \subset B^n$. Therefore,

$$\langle v^g, \nu\rangle_{gg} \cdot \Omega_{gg}^\partial = \pi^*(dg^\partial) \wedge \big(\langle v, \nu\rangle_g \cdot dg_{\mathsf{E}}(S_+^{n-1})\big) = \pi^*(dg^\partial) \wedge dg_{\mathsf{E}}(B^{n-1}).$$

Let $D^\partial(SM)$ denote the space of the unit tangent disk bundle over $\partial_1(SM)$. Again, by the Fubini theorem, applied to the fibration $\pi^\partial : D^\partial(SM) \to \partial M$, we get

$$vol_\Theta(\mathcal{T}(v^g)) = \int_{\partial M} \Big(\int_{(\pi^\partial)^{-1}(m)} dg_{\mathsf{E}}\Big)\, dg^\partial = vol_{\mathsf{E}}(B^{n-1}) \cdot vol_g(\partial M).$$

This proves formula (4.22).

Thus formula (4.24) follows from formula (4.26):

$$\mathsf{av}(\ell_{gg}) = \frac{\int_{SM} \beta_g \wedge \omega_g^{n-1}}{\int_{\partial_1^+(SM)} \omega_g^{n-1}} = \frac{vol_{gg}(SM)}{\int_{\partial_1^+(SM)} \omega_g^{n-1}} = \frac{vol_{\mathsf{E}}(S^{n-1})}{vol_{\mathsf{E}}(B^{n-1})} \cdot \frac{vol_g(M)}{vol_g(\partial M)}.$$

Finally, (4.25) follows from Theorem 4.18. □

Definition 4.16. Let (M, g) be a compact Riemannian manifold with boundary, where g is a non-trapping metric. We denote by $\gamma_{x,y}$ a free geodesic arc that connects a pair of points $x, y \in M$ and whose interior belongs to the interior of M, and by $\ell_g(\gamma_{x,y})$ its length. (Due to possible concavity of ∂M, not any pair $x, y \in M$ can be connected by such a geodesic!)

Consider the size of the "longest" *free* geodesic segment in M:

$$\mathsf{gd}(M, g) =_{\mathsf{def}} \sup_{\{x,y \in M\}} \{\ell_g(\gamma_{x,y})\} = \sup_{\{x,y \in \partial M\}} \{\ell_g(\gamma_{x,y})\},$$

and call it the geodesic diameter of M. ◇

The relation between the geodesic diameter $\mathsf{gd}(M, g)$ and the ordinary diameter $\mathsf{d}(M, g)$ is subtle due to the boundary effects, as shown in Fig. 4.5.

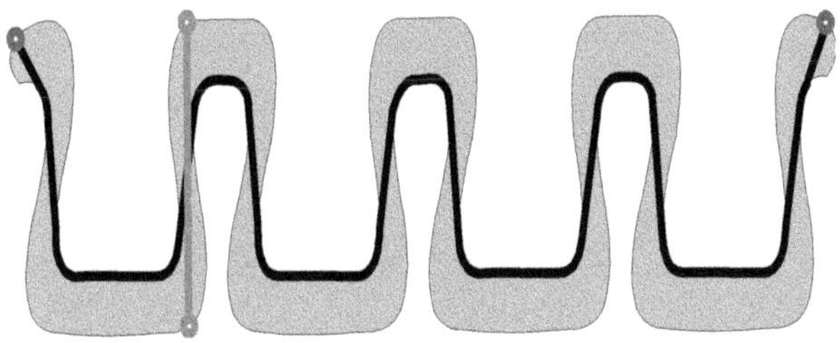

Fig. 4.5 The intrinsic diameter $\mathsf{d}(M, g)$ (the length of the wavy black curve) and the geodesic diameter $\mathsf{gd}(M, g)$ (the length of the vertical red segment) of a Riemannian manifold $M \subset \mathbb{E}^2$ with boundary. The Riemannian metric g on M is Euclidean.

If M is compact and g is trapping, then $\mathsf{gd}(M, g) = +\infty$, while $\mathsf{d}(M, g) < \infty$. At the same time, if (M, g) is geodesically convex and non-trapping (a rare occasion), then $\mathsf{gd}(M, g) = \mathsf{d}(M, g)$.

Using formula (4.24), we get the following inequality.

Corollary 4.19. *For any non-trapping metric g on a compact connected manifold M with boundary,*

$$vol_g(M) \leq \frac{vol_{\mathsf{E}}(B^{n-1})}{vol_{\mathsf{E}}(S^{n-1})} \cdot \mathsf{gd}(M, g) \cdot vol_{g|}(\partial M). \tag{4.27}$$

◇

Corollary 4.19 should be compared with the following classical result of Christopher Croke [Cr]. At the first glance, Theorem 4.22 below looks very similar to Corollary 4.19, but it is based on a quantity $\mathsf{d}(M, g|_M)$ that behaves very differently from $\mathsf{gd}(M, g)$, as illustrated in Fig. 4.5.

Theorem 4.22. [*Cr*] *Let M be a codimension zero compact smooth submanifold of a closed Riemannian n-manifold (N, g). Assume that the diameter* $\mathsf{d}(M, g|_M) < \mathsf{inj\,rad}(N, g)$, *the injectivity radius of* (N, g). *Then*

$$vol_g(M) \;\leq\; \frac{vol_{\mathsf{E}}(S^n)}{2\pi\, vol_{\mathsf{E}}(S^{n-1})} \cdot \mathsf{d}(M, g|_M) \cdot vol_{g|}(\partial M). \qquad \diamondsuit$$

Definition 4.17. Given a compact Riemannian manifold (N, g), consider the number

$$\Lambda(N, g) =_{\mathsf{def}} \sup_{\{(M,\, g|_M) \subset (N,\, g) \,\big|\, g|_M \text{ being non-trapping}\}} \big\{vol_{g|_M}(M)\big\} \qquad (4.28)$$

and call it the Lyapunov volume of (N, g). Evidently, $\Lambda(N, g) \leq vol_g(N)$. $\diamondsuit$

The two conjectures below, which accompany the Definitions 4.17-4.19, are quite speculative. Nevertheless, based on the example of a flat 2-torus and, to a lesser degree, on the proof of Theorem 3.12, being applied to the geodesic vector field v^g, we formulate the following conjecture.

Conjecture 4.5. *For any compact Riemannian manifold* (N, g),

$$\Lambda(N, g) = vol_g(N).$$

$\diamondsuit$

Definition 4.18. For a given closed Riemannian manifold (N, g), keeping in mind the inequality (4.27), consider the scale-invariant quantity:

$$\mathsf{A}(N, g) =_{\mathsf{def}}$$

$$\sup_{\{(M,\, g|_M) \subset (N,\, g) \,\big|\, g|_M \text{ being non-trapping}\}} \Big\{\frac{vol_{g|_M}(M)}{vol_{g|_{\partial M}}(\partial M) \cdot \mathsf{gd}(M, g|_M)}\Big\}.$$

Note that, for a very small and convex $M \subset N$, the expression in the brackets $\{\sim\}$ approaches a positive constant, universal for each dimension.

We also introduce a function of the numerical parameter $V \in [0, \Lambda(N, g))$ by the formula

$$\mathsf{B}(N, g, V) =_{\mathsf{def}} \inf_{\{(M, g|_M) \subset (N, g) \mid vol_{g|_M}(M) = V\}} \big\{\mathsf{gd}(M, g|_M) \cdot vol_{g|_{\partial M}}(\partial M)\big\},$$

where $(M, g|_M)$ runs over all codimension zero smooth compact submanifolds $M \subset N$ such that $g|_M$ is non-trapping, and the volume of M is a given number V.

The function $\mathsf{B}(N, g, V)$ measures how "tightly" one can embed *isometrically* non-trapping domains $(M, g|_M)$ of volume V into a given (N, g). $\diamondsuit$

With Definition 4.18 in place, formula (4.27) leads instantly to the following estimate.

Corollary 4.20. *Let (N, g) be a closed smooth Riemannian manifold. Then, for each number $V \in [0, \Lambda(N, g))$,*

$$\mathsf{A}(N,g) \leq \frac{1}{2\sqrt{\pi}} \cdot \frac{\Gamma(\frac{n}{2})}{\Gamma(\frac{n+1}{2})}, \qquad \mathsf{B}(N,g,V) \geq 2\sqrt{\pi} \cdot \frac{\Gamma(\frac{n+1}{2})}{\Gamma(\frac{n}{2})} \cdot V. \qquad \diamondsuit$$

With the same mindset, we propose the following definition.

Definition 4.19. Consider a compact n-dimensional Riemannian manifold (N, g) and a finite collection $\mathcal{B} = \{B_\alpha\}_\alpha$ of *disjoint* closed smooth n-balls B_α such that:

- each ball B_α is geodesically *strictly convex* in N,
- each connected component of the intersection of any geodesic curve γ in N with the complement $M_\mathcal{B} =_{\mathsf{def}} N \setminus \big(\bigcup_\alpha \mathsf{int}(B_\alpha)\big)$ is a closed segment or a singleton (so that $g|_{M_\mathcal{B}}$ is non-trapping).

We call the Riemannian manifold $(M_\mathcal{B}, g|)$ the non-trapping geodesic Swiss cheese model of (N, g) and the number

$$\mathsf{SC}(N,g) =_{\mathsf{def}} \sup_{\{M_\mathcal{B}\}} \{vol_g(M_\mathcal{B})\}$$

the Swiss cheese volume of (N, g). $\diamondsuit$

Evidently, $\mathsf{SC}(N, g) \leq \Lambda(N, g)$. We do not anticipate the two quantities to be equal.

Conjecture 4.6. *Any compact Riemannian manifold (N, g) admits a non-trapping geodesic Swiss Cheese Model.* $\diamondsuit$

For $(N, \mathcal{B}, g)$ as in Definition 4.19, the geodesic vector field v^g on $SM_\mathcal{B} =_{\mathsf{def}} \pi^{-1}(M_\mathcal{B})$ is traversing and boundary concave. Thus, v^g is boundary generic with respect to the union of the tori $\partial_1(SM_\mathcal{B}) =$

$S^{n-1} \times \left(\bigcup_\alpha \partial B_\alpha\right)$ and admits Lyapunov functions both in $SM_{\mathcal{B}}$ and in $S\mathcal{B} =_{\mathsf{def}} \pi^{-1}(\mathcal{B}) \approx S^{n-1} \times \mathcal{B}$. By [Si], [Si1], [CM], and [BFK], the billiard on $M_{\mathcal{B}}$ is dispersing and thus ergodic, provided that g is Euclidean. Therefore, Corollary 4.21 below is applicable to such a Swiss cheese billiard.

Let us revisit ergodic billiards. Assuming that we know the g-induced volume of ∂M and the *time record* of the billiard ball hitting the boundary ∂M (say, as an infinite sequence of bell rings that broadcast each collision of the billiard ball with the boundary of a curved billiard table (M, g)), for an ergodic billiard and a non-trapping metric g, we can "hear" the volume $vol_g(M)$!

Corollary 4.21. *For a non-trapping boundary generic metric g on M and an ergodic billiard map B_{v^g}, knowing the average length of free geodesic arcs in M, given by the limit*

$$\mathsf{av}(\ell_{gg}) = \lim_{m \to \infty} \frac{1}{m} \sum_{k=0}^{m-1} \ell_{gg}\big((B_{v^g})^{\circ k}(z)\big)$$

for almost any $z \in \partial_1^+(SM)$, allows to determine the isoperimetric proportion $\frac{vol_g(M)}{vol_g(\partial M)}$.

The time intervals $\big\{\ell_{gg}\big((B_{v^g})^{\circ k}(z)\big)\big\}_{k \in \mathbb{Z}_+}$ and thus the quantity $\mathsf{av}(\ell_{gg})$ are data accessible to an observer confined to the boundary ∂M. ◇

Question 4.5. *What are other metric quantities of (M, g) that can be recovered from the sequence $\big\{\ell_{gg}\big((B_{v^g})^{\circ k}(z)\big)\big\}_{k \in \mathbb{Z}_+}$ of time intervals (these data may be registered along ∂M) for ergodic non-trapping billiards?* ◇

Combining Theorem 4.20 with Theorem 4.21 leads to the following claim.

Corollary 4.22. *For a balanced Lyapunov function $F : SM \to \mathbb{R}$ and a v^g-harmonizing metric $g^\bullet$ on SM such that $*_{g^\bullet}(dF) = \omega_g^{n-1}$, we get*

$$vol_{g^\bullet}(SM) = vol_{gg}(SM) = vol_{\mathsf{E}}(S^{n-1}) \cdot vol_g(M).$$

Thus, for an ergodic billiard map B_{v^g} and for almost any $z \in \partial_1^+(SM)$, according to Corollary 4.21, the numerical sequence $\big\{\Delta_F\big((B_{v^g})^{\circ k}(z)\big)\big\}_{k \in \mathbb{Z}_+}$ together with $vol_g(\partial M)$ allow to reconstruct $vol_g(M)$. ◇

Theorem 4.23. *Let (M, g) be a compact connected Riemannian n-manifold with boundary, such that the metric g is non-trapping and bound-*

ary generic. Let $F : SM \to \mathbb{R}$ be a Lyapunov function for v^g, and let

$$var(F) =_{\mathsf{def}} \max_{SM} F - \min_{SM} F.$$

We denote by $g^\bullet$ a v^g-harmonizing metric on SM such that $*_{g^\bullet}(dF) = \omega_g^{n-1}$, where ω_g is the restriction of the symplectic form on TM to SM.

For each $t \in \mathbb{R}$, consider the $g^\bullet$-minimal hypersurface $F^{-1}(t) \subset SM$ and its $g^\bullet$-induced $(2n-2)$-volume $A_{g^\bullet}(t)$. This construction produces a measurable function $A_{g^\bullet} : F(SM) \to \mathbb{R}_+$ on the segment $F(SM) \subset \mathbb{R}$.

With these ingredients in place, the following statements hold:

-
$$A_{g^\bullet}(t) = \int_{F^{-1}(t)} \omega_g^{n-1} = \int_{F^{-1}(t)\cap\partial_1(SM)} \beta_g \wedge \omega_g^{n-2} \\ \leq vol_{\{\omega_g^{n-1}\}}\big(\mathcal{T}(v^g)\big). \quad (4.29)$$

- *The average* $\mathsf{av}(A_{g^\bullet})$ *of the $g^\bullet$-induced volumes of the F-constant lever slices can be calculated via the formula*

$$\mathsf{av}(A_{g^\bullet}) =_{\mathsf{def}} \frac{\int_{F(SM)} \big(\int_{F^{-1}(t)} \omega_g^{n-1}\big)\, dt}{var(F)} \\ = \frac{\int_{SM} dF \wedge \omega_g^{n-1}}{var(F)} = \frac{vol_{g^\bullet}(SM)}{var(F^\partial)}. \quad (4.30)$$

Proof. For a v^g-harmonizing metric $g^\bullet$ that satisfies the hypotheses of the theorem, the restriction of the volume form $dg^\bullet|_{F^{-1}(t)} = \omega_g^{n-1}$. Thus $A_{g^\bullet}(t) = \int_{F^{-1}(t)} \omega_g^{n-1}$. Since $\omega_g^{n-1} = d(\beta_g \wedge \omega_g^{n-2})$, by Stokes' theorem,

$$\int_{F^{-1}(t)\cap\partial_1(SM)} \beta_g \wedge \omega_g^{n-2} = \int_{F^{-1}(t)} \omega_g^{n-1} = A_{g^\bullet}(t).$$

Consider the measurable set $\mathcal{X}(t)$ of v^g-trajectories that have a nonempty intersection with the compact locus $F^{-1}(t)$, where $t \in F(\partial_1(SM))$. Since $\mathcal{X}(t) \subset \mathcal{T}(v^g)$, its ω_g^{n-1}-induced measure does not exceed the ω_g^{n-1}-induced measure of $\mathcal{T}(v^g)$. Therefore, we get the isoperimetric inequality (4.29) for all $t \in F(\partial_1(SM))$ and the constant level hypersurfaces $F^{-1}(t)$:

$$A_{g^\bullet}(t) = \int_{F^{-1}(t)} \omega_g^{n-1} \leq \int_{\partial_1^+(SM)} \omega_g^{n-1} \\ =_{\mathsf{def}} vol_{\{\omega_g^{n-1}\}}\big(\mathcal{T}(v^g)\big).$$

This validates the inequality in (4.29).

Using that $F(v^g) > 0$, we conclude that $F(SM) = F(\partial_1(SM))$. Hence, $var(F) = var(F^\partial)$. Since F has no critical values and the critical values of F^∂ have zero measure, we may treat $F : SM \to \mathbb{R}$ as "almost a fibration", whose fibers are compact $(2n-2)$-manifolds. Thus the integration over the F-fibers is well-defined. Therefore, the Fubini formula holds:

$$\int_{F(SM)} A_{g^\bullet}(t)\, dt = \int_{F(SM)} \Big(\int_{F^{-1}(t)} \omega_g^{n-1}\Big)\, dt = \int_{SM} dF \wedge \omega_g^{n-1},$$

which proves (4.30). □

Corollary 4.23. *Under the hypotheses and notations of Theorem 4.23, the following claims are valid:*

- $vol_{g^\bullet}(SM) \leq var(F^\partial) \cdot \int_{\partial_1^+(SM)} \omega_g^{n-1} = var(F^\partial) \cdot vol_{\{\omega_g^{n-1}\}}(\mathcal{T}(v^g)).$
- *The* proportion

$$\frac{\mathsf{av}(A_{g^\bullet})}{\mathsf{av}(\ell_{g^\bullet})} = \frac{vol_{\omega_g^{n-1}}(\mathcal{T}(v^g))}{var(F^\partial)}, \tag{4.31}$$

  and thus does not depend on the choice of the v^g-harmonizing metric $g^\bullet$ that satisfies the hypotheses of Theorem 4.23. In fact, the proportion in (4.31) is "holographic": it depends only on the data that are confined to the boundary $\partial_1(SM)$.
- *If, in addition, F is balanced, then a $g^\bullet$-independent inequality is valid:*

$$vol_g(M) \leq \frac{1}{vol_{\mathsf{E}}(S^{n-1})} \cdot var(F^\partial) \cdot vol_{\{\omega_g^{n-1}\}}(\mathcal{T}(v^g)).$$

Proof. The claim in the first bullet of Corollary 4.23 follows from (4.29).

Formula (4.31) is obtained by combining (4.22) with (4.30).

For a balanced F and a v^g-harmonizing pair $(g^\bullet, dF)$, we have $\mathsf{av}(\ell_{gg}) = \mathsf{av}(\ell_{g^\bullet})$. Therefore, the inequality in the third bullet follows from the inequality in the first bullet and formula (4.24). □

Remark 4.11. Contemplating about the difference

$$vol_{\{\omega_g^{n-1}\}}(\mathcal{T}(v^g)) - \max_{t \in F(SM)} \{A_{g^\bullet}(t)\} \geq 0$$

between the volume of the trajectory space and the maximum of the volumes of the F-slices, we may view this quantity as measuring the complexity of the geodesic flow. More accurately, the difference measures "how slanted on average" the geodesic flow v^g is with respect to the locus $\partial_1^+(SM)$. ◇

4.9 Detecting intrinsic global geometry of an obstacle via a layered scattering

All the results of this section can be found in [BK], a joint work with Leonid Bunimovich. This section is about reconstructing a shape of a scatterer from scattering data. We place the scatterer within a container and send signals (rays) from its boundary to the interior of the container and see whether a signal returns to the boundary. We outline a novel approach to the problem of determining an inner global geometry (such as volume, curvature, etc.) of a scatterer (obstacle) K, residing in the interior of a Riemannian manifold (M, g) with the boundary, by the probing geodesic rays that originate in the boundary ∂M. In the previous research of this sort, K was a submanifold of M of codimension zero [S], [St], [NS], [NS1], [NS2]. We aim here to investigate manifold obstacles K of an arbitrary codimension. The goal of this section is to point out to an interesting interplay between intrinsic and extrinsic geometry of the obstacle K.

A mathematical setting that we employ here is actually a crude approximation of more realistic physical models, which are used in echolocation, linear optics, etc.

Our main results about the recovery of the intrinsic global invariants of an obstacle K from the scattering data, delivered by the maps $\{\mathcal{B}(K, \epsilon)\}_\epsilon$ in (4.32), are described in Theorem 4.26, Theorem 4.27, Corollary 4.27, and Theorem 4.28.

Let us describe our general setting more accurately. Let M be a compact domain in the Euclidean space $\mathbb{E}^n$. For $k \leq n - 2$, let K be a closed k-dimensional (not necessarily connected) submanifold of M.

We consider the problem of determining the intrinsic geometry of the obstacle K (such as volume, integral curvature) from the scattering data, produced by the reflections of geodesic trajectories from the boundary of a tubular ϵ-neighborhood $\mathsf{T}(K, \epsilon)$ of K in M. The geodesics that participate in this scattering emanate from the boundary ∂M and terminate there after a few reflections from the boundary $\partial\mathsf{T}(K, \epsilon)$. However, the major problem in this setting is that a ray (a billiard trajectory) may get stuck in the vicinity of K by entering some trap there so that this ray will have infinitely many reflections from $\partial\mathsf{T}(K, \epsilon)$. To rule out such a possibility, we modify the geometry of a tube $\mathsf{T}(K, \epsilon)$ by building it from spherical bubbles. We need to use $\lceil \dim(K)/2 \rceil$ many bubbling tubes $\{\mathsf{T}(K, \epsilon_j)\}_j$ for detecting certain global invariants of K, invariants that reflect its intrinsic geometry. Thus the words "layered scattering" are present in the title of this section.

These invariants were studied by Hermann Weyl in his classical theory of tubes $\mathsf{T}(K, \epsilon)$ and their volumes [We].

Clearly, the space geodesics in the ambient manifold M that intersect an obstacle K of codimension ≥ 2 has measure zero; therefore, the "direct" scattering from K is physically unobservable. However, thickening the obstacle K may solve the problem. A manifold obstacle K is a core of a tubular ϵ-neighborhood $\mathsf{T}(K, \epsilon)$ of K, residing in the interior of M, where $\epsilon > 0$ is so small that $\mathsf{T}(K, \epsilon)$ is the space of fibration over K with the fiber being an ϵ-ball of dimension $\dim M - \dim K$ (see Fig. 4.7). The probing geodesic rays generate scattering data. The ones, which are of interest to us, are produced by the reflections of geodesic trajectories in $M \setminus \mathsf{T}(K, \epsilon)$ from the boundary $\partial\mathsf{T}(K, \epsilon)$, where K is assumed to be a closed submanifold of the interior of M. In fact, we need to use *several* nested tubes $\{\mathsf{T}(K, \epsilon_j)\}_j$ for detecting certain intrinsic global invariants of K, studied by Hermann Weyl in terms of the volumes of the tubes, viewed as functions of ϵ [We]. The minimal number of such ϵ-dependent reflective layers is $\lceil k/2 \rceil$, where $k = \dim(K)$. One may think of the layered scattering as probing K by waves with sufficiently high frequencies so that the corresponding wavelengths were much smaller than the sickness ϵ of a tube. Waves of a frequency f_j are reflected from the boundary $\partial\mathsf{T}(K, \epsilon_j)$ of a tube $\mathsf{T}(K, \epsilon_j)$. Our ability to determine global inner invariants of K precisely depends on *a priori* assumptions that the shadow volumes $\mathcal{W}(\epsilon)$ (see formula (4.41)) of sufficiently narrow tubes $\partial\mathsf{T}(K, \epsilon)$ with the core K vanish.

Recall that, if M is a compact domain in the Euclidean space $\mathbb{E}^n$ or hyperbolic space $\mathbb{H}^n$, the induced metric g on M is non-trapping. Therefore, any geodesic in M originates and terminates at points of the boundary ∂M.

For a sufficiently small $\epsilon > 0$, the tube $\mathsf{T}(K, \epsilon)$ is a disjoint union of all normal to K geodesic ϵ-balls. Each ball is formed by geodesic segments γ of length ϵ, emanating from a point $m \in K$ so that γ is orthogonal to K at m. For a closed manifold K and a small $\epsilon > 0$, the tube $\mathsf{T}(K, \epsilon)$ coincides with the ϵ-neighborhood $\mathsf{U}(K, \epsilon)$ of K in M. For a manifold K with a boundary, the tube $\mathsf{T}(K, \epsilon)$ is contained in $\mathsf{U}(K, \epsilon)$, but is different from a $\mathsf{U}(K, \epsilon)$ in the vicinity of ∂K.

Let $N_\epsilon =_{\mathsf{def}} M \setminus \mathsf{int}(\mathsf{T}(K, \epsilon))$. Then $\partial N_\epsilon = \partial M \coprod \partial\mathsf{T}(K, \epsilon)$ for all sufficiently small $\epsilon > 0$.

Consider now the geodesic billiard trajectories γ in the smooth (or at least C^3-differentiable) Riemannian manifold N_ϵ that emanate from ∂M

and are reflected from $\partial \mathsf{T}(K, \epsilon)$ according to the laws of geometric optics [Tab].

We stress our convention: if such a billiard trajectory reaches ∂M in positive time, then it terminates there, thus providing us with the "scattering data" as in (4.32). Since the metric g in M is non-trapping, any geodesic curve in M originates and terminates at points of ∂M. However, some of the billiard trajectories γ that originate in ∂M may be "trapped" in $N_\epsilon \setminus \partial M$ without reaching ∂M again in a positive time. A trajectory γ of this type must have infinitely many reflections from the boundary $\partial \mathsf{T}(K, \epsilon)$. Such a phenomenon may occur, e.g., if the boundary of K is not smooth enough (see [H]). If an initial point $(m, v) \in \partial_1^+(SM)$ generates such a trapped trajectory, we call (m, v) trapping initial data. They form a subset $\mathsf{Trap}(\partial M \rightsquigarrow, SN_\epsilon)$ of $\partial_1^+(SM)$. Since g in M is non-trapping, any billiard trajectory γ in N_ϵ, which is determined by a point

$$x \in \partial_1^+(SM) \setminus \mathsf{Trap}(\partial M \rightsquigarrow, \; SN_\epsilon),$$

will reach ∂M again in a finite time. Its lift $\tilde{\gamma}$ to SN_ϵ, after a few reflections of the boundary $SN_\epsilon|_{\partial \mathsf{T}(K,\epsilon)}$, will terminate at a point of $\partial_1^-(SM)$ (see Fig. 4.6, the top diagram). A smooth involution $\tau : \partial_1^-(S\mathsf{T}(K, \epsilon)) \to \partial_1^+(S\mathsf{T}(K, \epsilon))$ represents the effect of an elastic reflection from the boundary $\partial \mathsf{T}(K, \epsilon)$.

As a result, we have the following scattering map

$$\mathcal{B}(K, \epsilon) : \partial_1^+(SM) \setminus \mathsf{Trap}(\partial M \rightsquigarrow, SN_\epsilon) \to \partial_1^-(SM), \tag{4.32}$$

which is our main probing instrument (see Fig. 4.8 and Fig. 4.6). We will adjust the map $\mathcal{B}(K, \epsilon)$ by varying $\epsilon > 0$.

Let ω_g be the symplectic 2-form on TN_ϵ. The map $\mathcal{B}(K, \epsilon)$ preserves the $(2n - 2)$-dimensional measure μ on $\partial_1(SN_\epsilon)$, defined by integrating the differential form $\pm\omega_g^{n-1}|_{\partial_1(SN_\epsilon)}$ against Lesbegue measurable sets in $\partial_1^\pm(SN_\epsilon)$ (see Theorem 4.21).

In summary, a physical model, which we explore here, is the one used in echolocation or in linear optics, where a probing signal originates at the boundary of a known region M, and the unknown obstruction $K \subset \mathsf{int}(M)$ is a submanifold, surrounded by several reflecting layers (e.g., with different reflecting properties with respect to different penetrating abilities).

4.9.1 *Probing an obstacle by remote scattering*

The locus $\mathsf{Trap}(\partial M \rightsquigarrow, \; SN_\epsilon)$ has the Lesbegue measure zero ([LP], Theorem 1.6.2). In contrast, the trapping locus $\mathsf{Trap}(SN_\epsilon) \subset SN_\epsilon$, consisting of

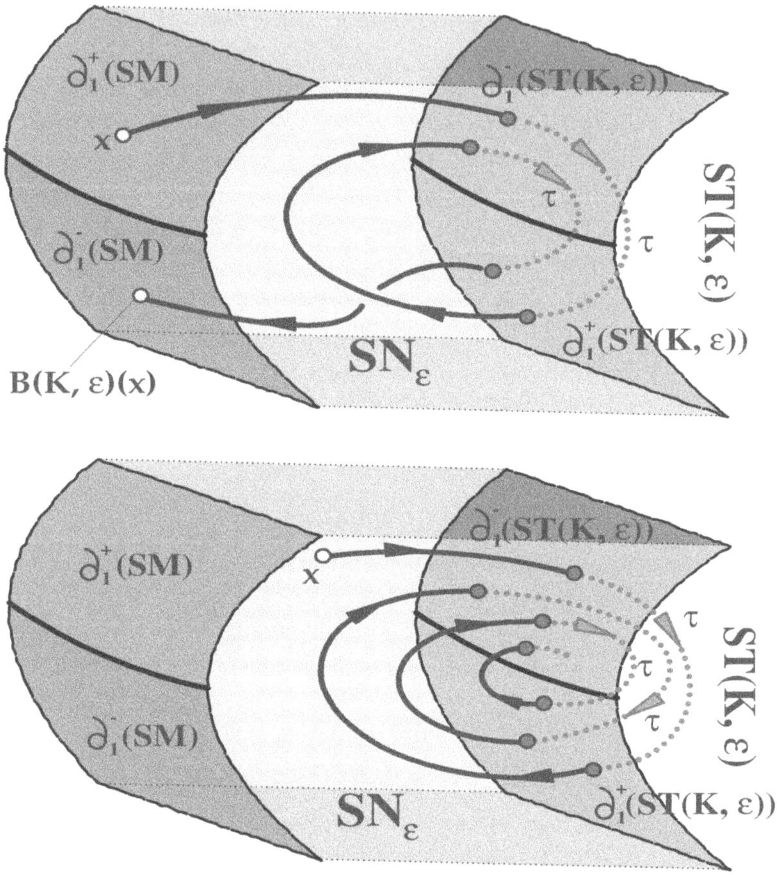

Fig. 4.6 Top picture: the lift $\tilde{\gamma}$ of a non-trapped billiard trajectory γ in N_ϵ to SN_ϵ, which originates at $x \in \partial_1^+(SM)$ and, after two "reflections" τ from $\partial_1^-(\mathsf{T}(K,\epsilon))$, terminates at a point $\mathcal{B}(K,\epsilon)(x) \in \partial_1^-(SM)$. Bottom picture: the lift $\tilde{\gamma}$ to SN_ϵ of a trapped billiard trajectory γ in N_ϵ, which originates at $x \in SN_\epsilon$ and does not reach $\partial_1(SM)$ again.

initial data $(m, v) \in SN_\epsilon$ for which the billiard trajectory in N_ϵ will never reach ∂M, may have a *positive* Lesbegue measure μ [Pe] (see Fig. 4.6, the bottom diagram). This possibility significantly complicates our efforts. We conjecture though that, for a non-trapping metric g on M and a generic obstacle K whose dimension is less than $n-1$, the locus $\mathsf{Trap}(SN_\epsilon)$ has zero measure μ for all sufficiently small $\epsilon > 0$ (see Conjecture 4.7).

It is known that if M is a ball in the Euclidean space $\mathbb{E}^n$ and the tube $\mathsf{T}(K, \epsilon)$ is replaced by any disjoint union $\tilde{K}$ of convex domains, then $\mu(\mathsf{Trap}(\partial M \rightsquigarrow, \tilde{K})) = 0$ by [NS], [NS2]. However, the intrinsic geometry and topology of the manifolds $\tilde{K}$'s that are unions of convex domains are both trivial... Therefore, in the case of K being a finite set in $\mathbb{E}^n$, one has $\mu(\mathsf{Trap}(SN_\epsilon)) = 0$. On the other hand, if K is not of a homotopy type of a finite set, then the tube $\mathsf{T}(K, \epsilon)$ cannot be a disjoint union of convex domains for all sufficiently small $\epsilon > 0$.

We denote by $L_\epsilon(x)$ the length of the billiard trajectory (which may reflect from $\partial\mathsf{T}(K, \epsilon)$ several times) in N_ϵ, determined by the initial data $x \in \partial_1^+(SM) \setminus \mathsf{Trap}(\partial M \rightsquigarrow, SN_\epsilon)$. It is known that $L(x) < \infty$ almost everywhere in $x \in \partial_1^+(SM)$ [LP].

Our aim is to probe K by means of the scattering maps $\{\mathcal{B}(K, \epsilon)\}_\epsilon$ (see (4.32)). In order to detect some global intrinsic invariants of the Riemannian manifold K (to be described in the next subsection), we will use a nested family of tubes

$$\mathsf{T}(K, \epsilon_0) \supset \mathsf{T}(K, \epsilon_1) \supset \cdots \supset \mathsf{T}(K, \epsilon_k)$$

and the family of the scattering maps $\{\mathcal{B}(K, \epsilon_i)\}_{0 \leq i \leq k}$ that are associated with them. For a "clean" detection, we need to assume that $\mu(\mathsf{Trap}(SN_{\epsilon_i}) = 0$. It worthwhile to mention that we do not require $\mathsf{T}(K, \epsilon_i)$ to be very narrow.

4.9.2 *Weyl's theory of tubes*

To formulate the Hermann Weyl theory of tubes [We], we need to introduce several geometric ingredients.

Let K be a compact smooth k-dimensional submanifold of the Euclidean space $\mathbb{E}^n$. We also consider K as a Riemannian manifold with a symmetric connection $\{\Gamma^\kappa_{\ \alpha\beta}\}$.

Following [We], we introduce the Riemann curvature tensor $\{R^\kappa_{\ \lambda\alpha\beta}\}$ on K as:

$$R^\kappa_{\ \lambda\alpha\beta} = \Big(\frac{\partial\Gamma^\kappa_{\ \lambda\beta}}{\partial u^\alpha} - \frac{\partial\Gamma^\kappa_{\ \lambda\alpha}}{\partial u^\beta}\Big) + \sum_\rho \big(\Gamma^\kappa_{\ \rho\alpha}\,\Gamma^\rho_{\ \lambda\beta} - \Gamma^\kappa_{\ \rho\beta}\,\Gamma^\rho_{\ \lambda\alpha}\big), \qquad (4.33)$$

and the skew-symmetric in α, β and in κ, λ the Weyl tensor $\{R^{\kappa\mu}_{\alpha\beta}\}$ by the formula:

$$R^\kappa_{\ \lambda\alpha\beta} = \sum_\mu g_{\lambda\mu} R^{\kappa\mu}_{\alpha\beta} \qquad (4.34)$$

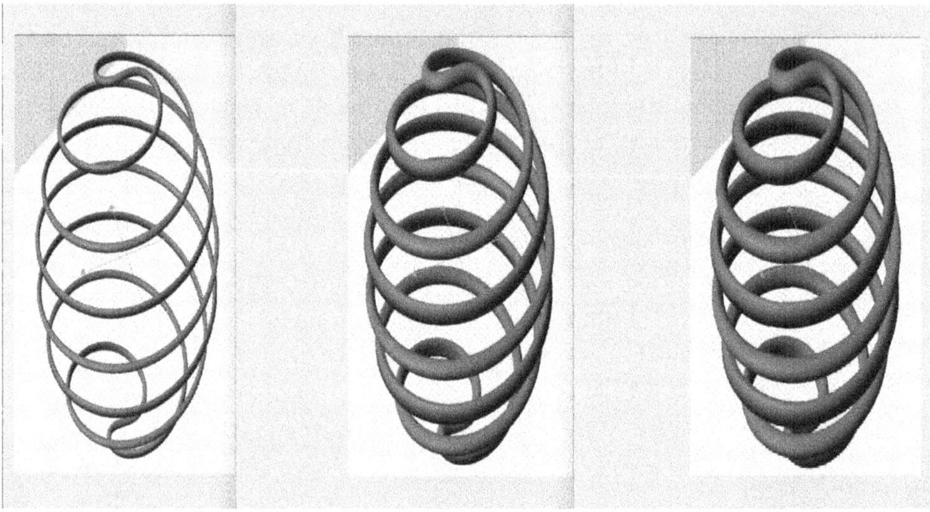

Fig. 4.7 Three Weyl tubes $\{\mathsf{T}(K, \epsilon_i)\}_{i=1,2,3}$ with the same 1-dimensional core loop $K \subset \mathbb{E}^3$.

Theorem 4.24. [*We*] *For any compact smooth k-dimensional submanifold $K \subset \mathbb{E}^n$, the volume of the tube $\mathsf{T}(K, \epsilon)$, for all sufficiently small $\epsilon > 0$, is a polynomial $P_K(\epsilon)$ of degree n in the variable ϵ, divisible by the monomial ϵ^{n-k}. It is given by the formula*

$$vol(\mathsf{T}(K,\epsilon)) = \tag{4.35}$$
$$vol(B^{n-k}) \cdot \sum\nolimits_{\{\ell \in [0,k],\ \ell \equiv 0\, mod\, 2\}} \frac{\mathsf{Q}_\ell(K)}{(n-k+2)(n-k+4)\dots(n-k+\ell)}\ \epsilon^{n-k+\ell},$$

where $vol(B^{n-k})$ is the volume of the unit Euclidean ball, and

$$\mathsf{Q}_\ell(K) =_{\mathsf{def}} \int_K H_\ell\ \mu_K, \tag{4.36}$$

where the measure μ_K is the volume k-form on K, and the function $H_\ell : K \to \mathbb{R}$ can be written in terms the Weyl tensor $\{R^{\kappa\mu}_{\alpha\beta}\}$ of K as

$$H_\ell = \frac{1}{2^\ell (\ell/2)!} \sum_{1', \dots, \ell'} \Big\{ \pm \sum_{\alpha_1, \dots, \alpha_\ell} R^{\alpha_{1'}\alpha_{2'}}_{\alpha_1 \alpha_2}\ R^{\alpha_{3'}\alpha_{4'}}_{\alpha_3 \alpha_4}\ \dots\ R^{\alpha_{\ell'-1}\alpha_{\ell'}}_{\alpha_{\ell-1}\alpha_\ell} \Big\}. \tag{4.37}$$

◇

The first coefficient $\mathsf{Q}_0(K)$ in (4.35) is just the k-volume of K, while the second coefficient $\mathsf{Q}_2(K) = \frac{1}{2}\int_K \kappa\, d\mu$, where κ is the scalar curvature of K [We].

By their very construction, all the terms $\{\mathsf{Q}_\ell(K)\}_\ell$ do not depend on the *extrinsic geometry* of the embedding $K \subset \mathbb{E}^n$; i.e., $\{\mathsf{Q}_\ell(K)\}_\ell$ are the same for all *isometric* embeddings of a given Riemannian manifold K into the Euclidean space. In fact, these coefficients can be expressed in terms of the second fundamental form on K [We]. In case when K is closed and $\dim K = 2q$, the leading coefficient $\mathsf{Q}_{2q}(K) = (2\pi)^q \cdot \chi(K)$, where $\chi(K)$ is the Euler characteristic of K [Gray1]. For a compact surface $K \subset \mathbb{R}^3$, the formula (4.35) reduces to the following cubic polynomial in ϵ

$$vol(\mathsf{T}(K,\epsilon)) = 2\epsilon \cdot vol(K) + \frac{4\pi\epsilon^3}{3} \cdot \chi(K),$$

which is divisible by ϵ.

See [Gray1] for a comprehensive account of the analogues of these formulas, when the ϵ-tubes $\mathsf{T}(K,\epsilon)$ are considered in a general Riemannian manifold (M,g).

4.9.3 *Layered scattering by reflections from nested Weyl tubes*

Let β_g be the Lioville 1-form for the Euclidean metric g on SN_ϵ. We introduce the following quantities:

$$vol(SM) =_{\mathsf{def}} \int_{SM} \beta_g \wedge \omega_g^{n-1}, \; vol(SN_\epsilon) =_{\mathsf{def}} \int_{SN_\epsilon} \beta_g \wedge \omega_g^{n-1},$$

$$vol(S\mathsf{T}(K,\epsilon)) =_{\mathsf{def}} \int_{S\mathsf{T}(K,\epsilon)} \beta_g \wedge \omega_g^{n-1}.$$

Let $L_\epsilon(x)$ be the combined length of several trajectories of the geodesic flow $\phi_g^t : SN_\epsilon \to SN_\epsilon$, the first of which starts at a point

$$x \in \partial_1^+(SM) \setminus \mathsf{Trap}(\partial M \rightsquigarrow, SN_\epsilon)$$

and the last of which terminates at a point $y \in \partial_1^-(SM)$; the other ends of these trajectories reside in $\partial_1(S\mathsf{T}(K,\epsilon)) \subset \partial_1(SN_\epsilon)$ (see Fig. 4.6, the upper diagram). Recall that the length of a v^g-trajectory $\tilde\gamma \subset SN_\epsilon$ in the Sasaki metric gg on SN_ϵ equals the length of its projection $\gamma \subset N_\epsilon$ in the metric g.

The images of these trajectories, under the projection $\pi : SN_\epsilon \to N_\epsilon$, give rise to the broken billiard trajectory in N_ϵ that starts at $\pi(x)$ and, after several reflections from $\partial \mathsf{T}(K, \epsilon)$, terminates at $\pi(y)$.

Let us formulate now an important claim from the Stoyanov paper [St], as it applies to our setting:

Theorem 4.25. [*St*] *Let M be a compact smooth n-dimensional manifold, equipped with a non-trapping metric g, and $N_\epsilon =_{\mathsf{def}} M \setminus \mathsf{T}(K, \epsilon)$, where $\epsilon > 0$ is so small that the ϵ-tubes, whose cores are the connected components of K, do not intersect each other and $\mathsf{T}(K, \epsilon) \cap \partial M = \emptyset$.*

Then the volume of the non-trapping region in SN_ϵ can be calculated as

$$\int_{\{SN_\epsilon \setminus \mathsf{Trap}(SN_\epsilon)\}} \beta_g \wedge \omega_g^{n-1} =$$

$$\int_{\{\partial_1^+(SM) \setminus \mathsf{Trap}(\partial M \rightsquigarrow,\, SN_\epsilon)\}} L_\epsilon \cdot \omega_g^{n-1}. \tag{4.38}$$

◇

The relation in (4.38) uses that $\beta_g(v^g) = 1$, where v^g is the geodesic vector field on SN_ϵ.

Let us introduce a variant of volume by

$$\mathcal{V}(\epsilon) =_{\mathsf{def}} \int_{\{SN_\epsilon \setminus \mathsf{Trap}(SN_\epsilon)\}} \beta_g \wedge \omega_g^{n-1}. \tag{4.39}$$

Stoyanov's theorem says that the volume $\mathcal{V}(\epsilon)$ is "observable" and may be computed from the scattering information, encoded in the following integrable function:

$$L_\epsilon : \partial_1^+(SM) \setminus \mathsf{Trap}(\partial M \rightsquigarrow, SN_\epsilon) \to \mathbb{R}_+. \tag{4.40}$$

In other words, a value $\mathcal{V}(\epsilon)$ is detectable by the scattering data that include traveling times of the probing signal (of an appropriate frequency f_ϵ). Note that any two functions L_ϵ and $\tilde{L}_\epsilon$, which differ on a set of measure zero, both make equal contributions to the integral in the LHS of (4.38).

We will call the volume of the trapping portion $\mathsf{Trap}(SN_\epsilon)$ of SN_ϵ the shadow ϵ-volume of K. It is given by the formula

$$\mathcal{W}(\epsilon) =_{\mathsf{def}} vol(\mathsf{Trap}(SN_\epsilon)) = \int_{\{\mathsf{Trap}(SN_\epsilon)\}} \beta_g \wedge \omega_g^{n-1} \tag{4.41}$$

At first glance, $\mathcal{W}(\epsilon)$ looks as an *extrinsic* invariant of the obstacle K, which, among other things, depends on M and the inclusion $K \hookrightarrow M$.

By definition,

$$\begin{aligned} vol(SM) &= vol(SN_\epsilon) + vol(S\mathsf{T}(K,\epsilon)) \\ &= \mathcal{V}(\epsilon) + \mathcal{W}(\epsilon) + vol(S\mathsf{T}(K,\epsilon)). \end{aligned} \tag{4.42}$$

Thus, by Theorem 4.25, the shadow ϵ-volume of K equals

$$\begin{aligned} \mathcal{W}(\epsilon) &= vol(SM) - vol(S\mathsf{T}(K,\epsilon)) - \\ &- \int_{\{\partial_1^+(SM)\setminus \mathsf{Trap}(\partial M^{\leadsto}, SN_\epsilon)\}} L_\epsilon \cdot \omega_g^{n-1}. \end{aligned} \tag{4.43}$$

The volume $vol(S\mathsf{T}(K,\epsilon))$ will turn out to be an invariant that is shared by all isometric embeddings $\alpha : K \hookrightarrow M$, subject to the constraints $\mathsf{dist}_g(\partial M, \alpha(K)) \geq \epsilon$. The integral $\mathcal{V}(\epsilon)$ in (4.42) is an observable quantity.

Generally, the shadow ϵ-volume $\mathcal{W}(\epsilon)$ is "a black hole" or "a known unknown" of the scattering enterprise here. However, it has some good features too: i.e., by Theorem 1.2 from [St], $\mathcal{W}(\epsilon)$ is a continuous function of the smooth embedding $\partial\mathsf{T}(K,\epsilon) \hookrightarrow M$, and thus of the parameter $\epsilon > 0$ and of the smooth regular embedding $\alpha : K \hookrightarrow M$.

• Hypothesis W. *For all sufficiently small $\epsilon > 0$, the volume $\mathcal{W}(\epsilon)$ of the trapping locus $\mathsf{Trap}(SN_\epsilon)$ is zero.* ♢

Conjecture 4.7. *Hypothesis* W *is valid for any regular imbedding $K \subset \mathbb{E}^n$ of a smooth manifold K of codimension ≥ 2.* ♢

Our main result in this section is obtained by combining Theorem 4.24 and Theorem 4.25:

Theorem 4.26. *Let K be a closed smooth k-dimensional submanifold of the Euclidean space $\mathbb{E}^n$. Let $\mathsf{T}(K,\epsilon)$ be the ϵ-tube, properly contained in a smooth compact domain $M \subset \mathbb{E}^n$. Let $N_\epsilon =_{\mathsf{def}} M \setminus \mathsf{int}(\mathsf{T}(K,\epsilon))$.*

If Hypothesis W *is valid, then, for any sequence*

$$\epsilon_0 > \epsilon_1 > \dots > \epsilon_{\lceil k/2 \rceil} > 0,$$

where $\epsilon_0 > 0$ is sufficiently small, the lengths (travel times) functions $\{L_{\epsilon_j} : \partial_1^+(SM) \to \mathbb{R}\}_{0 \leq j \leq \lceil k/2 \rceil}$ of billiard trajectories in the domains $\{N_{\epsilon_j}\}_{0 \leq j \leq \lceil k/2 \rceil}$ determine the intrinsic global invariants

$$\Big\{\mathsf{Q}_\ell =_{\mathsf{def}} \int_K H_\ell\, \mu_K\Big\}_{0 \leq \ell \leq \lceil k/2 \rceil}$$

of the Riemannian manifold K, defined in formulas (4.36)-(4.37).

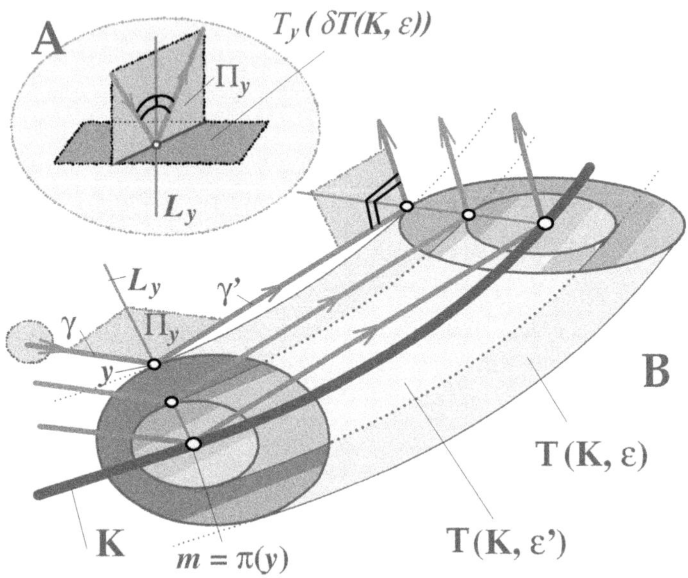

Fig. 4.8 Diagram A : A single reflection from the boundary of the tube $\mathsf{T}(K, \epsilon)$. Diagram B : Double reflections from the boundaries of two tubes, $\mathsf{T}(K, \epsilon)$ and $\mathsf{T}(K, \epsilon')$, and from their core K.

Proof. For the time being, let $\epsilon = \epsilon_i$. By equation (4.43) and the Fubini theorem,

$$vol(\mathsf{Trap}(SN_\epsilon)) = vol(SN_\epsilon) - \mathcal{V}(\epsilon) = vol(SM) - vol(S\mathsf{T}(K, \epsilon)) - \mathcal{V}(\epsilon)$$

$$\overset{Fubini}{=} \sigma_{n-1} \cdot \big[vol(M) - vol(\mathsf{T}(K, \epsilon))\big] - \mathcal{V}(\epsilon),$$

where σ_{n-1} denotes the volume of the standard unit $(n-1)$-sphere in $\mathbb{E}^n$.

By (4.35), $vol(T(K, \epsilon))$ is given by a polynomial $P_K(\epsilon)$ of the form $\epsilon^{n-k} \cdot Q_K(\epsilon^2)$, where $Q_K(\epsilon^2)$ is polynomial of degree $\lceil k/2 \rceil$ in ϵ^2. Therefore,

$$vol(\mathsf{Trap}(SN_\epsilon)) = \sigma_{n-1} \cdot \big[vol(M) - \epsilon^{n-k} \cdot Q_K(\epsilon^2))\big] - \mathcal{V}(\epsilon). \quad (4.44)$$

Assuming that $vol(\mathsf{Trap}(SN_\epsilon)) = 0$, we get

$$Q_K(\epsilon^2)) = \frac{1}{\epsilon^{n-k}}\Big[vol(M) - \frac{1}{\sigma_{n-1}}\mathcal{V}(\epsilon)\Big], \quad (4.45)$$

where $\mathcal{V}(\epsilon)$ is defined by formula (4.39) and is determined by an integrable function $L_\epsilon : \partial_1^+(SM) \setminus \big(\mathsf{Trap}(SN_\epsilon, \partial M) \to \mathbb{R}$.

Since Q_K is a polynomial of degree $\lceil k/2 \rceil$ in ϵ^2, by choosing any sequence $\epsilon_0 > \epsilon_1 > \cdots > \epsilon_{\lceil k/2 \rceil} > 0$, and applying (4.45) to the members of the sequence, we conclude that the collection of integrable functions

$$\{L_{\epsilon_i} : \partial_1^+(SM) \setminus \mathsf{Trap}(\partial M \rightsquigarrow SN_{\epsilon_i}) \to \mathbb{R}\}_{0 \le i \le \lceil k/2 \rceil}$$

determines the polynomial Q_K. Thus, the polynomial $P_K = \epsilon^{n-k} \cdot Q_K$ is the one in (4.35). As a result, the intrinsic integral invariants $\{\mathsf{Q}_\ell(K)\}$ from formula (4.36) are determined by the scattering data $\{L_{\epsilon_i}\}_{0 \le i \le \lceil k/2 \rceil}$, provided that $\{vol(\mathsf{Trap}(SN_{\epsilon_i})) = 0\}_{0 \le i \le \lceil k/2 \rceil}$. □

Corollary 4.24. *Under Hypotheses* W, *the length/travel time integrable functions* $\{L_{\epsilon_j} : \partial_1^+(SM) \to \mathbb{R}\}_{0 \le j \le \lceil k/2 \rceil}$ *detect the following quantities:*

(1) *the* k*-dimensional volume of the obstacle* K,

(2) *the integral* $\mathsf{Q}_2(K) = \frac{1}{2}\int_K \kappa \, d\mu$, *where* κ *is the scalar curvature of* K,

(3) *for an even* k, *the invariant*

$$\mathsf{Q}_k(K) = (2\pi)^{k/2}\chi(K) = (2\pi)^{k/2} \deg\big(\mathsf{Gauss}(\partial T(K, \epsilon))\big),$$

where $\chi(K)$ *is the Euler number of* K *and* $\deg\big(\mathsf{Gauss}(\partial T(K,\epsilon))\big)$ *is the degree of the Gaussian map* $\mathsf{Gauss} : \partial \mathsf{T}(K, \epsilon) \to S^{n-1}$ *(defined by the exterior normals to* $\partial \mathsf{T}(K, \epsilon)$*).*

Proof. All the claims follow from the geometric interpretations of the coefficients $\{\mathsf{Q}_\ell(K)\}_\ell$ in [We]. In particular, in the case when K is closed and $\dim K = k = 2q$, the leading coefficient $\mathsf{Q}_{2q}(K) = (2\pi)^q \chi(K)$, where $\chi(K)$ is the Euler characteristic of K [Gray]. This number coincides with the degree $\deg(\mathsf{Gauss}(\partial \mathsf{T}(K, \epsilon))$ of the Gaussian normal map times $(2\pi)^q$. □

Since the Euler class $\chi(K)$ determines the genus of the closed connected orientable surface K, Corollary 4.24 implies the following claim.

Corollary 4.25. *Let* M *be a compact domain in* $\mathbb{E}^n$ *with a smooth boundary, and* K *is a closed smooth connected two-dimensional surface in the interior of* M. *Then the scattering data* $\{L_{\epsilon_j} : \partial_1^+(SM) \to \mathbb{R}_+\}_{0 \le j \le 1}$ *detect the genus of* K, *provided that the Hypothesis* W *holds.* ◇

We define now the average length (travel time) of non-trapped billiard trajectories in N_ϵ (these are the trajectories that originate and terminate in ∂M, with possible multiple reflections from $\partial T(K, \epsilon)$ in-between) by

$$\mathcal{L}^{\mathsf{av}}_\epsilon(M, K) = \frac{\int_{\{\partial_1^+(SM) \setminus \mathsf{Trap}(\partial M \rightsquigarrow, \, SN_\epsilon)\}} L_\epsilon \cdot \omega_g^{n-1}}{\int_{\{\partial_1^+(SM) \setminus \mathsf{Trap}(\partial M \rightsquigarrow, \, SN_\epsilon)\}} \omega_g^{n-1}}. \tag{4.46}$$

Theorem 4.27. *Let K be a closed smooth k-dimensional Riemannian manifold, and M is a smooth compact domain in the Euclidean space $\mathbb{E}^n$. Then, for any* isometric *imbedding $\alpha : K \hookrightarrow M \setminus T_\epsilon(\partial M)$, the trapping volume, defined by*

$$vol(\mathsf{Trap}(SN_\epsilon)) = \int_{\{\mathsf{Trap}(SN_\epsilon)\}} \beta_g \wedge \omega_g^{n-1},$$

depends continuously on $\epsilon > 0$, on the average length $\mathcal{L}_\epsilon^{\mathsf{av}}(M, \alpha(K))$ of the non-trapped billiard trajectories in $N_\epsilon(\alpha) =_{\mathsf{def}} M \setminus T(\alpha(K), \epsilon)$, and on the following three α-independent[19] *volumes:*

(i) $\int_{S\mathsf{T}(\alpha(K),\, \epsilon)} \beta_g \wedge \omega_g^{n-1}$, (ii) $\int_{\partial_1^+(SM)} \omega_g^{n-1}$, *and* (iii) $\int_{SM} \beta_g \wedge \omega_g^{n-1}$.

Moreover, all these quantities together determine $vol(\mathsf{Trap}(SN_\epsilon))$.

Proof. By Theorem 1.6.2, [LP], the locus $\mathsf{Trap}(\partial M \rightsquigarrow, SN_\epsilon)$ has the Lesbegue measure zero. Thus, the denominator of (4.46) equals

$$\int_{\{\partial_1^+(SM)\setminus \mathsf{Trap}(\partial M \rightsquigarrow,\, SN_\epsilon)\}} \omega_g^{n-1} = \int_{\{\partial_1^+(SM)\}} \omega_g^{n-1},$$

and it does not depend on α and K. Therefore, the numerator in (4.46) depends only on the average length $\mathcal{L}_\epsilon(\alpha(K))$.

On the other hand, by Theorem 4.24, the volume of an ϵ-tube $\mathsf{T}(\alpha(K), \epsilon)$ depends only on the intrinsic geometry of K. Thus, by the Fubini theorem, the volume of $S\mathsf{T}(\alpha(K), \epsilon) \subset SM$ depends only on the intrinsic geometry of K as well. Here $S\mathsf{T}(\alpha(K), \epsilon)$ denotes the space of the spherical fibration $SM \to M$ restricted to the tube $\mathsf{T}(\alpha(K), \epsilon) \subset M$.

Therefore, formula (4.42) implies that the trapping volume $vol(\mathsf{Trap}(SN_\epsilon))$ depends only on the following four quantities:

(1) $\mathcal{L}_\epsilon^{\mathsf{av}}(M, \alpha(K))$,
(2) $\int_{S\mathsf{T}(\alpha(K),\epsilon)} \beta_g \wedge \omega_g^{n-1}$,
(3) $\int_{\partial_1^+(SM)} \omega_g^{n-1}$,
(4) $\int_{SM} \beta_g \wedge \omega_g^{n-1}$.

Here the first quantity is *observable* and *extrinsic* to K, the second is *intrinsic* to K, and the last two quantities are α- and K-*independent.*

Finally, by [St], Theorem 1.2, the volume $vol(\mathsf{Trap}(SN_\epsilon))$ of the trapping region depends continuously on the embedding $\partial\mathsf{T}(K, \epsilon) \hookrightarrow M$ and, therefore, on the embedding $K \hookrightarrow M$ and on ϵ, provided that $\epsilon > 0$ is sufficiently small. □

[19]The quantity in (i) is K-dependent, but, by Theorem 4.24, it is α-independent.

Example 4.10. Let $\alpha : K \hookrightarrow \mathbb{E}^3$ be a link or a knot. Of course, in such cases, K has no interesting intrinsic geometry, but it has its total length $\ell(K)$. Then the polynomial

$$Q_K(\epsilon^2) = \frac{1}{\epsilon^2}\Big[vol(M) - \frac{1}{\sigma_2}\mathcal{V}(\epsilon)\Big].$$

equals the constant $\pi \cdot \ell(K)$, where where $\sigma_2 = 4\pi$, the volume of the standard unit 2-sphere. This leads to a representation of the ϵ-trapping volume of K as a sum of one observable and two intrinsic quantities:

$$\mathcal{W}(\epsilon) = -\Big(\int_{\{\partial_1^+(SM)\setminus \mathsf{Trap}(\partial M^{\rightsquigarrow}, SN_\epsilon)\}} L_\epsilon \cdot \omega_g \wedge \omega_g\Big) - 4\pi \cdot vol(M) - 4\pi^2\epsilon^2\ell(K).$$

It seems that one can construct an example of a loop $K \subset \mathbb{E}^3$ such that $\mathcal{W}(\epsilon) > 0$ for some "sufficiently big" $\epsilon > 0$ (see Fig. 4.7). However, if $\mathcal{W}(\epsilon) = 0$ for a small $\epsilon > 0$, then $\ell(K)$ is determined by the one-layered scattering:

$$\ell(K) = \frac{1}{4\pi^2\epsilon^2}\{4\pi \cdot vol(M) - \mathcal{V}(\epsilon)\}.$$

$\diamondsuit$

Corollary 4.26. *Let M be an n-dimensional smooth domain in $\mathbb{E}^n$, and let K be a compact closed smooth k-dimensional submanifold of M. We consider Weyl tubes $\mathsf{T}(K, \epsilon)$ as in Theorem 4.26.*

Then there exists a constant $c(K) > 0$, which depends only on the intrinsic geometry of K (in particular, $c(K)$ is ϵ-independent and M-independent) and such that for all sufficiently small $\epsilon > 0$, the average length/travel time $\mathcal{L}^{\mathsf{av}}_\epsilon(M, K)$ of billiard trajectories in N_ϵ (see (4.46)) can detect the volume

$$vol(\mathsf{Trap}(SN_\epsilon)) =_{\mathsf{def}} \int_{\mathsf{Trap}(SN_\epsilon)} \beta_g \wedge (\omega_g)^{n-1}$$

of the trapping region $\mathsf{Trap}(SN_\epsilon)$ with the precision $c(K) \cdot \epsilon^{n-k}$.

Proof. The claim follows from the formula (4.44) by bounding from above the absolute value $|Q_K(\epsilon^2)|$, say, in the interval $[0, 1]$, provided $\epsilon^2 < 1$. Note that $c(K)$ depends only on the inner geometry of K, and thus, it is shared by all isometric embeddings $\alpha : K \hookrightarrow M$. $\square$

The next corollary shows that only the knowledge of the average lengths/travel times $\{\mathcal{L}^{\mathsf{av}}_{\epsilon_j}(M, K)\}_{0 \leq j \leq \lceil k/2 \rceil}$ is needed in order to determine the global Weyl invariants of K, provided that Hypothesis W is valid.

Corollary 4.27. *Let K be a closed smooth k-dimensional Riemanian manifold and M is a smooth compact domain in the Euclidean space $\mathbb{E}^n$.*

Then, assuming the Hypothesis W*, the average length (travel time) $\mathcal{L}^{\mathsf{av}}_\epsilon(M,K)$ of the non-trapped billiard trajectories in N_ϵ which originate and terminate in ∂M* [20] *is shared by all* isometric *embeddings $\alpha : K \hookrightarrow M \setminus \mathsf{T}_\epsilon(\partial M)$. In fact,*

$$\mathcal{L}^{\mathsf{av}}_\epsilon(M,K) = \frac{\int_{SM} \beta_g \wedge \omega_g^{n-1} - \int_{S\mathsf{T}(\alpha(K),\,\epsilon)} \beta_g \wedge \omega_g^{n-1}}{\int_{\partial_1^+(SM)} \omega_g^{n-1}}, \tag{4.47}$$

where the quantity $\int_{S\mathsf{T}(\alpha(K),\,\epsilon)} \beta_g \wedge \omega_g^{n-1}$ depends only on the intrinsic geometry of K (and not on its isometric embedding α).

Moreover, for any sequence $\epsilon_0 > \epsilon_1 > \cdots > \epsilon_{\lceil k/2 \rceil} > 0$, where ϵ_0 is sufficiently small, the average lengths/travel times $\{\mathcal{L}^{\mathsf{av}}_{\epsilon_j}(M,K)\}_{0 \leq j \leq \lceil k/2 \rceil}$ of the non-trapped billiard trajectories in $\{N_{\epsilon_j}\}_{0 \leq j \leq \lceil k/2 \rceil}$ determine the intrinsic global invariants

$$\Big\{\mathsf{Q}_\ell = \int_K H_\ell\, \mu_K\Big\}_{0 \leq \ell \leq \lceil k/2 \rceil}$$

of the Riemannian manifold K, defined by the formulae (4.36)-(4.37).

Proof. By Theorem 1.6.2 in [LP], the Lesbegue measure of the locus $\mathsf{Trap}(\partial M \rightsquigarrow, SN_\epsilon)$ is zero. Thus the denominator of (4.46) equals

$$\int_{\{\partial_1^+(SM) \setminus \mathsf{Trap}(\partial M \rightsquigarrow,\, SN_\epsilon)\}} \omega_g^{n-1} = \int_{\{\partial_1^+(SM)\}} \omega_g^{n-1}.$$

It is a quantity that does not depend on α and K. In fact, by the Fubini theorem, $\int_{\{\partial_1^+(SM)\}} \omega_g^{n-1} = \theta_{n-1} \cdot vol(\partial M)$, where θ_{n-1} denotes the volume of the unit $(n-1)$-ball. Therefore, the numerator in (4.47) depends on the average length $\mathcal{L}^{\mathsf{av}}_\epsilon(\alpha(K))$ and $vol(\partial M)$.

On the other hand, by Theorem 4.24, the volume of the ϵ-tube $\mathsf{T}(\alpha(K), \epsilon)$ depends only on the intrinsic geometry of K. Thus, by the Fubini theorem, the volume of $S\mathsf{T}(\alpha(K), \epsilon) \subset SM$ depends only on the intrinsic geometry of K as well.

Therefore, the formula (4.42) implies (4.47). Note that $\mathcal{L}^{\mathsf{av}}_\epsilon(M,K)$ is an *observable* quantity.

The last claim follows from Theorem 4.26. □

[20]with possible multiple reflections from $\partial \mathsf{T}(K, \epsilon)$ in-between

4.9.4 *Approximating the Weyl tubes by tamed bubbling tubes*

As before, we assume that the obstacle $K \subset M \subset \mathbb{E}^n$ is a closed smooth k-dimensional manifold and M is a compact domain with a smooth boundary ∂M. We assume that $\mathsf{T}(K, \epsilon) \to K$ is a fibration whose fibers are the $(n-k)$-dimensional ϵ-balls, orthogonal to K. Thus the Weyl tube $\mathsf{T}(K, \epsilon) \subset M$ is the union of all ϵ-balls, whose centers belong to K. We assume also that all ϵ are such that $\mathsf{dist}(K, \partial M) > \epsilon$.

We do not know whether Hypothesis W generally holds. Therefore, we need to change the notion of a tube so that this hypothesis will become valid for the modified tubes.

Let us adopt a definition from the paper of Burago, Ferleger, and Kononenko [BFK].

Definition 4.20. Let $\mathcal{U}$ be a finite collection of convex domains $\{B_i\}_{i \in [1,m]}$ in a smooth Riemannian manifold (M, g). Let $B =_{\mathsf{def}} \bigcup_{i=1}^m B_i$.

Then $\mathcal{U}$ is called non-degenerate in an open set $U \subset M$, if for any multi-index $I \subset \{1, 2, \dots, m\}$ and any $y \in U \cap (B \setminus \bigcap_{i \in I} B_i)$,

$$\max_{k \in I} \left\{ \frac{\mathsf{dist}(y, B_k)}{\mathsf{dist}(y, \bigcap_{j \in I} B_j)} \right\} \geq c, \quad \text{for a constant } c > 0.$$

A collection $\mathcal{U}$ is called non-degenerate, if there exist $\delta > 0$ and constant $c > 0$ so that $\mathcal{U}$ is non-degenerate with the constant $c > 0$ in any δ-ball. ◇

Speaking informally, this means that if a point in $U \cap (M \setminus B)$ is d-close to all the walls $\{\partial B_i\}_{i \in I}$ then it is (d/c)-close to their intersection.

Definition 4.21. A tamed bubbling tube $\mathbf{T}(K, \epsilon) \subset \mathsf{T}(K, \epsilon)$ is a finite union $\bigcup_{j=1}^q B_\epsilon(x_j)$ of closed ϵ-balls, whose distinct centers $x_j \in K$ and which satisfy the following properties:

- $\mathbf{T}(K, \epsilon) \cap K = K$,
- the system of balls $\{B_\epsilon(x_j)\}_{j \in [1,q]}$ is nondegenerate in the sense of Definition 4.20,
- any two intersecting spheres $\partial B_\epsilon(x_j)$ and $\partial B_\epsilon(x_l)$, $j \neq l$, intersect transversally.

We call the number

$$\rho(\mathbf{T}(K, \epsilon)) =_{\mathsf{def}} 1 - \frac{vol(\mathbf{T}(K, \epsilon))}{vol(\mathsf{T}(K, \epsilon))} \in [0, 1) \tag{4.48}$$

the roughness of the tamed bubbling tube $\mathbf{T}(K, \epsilon)$. The bigger $\rho(\mathbf{T}(K, \epsilon))$ is, the "rougher" is the tube. ◇

Clearly, by adding new ϵ-balls with the centers in K, one can increase the volume $vol(\mathbf{T}(K, \epsilon)) \leq vol(\mathsf{T}(K, \epsilon))$, while keeping new system of balls nondegenerate (see Fig. 4.9).

In what follows, we replace the original scattering problem for the smooth manifold $N_\epsilon = M \setminus \mathsf{T}(K, \epsilon)$ with a similar scattering problem for the manifold $\mathbf{N}_\epsilon =_{\mathsf{def}} M \setminus \mathbf{T}(K, \epsilon)$ with *corners*. The main advantage of this replacement is that the billiard in $\mathbf{N}_\epsilon$ is dispersing and, therefore, has zero non-trapping volume (see Lemma 4.11). This conclusion is in line with the results of, i.g., [BFK] which claim, in particular, that the trajectories of nondegenerate (see Definition 4.20) semi-dispersing billiards have a finite number of collisions with the boundary, before escaping to infinity.

At the same time, we can control the volume $vol(\mathbf{T}(K, \epsilon))$, in terms of the intrinsic geometry of K, with accuracy that depends on the roughness $\rho(\mathbf{T}(K, \epsilon))$ from (4.48).

Following [BFK], we declare any point of the intersection $\partial B_\epsilon(x_i) \cap \partial B_\epsilon(x_j)$, $i \neq j$, reached by a billiard trajectory γ in $\mathbf{N}_\epsilon$, to be *the terminal point* on γ. Since such multiple intersections form a measure zero set in the boundary $\partial \mathbf{T}(K, \epsilon)$, the measure of the points in $S\mathbf{N}_\epsilon$, visited by the lifts $\tilde{\gamma}$ of such γ, vanishes as well. Therefore, their contributions vanish in all the integral formulas to follow.

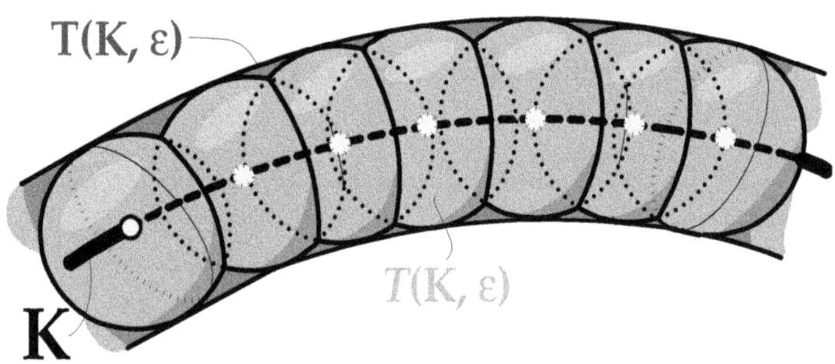

Fig. 4.9 Approximating a Weyl tube $\mathsf{T}(K, \epsilon)$ by a tame bubbling tube $\mathbf{T}(K, \epsilon)$.

Lemma 4.11. *Let* $\mathbf{T}(K, \epsilon) \subset M$ *be a tame bubbling tube. Consider the billiard with the table* $\mathbf{N}_\epsilon =_{\mathsf{def}} M \setminus \mathbf{T}(K, \epsilon)$. *Then the measure of the trapping set* $\mathsf{Trap}(S\mathbf{N}_\epsilon) \subset S\mathbf{N}_\epsilon$ *is zero.*

Proof. The set $\partial \mathbf{T}(K, \epsilon) \subset \mathbb{E}^n$ consists of pieces of $(n-1)$-spheres in $\mathbb{E}^n$. Each such piece is a manifold with corners. Indeed, it is possible to choose a tame bubbling tube so that any group of spheres intersects transversally (so that Definition 4.20 is satisfied).

Without loss of generality, we may also assume that $\mathbf{T}(K, \epsilon)$ is contained inside of a n-cube $Q \subset \mathbb{E}^n$, whose interior contains M. We view this cube as the fundamental domain for a flat n-torus T^n. Thus, we may assume that $\mathbf{T}(K, \epsilon)$ resides in the torus T^n.

Consider now the billiard on the table $\mathbf{P}(K, \epsilon) =_{\mathsf{def}} T^n \setminus \mathsf{int}(\mathbf{T}(K, \epsilon))$. The billiard $\mathbf{P}(K, \epsilon)$ is dispersing: i.e., each boundary component is convex inward. Indeed, $\partial\mathbf{P}(K, \epsilon)$ consists of pieces of spheres. However, it is not a Sinai billiard, because the boundary has singularities, which consists of the sets where at least two spheres intersect. Such singularities in the theory of billiards are called corner singularities. Thanks to transversality of such intersections and to the fact that the number of spheres in $\partial\mathbf{T}(K, \epsilon)$ is finite, all the angles between the intersecting spheres exceed some value $\alpha > 0$.

Actually, there are two types of singularities in our billiards $\mathbf{P}(K, \epsilon)$. Only one type of singularities, $\Sigma^{\mathsf{tan}} \subset \partial_1 S\mathbf{P}(K, \epsilon)$, the tangent singularities, is present in Sinai billiards, where the boundary is dispersing (convex inward) and smooth. Note that $\dim \Sigma^{\mathsf{tan}} = 2n - 3$. However, in billiards we consider, the boundary itself has the corner singularities $\Sigma^{\vee} \subset \partial_1 S\mathbf{P}(K, \epsilon)$, also of dimension $2n - 3$. It is well known that ergodic properties of billiards are essentially determined by the time evolution of the singular set $\Sigma =_{\mathsf{def}} \Sigma^{\mathsf{tan}} \bigcup \Sigma^{\vee}$ under the iterations of the billiard map

$$\mathcal{B} =_{\mathsf{def}} \mathcal{B}(K, \epsilon) : \partial_1^+ S\mathbf{P}(K, \epsilon) \to \partial_1^+ S\mathbf{P}(K, \epsilon).$$

Thus, the $\mathcal{B}$-invariant set $\bigcup_{i=-\infty}^{+\infty} \mathcal{B}^{(i)}(\Sigma)$ is everywhere dense in $\partial_1^+ S\mathbf{P}(K, \epsilon)$ [CM].

Recall that Sinai billiards are ergodic [Si], [KSS]. For dispersing billiards, their ergodicity can be derived from, e.g., [KSS]. We will make several comments in order to explain why the billiards in the class $\mathbf{P}(K, \epsilon)$ are ergodic.

The main fact is that the images under positive and negative iterations of the billiard map $\mathcal{B}$ of any smooth piece of the singularity set Σ do not locally coincide with any other piece of the singularity set [CM], [SiC], [KSS]. The last means that corresponding sets intersect transversally. Indeed, consider the, so-called, wave fronts (see [CM]), i.e., local (small) manifolds, orthogonal to narrow beams of rays (billiard orbits).

A smooth portion of the complement to the singularity set Σ can be smoothly foliated by such strictly convex future wave fronts (i.e., having a strictly positive curvature tensor). One just needs to perturb a unit velocity vector arbitrarily (the perturbation is again a unitary vector field), so that the foot points of the perturbed vectors remain the same. In other words, we consider a beam of rays emanating from a point in a billiard table. Because of convexity of the boundary, these strictly convex wave fronts remain strictly convex under billiard dynamics [CM].

Tamed bubbling tubes have strictly dispersing billiards with only corner-type singularities. In addition to that, at the corners, the opening spatial angles between the spheres forming such tubes are always positive. There are also singularities arising because of tangencies of billiard orbits to the boundary of the tamed bubbling tubes. However, such singularities are not dangerous for our purposes [BS]. Recall also that in dispersing billiards, the set of orbits hitting (going through) singularities has a zero measure. However, the corner singularities must be studied separately. One need to show that a future image of a singularity does not exactly coincide (not even locally) with a singularity. The reason is that these singularities can be smoothly foliated with strictly convex future wave fronts, i.e., with a strictly positive curvature tensor. The construction of these wave fronts is the so-called candle construction ([KSS]). One just needs to perturb only the unit velocity vector arbitrarily, but so that it stays unit, while the footpoint of the velocity vector stays fixed. Because the billiard is strictly dispersing, these strictly convex wave fronts stay strictly convex in the future. Similarly, the backward images of singularities can be foliated by the "inverse candles", i.e., strictly concave wave fronts with a strictly negative definite curvature form. These strictly convex and concave fronts are transversal to each other at any intersection point, and therefore, ergodicity follows ([SiC]).

The ergodicity of the billiard maps $\mathcal{B}(K, \epsilon)$ follows, since, except for a set of zero measure and topological codimension at least two, there is a hyperbolic structure in the complement to the singular set and its iterations in $\partial_1^+ S\mathbf{P}(K, \epsilon)$; that is, local stable and unstable manifolds do exist almost everywhere. For such points, the conditions of the theorem on local ergodicity [SiC] are satisfied. Recall that a dynamical system is called locally ergodic if for almost every point there exists a neighborhood which belongs to one ergodic component.

A global ergodicity, i.e., the existence of a single ergodic component of full measure, also follows. Indeed, observe that the set of points, which satisfy the local ergodicity, is linearly connected and is at least of codimension two.

Finally, contrary to the lemma claim, assume that the invariant set $A =_{\mathsf{def}} \mathsf{Trap}(S\mathbf{N}_\epsilon)$ has a positive measure. Then A is an invariant positive measure set in the spherical fibration $SQ \to Q$ over the cube $Q \supset M \supset \mathbf{N}_\epsilon$. The set A is invariant under the factor map $S\mathbb{E}^n \to ST^n$, and clearly, it is not of a full measure, since there are open families of lines through points of M that miss $\mathbf{P}(K,\epsilon) \subset \mathsf{int}(M)$. On the other hand, a strictly dispersing billiard $\mathbf{P}(K,\epsilon)$ on a torus T^n with scatterers $\partial\mathbf{T}(K,\epsilon)$ is ergodic [Si], [SiC]. Thus we came to a contradiction, which proves lemma. □

Remark 4.12. In our billiard $\mathbf{P}(K,\epsilon)$, the scatters are pieces of spheres. Therefore, they are real semi-algebraic sets, and thus the results of [SiC], [KSS] apply directly. ◇

Theorem 4.28. *Let K be a closed smooth k-dimensional submanifold of the Euclidean space $\mathbb{E}^n$. Consider a sequence $\epsilon_0 > \epsilon_1 > \cdots > \epsilon_{\lceil k/2\rceil} > 0$, where ϵ_0 is sufficiently small. Let $\{\mathbf{T}(K,\epsilon_j)\}_j$ be tame bubbling tubes, contained in a smooth compact domain $M \subset \mathbb{E}^n$ so that K is at least ϵ_0-away from ∂M. Consider the domains $\{\mathbf{N}_{\epsilon_j} =_{\mathsf{def}} M \setminus \mathsf{int}(\mathbf{T}(K,\epsilon_j))\}_j$.*

Then, the averages $\{\mathbf{L}^{\mathsf{av}}_{\epsilon_j}\}_{0\le j\le\lceil k/2\rceil}$ of the lengths (travel times) functions

$$\left\{\mathbf{L}_{\epsilon_j} : \partial_1^+(SM) \to \mathbb{R}\right\}_{0\le j\le\lceil k/2\rceil}$$

of the billiard trajectories (which originate and terminate in ∂M) in the domains $\{\mathbf{N}_{\epsilon_j}\}_{0\le j\le\lceil k/2\rceil}$, together with the roughness coefficients $\{\rho_j =_{\mathsf{def}} \rho(\mathbf{T}(K,\epsilon_j))\}_{0\le j\le\lceil k/2\rceil}$ of the tame bubbling tubes (see (4.48)), determine the intrinsic global invariants

$$\left\{\mathsf{Q}_\ell = \int_K H_\ell\, \mu_K\right\}_{0\le \ell\le\lceil k/2\rceil}$$

of the Riemannian manifold K, defined by the formulas (4.36)-(4.37).

Proof. Since, by Lemma 4.11, the measure of $\mathsf{Trap}(S\mathbf{N}_{\epsilon_j})$ is zero, we get $vol(\mathsf{Trap}(S\mathbf{N}_{\epsilon_j})) = 0$. Therefore,

$$\int_{\partial_1^+(SM)} \mathbf{L}_{\epsilon_j}\, \omega_g^{n-1} = \int_{S\mathbf{N}_{\epsilon_j}} \beta_g \wedge \omega_g^{n-1} = vol(SM) - vol(S\mathbf{T}(K,\epsilon_j))$$

By the definition of $\mathbf{L}^{\mathsf{av}}_{\epsilon_j}$, we get

$$\mathbf{L}^{\mathsf{av}}_{\epsilon_j} \cdot \int_{\partial_1^+(SM)} \omega_g^{n-1} = vol(SM) - vol(S\mathbf{T}(K,\epsilon_j))$$

By the Fubini theorem, the formula above transforms as

$$\mathbf{L}^{\mathsf{av}}_{\epsilon_j} \cdot \theta_{n-1} \cdot vol(\partial M) = \sigma_{n-1} \cdot \left[vol(M) - vol(\mathbf{T}(K,\epsilon_j))\right],$$

where σ_{n-1} denotes the volume of the unit $(n-1)$-sphere in $\mathbb{E}^n$ and θ_{n-1} the volume of the unit $(n-1)$-ball in $\mathbb{E}^{n-1}$.

By the definition of $\rho_j = \rho(\mathbf{T}(K, \epsilon_j))$ and putting $\hat{\rho}_j =_{\mathsf{def}} 1 - \rho_j$, we get

$$\mathbf{L}^{\mathsf{av}}_{\epsilon_j} \cdot \theta_{n-1} \cdot vol(\partial M) = \sigma_{n-1} \cdot \big[vol(M) - \hat{\rho}_j \cdot vol(\mathsf{T}(K, \epsilon_j))\big].$$

Solving for $vol(\mathsf{T}(K, \epsilon_j))$ leads to

$$vol(\mathsf{T}(K, \epsilon_j)) = \hat{\rho}_j^{-1}\big\{vol(M) - [\theta_{n-1}/\sigma_{n-1}] \cdot vol(\partial M) \cdot \mathbf{L}^{\mathsf{av}}_{\epsilon_j}\big\}. \quad (4.49)$$

According to (4.35), $vol(\mathsf{T}(K, \epsilon))$ is given by a polynomial $P_K(\epsilon)$ of the form $\epsilon^{n-k} \cdot Q_K(\epsilon^2)$, where $Q_K(x)$ is a polynomial of degree $\lceil k/2 \rceil$ in x. Therefore, using (4.49),

$$Q_K(\epsilon_j^2) = \hat{\rho}_j^{-1} \epsilon_j^{k-n} \Big\{vol(M) - [\theta_{n-1}/\sigma_{n-1}] \cdot vol(\partial M) \cdot \mathbf{L}^{\mathsf{av}}_{\epsilon_j}\Big\}. \quad (4.50)$$

Note that for a relatively smooth bubbling tube $\mathbf{T}(K, \epsilon_j)$ the "smoothness" coefficient $\hat{\rho}_j^{-1} =_{\mathsf{def}} \frac{vol(\mathbf{T}(K,\epsilon))}{vol(\mathsf{T}(K,\epsilon))}$ is close to 1. Assuming that all these roughness coefficients $\{\hat{\rho}_j\}_j$ are equal, i.e., j-independent, will simplify our computations.

Next, we fix a sequence $\epsilon_0 > \epsilon_1 > \cdots > \epsilon_{\lceil k/2 \rceil} > 0$ and apply the Lagrange interpolation formula

$$Q_K(u^2) = \sum_{i=0}^{\lceil k/2 \rceil} \Big(\prod_{0 \leq j \leq \lceil k/2 \rceil,\, j \neq i} \frac{u^2 - \epsilon_j^2}{\epsilon_i^2 - \epsilon_j^2} \Big) Q_K(\epsilon_i^2)$$

to reconstruct the polynomial

$$Q_K(u^2) =_{\mathsf{def}}$$

$$\theta_{n-k} \times \sum_{\{\ell \in [0,k],\, \ell \equiv 0 \mod 2\}} \frac{\mathsf{Q}_\ell(K)}{(n-k+2)(n-k+4)\dots(n-k+\ell)}\, u^\ell$$

from the "observables" $\{Q_K(\epsilon_j^2)\}_j$ in (4.50).

Now, we can conclude that the collection of numbers $\{\mathbf{L}^{\mathsf{av}}_{\epsilon_j}\}_{0 \leq i \leq \lceil k/2 \rceil}$ determines a polynomial $Q_K(\epsilon^2)$, and thus, the polynomial $P_K = \epsilon^{n-k} \cdot Q_K$, given by the Weyl formula (4.35). As a result, the intrinsic integral invariants $\{\mathsf{Q}_\ell(K)\}$ in (4.36) are determined by $\{\mathbf{L}^{\mathsf{av}}_{\epsilon_j}\}_{0 \leq i \leq \lceil k/2 \rceil}$. □

Conjecture 4.8. *Let K be a closed smooth k-dimensional submanifold of the flat torus T^n, $k \leq n-2$. Consider a sequence of numbers $\{\epsilon_0 > \epsilon_1 > \cdots > \epsilon_{\lceil k/2 \rceil} > 0\}$ such that $\{\mathsf{T}(K, \epsilon_j)\}_j$ are the Weyl tubes, properly contained in a smooth compact connected domain $M \subset T^n$. Assume that* *the natural homomorphism $\pi_1(M) \to \pi_1(T^n)$ of the fundamental groups is trivial*[21] *and the billiards $P(K, \epsilon_j) =_{\mathsf{def}} T^n \setminus \mathsf{int}(\mathsf{T}(K, \epsilon_j))$ are ergodic.*

Then the averages $\{\mathcal{L}^{\mathsf{av}}_{\epsilon_j}\}_{0 \leq j \leq \lceil k/2 \rceil}$ of lengths (travel times) functions

$$\left\{\mathcal{L}_{\epsilon_j} : \partial_1^+(SM) \to \mathbb{R}\right\}_{0 \leq j \leq \lceil k/2 \rceil}$$

of the billiard trajectories (which originate and terminate in ∂M) of the billiards $\left\{N(K, \epsilon_j) =_{\mathsf{def}} M \setminus \mathsf{T}(K, \epsilon_j)\right\}_{0 \leq j \leq \lceil k/2 \rceil}$ determine intrinsic global invariants

$$\left\{\mathsf{Q}_\ell =_{\mathsf{def}} \int_K H_\ell \, \mu_K\right\}_{0 \leq \ell \leq \lceil k/2 \rceil}$$

of the Riemannian manifold K, defined by the formulas (4.36)-(4.37). $\diamondsuit$

4.9.5 *Some concluding remarks and problems*

Weyl's Theorem 4.24 has versions for the volumes of tubes in the *spherical* and *projective* spaces, both real and complex. For domains M in these spaces, such that the metric on M is non-trapping, many of our arguments seem to hold.

In contrast, for compact domains M in the hyperbolic spaces $\mathbb{H}^n$, the Weyl equations for the volumes of ϵ-tubes turn into less informative *inequalities* [Gray]. As a result, for domains in $\mathbb{H}^n$, one may hope just to get some *upper estimates* of the volume $vol(\mathsf{T}(K, \epsilon))$ and of the trapping volume $vol(\mathsf{Trap}(SN_\epsilon))$ in terms of the scattering data, as it is done in Theorem 4.27. Since by [We] isometric embeddings of K produce ϵ-tubes of the same volume, we conjecture that these estimates may be shared by all isometric embeddings $\alpha : K \hookrightarrow M \setminus \mathsf{T}(\partial M, \epsilon)$.

The next problem seems to be quite challenging.

Problem 4.1. Let M be a compact Riemannian manifold with the boundary, equipped with a non-trapping metric g. Let K be a closed Riemannian manifold which admits an isometric imbedding in M. For a given (small) $\epsilon > 0$, find the isometric embeddings $\alpha_\star : K \hookrightarrow M$, for which the volume of the trapping set $\mathsf{Trap}(SN_\epsilon(\alpha_\star(K)))$ attains its infimum/minimum.

Clearly, the isometry group of M acts on such optimal embeddings $\alpha_\star$. Thus, the optimal isometric embedding may not be unique.

Note that the problem of finding such optimal $\alpha_\star$'s is equivalent to the problem of finding an isometric embeddings $\alpha_\star : K \hookrightarrow M$ for which the

[21]This assumption implies that the flat metric on M, induced by the metric on the ambient torus, is non-trapping.

average length of a billiard trajectory $\mathcal{L}^{\mathsf{av}}\big(N_\epsilon(\alpha(K))\big)$ in (4.47) attains a supremum. $\diamondsuit$

4.10 Applying Gromov's amenable localization to the traversally generic geodesic flows

This section bridges between the two parts of this book: its results deal with traversing geodesic flows, the main actors of Part I, and homological methods of Part II. The contents of this section can be found in [K10]. In turn, the latter work is based heavily on [AK] and [K6].

We are now in position to apply Gromov's amenable localization techniques ([Gr], [Gr1]) to the study of traversally generic Riemannian metrics (see Definition 4.3) on smooth compact manifolds M with boundary. As in [K6] and [K10], we will focus on the combinatorics of tangency patterns, exhibited by the geodesic curves with respect to the boundary ∂M. We will employ Gromov's simplicial semi-norm (see Definition 4.8) to get some lower bounds of the combinatorial complexity of geodesic flows.

This section is a version of Section 8.3 from the book [K7], being adjusted and applied to the traversally generic geodesic flows.

For a smooth compact n-dimensional manifold M with boundary, we form its double $DM =_{\mathsf{def}} M \cup_{\partial M} M$. We denote by $\rho_M : DM \to DM$ a smooth involution such that the orbit-space $DM/\{\rho_M\} = M$ and $(DM)^\rho = \partial M$. If M is oriented, then the fundamental cycle $[DM]$ of DM is represented by $[M] + (\rho_M)_*[-M]$.

Let $D(SM) =_{\mathsf{def}} SM \cup_{(SM|_{\partial M})} SM$ be the double of SM. Note that $D(SM)$ fibers over DM with the spherical fiber S^{n-1}; however, this spherical fibration is *not* the tangent spherical fibration of the manifold DM! For instance, $D(SM) \to DM$ admits a continuous section.

Evidently, ρ_M generates an involution $\rho_{SM} : D(SM) \to D(SM)$ such that $D(SM)/\{\rho_{SM}\} = SM$.

Let M be a compact connected smooth Riemannian n-manifold with boundary, equipped with a *traversally generic* metric g. Its traversally generic geodesic flow v^g (see Definition 2.6), defines a stratification of the space of the tangent spherical fibration $SM \to M$ by the pure strata $\{SM(v^g, \omega)\}_\omega$, where the combinatorial pattern ω from the poset $\mathbf{\Omega}^\bullet_{|\sim|' \leq 2n-2}$ (see Definition 2.5) labels the combinatorial types of tangency of the v^g-flow to the boundary $\partial_1(SM)$. Let

$$SM^\circ(v^g, \omega) =_{\mathsf{def}} SM(v^g, \omega) \cap \mathsf{int}(SM).$$

In turn, these pure strata generate the filtration

$$\Big\{SM^\circ_{-(k+1)}(g) =_{\mathsf{def}} \bigcup_{\omega \in \mathbf{\Omega}^\bullet \,\big|\, |\omega|' \geq k+1} SM^\circ(v^g, \omega)\Big\}_k \tag{4.51}$$

of $SM \setminus \partial_1(SM)$ and the filtration

$$\Big\{SM_{-(k+1)}(g) =_{\mathsf{def}} \Big(\bigcup_{\omega \in \mathbf{\Omega}^\bullet \,\big|\, |\omega|' \geq k+1} SM^\circ(v^g, \omega)\Big)$$
$$\bigcup \Big(\bigcup_{\omega \in \mathbf{\Omega}^\bullet \,\big|\, |\omega|' \geq k} \partial_1(SM) \cap SM(v^g, \omega)\Big)\Big\}_k \tag{4.52}$$

of SM. Both filtrations are produced by the ω-labeled strata of a particular dimension.

The filtration $\{SM_{-(k+1)}(g)\}$ from (4.52) induces the ρ_{SM}-equivariant filtration $\{D(SM)_{-(k+1)}(g)\}$ of the double $D(SM)$.

Given a commutative ring R, we consider the free R-modules:

$$\mathsf{C}^{2n-1-j}_{\mho}(D(SM), g;\, \mathsf{R}) =_{\mathsf{def}} H^{2n-1-j}\big(D(SM)_{-j}(g),\, D(SM)_{-(j+1)}(g);\, \mathsf{R}\big)$$

and

$$\mathsf{C}^{2n-1-j}_{\mho}(SM^\circ, g;\, \mathsf{R}) =_{\mathsf{def}}$$
$$H^{2n-1-j}\Big(SM_{-j}(g),\; SM_{-(j+1)}(g) \cup \big(SM_{-j}(g) \cap \partial_1(SM)\big);\, \mathsf{R}\Big).$$

Therefore, each traversally generic metric g on M gives rise to a differential complex

$$\mathbf{C}^*_{\mho}(D(SM), g;\, \mathsf{R}) =_{\mathsf{def}} \tag{4.53}$$
$$\Big\{0 \to \mathsf{C}^0_{\mho}(D(SM), g;\, \mathsf{R}) \xrightarrow{\delta_0} \mathsf{C}^1_{\mho}(D(SM), g;\, \mathsf{R}) \xrightarrow{\delta_1} \dots$$
$$\dots \xrightarrow{\delta_{2n-2}} \mathsf{C}^{2n-1}_{\mho}(D(SM), g;\, \mathsf{R}) \to 0\Big\},$$

where the differentials $\{\delta_j\}$ are the boundary homomorphisms from the long exact cohomology sequences of the triples

$$\big\{D(SM)_{-(j-1)}(g) \supset D(SM)_{-j}(g) \supset D(SM)_{-(j+1)}(g)\big\}_j.$$

Similarly, the traversally generic g produces the differential complex

$$\mathbf{C}^*_{\mho}(SM^\circ, g;\, \mathsf{R}) =_{\mathsf{def}} \tag{4.54}$$
$$\Big\{0 \to \mathsf{C}^0_{\mho}(SM^\circ, g;\, \mathsf{R}) \xrightarrow{\delta_0} \mathsf{C}^1_{\mho}(SM^\circ, g;\, \mathsf{R}) \xrightarrow{\delta_1} \dots$$
$$\dots \xrightarrow{\delta_{2n-2}} \mathsf{C}^{2n-1}_{\mho}(SM^\circ, g;\, \mathsf{R}) \to 0\Big\},$$

where the differentials $\{\delta_j\}$ are the boundary homomorphisms from the long exact homology sequences of the triples

$$\big\{SM_{-(j-1)}(g) \cup \partial_1(SM) \supset SM_{-j}(g) \cup \partial_1(SM) \supset SM_{-(j+1)}(g) \cup \partial_1(SM)\big\}_j.$$

We denote by

$$\mathsf{B}^{2n-1-j}_{\mho}(SM^\circ, g;\, \mathsf{R}) \subset \mathsf{C}^{2n-1-j}_{\mho}(SM^\circ, g;\, \mathsf{R})$$

the image of the differential δ_{2n-2-j} from (4.54). Similarly, let

$$\mathsf{B}^{2n-1-j}_{\mho}(D(SM), g;\, \mathsf{R}) \subset \mathsf{C}^{2n-1-j}_{\mho}(D(SM), g;\, \mathsf{R})$$

stand for the image of the differential δ_{2n-2-j} from (4.53).

Thus, for any commutative ring R, any traversally generic Riemannian metric g on a compact connected smooth n-manifold M with boundary generates the differential complexes $\mathbf{C}^*_{\mho}(SM^\circ, g; \mathsf{R})$ and $\mathbf{C}^*_{\mho}(D(SM), g;\, \mathsf{R})$ of free R-modules.

Conjecture 4.9. *The homology groups of the differential complexes* $\mathbf{C}^*_{\mho}(SM^\circ, g;\, \mathsf{R})$ *and* $\mathbf{C}^*_{\mho}(D(SM), g;\, \mathsf{R})$ *depend only on the connected component of the space of traversally generic metrics on* M *to which* g *belongs.* ◇

Theorem 4.29. *Let* g *be a traversally generic Riemmanian metric on a compact connected smooth* n*-manifold* M *with boundary.*

The differential complexes $\mathbf{C}^*_{\mho}(SM^\circ, g;\, \mathsf{R})$ *and* $\mathbf{C}^*_{\mho}(D(SM), g;\, \mathsf{R})$ *in (4.54) and (4.53) are "holographic": i.e., they can be reconstructed from the geodesic scattering map* $C_{v^g} : \partial_1^+(SM) \to \partial_1^-(SM)$.

Proof. Any traversally generic geodesic vector field v^g is boundary generic and traversing. If the geodesic field v^g is boundary generic and traversing, then by Corollary 4.2, the scattering map $C_{v^g} : \partial_1^+(SM) \to \partial_1^-(SM)$ allows for a reconstruction of the pair $(SM, \mathcal{F}(v^g))$, up to a homeomorphism $\Phi : SM \to SM$ which is the identity on the boundary $\partial_1(SM)$.

The homeomorphism Φ preserves the refined stratification $\mathcal{S}^\bullet_{v^g}(SM)$, whose strata are the connected components of the stratification

$$\mathcal{S}_{v^g}(SM) = \big\{\partial_1(SM)(v^g, \omega),\ SM^\circ(v^g, \omega)\big\}_\omega.$$

Similarly, the connected components of the stratification $\mathcal{S}_{v^g}(D(SM))$ give rise to the more refined stratification $\mathcal{S}^\bullet_{v^g}(D(SM))$ of $D(SM)$.

Therefore, the *topological types* of the stratifications $\mathcal{S}^\bullet_{v^g}(SM^\circ)$ and $\mathcal{S}^\bullet_{v^g}(D(SM))$ are determined by C_{v^g}. As a result, the differential complexes

$\mathbf{C}^*_{\mho}(SM^\circ, g; \mathsf{R})$ and $\mathbf{C}^*_{\mho}(D(SM), g; \mathsf{R})$, whose construction depends only on the *stratified* topological types $\mathcal{S}^\bullet_{v^g}(SM)$ and $\mathcal{S}^\bullet_{v^g}(D(SM))$ of the spaces SM and $D(SM)$, respectively, can be reconstructed from the geodesic scattering map C_{v^g}. □

Abusing notations, we denote by $C^j_{\mho}(D(SM), g; \mathbb{Z})$ the image of the $\mathbb{Z}$-module $C^j_{\mho}(D(SM), g; \mathbb{Z})$ in the vector space $C^j_{\mho}(D(SM), g; \mathbb{R})$. It is an integral lattice. Similarly, we consider the integral lattice $C^j_{\mho}(SM^\circ, g; \mathbb{Z}) \subset C^j_{\mho}(SM^\circ, g; \mathbb{R})$.

The lattice $C^j_{\mho}(SM, g; \mathbb{Z})$ comes equipped with a $\mathbb{Z}$-basis whose vectors correspond to the connected components of the strata

$$\{SM(v^g, \omega) \cap \text{int}(SM)\}_{\omega \in \mathbf{\Omega}^\bullet \,\big|\, |\omega|' = j}.$$

Similarly, the lattice $C^j_{\mho}(D(SM), g; \mathbb{Z})$ comes equipped with a $\mathbb{Z}$-basis whose vectors correspond to the connected components of the strata $\{SM(v^g, \omega)\}_{\omega \in \mathbf{\Omega}^\bullet \,\big|\, |\omega|' = j}$, each stratum being considered *twice*, together with the connected components of the strata

$$\{SM(v^g, \omega) \cap \partial(SM)\}_{\omega \in \mathbf{\Omega}^\bullet \,\big|\, |\omega|' = j-1}.$$

Using these bases, we introduce the l_1-norms $\| \sim \|_{\mho}$ in the vector spaces $C^j_{\mho}(SM^\circ, g; \mathbb{R})$ and $C^j_{\mho}(D(SM), g; \mathbb{R})$ so that the basic vectors, which belong to the lattices $C^j_{\mho}(SM^\circ, g; \mathbb{Z})$ and $C^j_{\mho}(D(SM), g; \mathbb{Z})$, all have unit length. The unit balls in these norms are the perfect polyhedra, dual to the hypercubes in the corresponding spaces.

Recall that the "quotient norm" of a given vector $\vec{V}$ in the quotient space is defined to be the infimum of the norms of all the vectors (in the original space) that represent $\vec{V}$.

For each j, the norms $\| \sim \|_{\mho}$ induce norms $\| \sim \|^\bullet_{\mho}$ in the quotient spaces

$$C^j_{\mho}(SM^\circ, g; \mathbb{R}) / B^j_{\mho}(SM^\circ, g; \mathbb{R}) \text{ and } C^j_{\mho}(D(SM), g; \mathbb{R}) / B^j_{\mho}(D(SM), g; \mathbb{R}),$$

respectively.

We denote by $\diamondsuit^j_{\mho}(SM^\circ, g)$ and $\diamondsuit^j_{\mho}(D(SM), g)$ the unit balls in these quotient norms $\| \sim \|^\bullet_{\mho}$. They are convex hulls of the images of the vertices of the perfect polyhedron $\{\| \sim \|_{\mho} = 1\}$ under the quotient maps.

The manifolds TM and $T(DM)$ are orientable; thus so are the manifolds SM and $S(DM)$. Therefore, the Poincaré Duality is available for their homology and cohomology with coefficients in $\mathbb{R}$ or $\mathbb{Z}$. As in [K6] (where we dealt with arbitrary traversally generic flows) or in [K10], for each j, we consider two localized Poincaré Duality operators over the coefficient rings $\mathsf{R} = \mathbb{Z}, \mathbb{R}$ (in the formulas below, we suppress the dependence of homology and cohomology on the coefficient ring R):

$$\begin{aligned}
\mathcal{L}_j(g) : H_j(D(SM)) &\stackrel{\approx\mathcal{D}}{\longrightarrow} H^{2n-1-j}(D(SM)) \stackrel{i^*_{\mathsf{loc}}}{\longrightarrow} H^{2n-1-j}\big(D(SM)_{-j}(g)\big) \\
&\approx C^{2n-1-j}_{\mho}(D(SM), g)\Big/ B^{2n-1-j}_{\mho}(D(SM), g), \quad \text{and} \\
\mathcal{M}_j(g) : H_j(SM) &\stackrel{\approx\mathcal{D}}{\longrightarrow} H^{2n-1-j}(SM, \partial_1(SM)) \\
&\stackrel{i^*_{\mathsf{loc}}}{\longrightarrow} H^{2n-1-j}\big(SM_{-j}(g),\, SM_{-j}(g) \cap \partial_1(SM)\big) \\
&\approx C^{2n-1-j}_{\mho}(SM^\circ, g)\Big/ B^{2n-1-j}_{\mho}(SM^\circ, g).
\end{aligned} \tag{4.55}$$

Here the natural homomorphisms i^*_{loc} are induced by the inclusions of the strata

$$SM^\circ_{-j}(g) \subset SM \text{ and } D(SM)_{-j}(g) \subset D(SM)$$

—thus the term "localized" in the names of the two operators.

The RHS isomorphisms "$\approx$" in (4.55) can be justified by considering the dimensions of the relevant strata (exactly by the same homological argument as in [K7], page 446; just replace an arbitrary traversally generic vector fields v on X with the geodesic traversally generic vector fields v^g on SM).

Note that $\mathcal{L}_j(g; \mathbb{R})$ maps the integral lattice $H_j(D(SM); \mathbb{Z})$ to the integral lattice $H^{2n-1-j}\big(D(SM)_{-j}(g); \mathbb{Z}\big)$. Similarly, $\mathcal{M}_j(g; \mathbb{R})$ maps the lattice $H_j(SM; \mathbb{Z})$ to the lattice $H^{2n-1-j}\big(SM^\circ_{-j}(g); \mathbb{Z}\big)$.

We denote by $\tilde{\mathsf{B}}^{\mathbf{\Delta}}_j(D(SM)) \subset H_j(D(SM); \mathbb{R})$ and by $\tilde{\mathsf{B}}^{\mathbf{\Delta}}_j(SM) \subset H_j(SM; \mathbb{R})$ the unit balls, i.e., the set of vectors whose simplicial semi-norms $|\sim|_{\mathbf{\Delta}}$ do not exceed 1. Let $\partial\tilde{\mathsf{B}}^{\mathbf{\Delta}}_j(D(SM))$ and $\partial\tilde{\mathsf{B}}^{\mathbf{\Delta}}_j(SM)$ denote the sets of vectors whose semi-norms are equal 1; these "spheres" may not be compact.

Note that, for any pair of vectors V and W such that $|W|_{\mathbf{\Delta}} = 0$, we get $|V + W|_{\mathbf{\Delta}} = |V|_{\mathbf{\Delta}}$. Therefore, $|\sim|_{\mathbf{\Delta}}$ becomes a *norm* on the quotient $H^{\mathbf{\Delta}}_*(\sim; \mathbb{R})$ of the homology space $H_*(\sim; \mathbb{R})$ by the vector subspace $H^{\{\|\sim\|_{\mathbf{\Delta}}=0\}}_*(\sim; \mathbb{R})$ of vectors whose simplicial semi-norms vanish.

Thus, we may form the compact convex ball $\mathsf{B}^{\Delta}_j(D(SM)) \subset H^{\Delta}_j(D(SM);\mathbb{R})$—the image of the unite ball $\tilde{\mathsf{B}}^{\Delta}_j(D(SM))$ under the quotient map $H_j(D(SM);\mathbb{R}) \to H^{\Delta}_j(D(SM);\mathbb{R})$. Similarly, we may form the compact convex ball $\mathsf{B}^{\Delta}_j(SM) \subset H^{\Delta}_j(SM;\mathbb{R})$—the image of the unite ball $\tilde{\mathsf{B}}^{\Delta}_j(D(SM))$ under the quotient map $H_j(SM;\mathbb{R}) \to H^{\Delta}_j(SM;\mathbb{R})$.

We will use the localized Poincaré duality operators $\mathcal{L}_j(g;\mathbb{R})$ and $\mathcal{M}_j(g;\mathbb{R})$ from (4.55) to project linearly the unit balls $\tilde{\mathsf{B}}^{\Delta}_j(D(SM))$ and $\tilde{\mathsf{B}}^{\Delta}_j(SM)$ on the g-dependent "screens"

$$C^{2n-1-j}_{\mho}(D(SM), g;\mathbb{R})/B^{2n-1-j}_{\mho}(D(SM), g;\mathbb{R})$$

and

$$C^{2n-1-j}_{\mho}(SM^{\circ}, g;\mathbb{R})/B^{2n-1-j}_{\mho}(SM^{\circ}, g;\mathbb{R}),$$

respectively. The screens are manufactured by various metrics $g \in \mathcal{NT}^{\ddagger}(M)$.

Theorem 4.30 below makes a few claims about the geometry of these projections.

Definition 4.22. A discrete group G is amenable if, for every finite subset $S \subset \mathsf{G}$ and every $\epsilon > 0$, there exists a finite nonempty set $A = A(S, \epsilon) \subset \mathsf{G}$ such that the proportion of cardinalities

$$\frac{|(gA \cup A) \setminus (gA \cap A)|}{|A|} < \epsilon$$

for all $g \in S$. $\diamondsuit$

Theorem 4.30. *Let M be a compact connected smooth n-manifold with boundary, where $n \geq 3$.*

- *Let $j \in [0, n]$. Assume that, for each connected component of the boundary ∂M, the image of its fundamental group in $\pi_1(DM)$ is amenable.*

  Then there is a universal constant $\lambda = \lambda(n, j) \geq 1$ such that, for every M and every traversally generic Riemannian metric g on M, the image of the unit (in the simplicial semi-norm) sphere $\partial\tilde{\mathsf{B}}^{\Delta}_j(D(SM))$ under the localized Poincaré Duality operator $\mathcal{L}_j(g;\mathbb{R})$ is contained in the complement to the ball

  $$\lambda^{-1} \cdot \diamondsuit^{2n-1-j}_{\mho}(D(SM), g)$$

  of radius λ^{-1}.

- *Let $j \in [0, n-1]$. Assume that, for each connected component of the boundary ∂M, the image of its fundamental group in $\pi_1(M)$ is amenable.*

  Then there is a universal constant $\mu = \mu(n, j) \geq 1$ such that, for every M and every traversally generic Riemannian metric g on M, the image of the unit sphere $\partial\tilde{\mathsf{B}}_j^{\mathbf{\Delta}}(SM)$ under the localized Poincaré Duality operator $\mathcal{M}_j(g;\mathbb{R})$ is contained in the complement to the ball

$$\mu^{-1} \cdot \Diamond_{\mho}^{2n-1-j}(SM, g)$$

  of radius μ^{-1}.

Proof. Consider the $(n-1)$-dimensional spherical fibrations $p : SM \to M$ and $q : D(SM) \to DM$. Let $\partial_\alpha M$ be a typical connected component of ∂M. Using the long homotopy sequences of the fibration p and of its restriction to $\partial_\alpha M$, we notice that, for $n \geq 3$, if the natural image of $\pi_1(\partial_\alpha M) \to \pi_1(M)$ is an amenable group, then so is the natural image of $\pi_1(p^{-1}(\partial_\alpha M)) \to \pi_1(SM)$. Similarly, if the image of $\pi_1(\partial_\alpha M) \to \pi_1(DM)$ is an amenable group, then so is the image of $\pi_1(q^{-1}(\partial_\alpha M)) \to \pi_1(D(SM))$.

As a default, we consider homology with coefficient in $\mathbb{R}$.

For dimensional reasons, the spectral sequences of the fibratios p and q collapse since p admits a section σ and q a admits a section $D\sigma$, the double of σ. Thus $H_*(SM) \approx H_*(M) \oplus H_{*-n+1}(M)$, and $H_*(D(SM)) \approx H_*(DM) \oplus H_{*-n+1}(DM)$.

We claim that the simplicial semi-norms of all elements of the second factor of the direct sums vanish. Let us justify the claim.

Recall that an odd multiple $[f]$ of every integral homology class $[h] \in H_k(M;\mathbb{Z})$ can be realized by a continuous map $f : N \to M$, where N is a closed oriented smooth k-manifold [CF]. Similarly, odd multiples $[F]$ of the elements $[H] \in H_{k+n-1}(SM;\mathbb{Z})$ that form the second summand are realizable by cycles of the form $F : K \to SM$, where K is the space of the spherical fibration $\pi : K \to N$, induced by f from the fibration $SM \to M$.

For $n \geq 3$, the projection p induces an isomorphism $p_* : \pi_1(SM) \to \pi_1(M)$. Let $\alpha : M \to K(\pi_1(M), 1)$ be the classifying map (that induces an isomorphism of the fundamental groups). Then $\alpha \circ p : SM \to K(\pi_1(M), 1)$ is also the classifying map. Note that $(\alpha \circ \pi)_*([F]) = 0$ in $H_{k+n-1}(K(\pi_1(M), 1))$ since K is the boundary of the disk bundle $\Pi : Cyl(\pi) \to N$, associated with π, and $\alpha \circ \pi$ extends to $\alpha \circ \Pi$. Using that continuous maps of spaces, which induce isomorphisms of their fundamental groups, are *isometries* in the semi-norms $\| \sim \|_{\mathbf{\Delta}}$ ([Gr]), we conclude that $\| [F] \|_{\mathbf{\Delta}} = 0$ and thus $\| [H] \|_{\mathbf{\Delta}} = 0$.

Therefore, we get the canonical isomorphism $p_* : H_*^{\mathbf{\Delta}}(SM) \approx H_*^{\mathbf{\Delta}}(M)$ of the two quotient spaces, viewed as an *isometry of normed spaces.*

Similar arguments, applied to the spherical fibration $q : D(SM) \to DM$, lead to the isometry $q_* : H_*^{\mathbf{\Delta}}(D(SM)) \approx H_*^{\mathbf{\Delta}}(DM)$.

Thus we proved that the hypotheses of Theorem 4.30 imply the validity of the hypotheses of Theorems 3.1, 3.2 from [K6] (see also [K7], Theorems 8.3 and 8.4). The latter theorems apply to any traversally generic vector field v on a connected compact smooth manifold X with boundary. In particular, they apply to any traversally generic geodesic vector field v^g on SM. Note that Theorems 3.1, 3.2 from [K6] are stated in a slightly different and less geometrical form than Theorem 4.30 here: they claims the validity of the inequality $\|h\|_\Delta \leq \lambda \cdot \|\mathcal{L}_j(g)(h)\|_{\mho}^{\bullet}$ for every class $h \in H_j(D(SM); \mathbb{R})$, and of the inequality $\|h\|_\Delta \leq \mu \cdot \|\mathcal{M}_j(g)(h)\|_{\mho}^{\bullet}$ for every class $h \in H_j(SM; \mathbb{R})$, where $\lambda \geq \mu \geq 1$ being some universal constants. The constants depend only on $\dim(M)$ and the index $j \in [0, n]$.

The inequality $\|h\|_\Delta \leq \lambda \cdot \|\mathcal{L}_j(g)(h)\|_{\mho}^{\bullet}$ may be interpreted as claiming that the $\mathcal{L}_j(g; \mathbb{R})$-image of the unit sphere $\partial\tilde{\mathsf{B}}_j^{\mathbf{\Delta}}(D(SM))$ (in the simplicial semi-norm) is contained in the complement to the ball $\lambda^{-1} \cdot \diamondsuit_{\mho}^{2n-1-j}(D(SM), g)$ of radius λ^{-1} in the target space. A similar interpretation is available for $\mathcal{M}_j(g; \mathbb{R})$.

Let us describe briefly the source of the universal constants λ, μ, participating in these inequalities[22]. In Chapter 7, formula (7.4), we will introduce the model space $\mathcal{E}_\omega \subset \mathbb{R} \times \mathbb{R}^{|\omega|'}$, given by the polynomial inequality $\wp(u, \vec{x}) \leq 0$, where $\wp$ is defined by the LHS of (2.12). Consider the obvious projection $p : \mathbb{R} \times \mathbb{R}^{|\omega|'} \to \mathbb{R}^{|\omega|'}$. The fibers of the projection $p : \mathcal{E}_\omega \to \mathbb{R}^{|\omega|'}$ are unions the ∂_u-trajectories in $\mathcal{E}_\omega$. Therefore, $\mathcal{E}_\omega$ acquires a stratification $\{\mathcal{E}_\omega(\partial_u, \hat{\omega})\}_{\hat{\omega}}$, labeled by the elements $\hat{\omega} \succeq \omega$. They form the subposet $\omega_{\preceq} \subset \mathbf{\Omega}^{\bullet}_{|\sim|' \leq 2n-2}$.

Put $\partial\mathcal{E}_\omega(\partial_u, \hat{\omega}) =_{\mathsf{def}} \mathcal{E}_\omega(\partial_u, \hat{\omega}) \cap \partial\mathcal{E}_\omega$ and $\mathcal{E}_\omega^\circ(\partial_u, \hat{\omega}) =_{\mathsf{def}} \mathcal{E}_\omega(\partial_u, \hat{\omega}) \cap \mathrm{int}(\mathcal{E}_\omega)$. We denote $\mathcal{S}^\bullet(\mathcal{E}_\omega^\circ)$ the stratification of the space $\mathcal{E}_\omega^\circ$ by the *connected components* of these strata.

The doubling $D\mathcal{E}_\omega$ is stratified by the connected components of $\{\partial\mathcal{E}_\omega(\partial_u, \hat{\omega})\}_{\hat{\omega} \in \omega_{\preceq}}$, together with the connected components of $\{\mathcal{E}_\omega^\circ(\partial_u, \hat{\omega})\}_{\hat{\omega} \in \omega_{\preceq}}$, and their images under the involution $\tau : D(\mathcal{E}_\omega) \to D(\mathcal{E}_\omega)$ affiliated with the doubles. We denote by $\mathcal{S}^\bullet(D(\mathcal{E}_\omega))$ this stratification.

---

[22]This is the only place where the hypotheses that v^g is traversally generic (and not only boundary generic and traversing) seems to be important.

See [Gr1] and [AK] for an accurate definition of the stratified simplicial norm. This definition puts some natural constraints on the singular simplexes in relation to the strata.

The universal constant μ is the maximum of the $\mathcal{S}^\bullet(\mathcal{E}^\circ_\omega)$-stratified simplicial norms of small j-disks, relative to their boundaries. Each disk is normal to a particular connected component of the $(2n+1-j)$-dimensional strata $\{\mathcal{E}^\circ_\omega(\partial_u, \hat{\omega})\}_{\omega,\, \hat{\omega} \in \omega_{\preceq}}$. The maximum is taken over all pairs $\omega \preceq \hat{\omega}$, where $\omega \in \mathbf{\Omega}^\bullet_{|\sim|' \leq 2n-2}$.

The universal constant λ is the maximum of the $\mathcal{S}^\bullet(D(\mathcal{E}_\omega))$-stratified simplicial norms of small j-disks, relative to their boundaries. Each disk is normal to a particular connected component of the $(2n+1-j)$-dimensional strata $\big\{\mathcal{E}^\circ_\omega(\partial_u, \hat{\omega}),\ \tau(\mathcal{E}^\circ_\omega(\partial_u, \hat{\omega})),\ \partial\mathcal{E}_\omega(\partial_u, \hat{\omega})\big\}$ in the double $D(\mathcal{E}_\omega)$. The maximum is taken over all pairs $\omega \preceq \hat{\omega}$, where $\omega \in \mathbf{\Omega}^\bullet_{|\sim|' \leq 2n-2}$. □

Example 4.11. Let (N, g) be a closed connected smooth n-manifold, $n \geq 3$. Let $U \subset N$ be a codimension zero submanifold with a smooth boundary. Assume that U is contained in a topological n-ball.

Put $M =_{\mathsf{def}} N \setminus \mathsf{int}(U)$ and let us assume that v^g is traversally generic on SM. Then, by Theorem 4.30, the number of connected codimension n components of the $\mathcal{S}^\bullet_{v^g}(D(SM))$-stratification of $D(SM)$ exceeds the product $\lambda^{-1} \cdot \|[DM]\|_\Delta$, where $\lambda \geq 1$ depends only on n. In particular, when $\|[DM]\|_\Delta \neq 0$, there exists a $(n-1)$-dimensional family of geodesics γ in M such that their reduced multiplicity $m'(\gamma) = n$.

When U is a smooth n-ball, then $DM = N\#N$, where "#" stands for the connected sum. For $n \geq 3$, we get $\|[DM]\|_\Delta = 2 \cdot \|[N]\|_\Delta$ ([Gr]).

If N is a closed hyperbolic manifold, then $\|[N]\|_\Delta$ can be expressed as $vol_{\mathsf{hyp}}(N)/vol_{\mathsf{hyp}}(\Delta^n)$, the normalized hyperbolic volume of N ([Gr]). Here $vol_{\mathsf{hyp}}(\Delta^n)$ denotes the hyperbolic volume of an ideal n-simplex in the hyperbolic space $\mathbb{H}^n$. In such case, the number of codimension n connected components of the $\mathcal{S}^\bullet_{v^g}(D(SM))$-stratification of $D(SM)$ exceeds

$$2\big(\lambda \cdot vol_{\mathsf{hyp}}(\Delta^n)\big)^{-1} \cdot vol_{\mathsf{hyp}}(N) > 0.$$

Thus, for a given family of closed hyperbolic n-manifolds $\{N_k\}_{k \to \infty}$ whose volumes approach infinity, the number of codimension n connected components of the $\mathcal{S}^\bullet_{v^g}(D(SM))$-stratification of $D(SM_k)$ grows at least as fast as $\{vol_{\mathsf{hyp}}(N_k)\}_{k \to \infty}$

Note that $\|[D(SM)]\|_\Delta = 0$, since $D(SM) \to DM$ is a spherical fibration. Therefore, Theorem 4.30 does not tell anything about the number of 0-dimensional components of the $\mathcal{S}^\bullet_{v^g}(D(SM))$-stratification of $D(SM)$.

These 0-dimensional components arise from the finitely many geodesics γ that have the maximal reduced multiplicity $m'(\gamma) = 2n-2$ to the boundary ∂M. $\diamondsuit$

Theorem 4.31. *For any metric $g \in \mathcal{NT}^\ddagger(M)$, the localized Poincaré Duality operators $\mathcal{L}_j(g;\mathbb{R})$ and $\mathcal{M}_j(g;\mathbb{R})$ in (4.55), as well as the simplicial semi-norms $\| \sim \|_\Delta$ in their source spaces, and the norms $\| \sim \|^\bullet_{\mho}$ in their target spaces, can be reconstructed from the geodesic scattering map $C_{v^g} : \partial_1^+(SM) \to \partial_1^-(SM)$.*

Proof. By Corollary 4.2, the scattering map C_{v^g} allows for a reconstruction of the pair $(SM, \mathcal{F}(v^g))$, up to a stratification-preserving homeomorphism of SM which is the identity on $\partial_1(SM)$. Therefore, with the help of the involution $\tau : D(SM) \to D(SM)$ (an ingredient of the doubling construction), we get also a reconstruction of the stratified topological type of the double $D(SM)$.

The localized Poincaré Duality operators $\mathcal{L}_j(g;\mathbb{R})$ and $\mathcal{M}_j(g;\mathbb{R})$ also depend only on the $\mathcal{S}^\bullet_{v^g}(D(SM))$-stratified and $\mathcal{S}^\bullet_{v^g}(SM^\circ)$-stratified topological types of $D(SM)$ and SM°, respectively. Thus $\mathcal{L}_j(g;\mathbb{R})$ and $\mathcal{M}_j(g;\mathbb{R})$ can be recovered from C_{v^g}.

The semi-norm $\| \sim \|_\Delta$ of a given homology class is an invariant of the topological type of the underlying space. Therefore, in accordance with Theorem 4.30 (also Theorem 3.5 in [K7]), the semi-norms $\| \sim \|_\Delta$ on the source spaces of the maps $\mathcal{L}_j(g;\mathbb{R})$ and $\mathcal{M}_j(g;\mathbb{R})$ can be recovered from C_{v^g}. The norm $\| \sim \|^\bullet_{\mho}$ on the target spaces is defined also solely in terms of the stratified topological types of $D(SM)$ and SM°. Therefore, by Corollary 4.2, $\| \sim \|^\bullet_{\mho}$ depends on the scattering map C_{v^g} only. $\square$

Theorem 4.32. *Let M be a compact connected smooth n-manifold with boundary, where $n \geq 3$.*

- *Let $j \in [0, n-1]$. Assume that, for each connected component of the boundary ∂M, the image of its fundamental group in $\pi_1(M)$ is amenable.*

  Then, for any traversally generic Riemannian metric g on M, the number of $(2n-1-j)$-dimensional connected components in the stratification $\{SM^\circ(v^g, \omega)\}_\omega$ is greater than or equal to $\mathrm{rank}\big(H_j^\Delta(M)\big)$.
- *Let $j \in [0, n]$. Assume that, for each connected component of the boundary ∂M, the image of its fundamental group in $\pi_1(DM)$ is amenable.*

Then, for any traversally generic Riemannian metric g on M, the number of $(2n-1-j)$-dimensional connected components in the stratification $\{D(SM)(v^g, \omega)\}_\omega$ is greater than or equal to $\mathrm{rank}\big(H_j^{\mathbf{\Delta}}(DM)\big)$.

Proof. By Theorem 4.30, the kernel $\mathsf{ker}(\mathcal{L}_j(g))$ is contained in the subspace $H_j^{\{\|\sim\|=0\}}(D(SM);\mathbb{R}) \subset H_j(D(SM);\mathbb{R})$. Thus, employing (4.55), we get

$$\mathrm{rank}(H_j^{\mathbf{\Delta}}(D(SM);\mathbb{R}) \leq \mathrm{rank}\big(H_j(D(SM);\mathbb{R})/\mathsf{ker}(\mathcal{L}_j(g))\big) = \mathrm{rank}(\mathrm{im}\big(\mathcal{L}_j(g)\big)$$

$$\leq \mathrm{rank}\big(C_{\mho}^{2n-1-j}(D(SM), g)\big/B_{\mho}^{2n-1-j}(D(SM), g)\big)$$

$$\leq \mathrm{rank}\big(C_{\mho}^{2n-1-j}(D(SM), g)\big),$$

the number of codimension j connected components in the stratification $\{D(SM)(v^g, \omega)\}_\omega$. Similarly,

$$\mathrm{rank}(H_j^{\mathbf{\Delta}}(SM;\mathbb{R}) \leq \mathrm{rank}\big(C_{\mho}^{2n-1-j}(SM, g)\big),$$

the number of codimension j connected components in the stratification $\{SM^\circ(v^g, \omega)\}_\omega$. □

The next corollary spells out the claims of Theorem 4.32 by applying formula (3.30) in Corollary 3.5 from [K6], or better, formula on the bottom of page 537 from [K6] (see also the formula on the top of page 457 in [K7]) to the settings of that theorem.

Corollary 4.28. *Let M be a compact connected smooth n-manifold with boundary, and g a traversally generic metric on M. Let $n \geq 3$.*

- *Assume that, for each connected component of the boundary ∂M, the image of its fundamental group in $\pi_1(M)$ is amenable.*

  Then, for each $j \in [0, n-1]$, the space of geodesics $\mathcal{T}(v^g)$ has the property

$$\sum_{\omega \in \mathbf{\Omega}^\bullet \,\big|\, |\omega|' = j} \mathrm{sup}(\omega) \cdot \#\big(\pi_0\big(\mathcal{T}(v^g, \omega)\big)\big) \geq \mathrm{rank}\big(H_j^{\mathbf{\Delta}}(M)\big),$$

  where $\mathcal{T}(v^g, \omega)$ stands for the family of geodesics $\gamma \subset M$ whose intersections $\gamma \cap \partial M$ conform to the combinatorial tangency pattern ω, $\mathrm{sup}(\omega) =_{\mathsf{def}} |\omega| - |\omega|'$, *and $\pi_0(\sim)$ denotes the set of connected components of the appropriate space.*

- *Assume that, for each connected component of the boundary ∂M, the image of its fundamental group in $\pi_1(DM)$ is amenable.*

Then, for each $j \in [0, n]$, the space of geodesics $\mathcal{T}(v^g)$ has the property

$$\sum_{\omega \in \mathbf{\Omega}^\bullet \,\big|\, |\omega|' = j} \sup(\omega) \cdot \#\big(\pi_0\big(\mathcal{T}(v^g, \omega)\big)\big) +$$

$$+\, 2 \cdot \sum_{\hat\omega \in \mathbf{\Omega}^\bullet \,\big|\, |\hat\omega|' = j+1} (\sup(\hat\omega) - 1) \cdot \#\big(\pi_0\big(\mathcal{T}(v^g, \hat\omega)\big)\big) \geq \operatorname{rank}\big(H_j^{\mathbf{\Delta}}(DM)\big).$$

$\diamondsuit$

Example 4.12. Consider a collection $\{\Sigma_k\}_{k \in [1,q]}$ of closed surfaces of genera $\mathsf{g}_k \geq 2$. Let $N_k =_{\mathsf{def}} \Sigma_k \times S^2$. We denote by N the connected sum of all N_k's. Note that N admits the obvious map $p_k : N \to \Sigma_k \times S^2$, which collapses all N_j's, $j \neq k$, and their 1-handles to a point or to a pair of points in ∂N_k. It induces an epimorphism $(p_k)_* : \pi_1(N) \to \pi_1(\Sigma_k \times S^2)$ of the fundamental groups.

Let M be a compact 4-dimensional manifold, obtained from N by removing a number of smooth 4-balls D^4 and solid "tori" of the form $S^1 \times D^3$, residing in N. We assume that these domains do not intersect the surfaces $\{\tilde\Sigma_k =_{\mathsf{def}} \Sigma_k \times pt\}$, where $pt \in S^2$. This surgery on N may add new 2-cycles to $H_2(M)$, which all are represented by 2-spheres. Thus, the simplicial norm of the new 2-cycles vanish. This observation implies that the map $M \subset N$ induces an isometry $H_2^{\mathbf{\Delta}}(M) \approx H_2^{\mathbf{\Delta}}(N)$.

Let $[\tilde\Sigma_k]$ denote the image of the fundamental class $[\Sigma_k]$ in $H_2^{\mathbf{\Delta}}(M)$. Employing the maps $\{p_k\}_k$ and the isomorphism $H_2^{\mathbf{\Delta}}(M) \approx H_2^{\mathbf{\Delta}}(N)$, we see that the classes $\{[\tilde\Sigma_k]\}_k$ are linearly independent in $H_2^{\mathbf{\Delta}}(M)$.

Since $\|\tilde\Sigma_k\|_{\mathbf{\Delta}} \geq 2\mathsf{g}_k - 2 > 0$ and recycling the arguments in the proof of Theorem 4.30 that deal with the spherical fibrations $SM \to M$ and $S(DM) \to DM$, we get

$$\operatorname{rank}\big(H_2^{\mathbf{\Delta}}(SM)\big) = \operatorname{rank}\big(H_2^{\mathbf{\Delta}}(M)\big) \geq q$$

and

$$\operatorname{rank}\big(H_2^{\mathbf{\Delta}}(D(SM))\big) = \operatorname{rank}\big(H_2^{\mathbf{\Delta}}(DM)\big) \geq 2q.$$

By Theorem 4.32, for any traversally generic metric g on M, the number of 5-*dimensional* connected components in the stratification $\{SM^\circ(v^g, \omega)\}_\omega$ of SM° is q at least. These are connected components of the strata, indexed by ω's such that $|\omega|' = 2$. Such ω's belong to the list $\{(1221), (13), (31)\}$.

Let us rephrase these conclusions in terms of M. For any traversally generic metric g on M, there exists at least q *four-dimensional* families of geodesics γ in M, such that either γ is quadratically tangent to ∂M at a pair of distinct points (the case of $\omega = (1221)$), or γ has a single tangency to ∂M of order 3 (the case of $\omega = (13), (31)$). Each family is continuously parametrized by an open smooth 4-manifold. Different families do not share geodesics.

Now let us illustrate the claims of Theorem 4.30 in this setting. Take, for example, the 2-cycle $h = \sum_{k=1}^{q} [\tilde{\Sigma}_k] \in H_2^{\mathbf{\Delta}}(M)$. Its simplicial norm $\|h\|_\Delta = \sum_{k=1}^{q}(2\mathsf{g}_k - 2)$.

For an universal constant $\mu \geq 1$ and any traversally generic metric g on M, the number of 5-dimensional connected components in the stratification $\{SM^\circ(v^g, \omega)\}_\omega$ of SM, where $\omega = (1221), (13), (31)$, is greater than or equal to $\mu^{-1} \cdot \sum_{k=1}^{q}(2\mathsf{g}_k - 2)$. Although estimating μ from above may be technically challenging, at least we know the rate of growth of the number of 5-dimensional connected components in the $\mathcal{S}_{v^g}(SM^\circ)$-stratification, as $k \to \infty$ or as individual genus $\mathsf{g}_k \to \infty$.

Finally, let M' be any smooth compact 4-dimensional manifold which is *homotopy equivalent* to M and such that $\partial M' = \partial M$. Let $H : M' \to M$ denote this homotopy equivalence. Since a homotopy equivalence induces an isometry in the normed homology $H_*^{\mathbf{\Delta}}(\sim)$, we see that the same conclusions are valid for any traversally generic metric g' on M'. $\diamondsuit$

The formulas from Corollary 4.28 have an instant implication.

Corollary 4.29. *Under the hypotheses of Theorem 4.32, the non-triviality of the groups $H_j^{\mathbf{\Delta}}(M)$ and/or $H_j^{\mathbf{\Delta}}(DM)$ represents an obstruction to the existence of a globally j-convex (see Definition 4.6) and traversally generic metric on M.* $\diamondsuit$

The assumption that a metric g on a given manifold M is traversally generic, perhaps, could be relaxed. To extend all the results of this section to boundary generic and traversing geodesic flows, we need to consider only such geodesic flows that match a *finite list* of a priori fixed semi-local models (in the spirit of formula (2.12)) of the vicinity of every v^g-trajectory γ. However, these models could not be determined by the combinatorics of tangency (like ω) of γ to $\partial_1(SM)$ alone.

4.11 Complexity of fundamental groups and the minimal families of geodesics

Let (M, g) be a compact connected smooth Riemannian manifold. We just have internalized the crucial role that the non-amenable fundamental groups $\pi_1(M)$ and $\pi_1(DM)$ play in delivering some lower bounds of the numbers of connected components in the $\mathbf{\Omega}^\bullet$-stratifications of the space of geodesics $\mathcal{T}(v^g)$.

The following considerations are quite elementary; in particular, the amenability properties do not play any role here. Our arguments are a replay of Section 8.6 in [K7], being adjusted for geodesic flows (instead of general traversing flows).

Definition 4.23. Given a finitely presentable group π, we denote by $c_{\mathsf{gen}}(\pi)$ the minimal number of its generators, the minimum is taken over all finite presentations of the group π. $\diamondsuit$

For $\dim M \geq 3$, we will connect the complexity $c_{\mathsf{gen}}(\pi_1(M/\partial M))$ of the fundamental group $\pi_1(M/\partial M)$ with the number of *minimal* connected components in the $\mathbf{\Omega}^\bullet$-stratification $\{SM(v^g, \omega)\}_\omega$ of SM.

Here is the key geometrical observation in the basis of a few propositions below. For a traversing geodesic flow, each v^g-trajectory γ of the combinatorial type ω defines a loop or a bouquet of loops in the quotient space $SM/\partial_1(SM)$. Similarly, each γ defines its double $D(\gamma) \subset D(SM)$. The double $D(\gamma)$ is a chain of touching loops (like "○", or "○○", or "○○○", etc), the number of loops in the chain being equal to $\#(\gamma \cap \partial_1(SM)) - 1$.

Therefore, each v^g-trajectory γ produces an element $[\gamma] \in \pi_1(SM/\partial_1(SM))$, represented by the bouquet $\gamma/(\gamma \cap \partial_1(SM))$. We denote by $[[\gamma]]$ the subgroup of $\pi_1(SM/\partial_1(SM))$, generated by the loops from the bouquet $\gamma/(\gamma \cap \partial X)$. The group $[[\gamma]]$ is equipped with the ordered set of $(|\sup(\omega)| - 1)$ generators—the v^g-ordered loops of the bouquet $\gamma/(\gamma \cap \partial_1(SM))$. Recall that, for a traversally generic v^g, $|\sup(\omega)|$ is the cardinality of the set $\gamma \cap \partial_1(SM)$. The element $[\gamma]$ and the subgroup $[[\gamma]]$ are constant within each connected component $SM(v^g, \sigma)$ of the pure stratum $SM(v^g, \omega)$.

Let us denote by $\mathcal{S}^\bullet(v^g)$ the poset whose elements are the connected components $\{SM(v^g, \sigma)\}_\sigma$. Let $\mathcal{S}^\bullet_{\min}(v^g)$ denote the set of minimal elements in the poset $\mathcal{S}^\bullet(v^g)$. Equivalently, $S^\bullet_{\min}(v^g)$ is the set of *minimal* connected components of the strata $\{\mathcal{T}(v^g, \omega)\}_{\omega \in \mathbf{\Omega}^\bullet_{|\sim|' \leq 2n-2}}$ of the space of geodesics $\mathcal{T}(v^g)$. The term "minimal" means that the closure of the relevant component of $\mathcal{T}(v^g, \omega)$ does not contain a component of a smaller stratum $\mathcal{T}(v^g, \omega')$, where $\omega' \prec \omega$.

Thus, we get a system of groups $\{[[\gamma_\sigma]]\}_{\sigma \in \mathcal{S}^\bullet(v^g)}$. By Lemma 8.5 in [K7], this construction leads to the following claim.

Lemma 4.12. *For a traversally generic geodesic vector field v^g on SM, the "minimal" subgroups $\big\{[[\gamma_\sigma]]\big\}_{\sigma \in \mathcal{S}^\bullet_{\min}(v^g)}$ generate $\pi_1(SM/\partial_1(SM))$.* ◇

For $n \geq 3$, $\pi_1(SM/\partial_1(SM)) \approx \pi_1(M/\partial M)$. Therefore, as in Section 8.6 in [K7], Lemma 4.12 leads to the following assertion (c.f. [K7], Theorem 8.16).

Theorem 4.33. *Let (M, g) be a compact connected smooth Riemannian n-manifold with boundary, where $n \geq 3$. For a traversally generic geodesic vector field v^g on SM, the number of* minimal *connected components in the stratification $\{\mathcal{T}(v^g, \omega)\}_\omega$ is greater than or equal to the complexity $c_{\mathsf{gen}}\big(\pi_1(M/\partial M)\big)$ of the fundamental group $\pi_1(M/\partial M)$:*

$$\sum_{\sigma \in \mathcal{S}^\bullet_{\min}(v^g)} \Big(|\sup(\omega(\sigma))| - 1\Big) \geq c_{\mathsf{gen}}\big(\pi_1(M/\partial M)\big),$$

where $\omega(\sigma) \in \mathbf{\Omega}^\bullet_{|\sim|' \leq 2n-2}$ denotes the combinatorial type of a v^g-trajectory passing through the connected component labeled by σ. ◇

As in [K7], Section 8.6, Theorem 4.33 implies the following corollary.

Corollary 4.30. *Let (M, g) be a compact connected smooth Riemannian n-manifold with boundary, where $n \geq 3$. For a traversally generic geodesic vector field v^g on SM, the number of minimal connected components in the space of geodesics $\mathcal{T}(v)$ satisfies the inequality:*

$$\#\big(\mathcal{S}^\bullet_{\min}(v^g)\big) \geq c_{\mathsf{gen}}\big(\pi_1(M/\partial M)\big)/(2n-1).$$

In particular, if ∂M consists of m connected components, then

$$\#\big(\mathcal{S}^\bullet_{\min}(v^g)\big) \geq (m-1)/(2n-1).$$

◇

Remark 4.13. For a n-dimensional Riemannian manifold (M, g) such that v^g is traversally generic, the deepest filtration $\mathcal{T}(v^g)_{-(2n+2)}$ is represented by the finite set of geodesic trajectories of the combinatorial type ω such that $|\omega|' = 2n - 2$. Therefore, by definition, $\#\big(\mathcal{S}^\bullet_{\min}(v^g)\big) \geq \#\big(\mathcal{T}(v^g)_{-(2n-2)}\big)$. Unfortunately, even in the non-amenable case, since

$\|D(SM)\|_\Delta = 0$, we do not have any nontrivial lower estimate of the cardinality $\#\big(\mathcal{T}(v^g)_{-(2n-2)}\big)$.

For $n \geq 3$, by Corollary 4.30, we get

$$\#\big(\mathcal{S}^\bullet_{\min}(v^g)\big) \geq c_{\mathsf{gen}}\big(\pi_1(M)\big)/(2n-1).$$

Therefore,

$$\#\big(\mathcal{S}^\bullet_{\min}(v^g)\big) \geq \max\big\{\#\big(\mathcal{T}(v^g)_{-(2n-2)}\big),\ c_{\mathsf{gen}}\big(\pi_1(M)\big)/(2n-1)\big\}.$$

◇

PART 2

Homology of Traversing Flows

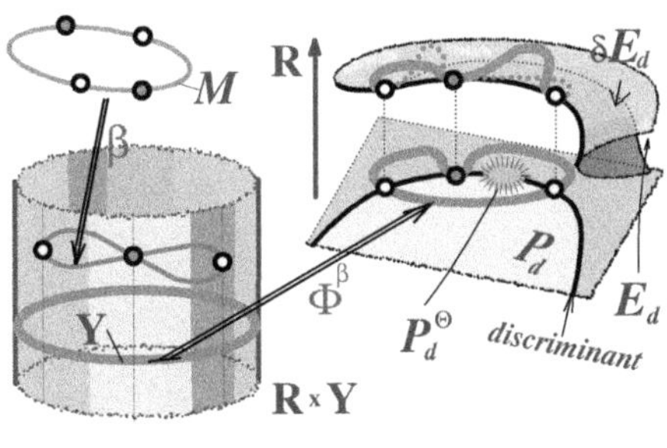

Chapter 5

Doodles and Blobs on a Ruled Page: Convex Quasi-envelops of Traversing Flows on Surfaces

"If there is one thing in mathematics that fascinates me more than anything else, it is neither number nor size, but always form."

Alexander Grothendieck

5.1 In this chapter...

The chapter illustrates the richness of traversing vector flows (see Definition 2.1) on surfaces with boundary. It also provides tools for constructing such flows (see Fig. 5.2). Some multi-dimensional versions of these ideas and constructions will appear in Chapters 7 and 8 (they also can be found in [K6], [K7], and [KSW1]). However, the case of, so called, 2-moderate one- and two-dimensional embeddings and immersions of curves and surfaces against the background of a fixed 1-dimensional foliation on a target surface A has its unique and pleasing features. One of which is the drastic simplification of the combinatorial considerations that characterize our treatment in Chapters 7 and 8 of similar multidimensional immersions.

Propositions 5.1 and 5.2 below illustrate the flavor of some of our results.

Consider the vector space $\mathbb{R}^3_{xyz}$ with coordinates x, y, z and the obvious projections $P_z : \mathbb{R}^3_{xyz} \to \mathbb{R}^2_{xy}$ and $p_z : \mathbb{R}^2_{xz} \to \mathbb{R}^1_x$. Let $\mathcal{C} \subset \mathbb{R}^2_{xz}$ be a smooth simple curve, the boundary of a compact domain $\mathcal{D} \subset \mathbb{R}^2_{xz}$, such that $p_z : \mathcal{C} \to \mathbb{R}^1_x$ has only singularities that are quadratic (folds). We say that $\mathcal{C}$ is overall positively (negatively) concave if the function $x : \mathcal{D} \to \mathbb{R}$ has more local maxima (minima) than minima (maxima) (see Fig. 5.5, (b)).

Proposition 5.1. *Let $\mathcal{C} \subset \mathbb{R}^2_{xz}$ be a simple smooth curve such that the projection $p_z : \mathcal{C} \to \mathbb{R}^1_x$ has only quadratic folding singularities and $\mathcal{C}$ is overall positively (negatively) concave.*

Then any smooth compact surface $\mathcal{S} \subset \mathbb{R}^3_{xyz} \cap \{y \geq 0\}$ *that bounds* $\mathcal{C}$ *must have at least cubic singularities (cusps) under the map* $P_z : \mathcal{S} \to \mathbb{R}^2_{xy}$. ◊

Proposition 5.1 is implied directly by Theorem 5.2. The key feature here is the interplay between the curve $\mathcal{C}$, surface $\mathcal{S}$, and the simple 1-dimensional foliation in $\mathbb{R}^3_{xyz}$ whose leaves are the fibers of the projection P_z.

A different, but related phenomenon is exemplified by the next proposition, which follows from the proof of Theorem 5.3.

Proposition 5.2. *Consider the immersed curve* $\mathcal{C}$ *in* $\mathbb{R}^2_{xz}$ *shown in Figure 5.9,* (a).

Then any compact immersed surface $\mathcal{S} \subset \mathbb{R}^3_{xyz} \cap \{y \geq 0\}$*, that bounds* $\mathcal{C}$ *must have two triple intersection points at least.* ◊

Although doodles on surfaces has been a well-traveled destination [CKS], [BFKK], doodles against the background of a given foliation on a compact surface are not. The same can be said about submersions $\alpha : X \to A$ of compact surfaces X on the cylinders or strips A, equipped with a product foliation $\mathcal{F}(\hat{v})$. This simple product foliation is responsible for the term "ruled" in the title of this chapter.

Our general inspiration comes from the pioneering works of V. I. Arnold [Ar], [Ar1], [Ar2], and V. A. Vassiliev [Va], [Va1].

The main problem we are dealing here is to classify such submersions $\alpha : X \to A$ (up to a kind of cobordism), while *controlling the tangency patterns* of the boundary ∂X, or rather of the curves $\alpha(\partial X)$ to the foliation $\mathcal{F}(\hat{v})$. As a byproduct, we getting some computable bordism-like relation among traversing vector fields on compact surfaces with boundary.

In this chapter, we consider two target spaces, the cylinder $\mathbb{R} \times S^1$ and and the strip $\mathbb{R} \times [0, 1]$. Some of the constructions will work for the cylinder, and some for the strip; for both, we use the same notation "A". The space A is equipped with a traversing vector field $\hat{v}$ and the 1-dimensional oriented foliation $\mathcal{F}(\hat{v})$ it generates. Its leaves are of the form $\{\mathbb{R} \times \theta\}_\theta$, where θ belongs either to the circle S^1 or to the interval $I = [0, 1]$.

We draw some "doodles" (finite collections of loops) on A (as in Fig. 5.1, diagram (a)) and pay close attention to the way they intersect the leaves of the foliation $\mathcal{F}(\hat{v})$, especially to the way they are tangent to the leaves. These interactions of doodles with the leaves are combinatorial in nature.

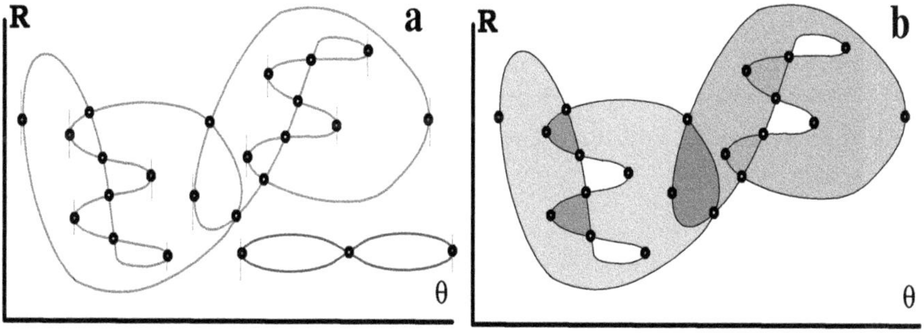

Fig. 5.1 Diagram (a) shows doodles—an immersion $\beta : \mathcal{C} \to A$ of 3 loops $\mathcal{C}$ in the surface $A = \mathbb{R} \times [0, 1]$. Diagram (b) shows blobs—an immersion $\alpha : X \to A$ of two disks X in A. The self-intersections of the curves $\beta(\mathcal{C})$ and of $\alpha(\partial X)$ and the points of tangency of $\beta(\mathcal{C})$ and of $\alpha(\partial X)$ to the vertical foliation $\mathcal{F}(\hat{v})$ on A are marked. Thanks to the presence of figure "∞", β does not extend to an immersion α of any compact surface X into A.

We will impose some a priori restrictions (called "2-moderate") on these combinatorial patterns and will classify the doodles that respect the restrictions.

In this chapter, we will also consider doodles that are the images of boundaries of compact surfaces X, *immersed* in A (as shown in Fig. 5.1, (b)). The images of X in A form overlapping "blobs".

Our main results about doodles and blobs on a ruled page $(A, \mathcal{F}(\hat{v}))$ are contained in Theorems 5.1-5.2 and Theorems 5.3-5.6. Perhaps, some new invariants of doodles and blobs that lead to these theorems will have an independent life.

Let us set the stage for these results in a more formal way. Let X be a compact surface with boundary and v a traversing vector field on X.

As a function of a point $x \in X$, the v-trajectory $\gamma_x \subset X$ through x exhibits a discontinuous behavior in the vicinity of any "concave" point (see Fig. 1.1, the right diagram) of the boundary In order to get around this fundamental difficulty, we "envelop" the pair (X, v) into a pair $(\hat{X}, \hat{v})$, where an ambient compact surface $\hat{X} \supset X$ with a traversing vector field $\hat{v}$, is such that:

(1) $\partial\hat{X}$ is convex (see Definition 1.3 and Fig. 1.1, the left diagram) with respect to the $\hat{v}$-flow on $\hat{X}$,

(2) $\hat{v}|_X = v$.

Without lost of generality, the reader may think of $\hat{X}$ as residing in the cylinder $\mathbb{R} \times S^1$ or in the strip $\mathbb{R} \times [0, 1]$ and the vector field $\hat{v}$ as being the unit vector field, tangent to the leaves of the product foliation $\mathcal{F}(\hat{v})$.

Not any pair (X, v) admits such convex envelop (see Lemma 5.1). However, when available, the convex envelop $(\hat{X}, \hat{v})$ simplifies the analysis of (X, v) greatly.

In this context, our goal is to study convex envelops (and their generalizations, the so called, quasi-envelops, shown in Fig. 5.2) together with the doodles or blobs they contain, up to some natural equivalence relations that we call in [K6], [K7] "quasitopies", a crossover between bordisms and pseudoisotopies of immersions. When $A = \mathbb{R} \times [0, 1]$, the quasitopy (bordism) equivalence classes can be organized into *groups.*

In Chapter 7 (see also [K6]), we compute these algebraic structures for a priori prescribed set of combinatorial tangency patterns of "n-dimensional doodles" to product-like 1-dimensional foliations. For $(n + 1)$-dimensional traversing convexly enveloped flows, this goal is achieved in full generality in Chapter 8. However, the case of *quasi*-enveloped traversing flows is far from being settled. Although in two dimensions all these structures are relatively primitive, as this chapter illustrates, they are not completely trivial ether.

Recall that in the study of manifolds and fibrations the universal classifying spaces like Grassmanians play a pivotal role. In the category of convex envelops, the role of universal objects (of "the new Grassmanians") is played by various spaces of smooth functions $f : \mathbb{R} \to \mathbb{R}$ whose zeros (considered with their multiplicities) are modeled after the real divisors of real polynomials. The topology of these functional spaces with constrained zero divisors is interesting on its own right (see Chapter 6, where it is described in detail, and [KSW1], [KSW2]). One particular kind of these functional spaces, called spaces of smooth functions/polynomials with k-moderate singularities, has been introduced and studied in depth by V. I. Arnold [Ar], [Ar1] and V. A. Vassiliev [Va], [Va1]. In Chapter 6, we compute the homology of similar functional spaces, based on real polynomials in one variable, in terms of appropriate *universal* combinatorics. This is reminiscent to the role played by the Schubert calculus in depicting the characteristic classes of classical Grassmanians.

Recall that, for boundary generic 2-dimensional traversing flows (see Definition 2.2), no tangencies of orders ≥ 3 to the boundary may occur [K2]. In light of what has been outlined above, we should anticipate a link between boundary generic flows on surfaces and the spaces of smooth functions $f : \mathbb{R} \to \mathbb{R}$ (or even real polynomials) that have no zeros of multiplicities ≥ 3. This connection and its generalizations are validated in Chapters 7 and 8.

Let $\mathcal{F}$ denote the space of smooths functions $f : \mathbb{R} \to \mathbb{R}$ that are identically 1 outside of a compact interval (the interval may depend on a particular function). The space $\mathcal{F}$ is considered in the C^∞-topology. Let $\mathcal{F}^{<k}$ be its subspace, formed by functions that have zeros only of the multiplicities less than k. Arnold calls such functions "functions with k-moderate singularities". The property of a function to have k-moderate zeros is an *open property* in the C^∞-topology; that is, the space $\mathcal{F}^{<k}$ is open in $\mathcal{F}$.

For $k = 2$, the combinatorial patterns ω of zero divisors of such functions are finite sequences of natural numbers, build of 1's and 2's only, as well as the empty sequence.

A fundamental theorem of Vassiliev ([Va], Corollary on page 81 and The First and Second Main Theorems on pages 78-79) identifies the weak homotopy types of the spaces $\mathcal{F}^{<k}$ (for all $k \geq 4$) and their integral homology types (for all $k \geq 3$) as ΩS^{k-1}, the space of loops on a $(k-1)$-sphere! In particular, the integral homology of the space $\mathcal{F}^{\leq 2} =_{\mathsf{def}} \mathcal{F}^{<3}$ is isomorphic to the homology of the loop space ΩS^2.

Arnold proved that the fundamental group $\pi_1(\mathcal{F}^{\leq 2}) \approx \mathbb{Z}$ ([Ar]), the result that we will use on a number of occasions. Thus, $H^1(\mathcal{F}^{\leq 2}; \mathbb{Z}) \approx \mathbb{Z}$ as well.

Therefore, for a 2-dimensional convex quasi-envelop $\alpha : (X, v) \subset (A, \hat{v})$ with no tangencies of $\alpha(\partial X)$ to the oriented foliation $\mathcal{F}(\hat{v})$ of order ≥ 3, this fact allows to define a characteristic class $J^*_\alpha \in H^1(A, \partial_\star A; \mathbb{Z}) \approx \mathbb{Z}$ of α. Here $\partial_\star A = \emptyset$ for $A = \mathbb{R} \times S^1$, and $\partial_\star A = \mathbb{R} \times \partial([0,1])$ for $A = \mathbb{R} \times [0,1]$.

- Let d be an even positive integer. We will employ the subspaces $\mathcal{F}^{\leq 2}_{\bar{d}} \subset \mathcal{F}^{\leq 2}$, formed by functions whose "degree"—the sum of multiplicities of all its zeros—is even and does not exceed d.

5.2 Convex quasi-envelops of traversing flows on surfaces and spaces of smooth functions with 2-moderate singularities

Let us revisit different kinds of vector fields, already familiar from Chapters 1 and 2 (see also [K1]-[K5]).

Let v be a smooth vector field on a compact connected surface X with boundary, such that $v \neq 0$ along the boundary $\partial_1 X$. Following [Mo], we consider the closed locus $\partial_1^+ X(v) \subset \partial_1 X$, where the field points inside of X or is tangent to $\partial_1 X$, and the closed locus $\partial_1^- X(v) \subset \partial_1 X$, where v points outside of X or is tangent to $\partial_1 X$. The intersection

$$\partial_2 X(v) =_{\mathsf{def}} \partial_1^+ X(v) \cap \partial_1^+ X(v)$$

is the locus where v is *tangent* to the boundary $\partial_1 X$.

Points $z \in \partial_2 X(v)$ come in two flavors: by definition, $z \in \partial_2^+ X(v)$ when $v(z)$ points inside of the locus $\partial_1^+ X(v)$, otherwise $z \in \partial_2^- X(v)$.
In particular, for a boundary generic v, the loci $\partial_1^\pm X(v)$ are finite unions of closed intervals and circles, residing in ∂X; the loci $\partial_2^\pm X(v)$ are finite unions of points, residing in $\partial_1 X$.

Each trajectory γ of a traversing vector field v must reach the boundary both in positive and negative times: otherwise γ is not homeomorphic to a closed interval.

Definition 5.1. A traversing vector field v on a compact surface X is called traversally generic, if the following properties hold:

(1) if a trajectory γ is tangent to the boundary $\partial_1 X$, then the tangency is simple,

(2) no v-trajectory γ contains more then one simple point of tangency to $\partial_1 X$. ♢

Definition 5.2. Let $\hat{v}$ be the standard traversing vector field on the surface A—a strip or a cylinder—, tangent to the fibers of the projections $\mathbb{R} \times [0, 1] \to [0, 1]$ or $\mathbb{R} \times S^1 \to S^1$. Let $\mathcal{C}$ be a closed 1-dimensional manifold (a finite collection of several circles). Consider an *immersion* $\beta : \mathcal{C} \to A$.

We say that such β is 2-moderately generic relatively to $\hat{v}$, if no $\hat{v}$-trajectory $\hat{\gamma}$ has order of tangency ≥ 3 to $\beta(\mathcal{C})$. Here the order of tangency is understood as the sum of tangency orders of local branches of $\beta(\mathcal{C})$ that pass through a given point of $\hat{\gamma}$. ♢

By standard techniques of the singularity theory, we can perturb any given immersion $\beta : \mathcal{C} \to A$ so that $\beta(\mathcal{C})$ will have only transversal self-intersections and will become boundary generic, and thus, 2-moderately generic.

In fact, any orientable connected surface X with boundary admits an immersion $\alpha : X \to A$ (or in the plane $\mathbb{R}^2$) (see Fig. 5.2). We will use this fact to pull-back to X the standard traversing vector field $\hat{v}$ on A.

Definition 5.3. Consider an *immersion* $\alpha : X \to A$ of a given compact (orientable) surface X into the surface A, equipped with the standard vector field $\hat{v}$ and the foliation $\mathcal{F}(\hat{v})$ it generates.

- Given a transversing vector field v on X, we call an immersion α a convex quasi-envelop of (X, v) if $v = \alpha^\dagger(\hat{v})$, the pull-back of $\hat{v}$ under α.
- Such α is called 2-moderately generic relative to $\hat{v}$, if the restriction $\alpha|_{\partial X}$ is 2-moderately generic with respect to $\hat{v}$ in the sense of Definition 5.2. ◇

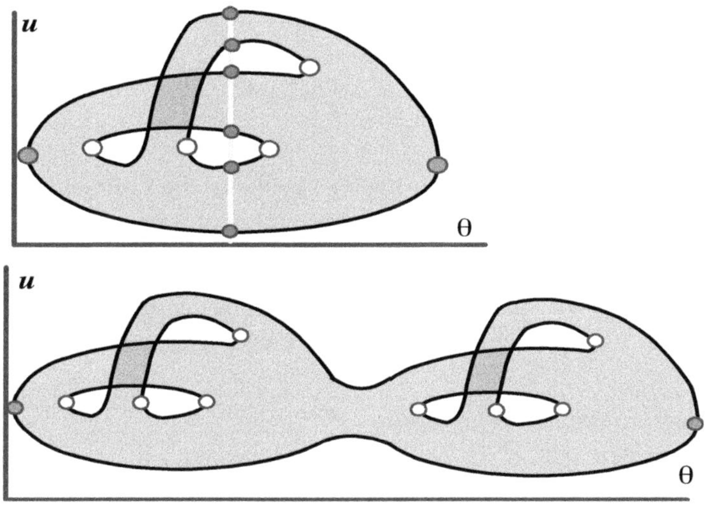

Fig. 5.2 A convex quasi-envelop $\alpha : X \to A$ of a 2-moderately generic (even traversally generic) vector field $\alpha^\dagger(\partial_u)$ on a compact surface X. The surface X is the torus from which an open smooth disk is removed (the top diagram), and X is the closed surface of genus 2 from which an open smooth disk is removed (the bottom diagram). In both examples, the cardinality of the fibers of the map $\theta \circ \alpha : \partial_1 X \to \mathcal{T}(\hat{v})$ does not exceed 6.

The next definition is a special case of Definition 5.3.

Definition 5.4. Let $\alpha : X \to A$ be a *regular embedding* of a given compact surface X into the surface A, carrying the standard vector field $\hat{v}$. We denote by v the pull-back $\alpha^\dagger(\hat{v})$ of $\hat{v}$ under α.

We say that the pair $(A, \hat{v})$ is a convex envelop of (X, v). If $\alpha|_{\partial X}$ is 2-moderately generic, then we call α a 2-moderately generic convex envelop. ◇

As the next lemma testifies, the existence of a convex envelop puts significant restrictions on the topology of orientable surfaces X: such X are disks with holes or unions of such (c.f., Theorem 1.1).

Lemma 5.1. *If a compact connected surface X with boundary has a pair of loops whose transversal intersection is a singleton, then no traversal flow on X admits a convex envelop. In other words, if a connected surface X with boundary has a handle, then no traversal flow on X can be convexly enveloped.*

Proof. By Theorem 1.1 ([K4]), the space $\hat{X}$ of a convex envelop is either a disk or an annulus, both surfaces residing in the plane. No two loops in the plane intersect transversally at a singleton. Thus, for surfaces with a handle, no convex envelops exist. □

To incorporate surfaces with handles into our constructions, in Definition 5.3, we introduced the less restrictive notion of a convex *quasi*-envelop.

Now we would like to explore closely a nice connection between:

(1) immersions $\alpha : (X, v) \to (A, \hat{v})$ of a compact surface X in the surface A, such that $v = \alpha^\dagger(\hat{v})$ and $\alpha(\partial_1 X)$ is 2-moderately generic with respect to $\hat{v}$, and

(2) loops in the functional spaces $\mathcal{F}^{\leq 2}$.

Let $\alpha(\partial_1 X)^\times$ denote the finite set of self-intersections of the curves forming the image $\alpha(\partial_1 X)$. Let $\alpha(\partial_1 X)^\circ$ denote the set $\alpha(\partial_1 X) \setminus \alpha(\partial_1 X)^\times$.

With the curve $\alpha(\partial_1 X) \subset A$ we associate an auxiliary smooth function $z_\alpha : A \to \mathbb{R}$, defined by the following properirs:

(5.1)

- $z_\alpha^{-1}(0) = \alpha(\partial X)$,
- 0 is the regular value of z_α at the points of $\alpha(\partial_1 X)^\circ$,
- in the vicinity of each transversal crossing point $a \in \alpha(\partial_1 X)^\times$, there exist locally defined smooth functions $x_1, x_2 : A \to \mathbb{R}$ such that: 0 is their regular value, $\{x_1^{-1}(0)\}$ and $\{x_2^{-1}(0)\}$ define the two local intersecting branches of $\alpha(\partial_1 X)$, and $z_\alpha = x_1 \cdot x_2$ in the vicinity of a.
- z_α approaches 1 at infinity in $A = \mathbb{R} \times S^1$ or in $A = \mathbb{R} \times [0, 1]$.

The sign of z_α changes to the opposite, as a path in A crosses an arc from $\alpha(\partial X)^\circ$ transversally, thus providing a "checker board" coloring of the domains in $A \setminus \alpha(\partial X)$. The properties in (5.1) do not determine a unique function z_α, but any such function will serve equally well in the forthcoming constructions.

Given a smooth traversing vector field v on a compact surface X, we denote by $\mathcal{T}(v)$ the space of its trajectories. For a boundary generic and traversing v, the space $\mathcal{T}(v)$ is a finite graph; for a traversally generic v, the space $\mathcal{T}(v)$ is a finite graph whose vertexes are only of valency 3 and 1 (see Fig. 1.2 and [K4]). For the standard vector field $\hat{v}$ on A, the trajectory space $\mathcal{T}(\hat{v})$ is either a circle S^1, or an interval $[0,1]$.

Let $u : A \to \mathbb{R}$ be the obvious projection on the first ("vertical") factor. In particular, $du(\hat{v}) > 0$ and $u(\hat{\gamma}) = \mathbb{R}$ for all $\hat{v}$–trajectories $\hat{\gamma}$ in A. Then, with the help of z_α and u, we get a map $J_{z_\alpha} : \mathcal{T}(\hat{v}) \to \mathcal{F}^{\leq 2}$ whose source, the trajectory space $\mathcal{T}(\hat{v})$, is either a circle or a closed interval. The image of J_{z_α} belongs to the space of smooth functions $f : \mathbb{R} \to \mathbb{R}$ such that the set $\{f \leq 0\}$ is compact in $\mathbb{R}$, f has no zeros of multiplicity ≥ 3, and $\lim_{u\to\pm\infty} f(u) = 1$. We define the map J_{z_α} by the formula

$$J_{z_\alpha}([\hat{\gamma}]) = (z_\alpha|_{\hat{\gamma}}) \circ (u|_{\hat{\gamma}})^{-1}, \tag{5.2}$$

where $\hat{\gamma}$ stands for a $\hat{v}$-trajectory in A, and $[\hat{\gamma}]$ for the corresponding point in the trajectory space $\mathcal{T}(\hat{v})$. If A is the strip, then $\mathcal{T}(\hat{v}) = [0,1]$. The two ends of the interval are mapped by J_{z_α} to the convex subspace $\mathcal{F}^+$ of $\mathcal{F}$ that is formed by *strictly positive* smooth functions. Therefore, we get a map of pairs

$$J_{z_\alpha} : ([0,1],\ \partial[0,1]) \to (\mathcal{F}^{\leq 2},\ \mathcal{F}^+).$$

The constant point-function $\mathbf{1}$ is a deformation retract of the space $\mathcal{F}^+$. Thus, homotopically, J_{z_α} can be regarded as a based loop

$$J_{z_\alpha} : (S^1, pt) \to (\mathcal{F}^{\leq 2},\ \mathcal{F}^+) \sim (\mathcal{F}^{\leq 2},\ \mathbf{1}).$$

For a fixed α, the homotopy class $[J_\alpha]$ of the map J_{z_α} does not depend on the choice of the auxiliary function z_α, subject to the four properties in (5.1). Indeed, consider the α_*-image $\alpha_*(\nu)$ of the outer normal vector field ν to the boundary $\partial_1 X$ in X. At the points of $a \in \alpha(\partial_1 X)^\times$, we get two vectors $v_1(a), v_2(a) \in \alpha_*(\nu)$, one for each branch of $\alpha(X)$.

If z_α and z'_α are any two functions that satisfy all the properties from the list (5.1), we get that the directional derivatives $\mathcal{L}_{\hat{v}} z_\alpha > 0$ and $\mathcal{L}_{\hat{v}} z'_\alpha > 0$ at each point of $\alpha(\partial X)^\circ$. Therefore, we get $\mathcal{L}_{\hat{v}}\{\tau z_\alpha + (1-\tau) z'_\alpha\} > 0$, where $\tau \in [0,1]$, at each point of $\alpha(\partial X)^\circ$. At the same time, for $i = 1, 2$, at each transversal crossing $a \in \alpha(\partial X)^\times$, we have $\mathcal{L}_{\hat{v}} x_i > 0$, and $\mathcal{L}_{\hat{v}} x'_i > 0$. Hence,

$\mathcal{L}_{\hat{v}}\{\tau x_i + (1-\tau)x'_i\} > 0$. The rest of properties from (5.1) are obviously "convex".

Therefore, the space of functions z_α that satisfy (5.1) is convex and thus contractible, which implies that the homotopy class $[J_\alpha]$ does not depend on the choice of z_α.

As a result, for $A = \mathbb{R} \times S^1$, any moderately generic convex quasi-envelop $\alpha : X \to A$, produces a homotopy class $[J_\alpha] \in [S^1, \mathcal{F}^{\leq 2}]$. For $A = \mathbb{R} \times [0,1]$, any moderately generic convex quasi-envelop $\alpha : X \to A$, produces a homotopy class $[J_\alpha] \in [(S^1, pt), (\mathcal{F}^{\leq 2}, \mathbf{1})]$.

Note that these homotopy classes do not depend on the orientation of the surface X.

We pick a generator $\kappa \in \pi_1(\mathcal{F}^{\leq 2}, \mathbf{1}) \approx \mathbb{Z}$ (see [Ar1] and Fig. 5.5, (a) or (b), for realistic portraits of κ). For $A = \mathbb{R} \times [0,1]$, we define an integer J^α by the formula $\mathsf{J}^\alpha \cdot \kappa = [J_\alpha]$.

The isomorphism $\pi_1(\mathcal{F}^{\leq 2}, \mathbf{1}) \approx \mathbb{Z}$ follows from [Ar1]. A slight modification of Arnold's arguments leads to Theorem 5.1 below. The main difference between our constructions and the ones from [Ar1] is that Arnold is concerned with immersions of 1-dimensional oriented "doodles" in A, while we deal also with immersions of "blobs" (compact orientable surfaces) in A (compare diagrams (a) and (b) in Fig. 5.1).

We denote by $\hat{v}^\bullet$ the vector field $(\hat{v}, 0)$ on the solid $A \times [0,1]$ and by $\pi : A \times [0,1] \to [0,1]$ the obvious projection.

We have seen how a 2-moderately generic immersion $\alpha : X \to A$ produces a homotopy class in $[S^1, \mathcal{F}^{\leq 2}]$ or in $[(D^1, \partial D^1), (\mathcal{F}^{\leq 2}, \mathbf{1})]$.

On the other hand, generic (oriented) loops in $\beta : S^1 \to \mathcal{F}^{\leq 2}$ have an interpretation as finite collections $\mathcal{C}$ of smooth *embedded* closed curves (embedded "doodles") in the surface A. These curves have no tangency to the $\hat{v}$-trajectories $\hat{\gamma}$ of order that exceeds two. In particular, any *inflections* of the curves with respect to the $\hat{\gamma}$-trajectories are forbidden. Furthermore, a generic homotopy between two maps $\beta : S^1 \to \mathcal{F}^{\leq 2}$ corresponds to some cobordism like relation between the corresponding plane curves $\mathcal{C}$. The cobordism also avoids the forbidden tangencies of orders ≥ 3 to the standard foliation $\mathcal{F}(\hat{v}^\bullet)$ on $A \times [0,1]$.

In order to define this cobordism, let us spell out more accurately the 2-moderate genericity requirements on the collections of doodles in A:

(5.3)

- $\beta : \mathcal{C} \to A$ is a smooth immersion of a finite collection of (oriented) circles,
- no intersections of $\beta(\mathcal{C})$ of multiplicities ≥ 3 are permitted,
- the order of tangency between each branch of $\beta(\mathcal{C})$ and each $\hat{v}$-trajectory $\hat{\gamma}$ does not exceed 2.
- if a branch of $\beta(\mathcal{C})$ is quadratically tangent to a $\hat{v}$-trajectory at a point x, then no other branch contains x. ♢

Definition 5.5. Given two immersions $\beta_0 : \mathcal{C}_0 \to A$ and $\beta_1 : \mathcal{C}_1 \to A$ as in (5.3), we say that they are (orientably) 2-moderately cobordant, if there is a compact (oriented) surface S and a smooth immersion $B : S \to A \times [0,1]$ such that:

(1) $\partial S = \mathcal{C}_0 \coprod -\mathcal{C}_1$,
(2) $B|_{\mathcal{C}_0 \coprod \mathcal{C}_1} = \beta_0 \coprod \beta_1$,
(3) the immersion $B| : S \to A \times [0,1]$ is 2-to-1 at most,
(4) if two local branches of $B(S)$ intersect at a point $z = a \times \{t\}$, then each of them is transversal to the trajectory $\hat{\gamma} \times \{t\}$ through z,
(5) each $\hat{v}^\bullet$-trajectory $\hat{\gamma} \times \{t\} \subset A \times [0,1]$ is tangent to each local branch of the surface $B(S)$ with the order of tangency that does not exceed 2,[1]
(6) the map $S \xrightarrow{B|} A \times [0,1] \to [0,1]$ is a Morse function with the regular values $\{0\}$ and $\{1\}$, ♢

We introduce the bordism set $\mathbf{C}^{\mathsf{imm}}_{\mathsf{moderate}\leq 2}(A)$ of immersions $\beta : \mathcal{C} \to A$ with 2-moderate tangencies to the foliation $\mathcal{F}(\hat{v})$. It is possible to verify that this cobordism between immersions is an *equivalence relation* (see [K6], [K7]).

Replacing "immersions" with "regular embeddings" in Definition 5.5, we get a modified notion of bordisms of (oriented) doodles. We denote the set of such bordisms by $\mathbf{C}^{\mathsf{emb}}_{\mathsf{moderate}\leq 2}(A)$. Depending on whether we consider the loops, forming $\mathcal{C}$, oriented or not, we get oriented versions $\mathbf{OC}^{\mathsf{imm}}_{\mathsf{moderate}\leq 2}(A)$ and $\mathbf{OC}^{\mathsf{emb}}_{\mathsf{moderate}\leq 2}(A)$ of the bordisms $\mathbf{C}^{\mathsf{imm}}_{\mathsf{moderate}\leq 2}(A)$ and $\mathbf{C}^{\mathsf{emb}}_{\mathsf{moderate}\leq 2}(A)$.

In the next definition, all surfaces and 3-folds that admit submersions in A or in $A \times [0,1]$ automatically must be orientable, but not necessarily oriented.

[1] In particular, the cubic tangencies are forbidden.

Definition 5.6. Given two (oriented) submersions $\alpha_0 : X_0 \to A$ and $\alpha_1 : X_1 \to A$ with $\alpha_0|_{\partial X_0}$ and $\alpha_1|_{\partial X_1}$ as in (5.3), we say that they are 2-moderately cobordant, if there is a compact (oriented) 3-fold W with corners $\partial X_0 \coprod \partial X_1$ and a submersion $B : W \to A \times [0,1]$ such that:

(1) $\partial W = X_0 \cup -X_1 \cup \delta W$, where $\delta W =_{\mathsf{def}} \partial W \setminus \mathsf{int}(X_0 \coprod X_1)$,
(2) $B|_{X_0 \coprod X_1} = \alpha_0 \coprod \alpha_1$,
(3) $B| : \delta W \to A \times [0,1]$ has at most double self-intersections,
(4) if two local branches of $B(\delta W)$ intersect at a point $z = a \times \{t\} \in A \times [0,1]$, then each of them is transversal to the trajectory $\hat{\gamma} \times \{t\}$ through z,
(5) every $\hat{v}^\bullet$-trajectory $\hat{\gamma} \times \{t\} \subset A \times [0,1]$ is tangent to each local branch of the surface $B(\delta W)$ with the order of tangency ≤ 2,
(6) the map $\delta W \xrightarrow{B|} A \times [0,1] \to [0,1]$ is a Morse function with the regular values $\{0\}$ and $\{1\}$. ◇

Thus, we can talk about the bordism set $\mathbf{B}^{\mathsf{imm}}_{\mathsf{moderate}\leq 2}(A)$ or $\mathbf{OB}^{\mathsf{imm}}_{\mathsf{moderate}\leq 2}(A)$ of immersions/embeddings $\alpha : X \to A$ with 2-moderate tangencies to the foliation $\mathcal{F}(\hat{v})$. It is possible to verify that this cobordism between immersions is an *equivalence relation* (see Chapter 8 and [K7]).

If in Definition 5.6 we replace all the "immersions" with the "regular embeddings" and drop vacuous constraints (2) and (3) of the definition, we will get a similar version of (oriented) bordisms of regular embeddings. We denote them by $\mathbf{B}^{\mathsf{emb}}_{\mathsf{moderate}\leq 2}(A)$ and $\mathbf{OB}^{\mathsf{emb}}_{\mathsf{moderate}\leq 2}(A)$.

If in Definition 5.6 we drop all the constraints related to how the boundary of embedded blobs and of the solid embedded cobordisms interact with the $\hat{v}$-trajectories in A and with the $\hat{v}^\bullet$-trajectories in $A \times [0,1]$, then the corresponding bordism groups $\mathbf{B}^{\mathsf{emb}}(A)$ and $\mathbf{OB}^{\mathsf{emb}}(A)$ of A are trivial. Indeed, by pushing (oriented) blobs $\alpha : X \hookrightarrow A$ inside $A \times [0,1]$, while keeping their boundaries $\alpha(\partial X)$ fixed in $A \times \{0\}$, we create a (oriented) solid that delivers the cobordism between $\alpha(X)$ and the empty blob.

The following obvious maps are available:

$$\begin{aligned}\mathcal{A}:\ \mathbf{B}^{\mathsf{emb}}_{\mathsf{moderate}\leq 2}(A) &\longrightarrow \mathbf{B}^{\mathsf{imm}}_{\mathsf{moderate}\leq 2}(A),\\ \mathcal{A}:\ \mathbf{OB}^{\mathsf{emb}}_{\mathsf{moderate}\leq 2}(A) &\longrightarrow \mathbf{OB}^{\mathsf{imm}}_{\mathsf{moderate}\leq 2}(A),\end{aligned} \tag{5.4}$$

$$\begin{aligned}\mathcal{A}:\ \mathbf{C}^{\mathsf{emb}}_{\mathsf{moderate}\leq 2}(A) &\longrightarrow \mathbf{C}^{\mathsf{imm}}_{\mathsf{moderate}\leq 2}(A),\\ \mathcal{A}:\ \mathbf{OC}^{\mathsf{emb}}_{\mathsf{moderate}\leq 2}(A) &\longrightarrow \mathbf{OC}^{\mathsf{imm}}_{\mathsf{moderate}\leq 2}(A).\end{aligned} \tag{5.5}$$

Also, taking the boundaries of (oriented) blobs, we get the obvious boundary maps:

$$\mathcal{B}^\partial : \mathbf{B}^{\mathsf{imm}}_{\mathsf{moderate}\leq 2}(A) \longrightarrow \mathbf{C}^{\mathsf{imm}}_{\mathsf{moderate}\leq 2}(A),$$
$$\mathcal{B}^\partial : \mathbf{OB}^{\mathsf{imm}}_{\mathsf{moderate}\leq 2}(A) \longrightarrow \mathbf{OC}^{\mathsf{imm}}_{\mathsf{moderate}\leq 2}(A), \tag{5.6}$$

$$\mathcal{B}^\partial : \mathbf{B}^{\mathsf{emb}}_{\mathsf{moderate}\leq 2}(A) \longrightarrow \mathbf{C}^{\mathsf{emb}}_{\mathsf{moderate}\leq 2}(A),$$
$$\mathcal{B}^\partial : \mathbf{OB}^{\mathsf{emb}}_{\mathsf{moderate}\leq 2}(A) \longrightarrow \mathbf{OC}^{\mathsf{emb}}_{\mathsf{moderate}\leq 2}(A). \tag{5.7}$$

In fact, the boundary maps $\mathcal{B}^\partial$ in (5.7) (i.e., for the category of embedding) are *bijections* (see Theorem 5.1 and [K6], [K7]).

Let us add one mild requirement to the list of the six properties in Definition 5.6:

- All the double intersections of $B| : \delta W \to A \times [0,1]$ are transversal. (5.8)

Combining the six properties from Definition 5.6 with the property in (5.8), we could, at the first glance, get a more rigid notion of cobordisms of immersions. Let us denote them temporarily by $\mathbf{B}^{\times\mathsf{imm}}_{\mathsf{moderate}\leq 2}(A)$.

Thus, we have the obvious map

$$\mathcal{A}^\times : \mathbf{B}^{\times\mathsf{imm}}_{\mathsf{moderate}\leq 2}(A) \to \mathbf{B}^{\mathsf{imm}}_{\mathsf{moderate}\leq 2}(A).$$

$$\mathcal{A}^\times : \mathbf{OB}^{\times\mathsf{imm}}_{\mathsf{moderate}\leq 2}(A) \to \mathbf{OB}^{\mathsf{imm}}_{\mathsf{moderate}\leq 2}(A).$$

By the general position argument, we may isotop an immersion B via immersions so that all its self-intersections become transversal [Th]. Therefore, the maps $\mathcal{A}^\times$ are actually bijections [K6].

In fact, the sets

$$\mathbf{B}^{\mathsf{imm}}_{\mathsf{moderate}\leq 2}(A),\ \mathbf{B}^{\mathsf{emb}}_{\mathsf{moderate}\leq 2}(A) \ \text{ and } \ \mathbf{OB}^{\mathsf{imm}}_{\mathsf{moderate}\leq 2}(A),\ \mathbf{OB}^{\mathsf{emb}}_{\mathsf{moderate}\leq 2}(A)$$

have a group structure. The group operation $\star$ takes a pair of 2-moderate submersions, $\alpha_1 : X_1 \to A$ and $\alpha_2 : X_2 \to A$, to a new 2-moderate submersion $\alpha_1 \star \alpha_2 : X_1 \coprod X_2 \to A$ by placing the image of α_2 in $\mathbb{R} \times S^1$ or in $\mathbb{R} \times [0,1]$ above the image of α_1 (see Fig. 5.5, (c) and (d)). At the first glance, the operation $\star$ seems to be non-commutative. At least for $A = \mathbb{R} \times [0,1]$, this first impression is wrong: by an appropriate isotopy one can switch the vertical order of $\alpha_1(X_1)$ and $\alpha_2(X_2)$ without violating the requirements that all the immersions are 2-moderate.

Note that the cardinality of the fibers of the map $\theta \circ (\alpha_1 \star \alpha_2)$ is less than or equal to the *sum* of cardinalities of the fibers of $\theta \circ \alpha_1$ and of $\theta \circ \alpha_2$, where $\theta : A \to \mathcal{T}(\hat{v})$ is the obvious map.

In the case of $A = \mathbb{R} \times [0,1]$, another *commutative* group structure is available in the sets $\mathbf{B}^{\mathsf{imm/emb}}_{\mathsf{moderate}\leq 2}(A)$ and $\mathbf{OB}^{\mathsf{imm/emb}}_{\mathsf{moderate}\leq 2}(A)$ (for more general than the 2-moderate combinatorial constraints on the tangency patterns, similar groups may not be commutative [K7], as described in Chapters 7 and 8). It is induced by the operation

$$\uplus : \mathbf{B}^{\mathsf{imm/emb}}_{\mathsf{moderate}\leq 2}(A) \times \mathbf{B}^{\mathsf{imm/emb}}_{\mathsf{moderate}\leq 2}(A) \to \mathbf{B}^{\mathsf{imm/emb}}_{\mathsf{moderate}\leq 2}(A)$$

$$\uplus : \mathbf{OB}^{\mathsf{imm/emb}}_{\mathsf{moderate}\leq 2}(A) \times \mathbf{OB}^{\mathsf{imm/emb}}_{\mathsf{moderate}\leq 2}(A) \to \mathbf{OB}^{\mathsf{imm/emb}}_{\mathsf{moderate}\leq 2}(A)$$

that places the image of α_2 to the right of the image of α_1 along the segment $[0,1]$; that is, $\alpha_1(X_1)$ is placed in the strip $\mathbb{R} \times (0,\, 0.5)$, while $\alpha_2(X_2)$ is placed in the strip $\mathbb{R} \times (0.5,\, 1)$ (see Fig. 5.2, lower diagram).

The zero element in $\mathbf{B}^{\mathsf{imm/emb}}_{\mathsf{moderate}\leq 2}(A)$ or in $\mathbf{OB}^{\mathsf{imm/emb}}_{\mathsf{moderate}\leq 2}(A)$ is represented by the empty surface X, and the minus of an immersion $\alpha : X \to A$ by a composition of α with a flip of A with respect to the trajectory $\mathbb{R} \times \{0.5\}$.

In contrast with the operation $\star$, the cardinality of the fibers of the map $\theta \circ (\alpha_1 \uplus \alpha_2)$ is less or equal to the *maximum* of cardinalities of the fibers of $\theta \circ \alpha_1$ and of $\theta \circ \alpha_2$. Due of this good feature of the operation $\uplus$, we choose to study its generalizations in Chapters 7 and 8.

Lemma 5.2. *For $A = \mathbb{R} \times [0,1]$, the operation $\uplus$ in $\mathbf{B}^{\mathsf{imm}}_{\mathsf{moderate}\leq 2}(A)$ or in $\mathbf{OB}^{\mathsf{imm}}_{\mathsf{moderate}\leq 2}(A)$ and in $\mathbf{B}^{\mathsf{emb}}_{\mathsf{moderate}\leq 2}(A)$ or in $\mathbf{OB}^{\mathsf{emb}}_{\mathsf{moderate}\leq 2}(A)$ is commutative.*

Proof. By a parallel transport in A of the images $\alpha_1(X_1) \subset \mathbb{R} \times (0,\, 0.5)$ and $\alpha_2(X_2) \subset \mathbb{R} \times (0.5,\, 1)$, we can switch their order along $[0,1]$. This can be done by sliding up $\alpha_2(X_2)$ so that it will reside in $(q, +\infty) \times (0.5,\, 1)$, then by sliding down $\alpha_1(X_1)$ so that it will reside in $(-\infty, -q) \times (0.5,\, 1)$. Here q is a positive number that exceeds the vertical size of both images. Then we slide horizontally the new $\alpha_2(X_2)$ and place it $(q, +\infty) \times (0,\, 0.5)$. Similarly, we slide horizontally the new $\alpha_1(X_1)$ and place it $(q, +\infty) \times (0.5,\, 1)$. Finally, we slide down $\alpha_1(X_1)$ and slide up $\alpha_2(X_2)$, thus completing the exchange.

Thanks to the nature of all these slides (parallel shifts), the maximal order of tangency of the $\hat{v}$-trajectories to the moving images $\alpha_1(\partial X_1)$ and $\alpha_2(\partial X_2)$ does not exceed 2.

Note that the maximal cardinality of the fibers of the projection $\alpha_1(\partial X_1) \coprod \alpha_2(\partial X_2) \to [0,1]$ did increase in the exchange process. □

Let $\mathcal{F}^{=2} \subset \mathcal{F}^{\leq 2}$ denote the discriminant hypersurface in the space $\mathcal{F}^{\leq 2}$, formed by the functions from $\mathcal{F}^{\leq 2}$ with at least one zero of multiplicity 2.

The following proposition is similar to Theorem from [Ar1]; however, our notion of cobordism of embedded blobs is different from the Arnold's more combinatorial notion of the cobordism of embedded doodles with no vertical inflection points.

Theorem 5.1. • *For $A = \mathbb{R} \times [0,1]$, the construction $\{\alpha \rightsquigarrow J_{z_\alpha}\}$ in (5.2), where the immersion $\alpha| : \partial X \to A$ has only 2-moderate tangencies to the foliation $\mathcal{F}(\hat{v})$, delivers a group homomorphism*

$$J^{\mathsf{imm}} : \mathbf{B}^{\mathsf{imm}}_{\mathsf{moderate}\leq 2}(A) \to \pi_1(\mathcal{F}^{\leq 2}, \mathbf{1}) \approx \mathbb{Z},$$

where the group addition in $\mathbf{B}^{\mathsf{imm}}_{\mathsf{moderate}\leq 2}(A)$ is the operation $\uplus$.

• *The same construction produces a group* isomorphism

$$J^{\mathsf{emb}} : \mathbf{B}^{\mathsf{emb}}_{\mathsf{moderate}\leq 2}(A) \approx \pi_1(\mathcal{F}^{\leq 2}, \mathbf{1}) \approx \mathbb{Z}.$$

The inverse map $(J^{\mathsf{emb}})^{-1}$ is delivered by the correspondence

$$K : \{\tau : ([0,1], \partial[0,1]) \to (\mathcal{F}^{\leq 2}, \mathbf{1})\} \Rightarrow \textstyle\bigcup_{\theta \in [0,1]} \big(\tau(\theta)^{-1}((-\infty, 0])\big),\ \theta\big) \\ \subset \mathbb{R}^1 \times [0,1],$$

where τ is a continuous path in $\mathcal{F}^{\leq 2}$, transversal to the discriminant hypersurface $\mathcal{F}^{=2}$.

• *For $A = \mathbb{R} \times S^1$, the construction $\{\alpha \rightsquigarrow J_{z_\alpha}\}$, where an embedding $\alpha : X \to A$ has only 2-moderate tangencies, delivers a 1-to-1 map*

$$J^{\mathsf{emb}} : \mathbf{B}^{\mathsf{emb}}_{\mathsf{moderate}\leq 2}(A) \approx [S^1, \mathcal{F}^{\leq 2}].$$

The inverse map $(J^{\mathsf{emb}})^{-1}$ is induced by the correspondence

$$K : \{\tau : S^1 \to \mathcal{F}^{\leq 2}\} \Rightarrow \bigcup_{\theta \in S^1} \big(\tau(\theta)^{-1}((-\infty, 0])\big),\ \theta\big) \subset \mathbb{R}^1 \times S^1.$$

Proof. The validation of this theorem is inspired by the *graphic calculus* in [Ar1] that converts homotopies of loops in the functional space $\mathcal{F}_{\leq 2}$ into cobordisms (surgeries) of doodles in A with 2-moderate tangencies to the foliation $\mathcal{F}(\hat{v})$. With the help of this calculus, the isomorphism $\pi_1(\mathcal{F}^{\leq 2}, \mathbf{1}) \approx \mathbb{Z}$ is established [Ar1]. In the present case, the loops are images of boundaries of oriented compact surfaces under their immersions/embeddings in A. Fig. 5.6 shows our modification of this graphic calculus in action. A generator of $\pi_1(\mathcal{F}_{\leq 2}, \mathbf{1}) \approx \mathbb{Z}$ is depicted in Fig. 5.5, diagram (b).

If two 2-moderate immersions, $\alpha_0 : X_0 \to A$ and $\alpha_1 : X_1 \to A$, are cobordant with the help of a 2-moderate immersion $B : W \to A \times [0,1]$ as in Definition 5.6, then extending the function

$$z_{\alpha_0} \coprod z_{\alpha_1} : A \times (\{0\} \coprod \{1\}) \to \mathbb{R}$$

to a smooth function $Z : A \times [0,1] \to \mathbb{R}$ with similar properties delivers a homotopy between the loops $J_{z_{\alpha_0}}$ and $J_{z_{\alpha_1}}$ (see (5.2)) in $\mathcal{F}_{\leq 2}$.

On the other hand, if a path $\tau : ([0,1], \partial[0,1]) \to (\mathcal{F}^{\leq 2}, \mathbf{1})$ is transversal to the discriminant hypersurface $\mathcal{F}^{=2} \subset \mathcal{F}^{\leq 2}$, then the locus

$$K(\tau) =_{\mathsf{def}} \bigcup_{\theta \in [0,1]} \big(\tau(\theta)^{-1}(0),\, \theta\big) \subset \mathbb{R}^1 \times [0,1]$$

is a smooth embedded curve in A (see [KSW1], Lemma 3.4, for validation of this claim and its generalizations). Since the set $f^{-1}((-\infty, 0]))$ is compact for any $f \in \mathcal{F}_{\leq 2}$, the curve $K(\tau)$ is the boundary of a compact surface

$$X(\tau) =_{\mathsf{def}} \bigcup_{\theta \in [0,1]} \big(\tau(\theta)^{-1}((-\infty, 0])),\, \theta\big) \subset A.$$

Similar arguments hold for a homotopy

$$B : \big([0,1],\, \partial[0,1]\big) \times [0,1] \to (\mathcal{F}^{\leq 2}, \mathbf{1})$$

between two such paths τ_0 and τ_1, where B is transversal to $\mathcal{F}^{=2}$. The 3-fold

$$W(B) =_{\mathsf{def}} \bigcup_{\theta \in [0,1],\, t \in [0,1]} \big(B(\theta)^{-1}((-\infty, 0])),\, \theta,\, t\big) \subset A \times [0,1]$$

delivers the cobordism (as in Definition 5.6) between $X(\tau_0)$ and $X(\tau_1)$. Therefore,

$$J^{\mathsf{emb}} : \mathbf{B}^{\mathsf{emb}}_{\mathsf{moderate}_{\leq 2}}(A) \approx \pi_1(\mathcal{F}^{\leq 2}, \mathbf{1})$$

is a bijection (actually, a group isomorphism with respect to the operation $\uplus$).

This fact has a curious implication for the map $\mathcal{A}$ from (5.6): since the map $J^{\mathsf{imm}} : \mathbf{B}^{\mathsf{imm}}_{\mathsf{moderate}_{\leq 2}}(A) \to \pi_1(\mathcal{F}^{\leq 2}, \mathbf{1})$ is available, we can compose it with the inverse of the bijection J^{emb} to get a surjective map

$$\mathcal{R}_J : \mathbf{B}^{\mathsf{imm}}_{\mathsf{moderate}_{\leq 2}}(A) \to \mathbf{B}^{\mathsf{emb}}_{\mathsf{moderate}_{\leq 2}}(A) \approx \mathbb{Z}, \tag{5.9}$$

which serves as the right inverse of the map (homomorphism) $\mathcal{A}$; that is, $\mathcal{R}_J \circ \mathcal{A} = \mathsf{id}$. In fact, for $A = \mathbb{R} \times [0,1]$, $\mathcal{R}_J$ is a group epimorphism.

The case $A = \mathbb{R} \times S^1$ is similar, but the operation $\uplus$ is not available. Instead, the opreration $\star$ is available, but its commutativity is in question. □

We orient the surface A so that the θ-coordinate, corresponding to the trajectory space $\mathcal{T}(\hat{v}) = [0, 1]$ or S^1, is the first, and the u-coordinate, corresponding to the multiplier $\mathbb{R}$, is the second. With this counterclockwise orientation of A being fixed, any immersion $\alpha : X \to A$ induces an orientation of the surface X, thus choosing orientations of each component of ∂X. However, this induced orientation, may differer from the original orientation of X, an ingredient in the definition of $\mathbf{OB}^{\mathsf{imm}}_{\mathsf{moderate}\leq 2}(A)$!

Given an immersion $\alpha : (X, v) \subset (A, \hat{v})$ such that $\alpha(\partial X)$ has the properties as in (5.3), we attach a new α-dependent polarity to each "concave" point $a \in \partial_2^+ X(v)$: by definition, the polarity of a is "$\oplus$" if $\alpha_*(\nu_a)$, where ν_a is the inner normal to ∂X at a, points in the direction of the coordinate θ. Otherwise, the new polarity of a is defined to be "$\ominus$" (see Fig. 5.5). Equivalently, a is of polarity "$\oplus$" if crossing the critical value $\theta(a)$ in the positive direction increases the cardinality of the fiber of the map $\theta : \alpha(\partial_1 X) \to \mathcal{T}(\hat{v})$ in the vicinity of a.

As a result, the loci $\{\partial_j^+ X(v)\}_j$ and $\{\partial_j^- X(v)\}_j$ acquire the additional polarities $\oplus$ and $\ominus$; all together, four flavors for the tangencies of $\alpha(\partial X)$ to $\mathcal{F}(\hat{v})$ are available: $\{(+, \oplus), (+, \ominus), (-, \oplus), (-, \ominus)\}$. These flavors are independent of the orientations of X.

When dealing with *oriented* 2-moderate blobs and doodles, it will be useful to introduce still another polarity (orientation) of tangency points, which will be denoted "$\uparrow$" and "$\downarrow$". Each point $a \in \partial_2^+ X(v)$ has polarity "$\uparrow$" if the vector field $\hat{v}(a)$ is consistent with the orientation of $\alpha(\partial_1 X)$ at a. Otherwise, the polarity of a is "$\downarrow$". Similar definition is applied to the points of the locus $\partial_2^- X(v)$. As a result, each point of the locus $\partial_2 X(v)$, with the help of α, acquires the following eight flavors:

$$\{(+, \oplus, \uparrow),\ (+, \ominus, \uparrow),\ (-, \oplus, \uparrow),\ (-, \ominus, \uparrow),$$
$$(+, \oplus, \downarrow),\ (+, \ominus, \downarrow),\ (-, \oplus, \downarrow),\ (-, \ominus, \downarrow)\}.$$

Four of these eight flavors will play an essential role:

$$\Big\{ \underbrace{(+, \oplus, \uparrow),\ (+, \ominus, \downarrow)}_{clockwise},\ \underbrace{(+, \oplus, \downarrow),\ (+, \ominus, \uparrow)}_{counterclockwise} \Big\}, \tag{5.10}$$

the first pair occurring in the blobs oriented clockwise, the second pair in the blobs oriented counterclockwise.

Thus, with any embedded *oriented* surface $\alpha : X \hookrightarrow A$, we associate two integers

$$\mathsf{K}^\alpha =_{\mathsf{def}} \#\{\partial_2^{+,\,\oplus,\,\downarrow} X(\hat{v})\} - \#\{\partial_2^{+,\,\ominus,\,\uparrow} X(\hat{v})\},$$
$$\mathsf{L}^\alpha =_{\mathsf{def}} \#\{\partial_2^{+,\,\oplus,\,\uparrow} X(\hat{v})\} - \#\{\partial_2^{+,\,\ominus,\,\downarrow} X(\hat{v})\} \tag{5.11}$$

that correspond to the blobs that are oriented counterclockwise and clockwise, respectively.

We introduce also an integer

$$\mathsf{J}^\alpha =_{\mathsf{def}} \#\{\partial_2^{+,\oplus} X(\hat{v})\} - \#\{\partial_2^{+,\ominus} X(\hat{v})\} \tag{5.12}$$
$$(\text{in the oriented case, } \mathsf{J}^\alpha = \mathsf{K}^\alpha + \mathsf{L}^\alpha).$$

As the next theorem testifies, this number plays a role as a bordism invariant of immersions of non-oriented blobs and as a tool for characterizing elements of the fundamental group $\pi_1(\mathcal{F}^{\leq 2}, \mathbf{1})$.

Theorem 5.2. *Let* $A = \mathbb{R} \times [0,1]$.

• *Any regular embedding* $\alpha : (X, v) \to (A, \hat{v})$ *with only* 2*-moderate tangencies of* $\alpha(\partial X)$ *to* $\mathcal{F}(\hat{v})$ *produces a map* J_{z_α} *as in Theorem 5.1. Its homotopy class* $[J_{z_\alpha}] = \mathsf{J}^\alpha \cdot \kappa$, *where* κ *denotes a generator of* $\pi_1(\mathcal{F}^{\leq 2}, \mathbf{1}) \approx \mathbb{Z}$ *(shown in Fig. 5.5,* (b)*), and* $\mathsf{J}^\alpha \in \mathbb{Z}$.

The integer J^α *can be computed by the formula (5.12).*

• *Any regular embedding* $\alpha : (X, v) \to (A, \hat{v})$ *of an* oriented *surface* X *with only* 2*-moderate tangencies produces two maps*

$$K_{z_\alpha} : ([0,1],\, \partial[0,1]) \to (\mathcal{F}^{\leq 2}, \mathbf{1}) \text{ and } L_{z_\alpha} : ([0,1],\, \partial[0,1]) \to (\mathcal{F}^{\leq 2}, \mathbf{1})$$

as in Theorem 5.1; the first map is generated by the counterclockwise oriented blobs, and the second one by the clockwise oriented blobs. The homotopy classes of these maps are: $[K_{z_\alpha}] = \mathsf{K}^\alpha \cdot \kappa$ *and* $[L_{z_\alpha}] = \mathsf{L}^\alpha \cdot \kappa$.

The integers K^α *and* L^α *can be computed by formula (5.11).*

• *Moreover,* K^α *and* L^α *deliver an isomorphism*

$$\mathsf{K} \times \mathsf{L} : \mathbf{OB}^{\mathsf{emb}}_{\mathsf{moderate}\leq 2}(A) \xrightarrow{\approx} \mathbb{Z} \times \mathbb{Z}. \tag{5.13}$$

Proof. Let $d =_{\mathsf{def}} \max_{\hat{\gamma}} \#\{\hat{\gamma} \cap \alpha(\partial_1 X)\}$ be the maximal cardinality of the intersections of the $\hat{v}$-trajectories $\hat{\gamma}$ with the loop pattern $\alpha(\partial_1 X)$. Since X bounds $\partial_1 X$, d must be even.

We start with the case of non-oriented blobs.

For any 2-moderate immersion $\alpha : X \to A$, we pick an auxiliary function $z_\alpha : A \to \mathbb{R}$, adjusted to α as in (5.1). By the previous arguments, this choice of z_α produces the relative loop $J_{z_\alpha} : ([0,1], \partial[0,1]) \to (F_d^{\leq 2}, \mathbf{1})$.

Consider a cobordism $B : W \to A \times [0,1]$ as in Definition 5.6 and its restriction $B^\delta =_{\mathsf{def}} B| : \delta W \to A \times [0,1]$, followed by the obvious projection $\phi : A \times [0,1] \to [0,1]$. By an arbitrary C^∞-small perturbation of B, we may assume that the composition $f =_{\mathsf{def}} \phi \circ B^\delta$ is a Morse function. Since

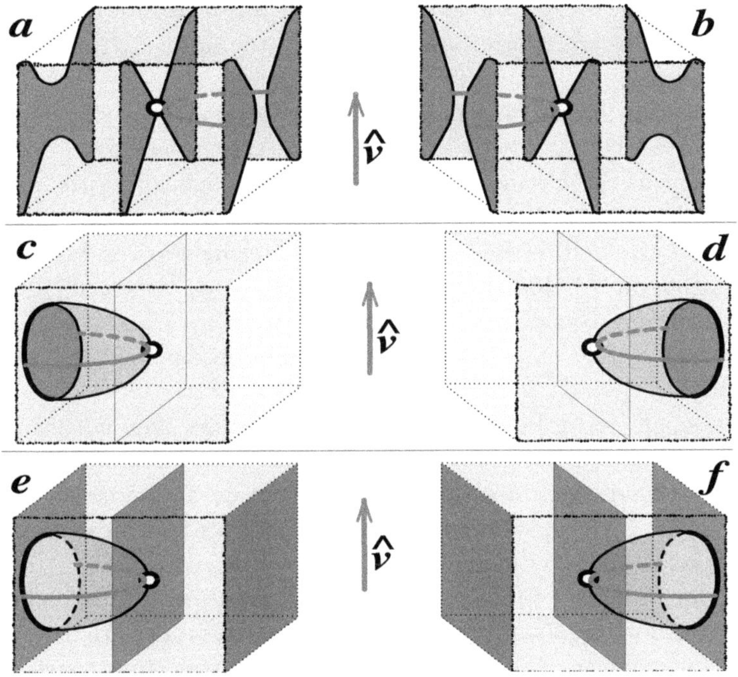

Fig. 5.3 Changing topology of slices $B^{-1}(A \times \{t\})$, as t crosses a critical value $t_\star$ of the Morse function $f : \delta W \to [0,1]$. Different shades correspond to different slices; each box is shown with 3 slices. In (a) and (b), the portion of $B(W)$ over a small interval $[t_\star - \epsilon, t_\star + \epsilon]$ is a pair of solid pants. In (c), (d), this portion is a solid half-ball. In (e) and (f), it is the complements to such half-balls in the solid cube. The figure does not show the complements to solid pants, depicted in (a) and (b). Note the "parabolic locus" (a dashed horizontal arc) in $B(\delta W)$, where the vector field $\hat{v}$ is quadratically tangent to the surface $B(\delta W)$.

the space $\mathcal{F}^{\leq 2}$ is open in $\mathcal{F}$, compactly supported perturbations of smooth 2-moderate functions or of their families remain 2-moderate. Therefore, we may assume that the cobordism B is such that $\phi \circ B^\delta$ is a Morse function.

Since, by property (6) from Definition 5.6, $B : W \to A \times [0,1]$ is a submersion and $f : \delta W \xrightarrow{B^\delta} A \times [0,1] \to [0,1]$ is a Morse function, as t ranges in $[0,1]$, the topology of a regular slice $B^{-1}(A \times \{t\}) \subset W$ may change via a relative, elementary surgery only when t crosses a critical value $t_\star$ of f. The topology of the slice $B^{-1}(A \times \{t\}) \cap \delta W$ changes via an elementary surgery inside δW as t crosses $t_\star$. Thus, there are four types of local surgery, shown in Fig. 5.3: two types of solid pants that correspond to f-critical points $(x_\star, t_\star) \in B(\delta W)$ of the Morse index 1 and two types of "indented" solids that correspond to f-critical points $(x_\star, t_\star) \in B(\delta W)$ of indexes 0 and 2.

It is crucial for our arguments that the geometry of the elementary surgery blocks $W \subset A \times [0, 1]$ is such that the surface $\delta W \subset \partial W$ has no tangency to the $\hat{v}^\bullet$-trajectories of order 3 or higher (see the parabolic tangency curves with dashed arcs from Fig. 5.3).

Now, let us describe an algorithm for the elementary moves (surgery) (see Fig. 5.6) that reduces a given pattern $X_0 = J_{z_\alpha}^{-1}((-\infty, 0]) \subset A$ to a pattern from *the canonical set of patterns* $\{n \cdot K\}_{n \in \mathbb{Z}}$ (as in Fig. 5.5) by a cobordism $B : W \to A \times [0, 1]$ (as in Definition 5.6) which is a regular embedding.

By a small perturbation of α_0, we may assume that no $\hat{v}$-trajectory has a combinatorial tangency pattern ω with two or more 2's. Then, we select the $\hat{v}$-trajectories $\hat{\gamma}_1, \dots \hat{\gamma}_s$ that contain the tangency patterns $\omega = (1, \dots, 1, 2, 1, \dots, 1)$, where the number of 1's that precede 2 is *odd* (which implies that the unique quadratic tangency point on $\hat{\gamma}_k$ resides in the *interior* of the set $\hat{\gamma}_k \cap \alpha_0(X_0)$). These trajectories correspond exactly to the "concave" points of $\partial_2^+(\alpha_0(X), \hat{v})$. The trajectories are listed by the order of their images $[\hat{\gamma}_1], \dots [\hat{\gamma}_s]$ in the oriented trajectory space $\mathcal{T}(\hat{v})$, a closed interval. We notice that such trajectories $\hat{\gamma}_k$ come in two flavors: "$\oplus$" and "$\ominus$", depending on the polarity of the tangency of $\hat{\gamma}_k$ to $\alpha_0(\partial X)$.

Then, for each adjacent pair $\hat{\gamma}_k, \hat{\gamma}_{k+1}$, we choose a $\hat{v}$-trajectory $\hat{\gamma}_k^\star$ in-between $\hat{\gamma}_k$ and $\hat{\gamma}_{k+1}$. Such $\hat{\gamma}_k^\star$ is traversal to $\alpha_0(X_0)$ and its combinatorial type $\omega(\hat{\gamma}_k^\star)$ is a sequence of 1's of a length $2q$. The intersection $\alpha_0(X_0) \cap \hat{\gamma}_k^\star$ is a disjoint union of closed intervals $I_{k,1}, \dots, I_{k,q}$. In the interior of each interval $I_{k,j}$ we pick a point $p_{k,j}^\star$ and will use it as the critical point for a surgery in $A \times [0, 1]$ on the solid $W_0 = \alpha_0(X_0) \times [0, 0.5)$ (see Fig. 5.3, (a) and (b)). If the polarity of $\hat{\gamma}_k$ is $\oplus$, we perform a surgery on W_0 as in Fig. 5.3, (a), if the polarity of $\hat{\gamma}_k$ is $\ominus$, we perform a surgery on W_0 as in Fig. 5.3, (b), by attaching solid pants Z_0 to W_0. We denote by W_1 the resulting solid $W_0 \cup Z_0$ in $A \times [0, 1]$ and by $F_1 : W_1 \to [0, 1]$ the obvious projection.

It is essential that, since F_1 is a Morse function, the attached pants Z_0 do not violate property (5) from Definition 5.6. Note that each elementary surgery of type (a) introduces a pair of new $\hat{v}^\bullet$-tangent points to the slice $\partial(F_1^{-1}(0.5))$; fortunately, none of new tangency pair is of the type $\omega = (1, \dots, 1, 2, 1, \dots, 1)$, where the number of 1's that precede 2 is odd (i.e., the pair belongs to $\partial_2^-(F_1^{-1}(0.5))$).

Now the preimage $F_1^{-1}(0.5)$ consists of several connected components $U_1, \dots U_s$, each of which is a domain in $A \times \{0.5\}$ which contains at most a single trajectory of the combinatorial type $(\dots, 1, 2, 1, \dots)$, where the number of 1's that precede 2 is odd and whose polarity is either $\oplus$ or $\ominus$.

If a connected component U is a ball with several holes, then applying the previous 1-surgery to a segment of trajectory in U that connects different connected components of ∂U, we will convert U into a topological 2-ball without introducing new trajectories of the combinatorial type $(\dots, 1, 2, 1, \dots)$ with the number of 1's that precede 2 being odd. Therefore, we may assume that all the connected components U are 2-balls.

Thus, we divide the balls U in three types:

(i) U intersects with a single trajectory of the polarity $\oplus$,

(ii) U intersects with a single trajectory of the polarity $\ominus$, and

(iii) U does not intersect with trajectories of the combinatorial type $(\dots, 1, 2, 1, \dots)$, where the number of 1's that precede 2 is odd.

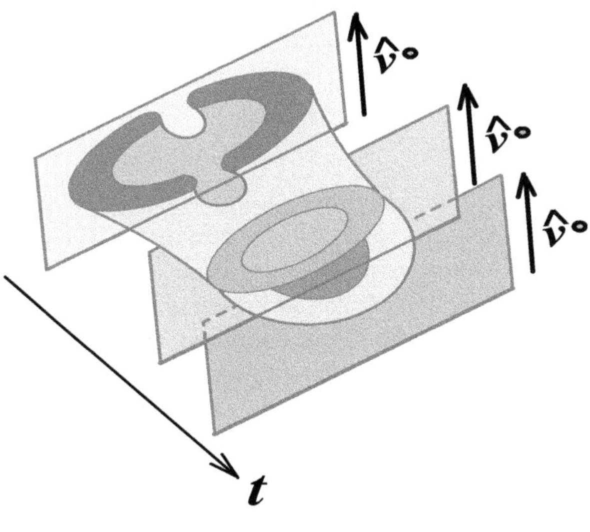

Fig. 5.4 Eliminating a pair of dark-shaded "kidneys", with their horns facing each other, by a surgery: the first surgery transforms the kidneys in the first slice into a ring in the second slice, and then the second surgery transforms the ring into an empty slice.

Next, we eliminate the type (iii) $\hat{v}^\bullet$-convex 2-balls by a 3-surgery on W_1 that amounts to attaching a few 3-balls to the slice $F_1^{-1}(0.5)$ as in

Fig. 5.3, (c). We denote by W_2 the resulting solid in $A \times [0,1]$, and by $F_2 : W_2 \to [0,1]$ the obvious projection. We may assume that the F_2-image of W_2 in $[0,1]$ resides in the interval $[0, 0.7)$. Because the balls of type (iii) are $\hat{v}^\bullet$-convex, we may assume that $\delta W_2 \setminus \delta W_1$ avoids $\hat{v}^\bullet$-tangencies of order ≥ 3. In particular, property (5) from Definition 5.6 is respected by these surgeries.

Now, we are left with the balls U in $F_2^{-1}(0.7)$ of types (i) and (ii) (with so called, Arnold's "kidneys"). Each pair of 2-balls of the distinct types (i) and (ii) can be eliminated by an isotopy in $A \times \{0.7\}$ (which involves scaling and parallel shifts), followed by a surgery, as shown in Fig. 5.4. Indeed, if the horns of a pair of kidneys U and U' of different polarities $\oplus$ and $\ominus$ are facing each other, then by attaching two solid pants we may surgery their union into a ring in a new slice. Then by attaching a relative handle $(D^2_+ \times S^1,\, D^1 \times S^1)$ to $A \times \{0\}$ in $A \times [0,1]$, we eliminate the ring. If the horns of U and U' of different polarities are facing in opposite directions, then by an isotopy in A (as in Lemma 5.2) we may switch their order along the $\mathcal{T}(\hat{v})$-direction. The switching will reduce the situation to the case when the horns face each other. Again, this surgery will not introduce new points of the combinatorial types $(\dots 121 \dots)$, where the number of 1's that precede 2 is odd. Thus, we will be left with a disjoint union of kidneys of the *same* polarity $\oplus$ or $\ominus$.

Let $\alpha : D^2 \to A$ be the regular imbedding realizing a $\oplus$-polarized kidney. By [Ar], under the map J^α, each kidney is mapped to a non-contractible loop Φ_α in $\mathcal{F}^{\leq 2}$, a generator of $\pi_1(\mathcal{F}^{\leq 2}, \mathbf{1}) \approx \mathbb{Z}$. Therefore, no further simplifications among the remaining kidneys are possible.

Thus, the difference between the numbers of kidneys of types (i) and (ii) is an invariant of the bordism class (introduced in Definition 5.6). This difference equals to the original difference $\#\{\partial_2^{+,\oplus} X(v)\} - \#\{\partial_2^{+,\ominus} X(v)\}$ between the numbers of $\hat{v}$-trajectories with polarities $\oplus$ and $\ominus$ and was preserved under all the surgeries that led to the final slice with the kidneys of the same polarity. Thus, the number $\mathsf{J}^\alpha = \#\{\partial_2^{+,\oplus} X(v)\} - \#\{\partial_2^{+,\ominus} X(v)\}$. did not change under all the surgeries above.

We are ready to investigate the 2-moderate bordisms $\mathbf{OB}^{\mathsf{emb}}_{\mathsf{moderate}\leq 2}(A)$ of embedded *oriented* blobs. Each connected blob can be oriented counter-clockwise, i.e., coherently with the fixed orientation of the ambient surface A, or clockwise, i.e., opposite to the fixed orientation of A. We aim to show that the numbers $\mathsf{K}^\alpha, \mathsf{L}^\alpha$ from (5.11) define an isomorphism (5.13).

It is possible to attach *embedded/immersed* 1-handles in A only to *similarly oriented* blobs: the "twisted 1-handles" cannot be immersed in A.

We notice (see Fig. 5.6, the lower diagram) that attaching a narrow 1-handle, whose core is transversal to the field $\hat{v}$, to similarly oriented blobs contributes a new pair of points of the polarities $(+, \oplus, \downarrow)$ and $(+, \ominus, \uparrow)$ (the counterclockwise oriented blobs), or a new pair of points of the polarities $(+, \oplus, \uparrow)$ and $(+, \ominus, \downarrow)$ (the clockwise oriented blobs). Deleting a 1-handle contributes a new pair of points of the polarities $(-, \ominus, \uparrow)$ and $(-, \oplus, \downarrow)$ (the counterclockwise oriented blobs), or $(-, \ominus, \downarrow)$ and $(-, \oplus, \uparrow)$ (the clockwise oriented blobs); however, these changes affect only the convex loci $\partial_2^-(\sim)$ which play secondary roles. Thus, all these surgeries do not change the values $\mathsf{K}^\alpha, \mathsf{L}^\alpha$. Similarly, adding or deleting a 2-ball D such that $\partial_2^+ D(\hat{v}) = \emptyset$, does not affect the values of $\mathsf{K}^\alpha, \mathsf{L}^\alpha$.

Thus, we will treat the counterclockwise and clockwise oriented blobs separately, recycling the our arguments in the non-oriented case. Each embedded connected blob $\alpha(X_0)$ is a disk or a disk with a number of holes. By deleting narrow 1-handles from $\alpha(X_0)$ as in Fig. 5.3, (a), or as in Fig. 5.6, we replace, via 2-moderate bordisms, each counterclockwise/clockwise oriented blob with holes by a similarly oriented topological 2-ball D_0. In the process, we keep the original values $\mathsf{K}^\alpha, \mathsf{L}^\alpha$. The tangency of ∂D_0 to $\hat{v}$ is still 2-moderate.

Now all the connected components of the new embedded surface $Y \subset A$ are 2-balls. As in the non-oriented case, by deleting 1-handles, each ball can be split into a number of balls, each of which has a single point from $\partial_2^+(\sim)$ at most. By a 2-surgery, we delete the balls which are convex with respect to $\hat{v}$ (i.e., they do not have singletons from $\partial_2^+(\sim)$). Now we a left with the balls that have a single point of the polarity from the list (5.10). Some of them are oriented counterclockwise, others clockwise.

Thus,

$$\mathsf{K} \times \mathsf{L} : \mathbf{OB}^{\mathsf{emb}}_{\mathsf{moderate}\leq 2}(A) \longrightarrow \mathbb{Z} \times \mathbb{Z}. \tag{5.14}$$

is an epimorphism.

Assuming that $\mathsf{K}^\alpha = 0 = \mathsf{L}^\alpha$ for some α, we see that this property allows to pair all the counterclockwise/clockwise oriented kidneys so that each pair by 1-surgery can be transformed into a ring and then eliminated as in Fig. 5.4. Therefore, $\mathsf{K} \times \mathsf{L}$ is a monomorphism. □

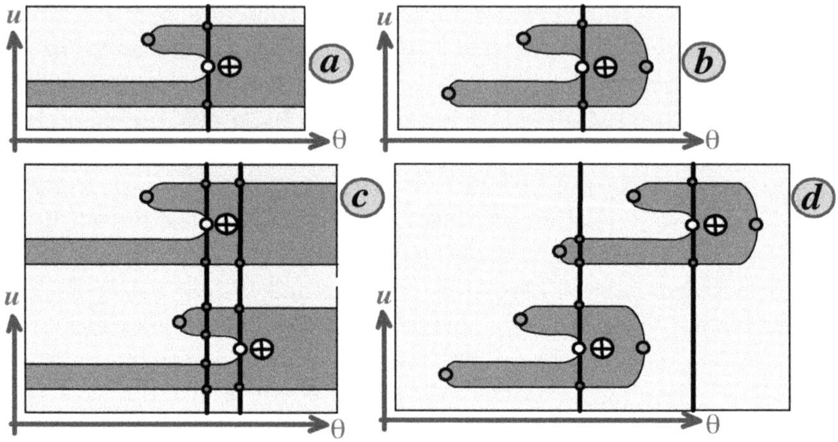

Fig. 5.5 Two portraits of a generator $\kappa \in \pi_1(\mathcal{F}^{\leq 2}, p)$; the base point $p \in \mathcal{F}^{\leq 2}$ is modeled after the polynomial $p(u) = u^4 - 1$ in diagram (a) and by $p(u) = u^4 + 1$ in diagram (b). Diagrams (c) and (d) portray 2κ. In diagrams (a) and (c) that represent the case $\mathcal{T}(\hat{v}) = S^1$, the left and the right edges of the rectangle should be identified so that the shaded regions match. Note the polarity $\oplus$ of the tangent $\hat{v}$-trajectories with the combinatorial pattern $\omega = (\dots 121 \dots)$, where the number of 1's that precede 2 is odd.

As in the case of non-oriented blobs, using formulas (5.11), we get a surjective map

$$\mathcal{R}_{\mathsf{K,L}} : \mathbf{OB}^{\mathsf{imm}}_{\mathsf{moderate}_{\leq 2}}(A) \to \mathbb{Z} \times \mathbb{Z} \approx \mathbf{OB}^{\mathsf{emb}}_{\mathsf{moderate}_{\leq 2}}(A), \tag{5.15}$$

which serves as the right inverse to the obvious homomorphism

$$\mathbf{OB}^{\mathsf{emb}}_{\mathsf{moderate}_{\leq 2}}(A) \to \mathbf{OB}^{\mathsf{imm}}_{\mathsf{moderate}_{\leq 2}}(A).$$

Therefore, $\mathbf{OB}^{\mathsf{imm}}_{\mathsf{moderate}_{\leq 2}}(A)$ contains canonically $\mathbb{Z} \times \mathbb{Z}$.

Remark 5.1. The number $c^+(v) =_{\mathsf{def}} \#\{\partial_2^{+,\oplus}(v)\} + \#\{\partial_2^{+,\ominus}(v)\}$ may be interpreted as the *complexity* of the vector field v on X [K4], [K7]. We notice that

$$c^+(v) \geq \quad |\mathsf{J}^\alpha| =_{\mathsf{def}} |\#\{\partial_2^{+,\oplus}(v)\} - \#\{\partial_2^{+,\ominus}(v)\}|.$$

It is somewhat surprising that the invariant $\mathsf{J}^\alpha = \#\{\partial_2^{+,\oplus}(v)\} - \#\{\partial_2^{+,\ominus}(v)\}$ reflects more the local topology of the embedding α (or of the field $v = \alpha^*(\hat{v})$) than the global topology of the embedded surface X: in fact, any integral value of J^α can be realized by a traversally generic field v on a 2-ball D which even admits a convex envelop! A portion of the boundary ∂D looks like a snake with respect to the field $\hat{v}$ of the envelop.

For any X, the effect of deforming a portion of the boundary $\partial_1 X \subset A$ into a snake is equivalent to adding several spikes (an edge and a pair of a univalent and a trivalent vertices) to the graph $\mathcal{T}(v)$, the space of v-trajectories. Evidently, these operations do not affect $H_1(\mathcal{T}(v);\mathbb{Z}) \approx H_1(X;\mathbb{Z})$.

For example, for α as in Fig. 5.2, $\mathsf{J}^\alpha = 0$. If we subject α to an isotopy in A that introduces a snake-like pattern of Fig. 5.5, (a), then for the new immersion α', the invariant $\mathsf{J}^{\alpha'} = 1$.

In contrast, the number $\#\{\partial_2^{+,\oplus}(v)\} + \#\{\partial_2^{+,\ominus}(v)\}$ has a topological significance for X. If X is the compliment to k disjoint balls in a closed orientable surface with g handles, then by [K7], Lemma 1.1,

$$\#\{\partial_2^{+,\oplus}(v)\} + \#\{\partial_2^{+,\ominus}(v)\} \geq 4g - 4 + 2k.$$

$\diamondsuit$

5.3 Invariants that distinguish between bordisms of immersions and embeddings

Recall that $\mathcal{A}^\times : \mathbf{B}^{\times \mathsf{imm}}_{\mathsf{moderate}\leq 2}(A) \to \mathbf{B}^{\mathsf{imm}}_{\mathsf{moderate}\leq 2}(A)$ is a bijection. Therefore, without lost of generality, we deal now with immersions under additional assumption (5.8): we require that their self-intersections are transversal. We adopt the notations from Definition 5.6.

Consider an immersion $\alpha : X \to A$ with transversal self-intersections of $\alpha(\partial_1 X)$ and no multiple self-intersections of more than two local branches. At each self-intersection $a \in \alpha(\partial_1 X)$, there are exactly two points $a', a'' \in \partial X$ such that $\alpha(a') = \alpha(a'') = a$. Then we take two small neighborhoods $D_+(a')$ and $D_+(a'')$ of a' and a'', respectively, both diffeomorphic to a half-disk. The linearizations of their α-images produces two preferred half-spaces, $H_{a'}$ and $H_{a''}$, in the tangent space $T_a A$. Put $K_{a',a''} = H_{a'} \cap H_{a''}$. We denote by $K^\vee_{a',a''}$ the image of $K_{a',a''}$ under the central symmetry in $T_a A$ (see Fig. 5.7).

Using the sectors $K_{a',a''}$ and $K^\vee_{a',a''}$, we will divide the self-intersections $\{a\}$ into four types:

• type **I** occurs when the vector $\hat{v}(a)$ points in the interior of the sector $K_{a',a''}$,

• type **III** occurs when the vector $\hat{v}(a)$ points in the interior of the sector $K^\vee_{a',a''}$;

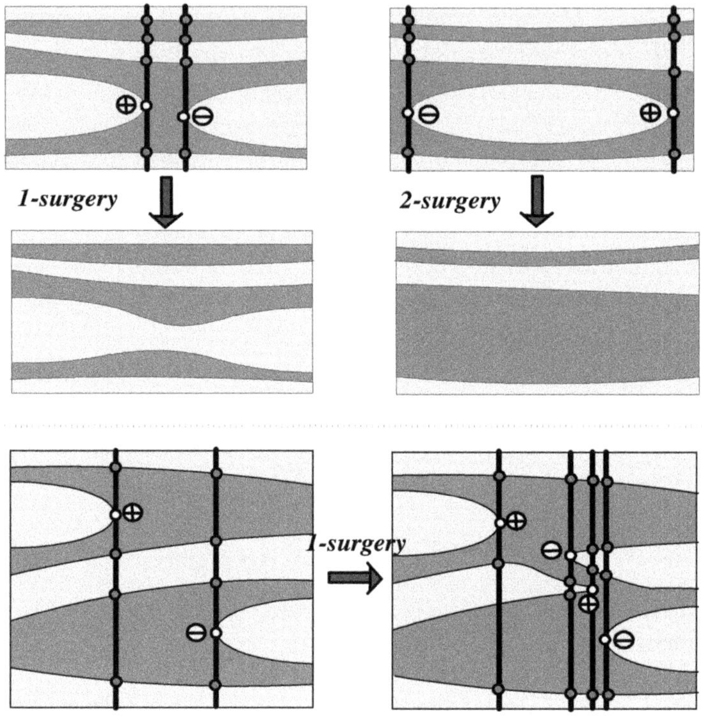

Fig. 5.6 The region $\alpha(X)$ in the cylinder $A = \mathbb{R} \times S^1$ is shaded, the vector field $\hat{v}$ is vertical. Elementary cancellations, via surgery in A, of a pair tangencies of $\alpha(\partial X)$ to $\hat{v}$ of opposite polarities $\oplus, \ominus$ (the upper diagrams). Increasing the local connectivity of the region $\alpha(X)$ between two tangent trajectories of opposite polarities by a 1-surgery in A (the lower diagram). This operation introduces a new pair of points of opposite polarities $\oplus, \ominus$.

• types **II** and **IV** arise when $\hat{v}(a)$ points in the interior of the complimentary to $K_{a',a''} \cup K^{\vee}_{a',a''}$ sectors. Type **II** corresponds to the case when the sector $K_{a',a''}$ is on the right of the line γ_a, while type **IV** corresponds to the case when the sector $K_{a',a''}$ is on the left of the line γ_a.

Note that the types **II** and **IV** are "crudely symmetric" with respect to a reflection in A that has the $\hat{v}$-trajectory γ_a through a as a line of symmetry.

Consider a small disk D, centered on a crossing point a. We consider a diffeotopy $\{\Psi_\phi : A \to A\}_\phi$ whose final stage is the turn on a given angle $\phi_\star$ inside the concentric disk $D' \subset D$ and the identity outside D. Note

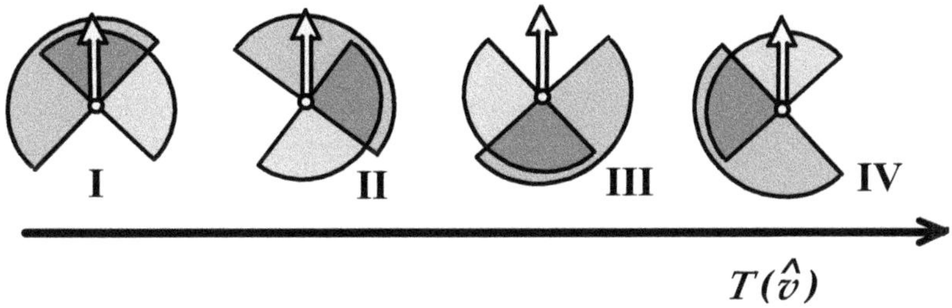

Fig. 5.7 Four configurations **I, II, III, IV** in the vicinity of a point $a \in A$ where two branches of $\alpha(\partial X)$ intersect transversally. The figure shows the α-images of neighborhoods in X of the two points $b_1, b_2 \in \alpha^{-1}(a) \cap \partial_1 X$. Note the position of the vector $\hat{v}(a)$ relative to the darkly shaded sector. The trajectory space $\mathcal{T}(\hat{v})$ is oriented.

that using Ψ_ϕ as a dial at a, we can change the type $\{$**I, II, III, IV**$\}$ of $\Psi_{\phi_\star}(\alpha(\partial_1 X))$ at a at will. However, $\alpha_{\phi_\star}$, the composition of α with $\Psi_{\phi_\star}$, is *not* 2-moderately cobordant to α: in the process, the curves $\alpha_\phi(\partial_1 X)$ may develop inflection points (cubic tangencies) to the foliation $\mathcal{F}(\hat{v})$. Moreover, the 2-moderate $\alpha_{\phi_\star}(\partial_1 X)$ may develop inside D new quadratic tangencies to $\mathcal{F}(\hat{v})$. Thus, there is a subtle interplay between the count of crossings of a particular type from $\{$**I, II, III, IV**$\}$ and the invariants of the type J^α.

Let us find out which of these four types $\{$**I, II, III, IV**$\}$ or their combinations are invariant under non-oriented cobordisms of immersions with 2-moderate tangencies.

We start with a key observation that guides us in this section.

Given a cobordism-immersion $B : W \to A \times [0,1]$ between two immersions, $\alpha_0 : X_0 \to A \times \{0\}$ and $\alpha_1 : X_1 \to A \times \{1\}$, with transversal self-intersection of $B(\delta W)$ of multiplicity 2 at most, consider the curve ℓ where the surface $B(\delta W) \subset A \times [0,1]$ self-intersects transversally. Either ℓ is:

(a) a simple segment in $A \times [0,1]$ that connects a pair of crossings of $\alpha_0(\partial X_0)$, or

(b) a pair of crossings of $\alpha_1(\partial X_1)$, or

(c) a pair of crossings, one of which belongs to $\alpha_0(\partial X_0)$ and the other one to $\alpha_1(\partial X_1)$, or

(d) a simple loop in $A \times (0,1)$.

The multiplicity of each intersection of the curve ℓ with a $\hat{v}^\bullet$-trajectory $\hat{\gamma} \subset A \times [0,1]$ is at least 2, and, according to Definition 5.6, cannot exceed 2. Therefore, each $\hat{v}^\bullet$-trajectory must be *transversal* to ℓ: otherwise, the multiplicity of the intersection $\ell \cap \hat{\gamma}$, where $\hat{\gamma}$ is tangent to ℓ, exceeds 2.

For each point $x \in \ell$, the two ℓ-localized branches, $B(\delta W_1)$ and $B(\delta W_2)$, of $B(\delta W)$ that intersect along ℓ divide the tubular neighborhood U_ℓ of ℓ into four chambers $S_x^{\mathbf{I}}, S_x^{\mathbf{II}}, S_x^{\mathbf{III}}, S_x^{\mathbf{IV}}$, where each point from the interior of $S_x^{\mathbf{I}}$ has two α-preimages in the vicinity of $\delta W_1 \cup \delta W_2$ in W, each point from the interior of $S_x^{\mathbf{III}}$ has no α-preimages in the vicinity of $\delta W_1 \cup \delta W_2$ in W, and each point from the interior of $S_x^{\mathbf{II}} \cup S_x^{\mathbf{IV}}$ has one α-preimage in the vicinity of $\delta W_1 \cup \delta W_2$ in W. Thanks to the transversality of the vector field $\hat{v}^\bullet$ to the curve ℓ, the vector field must point into the interior of one of the chambers; i.e., $\hat{v}^\bullet$ is not tangent to the branches $B(\delta W_1)$ and $B(\delta W_2)$. In $3D$, we cannot distinguish between the chambers $S_x^{\mathbf{II}}$ and $S_x^{\mathbf{IV}}$ without picking an orientation of ℓ. Thus, $S_x^{\mathbf{I}}$, $S_x^{\mathbf{III}}$, and $S_x^{\mathbf{II}} \cup S_x^{\mathbf{IV}}$ are geometrically distinct.

Now $\hat{v}^\bullet$ helps to define the three types of ℓ (not to be confused with the four types (a)-(d)): if $\hat{v}^\bullet(x)$ points into $S_x^{\mathbf{I}}$ or $S_x^{\mathbf{III}}$, we say that the curve ℓ is of types **I** or **III**, respectively. Otherwise, ℓ is of the "mixed type" **II** & **IV**.

In all cases (a)-(d), $\hat{v}^\bullet$ preserves the three champer types **I**, **III**, **II** & **IV** along ℓ. In fact, the case (d), in which ℓ is a loop, is irrelevant for what follows.

Let $x \cup y = \partial \ell$. Crucially, in cases (a)-(b), the type $S_x^{\mathbf{II}}$ turns into the type $S_y^{\mathbf{IV}}$, and $S_x^{\mathbf{IV}}$ turns into the type $S_y^{\mathbf{II}}$, while the types $S_x^{\mathbf{I}}$ and $S_y^{\mathbf{I}}$, $S_x^{\mathbf{III}}$ and $S_y^{\mathbf{III}}$ are the same.

Therefore, in cases (a)-(b) we get the following possible combinations: $S_x^{\mathbf{I}}$ and $S_y^{\mathbf{I}}$, or $S_x^{\mathbf{II}}$ and $S_y^{\mathbf{IV}}$, or $S_x^{\mathbf{III}}$ and $S_y^{\mathbf{III}}$, or $S_x^{\mathbf{IV}}$ and $S_y^{\mathbf{II}}$. In the case (c), all the types at x and y are the same.

Given an immersion $\alpha : X \to A$ as in Definition 5.6, let us denote by $\rho_{\mathbf{I}}(\alpha), \rho_{\mathbf{II}}(\alpha), \rho_{\mathbf{III}}(\alpha), \rho_{\mathbf{IV}}(\alpha)$ the number of crossings of $\alpha(\partial X)$ of the types **I**, **II**, **III**, **IV**, where the type of a crossing is determined by the configuration of the sector, formed by the two intersecting domains, and the vector $\hat{v}$ at the crossing.

For example, in Fig. 5.2, all the four crossings of $\alpha(\partial X)$ are of the different types **I**, **II**, **III**, **IV**; that is, $\rho_{\mathbf{I}}(\alpha) = 1$, $\rho_{\mathbf{II}}(\alpha) = 1$, $\rho_{\mathbf{III}}(\alpha) = 1$, $\rho_{\mathbf{IV}}(\alpha) = 1$. These pairings between different crossings of $\alpha_0(\partial X_0)$ and $\alpha_1(\partial X_1)$, which are delivered by the curves $\{\ell\}$ of types (a), (b), (c), lead directly to the following assertion.

Lemma 5.3. *If two 2-moderate immersions $\alpha_0 : X_0 \to A$ and $\alpha_1 : X_1 \to A$ are cobordant in $\mathbf{B}^{\times\mathsf{imm}}_{\mathsf{moderate}\leq 2}(A) \approx \mathbf{B}^{\mathsf{imm}}_{\mathsf{moderate}\leq 2}(A)$, then*

$$\rho_{\mathbf{I}}(\alpha_0) \equiv \rho_{\mathbf{I}}(\alpha_1) \mod 2, \qquad (5.16)$$
$$\rho_{\mathbf{III}}(\alpha_0) \equiv \rho_{\mathbf{III}}(\alpha_1) \mod 2,$$
$$\rho_{\mathbf{II}}(\alpha_0) - \rho_{\mathbf{IV}}(\alpha_0) = \rho_{\mathbf{II}}(\alpha_1) - \rho_{\mathbf{IV}}(\alpha_1).$$
◇

Thus, we produced three numerical invariants that have a potential to distinguish between the elements of the bordism $\mathbf{B}^{\mathsf{imm}}_{\mathsf{moderate}\leq 2}(A)$, modulo the ones that are in the image of the map $\mathbf{B}^{\mathsf{emb}}_{\mathsf{moderate}\leq 2}(A) \to \mathbf{B}^{\mathsf{imm}}_{\mathsf{moderate}\leq 2}(A)$.

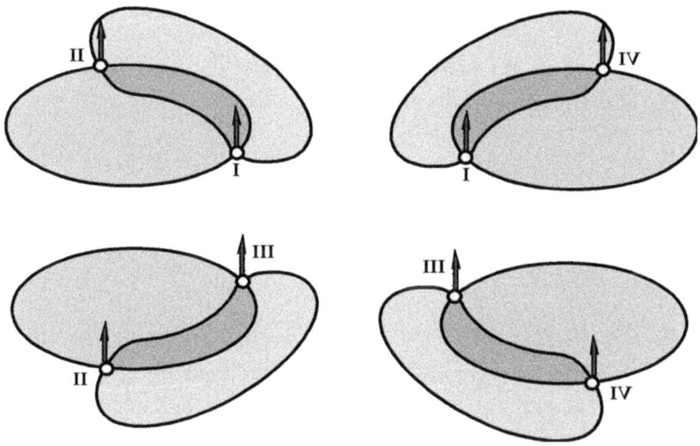

Fig. 5.8 Basic immersions of two disks in the surface A and the intersection patterns $\mathbf{I}, \mathbf{II}, \mathbf{III}, \mathbf{IV}$ from Fig. 5.7 they generate. As elements of $\mathbf{B}^{\mathsf{imm}}_{\mathsf{moderate}\leq 2}(A)$, the immersion on the left is minus the corresponding immersion on the right.

Proposition 5.3. *For $A = \mathbb{R} \times [0,1]$, the invariants $\rho_{\mathbf{I}}(\alpha)$, $\rho_{\mathbf{III}}(\alpha) \in \mathbb{Z}_2$, $\rho_{\mathbf{II}}(\alpha) - \rho_{\mathbf{IV}}(\alpha) \in \mathbb{Z}$ define a homomorphism*

$$\mathcal{I}\rho: \ \mathbf{B}^{\mathsf{imm}}_{\mathsf{moderate}\leq 2}(A)/\mathbf{B}^{\mathsf{emb}}_{\mathsf{moderate}\leq 2}(A) \to \mathbb{Z}_2 \times \mathbb{Z}_2 \times \mathbb{Z}$$

whose image is an abelian subgroup $\mathcal{M} \approx \mathbb{Z}_2 \times \mathbb{Z}$ of index 2. The monomorphism $\phi : \mathcal{M} \hookrightarrow \mathbb{Z}_2 \times \mathbb{Z}_2 \times \mathbb{Z}$ is delivered by the diagonal map $\mathbb{Z}_2 \to \mathbb{Z}_2 \times \mathbb{Z}_2$ on $\mathbb{Z}_2$ and $\phi(\mathbb{Z}) = 2\mathbb{Z}$.

Proof. The argument is based on the proof of Lemma 5.3 and on formulas (5.16).

Examining the types of crossings in the four diagrams from Fig. 5.8 that show the four immersions $\alpha_1, \alpha_2, \alpha_3, \alpha_4$, we get:

$$\begin{aligned}
&\rho_{\mathbf{I}}(\alpha_1) = 1,\ \ \rho_{\mathbf{III}}(\alpha_1) = 0,\ \ \rho_{\mathbf{II}}(\alpha_1) - \rho_{IV}(\alpha_1) = 1, \\
&\rho_{\mathbf{I}}(\alpha_2) = 1,\ \ \rho_{\mathbf{III}}(\alpha_2) = 0,\ \ \rho_{\mathbf{II}}(\alpha_2) - \rho_{\mathbf{IV}}(\alpha_2) = -1, \\
&\rho_{\mathbf{I}}(\alpha_3) = 0,\ \ \rho_{\mathbf{III}}(\alpha_3) = 1,\ \ \rho_{\mathbf{II}}(\alpha_3) - \rho_{\mathbf{IV}}(\alpha_3) = 1, \\
&\rho_{\mathbf{I}}(\alpha_4) = 0,\ \ \rho_{\mathbf{III}}(\alpha_4) = 1,\ \ \rho_{\mathbf{II}}(\alpha_4) - \rho_{\mathbf{IV}}(\alpha_4) = -1.
\end{aligned} \tag{5.17}$$

Note that in $\mathbf{B}^{\mathsf{imm}}_{\mathsf{moderate}\leq 2}(A)$, $\alpha_2 = -\alpha_1$ and $\alpha_4 = -\alpha_3$.

Also, for $\alpha : X \to A$ from Fig. 5.2, where X is a torus with a hole, we have $\rho_{\mathbf{I}}(\alpha) = 1$, $\rho_{\mathbf{III}}(\alpha) = 1$, and $\rho_{\mathbf{II}}(\alpha_4) - \rho_{\mathbf{IV}}(\alpha_4) = 0$.

Consider now just these three immersions: α_1, α_3, and α. By linear algebra, applied to the invariants $\rho_{\mathbf{I}}(\sim), \rho_{\mathbf{III}}(\sim)$, and $\rho_{\mathbf{II}}(\sim) - \rho_{\mathbf{IV}}(\sim)$ of α_1, α_3, and α, the image of the map

$$\mathcal{I}\rho :\ \mathbf{B}^{\mathsf{imm}}_{\mathsf{moderate}\leq 2}(A) \to \mathbb{Z}_2 \times \mathbb{Z}_2 \times \mathbb{Z}$$

is a subgroup of index 2 at most, which contains the $\mathbb{Z}$-module $\mathcal{M}$, spanned by the triples $(e, 0; 1), (0, e; 1), (e, e; 0)$, where $2e = 0$.

Lemma 5.4. *No loop $\mathcal{C}$ in $A = \mathbb{R} \times [0, 1]$ that bounds an immersed orientable surface Σ has an* odd *self-intersection number.*

Proof. Indeed, if an immersion $\alpha : \Sigma \to A$ with $\alpha(\partial\Sigma) = \mathcal{C}$ exists, then the pull-back vector field $v = \alpha^\dagger(\hat{v}) \neq 0$. Thus, its index $\mathsf{ind}(v) = 0$. Recall that *deg*(G), the degree of the Gauss normal map $G : \alpha(\partial\Sigma) \to S^1$ can be calculated by the Hopf-Gottlieb formula ([H], [Got]):

$$\deg(G) = \chi(\Sigma) - \mathsf{ind}(v) = \chi(\Sigma) = 1 - \beta_1(\Sigma),$$

where the first Betti number $\beta_1(\Sigma) = 2g$ is even. On the other hand, by the Whiney formula [W1],

$$\deg(G) = \mu + N^+ - N^-,$$

where $\mu = \pm 1$ and N^+, N^- count the positive and negative crossings of $\mathcal{C}$ (this polarity of crossings is based on "global" considerations). So, each "kink" of $\mathcal{C}$ adds ± 1 to $\deg(G)$. Odd number of self-crossings $N^+ + N^-$, by the Whiney formula, produces an even degree $\deg(G)$, which contradicts to the fact that $1 - \beta_1(\Sigma)$ is an odd number. In fact, the minimal number of self-intersections an orientable immersed surface of genus g with a circular boundary may have is $2g + 2$ [Gu]. □

Thus, by Lemma 5.4, none of the elements $(e, 0, 0), (0, e, 0), (0, 0, 1) \in \mathbb{Z}_2 \times \mathbb{Z}_2 \times \mathbb{Z}$ resides in the image of $\mathcal{I}\rho$. However, the elements $(e, 0, 1), (0, e, 1), (0, 0, 2) \in \mathbb{Z}_2 \times \mathbb{Z}_2 \times \mathbb{Z}$ do belong to the $\mathcal{I}\rho$-image. Therefore, the $\mathcal{I}\rho$-image is $\mathcal{M} =_{\mathsf{def}} \mathsf{span}_{\mathbb{Z}}\{(e, 0, 1), (0, e, 1), (0, 0, 2)\} \approx \mathbb{Z}_2 \times \mathbb{Z}$.

Evidently, $\mathbf{B}^{\mathsf{emb}}_{\mathsf{moderate}\leq 2}(A) \approx \mathbb{Z}$ is in the kernel of $\mathcal{I}\rho$. This completes the proof of Proposition 5.3. □

Note that *oriented* immersed doodles also generate patterns similar to the ones in Fig. 5.7 at each intersection point $a \in A$. Indeed, let $\beta : \mathcal{C} \to A$ be an immersion as in (5.3). Consider the two local branches $\mathcal{C}'_a$ and $\mathcal{C}''_a$ of $\beta(\mathcal{C})$ at a that are transversal and oriented. The pair of tangent vectors u'_a and u''_a to $\mathcal{C}'_a$ and to $\mathcal{C}''_a$ that are consistent with the orientations, generate a particular *sector* in T_aA, the convex hull of u'_a and u''_a (see Fig. 5.9, a).

By counting the four types of sectors in relation to $\hat{v}(x)$, which an oriented immersion β generates at its crossings $\{a\}$, we get the new quantities $\rho_{\mathsf{I}}(\beta)$, $\rho_{\mathsf{II}}(\beta)$, $\rho_{\mathsf{III}}(\beta)$, $\rho_{\mathsf{IV}}(\beta)$. The sector's number may be different by a permutation of four elements from the darkly shaded sectors in Fig. 5.7.

Therefore, the similar kind of invariants $\rho_{\mathsf{I}}(\beta) - \rho_{\mathsf{III}}(\beta)$, $\rho_{\mathsf{II}}(\beta) - \rho_{\mathsf{IV}}(\beta) \in \mathbb{Z}$ are available for the oriented doodles.

As in the case of blobs, these three quantities are invariants of the quotient $\mathbf{OC}^{\mathsf{imm}}_{\mathsf{moderate}\leq 2}(A)/\mathbf{OC}^{\mathsf{emb}}_{\mathsf{moderate}\leq 2}(A)$. The arguments that validate this claim are exactly the same as the ones that led to Lemma 5.3 for blobs. Moreover, for $A = \mathbb{R} \times [0, 1]$, these invariants are additive with respect to the connected sums $\uplus$ of oriented doodles.

Proposition 5.4. *For any oriented doodle* $\beta : \mathcal{C} \to A$, *the invariants* $\rho_{\mathsf{I}}(\beta) - \rho_{\mathsf{III}}(\beta)$, $\rho_{\mathsf{II}}(\beta) - \rho_{\mathsf{IV}}(\beta) \in \mathbb{Z}$ *define a surjective map*

$$\mathcal{I}\rho : \mathbf{OC}^{\mathsf{imm}}_{\mathsf{moderate}\leq 2}(A)/\mathbf{OC}^{\mathsf{emb}}_{\mathsf{moderate}\leq 2}(A) \to \mathbb{Z} \times \mathbb{Z}.$$

For $A = \mathbb{R} \times [0, 1]$, *the map* $\mathcal{I}\rho$ *is an epimorphism with respect to the group operation* $\uplus$.

Proof. We recycle the list of ρ-invariants for the oriented boundaries of submersions α_1, α_3, and α from the proof of Proposition 5.3 and add to them the ρ-invariants of new doodles that do not bound immersions of surfaces. Differently oriented figures "∞", placed in A horizontally with respect to the vector field $\hat{v}$ by immersions $\beta_+, \beta_- : S^1 \to A$, realize the values

$$\{\rho_{\mathsf{I}}(\beta_+) = 1,\ \rho_{\mathsf{III}}(\beta_+) = 0,\ \rho_{\mathsf{II}}(\beta_+) - \rho_{\mathsf{IV}}(\beta_+) = 0\}$$

and

$$\{\rho_{\mathsf{I}}(\beta_-) = 0,\ \rho_{\mathsf{III}}(\beta_-) = 1,\ \rho_{\mathsf{II}}(\beta_-) - \rho_{\mathsf{IV}}(\beta_-) = 0\}.$$

Consider another oriented doodle $\tilde{\beta} : S^1 \to A$ (not a boundary of an immersion) with a single self-intersection whose image is a "loop within a loop", symmetric with respect to the $\hat{v}$-trajectory through the self-intersection point. It realizes the values

$$\{\rho_{\mathsf{I}}(\tilde{\beta}) = 0,\ \rho_{\mathsf{III}}(\tilde{\beta}) = 0,\ \rho_{\mathsf{II}}(\tilde{\beta}) = 1, \rho_{\mathsf{IV}}(\tilde{\beta}) = 0\}.$$

Flipping the orientation of S^1, we realize the values

$$\{\rho_{\mathsf{I}}(\bar{\beta}) = 0,\ \rho_{\mathsf{III}}(\bar{\beta}) = 0,\ \rho_{\mathsf{II}}(\bar{\beta}) = 0, \rho_{\mathsf{IV}}(\bar{\beta}) = 1\}.$$

Therefore, by linear algebra, the map

$$\mathcal{I}\rho :\ \mathbf{OC}^{\mathsf{imm}}_{\mathsf{moderate}\leq 2}(A)/\mathbf{OC}^{\mathsf{emb}}_{\mathsf{moderate}\leq 2}(A) \to \mathbb{Z} \times \mathbb{Z}.$$

is surjective. For $A = \mathbb{R} \times [0,1]$, the map $\mathcal{I}\rho$ is an epimorphism with respect to the group operation $\uplus$. $\square$

Let us introduce one useful operation/notation. Given an (oriented) immersion $\beta : \mathcal{C} \to A$ (a doodle) and a nonnegative number n, we denote by $\beta(n)$ the new immersion, obtained by surrounding $\beta(\mathcal{C})$ by n counterclockwise oriented nested simple loops. If n is negative, $\beta(n)$ denotes an immersion in which $\beta(\mathcal{C})$ is surrounded by n clockwise oriented nested simple loops. When talking about 2-moderate immersions, we assume that the surrounding simple loops are 2-moderate as well. Similarly, given an (oriented) immersion $\alpha : X \to A$ (a blob) and a nonnegative n, $\alpha(n)$ denotes a submersion in which $\alpha(X)$ is surrounded by n nested (counterclockwise oriented) embedded disks whose boundaries are disjoint and 2-moderate. Again, for a negative n, we flip the orientations of the disks.

Let us list a few simple properties of these operations. We skip their straightforward validation. We consider all the immersions as elements of the appropriate (2-moderate) bordism groups, so that the following identities are understood as taking place in these groups:

$$\begin{aligned} \beta_1(n) \uplus \beta_2(n) &= (\beta_1 \uplus \beta_2)(n), \\ (\beta(n))(m) &= \beta(n+m); \end{aligned} \tag{5.18}$$

$$\begin{aligned} \alpha_1(n) \uplus \alpha_2(n) &= (\alpha_1 \uplus \alpha_2)(n), \\ (\alpha(n))(m) &= \alpha(n+m). \end{aligned} \tag{5.19}$$

We will use the operations $\odot_n : \beta \rightsquigarrow \beta(n)$ or $\odot_n : \alpha \rightsquigarrow \alpha(n)$, in combination with the addition $\uplus$, as main tools for generating new examples of immersions. In particular, we will apply $\odot_n$ to the immersed 2-moderate figures "∞" and "8", to get families of immersions "$\infty(n)$" and "$8(n)$" (see Fig. 5.9, (a), which depicts the immersion $\infty(2) \uplus \overline{\infty}$). We distinguish between "∞" and "8" by their position in relation to the vector field $\hat{v}$, so that "∞" stands for the crossing of types I, or III, while "8" for the crossing of types II, IV.

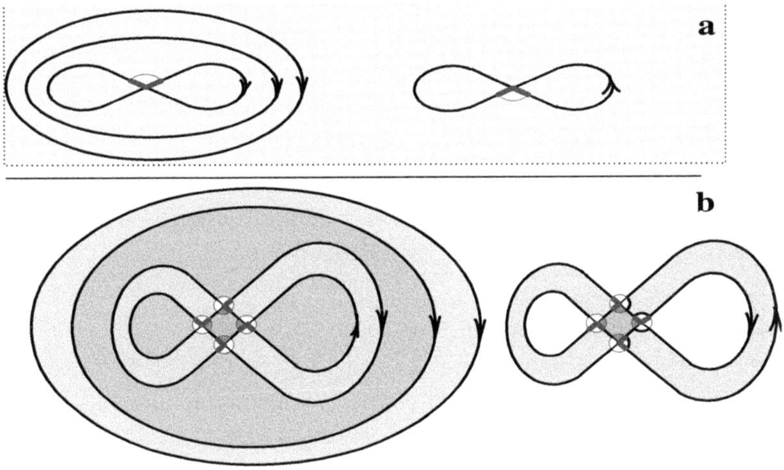

Fig. 5.9 (a) Oriented doodles $\infty(2) \uplus \overline{\infty}$; the two crossings are of the types I and III. (b) Oriented blobs ∞-ribbon(2) $\uplus\, \overline{\infty}$-ribbon; the four crossings on the left are of the types (**III**, II), (**IV**, III), (**I**, IV), (**II**, I), and on the right of the types (**III**, IV), (**IV**, I), (**I**, II), (**II**, III). The vector field $\hat{v}$ is vertical.

We can improve the claim of Proposition 5.4 by analyzing algebraically and geometrically the kernel of the epimorphism $\mathcal{I}\rho$. With this goal in mind, we need to take a *short detour* into combinatorics that mimics the geometric situation we are facing.

• Let Θ be a finite set of "colors". In the near future, for doodles, $\Theta = \{\mathsf{I}, \mathsf{II}, \mathsf{III}, \mathsf{IV}\}$ or $\Theta = \{\mathsf{I}, \mathsf{III}\}$; for blobs, $\Theta = \{\mathbf{I}, \mathbf{II}, \mathbf{III}, \mathbf{IV}\}$ or $\Theta = \{\mathbf{I}, \mathbf{III}\}$.

Consider a finite collection of smooth 2-moderately embedded oriented loops $\mathcal{E}$ in $A = \mathbb{R} \times [0, 1]$. We take a finite set of points Q in the complement to $\mathcal{E}$ (soon, Q will be the singular sets Σ_β of immersed doodles $\beta(\mathcal{C})$). In what follows, we consider the pairs $(Q, \mathcal{E})$ up to an ambient isotopy of A, or up to an ambient isotopy of A that preserves the foliation $\mathcal{F}(\hat{v})$.

We divide the set Q in complementary groups $Q_1, \dots, Q_s$, each group being painted with a different color from the pallet Θ. This coloring is a part of the Q-structure.

The elements of Q are considered in pairs. We impose some restrictions $\mathcal{R}$ on the pairings of $P : Q \to Q$, where P is a free involution. These restrictions are expressed in terms of colors from the pallet Θ. Some of the restrictions will require that the elements of a particular color may be paired only with other elements the same color, in which case we assume that the cardinality $|Q_i|$ of the corresponding Q_i is even; other restrictions from $\mathcal{R}$ will pair elements of a particular color only with elements of another color, in which case $|Q_j| = |Q_k|$ for the corresponding sets Q_j, Q_k.

We denote by $\wp(Q, \mathcal{R})$ the set of such pairings, i.e., the set of free involutions on Q, subject to the restrictions from $\mathcal{R}$.

• In one special case, we will pair elements of Q_{I} with elements of Q_{III}, elements of Q_{II} with elements of Q_{IV}. Let us denote this special set of rules by $\mathcal{R}^{\odot}$.

Using $\mathcal{E}$, with each pairing $P \in \wp(Q, \mathcal{R})$ we associate a function c_P on the set of ordered pairs $\{(q, P(q))\}_{q \in Q}$ with values in $\mathbb{Z}$. The integer $c(q, P) =_{\mathsf{def}} c_P(q, P(q))$ counts the *signed* transversal intersections of an oriented path $\gamma(q, P(q)) \subset A$, connecting q to $P(q)$, with the union of oriented loops that form the 1-cycle $\mathcal{E}$. For homological reasons, $c(q, P)$ does not depend on the choice of the path $\gamma(q, P(q))$, provided $A = \mathbb{R} \times [0, 1]$.

We may view $c(q, P)$ as the linking number of the pair $(q, P(q))$ with the cycle $\mathcal{E}$. Note that $c(q, P) = -c(P(q), P)$. This construction gives rise to a map

$$\mathbf{c}_{\mathcal{E},\, P} : \mathsf{Graph}(P) \to \mathbb{Z}, \tag{5.20}$$

where $\mathsf{Graph}(P) \subset Q \times Q$ denotes the graph of the free involution $P : Q \to Q$, subject to the $\mathcal{R}$-constraints.

For any $\mathcal{R}$-admissible pairing $P : Q \to Q$, we always can choose the connecting paths $\{\gamma(q, P(q))\}_{q \in Q}$ so that they do not intersect each other (recall that each $\gamma(q, P(q))$ is not a loop) and are transversal to $\mathcal{E}$. Without loss of generality, we may also assume that each path $\gamma(q, P(q))$ is 2-moderate (in particular, does not have vertical inflections).

Definition 5.7. • For a fix set of rules $\mathcal{R}$, we call the triple $\{Q, P, \mathcal{E})$ irreducible if there exist a set of paths $\{\gamma(q, P(q))\}_{q \in Q}$ such that all the paths are disjoint and each path $\gamma(q, P(q))$ has all its intersections with $\mathcal{E}$

of the same sign. By definition, if for some q, the intersection $\gamma(q, P(q)) \cap \mathcal{E}$ is empty, the collection $\{Q, P, \mathcal{E}\}$ is reducible.

• For a fix set of rules $\mathcal{R}$, we call the collection $\{Q, P, \mathcal{E}\}$ primitive if it is irreducible and each colored set $Q_i \subset Q$ consists of exactly two elements, when $P(Q_i) = Q_i$, and of a single element when $P(Q_i) = Q_j$ for a $Q_j \neq Q_i$. $\diamondsuit$

It is easy to see that, given two sets $(Q_1, P_1, \mathcal{E}_1)$ and $(Q_2, P_2, \mathcal{E}_2)$, their connected sum $\uplus$ in A produces a new triple $(Q_1 \uplus Q_2,\ P_1 \uplus P_2,\ \mathcal{E}_1 \uplus \mathcal{E}_2)$ by squeezing $(Q_1, \mathcal{E}_1)$ in $\mathbb{R} \times (0, 0.5)$ and $(Q_2, \mathcal{E}_2)$ in $\mathbb{R} \times (0.5, 1)$ and applying the pairing $P_1 \coprod P_2 \in \mathcal{R}$ to $Q_1 \coprod Q_2$.

For a fixed set of rules $\mathcal{R}$, let $\mathbf{D}(\mathcal{R})$ be the set of all triples $\{Q, P, \mathcal{E}\}$, being considered up to an ambient isotopy of A that preserves the foliation $\mathcal{F}(\hat{v})$. The operation $\uplus$ turns the set $\mathbf{D}(\mathcal{R})$ into an abelian semi-group. The associativity and commutativity of the operation $\uplus$ in $\mathbf{D}(\mathcal{R})$ can be validated as in Lemma 5.2.

For a given triple $(Q, \mathcal{R}, \mathcal{E})$, consider the ℓ_1-norm $\|\mathbf{c}_{\mathcal{E},\,P}\|_{\ell_1}$ of the discrete function $\mathbf{c}_{\mathcal{E},\,P} : \mathsf{Graph}(P) \to \mathbb{Z}$ in (5.20). It is independent of the choice of paths $\{\gamma(q, P(q))\}_{q \in Q}$. We introduce a non-negative integer by the formula

$$m(Q, \mathcal{R}, \mathcal{E}) =_{\mathsf{def}} \min_{P \in \wp(Q, \mathcal{R})} \big\{ \|\mathbf{c}_{\mathcal{E},\,P}\|_{\ell_1} \big\}, \tag{5.21}$$

the minimum is taken over all $\mathcal{R}$-admissible pairings in Q. Evidently,

$$m(Q_1 \uplus Q_2,\ \mathcal{R},\ \mathcal{E}_1 \uplus \mathcal{E}_2) \leq m(Q_1,\ \mathcal{R},\ \mathcal{E}_1) + m(Q_2,\ \mathcal{R},\ \mathcal{E}_2).$$

Thus, $m(Q, \mathcal{R}, \mathcal{E})$ resembles a semi-norm under the connected sum operation $\uplus$. However, in general, $m\big(\uplus_1^k\ \{Q,\ \mathcal{R},\ \mathcal{E}\}\big) \neq k \cdot m(Q, \mathcal{R}, \mathcal{E})$. At the same time, for a primitive $\{Q, \mathcal{R}, \mathcal{E}\}$,

$$m\big(\uplus_1^k\ \{Q,\ \mathcal{R},\ \mathcal{E}\}\big) = k \cdot m(Q, \mathcal{R}, \mathcal{E}).$$

• We denote by $\mathbf{D}_0(\mathcal{R})$ the sub-semigroup of $\mathbf{D}(\mathcal{R})$ consisting of the triples $\{Q, \mathcal{R}, \mathcal{E}\}$ with the property $\{m(Q, \mathcal{R}, \mathcal{E}) = 0\}$ and consider the quotients $\mathbf{D}(\mathcal{R})/\mathbf{D}_0(\mathcal{R})$.

Lemma 5.5. *$\mathbf{D}(\mathcal{R})/\mathbf{D}_0(\mathcal{R})$ is an abelian group with respect to the operation $\uplus$.*

If $\mathcal{R}$ allows for pairings within each given color, then any non-zero element of the group $\mathbf{D}(\mathcal{R})/\mathbf{D}_0(\mathcal{R})$ is of order 2.

Proof. The only property of groups that needs a clarification is the existence of the negation (minus element). Let $\tau : A \to A$ be a mirror involution whose fixed set is the line $\mathbb{R} \times \{0.5\}$. For any triple $\{Q, P, \mathcal{E}\}$ in $\mathbb{R} \times [0, 0.5)$, we form the triple $\{\bar{Q}, \bar{P}, \bar{\mathcal{E}}\}$ in $\mathbb{R} \times (0.5, 1]$, where $\bar{Q} = \tau(Q)$, $\bar{\mathcal{E}} = \tau(\mathcal{E})$. Each pair $(q, P(q))$ is replaced in $\{\bar{Q}, \bar{P}, \bar{\mathcal{E}}\}$ by the pair $(\tau(P(q)), \tau(q))$ so that $\tau(q)$ is given the color of $P(q)$ and $\tau(P(q))$ is given the color of q.

Consider the new triple $\{Q, P, \mathcal{E}\} \uplus \{\bar{Q}, \bar{P}, \bar{\mathcal{E}}\}$. With the new color assignment in place, the pairing $(q, \tau(q))$ is permissible. Since the orientations of $\bar{\mathcal{E}} = \tau(\mathcal{E})$ is opposite to the orientations of $\mathcal{E}$, the path γ_q that connects q and $\tau(q)$ has the property $\gamma_q \circ (\mathcal{E} \uplus \tau(\mathcal{E})) = 0$. Therefore, $\{Q, P, \mathcal{E}\} \uplus \{\bar{Q}, \bar{P}, \bar{\mathcal{E}}\} \in \mathbf{D}_0(\mathcal{R})$. Hence, $\{\bar{Q}, \bar{P}, \bar{\mathcal{E}}\}$ is the negative of $\{Q, P, \mathcal{E}\}$ in $\mathbf{D}(\mathcal{R})/\mathbf{D}_0(\mathcal{R})$. This validates the first claim.

Consider the triple $\{Q, P, \mathcal{E}\} \uplus \{Q, P, \mathcal{E}\}$ and a point $q \in Q$. Let $\gamma(q, s) \subset \mathbb{R} \times [0, 0.4]$ be any smooth path that is transversal to $\mathcal{E}$ and connects q to a point s whose $\mathbb{R}$-coordinate is bigger than h, where the box $[-h, h] \times [0, 0.4]$ contains Q and $\mathcal{E}$. We think of the second copy of $\{Q, P, \mathcal{E}\}$ as being obtained by the 0.5-shift $Sh : \mathbb{R} \times [0, 0.5] \to \mathbb{R} \times [0.5, 1]$ to the right. By the hypotheses about $\mathcal{R}$, we may pair q and $Sh(q)$.

We can connect s to $Sh(s)$ by a path ω that resides outside the box $[-h, h] \times [0, 1]$. Then the path

$$\tau = \gamma(q, s) \cup \omega \cup \overline{Sh(\gamma(q, s))}$$

(where $\overline{Sh(\gamma(q, s))}$ denotes the path $Sh(\gamma(q, s))$ with the reverse orientation) connects q to $Sh(q)$ and has a pair of adjacent intersections with $\mathcal{E} \cup Sh(\mathcal{E})$ of opposite signs. Therefore, $\{Q, P, \mathcal{E}\} \uplus \{Q, P, \mathcal{E}\} \in \mathbf{D}_0(\mathcal{R})$, and thus $\{Q, P, \mathcal{E}\}$ is an element of order 2 in $\mathbf{D}(\mathcal{R})/\mathbf{D}_0(\mathcal{R})$. □

Lemma 5.6. *The correspondence $\{Q, \mathcal{R}, \mathcal{E}\} \rightsquigarrow m(\{Q, \mathcal{R}, \mathcal{E}\})$ defines a norm-like map $m : \mathbf{D}(\mathcal{R})/\mathbf{D}_0(\mathcal{R}) \to \mathbb{Z}_+$ such that $m(a \uplus b) \leq m(a) + m(b)$ and $m(a) = 0$ implies $a = 0$, where $a, b \in \mathbf{D}(\mathcal{R})/\mathbf{D}_0(\mathcal{R})$.*

Proof. The verification of the claims is on the level of definitions. □

- This ends our combinatorial detour. ◇

We are in position to return to bordisms of 2-moderate immersions of oriented doodles and to state one of our main results.

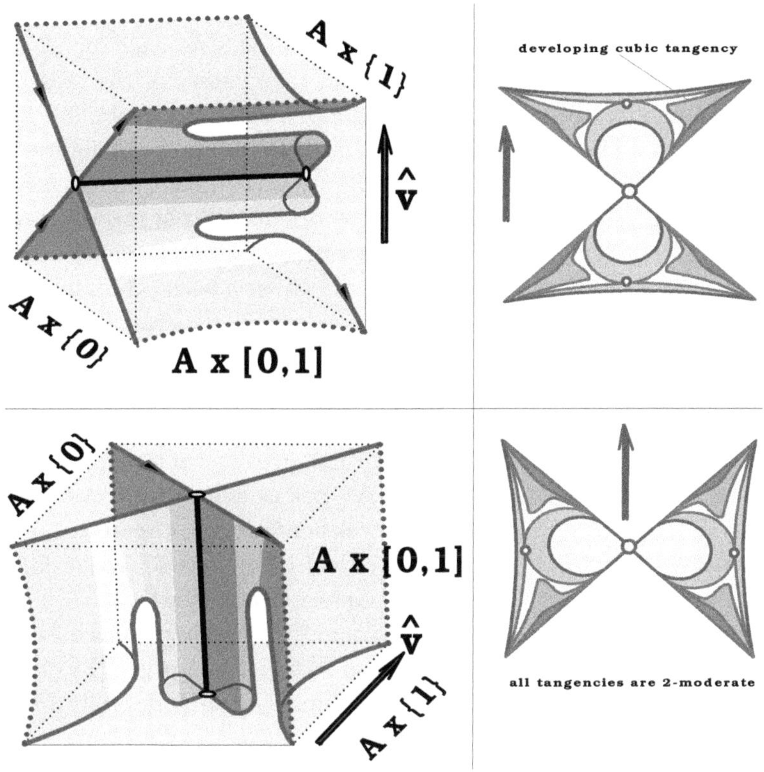

Fig. 5.10 Resolving crossings of types I or III into the figures "∞", "$\overline{\infty}$", and a pair of arcs via a 2-moderate cobordism (the lower figure). In contrast, resolving crossings of types II or IV into the figures "8", "$\bar{8}$", and a pair of arcs is impossible via a 2-moderate cobordism (the upper figure).

Theorem 5.3. *Let $A = \mathbb{R} \times [0,1]$ and $\mathcal{R} = \{\mathsf{I} \Leftrightarrow \mathsf{III}, \mathsf{II} \Leftrightarrow \mathsf{IV}\}$.*

- *There is an exact sequence of abelian groups*

$$0 \to \mathbf{K} \to \mathbf{OC}^{\mathsf{imm}}_{\mathsf{moderate}\leq 2}(A)/\mathbf{OC}^{\mathsf{emb}}_{\mathsf{moderate}\leq 2}(A) \xrightarrow{\mathcal{I}\rho} \mathbb{Z} \times \mathbb{Z} \to 0,$$

where the homomorphism $\mathcal{I}\rho$ is given by the two integral invariants, $\rho^{\mathsf{I}}(\sim) - \rho^{\mathsf{III}}(\sim)$ and $\rho^{\mathsf{II}}(\sim) - \rho^{\mathsf{IV}}(\sim)$.

- *There exists a monomorphism $\Theta : \mathbf{D}(\mathcal{R})/\mathbf{D}_0(\mathcal{R}) \hookrightarrow \mathbf{K}$ whose image $\mathbf{G}$ is spanned over $\mathbb{Z}$ by the doodles $\{\infty(n) \uplus \overline{\infty}\}$ and $\{8(n) \uplus \bar{8}\}$, where $n \in \mathbb{Z}$, $n \neq 0$. In fact, $\mathbf{G} \approx (\mathbb{Z})^{\infty}$.*
- $\mathbf{OC}^{\mathsf{emb}}_{\mathsf{moderate}\leq 2}(A) \approx \mathbb{Z} \times \mathbb{Z}$ *via an isomorphism as in (5.14).*

Proof. We start with a finite set of oriented loops $\mathcal{C}$ and their 2-moderately generic (relatively to $\hat{v}$) immersion $\beta : \mathcal{C} \to A$. At each point of self-intersection $a \in \beta(\mathcal{C})$ of types I and III, in the vicinity of the crossing, we perform a 2-moderate surgery on $\beta(\mathcal{C})$ whose result $\beta^\dagger$ is shown in Fig. 5.10, lower diagram. The resulting oriented curves contain small figures "∞_a" and "$\overline{\infty}_a$", whose self-intersection resides at a. The bar denotes the flip of the orientation of ∞_a, determined by the orientations of the two branches of $\beta(\mathcal{C})$ at a. The type of the self-intersection of ∞_a in relation to $\hat{v}(a)$ is the same as the type of the original crossing of $\beta(\mathcal{C})$ at a with values in $\{\mathsf{I}, \mathsf{III}\}$; so ∞_a occurs for type I, and $\overline{\infty}_a$ for type III crossings. The rest of the oriented curves $\mathcal{C}'$ are immersed in A by a map β' which has the crossings of the types $\{\mathsf{II}, \mathsf{IV}\}$ only. The immersions β and $\beta^\dagger$ are 2-moderately cobordant.

By a 2-surgery on $\beta(\mathcal{C}')$, we could replace the doodles $\mathcal{C}'$ by embedded doodles $\mathcal{C}''$ union several figures "8_a" for the crossings of types II and several figures "$\overline{8}_a$" for the crossings of types IV. Let us denote by $\beta^\ddagger$ the resulting immersion. Crucially, such a surgery, although being canonical, is not 2-moderate (see Fig. 5.10)! As a result, the immersions β and $\beta^\ddagger$ fail to be 2-moderately cobordant. This fact complicates our efforts.

In any case, via these canonical surgeries, we may replace any given immersion $\beta : \mathcal{C} \to A$ with the immersion $\beta^\ddagger$ which comprises the embedded doodles $\mathcal{C}'' \subset A$ disjoint union with a number of 2-moderately immersed figures "∞_a", "$\overline{\infty}_a$", and "8_a", "$\overline{8}_a$".

To summarize, for types I and III, we get the configurations "$\infty \uparrow \vee$" and "$\overline{\infty} \uparrow \wedge$"; for types II and IV, we either keep the original configurations "$\times \uparrow <$" and "$\times > \uparrow$" or, depending on the context, replace them with configurations "$8 \uparrow <$" and "$8 > \uparrow$". Here "$\uparrow$" mimics the direction of the vector field $\hat{v}$ in relation to the oriented figures "∞" and "$\times$", and the symbols "$\vee$, $\wedge$, $<$, $>$" indicate the preferred sector, determined by the orientations of the two branches at the crossing a, in relation to vector $\hat{v}(a)$, i.e., in relation to "$\uparrow$".

If the invariants $\rho_{\mathsf{I}}(\beta) - \rho_{\mathsf{III}}(\beta) \in \mathbb{Z}$, $\rho_{\mathsf{II}}(\beta) - \rho_{\mathsf{IV}}(\beta) \in \mathbb{Z}$ vanish, then we can pair figures "$8 \uparrow <$" of type II with figures "$8 > \uparrow$" of type IV, figures "$\infty \uparrow \vee$" of type I with figures "$\infty \uparrow \wedge$" of type III.

Consider immersions in the kernel of the homomorphism $\mathcal{I}\rho$ that consist of several 2-moderate figures ∞, $\overline{\infty}$, 8, $\overline{8}$ disjoint union with 2-moderate embeddings. Thus, the number of figures ∞ is equal to the number of figures $\overline{\infty}$, and the number of figures 8 is equal to the number of figures $\overline{8}$.

Let us denote by $\hat{\mathbf{G}}$ the set of such immersions, being considered up to the 2-moderate cobordisms. In particular, we focus on the subgroup

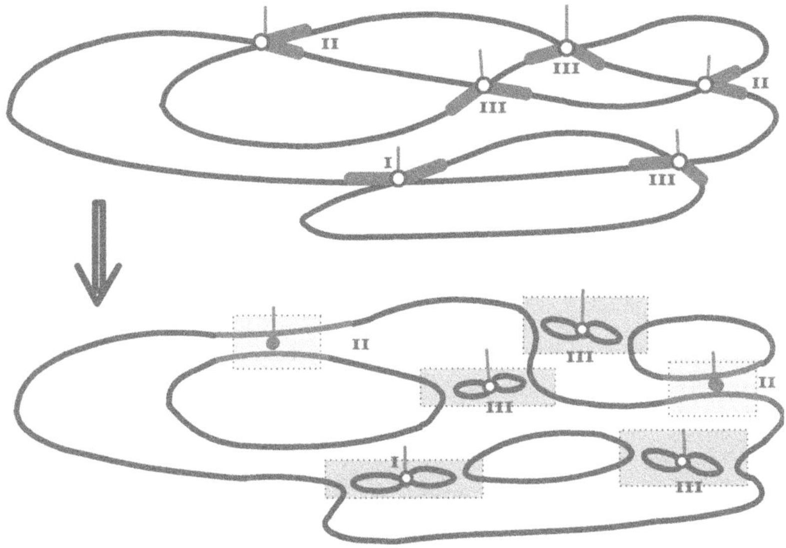

Fig. 5.11 Resolving oriented immersed doodles into several figures "∞" (type I), "$\overline{\infty}$" (type III), and number of 2-moderate embedded loops. The figure illustrates the homomorphism $\mathsf{res}^{\mathsf{II},\mathsf{IV}}$ from (5.22). The vertical vector field $\hat{v}$ points up.

$\mathbf{G} \subset \mathbf{K}$, spanned over $\mathbb{Z}$ by the elements $\{\infty(n) \uplus \overline{\infty}\}$ and $\{8(n) \uplus \overline{8}\}$, where $n \in \mathbb{Z}$, $n \neq 0$.

First we need to verify that these elements are nontrivial and distinct in $\mathbf{K}$. Moreover, we will show that they are elements of infinite order in $\mathbf{K}$.

Consider a 2-moderate immersion $B : W \to A \times [0,1]$ that delivers a cobordism between an immersion $\beta : \mathcal{C} \to A \times \{0\}$ and some embedding $\beta_1 : \mathcal{C}_1 \to A \times \{1\}$. Then the self-intersection curves $\{\ell\}$ of $B(W)$ define a pairing $\{P : a \rightsquigarrow P(a)\}_{a \in \Sigma_\beta}$ consistent with the set of rules $\mathcal{R} = \{\mathsf{I} \Leftrightarrow \mathsf{III}, \mathsf{II} \Leftrightarrow \mathsf{IV}\}$.

On the other hand, for $\beta : \mathcal{C} \to A$ which represents an element of $\hat{\mathbf{G}}$, any pairing $\{P : a \rightsquigarrow P(a)\}_{a \in \Sigma_\beta}$ from $\mathcal{R}$ allows us to attach in $A \times [0,1)$ 1-dimensional 2-moderate cross-handles $\infty_a \times I$ and $8_a \times I$ (the immersed images in $A \times [0,1)$ of the handles $S^1_a \times I$) to each pair $\infty_a \coprod \overline{\infty}_{P(a)} \subset A \times \{0\}$ or $8_a \coprod \overline{8}_{P(a)} \subset A \times \{0\}$. By a general position argument, these 2-moderate handles are disjoint. The union of these cross-handles with the embedded surface $\mathcal{C}_1 \times [0,1]$ delivers an oriented cobordism

$$B : \coprod_a \{S^1_a \times I\} \coprod \{\mathcal{C}_1 \times [0,1]\} \to A \times [0,1]$$

between β and β_1. It is important to attach the cross-handles with some care in order to satisfy the requirement for B to have only 2-moderate tangencies to the $\hat{v}^\bullet$-trajectories in $A \times [0,1]$.

First, we need to avoid the intersections of the 1-dimensional singular cores of the immersed surfaces $B(\coprod_a \{S^1_a \times I\})$ with the embedded surfaces $B(\mathcal{C}_1 \times [0,1])$ since such intersections would contribute at least cubic tangencies to the $\hat{v}^\bullet$-trajectories.

It turns out that there are new combinatorial obstructions to B being 2-moderate. These new obstructions are described in the next lemma.

Lemma 5.7. *Let* $A = \mathbb{R} \times [0,1]$ *and let* $\mathcal{R} = \{\mathsf{I} \Leftrightarrow \mathsf{III}, \mathsf{II} \Leftrightarrow \mathsf{IV}\}$ *or* $\mathcal{R} = \{\mathsf{I} \Leftrightarrow \mathsf{III}\}$. *Let* $\mathcal{C}$ *be a finite collection of oriented circles. Consider a* 2*-moderate immersion* $\beta : \mathcal{C} \to A$ *such that* $\rho_{\mathsf{I}}(\beta) = \rho_{\mathsf{III}}(\beta)$ *and* $\rho_{\mathsf{II}}(\beta) = \rho_{\mathsf{IV}}(\beta)$. *We denote by* Σ_β *the self-intersection locus of* β.

By replacing in $\beta(\mathcal{C})$ *each crossing with the figures "*∞_a*", "*$\overline{\infty}_a$*" or "*8_a*" or "*$\overline{8}_a$*" we get a new* 2*-moderate immersion* $\beta^\ddagger$ *that is a disjoint union of these figures with a* 2*-moderate embedding* $\beta'' : \mathcal{C}'' \to A$.

Then $\beta^\ddagger$ *is cobordant, via* 2*-moderate oriented immersions of surfaces, to an embedding* β_1 *if and only if* $m(\Sigma_\beta, \mathcal{R}, \mathcal{C}'') = 0$ *(see (5.21)), i.e., iff there exists a pairing* $P \in \wp(\Sigma_\beta, \mathcal{R})$ *whose function* $\mathbf{c}_{\mathcal{E},P} : \mathsf{Graph}(P) \to \mathbb{Z}$ *is identically zero.*

In the case $\mathcal{R} = \{\mathsf{I} \Leftrightarrow \mathsf{III}\}$, $\beta^\ddagger$ *is* 2*-moderately cobordant to the original* β.

Proof. We will prove the claim for $\mathcal{R} = \{\mathsf{I} \Leftrightarrow \mathsf{III}, \mathsf{II} \Leftrightarrow \mathsf{IV}\}$; the arguments in the case $\mathcal{R} = \{\mathsf{I} \Leftrightarrow \mathsf{III}\}$ are very similar and simpler. As Fig. 5.10 testifies, for this choice of $\mathcal{R}$, $\beta^\dagger$ is 2-moderately cobordant to the original β.

Let $B : W \to A \times [0,1]$ be a 2-moderate immersion that delivers the oriented cobordism between the immersion β and some embedding β_1. We have seen that the self-intersection locus $\Xi_B = \coprod \ell$ of $B(W)$ delivers a particular pairing $P_B : \Sigma_\beta \to \Sigma_\beta$ that is consistent with $\mathcal{R}$.

We denote by I a closed interval. We replace the singular surface $B(W)$ with the disjoint union of narrow 2-moderate cross-handles

$$\{B^\ddagger_a : S^1_a \times I \to A \times [0,1]\}_{a \in \Sigma_\beta / P_B}$$

whose images are $\infty_a \times I$, $\overline{\infty}_a \times I$, and $8_a \times I$, $\overline{8}_a \times I$, together with a number of narrow 2-moderate cross-tori

$$\{B^\ddagger_\kappa : S^1_\kappa \times S^1 \to A \times (0,1)\}_\kappa$$

(the images $B^\ddagger_\kappa(S^1_\kappa \times S^1)$ are diffeomorphic to $\infty_\kappa \times S^1$ or to $8_\kappa \times S^1$), and a 2-moderate regular embedding $B'' : W'' \hookrightarrow A \times [0,1]$ of an oriented surface W''. The trace of this cobordism in $A \times \{0\}$ produces the curves

$$\Big(\coprod_{a\in\Sigma^{\mathsf{I}}_\beta} \infty_a\Big)\coprod\Big(\coprod_{a\in\Sigma^{\mathsf{III}}_\beta} \overline{\infty}_a\Big)\coprod\Big(\coprod_{a\in\Sigma^{\mathsf{II}}_\beta} 8_a\Big)\coprod\Big(\coprod_{a\in\Sigma^{\mathsf{IV}}_\beta} \overline{8}_a\Big)$$

disjoint union with a *nonsingular* curve $\mathcal{C}'' = B''(W'') \cap (A \times \{0\})$ as has been described above.

Consider now the cores $\{\delta =_{\mathsf{def}} \delta_{a,P(a)}\}_{a\in\Sigma_\beta}$ of the cross-handles and of the cross-torii. We notice that $\delta \cap B''(W'')$ must be empty in order to insure that B is 2-moderate. Indeed, if $x \in \delta \cap B''(W'')$, then x is a triple-intersection point for $B(W)$, which violates the 2-moderate property of B.

Note that the algebraic (i.e., signed) intersection number $\delta_{a,P(a)} \circ B''(W'')$ is a homological invariant for the relative 1-homology class $[\delta_{a,P(a)}] \in H_1\big(A \times [0,1], a \cup P_B(a); \mathbb{Z}\big)$ of the path $\delta_{a,P(a)}$ and the relative 2-homology class

$$[B''(W'')] \in H_2\big(A \times [0,1],\ [A \setminus (a \cup P_B(a))] \times \{0\} \ \sqcup\ (A \times \{1\}); \mathbb{Z}\big)$$

of the relative 2-cycle $B''(W'')$. Since the relative homology class $[\delta_{a,P(a)}]$ and the relative homology class

$$[\gamma(a, P_B(a))] \in H_1(A \times [0,1],\ a \cup P_B(a);\ \mathbb{Z})$$

are equal (the two paths are even homotopic relative to $a \cup P_B(a)$ in $A \times [0,1]$), we conclude that the algebraic intersection number

$$\delta_{a,P(a)} \circ B''(W'') = \gamma(a, P_B(a)) \circ B''(W'') = \gamma(a, P_B(a)) \circ \mathcal{C}''.$$

Therefore, if $\gamma(a, P_B(a)) \circ \mathcal{C}_1 \neq 0$, we get a contradiction with the requirement $\delta_{a,P(a)} \cap B'(W') = \emptyset$. As a result, if $\gamma(a, P_B(a)) \circ \mathcal{C}'' \neq 0$ for some $a \in \Sigma_\beta$, then the cobordism B is not 2-moderate.

On the other hand, assume that, for some $\mathcal{R}$-amenable involution $P : \Sigma_\beta \to \Sigma_\beta$, the intersection $\gamma(a, P(a)) \circ \mathcal{C}'' = 0$ for all $a \in \Sigma_\beta$, where $\mathcal{C}''$ is a 2-moderate nonsingular component of the replacement $\beta^\ddagger$ of β. Then, let us show that β is 2-moderately cobordant to an embedding β_1.

Using the notations from the previous arguments, we start with the cobordism W which is the union $H \subset A\times[0,1]$ of 2-moderate cross-handles, attached to $A \times \{0\}$ at points of Σ_β according to the coupling P, with the non-singular surface $\mathcal{C}'' \times [0,1]$. Since $\hat{v}^\bullet$ is transversal to the cores of the narrow cross-handles, we may ensure that H is 2-moderate.

Although the surfaces H and $S = \mathcal{C}'' \times [0,1]$ individually are 2-moderate, their union $W = H \cup S$ may be not.

However, if we assume that all the cores δ of cross-handles from H have the property $\delta \circ S = 0$, then we can correct the failure of $W = H \cup S$ to be 2-moderate. Indeed, if two adjacent intersections x, y of δ with the oriented surface S are of opposite signs, then we can attach a 1-handle $S^1 \times [x, y]$ to S whose core is the segment $[x, y] \subset \delta$ and whose interior $D^2 \times [x, y]$ engulfs the cross-handles $\infty_a \times [x, y]$, or $\overline{\infty}_a \times [x, y]$, or $8_a \times [x, y]$, or $\overline{8}_a \times [x, y]$. The result is a new embedded oriented surface S' which intersects δ_a at fewer points: the intersections x and y has been eliminated. The rest of the cores δ_b keep their old intersections with S. Continuing this elimination of the adjacent along the cores intersections of opposite signs with evolving nonsingular surfaces, eventually we will get a nonsingular oriented surface $\tilde{S}$ which is disjoint from all the cross-handles. The union of $\tilde{S}$ with all the cross-handles delivers the desired 2-moderate cobordism between the immersion β and some embedding β_1.

By the previous arguments, this recipe for constructing β_1 works when, for some free involution $P : \Sigma_\beta \to \Sigma_\beta$, we have the property $\gamma(a, P(a)) \circ \mathcal{C}'' = 0$ for all $a \in \Sigma_\beta$. This means exactly that there exists a pairing $P \in \wp(\Sigma_\beta, \mathcal{R})$ whose function $\mathbf{c}_{\mathcal{C}'', P} : \mathsf{Graph}(P) \to \mathbb{Z}$ is identically zero. In other words, the semi-norm $m(\Sigma_\beta, \mathcal{R}, \mathcal{C}'')$ must vanish. □

We continue with the proof of Theorem 5.3. As before, in our notations, we suppress the choices of the paths $\{\gamma(q, P(q))\}_{q \in Q}$

Given an element $\{Q, P, \mathcal{E}\}$ of the group $\mathbf{D}(\mathcal{R})$, we associate with it a 2-moderate immersion $\beta : \mathcal{C} \to A$ by replacing each point $q \in Q$ with a small figure $\infty, \overline{\infty}$, or $8, \overline{8}$ (depending on the type $\{\mathsf{I}, \mathsf{II}, \mathsf{III}, \mathsf{IV}\}$ of q), and keeping the 2-moderate embedding $\mathcal{E}$ for the role of $\mathcal{C}''$. By Lemma 5.7, β is cobordant via a 2-moderately cobordism $B : W \to A \times [0,1]$ (whose self-intersection locus $\coprod \ell$ defines P) to an embedding if and only if the element $\{Q, P, \mathcal{E}\}$ is such that the function $\mathbf{c}_{\mathcal{E}, P} : \mathsf{Graph}(P) \to \mathbb{Z}$ is identically zero. Therefore, by the definition of $\mathbf{D}_0(\mathcal{R})$, this construction $\{Q, P, \mathcal{E}\} \rightsquigarrow \beta$ defines a momomorphism $\Theta : \mathbf{D}(\mathcal{R})/\mathbf{D}_0(\mathcal{R}) \hookrightarrow \mathbf{K}$. Its image is exactly the subgroup $\hat{\mathbf{G}}$ (by this group's definition).

Consider the subgroup $\mathbf{G} \subset \hat{\mathbf{G}}$ spanned over $\mathbb{Z}$ by the elements $\{\infty(n) \uplus \overline{\infty}\}$ and $\{8(n) \uplus \overline{8}\}$, where $n \in \mathbb{Z}$, $n \neq 0$. For any $\mathcal{R}$-admissible pairing P between the two crossing a and $P(a)$ of each figure, the intersection number $\gamma(a, P(a)) \circ C''$ of any path $\gamma(a, P(a))$ with the concentric loops C'' surrounding $\infty, \overline{\infty}$, or $8, \overline{8}$ is equal to $n \neq 0$. Thus, by Lemma 5.7, the elements $\{\infty(n) \uplus \overline{\infty}\}$ and $\{8(n) \uplus \overline{8}\}$ are nontrivial in the kernel $\mathbf{K}$ for all $n \neq 0$. Moreover, for the same reason, their k-multiple copies (with respect to $\uplus$) are not 2-moderately cobordant to an embedding. Therefore, the elements $\{\infty(n) \uplus \overline{\infty}\}$ and $\{8(n) \uplus \overline{8}\}$ are of the infinite order in $\mathbf{K}$. Along the same lines, for $n_1 \neq n_2 \neq 0$, no k-multiple of $\{\infty(n_1) \uplus \overline{\infty}\}$ is 2-moderately cobordant to $\{\infty(n_2) \uplus \overline{\infty}\}$. Since $\mathbf{K}$ is abelian, it follows that $\mathbf{G} \approx (\mathbb{Z})^\infty$.

In fact, $\hat{\mathbf{G}} = \mathbf{G}$. Indeed, consider a 2-moderate simple loop $\mathcal{L}$ that contains in its interior several figures $\infty, \overline{\infty}$, or $8, \overline{8}$ and does not contain any other simple loops. By a 2-moderate 1-surgery (the 1-handles must avoid the figures $\infty, \overline{\infty}, 8, \overline{8}$) on the domain $\mathcal{D}$ that bounds $\mathcal{L}$, we split $\mathcal{D}$ in several topological disks so that each of them contains a single figure. Let us call temporarily a nested group of simple loops which contain a single figure $\infty, \overline{\infty}$, or $8, \overline{8}$ a "block". If a block contains oppositely oriented simple loops, they can be eliminated by a 2-moderate surgery on the annulus that is bounded by them (see Fig. 5.4, the last two slices). Thus, we may assume that all the loops in a block are oriented coherently.

If a simple loop $\mathcal{L}'$ bounds a domain $\mathcal{D}'$ that contains several blocks, then again by a 2-moderate 1-surgery on $\mathcal{D}'$ we can split $\mathcal{D}'$ in a number of disks, each of which bounds a single new block. Proceeding this way, we transform any configuration of figures $\{\infty, \overline{\infty}, 8, \overline{8}\}$ and simple loops into 2-moderately cobordant disjoint union of coherently oriented blocks. This transformation does not affect the original figures $\{\infty, \overline{\infty}, 8, \overline{8}\}$ and their pairings. Therefore, $\hat{\mathbf{G}} = \mathbf{G} \approx (\mathbb{Z})^\infty$.

The last claim of the theorem follows from the isomorphism (5.14), since any oriented embedded doodle in A is the boundary of an oriented embedded blob. This completes the proof of Theorem 5.3. $\square$

Let $\beta : \mathcal{C} \to A$ be a 2-moderate oriented immersion. By canonical resolutions of the types II and IV singularities "$\times \uparrow <$" and "$\times > \uparrow$", we eliminate them in β' (see Fig. 5.11). The result, $\beta^\dagger$, is a new 2-moderate embedding $\beta'' : \mathcal{C}'' \to A$ disjoint union with a number of immersions of the type "∞_a" and "$\overline{\infty}_a$". The resolution $\mathsf{res}_{\mathsf{II},\mathsf{IV}} : \beta \rightsquigarrow \beta^\dagger$ may change the 2-moderate bordism invariant $\rho_{\mathsf{II}}(\beta) - \rho_{\mathsf{IV}}(\beta)$ and thus may change the bordism class of β. These canonical resolutions produce a homomorphism

$$\mathsf{res}^{\mathsf{II},\mathsf{IV}} : \mathbf{OC}^{\mathsf{imm}}_{\mathsf{moderate}\leq 2}(A) \to \mathbf{OC}^{\mathsf{imm},\ \mathsf{I\&III}}_{\mathsf{moderate}\leq 2}(A), \tag{5.22}$$

where $\mathbf{OC}^{\mathsf{imm},\ \mathsf{I\&III}}_{\mathsf{moderate}\leq 2}(A)$ stands for the 2-moderate bordisms of oriented immersions whose crossings are of the types I and III only. The map $\mathsf{res}^{\mathsf{II},\mathsf{IV}}$ is evidently the right inverse of the obvious homomorphism $\mathbf{OC}^{\mathsf{imm},\ \mathsf{I\&III}}_{\mathsf{moderate}\leq 2}(A) \to \mathbf{OC}^{\mathsf{imm}}_{\mathsf{moderate}\leq 2}(A)$.

The treatment of $\mathbf{OC}^{\mathsf{imm},\ \mathsf{I\&III}}_{\mathsf{moderate}\leq 2}(A)$ is very similar to our treatment of $\mathbf{OC}^{\mathsf{imm}}_{\mathsf{moderate}\leq 2}(A)$; in fact, the case of $\mathbf{OC}^{\mathsf{imm},\ \mathsf{I\&III}}_{\mathsf{moderate}\leq 2}(A)$ is easier to analyze.

Theorem 5.4. *Let* $A = \mathbb{R} \times [0,1]$ *and* $\mathcal{R}^\circ = \{\mathsf{I} \Leftrightarrow \mathsf{III}\}$.

- *There exists an exact sequence of abelian groups:*

$$0 \to \mathbf{K}^{\mathsf{I}\,\&\,\mathsf{III}} \to \mathbf{OC}^{\mathsf{imm},\ \mathsf{I}\,\&\,\mathsf{III}}_{\mathsf{moderate}\leq 2}(A)/\mathbf{OC}^{\mathsf{emb}}_{\mathsf{moderate}\leq 2}(A) \xrightarrow{\mathcal{I}\rho^{\mathsf{I}\,\&\,\mathsf{III}}} \mathbb{Z} \to 0,$$

where the epimorphism $\mathcal{I}\rho^{\mathsf{I}\,\&\,\mathsf{III}}$ *is given by the integral invariant* $\rho^{\mathsf{I}}(\sim) - \rho^{\mathsf{III}}(\sim)$.

- *The kernel* $\mathbf{K}^{\mathsf{I}\,\&\,\mathsf{III}}$ *is generated by the doodles* $\{\infty(n) \coprod \overline{\infty}\}_{n\in\mathbb{Z};\, n\neq 0}$ *and is isomorphic to the group* $\mathbf{D}(\mathcal{R}^\circ)/\mathbf{D}_0(\mathcal{R}^\circ) \approx (\mathbb{Z})^\infty$.

- *Thus, any* 2*-moderate oriented doodle* β *with the crossings of the types* I *and* III *only and such that* $\rho_{\mathsf{I}}(\beta) = \rho_{\mathsf{III}}(\beta)$, *is* 2*-moderately cobordant to an embedding if and only if its image* $\Theta(\beta) \in \mathbf{D}_0(\mathcal{R}^\circ)$ *(i.e., the "semi-norm"* $m(\Sigma_\beta, \mathcal{R}^\circ, \mathcal{C}') = 0$*).*

Proof. Since the oriented figure "∞" has the crossing of type I, and "$\overline{\infty}$" has the crossing of type III, the map $\mathcal{I}\rho^{\mathsf{I}\,\&\,\mathsf{III}} = \rho_{\mathsf{I}}(\sim) - \rho_{\mathsf{III}}(\sim)$ is an epimorphism.

By Lemma 5.7, any 2-moderate immersion $\beta : \mathcal{C} \to A$ with the crossings of the types I and III only and such that $\rho_{\mathsf{I}}(\beta) = \rho_{\mathsf{III}}(\beta)$ is cobordant in $\mathbf{OC}^{\mathsf{imm},\ \mathsf{I}\,\&\,\mathsf{III}}_{\mathsf{moderate}\leq 2}(A)$ to an embedding β_1 if and only if the singular set Σ_β admits a free involution $P \in \wp(\Sigma_\beta, \mathcal{R}^\circ)$ with the following property: for any $a \in \Sigma_\beta$, the algebraic intersection number $\gamma(a, P(a)) \circ \mathcal{C}' = 0$. Here the embedded doodle $\mathcal{C}'$ is the doodle $\mathcal{C}''$ from the proof of Theorem 5.3. Equivalently, $m(\Sigma_\beta, \mathcal{R}^\circ, \mathcal{C}') = 0$.

For any triple $\{Q, P, \mathcal{E}\}$, representing an element of $\mathbf{D}(\mathcal{R}^\circ)$, we construct a 2-moderate immersion $\beta(\{Q, P, \mathcal{E}\})$ by placing a small figure ∞ or $\overline{\infty}$ at each point $q \in Q$ (depending on whether q is of color I or III) and using the embedded blob $\mathcal{E}$ as a part of the immersion. The role of the involution $P : Q \to Q$ is tenuous: it makes sure that the colors of q and $P(q)$ are different. By Lemma 5.7, $\beta(\{Q, P, \mathcal{E}\})$ is cobordant to an embedding if and only if there is a triple $\{Q, \tilde{P}, \mathcal{E}\}$ for which $m(\{Q, \tilde{P}, \mathcal{E}\}) = 0$; in other words, when $\{Q, \tilde{P}, \mathcal{E}\} \in \mathbf{D}_0(\mathcal{R}^\circ)$.

Therefore, the correspondence $\{Q, P, \mathcal{E}\} \rightsquigarrow \beta(\{Q, P, \mathcal{E}\})$ gives rise to a monomorphism

$$\Theta : \mathbf{D}(\mathcal{R}^\circ)/\mathbf{D}_0(\mathcal{R}^\circ) \to \mathbf{K}^{\mathsf{I}\,\&\,\mathsf{III}}.$$

On the other hand, we have seen that, for any 2-moderate immersion $\beta \in \mathbf{K}^{\mathsf{I}\,\&\,\mathsf{III}}$, the immersion $\beta^\dagger$ (see Fig. 5.11) from Lemma 5.7 is 2-moderately cobordant to β. Evidently, $\beta^\dagger$ is in the image of Θ. As a result, Θ is an epimorphism, and thus, an isomorphism.

Exactly as in the proof of Theorem 5.3 and with the help of Lemma 5.7, we verify that all the oriented doodles $\{\infty(n) \coprod \overline{\infty}\}_{n\in\mathbb{Z}}$ are distinct elements of infinite order in $\mathbf{K}^{\mathsf{I}\,\&\,\mathsf{III}}$ and generate a subgroup $(\mathbb{Z})^\infty$. □

Let "∞-ribbon" denotes the immersion of a band $S^1 \times I$ in A, shaped as a fat figure ∞ in relation to $\hat{v}$ (see Fig. 5.9, (b)). The symbol "∞-ribbon(n)" stands for the ∞-ribbon surrounded by n concentric nested disks. Similarly, the "8-ribbon" is the immersion of a band $S^1 \times I$ in A, shaped as a fat figure 8 in relation to $\hat{v}$ and 8-ribbon(n) denotes the 8-ribbon surrounded by n concentric nested disks.

In the case of non-oriented blobs, we get the following result:

Theorem 5.5. *Let* $A = \mathbb{R} \times [0, 1]$.

- *There is an exact sequence of abelian groups:*

$$0 \to \mathbf{M} \to \mathbf{B}^{\mathsf{imm}}_{\mathsf{moderate}\leq 2}(A)/\mathbf{B}^{\mathsf{emb}}_{\mathsf{moderate}\leq 2}(A) \xrightarrow{\mathcal{I}\rho} \mathbb{Z}_2 \times \mathbb{Z} \to 0,$$

where $\mathbf{B}^{\mathsf{emb}}_{\mathsf{moderate}\leq 2}(A) \overset{J}{\approx} \mathbb{Z}$ *is a direct summand of* $\mathbf{B}^{\mathsf{imm}}_{\mathsf{moderate}\leq 2}(A)$.

- *The homomorphism* $\mathcal{I}\rho$ *is defined by the invariants*

$$\rho_{\mathbf{I}}(\sim), \rho_{\mathbf{III}}(\sim) \in \mathbb{Z}_2,\ \ \rho_{\mathbf{II}}(\sim) - \rho_{\mathbf{IV}}(\sim) \in \mathbb{Z}$$

(which realize the subgroup $\mathbb{Z}_2 \times \mathbb{Z} \hookrightarrow \mathbb{Z}_2 \times \mathbb{Z}_2 \times \mathbb{Z}$*).*

- *The kernel* $\mathbf{M}$ *contains the group* $(\mathbb{Z}_2)^\infty$ *generated by the blobs* $\{\infty\text{-ribbon}(n) \uplus \overline{\infty}\text{-ribbon}\}_{n\in\mathbb{Z}}$ *(*$n \neq 0$*) as in Fig. 5.9,* (b).

Proof. Thanks to Proposition 5.3, to prove the theorem we need only to show that the kernel $\mathbf{M}$ contains the group $(\mathbb{Z}_2)^\infty$. It suffices to exhibit infinitely many distinct elements of $\mathbf{M}$ of order 2.

Consider the immersion $\beta_\bullet : X_\bullet \to A$ of a band $X_\bullet = S^1 \times I$, whose image is shaped as ∞-ribbon (see Fig. 5.9, (b)). With the help of $\beta_\bullet$, the orientation of $X_\bullet$ (and thus of $\partial_1 X_\bullet$) is induced by the preferred orientation of A. The image $\beta_\bullet(\partial X_\bullet)$ has four self-intersections of distinct colors $\{\mathbf{I}, \mathbf{II}, \mathbf{III}, \mathbf{IV}\}$ as blobs (see Fig. 5.7) and of distinct colors $\{\mathsf{I}, \mathsf{II}, \mathsf{III}, \mathsf{IV}\}$ (as doodles). Alternatively, we could use the immersion $\beta_\bullet : X_\bullet = D^2 \coprod D^2 \to A$ whose image has also four self-intersections of distinct colors $\mathbf{I}, \mathbf{II}, \mathbf{III}, \mathbf{IV}$ and of distinct colors $\mathsf{I}, \mathsf{II}, \mathsf{III}, \mathsf{IV}$ (see Fig. 5.12).

We add to $\beta_\bullet$ the regular embeddings of n coherently oriented concentric disks so that the image of the smallest disk in A contains properly the ∞-ribbon $\beta_\bullet(X_\bullet)$.

We claim that $\infty\text{-ribbon}(n)$ is a nontrivial element of $\mathbf{B}^{\mathsf{imm}}_{\mathsf{moderate}\leq 2}(A)$. Assume to the contrary that the $\infty\text{-ribbon}(n)$ is the boundary of a solid 2-moderate cobordism $B : W \to A \times [0, 1)$. Then the self-intersection curves $\ell \subset B(\delta W)$ cannot connect the unique self-intersection point of $\beta_\bullet(\partial X_\bullet)$ of type **I** to itself. Thus, no pairing consistent with

$$\mathcal{R}^\bullet =_{\mathsf{def}} \{\mathbf{I} \Leftrightarrow \mathbf{I},\ \mathbf{III} \Leftrightarrow \mathbf{III},\ \mathbf{II} \Leftrightarrow \mathbf{IV}\}$$

is available, and no such 2-moderate cobordism B exists.

Moreover, for $n \neq m$,

$$\infty\text{-ribbon}(n) \neq \infty\text{-ribbon}(m) \text{ in } \mathbf{B}^{\mathsf{imm}}_{\mathsf{moderate}\leq 2}(A)/\mathbf{B}^{\mathsf{emb}}_{\mathsf{moderate}\leq 2}(A).$$

Indeed, if

$$\infty\text{-ribbon}(n) = \infty\text{-ribbon}(m) \uplus \alpha' \text{ in } \mathbf{B}^{\mathsf{imm}}_{\mathsf{moderate}\leq 2}(A), \tag{5.23}$$

where $\alpha' : X' \hookrightarrow A$ is a 2-moderate embedding whose image is separated from the rest of the blobs by a $\hat{v}$-trajectory, then the 2-moderate solid cobordism $B : W \to A \times [0, 1)$ between the two blobs in the RHS and LHS of equality (5.23) is available.

Consider the self-intersection curve $\ell \subset B(\delta W)$ that connects the unique self-intersection point a of type **I** in $\infty\text{-ribbon}(n)$ to the unique self-intersection point b of type **I** in $\infty\text{-ribbon}(m)$. Since B is an immersion, the preimage $B^{-1}(\ell)$ must be union of several arcs, so that $B : B^{-1}(\ell) \to \ell$ is an immersion. By the construction of ℓ, two of these arcs, ℓ_1 and ℓ_2, belong to the boundary δW, while the rest of the arcs start at n points in the interior of W and terminate at m points in the interior of W. For $m \neq n$, at least one of these arcs must hit δW at some point c. Therefore, the point $B(c) \in \ell$ belongs to triple intersection locus of $B(\delta W)$, a contradiction with the assumption that $B : \delta W \to A \times [0, 1]$ is 2-moderate.

Note that $\infty\text{-ribbon}(n)$ does not belong to the kernel of $\mathcal{I}\rho$ since its $\rho_{\mathsf{I}} = 1$. At the same time, all the $\mathcal{I}\rho$-invariants of $\infty\text{-ribbon}(n) \uplus \overline{\infty}\text{-ribbon}$ do vanish; thus, it belongs to the kernel $\mathcal{I}\rho$.

We recycle the previous arguments and examine the $\mathcal{R}^\bullet$-permissible pairings between the four self-intersections of type **I** of the ribbons $\infty\text{-ribbon}(n) \uplus \infty\text{-ribbon}$ and $\infty\text{-ribbon}(m) \uplus \infty\text{-ribbon}$. These pairings are delivered by a pair of curves $\ell \subset B(\delta W)$ and $\ell' \subset B(\delta W)$ that belong to the self-intersection locus of $B(\delta W)$. Thus, for $m \neq n$, $B(\delta W)$ must develop triple intersections in $A \times [0, 1]$. We conclude that

$$\infty\text{-ribbon}(n) \uplus \infty\text{-ribbon} \neq \infty\text{-ribbon}(m) \uplus \infty\text{-ribbon} \mod \mathbf{B}^{\mathsf{emb}}_{\mathsf{moderate}\leq 2}(A).$$

On the other hand, if the mirror image (with respect to a $\hat{v}$-trajectory) of a non-oriented blob $\alpha : X \to A$ is isotopic to the original blob α, then $\alpha \uplus \alpha = 0$ in $\mathbf{B}^{\mathsf{imm}}_{\mathsf{moderate}\leq 2}(A)$. Here the isotopy is assumed to preserve the oriented foliation $\mathcal{F}(\hat{v})$. Indeed, in such symmetric case, $\alpha \uplus \alpha$ is the boundary of the 2-moderate cobordism $B : W \to A \times [0, 1)$, traced by a rotation in $A \times [0, 1)$ of $\alpha(X) \subset \mathbb{R} \times [0, 0.5] \subset A$ on the angles $0 \leq \phi \leq \pi$ around the axis $\gamma = \mathbb{R} \times \{0.5\} \subset A \times \{0\} \subset A \times [0, 1)$. Thus,

$$\big(\infty\text{-}\mathsf{ribbon}(n) \uplus \infty\text{-}\mathsf{ribbon}\big) \uplus \big(\infty\text{-}\mathsf{ribbon}(n) \uplus \infty\text{-}\mathsf{ribbon}\big) = 0$$

in $\mathbf{B}^{\mathsf{imm}}_{\mathsf{moderate}\leq 2}(A)$ and hence in $\mathbf{M}$. Therefore, $\{\infty\text{-}\mathsf{ribbon}(n) \uplus \infty\text{-}\mathsf{ribbon}\}_n$ deliver distinct elements of order 2 in $\mathbf{M}$.

Thanks to the homomorphism $\mathcal{R}_J : \mathbf{B}^{\mathsf{imm}}_{\mathsf{moderate}\leq 2}(A) \to \mathbf{B}^{\mathsf{emb}}_{\mathsf{moderate}\leq 2}(A)$ from (5.9), $\mathbf{B}^{\mathsf{emb}}_{\mathsf{moderate}\leq 2}(A) \stackrel{J}{\approx} \mathbb{Z}$ is a direct summand of $\mathbf{B}^{\mathsf{imm}}_{\mathsf{moderate}\leq 2}(A)$. □

Remark 5.2. We do not know whether $\mathbf{M} \approx (\mathbb{Z}_2)^\infty$ and we do not have geometric models for all the generators of $\mathbf{M}$, beyond the ones that have been constructed above and the similar ones from (5.25), based on Fig. 5.12. The main difficulty is that we do not have *localized* resolutions of singularities, present in the boundaries of immersed blobs, into canonical singular (like a small ∞-ribbon) and nonsingular parts. This prevents us from establishing an analogue of Lemma 5.7 for blobs. ◇

In the case of *oriented* blobs, many new invariants, counting self-intersections of their boundaries, arise. In principle, there are 16 ways to combine the orientation independent types $\mathbf{I} - \mathbf{IV}$ with the orientation dependent types $\mathsf{I} - \mathsf{IV}$. As before, the curves ℓ that belong to the self-intersections of the δ-boundaries of solid cobordisms W, define 8 parings π from the set $\mathcal{R}^{\bullet\bullet}$ between these 16 types:

$$\begin{aligned}
(\mathbf{I}, \mathsf{I}) \Leftrightarrow (\mathbf{I}, \mathsf{III}), &\quad (\mathbf{I}, \mathsf{II}) \Leftrightarrow (\mathbf{I}, \mathsf{IV}),\\
(\mathbf{II}, \mathsf{I}) \Leftrightarrow (\mathbf{IV}, \mathsf{III}), &\quad (\mathbf{II}, \mathsf{II}) \Leftrightarrow (\mathbf{IV}, \mathsf{IV}),\\
(\mathbf{III}, \mathsf{I}) \Leftrightarrow (\mathbf{III}, \mathsf{III}), &\quad (\mathbf{III}, \mathsf{II}) \Leftrightarrow (\mathbf{III}, \mathsf{IV}),\\
(\mathbf{IV}, \mathsf{I}) \Leftrightarrow (\mathbf{II}, \mathsf{III}), &\quad (\mathbf{IV}, \mathsf{II}) \Leftrightarrow (\mathbf{II}, \mathsf{IV}).
\end{aligned} \tag{5.24}$$

These pairings from $\mathcal{R}^{\bullet\bullet}$ follow the two rules:

1) start with a bold Roman number, indexing a sector that belongs to a pair of intersecting blobs at a point a where their boundaries intersect (see Fig. 5.7); subject the sector to a reflection with respect to the vector $\hat{v}(a)$, then the new sector accquires new bold Roman number;

2) the second Roman number that indexes the orientations of two intersecting boundary curves, is changed to the Roman number that indexes the two flipped orientations of the boundary curves.

Consider two subsets, **A** and **B**, of the 16 patterns, each subset containing 8 patterns: **A** consists of patterns whose second index is I or II, while **B** consists of patterns whose second index is III or IV. We notice that the pairings from (5.24) couple only elements from **A** with elements from **B**.

For a given oriented blob $\alpha : X \to A$, counting the crossings of each of the 16 types κ from the list (5.24), produces a number ρ_κ. Each of the 8 pairings $\pi \in \mathcal{R}^{\bullet\bullet}$ from the list (5.24), with the help of the self-intersection curves $\ell \subset \delta W$, generates an oriented cobordism invariant $\rho_\kappa - \rho_{\pi(\kappa)} \in \mathbb{Z}$. Together, $\{\rho_\kappa - \rho_{\pi(\kappa)}\}_\kappa$ produce a homomorphism

$$\mathcal{I}\rho^{\bullet\bullet} : \mathbf{OB}^{\mathsf{imm}}_{\mathsf{moderate}\leq 2}(A) \longrightarrow (\mathbb{Z})^8.$$

Examining the pairings π from the list (5.24) between different types of intersection patterns, we notice that π couples crossings of two *similarly oriented* (clockwise or counterclockwise) blobs with crossings of two *similarly oriented* blobs, and couples crossings of two *oppositely oriented* blobs with crossings of two *oppositely oriented* blobs. Therefore, $\mathcal{I}\rho^{\bullet\bullet}$ is the direct product of two homomorphisms $\mathcal{I}\rho^{\bullet}$ and $\overline{\mathcal{I}\rho^{\bullet}}$.

Theorem 5.6. *For $A = \mathbb{R} \times [0,1]$, there is an exact sequence of abelian groups:*

$$0 \to \mathbf{OM} \to \mathbf{OB}^{\mathsf{imm}}_{\mathsf{moderate}\leq 2}(A)/\mathbf{OB}^{\mathsf{emb}}_{\mathsf{moderate}\leq 2}(A) \xrightarrow{\mathcal{I}\rho^{\bullet} \times \overline{\mathcal{I}\rho^{\bullet}}} (\mathbb{Z})^4 \times (\mathbb{Z})^4$$

$$\xrightarrow{\Sigma \times \Sigma} \mathbb{Z}_2 \times \mathbb{Z}_2 \to 0,$$

where $\mathbf{OB}^{\mathsf{emb}}_{\mathsf{moderate}\leq 2}(A) \overset{\mathsf{K}\times\mathsf{L}}{\approx} \mathbb{Z} \times \mathbb{Z}$ is a direct summand of $\mathbf{OB}^{\mathsf{imm}}_{\mathsf{moderate}\leq 2}(A)$. The homomorphism $\Sigma : (\mathbb{Z})^4 \to \mathbb{Z}_2$ takes each 4-vector to the sum of its components modulo 2.

The kernel **OM** *contains the subgroup $(\mathbb{Z})^\infty$, generated by the blobs $\{Y_{(\bar\mu,\nu)}(n,0)\}$, $\{Y_{(\bar\mu,\nu)}(0,m)\}$ from (5.25) (see Fig. 5.12) and indexed by the elements of the set*

$$\{m, n \in \mathbb{Z};\ m, n \neq 0;\ \ (\bar\mu, \nu) \in \tilde{\mathbf{A}} =_{\mathsf{def}} \{\mathbf{I}, \mathbf{II}\} \times \{\mathsf{I}, \mathsf{II}\}\}.$$

Proof. Consider two smooth simple loops (say, ellipcii), $\mathcal{C}_1$ and $\mathcal{C}_2$, in $A = \mathbb{R} \times [0,1]$ such that they intersect transversally only at a pair of points a and b, where the vectors $\hat v(a)$ and $\hat v(b)$ are transversal to both curves (see Fig. 5.12). We pick any pair of indexes $(\bar\mu(a), \nu(a))$, where $\bar\mu(a)$ from the list $\{\mathbf{I}, \mathbf{II}, \mathbf{III}, \mathbf{IV}\}$ and $\nu(a)$ from the list $\{\mathsf{I}, \mathsf{II}, \mathsf{III}, \mathsf{IV}\}$. This choice of $(\bar{\mu}(a), \nu(a))$ determines orientations (counterclockwise or clockwise) of the curves $\mathcal{C}_1$ and $\mathcal{C}_2$ as well as one of the four possible choices of the domains $\mathcal{D}_1$ and $\mathcal{D}_2$ in A that bound $\mathcal{C}_1$ and $\mathcal{C}_2$, respectivelly. Only one of the four choices will produce a pair of compact domains. To adress this complication, we trim the non-compact domains by bounding them with very big concentric circles, disjoint from $\mathcal{C}_1 \cup \mathcal{C}_2$.

With these choices in place, the intersection point b acquired a unique type

$$(\bar{\mu}(b), \nu(b)) \in \{\mathbf{I}, \mathbf{II}, \mathbf{III}, \mathbf{IV}\} \times \{\mathsf{I}, \mathsf{II}, \mathsf{III}, \mathsf{IV}\}.$$

Thus, with the help of oriented $\mathcal{D}_1$ and $\mathcal{D}_2$, we get a permutation map Π of the set $\{\mathbf{I}, \mathbf{II}, \mathbf{III}, \mathbf{IV}\} \times \{\mathsf{I}, \mathsf{II}, \mathsf{III}, \mathsf{IV}\}$ to itself.

By a direct inspection, we see that Π preserves the two subsets $\mathbf{A} = \{\mathbf{I}, \mathbf{II}, \mathbf{III}, \mathbf{IV}\} \times \{\mathsf{I}, \mathsf{II}\}$ and $\mathbf{B} = \{\mathbf{I}, \mathbf{II}, \mathbf{III}, \mathbf{IV}\} \times \{\mathsf{III}, \mathsf{IV}\}$. Moreover, Π is a *free* involution on $\mathbf{A}$ and on $\mathbf{B}$. In particular, the involution Π generates 4 pairs-orbits in $\mathbf{A}$ and 4 pairs-orbits in $\mathbf{B}$.

In contrast, the involutions

$$\pi : \{\mathbf{I}, \mathbf{II}, \mathbf{III}, \mathbf{IV}\} \times \{\mathsf{I}, \mathsf{II}, \mathsf{III}, \mathsf{IV}\} \to \{\mathbf{I}, \mathbf{II}, \mathbf{III}, \mathbf{IV}\} \times \{\mathsf{I}, \mathsf{II}, \mathsf{III}, \mathsf{IV}\}$$

from the list $\mathcal{R}^{\bullet\bullet}$ pair elements of $\mathbf{A}$ with elements of $\mathbf{B}$. As a result, the invariants $\{\rho_\kappa - \rho_{\pi(\kappa)}\}_{\kappa \in \mathbf{A}}$ do not vanish for the blobs $\mathcal{D}_1$ and $\mathcal{D}_2$ whose crossings are indexed by $(\bar{\mu}(a),\, \nu(a))$ and $(\bar{\mu}(b),\, \nu(b))$, both from $\mathbf{A}$ or $\mathbf{B}$. Moreover, $\{\rho_{(\bar{\mu}(a),\, \nu(a))} = 1 = \rho_{(\bar{\mu}(b),\, \nu(b))}\}$ for any of the 8 choices of $(\bar{\mu}(a),\, \nu(a)) \in \mathbf{A}$ and for any of the 8 choices of $(\bar{\mu}(a),\, \nu(a)) \in \mathbf{B}$. Therefore, each choice of $(\bar{\mu}(a),\, \nu(a)) \in \mathbf{A}$ or $(\bar{\mu}(a),\, \nu(a)) \in \mathbf{B}$ produces a vector $\vec{z}_{(\bar{\mu}(a),\, \nu(a))} \in (\mathbb{Z})^{16}$ with exactly two components equal to 1 and the rest of components equal to 0. The pairs of 1's in each vector are indexed by either elements of $\mathbf{A}$ or by elements of $\mathbf{B}$. Again, by a direct inspection, the 8×16 matrix T, formed by these vectors is of the rank 8. It consists of four 4×8 blocks $T_{11}, T_{12}, T_{21}, T_{22}$, where $T_{11} = T_{22}$ and $T_{12} = \mathbf{0} = T_{21}$. In turn, T_{11} consists of two 4×4 blocks, the first of which has 1's along the diagonal, and the second one has 1's along the anti-diagonal; the rest of entries are zeros.

Therefore, the image of

$$\mathbf{OB}^{\mathsf{imm}}_{\mathsf{moderate}\leq 2}(A)/\mathbf{OB}^{\mathsf{emb}}_{\mathsf{moderate}\leq 2}(A) \overset{\mathcal{I}\rho^\bullet \times \overline{\mathcal{I}\rho^\bullet}}{\longrightarrow} (\mathbb{Z})^4 \times (\mathbb{Z})^4$$

contains the lattice spanned by the 8 vectors $\{\vec{z}_{(\bar{\mu}(a),\, \nu(a))}\}_{(\bar{\mu}(a),\, \nu(a)) \in \mathbf{A} \cup \mathbf{B}}$.

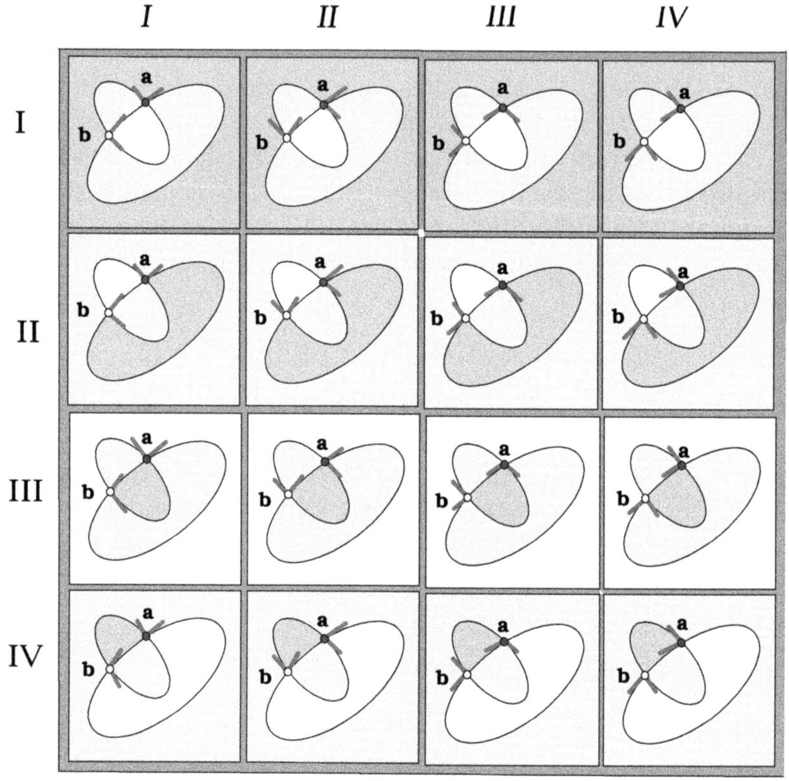

Fig. 5.12 The table lists 16 possible types at the crossing a of two blobs. Vertically are displayed blobs' types **I, II, III, IV**, horizontally doodles' types I, II, III, IV. The vertical vector field $\hat{v}$ points up. The corresponding types at the second crossing b define the pairing Π. The two bars at each crossing show the orientations of the intersecting curves. The shadings indicate the two choices of interior or exterior for each ellipse. Any non-compact domain is assumed to be bounded by a big circle, not shown.

On the other hand, by Lemma 5.4, no vector in $(\mathbb{Z})^4$ whose sum of components is odd (in particular, with a single component 1) can be realized by an immersion of a compact surface $\alpha : X \to A$, since the total number of crossings $\sum_\kappa(\rho_\kappa + \rho_{\pi(\kappa)})$ for $\alpha(\partial_1 X)$ must be even, and

$$\sum_\kappa(\rho_\kappa - \rho_{\pi(\kappa)}) \equiv \sum_\kappa(\rho_\kappa + \rho_{\pi(\kappa)}) \mod 2.$$

Thus, the image of $\mathcal{I}\rho^\bullet \times \overline{\mathcal{I}\rho^\bullet}$ is the kernel of the homomorphism

$$(\mathbb{Z})^4 \times (\mathbb{Z})^4 \xrightarrow{\Sigma\times\Sigma} \mathbb{Z}_2 \times \mathbb{Z}_2.$$

For any element $(\bar{\mu}, \nu) \in \mathbf{A}$, consider the two blobs $X_{(\bar{\mu},\nu)}$ in Fig. 5.12 for which the crossing at a has the type $(\bar{\mu}, \nu)$. Then the crossing at b has the type $\Pi(\bar{\mu}, \nu) \in \mathbf{A}$.

There are unique $\pi, \pi' \in \mathcal{R}^{\bullet\bullet}$ (see (5.24)) such that $\pi(\bar{\mu}, \nu) \in \mathbf{B}$ and $\pi'(\Pi(\bar{\mu}, \nu)) \in \mathbf{B}$. Moreover, $\pi'(\Pi(\bar{\mu}, \nu)) = \Pi(\pi(\bar{\mu}, \nu))$.

We denote by $X_{(\bar{\mu},\nu)}(n)$ the blob $X_{(\bar{\mu},\nu)}$ from Fig. 5.12, surrounded by n big concentric coherently oriented discs. Similarly, $X_{\pi(\bar{\mu},\nu)}(m)$ denotes the blob $X_{\pi(\bar{\mu},\nu)}$, surrounded by m big concentric coherently oriented discs.

Then the blob

$$Y_{(\bar{\mu},\nu)}(n, m) =_{\mathsf{def}} X_{(\bar{\mu},\nu)}(n) \uplus X_{\pi(\bar{\mu},\nu)}(m) \tag{5.25}$$

is in the kernel of $\mathcal{I}\rho^\bullet \times \overline{\mathcal{I}\rho^\bullet}$. Moreover, for $m \neq n$, $Y_{(\bar{\mu},\nu)}(n, m)$ is an element of infinite order in $\mathbf{OM}$. The validation of this claim is similar to the one used in the proof of Theorem 5.5. First, we show that $Y_{(\bar{\mu},\nu)}(n, m)$ is non-trivial element in $\mathbf{OM}$. Assume to the contrary that there is a solid cobordism $B : W \to A \times [0, 1)$ that bounds $Y_{(\bar{\mu},\nu)}(n, m)$. Consider the self-intersection curve $\ell \subset B(\delta W)$ that connects the unique self-intersection point a of type $(\bar{\mu}, \nu) \in X_{(\bar{\mu},\nu)}(n)$ to the unique self-intersection point a' of type $\pi(\bar{\mu}, \nu)$ in $X_{\pi(\bar{\mu},\nu)}(m)$. Since B is an immersion, the preimage $B^{-1}(\ell)$ must be union of several arcs. By the construction of ℓ, two of these arcs, ℓ_1 and ℓ_2, belong to the boundary δW, while the rest of the arcs start at n points in the interior of W and terminate at m points in the interior of W. For $m \neq n$, at least one of these arcs must hit δW at some point c. Therefore, the point $B(c) \in \ell$ belongs to triple intersection locus of $B(\delta W)$, a contradiction with the assumption that $B : \delta W \to A \times [0, 1]$ is 2-moderate. The demonstration that $Y_{(\bar{\mu},\nu)}(n, m)$ is an element of infinite order is similar. □

Conjecture 5.1. *For $A = \mathbb{R} \times [0, 1]$, consider the exact sequence of abelian groups:*

$$0 \to \mathbf{OM} \to \mathbf{OB}^{\mathsf{imm}}_{\mathsf{moderate}\leq 2}(A)/(\mathbb{Z}\times\mathbb{Z}) \xrightarrow{\mathcal{I}\rho^\bullet \times \overline{\mathcal{I}\rho^\bullet}} (\mathbb{Z})^4 \times (\mathbb{Z})^4 \xrightarrow{\Sigma\times\Sigma} \mathbb{Z}_2 \times \mathbb{Z}_2 \to 0$$

from Theorem 5.6. The kernel $\mathbf{OM}$ *is $\mathbb{Z}$-generated by the blobs $Y_{(\bar{\mu},\nu)}(n, m)$ from (5.25).*

For the set of rules $\mathcal{R}^{\bullet\bullet}$, there is an isomorphism

$$\Theta^{\bullet\bullet} : \mathbf{D}(\mathcal{R}^{\bullet\bullet})/\mathbf{D}_0(\mathcal{R}^{\bullet\bullet}) \hookrightarrow \mathbf{OM}.$$

With the help of $\Theta^{\bullet\bullet}$, the elements of $\mathbf{OM}$ *acquire "norms" $m(Q, \mathcal{R}^{\bullet\bullet}, \mathcal{E}) \in \mathbb{Z}_+$ as in (5.21).* ◇

To formulate the next conjecture, we introduce very informally a new topological space $\mathcal{G}_{\mathsf{moderate}\leq 2}$. Its points are the 2-moderate functions $f \in \mathcal{F}_{\mathsf{moderate}\leq 2}$ together with a "coupling" τ of their zeros. This τ may "link" either two simple zeros of f, or a double zero to itself, or a double zero to a pair of distinct simple zeros. One needs to describe also the deformation rules for τ, as a function f deforms within the space $\mathcal{F}_{\mathsf{moderate}\leq 2}$.

The space $\mathcal{G}_{\mathsf{moderate}\leq 2}$ admits a slightly different interpretation. Let $\mathcal{Q}_{\mathsf{moderate}\leq 2} \subset \mathcal{F}_{\mathsf{moderate}\leq 2}$ be the subspace of functions with either two simple zeros, or a single double zero, or with no zeros at all. Then any $f \in \mathcal{F}_{\mathsf{moderate}\leq 2}$ is a finite product $\prod_i f_i$, where $f_i \in \mathcal{Q}_{\mathsf{moderate}\leq 2}$ and the order of the multipliers is unimportant. Of course, the presentation $f = \prod_i f_i$ is far from being unique. For example, we always can replace $\prod_i f_i$ by a product $\prod_i h_i f_i$, where $\{h_i\}_i$ have no zeros and $\prod_i h_i = 1$.

We would like to think of points of the hypothetical space $\mathcal{G}_{\mathsf{moderate}\leq 2}$ as the presentations of functions $f \in \mathcal{F}_{\mathsf{moderate}\leq 2}$ as such unordered products, being considered up to the equivalence $\prod_i f_i \sim \prod_i h_i f_i$, where the positive functions $\{h_i\}$ are such that $\prod_i h_i = 1$.

Conjecture 5.2. *Let $A = \mathbb{R} \times [0,1]$. There exists a topological space $\mathcal{G}_{\mathsf{moderate}\leq 2}$ (whose construction is sketched above) and a surjective map $q : \mathcal{G}_{\mathsf{moderate}\leq 2} \to \mathcal{F}_{\mathsf{moderate}\leq 2}$, given by $q(f,\tau) = f$, with finite fibers and such that the fundamental group*

$$\pi_1(\mathcal{G}_{\mathsf{moderate}\leq 2}, pt) \approx \mathbf{B}^{\mathsf{imm}}_{\mathsf{moderate}\leq 2}(A).$$

The isomorphism is delivered by an analogue Q^{imm} of the map J^{imm} from Theorem 5.1 so that $q \circ Q^{\mathsf{imm}} = J^{\mathsf{imm}}$. ◇

5.4 A brief exploration of the combinatorics of tangency patterns via polynomial spaces

In this section, we intend to give the reader just a general feel for the circle of ideas and combinatorial structures that we will explore rigorously in Chapters 6-8. The section may be viewed as a gentile introduction into the combinatorics of Chapter 6.

First, let us glance at interpretations of Theorems 5.1 and 5.2 in terms of models, based on spaces of real monic polynomials of small even degrees.

Let $\mathcal{P}_d$ denote the space real monic polynomials in one variable of an even degree d, and let $\mathcal{P}_d^{\leq 2}$ be its open subspace, formed by polynomials with no real roots of multiplicity ≥ 3. The space $\mathcal{P}_d^{\leq 2}$ is a "natural approximation" of the functional space $\mathcal{F}_d^{\leq 2}$ of smooth functions with 2-moderate zeros whose total degree $\leq d$. Of course, an even degree polynomial from $\mathcal{P}_d^{\leq 2}$, $d \equiv 0 \mod 2$, is not approaching 1 in the vicinity of $\pm\infty$. However, there is an embedding

$$\mathcal{I}_d : \mathcal{P}_d^{\leq 2} \to \mathcal{F}^{\leq 2}$$

that, in the vicinity of $\pm\infty$, "levels down to 1" any real polynomial of an even degree. This embedding is described by an analytic formula (see [V]) as follows.

We fix an auxiliary smooth function $\chi : \mathbb{R} \to [0,1]$ such that $\chi(u) = 0$ for $|u| \leq 1$, $\chi(u) = 1$ for $|u| \geq 2$, and $\frac{\partial \chi}{\partial u} \neq 0$ for $1 < |u| < 2$. Let $\mu(P)$ denote the sum of absolute values of the coefficients of a monic polynomial P. Then we form a smooth function by the formula

$$\mathcal{I}_d(P)(u) =_{\mathsf{def}} P(u) + \big(1 - P(u)\big) \cdot \chi(u/\mu(p)).$$

In fact, the zeros of any polynomial P are in 1-to-1 correspondence with the zeros of the function $\mathcal{I}_d(P)$ and their multiplicities are preserved by $\mathcal{I}_d$.

Consider the "forbidden set" $\mathcal{F}_d^{\geq 3} \subset \mathcal{F}_d$ of functions f that have at least one zero of multiplicity ≥ 3. Among them, the functions f that have exactly one zero of multiplicity 3 form an open and dense subset $(\mathcal{F}_d^{\geq 3})^\circ$.

For each $f \in (\mathcal{F}_d^{\geq 3})^\circ$, let $u_f^\star \in \mathbb{R}$ be the unique zero of multiplicity 3.

The set $\mathcal{F}_d^{\geq 3}$ has codimension 2 in $\mathcal{F}_d$; so loops in $\mathcal{F}_d^{\leq 2}$ may be *linked* with the locus $\mathcal{F}_d^{\geq 3}$ in $\mathcal{F}_d$. Here is a model example of such a link.

Let $\theta \in S^1$. For any $u^* \in (-2, 2)$, consider the following θ-family of quartic u-polynomials (see Fig. 5.13)

$$\big\{P_{u^\star}(u,\theta) =_{\mathsf{def}} (u-2)\big[(u-u^\star)^3 + \cos(\theta)(u-u^\star) + \sin(\theta)\big]\big\}_{\theta \in [0,2\pi]}.$$

Each u-polynomial $P_{u^\star}(u,\theta)$ belongs to the space $\mathcal{P}_4^{\leq 2}$, and their θ-family forms a loop $S^1_{u^\star}$ in $\mathcal{P}_4^{\leq 2}$. The loop $S^1_{u^\star}$ bounds the 2-disk

$$D^2_{u^\star} =_{\mathsf{def}} \big\{(u-2)\big[(u-u^\star)^3 + x_1(u-u^\star) + x_0\big]\big\}_{\{x_0^2 + x_1^2 \leq 1\}}$$

in $\mathcal{P}_4$, which hits the forbidden subspace $\mathcal{P}_d^{\geq 3}$ at the singleton $o_\star = (u-2)(u-u^\star)^3$.

The $\mathcal{I}_4$-image of this family forms a loop $\mathcal{I}_4(S^1_{u^\star})$ in $\mathcal{F}_4^{\leq 2}$. The loop bounds a 2-disk $\mathcal{I}_4(D^2_{u^\star})$ in $\mathcal{F}^4$, which hits the subspace $\mathcal{F}^{\geq 3}$ at the singleton $\mathcal{I}_4(o_\star)$.

Similarly, for any $f \in (\mathcal{F}_d^{\geq 3})^\circ$ such that $u_f^\star$ is the zero of f of multiplicity 3 and $d \geq 4$, the loop

$$L_f =_{\mathsf{def}} \big\{f(u,\theta) = P_{u_f^\star}(u,\theta) \cdot \big[f(u)/(u-u_f^\star)^3\big]\big\}_{\theta \in [0,2\pi]}$$

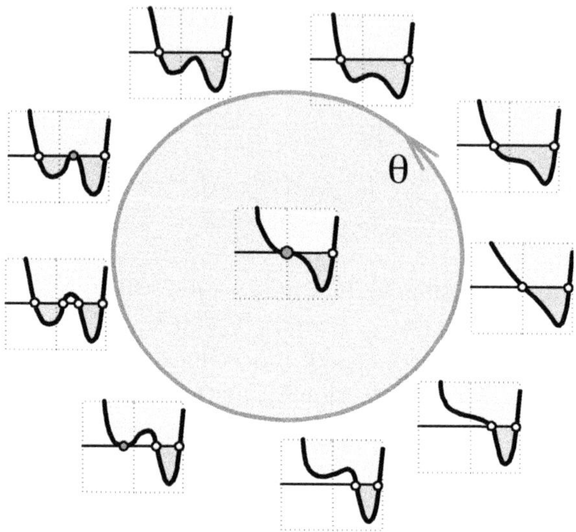

Fig. 5.13 A loop $\{P(\theta, u) =_{\mathsf{def}} (u-2)[u^3 + \cos(\theta)u + \sin(\theta)]\}_{\theta\in[0,2\pi]}$ in the space $\mathcal{P}^4_{\leq 2}$.

resides in $\mathcal{F}_d^{\leq 2}$ and is linked with the component of $(\mathcal{F}_d^{\geq 3})^\circ$ that contains the point f.

Since the correspondence $f \to u_f^\star$ is continuous for $f \in (\mathcal{F}_d^{\geq 3})^\circ$, the homotopy class $[L_f]$ of the loop L_f does not depend on the choice of f within each component of $(\mathcal{F}_d^{\geq 3})^\circ$.

Let us pick $p_0(u) = (u-2)(u^3+u)$ as the base point in $\mathcal{P}_4^{\leq 2}$ (until now the default choice of the base point was the polynomial $u^4 + 1 > 0$) and $\psi_0 = \mathcal{I}_4(p_0)$ be the base point in $\mathcal{F}_d^{\leq 2}$. The latter represents a function with a pair of distinct simple zeros. In fact, by Theorem 5.2, $[L_f]$ is the generator κ of $\pi_1(\mathcal{F}^d_{\leq 2}, \psi_0)$. In particular, the $\mathcal{I}_4$-image of the loop

$$\{P(\theta, u) =_{\mathsf{def}} (u-2)[u^3 + \cos(\theta)u + \sin(\theta)]\}_{\theta\in[0,2\pi]}$$

in $\mathcal{P}^4_{\leq 2}$ is a generator κ of $\pi_1(\mathcal{F}_{\leq 2}) \approx \mathbb{Z}$. Its zero set in the cylinder $A = \mathbb{R}\times S^1$ (equipped with the coordinates (u,θ)) is a union of two loops, similar to the ones shown in Fig. 5.5, (a), and bounding the shaded region.

Theorem 11 from [V], makes an important claim: the $\mathcal{I}_d$-induced map in homology

$$(\mathcal{I}_d)_* : H_j(\mathcal{P}_d^{\leq 2}; \mathbb{Z}) \to H_j(\mathcal{F}^{\leq 2}; \mathbb{Z})$$

is an isomorphism for all $j \leq d/3$. In particular,

$$(\mathcal{I}_4)_* : H_1(\mathcal{P}_4^{\leq 2}; \mathbb{Z}) \to H_1(\mathcal{F}^{\leq 2}; \mathbb{Z}) \approx \mathbb{Z}$$

is an isomorphism. Moreover,

$$(\mathcal{I}_4)_* : \pi_1(\mathcal{P}_4^{\leq 2}) \to \pi_1(\mathcal{F}^{\leq 2}) \approx \mathbb{Z}$$

is an isomorphism as well [Ar]. As we proceed with our polynomial models, let us keep these facts in mind.

Of course, it is much easier to visualize events in the 4-dimensional space $\mathcal{P}_4^{\leq 2}$ than their analogues in the infinite-dimensional $\mathcal{F}_4^{\leq 2} \subset \mathcal{F}^{\leq 2}$. This visualization is our next goal. It will lead us to explore the beautiful stratified geometry of the Swallow Tail discriminant hypersurface, whose 2-dimensional slice is shown in Fig. 5.14.

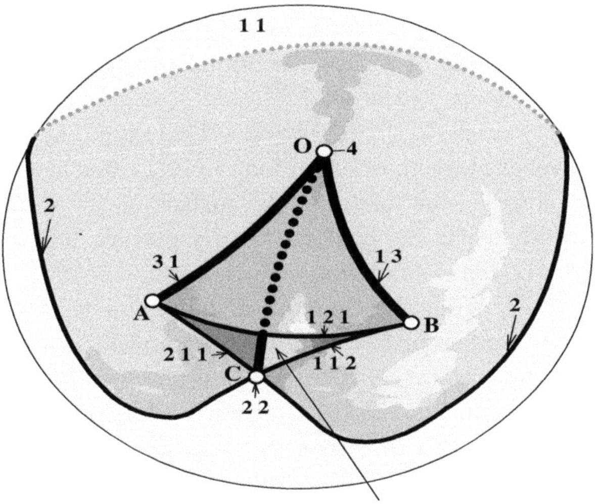

Fig. 5.14 The Swallow Tail Singularity is the critical locus of the Whitney projection of the hypersurface $\{P(u, x_0, x_1, x_2) = u^4 + x_2u^2 + x_1u + x_0 = 0\} \subset \mathbb{R}^4$ on the space $\mathbb{R}^3_{\mathsf{coef}}$ with the coordinates (x_0, x_1, x_2). The figure shows the portion of the projection inside of a unit ball, centered on the origin O. The strata in $\mathbb{R}^3_{\mathsf{coef}}$ are indexed by the combinatorial types $\{(1111), (11), \emptyset; (211), (121), (112), (2); (22), (13), (31); (4)\}$ of real divisors of the polynomial $P(u, \sim)$. They divide $\mathbb{R}^3_{\mathsf{coef}}$ into three 3-cells, four 2-cells, three 1-cells, and one 0-cell. The arrows in the picture point to the intersections of the strata with the unit sphere.

Consider the subspace $\tilde{\mathcal{P}}_4^{\leq 2} \subset \mathcal{P}_4^{\leq 2}$, formed by the monic **depressed** (i.e., with the zero coefficient next to u^3) quartic polynomials $P(u)$. Since $\tilde{\mathcal{P}}_4^{\leq 2}$ is a deformation retract of $\mathcal{P}_4^{\leq 2}$ ([Ar1]), these two spaces are homotopy equivalent. It is a bit easier to visualize the generator $\tilde{\kappa} \in \pi_1(\tilde{\mathcal{P}}_4^{\leq 2})$ (rather than $\kappa \in \pi_1(\mathcal{P}_4^{\leq 2})$): indeed, $\tilde{\mathcal{P}}_4^{\leq 2}$ is a domain in $\mathbb{R}^3_{\mathsf{coef}}$, the space of the coefficients x_0, x_1, x_2.

The polynomials from the space $\tilde{\mathcal{P}}_4^{\leq 2}$ with real roots of multiplicity ≥ 2 form a singular discriminant surface $\tilde{\mathcal{D}}_2 \subset \mathbb{R}^3_{\mathsf{coef}}$, the famous **Swallow Tail surface** (see Fig. 5.14). The point-polynomial u^4 (the strongest singularity O of $\tilde{\mathcal{D}}_2$), together with the polynomials that have one root of multiplicity 3, must be excluded from $\mathbb{R}^3_{\mathsf{coef}}$ to form $\tilde{\mathcal{P}}_4^{\leq 2}$. These excluded polynomials form the two branches of the curve $\mathcal{C}_\star \subset \tilde{\mathcal{D}}_2$, whose apex O is the origin and whose branches extend to infinity. One branch OB of $\mathcal{C}_\star$ correspond to polynomials with the smaller simple root, followed by the root of multiplicity 3; the other branch OA of $\mathcal{C}_\star$ correspond to polynomials with the smaller root of multiplicity 3, followed by the simple root. The curve OC, the self-intersection locus of $\tilde{\mathcal{D}}_2$, represents polynomials with two roots of multiplicity 2. It belongs to the set $\tilde{\mathcal{P}}_4^{\leq 2}$.

Thus, $\pi_1(\tilde{\mathcal{P}}_4^{\leq 2}) = \pi_1(\mathbb{R}^3_{\mathsf{coef}} \setminus \mathcal{C}_\star) \approx \mathbb{Z}$. Therefore, the generator $\tilde{\kappa} \in \pi_1(\tilde{\mathcal{P}}_4^{\leq 2})$ is represented by an oriented loop in $\mathbb{R}^3_{\mathsf{coef}}$ that is linked with the curve $\mathcal{C}_\star$. Such loop $\tilde{\kappa}$ must hit once the surface $\mathcal{D}^{(121)} \subset \tilde{\mathcal{D}}_2$, formed by polynomials whose real zeros conform to the pattern (121): indeed, the curve $\mathcal{C}_\star = OA \cup OB$ is the boundary of the surface $\mathcal{D}^{(121)}$.

Traversing $\mathcal{D}^{(121)}$ from the chamber of polynomials with 4 real roots to the chamber of polynomials with 2 real roots picks a normal to $\mathcal{D}^{(121)}$ orientation. Similar orientation rule can be applied to the stratum $\mathcal{D}^{(112)}$, bounded by the curves $OC \cup OB$, the stratum $\mathcal{D}^{(211)}$, bounded by the curves $OC \cup OA$, and the stratum $\mathcal{D}^{(2)}$ that separates the chamber with two real roots from the chamber with no real roots at all.

For any smooth loop $\beta : S^1 \to \tilde{\mathcal{P}}_4^{\leq 2}$ which is in general position to $\tilde{\mathcal{D}}_2$ and its strata, consider the zero set

$$\Delta(\beta) =_{\mathsf{def}} \big\{(u, \theta) |\ \beta(\theta)(u) = 0\big\} \subset \mathbb{R} \times S^1.$$

Thanks to the very definition of the target space $\tilde{\mathcal{P}}_4^{\leq 2}$, the set $\Delta(\beta)$ is a collection of closed curves with transversal self-intersections, no triple intersections, no θ-vertical self-tangencies, and no θ-vertical inflections. The cardinality of the fiber of $\theta : \Delta(\beta) \to S^1$ does not exceed 4.

Now the linking number $J^\beta =_{\mathsf{def}} \mathsf{lk}(\beta(S^1), \mathcal{C}_\star)$ equals the algebraic intersection of the loop $\beta(S^1)$ with the surface $\mathcal{D}^{(121)}$ that bounds $\mathcal{C}_\star$. Each time the loop $\beta(S^1)$ intersects the surface $\mathcal{D}^{(121)}$ transversally in the direction of the positive normal, a point from $\partial_2^{+,\oplus}\Delta(\beta)$ is generated, and each time $\beta(S^1)$ intersects $\mathcal{D}^{(121)}$ is in the direction of the negative normal, a point from $\partial_2^{+,\ominus}\Delta(\beta)$ is generated. In particular, the generator $\kappa : S^1 \to \tilde{\mathcal{P}}_4^{\leq 2}$ of $\pi_1(\tilde{\mathcal{P}}_4^{\leq 2}, pt)$ has the property $\mathsf{lk}(\beta(S^1), \mathcal{C}_\star) = 1$. Therefore, we get

$$\mathsf{lk}(\beta(S^1),\, \mathcal{C}_\star) = \beta(S^1) \circ \mathcal{D}^{(121)} = \#\{\partial_2^{+,\oplus}\Delta(\beta)\} - \#\{\partial_2^{+,\ominus}\Delta(\beta)\},$$

the baby model of the formula from Theorem 5.2. This number is an invariant of the homotopy class of the loop β.

Similarly, each time the loop $\beta(S^1)$ intersects the surface $\mathcal{D}^{(2)}$ transversally in the direction of the positive normal, a point from $\partial_2^{-,\oplus}\Delta(\beta)$ is generated, and each time $\beta(S^1)$ intersects $\mathcal{D}^{(2)}$ is in the direction of the negative normal, a point from $\partial_2^{-,\ominus}\Delta(\beta)$ is generated.

Note that, for any β that is transversal to the discriminant surface $\tilde{\mathcal{D}}_2$ and its strata, the set $\Delta(\beta)$ is the image $\alpha(\partial_1 X)$ of an embedding $\alpha : X \to S^1 \times \mathbb{R}$ for the orientable surface

$$X =_{\mathsf{def}} \big\{(u, \theta)|\ \beta(\theta)(u) \leq 0\big\} \subset \mathbb{R} \times S^1.$$

In particular, the generator κ in Fig. 5.5, (b), is.

Recall that, if we insist that the cardinality of the θ-fibers ≤ 4, then we cannot accommodate surfaces X with handles. According to Fig. 5.2, to accommodate them, we need to deal with polynomials/functions of degree 6 at least.

Let us describe briefly how this degree 6 polynomial model works. We see instantly that the increasingly complex combinatorics of tangency begins to play a significant role.

The combinatorial patterns ω of real divisors of monic degree 6 real polynomials are numerous:

- (111111), (1111), (11), $\emptyset$;
- (21111), (12111), (11211), (11121), (11112), (211), (121), (112), (2);
- (2211), (1221), (1122), (2121), (2112), (1212), (22);
- (3111), (1311), (1131), (1113), (31), (13);
- etc.

We denote by $\mathcal{P}_6^\omega$ the set of real monic polynomials of degree 6 whose real divisors conform to the combinatorial pattern $\omega = (\omega_1, \omega_2, \dots, \omega_s)$, where $\{\omega_i\}_{1\leq i\leq s}$ are natural numbers and $s \leq 6$. We denote by $|\omega| =_{\mathsf{def}} \sum_i \omega_i$, the l_1-norm of the vector ω. Evidently, $|\omega| \leq 6$.

The sets $\{\mathcal{P}_6^\omega\}_\omega$ form a partition of the space $\mathcal{P}_6 \approx \mathbb{R}^6$. In fact, each $\mathcal{P}_6^\omega$ is homeomorphic to an open ball of dimension $6 - |\omega|'$, where $|\omega|' = |\omega| - s$ ([K3]). The closure $\bar{\mathcal{P}}_6^\omega$ of $\mathcal{P}_6^\omega$ in $\mathbb{R}^6_{\mathsf{coef}}$ is an affine semi-algebraic variety. By resolving $\bar{\mathcal{P}}_6^\omega$ appropriately, one can show that the partition $\{\mathcal{P}_6^\omega\}_\omega$ defines a structure of a CW-complex on the space $\mathcal{P}_6$ (see Chapter 6 or [K3]). Therefore, we may think of $\{\mathcal{P}_6^\omega\}_\omega$ as being "cells" (although the closure $\bar{\mathcal{P}}_6^\omega$ may not be homeomorphic to an infinite cone over a closed ball).

Let $\mathcal{D} \subset \mathcal{P}_6$ denote the set of monic polynomials with multiple roots, the 5-dimensional discriminant variety. The first bullet in the list above presents the four 6-dimensional chambers-cells in which $\mathcal{D}$ divides $\mathcal{P}_6$. The second bullet lists all 5-dimensional strata in which $\mathcal{D}$ is divided by the strata of dimension 4. The first three bullets list the monic polynomials that form the space $\mathcal{P}_6^{\leq 2}$. The third and the fourth bullets list the 4-dimensional cells-strata. The forbidden locus $\mathcal{P}_6^{\geq 3}$ is the union of strata, labeled by the combinatorial types in the fourth bullet, and so on. Then $\mathcal{P}_6^{\geq 3}$ is the closure of the set

$$\mathcal{P}_6^{(3111)} \cup \mathcal{P}_6^{(1311)} \cup \mathcal{P}_6^{(1131)} \cup \mathcal{P}_6^{(1113)} \cup \mathcal{P}_6^{(31)} \cup \mathcal{P}_6^{(13)}.$$

We can orient each cell $\mathcal{P}_6^\omega$ so that their algebraic boundaries are:

$$\partial\bar{\mathcal{P}}_6^{(12111)} = \bar{\mathcal{P}}_6^{(3111)} - \bar{\mathcal{P}}_6^{(1311)} + \bar{\mathcal{P}}_6^{(1221)} - \bar{\mathcal{P}}_6^{(1212)},$$

$$\partial\bar{\mathcal{P}}_6^{(11121)} = \bar{\mathcal{P}}_6^{(1131)} - \bar{\mathcal{P}}_6^{(1113)} + \bar{\mathcal{P}}_6^{(2121)} - \bar{\mathcal{P}}_6^{(1221)},$$

$$\partial\bar{\mathcal{P}}_6^{(121)} = \bar{\mathcal{P}}_6^{(31)} - \bar{\mathcal{P}}_6^{(13)} + \bar{\mathcal{P}}_6^{(2121)} - \bar{\mathcal{P}}_6^{(1212)}.$$

The operator ∂ in these formulas should be understood in the spirit of algebraic topology as the boundary operator on cellular chains (and not as a topological boundary of the appropriate sets). See formulas (6.10) for the formal definitions and generalizations of this boundary operator.

Adding the three formulas above, we get that the forbidden set $\mathcal{P}_6^{\geq 3}$, viewed as a 4-chain, is an algebraic boundary of the 5-chain:

$$\mathcal{P}_6^{\geq 3} =_{\mathsf{def}} \bar{\mathcal{P}}_6^{(3111)} - \bar{\mathcal{P}}_6^{(1311)} + \bar{\mathcal{P}}_6^{(1131)} - \bar{\mathcal{P}}_6^{(1113)} + \bar{\mathcal{P}}_6^{(31)} - \bar{\mathcal{P}}_6^{(13)}$$

$$= \partial\big(\bar{\mathcal{P}}_6^{(12111)} + \bar{\mathcal{P}}_6^{(11121)} + \bar{\mathcal{P}}_6^{(121)}\big).$$

Now consider a smooth loop $\beta : S^1 \to \mathcal{P}_6^{\leq 2}$. By a small perturbation we may assume that $\beta(S^1)$ is transversal to the hypersurfaces $\mathcal{P}_6^{(12111)}, \mathcal{P}_6^{(11121)}, \mathcal{P}_6^{(121)}$ whose combination

$$\bar{\mathcal{P}}_6^{(12111)} + \bar{\mathcal{P}}_6^{(11121)} + \bar{\mathcal{P}}_6^{(121)}$$

bounds the cycle $\bar{\mathcal{P}}_6^{\geq 3}$. Therefore, the linking number

$$\mathsf{lk}\big(\beta(S^1), \mathcal{P}_6^{\geq 3}\big) = \beta(S^1) \circ \big(\mathcal{P}_6^{(12111)} \cup \mathcal{P}_6^{(11121)} \cup \mathcal{P}_6^{(121)}\big).$$

As in the case of polynomials of degree 4, we form the set

$$\Delta(\beta) =_{\mathsf{def}} \{(u, \theta) |\ \beta(\theta)(u) = 0\} \subset \mathbb{R} \times S^1.$$

(See Fig. 5.5 for the basic examples of such loops β and their loci $\Delta(\beta)$.)

Then

$$\beta(S^1) \circ \big(\mathcal{P}_6^{(12111)} \cup \mathcal{P}_6^{(11121)} \cup \mathcal{P}_6^{(121)}\big) = \#\{\partial_2^{+,\oplus}\Delta(\beta)\} - \#\{\partial_2^{+,\ominus}\Delta(\beta)\}.$$

Therefore, we get that the linking number

$$\mathsf{lk}\big(\beta(S^1), \mathcal{P}_6^{\geq 3}\big) = \#\{\partial_2^{+,\oplus}\Delta(\beta)\} - \#\{\partial_2^{+,\ominus}\Delta(\beta)\},$$

a version of the formula from Theorem 5.2, being applied to the loop $\beta = J_{z_\alpha}$. The loop is produced, with the help of an auxiliary function $z_{\alpha(\partial X)}$, by a 2-moderate with respect to $\hat{v}$ embedding $\alpha : X \to A$ of a compact surface X, where $\alpha(X)$ bounds $\Delta(\beta)$ in A. The cardinality of the fibers of $\theta : \alpha(\partial_1 X) \to S^1$ does not exceed 6. Recall that this constraint accommodates any compact connected oriented surface X with boundary.

In fact, the constructions of this kind are not restricted to polynomials/smooth functions of degree 6: they apply to any even degree d. The application requires a deeper dive into the combinatorics of real polynomial divisors and their modifications, as is done in Chapter 6.

In dimensions higher than two, similar considerations will produce characteristic classes of immersions and embeddings of compact $(n+1)$-manifolds X into a space $A = \mathbb{R} \times Y$, where Y is a compact n-dimensional manifold. They are based on computations in Chapter 6 of homology of spaces of real monic polynomials with *restricted combinatorics* of their real divisors. In Chapters 7 and 8, we will show that the topology of n-dimensional doodles on a ruled page $\mathbb{R}^{n+1}$ or the topology of high-dimensional convex envelops/quasi-envelops of traversing flows is at least as intricate as the homotopy groups of spheres!

Chapter 6

Topology of Spaces of Polynomials with Constrained Real Zero Divisors

"In these days (1939) the angel of topology and the devil of abstract algebra fight for the soul of each individual mathematical domain."

Hermann Weyl

"In our days (2025) the exhausted angel is taking a brake ..."

G. K.

6.1 In this chapter...

The results of this chapter will play a key technical role in the next two chapters which are devoted to a study of immersions and embeddings of compact n-manifolds into $(n+1)$-manifolds of the product type $\mathbb{R} \times Y$, where Y is a compact n-manifold. In parallel, we will study submersions of compact $(n+1)$-manifolds on compact $(n+1)$-manifolds, equipped with traversing and convex vector flows.

Some readers may choose to proceed to these applications, after familiarizing themselves with the notations and basic results of this chapter. Others may find the constructions and conclusions of this chapter to be of an independent interest.

In the late 80s, V. Arnold and V. Vassiliev initiated the topological study of the space of real univariate polynomials of a given degree, which have no real roots of multiplicity exceeding a given positive integer [Ar], [Va], [Va1]-[Va3]. Expanding their studies, we consider the spaces $\mathcal{P}_d^{\mathbf{c}\Theta}$ of real monic univariate polynomials of degree d whose real divisors avoid some given sequences ω of root multiplicities. These forbidden ω's are taken from an arbitrary poset Θ of compositions that are closed under the natural combinatorial operations of merges and inserts as in (2.2), (2.3).

All the constructions and results of this chapter originate in [KSW2] and [KSW1], our joint works with Boris Shapiro and Volkmar Welker. In particular, in [KSW2], we reduced the computation of the homology $H_*(\mathcal{P}_d^{\mathbf{c}\Theta}; \mathbb{Z})$ to the computation of the homology of a differential complex, defined purely combinatorially in terms of the given closed poset Θ. The results of *numeric computations* of these homology groups are organized into several tables. The computations are performed with the help of a software program developed by Volkmar Welker.

We also obtain the stabilization results about the homology $H_*(\mathcal{P}_d^{\mathbf{c}\Theta}; \mathbb{Z})$, as $d \to \infty$.

In fact, these results about the topology of spaces of polynomials with constrained real divisors are deduced from our description of the homology of related spaces $\mathcal{B}_d^{\mathbf{c}\Theta}$, whose points are binary real homogeneous forms, considered up to projective equivalence, with similarly Θ-constrained real divisors. In particular, we exhibit algebraic differential complexes that calculate the (co)homology of the spaces $\mathcal{B}_d^{\mathbf{c}\Theta}$.

As an example of our general approach, we compute the homology of the discriminants of projectivized binary real forms and of their complements in $\mathcal{B}_d \approx \mathbb{RP}^d$.

Let us describe the structure of Chapter 5 in some detail. In Section 6.3, we study the cellular structure of $\mathcal{P}_d$, given by the cells $\{\mathsf{R}_d^\omega\}_\omega$, labeled by the combinatorial types $\omega \in \mathbf{\Omega}_{\langle d]}$. In parallel, we investigate a cellular structure on the space $\mathcal{B}_d$ of projective classes of binary homogeneous forms of degree d with real coefficients. The cells of $\mathcal{B}_d$ are indexed by *pairs* consisting of a root pattern ω and a non-negative integer κ. The integer κ represents the root multiplicity at ∞. Similarly to the polynomial case, we consider a set Θ of pairs (ω, κ) and the space $\mathcal{B}_d^\Theta$ of forms, whose root multiplicity patterns belong to Θ, as well as its complement $\mathcal{B}_d^{\mathbf{c}\Theta} := \mathcal{B}_d \setminus \mathcal{B}_d^\Theta$.

In Section 6.4, we turn to (co)homology of the compact spaces $\bar{\mathcal{P}}_d^\Theta$, $\mathcal{B}_d^\Theta$, and their complements $\mathcal{P}_d^{\mathbf{c}\Theta}$, $\mathcal{B}_d^{\mathbf{c}\Theta}$. Here $\bar{\mathcal{P}}_d^\Theta$ denotes the one-point compactification of the space $\mathcal{P}_d^\Theta$. In Theorem 6.3, for a given closed Θ, we construct a combinatorial differential complex of free $\mathbb{Z}$-modules that calculates the homology of $\mathcal{B}_d^\Theta$.

Further, we can identify the univariate polynomial

$$a_0 + \cdots + a_{d-1}x^{d-1} + x^d$$

of degree d with the class of the homogeneous binary form

$$a_0 y^d + \cdots + a_{d-1}x^{d-1}y + x^d$$

of degree d with no (real) roots at infinity. Hence we can identify $\bar{\mathcal{P}}_d$ with a quotient of $\mathcal{B}_d$ by the closed subspace $\mathcal{B}_{d-1}$. Based on this identification, for a given closed Θ, Corollary 6.5 introduces the combinatorial complex that calculates the homology of the space of $\bar{\mathcal{P}}_d^{\Theta}$. The latter complex is the restriction of the combinatorial complex from Theorem 6.3 (which computes the homology of $\mathcal{B}_d^{\Theta}$) to the cells corresponding to univariate degree d polynomials. Via Alexander duality, this complex also computes the cohomology groups of $\mathcal{P}_d^{\mathbf{c}\Theta}$.

In Section 6.5, we discuss what happens to the homology of $\mathcal{B}_d^{\mathbf{c}\Theta}$ and $\mathcal{P}_d^{\mathbf{c}\Theta}$ when Θ is fixed and $d \to \infty$. Our main stabilization results are Theorems 6.7, 6.8, 6.9, and 6.10. For a closed Θ and a k, bounded by a certain function of d, it claims the stabilization of the kth homology groups $H_k(\mathcal{B}_d^{\mathbf{c}\Theta};\mathbb{Z})$ and $H_k(\mathcal{P}_d^{\mathbf{c}\Theta};\mathbb{Z})$. For, so called, profinite Θ the result holds for any k and d large enough.

Finally, in Section 6.5.1, we present the results of computer experiments calculating the homology of $\bar{\mathcal{P}}_d^{\Theta}$ for various Θ and d.

6.2 Background

In [Ar1], V. Arnold proved Theorems 6.1 and 6.2, stated below. Later, these results were generalized by V. Vassiliev. In particular, he described the multiplication in the cohomology of the relevant spaces, see [Va], [Va1]–[Va3]. All those papers study the topology of spaces of smooth functions/polynomials, which Arnold calls functions/polynomials with "moderate singularities." These works are the main sources of inspiration for our study of functions/polynomials with "*less moderate singularities.*"

In the following formulation of Arnold's theorems, we keep his original notation, which we will abandon later on.

For $1 \leq k \leq d$, let G_k^d be the space of real monic polynomials $x^d + a_{d-1}x^{d-1} + \cdots + a_0 \in \mathbb{R}[x]$ with no real roots of multiplicity $\geq k$.

Theorem 6.1. *[Ar1] If $k \leq d < 2k-1$, then G_k^d is diffeomorphic to the product of a sphere S^{k-2} by an Euclidean space. In particular, for all i, one has isomorphisms*

$$\pi_i(G_k^d) \cong \pi_i(S^{k-2})$$

of a homotopy groups. ◇

An analogous result holds for the space of polynomials whose sum of roots vanishes, i.e., the polynomials with the vanishing coefficient a_{d-1}.

Theorem 6.2. [*Ar1*] *The homology groups with integer coefficients of the space G_k^d are nonzero only for dimensions which are the multiples of $k-2$ and less than or equal to d. For $(k-2)r \leq d$, we have*

$$H_{r(k-2)}(G_k^d) \cong \mathbb{Z}. \quad \diamondsuit$$

6.2.1 *Our set-up*

Let $\mathcal{P}_d$ denote the space of real monic univariate polynomials of degree d. Given a polynomial $P(x) = x^d + a_{d-1}x^{d-1} + \cdots + a_0 \in \mathcal{P}_d$, we define its real divisor $D_{\mathbb{R}}(P)$ as the multiset

$$\underbrace{x_1 = \cdots = x_{i_1}}_{\omega_1 = i_1} < \underbrace{x_{i_1+1} = \cdots = x_{i_1+i_2}}_{\omega_2 = i_2 - i_1} < \cdots < \underbrace{x_{i_{\ell-1}+1} = \cdots = x_{i_\ell}}_{\omega_\ell = i_\ell - i_{\ell-1}}$$

of the real roots of $P(x)$. The ordered ℓ-tuple $\omega = (\omega_1, \ldots, \omega_\ell)$ of natural numbers describes the real root multiplicity pattern of $P(x)$.

Let $\mathring{\mathsf{R}}_d^\omega$ be the set of all real polynomials with a real root multiplicity pattern ω, and let R_d^ω be the closure of $\mathring{\mathsf{R}}_d^\omega$ in $\mathcal{P}_d$.

For a given collection Θ of root multiplicity patterns, we consider the union $\mathcal{P}_d^\Theta \subset \mathcal{P}_d$ of the subspaces $\mathring{\mathsf{R}}_d^\omega$, taken over all $\omega \in \Theta$. We denote by $\mathcal{P}_d^{\mathbf{c}\Theta} := \mathcal{P}_d \setminus \mathcal{P}_d^\Theta$ its complement. Our studies are restricted to the case when $\mathcal{P}_d^\Theta$ is closed in $\mathcal{P}_d$ and we call such collections Θ closed.

As has been observed in [KSW1], for any collection Θ of root multiplicity patterns which contains the pattern (d), the space $\mathcal{P}_d^\Theta$ is *contractible.* Thus topologically interesting situation occurs if one considers the one-point compactification $\bar{\mathcal{P}}_d^\Theta$ of $\mathcal{P}_d^\Theta$. For a closed Θ, the latter is the union of the one-point compactifications $\bar{\mathsf{R}}_d^\omega$ of the spaces R_d^ω, taken over $\omega \in \Theta$ with points at infinity identified. By Alexander duality in $\bar{\mathcal{P}}_d \cong S^d$, we get the relation

$$\bar{H}^j(\mathcal{P}_d^{\mathbf{c}\Theta}; \mathbb{Z}) \cong \bar{H}_{d-j-1}(\bar{\mathcal{P}}_d^\Theta; \mathbb{Z})$$

in reduced (co)homology. It implies that the spaces $\mathcal{P}_d^{\mathbf{c}\Theta}$ and $\bar{\mathcal{P}}_d^\Theta$ carry equivalent (co)homological information.

In particular, for Θ comprising all ω's such that at least one entry of ω is greater than or equal to k, we have that Θ is closed and $\mathcal{P}_d^{\mathbf{c}\Theta} \cong G_k^d$ (see [KSW1]).

6.3 Geometry and combinatorics of the cell structure on spaces of real polynomials and real binary forms

Let $\mathcal{B}_d$ denote the space of non-zero real bivariate homogeneous polynomials of the form

$$a_0 x^0 y^d + \cdots + a_d x^d y^0,$$

being considered up to a non-zero real scalar factor.

We have already associated to a polynomial $P(x) \in \mathbb{R}[x]$ its real root multiplicity pattern $(\omega_1, \ldots, \omega_\ell)$. Next, we define the combinatorial structure which will govern the cell decomposition of the space $\mathcal{B}_d$. Topologically, $\mathcal{P}_d$ is a real d-dimensional affine space, while $\mathcal{B}_d$ is the real d-dimensional projective space $\mathbb{RP}^d$. For $d' \leq d$, the map $\wp$, given by

$$\mathcal{P}_{d'} \ni a_0 + \cdots + a_{d'-1}x^{d'-1} + x^{d'} \xrightarrow{\wp} \\ a_0 x^0 y^d + \cdots + a_{d'-1}x^{d'-1}y^{d-d'+1} + x^{d'}y^{d-d'} \in \mathcal{B}_d, \tag{6.1}$$

is a homeomorphism of $\mathcal{P}_{d'}$ with a subspace of $\mathcal{B}_d$. The $\wp$-images for different d' are disjoint. We obtain the standard decomposition $\mathcal{B}_d = \mathcal{P}_d \sqcup \mathcal{P}_{d-1} \sqcup \mathcal{P}_{d-2} \sqcup \cdots \sqcup \mathcal{P}_0$. Here $\mathcal{P}_0$ can be identified with the one-point space of constant non-zero polynomials up to a non-zero factor.

Let us define the combinatorial structures which capture and refine this natural stratification.

The space $\mathcal{P}_d$ may be also identified with the space of positive degree d divisors on the complex plane $\mathbb{C}$, invariant under the complex conjugation $J : z \to \bar{z}$. At the same time, $\mathcal{B}_d$ may be identified with the space of positive degree d divisors on the Riemann sphere $\mathbb{CP}^1$ that are invariant under the complex conjugation $\hat{J} : [z : w] \to [\bar{z} : \bar{w}]$, where $[z : w]$ are homogeneous coordinates on $\mathbb{CP}^1$.

Definition 6.1. An arbitrary sequence $\omega = (\omega_1, \ldots, \omega_\ell)$ of positive integers is called a *composition*. We say that ω is a composition of the number $|\omega| := \omega_1 + \cdots + \omega_\ell$ (Cf. (2.5)). We also allow the *empty composition* $\omega = ()$ of the number $|()| = 0$. We call the ω_i, $i = 1, \ldots, \ell$, the parts of the composition ω.

Similarly, a pair (ω, κ) is called a marked composition, where ω is a composition and κ is a non-negative integer. For $\omega = (\omega_1, \ldots, \omega_\ell)$, we say that (ω, κ) is a marked composition of the number $|(\omega, \kappa)| := |\omega| + \kappa$.

For a composition ω, (resp., a marked composition (ω, κ),) we call $|\omega|$ (resp., $|(\omega, \kappa)|$) its norm and $|\omega|' := |\omega| - \ell$ (resp., $|(\omega, \kappa)|' = |\omega|' + \kappa$) its reduced norm.

If $\kappa = 0$, then we often identify the marked composition (ω, κ) with ω and speak of the norm $|\omega| = |(\omega, 0)|$ and the reduced norm $|\omega|' = |(\omega, 0)|'$ of ω. $\diamondsuit$

For a bivariant form $f = a_0x^0y^d + \cdots + a_dx^dy^0 \in \mathcal{B}_d$, we define its real degree d' as $\max\{i \mid a_i \neq 0\}$. As in (6.1), f can be identified with a polynomial $f^\dagger$ in $\mathcal{P}_{d'}$ such that $\wp(f^\dagger) = f$.

Let $[x : y] \in \mathbb{RP}^1$ be a real root of f. If $y \neq 0$, then we can choose $y = 1$ and x as a real root of $f^\dagger$. If $y = 0$, then we can choose $x = 1$ and identify $(1 : 0)$ with the point at infinity in $\mathbb{RP}^1$. Thus we can describe the root multiplicities of f by a marked composition (ω, κ), where ω is a composition with $|\omega| \leq d' = d - \kappa$ and κ is the multiplicity of the root at infinity. Any polynomial in $\mathcal{P}_0 \subseteq \mathcal{B}_d$ is associated with the marked composition $((), d)$.

Let (ω, κ) be a marked composition for which $d - |(\omega, \kappa)|$ is even and non-negative. Using the map $\wp$ from (6.1), we set

$$\mathring{\mathsf{R}}_d^{(\omega,\kappa)} =_{\mathsf{def}} \mathring{\mathsf{R}}_{d'}^{\omega} \subseteq \mathcal{P}_{d'} \subseteq \mathcal{B}_d,$$

where $d' = d - \kappa$. Analogously to the univariate case, we write $\mathsf{R}_d^{(\omega,\kappa)}$ for the closure of $\mathring{\mathsf{R}}_d^{(\omega,\kappa)}$ in $\mathcal{B}_d$. Evidently, for a given composition ω, the stratum R_d^{ω} is empty if and only if either $|\omega| > d$, or $|\omega| \leq d$ and $d - |\omega|$ is odd. Thus if (ω, κ) is a marked composition, then $\mathsf{R}_d^{(\omega,\kappa)}$ is empty if and only if either $|(\omega, \kappa)| > d$ or $|(\omega, \kappa)| \leq d$ and $d - |(\omega, \kappa)|$ is odd.

Note that for a composition ω, we have $\mathring{\mathsf{R}}_d^{\omega} = \mathring{\mathsf{R}}_d^{(\omega,0)}$. However, in general, $\mathsf{R}_d^{\omega} \neq \mathsf{R}_d^{(\omega,0)}$ since the closures of $\mathring{\mathsf{R}}_d^{\omega}$ and of $\mathring{\mathsf{R}}_d^{(\omega,0)}$ are taken in different spaces.

For a given collection Θ of marked compositions, consider the union $\mathcal{B}_d^{\Theta}$ of the subspaces $\mathring{\mathsf{R}}_d^{(\omega,\kappa)}$ over all $(\omega, \kappa) \in \Theta$. We denote its complement by $\mathcal{B}_d^{\mathbf{c}\Theta} := \mathcal{B}_d \setminus \mathcal{B}_d^{\Theta}$. Similarly, for a given collection of compositions Θ, we write $\mathcal{P}_d^{\Theta}$ for the union of the subspaces $\mathring{\mathsf{R}}_d^{\omega}$ over all $\omega \in \Theta$. Further, set $\mathcal{P}_d^{\mathbf{c}\Theta} := \mathcal{P}_d \setminus \mathcal{P}_d^{\Theta} = \bar{\mathcal{P}}_d \setminus \bar{\mathcal{P}}_d^{\Theta}$.

We restrict our studies to the case when either $\mathcal{B}_d^{\Theta}$ is closed in $\mathcal{B}_d$ or when $\mathcal{P}_d^{\Theta}$ is closed in $\mathcal{P}_d$. In such situations, we call the respective collection Θ closed.

As observed in [KSW1] , for any collection Θ of compositions containing (d), the space $\mathcal{P}_d^{\Theta}$ is contractible. Thus $\mathcal{P}_d^{\Theta}$ is contractible for any closed Θ. As a consequence, we concentrate on the *one-point compactification* $\bar{\mathcal{P}}_d^{\Theta} = \mathcal{P}_d^{\Theta} \sqcup \infty$ of $\mathcal{P}_d^{\Theta}$. It has a non-trivial topology related to $\mathcal{P}_d^{\mathbf{c}\Theta}$.

Note that if $\mathcal{P}_d^\Theta$ is closed in $\mathcal{P}_d$, then usually neither $\mathcal{P}_d^\Theta$ nor $\bar{\mathcal{P}}_d^\Theta$ are closed in $\mathcal{B}_d$.

We denote by $\mathbf{\Omega}^\infty$ the set of all marked compositions (ω, κ). For a given positive integer d, we denote by $\mathbf{\Omega}^\infty_{\langle d]}$ (resp., $\mathbf{\Omega}^\infty_d$) the set of all marked compositions (ω, κ), such that $|(\omega, \kappa)| \leq d$ and $|(\omega, \kappa)| \equiv d \mod 2$ (resp., $|(\omega, \kappa)| = d$). Analogously we write $\mathbf{\Omega}$ for the set of all compositions, $\mathbf{\Omega}_{\langle d]}$ for the set of all compositions ω for which $|\omega| \leq d$ and $|\omega| \equiv d \mod 2$, and $\mathbf{\Omega}_d$ for the set of all compositions ω with $|\omega| = d$.

Following (2.2) and (2.3), let us define two kinds operations on $\mathbf{\Omega}^\infty$, the merge and the insert operations, which will be instrumental in what follows.

For a marked composition (ω, κ), we set

$$\begin{cases} \mathsf{M}_j^\infty((\omega, \kappa)) = (\mathsf{M}_j(\omega), \kappa) \text{ for } 1 \leq j \leq \ell - 1, \\ \mathsf{M}_0^\infty((\omega, \kappa)) = ((\omega_2, \dots, \omega_\ell), \kappa + \omega_1), \\ \mathsf{M}_\ell^\infty((\omega, \kappa)) = ((\omega_1, \dots, \omega_{\ell-1}), \kappa + \omega_\ell). \end{cases}$$

We call $\mathsf{M}_j, \mathsf{M}_j^\infty$ the **merge operations** on compositions and marked compositions. In parallel, we extend the insert operations from the formula (2.3) define the insert operations I_j^∞ via the formula

$$\mathsf{I}_j^\infty((\omega, \kappa)) = (\mathsf{I}_j(\omega), \kappa). \tag{6.2}$$

Example 6.1. For $(\omega, \kappa) = ((12243), 2)$ we have

$$\mathsf{M}_0^\infty((\omega, \kappa)) = ((2243), \tilde{3}), \quad \mathsf{M}_1^\infty(\hat{\omega}) = ((\tilde{3}243), 2), \quad \mathsf{M}_2^\infty(\hat{\omega}) = ((1\tilde{4}43), 2),$$

$$\mathsf{M}_3^\infty((\omega, \kappa)) = ((12\tilde{6}3), 2), \quad \mathsf{M}_4^\infty(\hat{\omega}) = ((122\tilde{7}), 2), \quad \mathsf{M}_5^\infty(\hat{\omega}) = ((1224), \tilde{5}),$$

$$\mathsf{I}_0^\infty((\omega, \kappa)) = ((\tilde{2}12243), 2), \quad \mathsf{I}_1^\infty(\hat{\omega}) = ((1\tilde{2}2243), 2), \quad \mathsf{I}_2^\infty(\hat{\omega}) = ((12\tilde{2}243), 2),$$

$$\mathsf{I}_3^\infty((\omega, \kappa)) = ((122\tilde{2}43), 2), \quad \mathsf{I}_4^\infty(\hat{\omega}) = ((1224\tilde{2}3), 2), \quad \mathsf{I}_5^\infty(\hat{\omega}) = ((12243\tilde{2}), 2).$$

For example, for $d = 14$, we have $|\mathsf{I}_j((\omega, \kappa))| = 16$. Hence, $\mathring{\mathsf{R}}_d^{\mathsf{I}_j((\omega,\kappa))} = \emptyset$ for all $1 \leq j \leq \ell + 1$, while for all $1 \leq j \leq \ell - 1$, we have $\mathring{\mathsf{R}}_d^{\mathsf{M}_j((\omega,\kappa))} \neq \emptyset$.

The next Proposition 6.1 collects some basic properties of the cells $\mathring{\mathsf{R}}_d^\omega$ and of their closures R_d^ω in $\mathcal{P}_d$, see [K3] for details. In the proposition, we denote the upper half-plane of the complex numbers with positive imaginary part by $\mathbb{H}$, the m-fold symmetric product of a space X by $\mathrm{Sym}^m(X)$, and the open cone $\{(x_1, \dots, x_m) \in \mathbb{R}^m \mid x_1 < \cdots < x_m\}$ by Π_m. For later use, we also will introduce $\mathrm{Conf}^m(X)$ as a notation for the configuration space of m distinct unordered points in a space X. Note that $\Pi_m \cong \mathrm{Conf}^m(\mathbb{R})$.

Proposition 6.1. *Take $d \geq 1$ and $\omega = (\omega_1, \dots, \omega_\ell) \in \Omega$ such that $|\omega| \leq d$ and $|\omega| \equiv d \mod 2$. Then $\mathring{\mathsf{R}}_d^\omega \subset \mathcal{P}_d$ is homeomorphic to $\Pi_\ell \times \mathrm{Sym}^{\frac{d-|\omega|}{2}}\mathbb{H}$. In particular, it is an open cell of codimension $|\omega|'$. Moreover, R_d^ω is the union of the cells $\mathring{\mathsf{R}}_d^{\omega'}$, taken over all ω' that are obtained from ω by a sequence of merge and insert operations. In particular,*

(a) The cell $\mathring{\mathsf{R}}_d^\omega$ has the maximal dimension d if and only if, for $0 \leq \ell \leq d$ and $\ell \equiv d \mod 2$, $\omega = (\underbrace{1, 1, \dots, 1}_{\ell})$.

(b) The cell $\mathring{\mathsf{R}}_d^\omega$ has the minimal possible dimension 1 if and only if $\omega = (d)$. In this case, $\mathring{\mathsf{R}}^{(d)} = \mathsf{R}^{(d)} = \{(x-a)^d \mid a \in \mathbb{R}\}$. ◇

Geometrically speaking, if a typical point in $\mathring{\mathsf{R}}_d^\omega$ approaches the boundary $\partial\mathsf{R}_d^\omega := \mathsf{R}_d^\omega \setminus \mathring{\mathsf{R}}_d^\omega$, then either there is at least one value of the index j such that the distance between the jth and $(j+1)^{\text{st}}$ largest root in the divisor $D_{\mathbb{R}}(P)$ goes to 0, or there are two complex-conjugate real roots that converge to a double real root, which then either is j^{th} largest or adds 2 to the multiplicity of the jth largest real root.

The first situation corresponds to the application of the merge operation M_j from (2.2) to ω, and the second one to the application of the insertion operation I_j from (2.3). Of course, there exist points in the boundary of $\partial\mathsf{R}_d^\omega$ that can only be reached from $\mathring{\mathsf{R}}_d^\omega$ by applying sequentially a number of inserts and merges.

For $d' \leq d$, under the embedding $\wp : \mathcal{P}_{d'} \hookrightarrow \mathcal{B}_d$ from (6.1), the cell $\mathring{\mathsf{R}}_{d'}^\omega$ is sent to $\mathring{\mathsf{R}}_d^{(\omega,\, d-d')}$. The image $\wp(\mathcal{P}_{d'})$ lies in the closed subspace $\mathcal{B}_{d'} \cong \mathbb{RP}^{d'}$ of $\mathcal{B}_d$ of codimension $d-d'$. As described in Proposition 6.1, in addition to the cells $\mathring{\mathsf{R}}_d^{(\omega',\, d-d')}$, which are the $\wp$-images of the cells $\mathring{\mathsf{R}}_{d'}^{\omega'}$, the boundary of $\mathring{\mathsf{R}}_{d'}^\omega$ additionally contains only the cells which lie in the $\wp$-image of $\mathcal{P}_{d''}$ for some $d'' < d'$. The latter arise when the smallest or the largest real root of some $f \in \mathcal{P}_{d'}$ approaches infinity. Since the latter situation is described by the merge operations M_0^∞ and M_ℓ^∞ acting on marked compositions $((\omega_1, \dots, \omega_\ell), \kappa)$, we obtain the following result.

Corollary 6.1. *Let $d \geq 1$, $\omega = (\omega_1, \dots, \omega_\ell) \in \Omega$, and let $(\omega, \kappa) \in \Omega^\infty$ be such that $|(\omega,\kappa)| \leq d$ and $|(\omega,\kappa)| \equiv d \mod 2$. Then $\mathring{\mathsf{R}}_d^{(\omega,\kappa)} \subset \mathcal{B}_d$ is homeomorphic to $\Pi_\ell \times \mathrm{Sym}^{\frac{d-|(\omega,\kappa)|}{2}}\mathbb{H}$. In particular, $\mathring{\mathsf{R}}_d^{(\omega,\kappa)}$ is an open cell of codimension $|(\omega,\kappa)|'$.*

Moreover, $\mathsf{R}_d^{(\omega,\kappa)}$ is the union of cells $\mathring{\mathsf{R}}_d^{(\omega',\kappa')}$, taken over all (ω',κ') that are obtained from (ω,κ) by a sequence of merge and insert operations. In *particular,*

(a) *the cell* $\mathring{\mathsf{R}}_d^{(\omega,\kappa)}$ *has the maximal dimension* d *if and only if* $\omega = \underbrace{(1,1,\dots,1)}_{\ell}$ *for some* $0 \leq \ell \leq d$, $\ell \equiv d \mod 2$ *and* $\kappa = 0$.

(b) *the cell* $\mathring{\mathsf{R}}_d^{(\omega,\kappa)}$ *has the minimal possible dimension* 0 *and only if* $\omega = ()$ *and* $\kappa = d$. *In this case,* $\mathring{\mathsf{R}}_d^{((),d)} = \mathcal{P}_0 = \{a \mid a \in \mathbb{R} \setminus \{0\}\}/\sim$. ♢

Note that the norms $|\omega|$ and $|(\omega,\kappa)|$ are preserved under the merge operations for cells both in $\mathcal{P}_d$ and in $\mathcal{B}_d$. In both cases, the insert operations increase the norm by 2 and thus preserve its parity.

The merge and insert operations can be used to define a natural partial order "$\prec$" on the sets $\mathbf{\Omega}$ and $\mathbf{\Omega}^\infty$. It reflects the cellular structure, described in Proposition 6.1 and Corollary 6.1.

Definition 6.2. For $\omega, \omega' \in \mathbf{\Omega}$, we set $\omega' \prec \omega$, if ω' can be obtained from ω by a sequence of merge operations, M_j, $j \geq 1$, and insert operations I_j, $j \geq 0$.

Analogously, for $(\omega,\kappa), (\omega',\kappa') \in \mathbf{\Omega}^\infty$, we set $(\omega',\kappa') \prec (\omega,\kappa)$, if (ω',κ') can be obtained from (ω,κ) by a sequence of merge operations, M_j^∞, $j \geq 0$, and insert operations I_j^∞, $j \geq 0$.

If $\omega' \prec \omega$ or $(\omega',\kappa') \prec (\omega,\kappa)$, then we say that ω' is smaller than ω or that (ω',κ') is smaller than (ω,κ). ♢

One can easily check that "$\prec$" defines partial orders on $\mathbf{\Omega}$ and $\mathbf{\Omega}^\infty$. From now on, we will consider a subset $\Theta \subseteq \mathbf{\Omega}$ or $\Theta \subseteq \mathbf{\Omega}^\infty$ as a partially ordered set, poset for short, ordered by "$\prec$." Proposition 6.1 and Corollary 6.1 imply instantly the following two statements.

Corollary 6.2. *For a subset* $\Theta \subseteq \mathbf{\Omega}_{\langle d]}$,

(i) $\mathcal{P}_d^\Theta$ *is closed in* $\mathcal{P}_d$ *if and only if, for any* $\omega \in \Theta$ *and* $\omega' \prec \omega$, *we have* $\omega' \in \Theta$,

(ii) *if* $\mathcal{P}_d^\Theta$ *is closed in* $\mathcal{P}_d$, *then* $\bar{\mathcal{P}}_d^\Theta$ *carries the structure of a compact CW-complex with open cells* $\mathring{\mathsf{R}}_d^\omega$, *labeled by* $\omega \in \Theta$, *and the unique* 0*-cell, represented by the point* ∞. ♢

Corollary 6.3. *For a subset* $\Theta \subseteq \mathbf{\Omega}_{\langle d]}^\infty$,

(i) $\mathcal{B}_d^\Theta$ *is closed in* $\mathcal{B}_d$ *if and only if, for any* $(\omega,\kappa) \in \Theta$ *and* $(\omega',\kappa') \prec (\omega,\kappa)$, *we have* $(\omega',\kappa') \in \Theta$,

(ii) if $\mathcal{B}_d^\Theta$ is closed in $\mathcal{B}_d$, then $\mathcal{B}_d^\Theta$ carries the structure of a compact CW-complex with open cells $\mathring{\mathsf{R}}_d^{(\omega,\kappa)}$, labeled by $(\omega,\kappa) \in \Theta$. ◇

Corollaries 6.2 and 6.3 motivate the following definition.

Definition 6.3. A subposet $\Theta \subseteq \mathbf{\Omega}$ is called closed if, for any $\omega' \prec \omega$ and $\omega \in \Theta$, we have $\omega' \in \Theta$.

Similarly, a subposet $\Theta \subseteq \mathbf{\Omega}^\infty$ is called closed if, for any $(\omega', \kappa') \prec (\omega, \kappa)$ and $(\omega,\kappa) \in \Theta$, we have $(\omega', \kappa') \in \Theta$. ◇

Revisiting the beginning of Section 6.3, the closed posets Θ yield exactly the spaces $\bar{\mathcal{P}}_d^\Theta$, $\mathcal{P}_d^{\mathsf{c}\Theta}$ and $\mathcal{B}_d^\Theta$, $\mathcal{B}_d^{\mathsf{c}\Theta}$ we intend to study.

In Section 6.4 below, given *any* closed poset $\Theta \subset \mathbf{\Omega}_{\langle d]}$ or $\Theta \subseteq \mathbf{\Omega}^\infty_{\langle d]}$, we introduce a *combinatorial model* for the cellular chain complex that calculates the reduced homology of $\bar{\mathcal{P}}_d^\Theta$ or $\mathcal{B}_d^\Theta$.

However, for a *general* closed poset $\Theta \subset \mathbf{\Omega}_{\langle d]}$, it is impossible to obtain a *closed formula* description of the homology $\tilde{H}_*(\bar{\mathcal{P}}_d^\Theta; \mathbb{Z})$! To justify this "metaphysical" claim, consider the closed subposet $\mathbf{\Omega}_d \subseteq \mathbf{\Omega}_{\langle d]}$ of all $\omega \in \mathbf{\Omega}_{\langle d]}$ with $|\omega| = d$. The space $\mathcal{P}_d^{\mathbf{\Omega}_d}$ is the space of all real monic polynomials of degree d, having only real roots. We can identify $\mathcal{P}_d^{\mathbf{\Omega}_d}$ with $\{(x_1, \dots, x_d) \in \mathbb{R}^d \mid x_1 < \dots < x_d\}$, the closure of Π_d.

As a result, the cellular structure of $\mathcal{P}_d^{\mathbf{\Omega}_d}$ is the facial structure of the product of a $(d-1)$-dimensional cone $\mathcal{C}\Delta^{n-2}$ over a closed simplicial base Δ^{n-2} with a copy of the real line $\mathbb{R}$. The compositions in $\mathbf{\Omega}_d$ are in an order-preserving bijection with the open faces of the simplicial cone $\mathcal{C}\Delta^{n-2}$, so that $\mathring{\mathsf{R}}^\omega$ is the face corresponding to ω. Since the face poset of $\mathcal{C}\Delta^{n-2}$ is the Boolean lattice of subsets of $\{1, \dots, d-1\}$, it follows that one can identify simplicial complexes K with vertex set contained in $\{1, \dots, d-1\}$ and closed subposets of $\mathbf{\Omega}_d$. Then, for a closed subposet $\Theta \subseteq \mathbf{\Omega}_d$, the one-point compactification $\bar{\mathcal{P}}_d^\Theta$ is a double suspension the corresponding simplicial complex K whose vertexes are among $\{1, \dots, d-1\}$. This situation is considered in [SW].

In particular, for arbitrary closed subposets $\Theta \subseteq \mathbf{\Omega}_d$, the spaces $\bar{\mathcal{P}}_d^\Theta$ can, up to a shift by 2 in the (co)homological dimension, carry any homology that a simplicial complex on $d-1$ vertices can carry. Since $S^d \cong \bar{\mathcal{P}}_d$, the Alexander duality shows that the cohomology of $\mathcal{P}^{\mathsf{c}\Theta}$ can have a similarly arbitrary complex structure! In particular, arbitrary torsion may occur.

However, let us stress that in this example, the dimension of the simplicial complex, corresponding to Θ, and the degree d of the polynomials under consideration are closely linked. When it is possible to loosen this link, then, as we shall see in Subsection 6.5, the (co)homological stabilization takes place, and quite "tame" answers for the limiting homology $\tilde{H}_*(\bar{\mathcal{P}}_d^\Theta, \mathbb{Z})$ or $\tilde{H}_*(\mathcal{B}_d^\Theta, \mathbb{Z})$ emerge, as $d \to \infty$.

Before we further pursue our topological investigation, we would like to present a few combinatorial facts about the number of cells in $\bar{\mathcal{P}}_d$ and $\mathcal{B}_d$. Let p_j^d be the number of cells in $\bar{\mathcal{P}}_d$ of dimension j, and let b_j^d be the number of cells in $\mathcal{B}_d$ of dimension j. By Proposition 6.1 and Corollary 6.1, we know that $p_0^d = 1$ and, for $j \geq 1$, p_j^d counts the compositions $\omega \in \mathbf{\Omega}_{\langle d]}$ with $d - |\omega|' = j$. Similarly, for $j \geq 0$, b_j^d counts the marked compositions $(\omega, \kappa) \in \mathbf{\Omega}_{\langle d]}^\infty$ with $d - |(\omega, \kappa)|' = j$.

We introduce the generating t-polynomials

$$G(\bar{\mathcal{P}}_d; t) := \sum_{j=0}^{d} p_j^d\, t^j; \quad \text{and} \quad G(\mathcal{B}_d; t) := \sum_{j=0}^{d} b_j^d\, t^j.$$

Interpreting the value of the generating functions at $t = -1$ as the Euler characteristics of $\bar{\mathcal{P}}_d \cong S^d$ and $\mathcal{B}_d \cong \mathbb{PR}^d$, we get that $G(\bar{\mathcal{P}}_d, -1) = 1+(-1)^d$ and $G(\mathcal{B}_d; -1) = \frac{1}{2}(1 + (-1)^d)$. Proposition 6.1, (a), implies that $p_d^d = \lfloor \frac{d}{2} \rfloor + 1$.

Since $\mathcal{B}_d = \mathcal{P}_d \sqcup \mathcal{B}_{d-1}$, for $d \geq 1$, we get the recurrence relation

$$G(\mathcal{B}_d; t) = G(\mathcal{B}_{d-1}; t) + (G(\bar{\mathcal{P}}_d, t) - 1).$$

This recurrence implies the following relation for the (t, x)-generating functions

$$\sum_{d \geq 0} G(\mathcal{B}_d, t) x^d = 1 + x \sum_{d \geq 0} G(\mathcal{B}_d, t)\, x^d + \sum_{d \geq 0} G(\bar{\mathcal{P}}_d, t) x^d - 2 - \frac{x}{1-x}.$$

Note that the 2 in the recursion comes from $\bar{\mathcal{P}}_0$ consisting of 2 points. In particular,

$$\sum_{d \geq 0} G(\mathcal{B}_d, t) x^d = \frac{1}{1-x}\Big(-1 + \sum_{d \geq 0} G(\bar{\mathcal{P}}_d, t) x^d - \frac{x}{1-x}\Big).$$

Now, for $j \geq 1$, p_j^d equals the number of compositions of a number $d' \leq d$ with even $d - d'$ and $j - (d - d')$ parts. Thus for $j \geq 1$,

$$p_j^d = \sum_{\substack{d-j \leq d' \leq d \\ d-d' \text{ even}}} \binom{d'-1}{j-d+d'-1} = \sum_{k=0}^{\min\{\lfloor \frac{j-1}{2} \rfloor, \lfloor \frac{d-1}{2} \rfloor\}} \binom{d-1-2k}{j-1-2k}.$$

The latter observations implies via some standard calculations of generating functions the following result.

Lemma 6.1. *The number of cells in $\bar{\mathcal{P}}_d$ and $\mathcal{B}_d$ is given by the following generating functions:*

(i)

$$\sum_{d=0}^{\infty} G(\bar{\mathcal{P}}_d; t)x^d = \frac{(t^3+t^2)x^3 - (t^2-1)x^2 - (t+3)x + 2}{(1-t^2x^2)\,(1-x-tx)\,(1-x)}$$

(ii)

$$\sum_{d=0}^{\infty} G(\mathcal{B}_d, t)x^d = \frac{1}{(1-t^2x^2)\,(1-x-tx)}$$

(iii) For any positive odd d, $G(\bar{\mathcal{P}}_d; t) - 1 = t\,G(\mathcal{B}_{d-1})$, while for any positive even d,

$$G(\bar{\mathcal{P}}_d; t) - 1 = t\,G(\mathcal{B}_{d-1}) + t^d \quad \diamondsuit$$

Example 6.2. $G(\mathcal{B}_0; t) = 1$, $G(\mathcal{B}_1; t) = 1 + t$, $G(\mathcal{B}_2; t) = 1 + 2t + 2t^2$, $G(\mathcal{B}_3; t) = 1 + 3t + 4t^2 + 2t^3$, $G(\mathcal{B}_4; t) = 1 + 4t + 7t^2 + 6t^3 + 3t^4$, $G(\mathcal{B}_5; t) = 1 + 5t + 11t^2 + 13t^3 + 9t^4 + 3t^5$.

$G(\bar{\mathcal{P}}_1; t) = 1 + t$, $G(\bar{\mathcal{P}}_2; t) = 1 + t + 2t^2$, $G(\bar{\mathcal{P}}_3; t) = 1 + t + 2t^2 + 2t^3$, $G(\bar{\mathcal{P}}_4; t) = 1 + t + 3t^2 + 4t^3 + 3t^4$, $G(\bar{\mathcal{P}}_5; t) = 1 + t + 4t^2 + 7t^3 + 6t^4 + 3t^5$, $G(\bar{\mathcal{P}}_6; t) = 1 + t + 5t^2 + 11t^3 + 13t^4 + 9t^5 + 4t^6$. $\diamondsuit$

The coefficient sequences of $G(\bar{\mathcal{P}}_d, t)$ and $G(\mathcal{B}_d, t)$, up to a shift in indexing and removals of trailing or leading 1s, also has appeared in other contexts (see A035317 in [Sl]).

6.4 Combinatorial differential complexes for computing $\tilde{H}_*(\mathcal{B}_d^{\Theta}; \mathbb{Z})$ and $\tilde{H}_*(\bar{\mathcal{P}}_d^{\Theta}; \mathbb{Z})$

In this section, we use the natural CW-structure on $\mathcal{B}_d$ and $\bar{\mathcal{P}}_d$, described in Corollary 6.1 and Proposition 6.1, to construct *combinatorial differential complexes* that calculate the homology of $\mathcal{B}_d^{\Theta}$ and $\bar{\mathcal{P}}_d^{\Theta}$ for any given closed subposet $\Theta \subset \Omega_{\langle d]}^{\infty}$ or $\Theta \subset \Omega_{\langle d]}$.

Recall that the cells $\mathring{\mathsf{R}}_d^{(\omega,\kappa)}$ of the CW-complex $\mathcal{B}_d$ are indexed by marked compositions $(\omega, \kappa) \in \Omega_{\langle d]}^{\infty}$ and that the dimension of the cell $\mathring{\mathsf{R}}_d^{(\omega,\kappa)}$ equals

$d - |(\omega, \kappa)|' = d - |\omega|' - \kappa$. We define the dimension $\mathsf{dim}_d\big((\omega,\kappa) \big)$ of (ω, κ) in $\Omega^\infty_{\langle d]}$ by

$$\mathsf{dim}_d\big((\omega,\kappa) \big) = d - |(\omega,\kappa)|'.$$

Analogously, if $\Theta \subset \Omega_{\langle d]}$ for a composition $\omega \in \Theta_{\langle d]}$ we define the dimension of ω in $\Omega_{\langle d]}$ by

$$\mathsf{dim}_d\big(\omega \big) = d - |\omega|'.$$

Next, we provide an explicit description of the cellular chain complex of the CW-complex $\mathcal{B}^\Theta_d$. From now on, for $\Theta \subseteq \Omega_{\langle d]}$, or $\Theta \subseteq \Omega^\infty_{\langle d]}$, we write $\mathbb{Z}[\Theta]$ for the $\mathbb{Z}$-module, freely generated by the elements of Θ. In either case we consider $\mathbb{Z}[\Theta]$ as a graded $\mathbb{Z}$-module with grading given by dim.

For $\Theta \subseteq \Omega^\infty$ we define $\Theta_{|\sim|' \leq k}$ (resp., $\Theta_{|\sim|'=k}$) as the set of all (marked) compositions $(\omega,\kappa) \in \Theta$ with $|(\omega,\kappa)|' \leq k$ (resp., $|(\omega,\kappa)|' = k$). We use analogous notation in case $\Theta \subseteq \Omega$. In particular, if $\Theta \subseteq \Omega_{\langle d]}$ then $\Theta_{|\sim|'=d-j}$ is the set of all (ω,κ) of dimension j in $\Omega_{\langle d]}$. Again an analogous statement holds if $\Theta \subseteq \Omega\langle d]$.

For a composition $\omega = (\omega_1, \ldots, \omega_\ell)$, we denote by s_ω the number of its parts, i.e., $s_\omega = \ell$. Using the merge operators M and the insert operators I, introduced in 6.3, we define two homomorphisms which act on $\mathbb{Z}[\Theta]$. They are given by

$$\begin{aligned}
\partial^\infty_{\mathsf{M}}\big((\omega,\kappa) \big) &=_{\mathsf{def}} -\sum_{k=0}^{s_\omega} (-1)^k \mathsf{M}^\infty_k\big((\omega,\kappa) \big) \text{ and} \\
\partial^\infty_{\mathsf{I}}\big((\omega,\kappa) \big) &=_{\mathsf{def}} \sum_{k=0}^{s_\omega} (-1)^k \mathsf{I}^\infty_k\big((\omega,\kappa) \big) \text{ if } |(\omega,\kappa)| < d \text{ and} \\
\partial^\infty_{\mathsf{I}}\big((\omega,\kappa) \big) &=_{\mathsf{def}} 0 \text{ if } |(\omega,\kappa)| = d.
\end{aligned} \tag{6.3}$$

We set $\partial^\infty =_{\mathsf{def}} \partial^\infty_{\mathsf{M}} + \partial^\infty_{\mathsf{I}}$.

Lemma 6.2. *For any closed $\Theta \subseteq \Omega^\infty_{\langle d]}$, the homomorphisms $\partial^\infty_{\mathsf{M}}, \partial^\infty_{\mathsf{I}} : \mathbb{Z}[\Theta] \to \mathbb{Z}[\Theta]$ are graded of degree -1 and anticommuting differentials, i.e.*

$$(\partial^\infty_{\mathsf{I}})^2 = (\partial^\infty_{\mathsf{M}})^2 = \partial^\infty_{\mathsf{I}}\partial^\infty_{\mathsf{M}} + \partial^\infty_{\mathsf{M}}\partial^\infty_{\mathsf{I}} = 0.$$

In particular, $\partial^\infty = \partial^\infty_{\mathsf{M}} + \partial^\infty_{\mathsf{I}}$ is a differential, and $(\mathbb{Z}[\Theta], \partial)$ is a graded differential complex.

Proof. The fact that $\partial^\infty_{\mathsf{M}}$, $\partial^\infty_{\mathsf{I}}$ and ∂^∞ lower the grading by 1 is a simple computation.

Next we show that $(\partial_{\mathsf{I}}^{\infty})^2 = (\partial_{\mathsf{M}}^{\infty})^2 = \partial_{\mathsf{I}}^{\infty}\partial_{\mathsf{M}}^{\infty} + \partial_{\mathsf{M}}^{\infty}\partial_{\mathsf{I}}^{\infty} = 0$. Consider a marked composition (ω, κ) with $\omega = (\omega_1, \dots, \omega_\ell)$. Then for $\ell \geq 3$, the expression for $\partial_{\mathsf{M}}^{\infty}(\,(\omega, \kappa)\,)$ will, in particular, include the terms of the form

$$(-1)^k\big(\,(\omega_1, \dots, \omega_{k-1}, \omega_k + \omega_{k+1}, \omega_{k+2}, \dots, \omega_\ell), \kappa\,\big) +$$

$$(-1)^{k+1}\big(\,(\omega_1, \dots, \omega_k, \omega_{k+1} + \omega_{k+2}, \omega_{k+3}, \dots, \omega_\ell), \kappa\,\big).$$

Thus $(\partial_{\mathsf{M}}^{\infty})^2(\,(\omega, \kappa)\,)$ contains the vanishing sum

$$(-1)^{2k}\big(\,(\omega_1, \dots, \omega_{k-1}, \omega_k + \omega_{k+1} + \omega_{k+2}, \omega_{k+3}, \dots, \omega_\ell), \kappa\,\big) +$$

$$(-1)^{2k+1}\big(\,(\omega_1, \dots, \omega_{k-1}, \omega_k + \omega_{k+1} + \omega_{k+2}, \omega_{k+3}, \dots, \omega_\ell), \kappa\,\big) = 0.$$

For $\ell \geq 2$, the expansion of $\partial_{\mathsf{M}}^{\infty}(\,(\omega, \kappa)\,)$ will contains the terms $\big(\,(\omega_2, \dots, \omega_\ell), \omega_1 + \kappa) - ((\omega_1 + \omega_2, \omega_3, \dots, \omega_\ell), \kappa\,\big)$, which yields

$$\big(\,(\omega_1, \dots, \omega_\ell), \omega_1 + \omega_2 + \kappa) - ((\omega_3, \dots, \omega_\ell), \omega_1 + \omega_2 + \kappa\,\big) = 0$$

in $(\partial_{\mathsf{M}}^{\infty})^2(\,(\omega, \kappa)\,)$. For $\ell = 1$, $\partial_{\mathsf{M}}^{\infty}((\omega_1), \kappa) = 0$ by definition.

To show that $(\partial_{\mathsf{I}}^{\infty})^2 = 0$, expand

$$\partial_{\mathsf{I}}^{\infty}(\,(\omega, \kappa)\,) = \;\cdots + (-1)^k\big(\,(\omega_1, \dots, \omega_k, 2, \omega_{k+1}, \dots, \omega_\ell), \kappa\,\big) + \dots.$$

Thus $(\partial_{\mathsf{I}}^{\infty})^2(\,(\omega, \kappa)\,)$ will only consist of vanishing sums of the form

$$(-1)^{2k}\big(\,(\omega_1, \dots, \omega_k, 2, 2, \omega_{k+1}, \dots, \omega_\ell), \kappa\,\big)$$

$$+ (-1)^{k+1}\big(\omega_1, \dots, \omega_k, 2, 2, \omega_{k+1}, \dots, \omega_\ell), \kappa\,\big) = 0.$$

Finally, let us compute $(\partial_{\mathsf{I}}^{\infty}\partial_{\mathsf{M}}^{\infty} + \partial_{\mathsf{M}}^{\infty}\partial_{\mathsf{I}}^{\infty})(\,(\omega, \kappa)\,)$. Observe that $\partial_{\mathsf{I}}^{\infty}(\,(\omega, \kappa)\,)$ consists of the sums of the form

$$(-1)^k\big(\,(\omega_1, \dots, \omega_k, 2, \omega_{k+1}, \dots, \omega_\ell), \kappa\,\big)$$

$$+ (-1)^{k+1}\big(\,(\omega_1, \dots, \omega_{k+1}, 2, \omega_{k+2}, \dots, \omega_\ell), \kappa\,\big)$$

$$+ (-1)^{k+2}\big(\,(\omega_1, \dots, \omega_{k+2}, 2, \omega_{k+3}, \dots, \omega_\ell), \kappa\,\big).$$

Cancelling terms in $\partial_{\mathsf{M}}^{\infty}\partial_{\mathsf{I}}^{\infty}(\,(\omega, \kappa)\,)$, we are left with expressions

$$(-1)^{2k+2}\big(\,(\omega_1, \dots, \omega_k, 2, \omega_{k+1} + \omega_{k+2}, \omega_{k+3}, \dots, \omega_\ell), \kappa)\,\big) +$$

$$(-1)^{2k+3}\big(\,(\omega_1, \dots, \omega_k, \omega_{k+1} + \omega_{k+2}, 2, \omega_{k+3}, \dots, \omega_\ell), \kappa\,\big).$$

In a similar computation for $\partial_{\mathsf{I}}^{\infty}\partial_{\mathsf{M}}^{\infty}(\,(\omega, \kappa)\,)$, we obtain the contributions

$$(-1)^{2k+1}\big(\,(\omega_1, \dots, \omega_k, 2, \omega_{k+1} + \omega_{k+2}, \omega_{k+3}, \dots, \omega_\ell), \kappa\,\big) +$$

$$(-1)^{2k+2}\big(\,(\omega_1, \dots, \omega_k, \omega_{k+1} + \omega_{k+2}, 2, \omega_{k+3}, \dots, \omega_\ell), \kappa\,\big).$$

Adding them together, we get that $(\partial_{\mathsf{I}}^{\infty}\partial_{\mathsf{M}}^{\infty} + \partial_{\mathsf{M}}^{\infty}\partial_{\mathsf{I}}^{\infty})(\,(\omega,\kappa)\,) = 0$.

The claim that $\mathbb{Z}[\Theta]$ is a graded differential complex now follows. □

Proposition 6.2. *For any closed subposet $\Theta \in \mathbf{\Omega}^{\infty}_{\langle d]}$, the graded differential complex $(\mathbb{Z}[\Theta],\, \partial^{\infty})$ coincides with the cellular chain complex of $\mathcal{B}^{\Theta}_d$.*

In particular, $(\mathbb{Z}[\Theta],\, \partial^{\infty})$ calculates the homology of $\mathcal{B}^{\Theta}_d$.

Proof. By Corollary 6.1, the *topological* boundary $\partial\mathsf{R}_d^{(\omega,\kappa)}$ of $\mathsf{R}_d^{(\omega,\kappa)}$ coincides with

$$\Big(\bigcup_{k=0}^{s_\omega} \mathsf{R}_d^{\mathsf{M}_k^{\infty}((\omega,\kappa))}\Big) \bigcup \Big(\bigcup_{k=0}^{s_\omega} \mathsf{R}_d^{\mathsf{I}_k^{\infty}((\omega,\kappa))}\Big). \tag{6.4}$$

Therefore, the algebraic boundary of $\mathsf{R}_d^{(\omega,\kappa)}$ in the cellular chain complex of $\mathcal{B}^{\Theta}_d$ is given by a sum of the form

$$\sum_{k=0}^{s_\omega} a_k\, \mathsf{R}_d^{\mathsf{M}_k^{\infty}((\omega,\kappa))} + \sum_{k=0}^{s_\omega} b_k\, \mathsf{R}_d^{\mathsf{I}_k^{\infty}((\omega,\kappa))}, \tag{6.5}$$

where a_k and b_k are some integer coefficients.

In order to determine the coefficients a_k and b_k, we proceed as follows. By (6.4), if a path $\gamma = \{P_t\}_{t\in[0,1]}$ is such that $\gamma \setminus \gamma(1) \subset \mathring{\mathsf{R}}_d^{(\omega,\kappa)}$ and $\gamma(1)$ belongs to the boundary of $\mathsf{R}_d^{(\omega,\kappa)}$, then, as $t \to 1$, at least either two consecutive real roots of the polynomials P_t approach each other, or two complex-conjugate roots approach each other at a point of $\mathbb{R}$. These situations are encoded in the merge and insert operations, respectively. Thus in order to calculate a_k and b_k, we need to understand what happens along paths in $\mathring{\mathsf{R}}_d^{(\omega,\kappa)}$ that approach the boundary cells $\mathring{\mathsf{R}}_d^{\mathsf{M}_k^{\infty}((\omega,\kappa))}$ and $\mathring{\mathsf{R}}_d^{\mathsf{I}_k^{\infty}((\omega,\kappa))}$, respectively. Note that we can face a situation when, for $k \neq k'$, $\mathsf{M}_k^{\infty}((\omega,\kappa)) = \mathsf{M}_{k'}^{\infty}((\omega,\kappa))$ or $\mathsf{I}_k^{\infty}((\omega,\kappa)) = \mathsf{I}_{k'}^{\infty}((\omega,\kappa))$. Nevertheless, we will show that each operation M_k or I_k corresponds to a unique local homotopy type of paths γ in the vicinity of $\gamma(1)$ and hence yields a path–independent contributions a_k and b_k, respectively.

The preferred orientation of $\mathsf{R}_d^{(\omega,\kappa)}$ is induced by the canonical orientation of the open cell $\mathring{\mathsf{R}}_d^{(\omega,\kappa)}$. Recall that we can identify the class of a binary form from $\mathring{\mathsf{R}}_d^{(\omega,\kappa)}$ with a univariate polynomial of degree $d-\kappa$. By Corollary 6.1, we know that $\mathring{\mathsf{R}}_d^{(\omega,\kappa)} \cong \Pi_{s_\omega} \times \mathrm{Sym}^{m_\omega}(\mathbb{H})$ for $\omega = (\omega_1,\dots,\omega_{s_\omega})$ and $m_\omega =_{\mathsf{def}} \frac{d-|(\omega,\kappa)|}{2}$.

The orientation of the open cell $\mathrm{Sym}^{m_\omega}(\mathbb{H})$ is canonically induced by its complex structure, while the orientation of the polyhedron Π_{s_ω} is induced from the embient $\mathbb{R}^{s_\omega}$. In other words, the orientation of $\mathring{\mathsf{R}}_d^{(\omega,\kappa)} \cong \mathring{\mathsf{R}}_{d-\kappa}^{\omega}$ is given by the volume form

$$\rho_\omega := (dx_1 \wedge \cdots \wedge dx_{s_\omega}) \wedge \left(\frac{i}{2}\right)^{m_\omega} (dz_1 \wedge d\bar{z}_1 \wedge \ldots \wedge dz_{m_\omega} \wedge d\bar{z}_{m_\omega}),$$

considered on the product $\Pi_{s_\omega} \times \mathrm{Sym}^{m_\omega}(\mathbb{H})$.

Claim 1: In the formula (6.5) for the boundary operator, we have $a_k = (-1)^{k+1}$.

Note that, if for some k and l, $\mathsf{M}_k^\infty((\omega,\kappa)) = \mathsf{M}_l^\infty((\omega,\kappa))$, then either $k = l$ or $\{k, l\} = \{0, s_\omega\}$ and all the entries of ω are *identical.* In the latter case, $\mathsf{M}_l^\infty((\omega,\kappa)) = \mathsf{M}_0^\infty((\omega,\kappa))$ and the corresponding cell $\mathring{\mathsf{R}}_d^{(\omega,\kappa)}$ is adjacent to the cell $\mathring{\mathsf{R}}_d^{\mathsf{M}_0^\infty((\omega,\kappa))}$ "on both sides."

Case: $k \neq s_\omega, 0$

In this situation, $\mathsf{M}_k^\infty((\omega,\kappa)) \neq \mathsf{M}_l^\infty((\omega,\kappa))$ for $l \neq k$. Let $P \in \mathring{\mathsf{R}}_d^{\mathsf{M}_k^\infty((\omega,\kappa))}$ be a polynomial of degree $d - \kappa$. Consider a one-parameter family $\{P_t \subset \mathring{\mathsf{R}}_d^{(\omega,\kappa)}\}_{t \geq 0}$ such that $\lim_{t\to 0} P_t = P$. Then, for all sufficiently small $t > 0$, the kth and $(k+1)^{\text{st}}$ largest real roots $x_k(P_t)$ and $x_{k+1}(P_t)$ of P_t converge to the kth largest real root $x_k(P)$ of P. Moreover, for any two paths $P_t, Q_t \subset \mathring{\mathsf{R}}_d^{(\omega,\kappa)}$ such that $\lim_{t\to 0} P_t = \lim_{t\to 0} Q_t = P$, their germs at P can be deformed into one another by a *small* homotopy in $\mathring{\mathsf{R}}^{(\omega,\kappa)}$. Let us clarify this claim.

We say that $\epsilon > 0$ is *small enough for* P if the ϵ-disks around distinct roots of P are disjoint in $\mathbb{C}$ and the ϵ-disks around the non-real roots do not intersect the real line. Let $U_\epsilon(P)$ denote the union of such ϵ-disks.

For any $\epsilon > 0$ which is small enough for P, there exists $t_\epsilon > 0$ such that for $0 < t < t_\epsilon$, each root of P_t and of Q_t resides in $U_\epsilon(P)$. This implies that, with the exception of the two merging roots $x_k(P_t)$ and $x_{k+1}(P_t)$ which share the same ϵ-ball centered at the kth root of P, any other real root of P_t resides in a single ϵ-ball contained in U_ϵ. The same conclusion is valid for Q_t, $t < t_\epsilon$. The non-real roots of P_t and Q_t require more careful treatment, since more than one non-real roots of P_t or Q_t may belong the same ϵ-disk $D_\epsilon(z)$ centered on a non-real root z of multiplicity $m(z) \geq 2$ of the limiting polynomial P. As a result, there is no natural correspondence between the roots of P_t and Q_t that reside in $D_\epsilon(z)$.

Now, we can deform P_t into Q_t in $\mathring{\mathsf{R}}_d^{(\omega,\kappa)}$ so that, in the process of deformation, the roots of the intermediate polynomials do not exit $U_\epsilon(P)$. This may by accomplished by the linear homotopy of the corresponding real roots of P_t and Q_t, being combined with the following deformation of the non-real roots of P_t and Q_t that reside in each disk $D_\epsilon(z)$. We first contract radially all the roots of P_t in $D_\epsilon(z)$ to z, counted with the multiplicity $m(z)$, and then expand the divisor $m(z)z$ radially into the divisor of Q_t that is supported in $D_\epsilon(z)$. Thus, the desired deformation of P_t into Q_t takes place not only in the open cell $\mathring{\mathsf{R}}_d^{(\omega,\kappa)}$, but also in the vicinity of P.

Therefore, the germ of $\mathring{\mathsf{R}}_d^{(\omega,\kappa)}$ at $P \in \mathring{\mathsf{R}}_d^{\mathsf{M}_k^\infty((\omega,\kappa))}$ has a single connected component, contained in $\mathring{\mathsf{R}}_d^{(\omega,\kappa)}$. As a result, the incidence index a_k of the cell $\mathring{\mathsf{R}}_d^{(\omega,\kappa)}$ with the cell $\mathring{\mathsf{R}}_d^{\mathsf{M}_k^\infty(\omega,\kappa)}$ equals ± 1.

In order to determine the sign, it suffices to consider a family of polynomials $P_t \subset \mathring{\mathsf{R}}_d^{\omega}$ such that:

- $\lim_{t\to 0} P(t) = P \in \mathring{\mathsf{R}}_d^{\mathsf{M}_k^\infty((\omega,\kappa))}$,
- the polynomials $P(t)$ share the roots with P, except for the roots $x_k(P_t), x_{k+1}(P_t)$ and $x_k(P)$,
- $x_k(P_t) = x_k(P) - t,\ \ x_{k+1}(P_t) = x_k(P) + t.$

Since the complex roots of P_t and P coincide, we can restrict our attention to Π_ω. The tangent vector to the path $P(t)$ at $t = 0$ is given by $w = (0, \dots, 0, -1, 1, 0, \dots)$. This vector is the inward normal to the face $\{x_k = x_{k+1}\}$ of the polyhedron $\Pi_\omega \subset \mathbb{R}^{s_\omega}$.

The inner product "$\rfloor$" of w with the volume form in $\mathbb{R}^{s_\omega}$ is given by

$$w \rfloor (dx_1 \wedge \cdots \wedge dx_k \wedge dx_{k+1} \wedge \cdots \wedge dx_{s_\omega})$$

$$= (-1)^{k+1}(dx_1 \wedge \cdots \wedge dx_k \wedge dx_{k+2} \wedge \cdots \wedge dx_{s_\omega}).$$

Therefore, the orientation of the boundary $\partial\Pi_\omega$, induced by the orientation of its interior Π_ω°, differs from the preferred orientation of $\Pi^\circ_{\mathsf{M}_k^\infty(\omega)}$ exactly by the factor $(-1)^{k+1}$.

Case: $k = s_\omega := |\omega| - |\omega|'$

If a real polynomial P of degree $d - \kappa$ has its largest real root $x_k(P)$ of multiplicity ω_k, then P produces, with the help of the map $\wp$ from (6.1), a bilinear form $\wp(P)$ of degree d in the homogeneous variables $[z : w]$. In fact, $\wp(P) = (z - x_\ell w)^{\omega_k}\, w^\kappa \times Q(z, w)$, where $Q(z, w)$ is a real binary form of degree $d - \omega_k - \kappa$ such that $Q(z, 1) = P(z) \cdot (z - x_k)^{-\omega_k}$.

Let us examine the result of the merge operation $\mathsf{M}_\ell^\infty((\omega, \kappa))$ on the real zero divisors of the binary form $\wp(P)$.

First, we choose an open interval $(\alpha, \beta) \subset \mathbb{R}$ that contains all the real roots $x_1(P), \dots, x_k(P)$ of P. Then we choose an affine chart $U_\alpha =_{\mathsf{def}} \mathbb{RP}^1 \setminus$

$[\alpha : 1]$ and a projective transformation $\mathcal{A} \in \mathsf{PGL}(2, \mathbb{R})$ which maps the ordered sequence $[x_1 : 1], \dots, [x_k : 1]$ in the chart $U_\infty := \{[z : 1]\}$ to an ordered sequence $\mathcal{A}([x_1 : 1]), \dots, \mathcal{A}([x_k : 1])$ in the chart U_α. In particular, $\mathcal{A}$ maps $\infty = [1 : 0]$ to a point $[a : 1]$, where $a > \mathcal{A}([\beta, 1])$. In the original affine chart U_∞, the projective transformation $\mathcal{A}$ is given by the rational function $F(z) = a - \frac{b}{z-\alpha}$, where $F(\beta) < a$. Now, as $x_k \to \infty$, $F(x_k) \to a$. The real zero locus $\omega_1 \cdot x_1 + \dots + \omega_k \cdot x_k + \kappa \cdot \infty$ of the binary form $\wp(P)$ is mapped by $\mathcal{A}$ to the real zero locus $\omega_1 \cdot \mathcal{A}(x_1) + \dots + \omega_k \cdot \mathcal{A}(x_k) + \kappa \cdot a$ of the bilinear form $\mathcal{A}^*(\wp(P))$. Under the transformation $\mathcal{A}$, the combinatorial pattern (ω, κ) of $\wp(P)$ is transformed into the combinatorial pattern $((\omega_1, \dots, \omega_k, \kappa), 0)$ of $\mathcal{A}^*(\wp(P))$ in U_α. Now set $\omega^\kappa := (\omega_1, \dots, \omega_k, \kappa)$. Notice that under $\mathcal{A}^*$, the merge operation $\mathsf{M}_k^\infty((\omega, \kappa))$ is transformed into the operation $\mathsf{M}_k(\omega^\kappa) = \mathsf{M}_k^\infty(\omega^\kappa, 0)$. The latter case does not involve ∞ and has been analyzed above.

Case: $k = 0$

This case can be treated analogously to the case $k = s_\omega$.

Claim 2: In the formula (6.5) for the boundary operator, we have $b_k = (-1)^k$.

This is a more delicate situation since multiple sheets of $\mathring{\mathsf{R}}_d^{(\omega,\kappa)}$ may appear in the neighborhood of $P \in \mathring{\mathsf{R}}_d^{\mathsf{I}_k(\omega)}$. The simplest example of such phenomenon occurs when P has two consecutive real roots, x_i and x_{i+1}, each of multiplicity 2. Denote by P_1 and P_2 two polynomials obtained by resolving the first and, respectively, the second double root into a pair of simple complex-conjugate roots. Then the combinatorial patterns of the real root multiplicities of P_1 and P_2 are identical. Note that, although P_1 and P_2 are close to each other in $\mathsf{R}_d^{(\omega,\kappa)}$, there is no *short* path in $\mathring{\mathsf{R}}_d^{(\omega,\kappa)}$ connecting them in a neighborhood of P.

Let $x_1(P) < \dots < x_{s_\omega+1}(P)$ be the distinct real roots of $P \in \mathring{\mathsf{R}}_d^{\mathsf{I}_k^\infty((\omega,\kappa))}$, and let $\{(z_l^\star, \bar{z}_l^\star)\}_l$ be the unordered collection of non-real roots of P. By our assumptions on P, we have that $x_{k+1}(P)$ is a root of multiplicity 2. Let $P_t \subset \mathring{\mathsf{R}}_d^{(\omega,\kappa)}$ be a path in $\mathring{\mathsf{R}}_d^{(\omega,\kappa)}$ such that $\lim_{t\to 0} P_t = P$. We may assume that

- $x_j(P_t) = x_j(P)$ for all $j \leq k$ and $x_j(P_t) = x_{j+1}(P)$ for all $j > k+1$;
- $(x_{k+1}(P) + \mathsf{i}\, t,\ x_{k+1}(P) - \mathsf{i}\, t)$ is a pair of simple complex-conjugate roots for P_t and all the other non-real roots of P_t and P coincide.

With this convention and choosing a fixed ordering of the complex roots with positive imaginary parts, we may consider P_t as a curve in the the ordered "root space" $\mathbb{R}^{s_\omega} \times \mathbb{C}^{m_\omega}$. Here the vector $w = (0, \dots, 0, \mathsf{i}, 0, \dots, 0)$ is tangent to the curve P_t at P.

The volume form ρ_ω can be written as $\rho_\omega^{\mathbb{R}} \wedge \rho_\omega^{\mathbb{C}}$, where

$$\rho_\omega^{\mathbb{R}} := dx_1 \wedge \cdots \wedge dx_{s_\omega}$$

and

$$\rho_\omega^{\mathbb{C}} := \left(\frac{\mathsf{i}}{2}\right)^{m_\omega} (dz_1 \wedge d\bar{z}_1) \wedge \cdots \wedge (dz_{m_\omega} \wedge d\bar{z}_{m_\omega}).$$

Since $\mathsf{i} \rfloor \left(\frac{\mathsf{i}}{2} dz \wedge d\bar{z}\right) = (0,1) \rfloor (dx \wedge dy) = -dx$, we get

$$w \rfloor (\rho_\omega^{\mathbb{R}} \wedge \rho_\omega^{\mathbb{C}}) = (-1)^{s_\omega} \rho_\omega^{\mathbb{R}} \wedge (w \rfloor \rho_\omega^{\mathbb{C}}) = (-1)^{s_\omega} \rho_\omega^{\mathbb{R}} \wedge dx_{k+1} \wedge \rho_{\mathsf{l}_k(\omega)}^{\mathbb{C}}$$

$$= (-1)^{2s_\omega + k} \rho_{\mathsf{l}_k(\omega)}^{\mathbb{R}} \wedge \rho_{\mathsf{l}_k(\omega)}^{\mathbb{C}} = (-1)^k \rho_{\mathsf{l}_k(\omega)}.$$

This calculation implies that the part $\mathring{\mathsf{R}}_d^{\mathsf{l}_k^\infty(\omega)}$ of the boundary $\delta^\infty(\mathsf{R}_\omega)$, being approached via the path $P_{t,k} := P_t \subset \mathsf{R}_d^\omega$ as above, acquires an orientation that differs from its $\rho_{\mathsf{l}_k(\omega)}$-induced orientation by the factor $b_k = (-1)^k$.

Thus, we have shown that the (E^2, d^2)-term of the $\tilde{H}_*$-homological spectral sequence, associated to the filtration of the space $\mathcal{B}_d^\Theta$ by its skeletae, coincides with the graded differential complex complex $\partial^\infty : \mathbb{Z}[\Theta] \to \mathbb{Z}[\Theta]$, defined by the formula (6.3). This completes the proof of Proposition 6.2. □

As an immediate consequence of Proposition 6.2 and of the previous discussions, we obtain one of our main results—the construction of the combinatorial differential complex that calculates the homology of $\mathcal{B}_d^\Theta$.

Theorem 6.3. *Let $\Theta \subset \mathbf{\Omega}_{\langle d]}^\infty$ be a closed subposet. For all $j \geq 0$, the homology of $\mathcal{B}_d^\Theta$ is given by*

$$H_j(\mathcal{B}_d^\Theta; \mathbb{Z}) \cong \frac{\mathsf{ker}\{\partial^\infty : \mathbb{Z}[\Theta_{|\sim|'=d-j}] \to \mathbb{Z}[\Theta_{|\sim|'=d-j+1}]\}}{\mathsf{im}\{\partial^\infty : \mathbb{Z}[\Theta_{|\sim|'=d-j-1}] \to \mathbb{Z}[\Theta_{|\sim|'=d-j}]\}}, \tag{6.6}$$

where the differential ∂^∞ is given by formula (6.3). ◇

For a later use, let us also consider a relative version of this differential complex. Namely, given a closed subposet $\Theta \subseteq \mathbf{\Omega}_{\langle d]}^\infty$, consider the exact sequence of differential complexes

$$0 \to (\mathbb{Z}[\Theta], \partial^\infty) \to (\mathbb{Z}[\mathbf{\Omega}_{\langle d]}^\infty], \partial^\infty) \to (\mathbb{Z}[\mathbf{\Omega}_{\langle d]} \setminus \Theta], \partial^\#) \to 0, \tag{6.7}$$

where the first homomorphism is the obvious inclusion and the last term in the exact sequence is, by definition, the differential quotient complex. Its differential $\partial^{\#}$ is still given by formulas (6.3) in which all terms $\mathsf{M}_k^\infty((\omega,\kappa))$ (resp. $\mathsf{I}_k^\infty((\omega,\kappa))$) with $\mathsf{M}_k^\infty((\omega,\kappa)) \in \Theta$ (resp. $\mathsf{I}_k^\infty((\omega,\kappa)) \in \Theta$) are replaced by zeros.

Since all the $\mathbb{Z}$-modules in (6.7) are free, we get a short exact sequence of *dual* complexes:

$$0 \to (\mathbb{Z}[\boldsymbol{\Omega}_{\langle d]} \setminus \Theta]^*, (\partial^{\#})^*) \to (\mathbb{Z}[\boldsymbol{\Omega}_{\langle d]}^\infty]^*, (\partial^\infty)^*) \to (\mathbb{Z}[\Theta]^*, (\partial^\infty)^*) \to 0, \tag{6.8}$$

where $\mathbb{Z}[\sim]^*$ denotes $\mathsf{Hom}_{\mathbb{Z}}(\mathbb{Z}[\sim], \mathbb{Z})$, and $(\partial^{\#})^*, (\partial^\infty)^*$ are the dual differentials.

For any closed subposet $\Theta \subseteq \boldsymbol{\Omega}_{\langle d]}^\infty$, the following claim is an immediate consequence of Theorem 6.3 and of the fact that $\mathcal{B}_d^\Theta$ is a subcomplex of the CW-complex $\mathcal{B}_d$.

Corollary 6.4. *Let $\Theta \subset \boldsymbol{\Omega}_{\langle d]}^\infty$ be a closed subposet. Then for all $j \geq 0$, the reduced homology of the quotient $\mathcal{B}_d/\mathcal{B}_d^\Theta$ is given by*

$$H_j(\mathcal{B}_d/\mathcal{B}_d^\Theta; \mathbb{Z}) \cong \frac{\mathsf{ker}\big\{\partial^{\#} : \mathbb{Z}[(\boldsymbol{\Omega}_{\langle d]} \setminus \Theta)_{|\sim|'=d-j}] \longrightarrow \mathbb{Z}[(\boldsymbol{\Omega}_{\langle d]} \setminus \Theta)_{|\sim|'=d-j+1}]\big\}}{\mathsf{im}\big\{\partial^{\#} : \mathbb{Z}[(\boldsymbol{\Omega}_{\langle d]} \setminus \Theta)_{|\sim|'=d-j-1}] \longrightarrow \mathbb{Z}[(\boldsymbol{\Omega}_{\langle d]} \setminus \Theta)_{|\sim|'=d-j}]\big\}}. \tag{6.9}$$

The reduced cohomology of the quotient space $\mathcal{B}_d/\mathcal{B}_d^\Theta$ is isomorphic to the homology of the dual differential complex $(\mathbb{Z}[\boldsymbol{\Omega}_{\langle d]} \setminus \Theta]^*, (\partial^{\#})^*)$ introduced in (6.8). ◇

Now, let us consider of the one-point compactification $\bar{\mathcal{P}}_d^\Theta$ for a closed subposet $\Theta \subseteq \boldsymbol{\Omega}_{\langle d]}$. Analogously to the case of a closed subposet $\Theta \subseteq \boldsymbol{\Omega}_{\langle d]}^\infty$, we introduce two homomorphisms of of the $\mathbb{Z}$-module $\mathbb{Z}[\Theta]$ by the formula

$$\partial_{\mathsf{M}}(\omega) =_{\mathsf{def}} -\sum_{k=1}^{s_\omega - 1} (-1)^k \mathsf{M}_k(\omega) \ \text{ and } \ \partial_{\mathsf{I}}(\omega) =_{\mathsf{def}} \sum_{k=0}^{s_\omega} (-1)^k \mathsf{I}_k(\omega).$$

Next, we define a homomorphism

$$\partial = \partial_{\mathsf{M}} + \partial_{\mathsf{I}} : \mathbb{Z}[\Theta] \to \mathbb{Z}[\Theta]$$

by

$$\partial(\omega) =_{\mathsf{def}} \begin{cases} -\sum_{k=1}^{s_\omega - 1} (-1)^k \, \mathsf{M}_k(\omega) + \sum_{k=0}^{s_\omega} (-1)^k \, \mathsf{I}_k(\omega), \\ \quad \text{for } \omega \neq \infty, |\omega| < d, \\ -\sum_{k=1}^{s_\omega - 1} (-1)^k \, \mathsf{M}_k(\omega), \quad \text{for } \omega \neq \infty, (d), |\omega| = d, \\ 0, \quad \text{for } \omega = (d). \end{cases} \tag{6.10}$$

Corollary 6.5. *Let $\Theta \subset \mathbf{\Omega}_{\langle d]}$ be a closed subposet. The for all $j \geq 0$, the reduced homology of the one-point compactification $\bar{\mathcal{P}}_d^{\Theta}$ of $\mathcal{P}_d^{\Theta}$ is given by*

$$\tilde{H}_j(\bar{\mathcal{P}}_d^{\Theta};\mathbb{Z}) \cong \frac{\mathsf{ker}\{\partial : \mathbb{Z}[\Theta_{|\sim|'=d-j}] \to \mathbb{Z}[\Theta_{|\sim|'=d-j+1}]\}}{\mathsf{im}\{\partial : \mathbb{Z}[\Theta_{|\sim|'=d-j-1}] \to \mathbb{Z}[\Theta_{|\sim|'=d-j}]\}}, \tag{6.11}$$

where the differential ∂ is given by formula (6.10).

Proof. Consider the decomposition $\mathcal{B}_d = \mathcal{P}_d \sqcup \cdots \sqcup \mathcal{P}_0 = \mathcal{P}_d \sqcup \mathcal{B}_{d-1}$. For a closed $\Theta \subseteq \mathbf{\Omega}_{\langle d]}$, consider the smallest closed subposet $\Theta^{\infty} \subseteq \mathbf{\Omega}^{\infty}_{\langle d]}$ that contains $\Theta \times \{0\}$. Let Θ_ℓ be the subset of all ω such that $(\omega, d-\ell) \in \Theta^{\infty}$. (In particular, $\Theta = \Theta_d$.) Then the above decomposition induces the decomposition $\mathcal{B}_d^{\Theta^{\infty}} = \mathcal{P}_d^{\Theta_d} \sqcup \cdots \sqcup \mathcal{P}_0^{\Theta_0}$.

Since $(\Theta_{d-1} \times \{1\}) \cup \cdots \cup (\Theta_0 \times \{d\})$ is closed, we get that $Y = \mathcal{P}_{d-1}^{\Theta_{d-1}} \sqcup \cdots \sqcup \mathcal{P}_0^{\Theta_0}$ is a closed CW-subcomplex of $\mathcal{B}_d$. Thus $\bar{\mathcal{P}}_d^{\Theta}$ and $\mathcal{B}_d^{\Theta^{\infty}}/Y$ are isomorphic as CW-complexes. In particular, the cellular differentials of $\bar{\mathcal{P}}_d^{\Theta}$ and $\mathcal{B}_d^{\Theta^{\infty}}/Y$ coincide. Now the assertion follows from Corollary 6.4 and our definition of the differential $\partial^{\#}$ in (6.7). □

For any closed sub-poset $\Theta_{\langle d]} \subset \mathbf{\Omega}_{\langle d]}$, let us revisit the differential complex $\partial : \mathbb{Z}[\Theta_{\langle d]}] \to \mathbb{Z}[\Theta_{\langle d]}]$ in (6.10), whose $(d-j)$-grading is defined by the module $\mathbb{Z}[\Theta_{\langle d],\, |\sim|'=j}]$. We denote by $\partial^* : \mathbb{Z}[\Theta_{\langle d]}]^* \to \mathbb{Z}[\Theta_{\langle d]}]^*$ its dual, where $\mathbb{Z}[\Theta_{\langle d]}]^* := \mathsf{Hom}(\mathbb{Z}[\Theta_{\langle d]}], \mathbb{Z})$ and the operator ∂^* is the dual of the boundary operator ∂. Then we consider the quotient set $\Theta^{\#}_{\langle d]} := \mathbf{\Omega}_{\langle d]}/\Theta_{\langle d]}$. For the closed subposet $\Theta_{\langle d]}$, the partial order in $\mathbf{\Omega}_{\langle d]}$ induces a partial order in the quotient $\Theta^{\#}_{\langle d]}$.

Next, we introduce a new differential complex $(\mathbb{Z}[\Theta^{\#}_{\langle d]}], \partial^{\#})$ by including it in the short exact sequence of differential complexes:

$$0 \to (\mathbb{Z}[\Theta_{\langle d]}], \partial) \to (\mathbb{Z}[\mathbf{\Omega}_{\langle d]}], \partial) \to (\mathbb{Z}[\Theta^{\#}_{\langle d]}], \partial^{\#}) \to 0. \tag{6.12}$$

Using (6.12), Corollary 6.5 allows us to compute the (co)homology of $\mathcal{P}_d^{\mathbf{c}\Theta} =_{\mathsf{def}} \mathcal{P}_d \setminus \mathcal{P}_d^{\Theta} = S^d \setminus \bar{\mathcal{P}}_d^{\Theta}$.

Corollary 6.6. *Let $\Theta \subset \mathbf{\Omega}_{\langle d]}$ be a closed subposet. Then, for any $j \in [0, d]$, we get group isomorphisms*

$$\begin{aligned} \tilde{H}^j(\mathcal{P}_d^{\mathbf{c}\Theta};\mathbb{Z}) &\overset{\mathcal{D}}{\approx} H_{d-j}(\bar{\mathcal{P}}_d, \bar{\mathcal{P}}_d^{\Theta};\mathbb{Z}) \\ &\approx H_{d-j}\big(\partial^{\#} : \mathbb{Z}[\Theta^{\#}] \to \mathbb{Z}[\Theta^{\#}]\big), \end{aligned} \tag{6.13}$$

$$\begin{aligned} \tilde{H}_j(\mathcal{P}_d^{\mathbf{c}\Theta};\mathbb{Z}) &\overset{\mathcal{D}}{\approx} H^{d-j}(\bar{\mathcal{P}}_d, \bar{\mathcal{P}}_d^{\Theta};\mathbb{Z}) \\ &\approx H_{d-j}\big((\partial^{\#})^* : (\mathbb{Z}[\Theta^{\#}])^* \to (\mathbb{Z}[\Theta^{\#}])^*\big), \end{aligned} \tag{6.14}$$

where $\mathcal{D}$ denote the Poincaré duality isomorphisms. ◇

With the combinatorial complex that calculates the cohomology of $\mathcal{P}_d^{\mathbf{c}\Theta}$ in place, it is natural to look for a similar complex that would calculate the cohomology of the complement $\mathcal{B}_d^{\mathbf{c}\Theta} =_{\mathsf{def}} \mathcal{B}_d \setminus \mathcal{B}_d^{\Theta}$ for any closed $\Theta \subseteq \mathbf{\Omega}^{\infty}_{\langle d]}$. Since $\mathcal{B}_d$ is homeomorphic to $\mathbb{RP}^d$, we need to consider the Alexander duality for the real projective spaces. With this goal in mind, let us remind the reader of a few standard constructions and notions regarding the Poincaré duality on non-simply-connected manifolds (as a general reference, see [Wa3]).

Let X be a d-dimensional compact connected manifold or, more generally, a Poincaré CW-complex. Let Λ be the group ring $\mathbb{Z}[\pi_1(X)]$ of the fundamental group $\pi_1(X)$. An element of Λ is a finite combination $\sum_g n_g g$, where $g \in \pi_1(X)$ and $n_g \in \mathbb{Z}$.

We denote by $\tilde{X}$ the universal cover of X and by $C_*(\tilde{X}, \mathbb{Z})$ the cellular chain complex of $\tilde{X}$, viewed as a right Λ-module under the free $\pi_1(X)$-action on $\tilde{X}$. We define the homology and cohomology of X with coefficients in a right Λ-module $\mathbb{A}$ (in a local system of coefficients) by

$$H^*(X; \mathbb{A}) := H(\mathsf{Hom}_\Lambda(C_*(\tilde{X}, \mathbb{Z}), \mathbb{A})), \quad H_*^t(X; \mathbb{A}) := H(C_*(\tilde{X}, \mathbb{Z}) \otimes_\Lambda \mathbb{A}^t).$$

The operation $\otimes_\Lambda$ relies on converting $\mathbb{A}$ into a left Λ-module $\mathbb{A}^t$ via the formula $\lambda a := a\bar{\lambda}$, where $a \in \mathbb{A}$ and $\lambda \in \Lambda$. Here, for any $\lambda = \sum n_g\, g$, we denote by $\bar{\lambda} := \sum w(g) n_g\, g^{-1}$, where the homomorphism $w : \pi_1(X) \to \mathbb{Z}_2$ is defined by the first Stiefel-Whitney class, an element of $H^1(X; \mathbb{Z}_2)$.

The special case $\mathbb{A} = \mathbb{Z}$ is central for us.

Similar definitions of $H^*(X, \partial X; \mathbb{A})$ and $H_*^t(X, \partial X; \mathbb{A})$ are available for a manifold X with boundary ∂X. They also make sense for any pair $X \supset Y$, where Y is a compact subcomplex of a Poincaré complex X (see [Wa3]). In such a case, the cap product with the fundamental cycle $[X] \in H_d(X; \mathbb{Z}^t)$ delivers the Poincaré duality isomorphism

$$[X]\cap :\ H^j(X, K; \mathbb{A}) \stackrel{\cong}{\to} H_{d-j}(X \setminus K; \mathbb{A}^t),$$

where $\mathbb{A}$ is a right $\mathbb{Z}[\pi_1(X)]$-module. As before, in order to convert $\mathbb{A}$ into a left $\mathbb{Z}[\pi]$-module $\mathbb{A}^t$, we use the Stiefel-Whitney map

$$w : \pi_1(X \setminus K) \to \pi_1(X) \to \mathbb{Z}_2,$$

which describes whether the local orientation of X is preserved or reversed along a loop in $X \setminus K$.

Now consider the natural homomorphism $H^j(K; \mathbb{A}) \stackrel{\delta^*}{\to} H^{j+1}(X, K; \mathbb{A})$ from the long exact sequence of the pair (X, K) and compose it with the

Poincaré duality isomorphism $\Delta_{X\setminus K} = [X]\cap \sim$. This composition produces the Alexander homomorphism

$$\mathcal{A}\ell^j : H^j(K;\mathbb{A}) \stackrel{\delta^*}{\to} H^{j+1}(X,K;\mathbb{A}) \stackrel{\cong \Delta_{X\setminus K}}{\longrightarrow} H_{d-j-1}(X\setminus K;\mathbb{A}^t). \quad (6.15)$$

Obviously, if δ^* is an isomorphism (monomorphism/epimorphism), then so is $\mathcal{A}\ell^j$. Similarly, we get a homomorphism

$$\mathcal{A}\ell_j : H^{d-j-1}(X\setminus K;\mathbb{A}) \stackrel{\cong \Delta^*_{X\setminus K}}{\longrightarrow} H_{j+1}(X,K;\mathbb{A}^t) \stackrel{\partial_*}{\to} H_j(K;\mathbb{A}^t). \quad (6.16)$$

Again, if ∂_* is an isomorphism (monomorphism/epimorphism), then so is $\mathcal{A}\ell_j$.

We will apply $\mathcal{A}\ell$ in the case when $X = \mathbb{RP}^d$, $K = \mathcal{B}_d^\Theta$, $\mathbb{A} = \mathbb{Z}$, and $\mathbb{A}^t = \mathbb{Z}^t$ is the local coefficient system, defined by the homomorphism $w : \pi_1(\mathbb{RP}^d \setminus \mathcal{B}_d^\Theta) \to \pi_1(\mathbb{RP}^d) \cong \mathbb{Z}_2$.

Theorem 6.4. *Given a closed subposet $\Theta \subset \mathbf{\Omega}^\infty_{\langle d]}$, set $\mathbf{c}\Theta := \mathbf{\Omega}^\infty_{\langle d]} \setminus \Theta$. Then for any $j \geq 0$, we get isomorphisms:*

$$H^j(\mathcal{B}_d^{\mathbf{c}\Theta};\mathbb{Z}) \cong \quad (6.17)$$
$$\frac{\mathsf{ker}\{\partial^{\#}_{d-j} : \mathbb{Z}[\mathbf{c}\Theta_{|\sim|'=d-j}] \longrightarrow \mathbb{Z}[\mathbf{c}\Theta_{|\sim|'=d-j-1}]\}}{\mathsf{im}\{\partial^{\#}_{d-j+1} : \mathbb{Z}[\mathbf{c}\Theta_{|\sim|'=d-j+1}] \longrightarrow \mathbb{Z}[\mathbf{c}\Theta_{|\sim|'=d-j}]\}},$$

$$H_j(\mathcal{B}_d^{\mathbf{c}\Theta};\mathbb{Z}^t) \cong \quad (6.18)$$
$$\frac{\mathsf{ker}\{(\partial^{\#}_{d-j+1})^* : \mathbb{Z}[\mathbf{c}\Theta_{|\sim|'=d-j}]^* \longrightarrow \mathbb{Z}[\mathbf{c}\Theta_{|\sim|'=d-j+1}]^*\}}{\mathsf{im}\{(\partial^{\#}_{d-j})^* : \mathbb{Z}[\mathbf{c}\Theta_{|\sim|'=d-j-1}]^* \longrightarrow \mathbb{Z}[\mathbf{c}\Theta_{|\sim|'=d-j}]^*\}}.$$

Proof. The above isomorphisms follow from Theorem 6.3 and Corollary 6.4, being combined with the Poincare dualities:

$$\mathcal{PD} : H^j(\mathcal{B}_d^{\mathbf{c}\Theta};\mathbb{Z}) \cong H_{d-j}(\mathcal{B}_d, \mathcal{B}_d^\Theta;\mathbb{Z}^t) \cong H_{d-j}(\mathcal{B}_d/\mathcal{B}_d^\Theta;\mathbb{Z}^t),$$

$$\mathcal{PD}^{-1} : H_j(\mathcal{B}_d^{\mathbf{c}\Theta};\mathbb{Z}^t) \cong H^{d-j}(\mathcal{B}_d, \mathcal{B}_d^\Theta;\mathbb{Z}) \cong H^{d-j}(\mathcal{B}_d/\mathcal{B}_d^\Theta;\mathbb{Z}).$$

□

Classical example: the discriminant of real binary forms. Proposition 6.4 describes the result of our computation.

To the best of our knowledge, only very few of the spaces $\mathcal{B}_d^\Theta$ or $\mathcal{B}_d^{\mathbf{c}\Theta}$ have been considered in the literature before. The most significant result, albeit in a different setting, is again due to Vassiliev. He considered Arnold's setting from Theorem 6.2 for *real binary forms with k-moderate singularities* [Va1], [Va2], [Va3]. Following Vassiliev's notation, let $\mathcal{HP}_d \cong \mathbb{R}^{d+1}$ denote the space of real binary homogeneous forms $Q(x, y)$ of degree d. Thus our $\mathcal{B}_d$ is the porjective space of $\mathcal{HP}_d$. For $k \geq 2$, let $\Sigma_k \subset \mathcal{HP}_d$ be the subset consisting of all forms Q that vanish with multiplicity at least k on some line in the plane $\mathbb{R}^2$. The main theorem of [Va2], formulated below, describes the reduced homology $\bar{H}^*(\mathcal{HP}_d \setminus \Sigma_k; \mathbb{Z})$ for any $2 \leq k \leq d$.

Theorem 6.5. [*Va2*] *Fix a number* $k \in [2, d]$.

(i) *For* k *even, the reduced cohomology group* $\bar{H}^*(\mathcal{HP}_d \setminus \Sigma_k; \mathbb{Z})$ *is a free abelian group of rank* $2[d/k] + 1$ *with generators in degrees*

$$k-2, k-1,\ 2(k-2), 2(k-2)+1,\ \dots\ ,\ [d/k](k-2), [d/k](k-2)+1,$$

and $d - 2[d/k]$.

(ii) *For* k *odd and* d *not divisible by* k, $\bar{H}^*(\mathcal{HP}_d \setminus \Sigma_k; \mathbb{Z})$ *is a direct product of the following groups:*
   (a) *for any* $p = 1, 2, \dots, [d/k]$ *such that* $d - pk$ *is odd,* $\mathbb{Z}$ *in dimension* $p(k-2)$ *and* $\mathbb{Z}$ *in dimension* $p(k-2)+1$;
   (b) *for any* $p = 1, 2, \dots, [d/k]$ *such that* $d - pk$ *is even,* $\mathbb{Z}/2\mathbb{Z}$ *in dimension* $p(k-2)+1$;
   (c) $\mathbb{Z}$ *in dimension* $d - 2[d/k]$.

(iii) *For* k *odd and* d *divisible by* k, *the answer is almost the same as in Case (ii), but the summand* $\mathbb{Z}/2\mathbb{Z}$ *in dimension* $d - 2(d/k) + 1$ *disappears.* $\diamondsuit$

As an example of application of Theorem 6.4 and Theorem 6.3, let us now compute the reduced homology of the discriminant variety and the homotopy type of its complement in our projective setting. Corollary 6.4 should be compared with Theorem 6.5 for $k = 2$, in which case, the latter simplifies greatly.

Let $\Delta_d \subset \mathcal{B}_d$ be the discriminant variety that consists of all classes of binary degree d forms with at least one real root of multiplicity ≥ 2. Note that $\Delta_d = \mathcal{B}_d^{\Theta_{\mathsf{disc}}}$ for the poset

$$\Theta_{\mathsf{disc}} =_{\mathsf{def}} \Big\{((\omega_1, \dots, \omega_\ell), \kappa) \in \mathbf{\Omega}^{\infty}_{\langle d]} \;\Big|\; \exists\ 1 \leq i \leq \ell :\ \omega_i \geq 2 \text{ or } \kappa \geq 2\Big\}.$$

The following result is contained in Example 1 and Lemma 1 of [Va2].

Proposition 6.3. *The connected components of* $\mathcal{B}_d^{\mathsf{c}\Theta_{\mathsf{disc}}} =_{\mathsf{def}} \mathcal{B}_d \setminus \Delta_d$ *are labeled by marked compositions* $((\underbrace{1, \dots, 1}_{\ell}), 0)$, *where* $\ell \geq 0$ *and* $d - \ell \geq 0$

is even. If d is odd, then each connected component of $\mathcal{B}_d^{\mathbf{c}\Theta_{\mathrm{disc}}}$ is homotopy equivalent to a circle. If d is even, then each connected component of $\mathcal{B}_d^{\mathbf{c}\Theta_{\mathrm{disc}}}$ is homotopy equivalent to a circle, except for the component labeled by $((),0)$, which is contractible.

Proof. Let ℓ be such that $d-\ell$ is even. Then, for $\ell \neq 0$ and $\hat{\omega}_1 := ((\underbrace{1,\ldots,1}_{\ell}),0)$, $\hat{\omega}_2 := (\mathsf{M}_\ell^\infty(\hat{\omega}_1)) = (\mathsf{M}_0^\infty(\hat{\omega}_1)) = ((\underbrace{1,\ldots,1}_{\ell-1}),1)$, the union $\mathcal{C}_{\ell,d} := \mathring{\mathsf{R}}_d^{\hat{\omega}_1} \cup \mathring{\mathsf{R}}_d^{\hat{\omega}_2}$ is a connected component of $\mathcal{B}_d \setminus \Delta_d$. If d is even and $\ell = 0$, then $\mathcal{C}_{1,d} = \mathring{\mathsf{R}}_d^{((),0)}$ is a connected component. Since every cell of $\mathcal{B}_d$ that is not contained in Δ_d is contained in some $\mathcal{C}_{\ell,d}$, these are all the connected components of $\mathcal{B}_d \setminus \Delta_d$.

The space $\mathcal{C}_{0,d} = \mathring{\mathsf{R}}_d^{((),0)}$ is an open cell and hence contractible. Further, for $\ell \neq 0$ and even $d-\ell$, the polynomials in the corresponding connected components have ℓ simple real roots (including the root at infinity) and $\frac{d-\ell}{2}$ complex-conjugate pairs of roots in $\mathbb{CP}^1 \setminus \mathbb{RP}^1$, counted with their multiplicities. Then $\mathcal{C}_{\ell,d}$ is the Cartesian product of the configuration space $\mathrm{Conf}^\ell(\mathbb{RP}^1)$ and $\mathrm{Sym}^{\frac{d-\ell}{2}}(\mathbb{H})$. Since the upper half-plane $\mathbb{H}$ is contractible, $\mathrm{Sym}^{\frac{d-\ell}{2}}(\mathbb{H})$ is contractible as well. It is well-known that $\mathrm{Conf}^\ell(\mathbb{RP}^1) = \mathrm{Conf}^\ell(S^1)$ is homotopy equivalent to a circle [Mort]. As in [Mort], there is a fibration $\mathrm{Conf}^\ell(S^1) \to S^1$ whose fiber is the open simplicial cone $\Pi_{\ell-1}$. Since the latter is contractible, we conclude that $\mathrm{Conf}^\ell(S^1)$ is homotopy equivalent to a circle, which completes the proof. $\square$

The following lemma is the crucial step in determining the homology of $\mathcal{D}_d$.

Lemma 6.3.

(i) $H_i(\mathcal{B}_d, \mathcal{D}_d; \mathbb{Z}) = 0$ *for* $i \leq d-2$.
(ii) If d is odd, then $H_d(\mathcal{B}_d, \mathcal{D}_d; \mathbb{Z}) \cong \mathbb{Z}^{(d+1)/2}$ *and* $H_{d-1}(\mathcal{B}_d, \mathcal{D}_d; \mathbb{Z}) \cong \mathbb{Z}^{(d+1)/2}$.
(iii) If d is even, then $H_d(\mathcal{B}_d, \mathcal{D}_d; \mathbb{Z}) \cong \mathbb{Z}$ *and* $H_{d-1}(\mathcal{B}_d, \mathcal{D}_d; \mathbb{Z}) \cong (\mathbb{Z}/2\mathbb{Z})^{d/2}$.

We provide *two* proofs of the lemma in order to demonstrate the applicability of both Theorem 6.4 and of Theorem 6.3.

Proof of Lemma 6.3, based on Theorem 6.4. By the Poincaré duality and Theorem 6.4,

$$H_j(\mathcal{B}_d, \mathcal{D}_d; \mathbb{Z}) \cong H^{d-j}(\mathcal{B}_d \setminus \mathcal{D}_d; \mathbb{Z}^t),$$

where, for d odd, the local system of coefficients $\mathbb{Z}^t$ is constant, and, for d even, the local coefficient system $\mathbb{Z}^t$ is twisted by the monodromy along the generator of $\pi_1(\mathcal{B}_d) \cong \mathbb{Z}/2\mathbb{Z}$.

For an odd d, by Lemma 6.3, the space $\mathcal{B}_d \setminus \mathcal{D}_d$ is homotopy equivalent to a disjoint union of $(d+1)/2$ circles. This implies that $H^0(\mathcal{B}_d \setminus \mathcal{D}_d; \mathbb{Z}) \cong \mathbb{Z}^{(d+1)/2}$, and $H^1(\mathcal{B}_d \setminus \mathcal{D}_d; \mathbb{Z}) \cong \mathbb{Z}^{(d+1)/2}$, and all other cohomology groups vanish. As a result, for d, odd we get $H_d(\mathcal{B}_d, \mathcal{D}_d; \mathbb{Z}) \cong \mathbb{Z}^{(d+1)/2}$, and $H_{d-1}(\mathcal{B}_d, \mathcal{D}_d; \mathbb{Z}) \cong \mathbb{Z}^{(d+1)/2}$, and all other homology groups vanish.

For an even d, by Lemma 6.3, the space $\mathcal{B}_d \setminus \mathcal{D}_d$ is homotopy equivalent to a disjoint union of a point and $d/2$ circles. By the twist in the local coefficient system $\mathbb{Z}^t$, we get $H^0(\mathcal{B}_d \setminus \mathcal{D}_d; \mathbb{Z}^t) \cong \mathbb{Z}$ and $H^1(\mathcal{B}_d \setminus \mathcal{D}_d; \mathbb{Z}^t) \cong (\mathbb{Z}/2\mathbb{Z})^{d/2}$. All other cohomology groups $H^*(\mathcal{B}_d \setminus \mathcal{D}_d; \mathbb{Z}^t)$ vanish.

Thus, for d even, again by the Poincaré duality and Theorem 6.4, we get $H_d(\mathcal{B}_d, \mathcal{D}_d; \mathbb{Z}) \cong \mathbb{Z}$, $H_{d-1}(\mathcal{B}_d, \mathcal{D}_d; \mathbb{Z}) \cong (\mathbb{Z}/2\mathbb{Z})^{d/2}$, and all other homology groups vanish. □

Proof of Lemma 6.3, based on Theorem 6.3. Since in our cellulation the cells in $\mathcal{B}_d \setminus \Delta_d$ are of dimension d and $d-1$, the cellular chain complex of the pair $(\mathcal{B}_d, \Delta_d)$, arising from Theorem 6.3, has trivial chain groups in dimensions $\neq d, d-1$. This instantly implies (i).

From the preceding arguments it also follows that the group of cycles in dimension $d-1$ of the cellular chain complex of $(\mathcal{B}_d, \mathcal{D}_d)$ is freely generated by the cells of dimension $d-1$ in $\mathcal{B}_d \setminus \mathcal{D}_d$.

For d odd, the d-cells in $\mathcal{B}_d \setminus \mathcal{D}_d$ are labeled by $((\underbrace{1, \dots, 1}_{2j+1}), 0)$, where $j \in [0, \frac{d-1}{2}]$, and the $(d-1)$-dimensional cells by $((\underbrace{1, \dots, 1}_{2j}), 1)$, where $j \in [1, \frac{d-1}{2}]$. By Theorem 6.3, the differential of the cells of dimension d is fully contained in $\mathcal{D}_d$. Hence $H_d(\mathcal{B}_d, \mathcal{D}_d; \mathbb{Z})$ is freely generated by the cells of dimension d, and $H_{d-1}(\mathcal{B}_d, \mathcal{D}_d; \mathbb{Z})$ is freely generated by the cells of dimension $d-1$. A simple counting argument then yields (ii).

For d even, the d-cells in $\mathcal{B}_d \setminus \mathcal{D}_d$ are labeled by $((\underbrace{1, \dots, 1}_{2j}), 0)$, where $j \in [0, \frac{d}{2}]$, and the $(d-1)$-dimensional cells by $((\underbrace{1, \dots, 1}_{2j-1}), 1)$, where $j \in [1, \frac{d}{2}]$. By Theorem 6.3, the differential of the cell, labeled by $((\underbrace{1, \dots, 1}_{2j}), 0)$, consists only of cells from $\mathcal{D}_d$ if $j = 0$, and twice the cell, labeled by

$((\underbrace{1,\ldots,1}_{2j-1}),1)$, plus some terms which correspond to cells from $\mathcal{D}_d$ when $j \in [1, \frac{d}{2}]$. By simple counting and linear algebra, we then get $H_d(\mathcal{B}_d, \mathcal{D}_d; \mathbb{Z}) = \mathbb{Z}$ and $H_{d-1}(\mathcal{B}_d, \mathcal{D}_d; \mathbb{Z}) \cong (\mathbb{Z}/2\mathbb{Z})^{d/2}$. This validates (iii). □

Finally, we are in position to compute $\bar{H}_*(\mathcal{D}_d; \mathbb{Z})$.

Proposition 6.4. *The non-zero reduced integral homology groups $\bar{H}_i(\mathcal{D}_d; \mathbb{Z})$ of the discriminant $\mathcal{D}_d \subset \mathcal{B}_d$ have the following description:*

(i) if d is odd, then

$$\bar{H}_{d-1}(\mathcal{D}_d; \mathbb{Z}) \cong \mathbb{Z}^{(d-1)/2},\ \bar{H}_{d-2}(\mathcal{D}_d; \mathbb{Z}) \cong \mathbb{Z}^{(d+1)/2}, \text{ and}$$

$$\bar{H}_{d-4}(\mathcal{D}_d; \mathbb{Z}) \cong \bar{H}_{d-6}(\mathcal{D}_d; \mathbb{Z}) \cong \cdots \cong \bar{H}_1(\mathcal{D}_d; \mathbb{Z}) \cong \mathbb{Z}/2\mathbb{Z}.$$

The rest of the homology vanishes.

(ii) if d is even, then

$$\bar{H}_{d-1}(\mathcal{D}_d; \mathbb{Z}) \cong \mathbb{Z},\ \bar{H}_{d-2}(\mathcal{D}_d; \mathbb{Z}) \cong (\mathbb{Z}/2\mathbb{Z})^{(d/2)-1},$$

$$\text{and } \bar{H}_{d-3}(\mathcal{D}_d; \mathbb{Z}) \cong \bar{H}_{d-5}(\mathcal{D}_d; \mathbb{Z}) \cong \cdots \cong \bar{H}_1(\mathcal{D}_d; \mathbb{Z}) \cong \mathbb{Z}/2\mathbb{Z}.$$

The rest of the homology vanishes.

Proof. Since $\mathcal{D}_d$ has no cells in dimension d, by Theorem 6.3, we have $H_d(\mathcal{D}_d; \mathbb{Z}) = 0$.

Consider the long exact homology sequence of the pair $\mathcal{D}_d \subset \mathcal{B}_d$.

$$\begin{array}{lllll}
0 \cong H_d(\mathcal{D}_d; \mathbb{Z}) & \longrightarrow & H_d(\mathcal{B}_d; \mathbb{Z}) & \longrightarrow & H_d(\mathcal{B}_d, \mathcal{D}_d; \mathbb{Z}) \\
\hookrightarrow H_{d-1}(\mathcal{D}_d; \mathbb{Z}) & \longrightarrow & H_{d-1}(\mathcal{B}_d; \mathbb{Z}) & \longrightarrow & H_{d-1}(\mathcal{B}_d, \mathcal{D}_d; \mathbb{Z}) \\
\hookrightarrow \cdots & & \cdots & & \cdots \\
\hookrightarrow H_0(\mathcal{D}_d; \mathbb{Z}) & \longrightarrow & H_0(\mathcal{B}_d; \mathbb{Z}) & \longrightarrow & H_0(\mathcal{B}_d, \mathcal{D}_d; \mathbb{Z}) \cong 0.
\end{array}$$

Fig. 6.1 Long exact homology sequence of the pair $(\mathcal{B}_d, \mathcal{D}_d)$.

Basic algebraic topology [Ha] tells us that the homology of $\mathbb{RP}^d \cong \mathcal{B}_d$ is given by

$$H_p(\mathbb{RP}^d;\mathbb{Z}) = \begin{cases} \mathbb{Z} \text{ for } p=0 \text{ and, when } d \equiv 1 \mod 2, \text{ for } p=d \\ \mathbb{Z}/2\mathbb{Z},\ p \text{ odd},\ 0<p<d \\ 0, \text{ otherwise.} \end{cases} \tag{6.19}$$

By Lemma 6.3, (i) and the long exact sequence of the pair $(\mathcal{B}_d, \mathcal{D}_d)$ from Fig. 6.1, we get $\bar{H}_i(\mathcal{D};\mathbb{Z}) \cong \bar{H}_i(\mathbb{RP}^2;\mathbb{Z})$ for $i < d-2$.

It remains to determine $\bar{H}_i(\mathcal{D};\mathbb{Z})$ for $i = d-1, d-2$. For d odd, by Lemma 6.3 and the homology long exact sequence from Fig. 6.1, we obtain two short exact sequences:

$$0 \to \quad \mathbb{Z} \to \mathbb{Z}^{(d+1)/2} \quad \to H_{d-1}(\mathcal{D}_d;\mathbb{Z}) \ \to 0, \quad \text{and} \tag{6.20}$$

$$0 \to \mathbb{Z}^{(d+1)/2} \to H_{d-2}(\mathcal{D}_d;\mathbb{Z}) \ \to \mathbb{Z}/2\mathbb{Z} \quad \to 0. \tag{6.21}$$

In (6.20), $H_{d-1}(\mathcal{D}_d;\mathbb{Z})$ is the homology of $\mathcal{D}_d$ in homological dimension $\dim(\mathcal{D}_d)$ and hence is torsion free. From the exactness of the sequence it then follows that $H_{d-1}(\mathcal{D}_d;\mathbb{Z}) \cong \mathbb{Z}^{(d+2)/2-1} = \mathbb{Z}^{(d-1)/2}$.

For $0 \leq \ell \leq d-1$ even, the boundary of the cell, labeled by $((\underbrace{1,\dots,1}_{\ell}),1)$, lies completely in $\mathcal{D}_d$. In the expansion γ_ℓ of the boundary in the cellular chain complex, the cell, labeled by $((\underbrace{1,\dots,1}_{\ell-1}),2)$, has coefficient -2. As a boundary, each γ_ℓ is a cycle in the cellular chain complex of $\mathcal{D}_d$. Since the cell, labelled by $((\underbrace{1,\dots,1}_{\ell-1}),2)$, does not appear in the boundary of any cell of $\mathcal{D}_d$, it follows that each γ_ℓ represents a non-zero homology class of $\mathcal{D}_d$. Again, by the fact that none of the cells, labeled by $((\underbrace{1,\dots,1}_{\ell-1}),2)$, appear in the boundary of any cell of $\mathcal{D}_d$, it follows that a linear combination of the γ_ℓ is zero in homology if and only if all cells, labelled $((\underbrace{1,\dots,1}_{\ell-1}),2)$, cancel. But the cell, labeled $((\underbrace{1,\dots,1}_{\ell-1}),2)$, appears only in γ_ℓ. It follows that, over $\mathbb{Z}$, the collection of all γ_ℓ forms $\frac{d+1}{2}$ linearly independent non-trivial homology cycles. As a consequence, they generate the group $\mathbb{Z}^{\frac{d+1}{2}}$ inside $H_{d-2}(\mathcal{D}_d;\mathbb{Z})$. A simple calculation shows that, in the sum $\gamma_0 + \cdots + \gamma_{d-1}$, all non-zero coefficients are ± 2. Hence, $\gamma_0 + \cdots + \gamma_{d-1}$ is twice a cycle γ'

in the cellular chain complex of $\mathcal{D}_d$. In γ' each cell, labelled $((\underbrace{1,\ldots,1}_{\ell-1}),2)$, has coefficient -1. From the proof Lemma 6.3 we know that the classes of the cells, labelled by $((\underbrace{1,\ldots,1}_{\ell}),1)$, generate $H_{d-1}(\mathcal{B}_d,\mathcal{D};\mathbb{Z})$. Hence, in (6.21), their boundaries γ_ℓ generate the image of $\mathbb{Z}^{\frac{d+1}{2}}$ in $H_{d-2}(\mathcal{D}_d;\mathbb{Z})$. The arguments above show that γ' does not lie in the boundary and hence it follows from (6.21) that $H_{d-2}(\mathcal{D}_d;\mathbb{Z})\cong\mathbb{Z}^{\frac{d+2}{2}}$.

For d even, by Lemma 6.3 and the homology long exact sequence in Fig. 6.1, we obtain the exact sequence.

$$0\to\mathbb{Z}\to H_{d-1}(\mathcal{D}_d;\mathbb{Z})\to\mathbb{Z}/2\mathbb{Z}\to(\mathbb{Z}/2\mathbb{Z})^{d/2}\to H_{d-2}(\mathcal{D}_d;\mathbb{Z})\to 0. \quad (6.22)$$

The copy of $\mathbb{Z}$ in (6.22) is the top homology of the pair $(\mathcal{B}_d,\mathcal{D}_d)$. By the proof of Lemma 6.3, in our cellulation, the cell labeled $((),0)$ represents a class that generates this homology group. The image of this class, under the connecting homomorphism, is the class of its boundary, which consists only of the cell labeled by $((2),0)$. This is a top-dimensional cell of $\mathcal{D}_d$. Its boundary in the cellular chain complex of $\mathcal{D}_d$ is 0. Since the cell, labeled by $((2),0)$, is a basis element of the top-dimensional cellular chain group of $\mathcal{D}_d$, it follows that its class generates a $\mathbb{Z}$-summand of the homology group. Since this $\mathbb{Z}$-summand is the image $H_d(\mathcal{B}_d,\mathcal{D}_d;\mathbb{Z})$ under the connecting homomorphism, it follows from the exactness of (6.22) that $H_{d-1}(\mathcal{D}_d;\mathbb{Z})\cong\mathbb{Z}$ and (6.22) reduces to

$$0\to\mathbb{Z}/2\mathbb{Z}\to(\mathbb{Z}/2\mathbb{Z})^{d/2}\to H_{d-2}(\mathcal{D}_d;\mathbb{Z})\to 0.$$

This exact sequence can be seen as an exact sequence of vector spaces. Hence the sequence splits and $H_{d-2}(\mathcal{D}_d;\mathbb{Z})=(\mathbb{Z}/2\mathbb{Z})^{(d/2)-1}$. □

Remark 6.1. At the moment, we are unable to extend Proposition 6.4 to the case of the discriminant of classes binary forms in $\mathcal{B}_d$ with at least one root of multiplicity $k\geq 3$. One should mention that the latter case does not immediately follow from Vassiliev's Theorem 6.5, where the case of actual (non-projectivized) binary forms has been settled; this is related to a non-trivial action of $\mathbb{Z}/2\mathbb{Z}$ on the homology, when taking the projectivization. To illustrate this phenomenon, consider the table in Fig. 6.3. It shows the results of our computer calculations for the projectivized discriminant in case $k=3$, using the cellular chain complex from Theorem 6.3. Note the appearance of the higher torsion group $\mathbb{Z}/4\mathbb{Z}$ in the table, while in Vassiliev's case, all the torsion groups are sums of copies of $\mathbb{Z}/2\mathbb{Z}$. ◇

d/i	1	2	3	4	5	6	7	8	9	10	11
2	$\mathbb{Z}$										
3	$\mathbb{Z}^2$	$\mathbb{Z}$									
4	$\mathbb{Z}/2\mathbb{Z}$	$\mathbb{Z}/2\mathbb{Z}$	$\mathbb{Z}$								
5	$\mathbb{Z}/2\mathbb{Z}$	0	$\mathbb{Z}^3$	$\mathbb{Z}^2$							
6	$\mathbb{Z}/2\mathbb{Z}$	0	$\mathbb{Z}/2\mathbb{Z}$	$\mathbb{Z}/2\mathbb{Z}^2$	$\mathbb{Z}$						
7	$\mathbb{Z}/2\mathbb{Z}$	0	$\mathbb{Z}/2\mathbb{Z}$	0	$\mathbb{Z}^4$	$\mathbb{Z}^3$					
8	$\mathbb{Z}/2\mathbb{Z}$	0	$\mathbb{Z}/2\mathbb{Z}$	0	$\mathbb{Z}/2\mathbb{Z}$	$\mathbb{Z}/2\mathbb{Z}^3$	$\mathbb{Z}$				
9	$\mathbb{Z}/2\mathbb{Z}$	0	$\mathbb{Z}/2\mathbb{Z}$	0	$\mathbb{Z}/2\mathbb{Z}$	0	$\mathbb{Z}^5$	$\mathbb{Z}^4$			
10	$\mathbb{Z}/2\mathbb{Z}$	0	$\mathbb{Z}/2\mathbb{Z}$	0	$\mathbb{Z}/2\mathbb{Z}$	0	$\mathbb{Z}/2\mathbb{Z}$	$\mathbb{Z}/2\mathbb{Z}^4$	$\mathbb{Z}$		
11	$\mathbb{Z}/2\mathbb{Z}$	0	$\mathbb{Z}/2\mathbb{Z}$	0	$\mathbb{Z}/2\mathbb{Z}$	0	$\mathbb{Z}/2\mathbb{Z}$	0	$\mathbb{Z}^6$	$\mathbb{Z}^5$	
12	$\mathbb{Z}/2\mathbb{Z}$	0	$\mathbb{Z}/2\mathbb{Z}$	0	$\mathbb{Z}/2\mathbb{Z}$	0	$\mathbb{Z}/2\mathbb{Z}$	0	$\mathbb{Z}/2\mathbb{Z}$	$\mathbb{Z}/2\mathbb{Z}^4$	$\mathbb{Z}$

Fig. 6.2 Homology $\widetilde{H}_i(\mathcal{D}_d; \mathbb{Z})$ of the discriminant of binary forms.

6.5 Homological stabilization of polynomial spaces by the degree *d*

In this section we prove stabilization results for the homology of $\mathcal{B}_d^{\Theta}$ and $\bar{\mathcal{P}}_d^{\Theta}$. We first need to define suitable sets Θ for which such stabilizations make sense.

We either consider closed subposets $\Theta \subset \Omega^\infty$ such that, for all $(\omega, \kappa) \in \Theta$, the norms $|(\omega, \kappa)|$ have the same parity, or we consider closed subposets $\Theta \subset \Omega$ such that, for all $\omega \in \Theta$, the norms $|\omega|$ have the same parity. We call these posets Θ **closed equal parity subposets.** We say that a closed equal parity poset Θ is of parity d or equivalently d is of parity Θ if the norm of all elements of Θ has the same parity as d. Note that posets of mixed parity do not have a natural geometric interpretation.

If Θ is a closed equal parity poset, then we denote $\Theta_{\langle d]} =_{\mathsf{def}} \Theta \cap \mathbf{\Omega}^\infty_{\langle d]}$ (resp., $\Theta_{\langle d]} =_{\mathsf{def}} \Theta \cap \mathbf{\Omega}_{\langle d]}$). We also introduce $\Theta_d =_{\mathsf{def}} \Theta \cap \mathbf{\Omega}^\infty_d$ (resp., $\Theta_d =_{\mathsf{def}} \Theta \cap \mathbf{\Omega}_d$). (These constructions can only lead to non-empty posets if Θ is not of parity d.)

Let Θ be a closed equal parity poset in either Ω or Ω^∞. For a number d of parity Θ, we introduce the homomorphism

$$\mathsf{trunc} : \mathbb{Z}[\Theta_{\langle d+2]}] \to \mathbb{Z}[\Theta_{\langle d]}]$$

that is the identity on the elements of $\Theta_{\langle d+2]}$ of norm $\leq d$ and sends all elements of norm $d+2$ to 0. Note that in the univariate case (i.e., $\Theta \subset \Omega$),

$d \setminus i$	1	2	3	4	5	6	7	8	9	10	11	12
3	$\mathbb{Z}$											
4	$\mathbb{Z}^2$	$\mathbb{Z}$										
5	$\mathbb{Z} \oplus \mathbb{Z}/2\mathbb{Z}$	$\mathbb{Z}/2\mathbb{Z}$	0									
6	$\mathbb{Z}/2\mathbb{Z}$	0	$\mathbb{Z}^2$	$\mathbb{Z}$								
7	$\mathbb{Z}/2\mathbb{Z}$	0	$\mathbb{Z}^2 \oplus \mathbb{Z}/2\mathbb{Z}$	$\mathbb{Z} \oplus \mathbb{Z}/2\mathbb{Z}$	0							
8	$\mathbb{Z}/2\mathbb{Z}$	0	$\mathbb{Z} \oplus \mathbb{Z}/2\mathbb{Z}$	$\mathbb{Z}/4\mathbb{Z}$	$\mathbb{Z}$	$\mathbb{Z}$						
9	$\mathbb{Z}/2\mathbb{Z}$	0	$\mathbb{Z}/2\mathbb{Z}$	0	$\mathbb{Z}^2 \oplus \mathbb{Z}/2\mathbb{Z}$	$\mathbb{Z} \oplus \mathbb{Z}/2\mathbb{Z}$	0					
10	$\mathbb{Z}/2\mathbb{Z}$	0	$\mathbb{Z}/2\mathbb{Z}$	0	$\mathbb{Z}^2 \oplus \mathbb{Z}/2\mathbb{Z}$	$\mathbb{Z} \oplus \mathbb{Z}/4\mathbb{Z}$	$\mathbb{Z}$	$\mathbb{Z}$				
11	$\mathbb{Z}/2\mathbb{Z}$	0	$\mathbb{Z}/2\mathbb{Z}$	0	$\mathbb{Z} \oplus \mathbb{Z}/2\mathbb{Z}$	$\mathbb{Z}/4\mathbb{Z}$	$\mathbb{Z} \oplus \mathbb{Z}/2\mathbb{Z}$	$\mathbb{Z} \oplus \mathbb{Z}/2\mathbb{Z}$	0			
12	$\mathbb{Z}/2\mathbb{Z}$	0	$\mathbb{Z}/2\mathbb{Z}$	0	$\mathbb{Z}/2\mathbb{Z}$	0	$\mathbb{Z}^2 \oplus \mathbb{Z}/2\mathbb{Z}$	$\mathbb{Z} \oplus \mathbb{Z}/4\mathbb{Z}$	$\mathbb{Z}$	$\mathbb{Z}$		
13	$\mathbb{Z}/2\mathbb{Z}$	0	$\mathbb{Z}/2\mathbb{Z}$	0	$\mathbb{Z}/2\mathbb{Z}$	0	$\mathbb{Z}^2 \oplus \mathbb{Z}/2\mathbb{Z}$	$\mathbb{Z} \oplus \mathbb{Z}/4\mathbb{Z}$	$\mathbb{Z} \oplus \mathbb{Z}/2\mathbb{Z}$	$\mathbb{Z} \oplus \mathbb{Z}/2\mathbb{Z}$	0	

Fig. 6.3 Reduced homology groups $\{\bar{H}_i(\mathcal{B}_d^{\max \omega \geq 3}; \mathbb{Z})\}$ of the projectivized space of binary forms with at least one root (i.e., a line in the plane) of multiplicity ≥ 3. The rows of the table correspond to the fixed values of d.

the strata that correspond to compositions in $\Omega_{\langle d+2]}$ with norm $d+2$ are exactly the cells that consists of real polynomials whose roots are all real.

For the formulation of the next lemma we write $(\mathbb{Z}[\Theta_{\langle d]}], \partial)[-2]$ (resp., $(\mathbb{Z}[\Theta_{\langle d]}], \partial^\infty)[-2]$ for the complex $(\mathbb{Z}[\Theta_{\langle d]}], \partial)$ (resp., $(\mathbb{Z}[\Theta_{\langle d]}], \partial^\infty)$) with all degrees *raised* by 2.

Lemma 6.4. *Let Θ be a closed equal parity poset in Ω or Ω^∞ and $d \geq 0$ a number of parity Θ. Then the map* trunc *induces short exact sequences of differential complexes*

$$0 \to (\mathbb{Z}[\Theta_{d+2}], \partial_{\mathsf{M}}) \hookrightarrow (\mathbb{Z}[\Theta_{\langle d+2]}], \partial) \stackrel{\mathsf{trunc}}{\longrightarrow} (\mathbb{Z}[\Theta_{\langle d]}], \partial)[-2] \to 0,$$

or

$$0 \to (\mathbb{Z}[\Theta_{d+2}], \partial^\infty_{\mathsf{M}}) \hookrightarrow (\mathbb{Z}[\Theta_{\langle d+2]}], \partial^\infty) \stackrel{\mathsf{trunc}}{\longrightarrow} (\mathbb{Z}[\Theta_{\langle d]}], \partial^\infty)[-2] \to 0.$$

Proof. A simple calculation shows that trunc commutes with the differentials ∂^∞ (resp., ∂) and thus defines a homomorphism of differential complexes $(\mathbb{Z}[\Theta_{\langle d+2]}], \partial^\infty)$ and $(\mathbb{Z}[\Theta_{\langle d]}], \partial^\infty)$ (resp., $(\mathbb{Z}[\Theta_{\langle d+2]}], \partial)$ and $(\mathbb{Z}[\Theta_{\langle d]}], \partial)$). It also lowers the homological degree by 2. □

We say that Θ is generated by a subset $\Theta' \subseteq \Theta$ if Θ is the smallest closed subposet of either $\mathbf{\Omega}^\infty$ or $\mathbf{\Omega}$ that contains Θ'. If Θ' can be chosen to be finite, then we say that Θ is finitely generated. It is easily seen that, for any $\Theta' \subseteq \mathbf{\Omega}^\infty$ or $\Theta' \subseteq \mathbf{\Omega}$, there exists a unique smallest closed subposet Θ that contains Θ'. Moreover, if Θ is generated by Θ', then one can always assume that Θ' consists of maximal elements of Θ with respect to the order "$\prec$". In particular, being finitely generated is equivalent to having only finitely many maximal elements.

Definition 6.4. We call a closed poset $\Theta \subseteq \mathbf{\Omega}$ (resp., $\mathbf{\Omega}^\infty$) profinite if, for all integers $q \geq 0$, there exist only finitely many elements $(\omega, \kappa) \in \Theta$ (resp. $\omega \in \Theta$)) such that $|(\omega, \kappa)|' \leq q$ (resp., $|\omega|' \leq q$). ◇

Obviously, every finite Θ is profinite. In particular, for any closed $\Theta \subseteq \mathbf{\Omega}$ (resp., $\Theta \subseteq \mathbf{\Omega}^\infty$), we have that $\Theta_{\langle d]} = \Theta \cap \mathbf{\Omega}_{\langle d]}$ (resp., $\Theta_{\langle d]} = \Theta \cap \mathbf{\Omega}^\infty_{\langle d]}$) is finite, and hence profinite.

We say that Θ is generated by a subset $\Theta' \subseteq \Theta$ if Θ is the smallest closed subposet of either $\mathbf{\Omega}^\infty$ or $\mathbf{\Omega}$ that contains Θ'. If Θ' can be chosen to be finite, then we say that Θ is finitely generated. It is easily seen that, for any $\Theta' \subseteq \mathbf{\Omega}^\infty$ or $\Theta' \subseteq \mathbf{\Omega}$, there exists a unique smallest closed subposet Θ that contains Θ'. Moreover, if Θ is generated by Θ', then one can always assume that Θ' consists of maximal elements of Θ with respect to "$\prec$". In particular, being finitely generated is equivalent to having only finitely many maximal elements.

Given a closed equal parity poset Θ, we introduced the notations $\Theta_{\langle d]} =_{\mathsf{def}} \Theta \cap \mathbf{\Omega}^\infty_{\langle d]}$ (resp., $\Theta_{\langle d]} =_{\mathsf{def}} \Theta \cap \mathbf{\Omega}_{\langle d]}$). We also introduce $\Theta_d =_{\mathsf{def}} \Theta \cap \mathbf{\Omega}^\infty_d$ (resp., $\Theta_d =_{\mathsf{def}} \Theta \cap \mathbf{\Omega}_d$). These constructions can only lead to non-empty posets if Θ is not of parity d.

Definition 6.5. We call a closed poset $\Theta \subseteq \mathbf{\Omega}$ (resp., $\mathbf{\Omega}^\infty$) profinite if, for all integers $q \geq 0$, there exist only finitely many elements $(\omega, \kappa) \in \Theta$ (resp. $\omega \in \Theta$)) such that $|(\omega, \kappa)|' \leq q$ (resp., $|\omega|' \leq q$). ♢

Obviously, every finite Θ is profinite. In particular, for any closed $\Theta \subseteq \Omega$ (resp., $\Theta \subseteq \Omega^\infty$), we have that $\Theta_{\langle d]} = \Theta \cap \Omega_{\langle d]}$ (resp., $\Theta_{\langle d]} = \Theta \cap \Omega^\infty_{\langle d]}$) is finite and hence profinite.

Lemma 6.5. *Let Θ be a closed finitely generated equal parity subposet of $\mathbf{\Omega}^\infty$ (resp., of $\mathbf{\Omega}$). Let n be the largest norm of the maximal elements in Θ.*

Then, for any $q \geq 0$, we have inclusions

$$\{(\omega, \kappa) \in \Theta \mid |(\omega, \kappa)|' \leq q\} \subseteq \Theta_{\langle n+2q]} \quad \textit{and} \quad \{\omega \in \Theta \mid |\omega|' \leq q\} \subseteq \Theta_{\langle n+2q]}.$$

In particular, every closed finitely generated equal parity subposet is profinite.

Proof. Consider the case $\Theta \subseteq \mathbf{\Omega}^\infty$ and assume that (ω, κ) is one of its maximal elements. Let (ω, κ) be such that $|(\omega, \kappa)|' \leq q$. By definition any $(\omega', \kappa') \preceq (\omega, \kappa)$ can be obtained from (ω, κ) by a sequence of merge and insert operations. Both operations increase $| \sim |'$ by one. In particular, at most $q - |(\omega, \kappa)|' \geq 0$ insertion operations can be applied to (ω, κ) without $| \sim |'$ exceeding q. Each merge operation preserves the norm and each insertion operation increases the norm by 2. In particular, if $(\omega', \kappa') \preceq (\omega, \kappa)$ and $|(\omega', \kappa)|' \leq q$ then

$$|(\omega', \kappa')| \leq |(\omega, \kappa)| + 2(q - |(\omega, \kappa)|') = |(\omega, \kappa)| - 2|(\omega, \kappa)|' + 2q \leq n + 2q.$$

The proof in case $\Theta \subseteq \mathbf{\Omega}$ is analogous.

The last assertion now follows from the fact that $\Theta_{\langle n+2q]}$ is finite. □

Let $\Theta \subseteq \mathbf{\Omega}^\infty$ (resp., $\Theta \subseteq \mathbf{\Omega}$) be a closed equal parity subposet of parity d. Since each $\Theta_{\langle d]}$ is finite, it has only finitely many maximal elements. Let $n_1, \ldots, n_{\ell(d)}$ be the norms of the maximal elements in $\Theta_{\langle d]}$ and $c_1, \ldots, c_{\ell(d)}$

be the corresponding reduced norms. Now we define for d of parity Θ the quantity

$$\eta_\Theta(d) =_{\mathsf{def}} \max_{i=1,\dots,\ell(d)} \{n_i - 2c_i\}. \tag{6.23}$$

Note that, if Θ is finitely generated, then $\eta_\Theta(d)$ is constant for all d greater than or equal to the largest norm of a maximal element in Θ. In general, if (ω, κ) is maximal in $\Theta_{\langle d]}$, then it is also maximal in $\Theta_{\langle \ell]}$ for any $\ell \geq d$ of parity Θ. This implies that η_Θ is a *weakly increasing* function of d of parity Θ.

Further, using (6.23), for any number d of the parity Θ, we define the function

$$\psi_\Theta(d) =_{\mathsf{def}} \frac{1}{2}\big(d + \eta_\Theta(d)\big) = \frac{1}{2}\big(d + \max_{i=1,\dots,\ell(d)} \{n_i - 2c_i\}\big). \tag{6.24}$$

Since the function $\eta_\Theta(d)$ is weakly increasing as a function for arguments d of parity Θ, it follows that ψ_Θ is strictly increasing. In particular, we get that $\lim_{d\to\infty} \psi_\Theta(d) = \infty$, the limit being taken over d of parity Θ. Nevertheless, the growth of the function ψ_Θ depends on whether Θ is a finite, a profinite, or a general subposet.

Lemma 6.6. *Let $\Theta \subseteq \mathbf{\Omega}^\infty$ (resp., $\mathbf{\Omega}$) be a closed profinite equal parity poset. Then* $\lim_{d\to+\infty} (d - \psi_\Theta(d)) = +\infty$.

Proof. If Θ is finitely generated, then $\eta_\Theta(d) = c' \in \mathbb{Z}$ is constant for $d \gg 0$. Elementary manipulations validate the claim.

Assume that Θ is not finitely generated and let $\{n_i\}_{i\in\mathbb{N}}, \{c_i\}_{i\in\mathbb{N}}$ be the sequences of norms and the corresponding reduced norms of the maximal elements of Θ ordered such that the norms are weakly increasing. Since Θ is profinite, for any $q \geq 0$ there are only finitely many c_i with $c_i \leq q$. It follows that $\lim_{i\to\infty} c_i = \infty$.

Thus, for $d \gg 0$, we have $\psi_\Theta(d) = \frac{1}{2}(d + n_i - 2c_i)$ for some i such that $n_i \leq d$. Note that i may depend on d.

It follows that $d - \psi_\Theta(d) = \frac{1}{2}d - \frac{1}{2}(n_i - 2c_i) \geq 2c_i$. The claim now follows from the property $\lim_{i\to\infty} c_i = \infty$. □

Example 6.3. Let $\ell \geq 1$ be a number and $\Theta(\ell) \subseteq \mathbf{\Omega}^\infty$ be generated by the multiplicity patterns $\big\{\gamma_k = \big((\underbrace{3,\dots,3}_{2k}, \underbrace{1,\dots,1}_{2\ell k}), 0\big)\big\}_{k\geq 1}$.

Then $|\gamma_k| = k\,(6 + 2\,\ell)$ and $|\gamma_k|' = 4\,k$. It follows that $\Theta(\ell)$ is profinite, but not finitely generated. If for some k we have $d = k\,(6 + 2\,\ell)$, then $\eta_{\Theta(\ell)}(d) = \max_{k'\leq k}\{2\,k'\,(\ell - 2)\} \leq \frac{1}{3}d$ and $\psi_{\Theta(\ell)}(d) \leq \frac{5}{6}d$. It follows that $d - \psi_{\Theta(\ell)}(d) \geq \frac{1}{6}d \to +\infty$. ◇

We now show that the function ψ_Θ plays a fundamental role in the *homological stabilization* of spaces $\mathcal{B}_d^\Theta$ and $\bar{\mathcal{P}}_d^\Theta$.

Let Θ be a closed equal parity poset in either $\mathbf{\Omega}$ or $\mathbf{\Omega}^\infty$. For a number d of the parity Θ, we introduce the $\mathbb{Z}$-module homomorphism

$$\mathsf{trunc} : \mathbb{Z}[\Theta_{\langle d+2]}] \to \mathbb{Z}[\Theta_{\langle d]}][-2]$$

that is the identity on the elements of $\Theta_{\langle d+2]}$ of norm $\leq d$ and sends all elements of norm $d+2$ to 0. The latter elements represent divisors in $\mathbb{C}$ or $\bar{\mathbb{C}}$ of degree $d+2$ that are supported in the real line $\mathbb{R} \subset \mathbb{C}$ or its completion $\bar{\mathbb{R}} \subset \bar{\mathbb{C}}$.

A simple calculation shows that trunc commutes with the differentials ∂^∞ (resp., ∂) and thus defines a homomorphism of graded differential complexes $(\mathbb{Z}[\Theta_{\langle d+2]}], \partial^\infty)$ and $(\mathbb{Z}[\Theta_{\langle d]}], \partial^\infty)[-2]$ (resp., $(\mathbb{Z}[\Theta_{\langle d+2]}], \partial)$ and $(\mathbb{Z}[\Theta_{\langle d]}], \partial)[-2]$). It also lowers the homological degree by 2. Note that in the univariate case (i.e., for $\Theta \subset \mathbf{\Omega}$), the strata that correspond to compositions in $\mathbf{\Omega}_{\langle d+2]}$ with norm $d+2$ are exactly the cells that consists of real polynomials whose roots are all real.

Proposition 6.5. *Let Θ be a closed equal parity poset in $\mathbf{\Omega}$ or $\mathbf{\Omega}^\infty$ and $d \geq 0$ a number of parity Θ. Then*

(i) the operation trunc *lowers the homological grading by 2 and generates the short exact sequences of graded differential complexes*

$$0 \to (\mathbb{Z}[\Theta_{d+2}], \partial_{\mathsf{M}}) \to (\mathbb{Z}[\Theta_{\langle d+2]}], \partial) \overset{\mathsf{trunc}}{\longrightarrow} (\mathbb{Z}[\Theta_{\langle d]}], \partial)[-2] \to 0,$$

$$0 \to (\mathbb{Z}[\Theta_{d+2}], \partial_{\mathsf{M}}^\infty) \to (\mathbb{Z}[\Theta_{\langle d+2]}], \partial^\infty) \overset{\mathsf{trunc}}{\longrightarrow} (\mathbb{Z}[\Theta_{\langle d]}], \partial^\infty)[-2] \to 0,$$

(ii) if, for some k, the complexes $(\mathbb{Z}[\Theta_{d+2}], \partial_{\mathsf{M}})$ and $(\mathbb{Z}[\Theta_d], \partial_{\mathsf{M}})$, or the complexes $(\mathbb{Z}[\Theta_{d+2}], \partial_{\mathsf{M}}^\infty)$ and $(\mathbb{Z}[\Theta_d], \partial_{\mathsf{M}}^\infty)$ are acyclic in homological dimensions $\geq k$, then for all $j \geq k$, the homomorphism trunc *induces homological isomorphisms*

$$H_{j+2}(\mathbb{Z}[\Theta_{\langle d+2]}], \partial) \overset{\mathsf{trunc}_*}{\longrightarrow} H_j(\mathbb{Z}[\Theta_{\langle d]}], \partial),$$

$$H_{j+2}(\mathbb{Z}[\Theta_{\langle d+2]}], \partial^\infty) \overset{\mathsf{trunc}_*}{\longrightarrow} H_j(\mathbb{Z}[\Theta_{\langle d]}], \partial^\infty).$$

Proof. First note that on $\mathbb{Z}[\Theta_{d+2}]$ the differential ∂ (resp., ∂^∞) coincides with ∂_{M} (resp., $\partial_{\mathsf{M}}^\infty$). Now the fact that the two three-term sequences of differential complexes from (i) are exact follows instantly from definition of trunc. The long exact sequences in their homology, induced by the short exact sequences of differential complexes, then imply (ii). □

In Theorem 6.7, we will find out which posets Θ and integers k satisfy the hypotheses of Proposition 6.5, (ii). Prior to that, we need to investigate the geometry that leads to the (co)homological stabilization for the spaces $\mathcal{B}_d^\Theta$, $\mathcal{B}_d^{\mathbf{c}\Theta}$ and $\bar{\mathcal{P}}_d^\Theta$, $\mathcal{P}_d^{\mathbf{c}\Theta}$, as d increases.

For each $(\omega, \kappa) \in \mathbf{\Omega}^\infty_{\langle d]}$, consider the set $\mathcal{K}_d^{(\omega,\kappa)} \subseteq \mathcal{B}_d$ of projective classes of binary forms of degree d with *all zeros being real* and with the root multiplicity patterns $(\omega', \kappa') \preceq (\omega, \kappa)$. More precisely, we set

$$\mathcal{K}_d^{(\omega,\kappa)} =_{\mathsf{def}} \bigcup_{\substack{(\omega',\kappa') \preceq (\omega,\kappa) \\ |(\omega,\kappa)| = d}} \mathsf{R}_d^{(\omega',\kappa')}.$$

Analogously, for a closed equal parity poset $\Theta \subseteq \mathbf{\Omega}^\infty_{\langle d]}$, we define the cellular complex

$$\mathcal{K}_d^\Theta =_{\mathsf{def}} \bigcup_{(\omega,\kappa)\in\Theta} \mathcal{K}_d^{(\omega,\kappa)}.$$

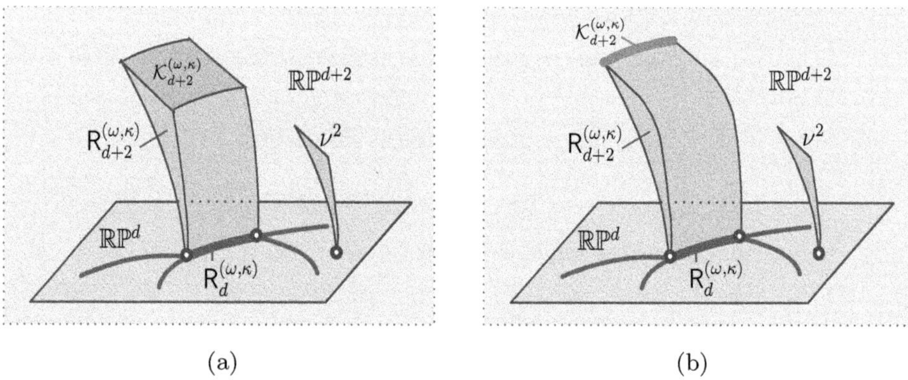

Fig. 6.4 A cell $\mathsf{R}_d^{(\omega,\kappa)}$ in $\mathcal{B}_d \cong \mathbb{RP}^d$ with $|(\omega,\kappa)| \leq d$ and the corresponding cell $\mathsf{R}_{d+2}^{(\omega,\kappa)}$ in $\mathcal{B}_{d+2} \cong \mathbb{RP}^{d+2}$. The set $\mathcal{K}_d^\Theta$ is shown as a graph in the plane $\mathbb{RP}^d$. The normal 2-bundle $\nu(\mathbb{RP}^d, \mathbb{RP}^{d+2})$ is denoted by ν^2. In (a), the portion $\mathcal{K}_{d+2}^{(\omega,\kappa)}$ of the boundary $\partial\mathsf{R}_{d+2}^{(\omega,\kappa)}$ has dimension $\dim(\mathsf{R}_{d+2}^{(\omega,\kappa)}) - 1$, while in (b), $\mathcal{K}_{d+2}^{(\omega,\kappa)}$ has dimension smaller than $\dim(\mathsf{R}_{d+2}^{(\omega,\kappa)}) - 1$.

Similarly, we consider the closed space $\mathcal{K}_{d+2}^\Theta$ that is the union of cells $\{\mathcal{K}_{d+2}^{(\omega,\kappa)}\}_{(\omega,\kappa)\in\Theta}$ indexed by elements (ω, κ) of the maximal norm $d + 2$.

Lemma 6.7.

(i) For $(\omega, \kappa) \in \mathbf{\Omega}^{\infty}_{\langle d]}$ or $\omega \in \mathbf{\Omega}_{\langle d]}$, we have

$$\dim(\mathcal{K}_d^{(\omega,\kappa)}) = \frac{1}{2}\Big(d + \big(\,|(\omega,\kappa)| - 2|(\omega,\kappa)|'\,\big)\Big)$$

or

$$\dim(\mathcal{K}_d^{\omega}) = \frac{1}{2}\Big(d + \big(\,|\omega| - 2|\omega|'\,\big)\Big).$$

(ii) For $d \geq 0$ and a closed equal d-parity subposet $\Theta \subseteq \mathbf{\Omega}^{\infty}_{\langle d]}$ or $\Theta \subseteq \mathbf{\Omega}$,

$$\dim(\mathcal{K}_d^{\Theta}) = \psi_\Theta(d).$$

Proof. It suffices to consider marked composition and $\Theta \subseteq \mathbf{\Omega}^{\infty}$. To settle item (i), we need to find the marked composition $(\omega', \kappa') \preceq (\omega, \kappa)$ with the minimal reduced norm $|(\omega', \kappa')|'$ and the norm $|(\omega', \kappa')| = d$. Since $d - |(\omega, \kappa)|$ is even, any sequence of $\frac{1}{2}(d - |(\omega, \kappa)|)$ insert operations, being applied to (ω, κ), does the job. The resulting (ω', κ') has the reduced norm $|(\omega', \kappa')|' = |(\omega, \kappa)|' + \frac{1}{2}(d - |(\omega, \kappa)|)$. Hence, the dimension of $\mathcal{K}_d^{(\omega',\kappa')}$ is given by

$$d - \frac{1}{2}(d - |(\omega,\kappa)| + 2|(\omega,\kappa)|') = \frac{1}{2}(d + (|(\omega,\kappa)| - 2|(\omega,\kappa)|')).$$

Item (ii) follows immediately from the definition of $\psi_\Theta(d)$ and (i). □

Let $\{\omega_i, \kappa_i\}_i$ be the finite list of all *maximal* elements in the poset $\Theta_{\langle d+2]}$.

Theorem 6.6. *The short exact sequences of differential complexes in Proposition 6.4 generate homology isomorphisms*

$$\begin{aligned} H_{j+2}(\mathcal{B}^{\Theta}_{d+2}) \approx H_{j+2}(\mathbb{Z}[\Theta_{\langle d+2]}], \partial^{\infty}) &\overset{\approx \mathsf{trunc}_*}{\longrightarrow} H_j(\mathbb{Z}[\Theta_{\langle d]}], \partial^{\infty}) \\ &\approx H_j(\mathcal{B}^{\Theta}_d), \end{aligned} \tag{6.25}$$

$$\begin{aligned} H_{j+2}(\mathcal{P}^{\Theta}_{d+2}) \approx H_{j+2}(\mathbb{Z}[\Theta_{\langle d+2]}], \partial) &\overset{\approx \mathsf{trunc}_*}{\longrightarrow} H_j(\mathbb{Z}[\Theta_{\langle d]}], \partial) \\ &\approx H_j(\mathcal{P}^{\Theta}_d) \end{aligned} \tag{6.26}$$

for all j subject to the constraint

$$j + 2 > \psi_\Theta(d+2). \tag{6.27}$$

imposed on the maximal generators $\{(\omega_i, \kappa_i)\}_i$ or $\{\omega_i\}_i$ of the posets $\Theta_{\langle d+2]}$ and $\Theta_{\langle d]}$. Under the constraint $j + 1 > \psi_\Theta(d+2)$, the degree lowering isomorphisms $\overset{\approx \mathsf{trunc}_*}{\longrightarrow}$ in (6.25) and (6.26) are monomorphisms.

Moreover, if

$$j + 2 > d + 2 - \min_i\{|(\omega_i, \kappa_i)|'\}, \tag{6.28}$$

then the groups $H_{j+2}(\mathbb{Z}[\Theta_{\langle d+2]}], \partial^\infty)$ *and* $H_{j+2}(\mathbb{Z}[\Theta_{\langle d+2]}], \partial)$ *vanish.*

Proof. By Lemma 6.7, the dimension of each cell $\mathcal{K}_{d+2}^{(\omega_i,\kappa_i)}$ is

$$\frac{1}{2}\big(d + 2 + |(\omega_i, \kappa_i)| - 2|(\omega_i, \kappa_i)|'\big).$$

Thus

$$\dim(\mathcal{K}_{d+2}^{\Theta}) = \frac{1}{2}\big(d + 2 + \max_i\{|(\omega_i, \kappa_i)| - 2|(\omega_i, \kappa_i)|'\}\big) =_{\mathsf{def}} \psi_\Theta(d+2).$$

As a result, $H_j(\mathcal{K}_{d+2}^{\Theta}) = 0$ for all $j > \psi_\Theta(d+2)$.

By Theorem 6.3, the group $H_j(\mathcal{K}_{d+2}^{\Theta})$ is isomorphic to the j-th homology of the differential complex $(\mathbb{Z}[\Theta_{d+2}], \partial_{\mathsf{M}}^\infty)$ from the short exact sequence in Proposition 6.5. As a result, for all j, k, we get

$$H_j(\mathcal{K}_{d+2}^{\Theta}) \approx H_j(\mathbb{Z}[\Theta_{d+2}], \partial_{\mathsf{M}}^\infty) \text{ and } H_k(\mathcal{K}_d^{\Theta}) \approx H_k(\mathbb{Z}[\Theta_d], \partial_{\mathsf{M}}^\infty).$$

Therefore, $H_j(\mathbb{Z}[\Theta_{d+2}], \partial_{\mathsf{M}}^\infty) = 0$ for all $j > \psi_\Theta(d+2)$, and $H_j(\mathbb{Z}[\Theta_d], \partial_{\mathsf{M}}^\infty) = 0$ for all $j > \psi_\Theta(d)$.

Therefore, by the long exact sequences, generated by short exact sequences of differential complexes from Proposition 6.5, we get isomorphisms

$$\underbrace{0 = H_j(\mathbb{Z}[\Theta_{d+2}], \partial_{\mathsf{M}}^\infty)} \to H_j(\mathbb{Z}[\Theta_{\langle d+2]}], \partial^\infty) \overset{\approx \mathsf{trunc}_*}{\longrightarrow}$$

$$\overset{\approx \mathsf{trunc}_*}{\longrightarrow} H_{j-2}(\mathbb{Z}[\Theta_{\langle d]}], \partial^\infty) \overset{\delta}{\longrightarrow} \underbrace{H_{j-1}(\mathbb{Z}[\Theta_d], \partial_{\mathsf{M}}^\infty) = 0}$$

for all $j - 1 > \psi_\Theta(d)$ and $j > \psi_\Theta(d+2)$. Thus,

$$H_j(\mathbb{Z}[\Theta_{\langle d+2]}], \partial^\infty) \overset{\approx \mathsf{trunc}_*}{\longrightarrow} H_{j-2}(\mathbb{Z}[\Theta_{\langle d]}], \partial^\infty)$$

for all $j > \max\{\psi_\Theta(d) + 1,\ \psi_\Theta(d+2)\}$. However, since $\psi_\Theta(d+2) > \psi_\Theta(d)$ (the ψ_Θ function is strictly increasing), we get $\max\{\psi_\Theta(d)+1,\ \psi_\Theta(d+2)\} = \psi_\Theta(d+2)$.

Equivalently, we get an isomorphism

$$H_{j+2}(\mathbb{Z}[\Theta_{\langle d+2]}], \partial^\infty) \overset{\approx \mathsf{trunc}_*}{\longrightarrow} H_j(\mathbb{Z}[\Theta_{\langle d]}], \partial^\infty)$$

for all $j + 2 > \psi_\Theta(d+2)$.

By Theorem 6.4, we get the desired isomorphisms

$$H_{j+2}(\mathcal{B}_{d+2}^{\Theta}) \approx H_{j+2}(\mathbb{Z}[\Theta_{\langle d+2]}], \partial^\infty) \overset{\approx \mathsf{trunc}}{\to} H_j(\mathbb{Z}[\Theta_{\langle d]}], \partial^\infty) \approx H_j(\mathcal{B}_d^{\Theta})$$

for all $j + 2 > \psi_\Theta(d + 2)$. By the same token, this composition $H_{j+2}(\mathcal{B}^\Theta_{d+2}) \to H_j(\mathcal{B}^\Theta_d)$ is a monomorphism for all $j + 3 > \psi_\Theta(d + 2)$.

At the same time, the dimension of each cell $\mathsf{R}^{(\omega,\kappa)}_{d+2}$ is $d + 2 - |(\omega,\kappa)|'$. Thus

$$\dim(\mathcal{B}^\Theta_{d+2}) = d + 2 - \min_i\{|(\omega_i,\kappa_i)|'\}_i =_{\mathsf{def}} \delta_\Theta(d+2) \tag{6.29}$$

Hence, $H_{j+2}(\mathcal{B}^\Theta_{d+2}) = 0$ for all $j + 2 > \delta_\Theta(d + 2)$.

Therefore, the range of $j + 2$ for which the stabilization $H_{j+2}(\mathcal{B}^\Theta_{d+2}) \approx H_j(\mathcal{B}^\Theta_d)$ occurs *and* the homology groups may be nontrivial is

$$\delta_\Theta(d+2) \geq j+2 > \psi_\Theta(d+2). \tag{6.30}$$

Using the definition in (6.24), this leads to the constraint on the *maximal* generators of the poset $\Theta_{\langle d+2]}$:

$$d + 2 > \max_i\{|(\omega_i,\kappa_i)| - 2|(\omega_i,\kappa_i)|'\} + 2\min_i\{|(\omega_i,\kappa_i)|'\}_i.$$

In particular, this constraint is satisfied when

$$d + 2 > \max_i\{|(\omega_i,\kappa_i)|\}.$$

Hence, if $\max_i\{|(\omega_i,\kappa_i)|\} < d+2$, then $H_{j+2}(\mathcal{B}^\Theta_{d+2}) \approx H_j(\mathcal{B}^\Theta_d)$ for all $j+2$ as in (6.30). □

Remark 6.2. If $(\omega,\kappa) \in \Theta_{[d]}$ (then $|(\omega,\kappa)| = d$) is a cycle, i.e., if $\partial_{\mathsf{M}}(\omega,\kappa) = 0$, but in $(\mathbb{Z}[\Theta_{\langle d+2]}], \partial)$, we have $\partial_{\mathsf{I}}(\omega,\kappa) \neq 0$, then (ω,κ) generates a non-stable cycle in $(\mathbb{Z}[\Theta_{\langle d]}], \partial)$ that dies in $(\mathbb{Z}[\Theta_{\langle d+2]}], \partial)$. On the other hand, any $(\omega,\kappa) \in \Theta_{[d+2]}$ (then $|(\omega,\kappa)| = d + 2$) such that $\partial_{\mathsf{M}}(\omega,\kappa) = 0$ generates a new cycle not present in $(\mathbb{Z}[\Theta_{\langle d]}], \partial)$. ◇

Now we are in position to establish our main stabilization results.

Theorem 6.7 (Short stabilization $\mathbf{d} \Rightarrow \mathbf{d+2}$, the projective case). *Let $\Theta \subseteq \mathbf{\Omega}^\infty$ be a closed equal parity poset and d of parity Θ.*

(i) For all $j + 2 > \psi_\Theta(d + 2)$ there is an isomorphism

$$T_j : H_j(\mathcal{B}^\Theta_d; \mathbb{Z}) \cong H_{j+2}(\mathcal{B}^\Theta_{d+2}; \mathbb{Z}).$$

(ii) For all $j + 1 > \psi_\Theta(d + 2)$ there is an isomorphism

$$T^\#_j : H_j(\mathcal{B}_d, \mathcal{B}^\Theta_d; \mathbb{Z}) \cong H_{j+2}(\mathcal{B}_{d+2}, \mathcal{B}^\Theta_{d+2}; \mathbb{Z}).$$

Both isomorphisms T_j and $T^\#_j$ are delivered by the inverses $(\mathsf{trunc}_*)^{-1}$ of truncations trunc_*.

Proof. Assertion (i) follows directly from Theorem 6.6.

For (ii), consider the new poset Θ', consisting of all the elements of Ω^∞ with the same parity as d. Of course, for that poset, we have $\mathcal{B}_d^{\Theta'_{\langle d]}} = \mathcal{B}_d$ and $\mathcal{B}_{d+2}^{\Theta'_{\langle d+2]}} = \mathcal{B}_{d+2}$. The maximal elements of $\Theta'_{\langle d+2]}$ are $((\underbrace{1,\dots,1}_{\ell}),0)$ for $0 \leq \ell \leq d+2$ of the same parity as d. The reduced norm of each of those marked compositions is 0. Hence $\psi_{\Theta'}(d+2)$ is either 1 or 0, depending on the parity of d. Thus by (i), trunc_* induces an isomorphism $H_{j+2}(\mathcal{B}_d;\mathbb{Z}) \cong H_j(\mathcal{B}_{d+2};\mathbb{Z})$ for all $j > 0$.

Consider the long exact homology sequences of the two pairs $(\mathcal{B}_{d+2}, \mathcal{B}_{d+2}^{\Theta})$ and $(\mathcal{B}_d, \mathcal{B}_d^{\Theta})$ and the homomorphisms trunc_* connecting them.

$$
\begin{array}{ccccccc}
\cdots \longrightarrow & H_{j+2}(\mathcal{B}_{d+2}^{\Theta}) & \longrightarrow & H_{j+2}(\mathcal{B}_{d+2}) & \longrightarrow & H_{j+2}(\mathcal{B}_{d+2}, \mathcal{B}_{d+2}^{\Theta}) & \to \\
 & \downarrow \mathsf{trunc}_* & & \downarrow \mathsf{trunc}_* & & \downarrow \mathsf{trunc}_* & \\
\cdots \longrightarrow & H_j(\mathcal{B}_d^{\Theta}) & \longrightarrow & H_j(\mathcal{B}_d) & \longrightarrow & H_j(\mathcal{B}_d, \mathcal{B}_d^{\Theta}) & \to \\
\to & H_{j+1}(\mathcal{B}_{d+2}^{\Theta}) & \longrightarrow & H_{j+1}(\mathcal{B}_{d+2}) & \longrightarrow & H_{j+1}(\mathcal{B}_{d+2}, \mathcal{B}_{d+2}^{\Theta}) & \\
 & \downarrow \mathsf{trunc}_* & & \downarrow \mathsf{trunc}_* & & \downarrow \mathsf{trunc}_* & \\
\to & H_{j-1}(\mathcal{B}_d^{\Theta}) & \longrightarrow & H_{j-1}(\mathcal{B}_d) & \longrightarrow & H_{j-1}(\mathcal{B}_d, \mathcal{B}_d^{\Theta}) &
\end{array}
$$

Fig. 6.5 Connecting the long exact homology sequences of $(\mathcal{B}_d^{\Theta}, \mathcal{B}_d)$ and $(\mathcal{B}_{d+2}^{\Theta}, \mathcal{B}_{d+2})$.

By (i), the two leftmost homomorphisms trunc_* in each row are isomorphisms for all $j+1 > \psi_\Theta(d+2)$. Thus, by the Five Lemma, the assertion (ii) follows. □

If true, the conjecture below would deliver a better geometric understanding of the spaces that appeared in the preceding proof.

Conjecture 6.1. *For any closed* Θ, *the quotient* $\mathcal{B}_{d+2}^{\Theta} / \mathcal{K}_{d+2}(\Theta)$ *is the Thom space of an orientable topological 2-dimensional microbundle over* $\mathcal{B}_d^{\Theta}$. ◇

The same proof, as the proof of Theorem 6.7, yields the following stabilization result for spaces of polynomials.

Theorem 6.8 (Short stabilization $\mathbf{d} \Rightarrow \mathbf{d}+\mathbf{2}$, the polynomial case). *Let $\Theta \subseteq \mathbf{\Omega}$ be a closed equal parity poset and d of parity Θ.*

(i) For all $j+2 > \psi_\Theta(d+2)$, there is an isomorphism

$$T_j : H_j(\bar{\mathcal{P}}_d^\Theta; \mathbb{Z}) \cong H_{j+2}(\bar{\mathcal{P}}_{d+2}^\Theta; \mathbb{Z}).$$

(ii) For all $j+1 > \psi_\Theta(d+2)$, there is an isomorphism

$$T_j^\# : H_j(\bar{\mathcal{P}}_d, \bar{\mathcal{P}}_d^\Theta; \mathbb{Z}) \cong H_{j+2}(\bar{\mathcal{P}}_{d+2}, \bar{\mathcal{P}}_{d+2}^\Theta; \mathbb{Z}).$$

Both isomorphisms T_j and $T_j^\#$ are delivered by the inverses $(\mathsf{trunc}_*)^{-1}$ *of the truncations* trunc_*. ♢

Next, we give two examples where the stabilization bound ψ_Θ from Theorem 6.7 and Theorem 6.8 can be explicitly determined.

Example 6.4. Let the closed poset $\Theta = \langle (4), (33) \rangle$ be generated by two elements: (4) and (33). Then $\eta_\Theta(d) = -2$, $\psi_\Theta(d) = \frac{1}{2}(d-2)$. Applying Theorem 6.8 to this Θ, we get isomorphisms

$$(\epsilon_{d,d+2})_* : H_j(\mathcal{P}_d^{\mathbf{c}\Theta}; \mathbb{Z}) \approx H_j(\mathcal{P}_{d+2}^{\mathbf{c}\Theta}; \mathbb{Z})$$

for all $j \leq \frac{1}{2}d + 2$. Moreover, since $\bar{\mathcal{P}}_d^\Theta$ is of codimension 3 in $\bar{\mathcal{P}}_d \approx S^d$, we get $\pi_1(\mathcal{P}_d^{\mathbf{c}\Theta}) = 1$ for all $d \geq 6$. Therefore, we get the homotopy groups isomorphisms

$$(\epsilon_{d,d+2})_* : \pi_j(\mathcal{P}_d^{\mathbf{c}\Theta}) \approx \pi_j(\mathcal{P}_{d+2}^{\mathbf{c}\Theta})$$

for all $j \leq \frac{1}{2}d + 2$, where $d \geq 6$. ♢

Example 6.5. In Section 6.5.1, we give a long list of homology computations for $\Theta \subseteq \mathbf{\Omega}$ that are generated by a single composition ω. They are organized in large tables. In these cases, for $d \geq |\omega|$, we have $\psi_\Theta(d) = \frac{1}{2}\big(d + |\omega| - 2|\omega|'\big)$. The reader is welcome to compare the results from Theorem 6.8, as they manifest themselves in these computer-generated tables. ♢

Interesting examples, where the stabilization bound from Theorem 6.7 is non-trivial can be derived from Example 6.3 or cases where Θ is *finitely generated.*

The next examples will not provide interesting applications of Theorem 6.7, but will show that the stabilization bound in it is *sharp*.

Example 6.6. For $\ell = 0$ or 1, the two closed equal parity posets given by

$$\Theta_{\ell,q} = \Omega^{\infty}_{|\sim|' \geq q} := \left\{ (\omega, \kappa) \in \Omega^{\infty} \,\middle|\, |(\omega,\kappa)|' \geq q,\ |(\omega,\kappa)| \equiv \ell \mod 2 \right\} \tag{6.31}$$

are easily seen to be non-profinite. Actually, the space $\mathcal{B}_d^{\Theta_{\ell,q}}$ is the $(d-q)$-skeleton of $\mathcal{B}_d$. By definition of $\eta_\Theta(d)$, we get $\eta_\Theta(d) = d - 2q$. Thus $\psi_\Theta(d) = d - q$ and $\psi_\Theta(d+2) = d + 2 - q$. Now, Theorem 6.7 shows that, if $j + 2 > d + 2 - q$ (or, equivalently, $j > d - q$), then $H_j(\mathcal{B}_d^{\Theta_{\ell,q}}; \mathbb{Z}) \cong H_{j+2}(\mathcal{B}_{d+2}^{\Theta_{\ell,q}}; \mathbb{Z})$.

For $q = 2$, the homology of $\mathcal{B}_d^{\Theta_{\ell,q}}$ is listed in Fig. 6.6 for small d and $\ell \equiv d \mod 2$. The list is the result of our computer-assisted calculations.

Note that the stabilization from Theorem 6.7 appears for $j \geq d-1$, which is just *right above* the top non-vanishing homology group. On the other hand, for $j = d-2$, Fig. 6.6 testifies that there is no stabilization. In fact, for $j < d-2$, the homology is stable, since here we are considering the skeletae of the real projective spaces. ◊

$d \setminus j$	1	2	3	4	5	6	7	8	9
4	$\mathbb{Z}/2\mathbb{Z}$	$\mathbb{Z}^3$							
5	$\mathbb{Z}/2\mathbb{Z}$	0	$\mathbb{Z}^7$						
6	$\mathbb{Z}/2\mathbb{Z}$	0	$\mathbb{Z}/2\mathbb{Z}$	$\mathbb{Z}^8$					
7	$\mathbb{Z}/2\mathbb{Z}$	0	$\mathbb{Z}/2\mathbb{Z}$	0	$\mathbb{Z}^{13}$				
8	$\mathbb{Z}/2\mathbb{Z}$	0	$\mathbb{Z}/2\mathbb{Z}$	0	$\mathbb{Z}/2\mathbb{Z}$	$\mathbb{Z}^{15}$			
9	$\mathbb{Z}/2\mathbb{Z}$	0	$\mathbb{Z}/2\mathbb{Z}$	0	$\mathbb{Z}/2\mathbb{Z}$	0	$\mathbb{Z}^{21}$		
10	$\mathbb{Z}/2\mathbb{Z}$	0	$\mathbb{Z}/2\mathbb{Z}$	0	$\mathbb{Z}/2\mathbb{Z}$	0	$\mathbb{Z}/2\mathbb{Z}$	$\mathbb{Z}^{24}$	
11	$\mathbb{Z}/2\mathbb{Z}$	0	$\mathbb{Z}/2\mathbb{Z}$	0	$\mathbb{Z}/2\mathbb{Z}$	0	$\mathbb{Z}/2\mathbb{Z}$	0	$\mathbb{Z}^{31}$

Fig. 6.6 Homology of the space $\mathcal{B}_d^{\Theta_{\ell,2}}$, the $(d-2)$-skeleton of $\mathcal{B}_d$. The poset $\Theta_{\ell,2}$, where $\ell \equiv d \mod 2$, is defined in (6.31).

Example 6.7. Let us fix a parity and a number $k \geq 2$. Consider the closed equal parity poset $\Theta = \Theta^{\max \geq k}$ containing all (ω, κ) of the chosen parity, for which there is a part of ω that is $\geq k$ or $\kappa \geq k$. Thus, $\mathcal{B}_d^\Theta$ is the space of classes of binary forms of degree d with at least one k^{fold} root (see Theorem 6.2 and Theorem 6.5 for some results on analogously defined spaces).

The maximal elements of Θ are the pairs the $(\omega, 0)$ where ω has one part of size k and all other parts being 1. It follows that $|(\omega, 0)|' = k - 1$. Hence $\eta_\Theta(d) = d - 2(k-1)$. Therefore $\psi_\Theta(d) = d - k + 1$ and $\psi_\Theta(d+2) = d - k + 3$.

Again, Theorem 6.7 shows that, for $j + 2 > d - k + 3$ (or, equivalently, for $j > d - k + 1$), $H_j(\mathcal{B}_d^\Theta; \mathbb{Z}) \cong H_{j+2}(\mathcal{B}_{d+2}^\Theta; \mathbb{Z})$. Looking at Fig. 6.3, one sees that the stabilization bound is again right about the top non-vanishing homology group. However, it seems that a stabilization below that bound is present here. $\diamondsuit$

Let us describe now, in terms of divisors, the effect of increasing the degree $d \to d + 2$ on the spaces of projectivized binary forms.

Let $\epsilon : \mathcal{B}_d \to \mathcal{B}_{d+2}$ be the map which sends the class of a binary form $f(x, y)$ in $\mathcal{B}_d$ to the class of the binary form $(x^2 + y^2) f(x, y)$ in $\mathcal{B}_{d+2}$. Since $x^2 + y^2$ has no real roots in $\mathbb{CP}^1$, for any $(\omega, \kappa) \in \mathbf{\Omega}_{\langle d]}^\infty$ the image of $\mathring{\mathsf{R}}_d^{(\omega,\kappa)}$ under ϵ is a subset of $\mathring{\mathsf{R}}_{d+2}^{(\omega,\kappa)}$.

For any $d' \geq d, d' \equiv d \mod 2$, we denote by $\epsilon_{d,d'}$ the $[(d'-d)/2]$-fold iteration of ϵ.

Similarly, for any $d' \geq d, d' \equiv d \mod 2$, we consider a natural map

$$\epsilon_{d,d'} : \mathcal{P}_d \to \mathcal{P}_{d'}, \text{ defined by}$$
$$(\epsilon_{d,d'}(P))(u) =_{\mathsf{def}} (u^2 + 1)^{\frac{d'-d}{2}} \cdot P(u). \tag{6.32}$$

We abbreviate again $\epsilon_{d,d+2}$ as ϵ.

Consider the 2-sphere $\mathbb{CP}^1$ and the involution $J : \mathbb{CP}^1 \to \mathbb{CP}^1$, given by the complex conjugation $J[z : w] = [\bar{z} : \bar{w}]$. The fixed point set of J is the circle $\bar{\mathbb{R}} = \mathbb{RP}^1$. We can identify $\mathcal{B}_d$ with the space positive J-$\mathsf{symmetric}$ divisors on $\mathbb{CP}^1$ of degree d. Then ϵ translates into a map

$$\epsilon^\bullet : [Sym^d(\mathbb{CP}^1)]^J \to [Sym^{d+2}(\mathbb{CP}^1)]^J$$

that takes each divisor D to the divisor $D + [\mathsf{i} : 1] + [-\mathsf{i} : 1]$. Thus the real divisors $\epsilon^\bullet(D)_{\bar{\mathbb{R}}}$ and $D_{\bar{\mathbb{R}}}$ share the same combinatorial type (ω, κ).

Our next goal is to give alternative pure *homological* definitions of the maps $(\mathsf{trunc})^*$, $(\mathsf{trunc})_*$. The new definitions will allow us to study stabilization in cohomology of the spaces $\mathcal{B}_d^{\mathsf{c}\Theta}$. With this goal in mind, let us describe a setting, which we call "Hypotheses νA", needed for such definitions.

• Hypotheses $\nu\mathbf{A}$. Let M be a closed n-dimensional manifold M, equipped with the structure of a finite CW-complex, and $N \subseteq M$ a closed $(n-k)$-dimensional submanifold, whose normal bundle is orientable. Assume that the intersections of N with the cells of M induce a structure of a cellular complex on N.

Let $K \subset M$ be a compact CW-subcomplex of M. Put $L =_{\mathsf{def}} K \cap N$.

Assume that, for each open cell e of M, the intersection $e \cap N$ is either a single cell such that $\dim(e) = \dim(e \cap N) + k$, or $e \cap N = \emptyset$. We assume also that M admits a triangulation T that is consistent with its cellular structure so that K, L, N and their cellular skeletons are closed subcomplexes of this triangulation. ◇

Put $\Lambda =_{\mathsf{def}} \mathbb{Z}[\pi_1(M)]$, where $\mathbb{Z}[\pi_1(M)]$ denotes the group ring of the fundamental group $\pi_1(M)$. Let $\mathbb{A}$ be a right Λ-module.

The embeddings $\epsilon : N \setminus L \subset M \setminus K \subset M$, with the help of the natural homomorphism $\epsilon_* : \pi_1(N) \to \pi_1(M)$, convert the Λ-module $\mathbb{A}$ into a $\mathbb{Z}[\pi_1(N)]$-module $\epsilon^*(\mathbb{A})$.

Let $[M] \in H_n(M; \mathbb{Z}^t)$ and $[N] \in H_{n-k}(M; \mathbb{Z}^t)$ be the fundamental classes of M and N, respectively.

We denote by $\mathcal{C}_\bullet(\sim)$ and $\mathcal{C}^\bullet(\sim)$ the $\mathbb{Z}$-modules of cellular chains of a cellular complex.

Lemma 6.8. *Under* Hypotheses $\nu\mathbf{A}$, *the following diagram is commutative:*

$$\begin{array}{ccc} H^j(N, L;\, \epsilon^*(\mathbb{A})) & \stackrel{(\mathsf{trunc})^*}{\longrightarrow} & H^{j+k}(M, K;\, \mathbb{A}) \\ \cong\downarrow [N]\cap & & \cong\downarrow [M]\cap \\ H_{n-j}(N \setminus L;\, \epsilon^*(\mathbb{A}^t)) & \stackrel{\epsilon_*}{\longrightarrow} & H_{n-j}(M \setminus K;\, \mathbb{A}^t), \end{array} \tag{6.33}$$

where the vertical isomorphisms are the Poincaré duality maps, $\mathcal{D}_N = [N]\cap \sim$ *and* $\mathcal{D}_M = [M]\cap \sim$. *The horizontal homomorphisms* $(\mathsf{trunc})^*$ *are induced by the duals of the chain homomorphisms* $\mathsf{trunc} : \mathcal{C}_{*+k}(M, K;\, \mathbb{A}) \to \mathcal{C}_*(N, L;\, \epsilon^*(\mathbb{A}))$*; the latter ones are produced by the intersection of cells* $\{e\}$ *in* M *with the submanifold* N.

The proof below contains an exact definition of the cohomological transfer $(\mathsf{trunc})^*$ *from diagram (6.33).*

Proof. We use the second barycentric subdivision of the triangulation T to form regular neighborhoods of closed subcomplexes of T.

Since, for each open cell e of M, the intersection $e \cap N$ is either a single cell such that $\dim(e) = \dim(e \cap N) + k$, or $e \cap N = \emptyset$, the restriction of the cellular structure on M to N gives rise to a cellular structure on N and produces the geometric correspondence $\mathsf{trunc} : e \rightsquigarrow e \cap N$. Since $\nu(N, M)$ is orientable, its orientation helps to use the orientation of e to orient $e \cap N$.

The correspondence trunc leads to a well-defined homomorphisms on the *cellular* chain level:

$$\mathsf{trunc} : \mathcal{C}_{j+k}(K;\, \mathbb{A}) \to \mathcal{C}_j(L;\, \epsilon^*(\mathbb{A})),$$

$$\mathsf{trunc} : \mathcal{C}_{j+k}(M, K;\, \mathbb{A}) \to \mathcal{C}_j(N, L;\, \epsilon^*(\mathbb{A}))$$

We use the identity $[(\mathsf{trunc})^*(\alpha)](e) =_{\mathsf{def}} \alpha(e \cap N)$ to introduce the dual operator $(\mathsf{trunc})^*$ for any cochain $\alpha \in \mathcal{C}^j(L;\, \epsilon^*(\mathbb{A}))$ and a $(j+k)$-dimensional cell $e \subset K$.

Our next goal is to define the truncation

$$H^j(N, L;\, \epsilon^*(\mathbb{A})) \stackrel{(\mathsf{trunc})^*}{\longrightarrow} H^{j+k}(M, K;\, \mathbb{A})$$

in pure homological terms, since the cap products in (6.33) are not directly amenable to the cellular structure $\{e\}$ on M.

Let M_p, K_p denote the p-skeletons of M, K in the cellular structure $\{e\}$, and N_p, L_p denote the p-skeletons of N, L in the cellular structure $\{N \cap e\}$. To simplify the notations, put

$$K_p^\sharp =_{\mathsf{def}} K_p \cup L, \quad M_p^\sharp =_{\mathsf{def}} M_p \cup K, \quad N_p^\sharp =_{\mathsf{def}} N_p \cup L.$$

We omit the relevant coefficient systems $\mathbb{A}$, $\mathbb{A}^t$ and $\epsilon^*(\mathbb{A})$, $\epsilon^*(\mathbb{A}^t)$ in our notations.

Consider the following filtrations by closed subsets:

$$N =_{\mathsf{def}} N_n^\sharp \supset \ldots \supset N_p^\sharp \supset N_{p-1}^\sharp \supset \ldots \supset N_0^\sharp,$$

$$M =_{\mathsf{def}} M_{n+k}^\sharp \supset \ldots \supset M_p^\sharp \supset M_{p-1}^\sharp \supset \ldots \supset M_0^\sharp,$$

These filtrations give rise to two spectral sequences, with E_1-pages

$$E_1^{p,q}(N, L) = H^{p+q}(N_p^\sharp, N_{p-1}^\sharp) \Rightarrow \text{ converges to } H^*(N, L),$$

$$E_1^{p,q}(M, K) = H^{p+q}(M_p^\sharp, M_{p-1}^\sharp) \Rightarrow \text{ converges to } H^*(M, K),$$

Therefore, in terms of the given cellular structures $\{e \cap N\}$ on N and $\{e\}$ on M, we get $E_1^{p,0}(N, L) \cong \mathcal{C}^p(N, L)$, and $E_1^{p,q}(N, L) = 0$ for all $q > 0$, since the cohomology with compact support of each open p-cell arise only in dimension p. Similarly, $E_1^{p,0}(M,K) \cong \mathcal{C}^p(M,K)$, and $E_1^{p,q}(M,K) = 0$ for all $q > 0$.

Note that

$$N_p^\sharp \setminus N_{p-1}^\sharp = N_p \setminus (N_{p-1} \cup L_p)),\ M_{p+k}^\sharp \setminus M_{p+k-1}^\sharp = M_{p+k} \setminus (M_{p+k-1} \cup K_{p+k}))$$

are unions of open p-cells and $(p+k)$-cells, respectively. Thus we have a homomorphism $(\mathsf{trunc}^*)^{p,q} : E_1^{p,q}(N,L) \to E_1^{p+k,q}(M,K)$, defined by:

$$H^p(N_p^\sharp, N_{p-1}^\sharp) \cong \mathcal{C}^p(N,L) \overset{(\mathsf{trunc})^*}{\longrightarrow}$$

$$\mathcal{C}^{p+k}(M,K) \cong H^{p+k}(M_{p+k}^\sharp, M_{p+k-1}^\sharp). \tag{6.34}$$

It relies on the coherent orientations of the pairs $(e, e \cap N)$ of cells and amounts to attaching the value $\alpha(e \cap N)$ to each $(p+k)$-cell e in M and each p-cochain α that represents an element of $H^p(N_p^\sharp, N_{p-1}^\sharp)$.

It remains to verify that $\mathsf{trunc}^* : E_1^{p,q}(N,L) \to E_1^{p+k,q}(M,K)$ commutes with the differentials of the two spectral sequences. This verification is in line with the standard arguments that show the equivalence of singular and cellular (co)homology. We incorporate the differentials from both spectral sequences in the commutative diagrams as on page 227 of [Mu] (with homology being replaced with cohomology and the arrows being reversed). The top of the two diagrams is based on the relative cohomology of different pairs from the quadruple

$$M_{p+k+1}^\sharp \supset M_{p+k}^\sharp \supset M_{p+k-1}^\sharp \supset M_{p+k-2}^\sharp,$$

and the bottom one on on the relative cohomology of different pairs from the quadruple

$$N_{p+1}^\sharp \supset N_p^\sharp \supset N_{p-1}^\sharp \supset N_{p-2}^\sharp.$$

The degree raising differentials

$$\delta_N^{p-1} : H^{p-1}(N_{p-1}^\sharp, N_{p-2}^\sharp) \to H^p(N_p^\sharp, N_{p-1}^\sharp),$$

$$\delta_N^p : H^p(N_p^\sharp, N_{p-1}^\sharp) \to H^{p+1}(N_{p+1}^\sharp, N_p^\sharp)$$

are taken from the long exact cohomology sequence of the triples

$$\{N_p^\sharp \supset N_{p-1}^\sharp \supset N_{p-2}^\sharp\} \text{ and } \{N_{p+1}^\sharp \supset N_p^\sharp \supset N_{p-1}^\sharp\},$$

respectively. Similarly, we get the differentials

$$\delta_M^{p+k-1} : H^{p+k-1}(M^\sharp_{p+k-1}, M^\sharp_{p+k-2}) \to H^{p+k}(M^\sharp_{p+k}, M^\sharp_{p+k-1})$$

$$\delta_M^{p+k} : H^{p+k}(M^\sharp_{p+k}, M^\sharp_{p+k-1}) \to H^{p+k+1}(M^\sharp_{p+k+1}, M^\sharp_{p+k}).$$

By the lemma hypotheses "$\partial(e)\cap N = \partial(e\cap N)$", these differentials coincide with the coboundary operators on the cochain level. By the diagrams as on page 227 of [Mu], the differentials

$$\delta_N^{p-1} = \partial^{p-1} : \mathcal{C}^{p-1}(N, L) \to \mathcal{C}^p(N, L), \quad \delta_N^p = \partial^p : \mathcal{C}^p(N, L) \to \mathcal{C}^{p+1}(N, L).$$

$$\delta_M^{p+k-1} = \partial^{p+k-1} : \mathcal{C}^{p+k-1}(M, K) \to \mathcal{C}^{p+k}(M, K),$$

$$\delta_M^{p+k} = \partial^{p+k} : \mathcal{C}^{p+k}(M, K) \to \mathcal{C}^{p+k+1}(M, K)$$

commute with the trunc^* maps: that is, $\delta_N^{p-1} \circ \mathsf{trunc}^* = \mathsf{trunc}^* \circ \delta_M^{p+k-1}$ and $\delta_N^p \circ \mathsf{trunc}^* = \mathsf{trunc}^* \circ \delta_M^{p+k}$. Therefore (see [Mu], Theorems 39.4 and 39.5) the map from (6.34) gives rise to a homomorphism

$$\mathsf{trunc}^* : H^p(N, L) = \frac{\mathsf{ker}(\delta_N^p)}{\mathsf{im}(\delta_N^{p-1})} \longrightarrow \frac{\mathsf{ker}(\delta_M^{p+k})}{\mathsf{im}(\delta_M^{p+k-1})} = H^{p+k}(M, K),$$

which is now well-defined on the level of cohomology (and not only on the cellular level). As a result, all the maps in the diagram (6.33) are well-defined homologically.

Thus, to validate the commutativity of (6.33), we seek to verify that, for a given element $\alpha \in H^p(N, L; \epsilon^*(\mathbb{A}))$, we get the equality

$$[M] \cap \mathsf{trunc}^*(\alpha) = \epsilon_*([N] \cap \alpha).$$

Let $U = U(N) \subset M$ be a regular closed neighborhood of N in M (defined via the second barycentric subdivision of T) and $\pi : U \to N$ its canonical retraction; thanks to the properties of the triangulation T, we get $\pi^{-1}(L) = K \cap U$.

Let $u \in H^k(U, \partial U; \mathbb{Z})$ be a relative cocycle, Poincaré-dual in U to the fundamental cycle $\epsilon_*([N]) \in H_n(U; \mathbb{Z}^t) \cong H_n(N; \mathbb{Z}^t)$. Then $[M] \cap u = \epsilon_*([N])$, and $\pi^*(\alpha) \cup u$ is a relative cocycle of dimension $p + k$ in $(U, \partial U)$. Since α vanishes on singular chains in L, the cocycle $\pi^*(\alpha) \cup u$ vanishes on singular chains on $\pi^{-1}(L) \cup \partial U = (K \cap U) \cup \partial U$. Therefore, $\pi^*(\alpha) \cup u$

extends to a relative cocycle on (M, K) that vanishes on $K \cup (M \setminus \mathsf{int}(U))$. Abusing notations, we denote this extension by $\pi^*(\alpha) \cup u$ as well.

For any $(p+k)$-cell e, we denote by e_U the intersection $e \cap U$. Note that the interior of e_U is an open $(p+k)$-cell, and e_U is a regular neighborhood of $e \cap N$ in e. Thanks to the constructions of T and U, it is equipped with the canonical projection $\pi : e_U \to e \cap N$ whose fiber is a k-dimensional PL-ball. Then, for any relative p-cocycle α that vanishes on the boundary $\partial(e \cap N)$, we get $(u \cup \pi^*(\alpha))(e_U) = \alpha(e \cap N)$. Thus, we conclude that the restriction $(u \cup \pi^*(\alpha))|_{e_U} = \mathsf{trunc}^*(\alpha)|_{e_U}$.

Next, we represent the fundamental class $[M]$ as a $(n+k)$-chain $\mathcal{M} = \sum_{\{e:\, \dim(e)=n+k\}} e$, where each $(n+k)$-cell e is viewed, with the help of the triangulation T, as a singular $(n+k)$-chain, a sum of singular $(n+k)$-simplexes.

Then, using the basic properties of cap and cup products ($\cap$ and $\cup$) (see [Sp]), we get

$$\mathcal{M} \cap \mathsf{trunc}^*(\alpha) = \sum_{\{e:\, \dim(e)=n+k\}} e \cap (\pi^*(\alpha) \cup u)$$

$$= \sum_{\{e:\, \dim(e)=n+k\}} (e \cap u) \cap \pi^*(\alpha) = \sum_{\{e:\, \dim(e)=n+k\}} e_u \cap \pi^*(\alpha),$$

where the n-chain $e_u =_{\mathsf{def}} e \cap u$. Applying π_* to the last expression, we get

$$\pi_*\big(\sum_e e_u \cap \pi^*(\alpha)\big) = \sum_e \pi_*(e_u \cap \pi^*(\alpha)) = \sum_e (\pi_*(e_u) \cap \alpha)$$

$$= \big(\sum_e \pi_*(e_u)\big) \cap \alpha = [N] \cap \alpha.$$

Since $\pi : U \to N$ is a homotopy and thus homology equivalence, the cycles $[M] \cap \mathsf{trunc}^*(\alpha)$ and $\epsilon_*([N] \cap \alpha)$ are homologous in $H_{n-j}(M \setminus K;\, \mathbb{A}^t)$. This completes the proof of the lemma. □

The commutativity of diagram (6.33) implies instantly the following claim.

Corollary 6.7. *Under the* Hypotheses $\nu\mathbf{A}$ *and notations of Lemma 6.8, the transfer* $H^j(N, L;\, \mathbb{A}) \stackrel{(\mathsf{trunc})^*}{\longrightarrow} H^{j+k}(M, K;\, \mathbb{A})$ *is an isomorphism if and only if the natural homomorphism* $H_{n-j}(N \setminus L;\, \epsilon^*(\mathbb{A}^t)) \stackrel{\epsilon_*}{\longrightarrow} H_{n-j}(M \setminus K;\, \mathbb{A}^t)$ *is.* ◇

Now we are in position to consider stabilization in cohomology for the spaces $\mathcal{B}_d^{\mathbf{c}\Theta} =_{\mathsf{def}} \mathcal{B}_d \setminus \mathcal{B}_d^{\Theta}$.

Theorem 6.9 (Dual short projective stabilization $\mathbf{d} \Rightarrow \mathbf{d+2}$). *Let $\Theta \subseteq \mathbf{\Omega}^{\infty}$ be a closed equal parity poset.*

Then for d of parity Θ, the embedding $\epsilon : \mathcal{B}_d^{\mathbf{c}\Theta} \subset \mathcal{B}_{d+2}^{\mathbf{c}\Theta}$, induces an isomorphism

$$\epsilon_* : H_k(\mathcal{B}_d^{\mathbf{c}\Theta};\, \epsilon^*(\mathbb{Z}^t)) \cong H_k(\mathcal{B}_{d+2}^{\mathbf{c}\Theta};\, \mathbb{Z}^t) \tag{6.35}$$

for all $k \leq d + 2 - \psi_\Theta(d+2)$.

If $d \equiv 1 \mod 2$, then one can replace in (6.35) the local coefficient system $\mathbb{Z}^t$ with the trivial $\mathbb{Z}[\pi_1(\mathcal{B}_{d+2}^{\mathbf{c}\Theta})]$-module $\mathbb{Z}$.

Proof. We apply Lemma 6.8 and Corollary 6.7 to the spaces of $\mathcal{B}_d$ and $\mathcal{B}_d^{\Theta}$, where $N = \mathcal{B}_d$, $M = \mathcal{B}_{d+2}$, $K = \mathcal{B}_{d+2}^{\Theta}$, and $L = \mathcal{B}_d^{\Theta}$. In order to apply Lemma 6.8, we use the fundamental fact that naturally stratified algebraic sets admit triangulations [Jo], [Har1], [Har2], consistent with their stratifications.

Let $\mathbb{A} = \mathbb{Z}$ in Corollary 6.7. We then get that

$$\epsilon_* : H_{d-j}(\mathcal{B}_d^{\mathbf{c}\Theta}; \epsilon^*(\mathbb{Z}^t)) \to H_{d-j}(\mathcal{B}_{d+2}^{\mathbf{c}\Theta}; \mathbb{Z}^t)$$

is an isomorphism if and only if

$$\mathsf{trunc}^* : H^j(\mathcal{B}_d, \mathcal{B}_d^{\Theta};\, \mathbb{Z}) \to H^{j+2}(\mathcal{B}_{d+2}, \mathcal{B}_{d+2}^{\Theta};\, \mathbb{Z})$$

is an isomorphism. By Theorem 6.7,

$$H_{j+2}(\mathcal{B}_{d+2}, \mathcal{B}_{d+2}^{\Theta}; \mathbb{Z}) \stackrel{(\mathsf{trunc})_*}{\longrightarrow} H_j(\mathcal{B}_d, \mathcal{B}_d^{\Theta}; \mathbb{Z})$$

is an isomorphism for all $j + 1 > \psi_\Theta(d+2)$. By the universal coefficient theorem,

$$H^j(\mathcal{B}_d, \mathcal{B}_d^{\Theta};\, \mathbb{Z}) \stackrel{(\mathsf{trunc})^*}{\longrightarrow} H^{j+2}(\mathcal{B}_{d+2}, \mathcal{B}_{d+2}^{\Theta};\, \mathbb{Z})$$

is an isomorphism for all $j + 1 > \psi_\Theta(d+2)$. Therefore, by Corollary 6.7,

$$\epsilon_* : H_k(\mathcal{B}_d^{\mathbf{c}\Theta}; \epsilon^*(\mathbb{Z}^t)) \cong H_k(\mathcal{B}_{d+2}^{\mathbf{c}\Theta}; \mathbb{Z}^t)$$

is an isomorphism for all $k = d+2-j$ and $j+1 > \psi_\Theta(d+2)$, which yields

$$k \leq d+2-\psi(d+2).$$

For $d > 1$, $\epsilon_* : \pi_1(\mathbb{RP}^d) \to \pi_1(\mathbb{RP}^{d+2})$ is an isomorphism between two copies of $\mathbb{Z}_2$. Note that the Stiefel-Whitney homomorphism $w : \pi_1(\mathcal{B}_d^{\mathbf{c}\Theta}) \to \mathbb{Z}_2$ factors through the isomorphism $\tilde{w} : \pi_1(\mathbb{RP}^d) \to \mathbb{Z}_2$, and the Stiefel-Whitney homomorphism $w^\# : \pi_1(\mathcal{B}_{d+2}^{\mathbf{c}\Theta}) \to \mathbb{Z}_2$ factors through the isomorphism $\tilde{w}^\# : \pi_1(\mathbb{RP}^{d+2}) \to \mathbb{Z}_2$. Therefore, $\epsilon^*(\mathbb{Z}^t) \cong \mathbb{Z}^t$ as $\mathbb{Z}[\pi_1(\mathcal{B}_d^{\mathbf{c}\Theta})]$-modules. Hence, if $d \equiv 1 \mod 2$, then one can replace in (6.35) the local coefficient system $\mathbb{Z}^t$ with the trivial $\mathbb{Z}[\pi_1(\mathcal{B}_{d+2}^{\mathbf{c}\Theta})]$-module $\mathbb{Z}$. □

Corollary 6.8 (Dual long stabilization d $\Rightarrow \infty$, the projective case). *Let $\Theta \subseteq \mathbf{\Omega}^\infty$ be a profinite (see Definition 6.5) closed equal parity poset.*

Then, for each $j > 0$, there is a sufficiently big number d of parity Θ such that the homomorphism

$$(\epsilon_{d,d'})_* : H_j(\mathcal{B}_d^{\mathbf{c}\Theta}; \epsilon^*(\mathbb{Z}^t)) \cong H_j(\mathcal{B}_{d'}^{\mathbf{c}\Theta}; \mathbb{Z}^t)$$

is an isomorphism for all $d' \geq d$ of parity Θ.

Proof. The assertion follows immediately from Lemma 6.6 and Theorem 6.9. □

Next, we derive the analogues of these stabilizations for the spaces of polynomials.

Theorem 6.10 (Dual short polynomial stabilization d $\Rightarrow$ d + 2). *Let Θ be a closed equal parity subposet of $\mathbf{\Omega}$. Then for d of parity Θ the embedding $\epsilon : \mathcal{P}_d^{\mathbf{c}\Theta} \subset \mathcal{P}_{d+2}^{\mathbf{c}\Theta}$, induces an isomorphism*

$$\epsilon_* : H_k(\mathcal{P}_d^{\mathbf{c}\Theta}; \mathbb{Z}) \cong H_k(\mathcal{P}_{d+2}^{\mathbf{c}\Theta}; \mathbb{Z})$$

for all $k \leq d+2-\psi_\Theta(d+2)$.

Proof. The long exact cohomology sequences of the pairs $(\bar{\mathcal{P}}_{d+2}, \bar{\mathcal{P}}_{d+2}^{\Theta})$ and $(\bar{\mathcal{P}}_d, \bar{\mathcal{P}}_d^{\Theta})$, linked by the vertical $(\mathsf{trunc})^*$ homomorphisms, lead to a commutative diagram analogous to the one in Fig. 6.5. Then combining Corollary 6.6 with the Five Lemma validates the first claim. Applying the diagram (6.33) from Lemma 6.8, proves the second claim. □

Example 6.8. Let the closed poset $\Theta = \langle (4), (33) \rangle$ be generated by two elements: (4) and (33). Then $\eta_\Theta(d) = -2$, $\psi_\Theta(d) = \frac{1}{2}(d-2)$. Applying Theorem 6.10 to this Θ, we get isomorphisms

$$(\epsilon_{d,d+2})_* : H_j(\mathcal{P}_d^{\mathbf{c}\Theta}; \mathbb{Z}) \approx H_j(\mathcal{P}_{d+2}^{\mathbf{c}\Theta}; \mathbb{Z})$$

for all $j \leq \frac{1}{2}d + 2$. Moreover, since $\bar{\mathcal{P}}_d^{\Theta}$ is of codimension 3 in $\bar{\mathcal{P}}_d \approx S^d$, we get $\pi_1(\mathcal{P}_d^{\mathbf{c}\Theta}) = 1$ for all $d \geq 6$. Therefore, we get the homotopy groups isomorpisms

$$(\epsilon_{d,d+2})_* : \pi_j(\mathcal{P}_d^{\mathbf{c}\Theta}) \approx \pi_j(\mathcal{P}_{d+2}^{\mathbf{c}\Theta})$$

for all $j \leq \frac{1}{2}d + 2$, where $d \geq 6$. ◇

Corollary 6.9 (Dual long stabilization d $\Rightarrow \infty$, the polynomial case).

For any profinite *closed* Θ *and for each* $j > 0$, *there is a sufficiently big number* d *of parity* Θ *such that the homomorphism*

$$(\epsilon_{d,d'})_* : H_j(\mathcal{P}_d^{\mathbf{c}\Theta}; \mathbb{Z}) \to H_j(\mathcal{P}_{d'}^{\mathbf{c}\Theta}; \mathbb{Z})$$

is an isomorphism for all $d' \geq d$ *of parity* Θ.

Proof. The assertion follows immediately from Lemma 6.6 and Theorem 6.10. □

6.5.1 *Software-assisted results for* $H_*((\bar{\mathcal{P}}_d^{\langle\omega\rangle}; \mathbb{Z}))$

Now, let us present somewhat surprising results of computer-assisted computations from [KSW2], dealing with the simplest possible poset Θ's. They are organized in tables displayed in Figures 6.7, 6.8, and 6.9.

For a given $\omega \in \mathbf{\Omega}$, we denote by $\langle\omega\rangle$ the minimal closed poset that contains ω. Below, for $d \leq 13$, we list all compositions ω for which the space $\mathcal{P}_d^{\langle\omega\rangle}$ is *homologically nontrivial.* In fact, for all $d \leq 13$, every homologically nontrivial $\mathcal{P}_d^{\langle\omega\rangle}$ has its homology concentrated in one degree, where it is free of rank one. By [KSW2], for ω's from the tables with $|\omega|' > 2$, the spaces $\mathcal{P}_d^{\mathbf{c}\langle\omega\rangle}$ are homotopy spheres! While we have theoretical arguments for some cases, the reason for this general phenomenon is unknown to us.

• To make the writing of ω's more compact, we adopt the following notation: for example, we abbreviate $\omega = (1132555)$ as $(1^2 325^3)$.

d	Codimension	ω	i with $\widetilde{H}_i = \mathbb{Z}$	ω	i with $\widetilde{H}_i = \mathbb{Z}$
4	0	(1^2)	3		
	1	(2)	3		
	3	(4)	1		
5	0	(1^3)	4		
	4	(5)	1		
6	0	(1^4)	5	(1^2)	5
	1	(2)	5		
	5	(6)	1		
7	0	(1^5)	6	(1^3)	6
	6	(7)	1		
8	0	(1^6)	7	(1^4)	7
		(1^2)	7		
	1	$(1, 2, 1)$	3	(2)	7
	2	$(1, 3)$	3	$(3, 1)$	3
	3	(4)	3		
	7	(8)	1		
9	0	(1^7)	8	(1^5)	8
		(1^3)	8		
	1	$(1^2, 2, 1)$	4	$(1, 2, 1^2)$	4
	2	$(1^2, 3)$	4	$(1, 3, 1)$	4
		$(3, 1^2)$	4		
	8	(9)	1		
10	0	(1^8)	9	(1^6)	9
		(1^4)	9	(1^2)	9
	1	$(1^3, 2, 1)$	5	$(1^2, 2, 1^2)$	5
		$(1, 2, 1^3)$	5	(2)	9
	2	$(1^3, 3)$	5	$(1^2, 3, 1)$	5
		$(1, 3, 1^2)$	5	$(3, 1^3)$	5
	9	(10)	1		

Fig. 6.7 The list of the ω's and of the corresponding unique homological degrees $i = i(\omega)$ for which $\widetilde{H}_i(\bar{\mathcal{P}}_d^{\langle\omega\rangle}; \mathbb{Z}) = \mathbb{Z}$ in case $d \leq 10$. For ω's and i's not listed above the homology vanishes.

d	Codimension	ω	i with $\widetilde{H}_i = \mathbb{Z}$	ω	i with $\widetilde{H}_i = \mathbb{Z}$
11	0	(1^9)	10	(1^7)	10
		(1^5)	10	(1^3)	10
	1	$(1^4, 2, 1)$	6	$(1^3, 2, 1^2)$	6
		$(1^2, 2, 1^3)$	6	$(1, 2, 1^4)$	6
	2	$(1^4, 3)$	6	$(1^3, 3, 1)$	6
		$(1^2, 3, 1^2)$	6	$(1, 3, 1^3)$	6
		$(3, 1^4)$	6	$(1, 2, 1, 2, 1)$	4
	3	$(1, 2, 1, 3)$	4	$(3, 1, 2, 1)$	4
	4	$(3, 1, 3)$	4		
	10	(11)	1		
12	0	(1^{10})	11	(1^8)	11
		(1^6)	11	(1^4)	11
		(1^2)	11		
	1	$(1^5, 2, 1)$	7	$(1^4, 2, 1^2)$	7
		$(1^3, 2, 1^3)$	7	$(1^2, 2, 1^4)$	7
		$(1, 2, 1^5)$	7	$(1, 2, 1)$	5
		(2)	11		
	2	$(1^5, 3)$	7	$(1^4, 3, 1)$	7
		$(1^3, 3, 1^2)$	7	$(1^2, 3, 1^3)$	7
		$(1, 3, 1^4)$	7	$(3, 1^5)$	7
		$(1, 2, 1^2, 2, 1)$	5	$(1, 2, 1, 2, 1^2)$	5
		$(1, 2^2, 1)$	3	$(1, 3)$	5
		$(3, 1)$	5		
	3	$(1^2, 2, 1, 3)$	5	$(1, 2, 1^2, 3)$	5
		$(1, 2, 1, 3, 1)$	5	$(1, 3, 1, 2, 1)$	5
		$(3, 1^2, 2, 1)$	5	$(3, 1, 2, 1^2)$	5
		$(1, 2, 3)$	3	$(3, 2, 1)$	3
		$(1, 4, 1)$	3	(4)	5
	4	$(1, 3, 1, 3)$	5	$(3, 1, 3, 1)$	5
		$(1, 5)$	3	$(5, 1)$	3
		(3^2)	3		
	5	(6)	3		
	11	(12)	1		

Fig. 6.8 The list of the ω's and of the corresponding unique homological degrees $i = i(\omega)$ for which $\widetilde{H}_i(\bar{\mathcal{P}}_d^{\langle\omega\rangle}; \mathbb{Z}) = \mathbb{Z}$ for $d = 11, 12$. For ω's and i's not listed above, the homology vanishes.

d	Codimension	ω	i with $\widetilde{H}_i = \mathbb{Z}$	ω	i with $\widetilde{H}_i = \mathbb{Z}$
13	0	(1^{11})	12	(1^9)	12
		(1^7)	12	(1^5)	12
		(1^3)	12		
	1	$(1^6, 2, 1)$	8	$(1^5, 2, 1^2)$	8
		$(1^4, 2, 1^3)$	8	$(1^3, 2, 1^4)$	8
		$(1^2, 2, 1^5)$	8	$(1, 2, 1^6)$	8
		$(1^2, 2, 1)$	6	$(1, 2, 1^2)$	6
	2	$(1^6, 3)$	8	$(1^5, 3, 1)$	8
		$(1^4, 3, 1^2)$	8	$(1^3, 3, 1^3)$	8
		$(1^2, 3, 1^4)$	8	$(1, 3, 1^5)$	8
		$(3, 1^6)$	8	$(1^2, 3)$	6
		$(1, 3, 1)$	6	$(3, 1^2)$	6
		$(1^3, 2, 1, 2, 1)$	6	$(1^2, 2, 1^2, 2, 1)$	6
		$(1^2, 2, 1, 2, 1^2)$	6	$(1, 2, 1^3, 2, 1)$	6
		$(1, 2, 1^2, 2, 1^2)$	6	$(1, 2, 1, 2, 1^3)$	6
		$(1^2, 2^2, 1)$	4	$(1, 2^2, 1^2)$	4
	3	$(1^3, 2, 1, 3)$	6	$(1^2, 2, 1^2, 3)$	6
		$(1^2, 2, 1, 3, 1)$	6	$(1^2, 3, 1, 2, 1)$	6
		$(1, 2, 1^2, 3, 1)$	6	$(1, 2, 1, 3, 1^2)$	6
		$(1, 3, 1^2, 2, 1)$	6	$(1, 3, 1, 2, 1^2)$	6
		$(3, 1^2, 2, 1^2)$	6	$(3, 1, 2, 1^3)$	6
		$(1^2, 2, 3)$	4	$(1, 2, 3, 1)$	4
		$(1, 3, 2, 1)$	4	$(3, 2, 1^2)$	4
		$(1^2, 4, 1)$	4	$(1, 4, 1^2)$	4
	4	$(1^2, 3, 1, 3)$	6	$(1, 3, 1, 3, 1)$	6
		$(3, 1, 3, 1^2)$	6	$(1, 3^2)$	4
		$(3^2, 1)$	4	$(1^2, 5)$	4
		$(1, 5, 1)$	4	$(5, 1^2)$	4
	12	(13)	1		

Fig. 6.9 The list of the ω's and of the corresponding unique homological degrees $i = i(\omega)$ for which $\widetilde{H}_i(\bar{\mathcal{P}}_d^{\langle\omega\rangle}; \mathbb{Z}) = \mathbb{Z}$ for $d = 13$. For the rest of ω's and i's, the homology vanishes.

6.5.2 *Topology of* $\mathcal{P}_d^{\mathbf{c}\langle\omega\rangle}$ *for small d*

The list below is produced from the previous three tables with the help of the Alexander duality. In (6.36), for $d \leq 13$, we list all compositions ω for which the space $\mathcal{P}_d^{\mathbf{c}\langle\omega\rangle}$ is *homologically nontrivial*. In fact, for $d \leq 13$, every homologically nontrivial $\mathcal{P}_d^{\mathbf{c}\langle\omega\rangle}$ is a *homology* sphere. Moreover, at least for ω with $|\omega|' > 2$, all such spaces $\mathcal{P}_d^{\mathbf{c}\langle\omega\rangle}$ are *homotopy spheres.*

In the list (6.36) below, we denote by "$\mathbf{S^k}$" the homology k-spheres. We use the notation "$\mathbf{S^k_\star}$" for the homotopy k-spheres.

Nontrivial $\mathbb{Z}$-homology types of $\mathcal{P}_d^{\mathbf{c}\langle\omega\rangle}$ for $4 \leq d \leq 13$: (6.36)

d = 4.

$|\omega|' = 0$: $\omega = (1^2) \Rightarrow \mathbf{S^0}$;
$|\omega|' = 1$: $\omega = (2) \Rightarrow \mathbf{S^0}$;
$|\omega|' = 3$: $\omega = (4) \Rightarrow \mathbf{S^2_\star}$;

d = 5.

$|\omega|' = 0$: $\omega = (1^3) \Rightarrow \mathbf{S^0}$;
$|\omega|' = 4$: $\omega = (5) \Rightarrow \mathbf{S^3_\star}$;

d = 6.

$|\omega|' = 0$: $\omega = (1^4), (1^2) \Rightarrow \mathbf{S^0}$;
$|\omega|' = 1$: $\omega = (2) \Rightarrow \mathbf{S^0}$;
$|\omega|' = 5$: $\omega = (6) \Rightarrow \mathbf{S^4_\star}$;

d = 7.

$|\omega|' = 0$: $\omega = (1^5), (1^3) \Rightarrow \mathbf{S^0}$;
$|\omega|' = 6$: $\omega = (7) \Rightarrow \mathbf{S^5_\star}$;

d = 8.

$|\omega|' = 0$: $\omega = (1^6), (1^4), (1^2) \Rightarrow \mathbf{S^0}$;
$|\omega|' = 1$: $\omega = (1, 2, 1) \Rightarrow \mathbf{S^4}$ $\quad \omega = (2) \Rightarrow \mathbf{S^0}$;
$|\omega|' = 2$: $\omega = (1, 3), (3, 1) \Rightarrow \mathbf{S^4}$;
$|\omega|' = 3$: $\omega = (4) \Rightarrow \mathbf{S^4_\star}$;
$|\omega|' = 7$: $\omega = (8) \Rightarrow \mathbf{S^6_\star}$.

d = 9.

$|\omega|' = 0$: $\omega = (1^7), (1^5), (1^3) \Rightarrow \mathbf{S^0}$;
$|\omega|' = 1$: $\omega = (1^2, 2, 1), (1, 2, 1^2), \Rightarrow \mathbf{S^4}$;
$|\omega|' = 2$: $\omega = (1^2, 3), (1, 3, 1), (3, 1^2) \Rightarrow \mathbf{S^4}$;
$|\omega|' = 8$: $\omega = (9) \Rightarrow \mathbf{S^7_\star}$.

d = 10.

$|\omega|' = 0$: $\omega = (1^8), (1^6), (1^4), (1^2) \Rightarrow \mathbf{S^0}$;
$|\omega|' = 1$: $\omega = (1^3, 2, 1), (1^2, 2, 1^2), (1, 2, 1^3), \Rightarrow \mathbf{S^4}$; $\quad \omega = (2) \Rightarrow \mathbf{S^0}$;
$|\omega|' = 2$: $\omega = (1^3, 3), (1^2, 3, 1), (1, 3, 1^2), (3, 1^3) \Rightarrow \mathbf{S^4}$;
$|\omega|' = 9$: $\omega = (10) \Rightarrow \mathbf{S^8_\star}$.

d = 11.

$|\omega|' = 0$: $\omega = (1^9), (1^7), (1^5), (1^3) \Rightarrow \mathbf{S^0}$;
$|\omega|' = 1$: $\omega = (1^4, 2, 1), (1^3, 2, 1^2), (1^2, 2, 1^3), (1, 2, 1^4) \Rightarrow \mathbf{S^4}$;
$|\omega|' = 2$: $\omega = (1^4, 3), (1^3, 3, 1), (1^2, 3, 1^2), (1, 3, 1^3), (3, 1^4) \Rightarrow \mathbf{S^4}$;
$\omega = (1, 2, 1, 2, 1) \Rightarrow \mathbf{S^6}$;
$|\omega|' = 3$: $\omega = (1, 2, 1, 3), (3, 1, 2, 1) \Rightarrow \mathbf{S^6_\star}$;
$|\omega|' = 4$: $\omega = (3, 1, 3) \Rightarrow \mathbf{S^6_\star}$;
$|\omega|' = 10$: $\omega = (11) \Rightarrow \mathbf{S^9_\star}$.

d = 12.

$|\omega|' = 0$: $\omega = (1^{10}), (1^8), (1^6), (1^4), (1^2) \Rightarrow \mathbf{S^0}$;
$|\omega|' = 1$: $\omega = (1^5, 2, 1), (1^4, 2, 1^2), (1^3, 2, 1^3), (1^2, 2, 1^4), (1, 2, 1^5) \Rightarrow \mathbf{S^4}$;
$\omega = (1, 2, 1) \Rightarrow \mathbf{S^6}$;
$\omega = (2) \Rightarrow \mathbf{S^0}$;
$|\omega|' = 2$: $\omega = (1^5, 3), (1^4, 3, 1), (1^3, 3, 1^2), (1^2, 3, 1^3), (1, 3, 1^4), (3, 1^5) \Rightarrow \mathbf{S^4}$;
$\omega = (1^2, 2, 1, 2, 1), (1, 2, 1^2, 2, 1), (1, 2, 1, 2, 1^2), (1, 3), (3, 1) \Rightarrow \mathbf{S^6}$;
$\omega = (1, 2^2, 1) \Rightarrow \mathbf{S^8}$;
$|\omega|' = 3$: $\omega = (1^2, 2, 1, 3), (1, 2, 1^2, 3), (1, 2, 1, 3, 1), (1, 3, 1, 2, 1), (3, 1^2, 2, 1),$
$(3, 1, 2, 1^2), (4) \Rightarrow \mathbf{S^6_\star}$;
$\omega = (1, 2, 3), (3, 2, 1), (1, 4, 1) \Rightarrow \mathbf{S^8_\star}$;
$|\omega|' = 4$: $\omega = (1, 3, 1, 3), (3, 1, 3, 1) \Rightarrow \mathbf{S^6_\star}$;
$\omega = (1, 5), (5, 1), (3^2) \Rightarrow \mathbf{S^8_\star}$;
$|\omega|' = 5$: $\omega = (6) \Rightarrow \mathbf{S^8_\star}$;
$|\omega|' = 11$: $\omega = (12) \Rightarrow \mathbf{S^{10}_\star}$.

d = 13.

$|\omega|' = 0$: $\omega = (1^{11}), (1^9), (1^7), (1^5), (1^3) \Rightarrow \mathbf{S^0}$;
$|\omega|' = 1$: $\omega = (1^6, 2, 1), (1^5, 2, 1^2), (1^4, 2, 1^3), (1^3, 2, 1^4), (1^2, 2, 1^5), (1, 2, 1^6) \Rightarrow \mathbf{S^4}$;
$\omega = (1^2, 2, 1), (1, 2, 1^2) \Rightarrow \mathbf{S^6}$;
$|\omega|' = 2$: $\omega = (1^6, 3), (1^5, 3, 1), (1^4, 3, 1^2), (1^2, 1, 3, 1^3), (1^2, 3, 1^4), (1, 3, 1^5),$
$(3, 1^6) \Rightarrow \mathbf{S^4}$;
$\omega = (1^2, 3), (1, 3, 1), (3, 1^2), (1^3, 2, 1, 2, 1), (1^2, 2, 1^2, 2, 1), (1^2, 2, 1, 2, 1^2),$
$(1, 2, 1^3, 2, 1), (1, 2, 1^2, 2, 1^2), (1, 2, 1, 2, 1^3) \Rightarrow \mathbf{S^6}$;
$\omega = (1^2, 2^2, 1), (1, 2^2, 1^2) \Rightarrow \mathbf{S^8}$;
$|\omega|' = 3$: $\omega = (1^3, 2, 1, 3), (1^2, 2, 1^2, 3), (1^2, 2, 1, 3, 1), (1^2, 3, 1, 2, 1), (1, 2, 1^2, 3, 1),$
$(1, 2, 1, 3, 1^2), (1, 3, 1^2, 2, 1), (1, 3, 1, 2, 1^2), (3, 1^2, 2, 1^2), (3, 1, 2, 1^3) \Rightarrow \mathbf{S^6_\star}$;
$\omega = (1^2, 2, 3), (1, 2, 3, 1), (1, 3, 2, 1), (3, 2, 1^2), (1^2, 4, 1), (1, 4, 1^2) \Rightarrow \mathbf{S^8_\star}$;
$|\omega|' = 4$: $\omega = (1^2, 3, 1, 3), (1, 3, 1, 3, 1), (3, 1, 3, 1^2), \Rightarrow \mathbf{S^6_\star}$;
$\omega = (1, 3^2), (3^2, 1), (1^2, 5), (1, 5, 1), (5, 1^2) \Rightarrow \mathbf{S^8_\star}$;
$|\omega|' = 12$: $\omega = (13) \Rightarrow \mathbf{S^{11}_\star}$.

◇

6.6 On the homotopy types of $\mathcal{P}_d^{c\Theta}$

We have seen on many occasions that, in contrast to $\mathcal{P}_d^{\Theta}$, its one-point compactification $\bar{\mathcal{P}}_d^{\Theta}$ often has non-trivial topology for closed posets Θ. The spaces $\mathcal{P}_d^{c\Theta}$ also exhibit a non-trivial topology.

In what follows, we use the map $q : \mathcal{P}_d \times [0, \infty) \to \mathcal{P}_d$ which sends each pair $(P(x), \lambda)$, where $P(x) = x^d + a_{d-1}x^{d-1} + \cdots + a_0$ and $\lambda \in \mathbb{R}$, to the polynomial

$$x^d + a_{d-1}\,\lambda\, x^{d-1} + \cdots + a_0\,\lambda^d.$$

Hence, this transformation amounts to the multiplication of all roots of $P(x)$ by λ.

The simple statement below gives us a start on the homotopy type of the polynomial spaces under consideration.

Lemma 6.9. *For any poset $\Theta \subset \mathbf{\Omega}_{\langle d]}$ that contains the element (d), the space $\mathcal{P}_d^{\Theta} \subset \mathcal{P}_d$ is contractible. In particular, for any closed poset $\Theta \subset \mathbf{\Omega}_{\langle d]}$, the space $\mathcal{P}_d^{\Theta}$ is contractible.*

Proof. For $P(x) \in \mathcal{P}_d^{\Theta}$ and $\lambda \geq 0$, the roots of $q(P(x), \lambda)$ are the roots of $P(x)$, being multiplied by λ. In particular, we have $q(P(x), 0) = x^d \in \mathcal{P}_d^{(d)}$. Since $(d) \in \Theta$, we have $q(P(x), \lambda) \in \mathcal{P}_d^{\Theta}$ for $P(x) \in \mathcal{P}_d^{\Theta}$. Then the restriction of q to $[0, 1] \times \mathcal{P}_d^{\Theta}$ is a well-defined homotopy between the identity map and the constant map that sends $\mathcal{P}_d^{\Theta}$ to x^d. The assertion now follows. □

Lemma 6.10. *Let ΣX denotes the suspension and $\tilde{\Sigma} X$ denotes the reduced suspension of a space X. Let $\mathcal{P}_{d,0}^{\Theta}$ consists of all polynomials in $\mathcal{P}_d^{\Theta}$ with the vanishing coefficient at x^{d-1}.*

Then, for any poset Θ, we have two homeomorphisms:

$$\bar{\mathcal{P}}_d^{\Theta} \cong \tilde{\Sigma}\bar{\mathcal{P}}_{d,0}^{\Theta}.$$

$$\bar{\mathcal{P}}_{d,0}^{\Theta} \cong \Sigma(S^{d-1} \cap \mathcal{P}_{d,0}^{\Theta}). \tag{6.37}$$

Proof. Besides the map q, the following map p has been frequently used in the literature on the topology of spaces of univariate polynomials. The map $p : \mathcal{P}_d \to \mathcal{P}_d$ sends

$$P(x) = x^d + a_{d-1}x^{d-1} + \cdots + a_0 \in \mathbb{R}[x]$$

to $P(x - \frac{a_{d-1}}{d})$. The map p, preserving the stratification $\{\mathring{\mathsf{R}}_d^{\omega}\}_{\omega}$, is a fibration with the fiber $\mathbb{R}$. Thus, for a closed poset $\Theta \subseteq \mathbf{\Omega}_{\langle d]}$, the restriction

$p|_{\mathcal{P}_d^\Theta} : \mathcal{P}_d^\Theta \to \mathcal{P}_d^\Theta$ is also a fibration with fiber $\mathbb{R}$. Its image $\mathcal{P}_{d,0}^\Theta$ consists of all polynomials in $\mathcal{P}_d^\Theta$ with the vanishing coefficient at x^{d-1}, i.e., with vanishing root sum. Therefore, we get a *homeomorphism* $\bar{\mathcal{P}}_d^\Theta \cong \tilde{\Sigma}\bar{\mathcal{P}}_{d,0}^\Theta$. Note, that we consider $\Sigma\,\emptyset$ as the discrete two-point space.

Since $q(P(x), \lambda)$ amounts to multiplying the roots of $P(x)$ by λ, their sum is multiplied by λ as well. In particular, the map q preserves $\mathcal{P}_{d,0}^\Theta$. Using the map

$$a_0 + \cdots + a_{d-1}x^{d-1} + x^d \mapsto (a_0, \ldots, a_{d-1}),$$

we identify $\mathcal{P}_d$ with Euclidean d-space, equipped with its metric structure. Clearly, for $\lambda \geq 0$ and fixed $0 \neq P(x) \in \mathcal{P}_d$, the norm $\|q(P(x), \lambda)\|$ is strictly monotone in λ. Since $\|q(P(x), 0)\| = 0$, there is a unique $\lambda_P > 0$ for which $\|q(P(x), \lambda_P)\| = 1$. Let $S^{d-1} \subset \mathcal{P}_d$ be the unit sphere. Therefore, on $\mathcal{P}_{d,0}^\Theta \setminus \{0\}$ the map $P(x) \mapsto q(P(x), \lambda_P)$ is a deformation retraction to the closed subspace $S^{d-1} \cap \mathcal{P}_{d,0}^\Theta$ of S^{d-1}. Thus, we get a *homeomorphism* $\bar{\mathcal{P}}_{d,0}^\Theta \cong \Sigma(S^{d-1} \cap \mathcal{P}_{d,0}^\Theta)$. □

Based on our homological investigations, we present special cases of Θ where we can describe the homotopy type of $\mathcal{P}_d^{\mathbf{c}\Theta}$.

For $k \in [0, d-1]$, $q \in [0, d]$, and $q \equiv d \mod 2$, let us consider the closed poset

$$\mathbf{\Omega}^{(q)}_{|\sim|' \geq k} =_{\mathsf{def}} \big\{\omega \in \mathbf{\Omega}_{\langle d]} \text{ such that } |\omega|' \geq k \text{ and } |\omega| \geq q\big\}. \tag{6.38}$$

Note that, for $\Theta = \mathbf{\Omega}^{(0)}_{|\sim|' \geq k} =_{\mathsf{def}} \mathbf{\Omega}_{|\sim|' \geq k}$, the space $\bar{\mathcal{P}}_d^\Theta$ is the entire $(d-k)$-skeleton of $\bar{\mathcal{P}}_d$. Generalizing this observation, let

$$A(d, k, q) =_{\mathsf{def}} \big|\chi\big((\mathbb{Z}\big[\mathbf{\Omega}^{(q)}_{|\sim|' \geq k}\big], \partial)\big)\big|, \tag{6.39}$$

the absolute value of the Euler number of the differential complex $\big(\mathbb{Z}\big[\mathbf{\Omega}^{(q)}_{|\sim|' \geq k}\big], \partial\big)$.

Proposition 6.6. *For $k \in [3, d)$ and $q \in [0, d]$, $q \equiv d \mod 2$, let $\Theta = \mathbf{\Omega}^{(q)}_{|\sim|' \geq k}$ be the closed poset from (6.38).*

Then $\mathcal{P}_d^{\mathbf{c}\Theta}$ has the homotopy type of a bouquet of $(k-1)$-spheres. The number $A(d, k, q)$ of spheres in the bouquet equals the absolute value of the reduced by one Euler characteristic of $\bar{\mathcal{P}}_d^\Theta$, given by (6.39).

Proof. Since, by the definition of Θ, $\bar{\mathcal{P}}_d^\Theta$ is the $(d-k)$-skeleton of $\bar{\mathcal{P}}_d \cong S^d$, the cohomology of $\bar{\mathcal{P}}_d^\Theta$ is torsion-free and is concentrated in a single dimension $d-k$. By the Alexander duality, the homology of $\mathcal{P}_d^{\mathbf{c}\Theta}$ is torsion-free and is concentrated in a single dimension $k-1$. By a general position argument, the space $\mathcal{P}_d^{\mathbf{c}\Theta}$ is simply-connected for $k>2$. By [Ha], Theorem 4C.1, this implies that $\mathcal{P}_d^{\mathbf{c}\Theta}$ has the homotopy type of a bouquet of $(k-1)$-spheres. By Alexander duality, the number of spheres in the bouquet equals the absolute value of the reduced Euler characteristic of the CW-complex $\bar{\mathcal{P}}_d^{\Theta}$. By Corollary 6.6, the latter can be computed from the combinatorial differential complex $(\mathbb{Z}[\Theta], \partial)$. □

6.7 The fundamental groups $\pi_1(\mathcal{P}_d^{\mathbf{c}\Theta})$

Next, we concentrate on the fundamental groups of the spaces $\mathcal{P}_d^{\mathbf{c}\Theta}$ and of some related topological spaces. All the results in the section are contained in [KSW1].

We will find explicit presentations for the groups $\pi_1(\mathcal{P}_d^{\mathbf{c}\Theta})$ in terms of generators and relations and will show that in a number of interesting cases they are free with rank bounded from above by a quadratic function in d. We also will show that the fundamental groups $\{\pi_1(\mathcal{P}_d^{\mathbf{c}\Theta})\}_d$ stabilize for large d.

The mechanism by which the fundamental groups $\pi_1(\mathcal{P}_d^{\mathbf{c}\Theta})$ are generated is, in some sense, similar to the one that produces the braid groups as the fundamental groups of spaces of complex degree d monic polynomials with no multiple roots. This mechanism is also related to Khovanov's and Thomas' paper [Kh], which studies the topology of spaces K_d, obtained from the space $\mathbb{R}^d$ by removing vectors $(x_1, \dots, x_d)$ such that $x_j = x_k = x_l$ for some distinct j, k, l, or/and vectors of the form $x_j = x_k$ and $x_l = x_m$ for some distinct j, k, l, m.

This space is an ordered analog of the space of real polynomials containing the polynomials with only real roots, no roots of multiplicity 3 or higher, and no two different roots both of multiplicity 2 or higher. Fig. 6.10 shows

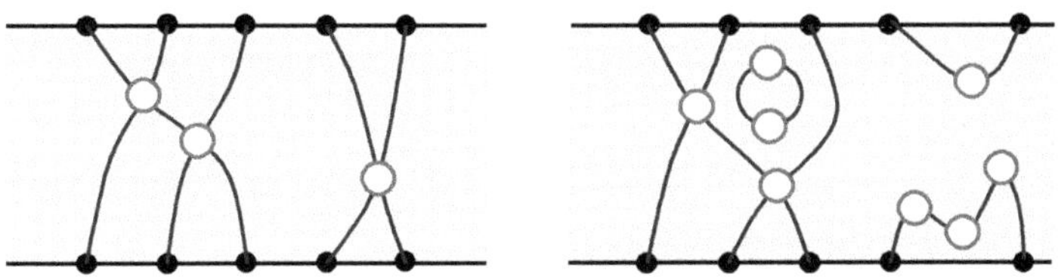

Fig. 6.10 (a) A graphic representation of a loop in the Khovanov's space K_5, and (b) in the space $\mathcal{P}_7^{\mathbf{c}\Theta}$ of real degree 7 polynomials with no real roots of multiplicity ≥ 3 and no pairs of real roots of multiplicity ≥ 2.

two contrasting images of objects from [Kh] and from the present section, each representing a *loop* in the relevant configuration space.

In Chapter 7, we will show that the groups $\pi_1(\mathcal{P}_d^{\mathbf{c}\Theta})$ admit an interpretation as special *bordisms of immersions* of 1-manifolds into the cylinder $S^1 \times \mathbb{R}$, immersions whose images avoid the tangency patterns from the poset Θ with respect to the generators of the cylinder.

Theorem 6.11. [*Ar1*] *The fundamental group of the space of smooth functions $f : S^1 \to \mathbb{R}$ without critical points of multiplicity ≥ 3 on a circle S^1 is isomorphic to the group of integers $\mathbb{Z}$.*

The space of smooth functions $f : \mathbb{R} \to \mathbb{R}$ without critical points of multiplicity ≥ 3 and which, for arguments $|x| > 1$, coincide either with x or with x^2 also have the fundamental group $\mathbb{Z}$. ◇

Theorem 6.12. [*Ar1*] *The latter fundamental group is naturally isomorphic to the group $\mathcal{B}$ of A_3-cobordism classes of embedded closed plane curves without vertical tangential inflections.* ◇

The generator of $\mathcal{B}$ is shown as the kidney-shaped loop in Fig. 7.4, (a).

6.8 Computing $\pi_1(\bar{\mathcal{P}}_d^{\Theta})$ and $\pi_1(\mathcal{P}_d^{\mathbf{c}\Theta})$

Lemmas 6.9 and 6.10 imply the following claim.

Proposition 6.7. *For any closed poset $\Theta \subseteq \mathbf{\Omega}_{\langle d]}$, we get $\pi_1(\bar{\mathcal{P}}_d^{\Theta}) = 0$, unless $\Theta = \{(d)\}$. If $\Theta = \{(d)\}$ then $\bar{\mathcal{P}}_d^{\Theta} \cong S^1$.*

Proof. By Lemma 6.10, we have the homeomorphism $\bar{\mathcal{P}}_d^{\Theta} \cong \tilde{\Sigma}\Sigma(S^{d-1} \cap \mathcal{P}_{d,0}^{\Theta})$. Therefore, $\bar{\mathcal{P}}_d^{\Theta}$ is simply connected, unless $S^{d-1} \cap P_{d,0}^{\Theta}$ is empty. But this can only happen if $\Theta = \{(d)\}$. It is easily seen that $\mathring{\mathsf{R}}_d^{(d)} = \{(x-\alpha)^d \mid \alpha \in \mathbb{R}\} \cong \mathbb{R}$. Hence its one-point compactification is S^1. □

Note, that the argument, employed in the proof of Proposition 6.7, also implies an isomorphism in homology

$$\tilde{H}_i(S^{d-1} \cap P_{d,0}^{\Theta}; \mathbb{Z}) \approx \tilde{H}_{i+2}(\bar{\mathcal{P}}_d^{\Theta}; \mathbb{Z})$$

for all $i \geq 0$.

6.8.1 The fundamental group of the complement of the codimension two skeleton of the space $\mathcal{P}_d$

In this subsection, we describe the fundamental group $\pi_1(\mathcal{P}_d^{\mathbf{c\Omega}_{\langle d],\,|\sim|'\geq 2}})$. Recall, that we collect in $\mathbf{\Omega}_{\langle d],\,|\sim|'\geq 2}$ all $\omega \in \mathbf{\Omega}_{\langle d]}$ such that $|\omega|' \geq 2$. Thus $\mathcal{P}_d^{\mathbf{c\Omega}_{\langle d],\,|\sim|'\geq 2}}$ is the complement of the codimension 2-skeleton of $\mathcal{P}_d$ in its ω-labelled cellulation.

First, let us outline the arguments below. The space $\mathcal{P}_d^{\mathbf{c}\Theta}$ is an open d-dimensional manifold without boundary, stratified into cells of dimensions d and $d-1$. Moreover, for any pair of top-dimensional cells, there exists at most one $(d-1)$-dimensional cell which separates them. Any such manifold is homotopy equivalent to a graph whose vertices are the top-dimensional cells and whose edges connect pairs of adjacent top-dimensional cells.

We associate a graph $\mathfrak{G}_d$ with the space $\mathcal{P}_d^{\mathbf{c\Omega}_{\langle d],\,|\sim|'\geq 2}}$, subdivided into open d-cells by the $(d-1)$-cells/walls. The set of vertices of $\mathfrak{G}_d$ is the union of the sets $\mathbf{\Omega}_{\langle d],\,|\sim|'=1}$ and $\mathbf{\Omega}_{\langle d],\,|\sim|'=0}$. We connect vertices $\omega \in \mathbf{\Omega}_{\langle d],\,|\sim|'=1}$ and $\omega' \in \mathbf{\Omega}_{\langle d],\,|\sim|'=0}$ by an edge $\{\omega, \omega'\}$, if the $(d-1)$-cell $\mathring{\mathsf{R}}_d^{\omega}$ lies in the boundary of the closure $\mathsf{R}_d^{\omega'}$ of $\mathring{\mathsf{R}}_d^{\omega'}$. In particular, the edges of $\mathfrak{G}_d$ correspond to single insertion or merging operations, applied to compositions from $\mathbf{\Omega}_{\langle d],\,|\sim|'=0}$. As usual, we identify the graph $\mathfrak{G}_d$ with the 1-dimensional simplicial complex, defined by its vertices and edges (see Fig. 6.11 for the example of $\mathfrak{G}_6$).

We embed the graph $\mathfrak{G}_d$ in the space $\mathcal{P}_d^{\mathbf{c\Omega}_{\langle d],\,|\sim|'\geq 2}}$ by mapping the vertex of $\mathfrak{G}_d$, labeled by $\omega \in \mathbf{\Omega}_{\langle d],\,|\sim|'=1}$, to a preferred point w_ω in the $(d-1)$-cell $\mathring{\mathsf{R}}_d^{\omega}$ and the vertex of $\mathfrak{G}_d$, labeled by $\omega' \in \mathbf{\Omega}_{\langle d],\,|\sim|'=0}$, to a preferred point $e_{\omega'}$ in the d-cell $\mathring{\mathsf{R}}_d^{\omega'}$. Then we identify each edge $\{\omega, \omega'\}$ of $\mathfrak{G}_d$, where $\omega \in \mathbf{\Omega}_{\langle d],\,|\sim|'=1}$ and $\omega' \in \mathbf{\Omega}_{\langle d],\,|\sim|'=0}$, with a smooth path $[w_\omega, e_{\omega'}]$ such that the semi-open segment $(w_\omega, e_{\omega'}] \subset \mathring{\mathsf{R}}_d^{\omega'}$. In addition, we can choose the paths so that $[w_{\omega_1}, e_{\omega'}] \cap [w_{\omega_2}, e_{\omega'}] = e_{\omega'}$ for any pair $\omega_1, \omega_2 \prec \omega'$ and $\omega_1 \neq \omega_2$. Moreover, for each $\omega \in \mathbf{\Omega}_{\langle d],\,|\sim|'=1}$, we may arrange for the two paths, $[w_\omega, e_{\omega_1'}]$ and $[w_\omega, e_{\omega_2'}]$, to share the tangent vector at their common end w_ω, so that the path $[e_{\omega_1'}, e_{\omega_2'}]$ is transversal to the hypersurface $\mathring{\mathsf{R}}_d^{\omega}$ at w_ω (the transversality is needed to insure the stability under perturbations of the map $\mathcal{E}$ (defined below) with respect to the hypersurfaces $\{\mathring{\mathsf{R}}_d^{\omega}\}$). This construction produces an embedding $\mathcal{E} : \mathfrak{G}_d \to \mathcal{P}_d^{\mathbf{c\Omega}_{\langle d],\,|\sim|'\geq 2}}$. In what follows, we do not distinguish between $\mathfrak{G}_d$ and its image $\mathcal{E}(\mathfrak{G}_d) \subset \mathcal{P}_d^{\mathbf{c\Omega}_{\langle d],\,|\sim|'\geq 2}}$.

Lemma 6.11. *The graph $\mathfrak{G}_d$ is homotopy equivalent to a wedge of $\frac{d(d-2)}{4}$ circles if d is even, and of $\frac{(d-1)^2}{4}$ circles if d is odd.*

Proof. A simple calculation shows that $|\mathbf{\Omega}_{\langle d],\,|\sim|'=0}| = \lfloor \frac{d}{2} \rfloor + 1$, and

$$|\mathbf{\Omega}_{\langle d],\,|\sim|'=1}| = \sum_{k=1}^{\lfloor \frac{d}{2} \rfloor}(2k-1) = \left\lfloor \frac{d}{2} \right\rfloor^2.$$

For each $k \in [2, d]$ with the same parity as d, the vertex $\underbrace{(1, \dots, 1)}_{k} \in \mathbf{\Omega}_{\langle d],\,|\sim|'=0}$ is contained in the $k-1$ edges that lead to the vertices

$$(\underbrace{1, \dots, 1}_{s}, 2, \underbrace{1, \dots, 1}_{k-2-s}) \in \mathbf{\Omega}_{\langle d],\,|\sim|'=1},$$

where $0 \leq s \leq k-2 \leq d-2$, and to the $k+1$ edges that lead to the vertices

$$(\underbrace{1, \dots, 1}_{s}, 2, \underbrace{1, \dots, 1}_{k-s}) \in \mathbf{\Omega}_{\langle d],\,|\sim|'=1},$$

where $0 \leq s \leq k \leq d-1$. For d even, each case yields $\frac{d^2}{4}$ edges. It is easily seen that the graph $\mathfrak{G}_d$ is connected and hence it is homotopy equivalent to a wedge of circles. Now a simple calculation of the Euler characteristic $\chi(\mathfrak{G}_d)$ yields

$$1 - (\frac{d}{2} + 1 + \frac{d^2}{4}) + 2\frac{d^2}{4} = \frac{d(d-2)}{4}$$

circles. The calculation for odd d is analogous. □

Our next result is inspired by Arnold's Theorem 6.11. Fig. 6.11, for $d = 6$, illustrates it by exhibiting the cell structure in $\mathcal{P}_6^{\mathbf{c\Omega}_{\langle 6],\,|\sim|'\geq 2}}$ and its graph $\mathfrak{G}_6$.

Theorem 6.13. *The space $\mathcal{P}_d^{\mathbf{c\Omega}_{\langle d],\,|\sim|'\geq 2}}$ is homotopy equivalent to a wedge of $\frac{d(d-2)}{4}$ circles for d even, and to a wedge of $\frac{(d-1)^2}{4}$ circles for d odd.*

In particular, the fundamental group $\pi_1(\mathcal{P}_d^{\mathbf{c\Omega}_{\langle d],\,|\sim|'\geq 2}})$ is the free group *on $\frac{d(d-2)}{4}$ generators for even d and on $\frac{(d-1)^2}{4}$ generators for odd d.*

Moreover, $\mathcal{P}_d^{\mathbf{c\Omega}_{\langle d],\,|\sim|'\geq 2}}$ is the corresponding $K(\pi, 1)$-space.

Proof. As an open subset of $\mathbb{R}^d$, the space $\mathcal{P}_d^{\mathbf{c\Omega}_{\langle d],\,|\sim|'\geq 2}}$ is paracompact. Now consider a finite open cover $\mathcal{X} = \{X_\omega\}_{\omega \in \mathbf{\Omega}_{\langle d],\,|\sim|'=1}}$ of the space $\mathcal{P}_d^{\mathbf{c\Omega}_{\langle d],\,|\sim|'\geq 2}}$, where X_ω is the union of the $(d-1)$-cell $\mathring{\mathsf{R}}_d^\omega$ with the two adjacent d-cells that contain $\mathring{\mathsf{R}}_d^\omega$ in their boundary.

Each X_ω is open in $\mathcal{P}_d^{\mathbf{c\Omega}_{\langle d],\, |\sim|' \geq 2}}$. Indeed, any point $x \in X_\omega$ either (1) lies in one of the two d-cells and thus has an open neighborhood in $\mathcal{P}_d^{\mathbf{c\Omega}_{\langle d],\, |\sim|' \geq 2}}$ (contained in that d-cell), or (2) $x \in \mathring{\mathsf{R}}_d^\omega$, in which case, it has on open neighborhood X_ω in $\mathcal{P}_d^{\mathbf{c\Omega}_{\langle d],\, |\sim|' \geq 2}}$.

By [K3], Lemma 4.2, the attaching maps $\phi : D^d \to \mathcal{P}_d$ of the d-cells D^d are injective on the ϕ-preimage of each open $(d-1)$ cell in $\mathcal{P}_d$. This implies that X_ω retracts to the $\mathring{\mathsf{R}}_d^\omega$ which, in turn, is contractible. For $i \geq 2$ and for pairwise distinct compositions $\omega_1, \dots, \omega_i \in \mathbf{\Omega}_{\langle d],\, |\sim|'=1}$, the intersection $\bigcap_{j=1}^i X_{\omega_i}$ is either empty, or is one of the open d-cells $\mathring{\mathsf{R}}_d^\omega$ for some $\omega \in \mathbf{\Omega}_{\langle d],\, |\sim|'=0}$. It follows that, for $i \geq 1$ and for compositions $\omega_1, \dots, \omega_i \in \mathbf{\Omega}_{\langle d],\, |\sim|'=1}$, the intersection $\bigcap_{j=1}^i X_{\omega_i}$ is either empty or contractible.

The preceding arguments show that the assumptions of Corollary 4G.3 in [Ha] are satisfied for the open covering $\mathcal{X}$ of $\mathcal{P}_d^{\mathbf{c\Omega}_{\langle d],\, |\sim|' \geq 2}}$. Hence $\mathcal{P}_d^{\mathbf{c\Omega}_{\langle d],\, |\sim|' \geq 2}}$ is homotopy equivalent to the nerve $N_\mathcal{X}$ of the covering $\mathcal{X}$. We can identify $N_\mathcal{X}$ with the simplicial complex, whose simplices are the non-empty subsets A of $\mathbf{\Omega}_{\langle d],\, |\sim|'=1}$ such that $\bigcap_{\omega \in A} X_\omega$ is non-empty. So the maximal simplices of the nerve $N_\mathcal{X}$ are in bijection with the elements of $\mathbf{\Omega}_{\langle d],\, |\sim|'=0}$.

The maximal simplex, corresponding to $\omega' \in \mathbf{\Omega}_{\langle d],\, |\sim|'=0}$, contains all $\omega \in \mathbf{\Omega}_{\langle d],\, |\sim|'=1}$ for which $\mathring{\mathsf{R}}_d^{\omega'} \subseteq X_\omega$. The intersection of the two maximal simplices, corresponding to $\omega', \omega'' \in \mathbf{\Omega}_{\langle d],\, |\sim|'=0}$, is labeled by all $\omega \in \mathbf{\Omega}_{\langle d],\, |\sim|'=1}$ for which there are edges from ω' to ω and from ω'' to ω in $\mathfrak{G}_d$ (see Fig. 6.12).

The graph $\mathfrak{G}_d$ can be covered by 1-dimensional subcomplexes Y_ω , $\omega \in \mathbf{\Omega}_{\langle d],\, |\sim|'=1}$, where Y_ω is the union of the two edges in $\mathfrak{G}_d$, containing ω. It is easy to check that the nerve of this covering $\mathcal{Y} =_{\mathsf{def}} \{Y_\omega\}_{\omega \in \mathbf{\Omega}_{\langle d],\, |\sim|'=1}}$ is again $N_\mathcal{X}$. In fact, under the embedding $\mathcal{E} : \mathfrak{G}_d \to \mathcal{P}_d^{\mathbf{c\Omega}_{\langle d],\, |\sim|' \geq 2}}$, one has $Y_\omega = X_\omega \cap \mathcal{E}(\mathfrak{G}_d)$.

By [Bj], Theorem 10.6, the nerve $N_\mathcal{Y}$ and the graph $\mathfrak{G}_d$ are homotopy equivalent.

Moreover, by the proof of Corollary 4G.3 in [Ha], the following claim is valid. Consider an embedding $Y \hookrightarrow X$ of a paracompact space Y into a paracompact space X and a locally finite open covering $\mathcal{X} = \{X_\alpha\}_\alpha$ of X. Put $\mathcal{Y} = \{Y_\alpha =_{\mathsf{def}} X_\alpha \cap Y\}_\alpha$. If, for any nonempty intersection $\cap_i X_{\alpha_i}$, the intersection $\cap_i Y_{\alpha_i} \neq \emptyset$, and both intersections are contractible, then the nerves $N_\mathcal{X}$ and $N_\mathcal{Y}$ are naturally isomorphic as simplicial complexes, and $Y \hookrightarrow X$ is a homotopy equivalence.

We conclude that the embedding $\mathcal{E} : \mathfrak{G}_d \to \mathcal{P}_d^{\mathbf{c\Omega}_{\langle d],\, |\sim|' \geq 2}}$ is a homotopy equivalence. The result now follows from Lemma 6.11. □

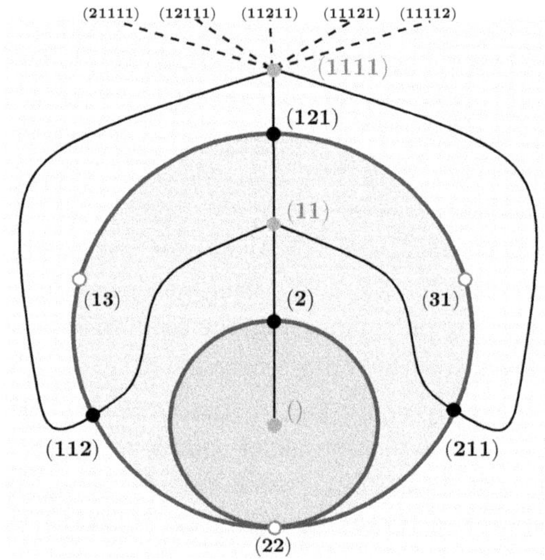

Fig. 6.11 A slice through the celluation of $\mathcal{P}_6^{\mathbf{c\Omega}_{\langle 6],\, |\sim|' \geq 2}}$ together with the graph $\mathfrak{G}_6$, dual to the celluation (shown by black curves). Patterns $(22), (13), (31)$ label the strata of codimension 2, while the arcs, labeled by $(2), (211), (121), (112)$, represent the strata of codimension 1.

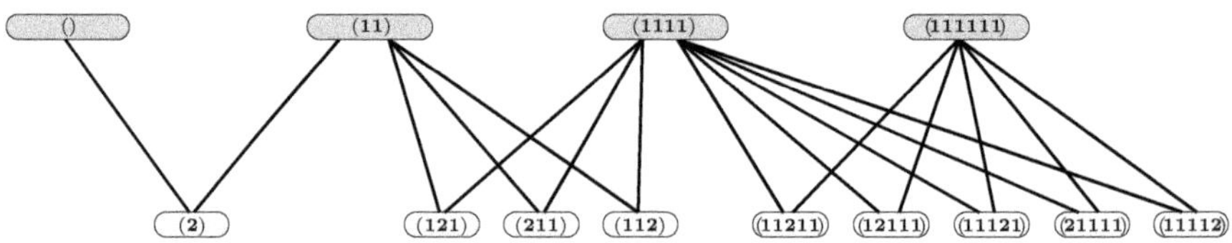

Fig. 6.12 The graph $\mathfrak{G}_6$ as part of the poset $\mathbf{\Omega}_{\langle 6]}$

Instead of $\mathfrak{G}_d$, let us also consider a related graph $\mathfrak{G}'_d$. In the new graph, any pair of edges e, e', forming a path that joins two different nodes that are labeled by elements from $\mathbf{\Omega}_{\langle d],\, |\sim|'=0}$, is replaced by a single edge, labeled by the unique $\omega \in \mathbf{\Omega}_{\langle d],\, |\sim|'=1}$ in the intersection of e and e' (see Fig. 6.13). Thus $\mathfrak{G}_d$ is the graph subdivision of $\mathfrak{G}'_d$ and in particular, we have $\pi_1(\mathfrak{G}_d) \cong \pi_1(\mathfrak{G}'_d)$ when both are considered as 1-dimensional simplicial complexes.

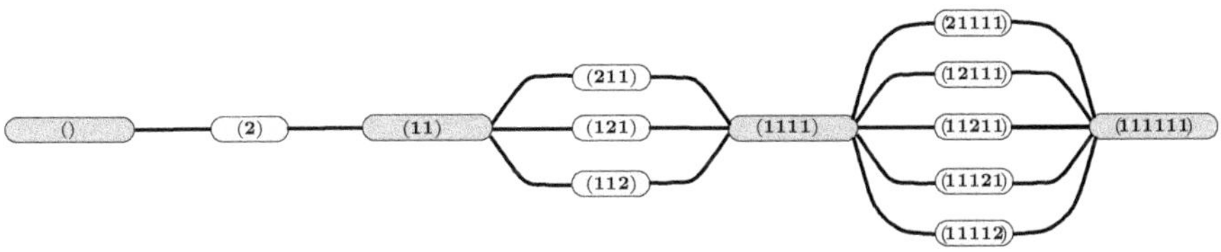

Fig. 6.13 The graph $\mathfrak{G}_6$, drawn as it is embedded in $\mathcal{P}_6^{\mathbf{c\Omega}_{\langle 6], |\sim|' \leq 2}}$

We will use the graph $\mathfrak{G}'_d$ in a second slightly different and perhaps more natural approach to the calculation of $\pi_1(\mathcal{P}_d^{\mathbf{c\Omega}_{\langle d], |\sim|' \geq 2}})$. For that, we introduce the alphabet A with letters

$$\omega_{ij} = (\underbrace{1, \ldots, 1}_{i}, 2, \underbrace{1, \ldots, 1}_{j}),$$

where $0 \leq i, j,\ i + j \leq d - 2$ and $i + j \equiv d \mod 2$. (See illustrations in Figures 6.11, 6.12 and 6.13).

We consider now the free group, generated by the letters from A. The letters are in obvious bijection with the edges of the graph $\mathfrak{G}'_d$. For a letter $\omega \in \mathsf{A}$, we use $\omega^+ = \omega$ or its inverse $\omega^- = \omega^{-1}$ to capture the orientations of the corresponding edges and to represent elements of the free group generated by A by words. We write $\mathsf{A}^{\pm}$ for the set of letters $\omega^{\pm}$ where $\omega \in \mathsf{A}$. We use the convention that the sign "+" corresponds to the orientation from the vertex with less ones towards the vertex with more ones, and the sign "−" corresponds to the opposite orientation.

We choose () as the basepoint of $\mathfrak{G}'_d$ if d is even, and (1) as the basepoint if d is odd. In particular, we can consider $\pi_1(\mathfrak{G}'_d)$ as a subgroup of the free group over A.

Next, we introduce a class of words which will be used to define canonical representatives of elements from $\pi_1(\mathfrak{G}'_d)$.

Definition 6.6. For $d \geq 1$, we say that a word w in the alphabet $\mathsf{A}^{\pm}$ is admissible if it is either empty or satisfies the following two conditions:

(a_e) if d is even, then w starts with the letter $(2)^+$ and ends with the letter $(2)^-$;

(a_o) if d is odd, then w starts with either the letter $(12)^+$ or with the letter $(21)^+$ and ends either with $(12)^-$ or with $(21)^-$.

(b) any pair $(\omega_1^{\pm}, \omega_2^{\pm})$ of two consecutive letters has the property that the number of 1's in ω_1 and ω_2 either coincide or differs by two. In the former case, the signs of the letters are different, and in the latter case, the signs should be as follows. If ω_1 has less ones than ω_2, then we only allow the pair (ω_1^+, ω_2^+); if ω_1 has more ones than ω_2, then we only allow the pair (ω_1^-, ω_2^-). ♢

Clearly, admissible words are in bijection with based loops in the graph $\mathfrak{G}'_d$. Here are two examples of admissible words representing based loops, the first one for even $d \geq 6$ and second one for odd $d \geq 7$:

$$w_1 = (2)^+ (112)^+ (121)^- (211)^+ (11112)^+ (12111)^- (121)^- (2)^-,$$

$$w_2 = (21)^+ (1112)^+ (112111)^+ (111211)^- (1211)^- (12)^-.$$

We observe that a word is admissible if and only if its reduction (in the sense of combinatorial group theory) is admissible.

Let $\mathcal{G}_d$ be the set of reduced admissible words over the alphabet $\mathsf{A}^\pm$. To each letter $\omega_{ij}^+ \in \mathsf{A}^\pm$, we associate the word γ_{ij} given by

- $$\gamma_{ij} = \omega_{0,0}^+\, \omega_{0,2}^+ \cdots \omega_{0,i+j-2}^+\, \omega_{ij}^+\, \omega_{0,i+j}^- \cdots \omega_{0,2}^-\, \omega_{0,0}^-$$
  if d is even and
- $$\gamma_{ij} = \omega_{0,1}^+\, \omega_{0,3}^+ \cdots \omega_{0,i+j-2}^+\, \omega_{ij}^+\, \omega_{0,i+j}^- \cdots \omega_{0,3}^-\, \omega_{0,1}^-$$
  if d is odd.

All γ_{ij} are admissible, and if $i \neq 0$, then γ_{ij} is reduced. We notice that each γ_{ij} represents a single *loop* in $\mathfrak{G}'_d$.

The following is then a standard fact about graphs considered as 1-dimensional CW-complexes.

Lemma 6.12. *The set $\mathcal{G}_d$ is a subgroup of the free group over the alphabet* A, *isomorphic to $\pi_1(\mathfrak{G}'_d)$. In particular, $\mathcal{G}_d$ is a free group itself. The words $\{\gamma_{ij}\}$ for $0 \leq i, j$, $i \neq 0$ and $i + j \leq d - 2$ form a minimal generating set of $\mathcal{G}_d$.* ♢

Note each $\gamma_{0,j}$ is 1 in $\mathcal{G}_d$. But for our geometric picture and for the sake of a less technical presentation, we will continue working with the $\gamma_{0,j}$.

Now we would like to relate this representation of $\pi_1(\mathfrak{G}'_d)$ to $\pi_1(\mathcal{P}_d^{\mathbf{c\Omega}_{\langle d],\, |\sim|' \geq 2}})$. We use the notations from the proof of Theorem 6.13.

For $\omega_{ij} \in \mathsf{A}$, consider the codimension one "wall" $\mathring{\mathsf{R}}_d^{\omega_{ij}}$. The walls divide $\mathcal{P}_d^{\mathbf{c\Omega}_{\langle d],\,|\sim|'\geq 2}}$ into a set of open d-cells $\{\mathring{\mathsf{R}}_d^{\omega}\}$ for $\omega = \underbrace{(1,\ldots,1)}_{i} \in \mathbf{\Omega}_{\langle d],\,|\sim|'=0}$, where $0 \leq i \leq d$ is even when d is even and odd when d is odd.

We orient each wall $\mathring{\mathsf{R}}_d^{\omega_{ij}}$ in such a way that crossing it in the preferred direction increases the number of simple real roots by 2. We consider this as a "+"-crossing and the crossing in the opposite direction as a "−"-crossing. This notation goes along with our convention for orienting the edges in $\mathfrak{G}'_d$.

Consider an oriented loop $\gamma : S^1 \to \mathcal{P}_d^{\mathbf{c\Omega}_{\langle d],\,|\sim|'\geq 2}}$, with the base point in the cell $\mathring{\mathsf{R}}_d^{()}$ if d is even and in the cell $\mathring{\mathsf{R}}_d^{(1)}$ if d is odd. By the general position arguments, we may assume that γ is smooth and transversal to each wall. In particular, since S^1 is compact, the (transversal) intersection of $\gamma(S^1)$ with each wall is a finite set. As we move along γ, we record each transversal crossing $\gamma \cap \mathring{\mathsf{R}}_d^{\omega_{ij}}$ and the direction of the crossing by the corresponding letter in $\mathsf{A}^{\pm}$.

Proposition 6.8. *The homotopy classes of based loops $\gamma \subset \mathcal{P}_d^{\mathbf{c\Omega}_{\langle d],\,|\sim|'\geq 2}}$ and the homotopy classes of based loops in $\mathfrak{G}'_d$ are in one-to-one correspondence with the reduced admissible words over the alphabet $\mathsf{A}^{\pm}$. In addition, concatenation of loops corresponds to the concatenation and reduction of words.*

In particular, $\pi_1(\mathcal{P}_d^{\mathbf{c\Omega}_{\langle d],\,|\sim|'\geq 2}}) \simeq \mathcal{G}_d \simeq \pi_1(\mathfrak{G}'_d)$ is a free group.

Sketch of proof: For $\pi_1(\mathfrak{G}'_d)$ the assertion already follows from Lemma 6.12 and the arguments preceding the lemma.

Our choice of basepoint in $\mathcal{P}_d^{\mathbf{c\Omega}_{\langle d],\,|\sim|'\geq 2}}$ corresponds exactly to the choice of the base vertices () for d even and (1) for d odd for admissible words in $\mathfrak{G}'_d$. The above arguments and the conditions for admissibility of words over the alphabet $\mathsf{A}^{\pm}$ show that based loops in $\mathcal{P}_d^{\mathbf{c\Omega}_{\langle d],\,|\sim|'\geq 2}}$ are represented by admissible words. Reduction is easily seen to be a homotopy equivalence. Moreover, loops corresponding to different reduced admissible words are not homotopy equivalent. Since the concatenation of admissible words is admissible, the remaining assertion now follows. □

Remark 6.3. Let X be a compact path-connected Hausdorff space and $C(X)$ the space of continuous functions on X. Let

$$P(u, x) = u^d + a_{d-1}u^{d-1} + \ldots + a_1 u + a_0$$

be a polynomial in u with coefficients $a_i \in C(X)$. Assume, that for each $x \in X$, $P(u,x) \in \mathcal{P}_d^{\mathbf{c\Omega}_{\langle d],\, |\sim|' \geq 2}}$; that is, $P(u,x)$ has only simple real roots and no more than a single root of multiplicity 2.

Thus every such family of polynomials $\{P(u,x)\}_{x\in X}$ generates a continuous map $\alpha_P : X \to \mathcal{P}_d^{\mathbf{c\Omega}_{\langle d],\, |\sim|' \geq 2}}$. Since a subgroup of a free group is free or trivial, it follows that when $\pi_1(X)$ does not admit an epimorphism with the free target group, then the induced map

$$(\alpha_P)_* : \pi_1(X) \to \pi_1(\mathcal{P}_d^{\mathbf{c\Omega}_{\langle d],\, |\sim|' \geq 2}})$$

must be trivial. For example, this is the case for any X with a finite fundamental group.

Here is one implication of the triviality of $(\alpha_P)_*$: for any loop $\delta : S^1 \to X$, the loop $(\alpha_P)_*(\delta)$ hits the non-singular part of the discriminant variety $\mathcal{D}_d \subset \mathcal{P}_d$ so that the admissible word in the alphabet $\mathsf{A}^{\pm}$, generated by $(\alpha_P)_*(\delta)$, can be reduced to the identity 1. In particular, the number of positive and negative crossings of $(\alpha_P)_*(\delta)$ are equal. ◊

6.8.2 *The fundamental group $\pi_1(\mathcal{P}_d^{\mathbf{c}\Theta})$ for an arbitrary closed poset Θ with the locus $\mathcal{P}_d^{\Theta} \subset \mathcal{P}_d$ of codimension ≥ 2*

This section describes how the fundamental group changes if we add to the space $\mathcal{P}_d^{\mathbf{c\Omega}_{\langle d],\, |\sim|' \geq 2}}$ some strata of codimension 2. Each of these strata will create a relation in the fundamental group of the new space under construction. We provide an explicit description of these relations and show that, in many cases, the resulting quotient is still a *free* group.

We start with the following simple statement.

Lemma 6.13. *Let $\Theta \subset \mathbf{\Omega}_{\langle d]}$ be a closed poset such that $\Theta \subset \mathbf{\Omega}_{\langle d],\, |\sim|' \geq k}$.*

Then the homotopy groups $\pi_i(\mathcal{P}_d^{\mathbf{c}\Theta})$ vanish for all $i < k-1$. In particular, $\pi_1(\mathcal{P}_d^{\mathbf{c}\Theta})$ vanishes, provided $\Theta \subset \mathbf{\Omega}_{\langle d],\, |\sim|' \geq 3}$. As a special case, we have $\pi_1(\mathcal{P}_d^{\mathbf{c\Omega}_{\langle d],\, |\sim|' \geq 3}}) = 0$.

Proof. We observe that if $\Theta \subset \mathbf{\Omega}_{\langle d],\, |\sim|' \geq k}$, then $\mathrm{codim}(\bar{\mathcal{P}}_d^{\Theta}, \bar{\mathcal{P}}_d) \geq k$. Therefore, by the general position argument, $\pi_i(\mathcal{P}_d^{\mathbf{c}\Theta}) = 0$ for all $i < k-1$. In particular, $\pi_1(\mathcal{P}_d^{\mathbf{c}\Theta}) = 0$, provided that $\Theta \subset \mathbf{\Omega}_{\langle d],\, |\sim|' \geq 3}$. □

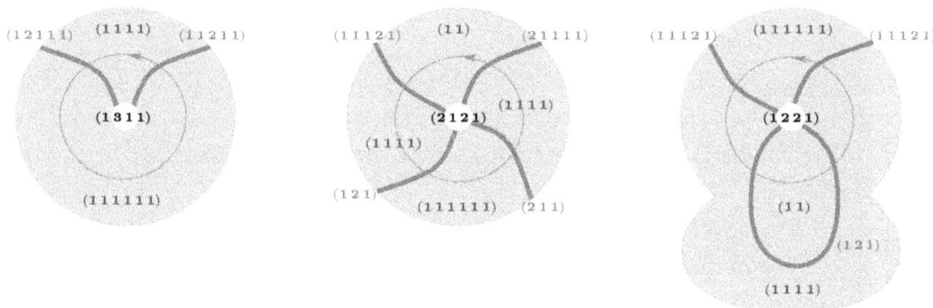

Fig. 6.14 The normal disks to the strata $\mathcal{P}_6^{(1311)}$ (left), $\mathcal{P}_6^{(2121)}$ (middle), and $\mathcal{P}_6^{(1221)}$ (right) and the loops bounding them (in red).

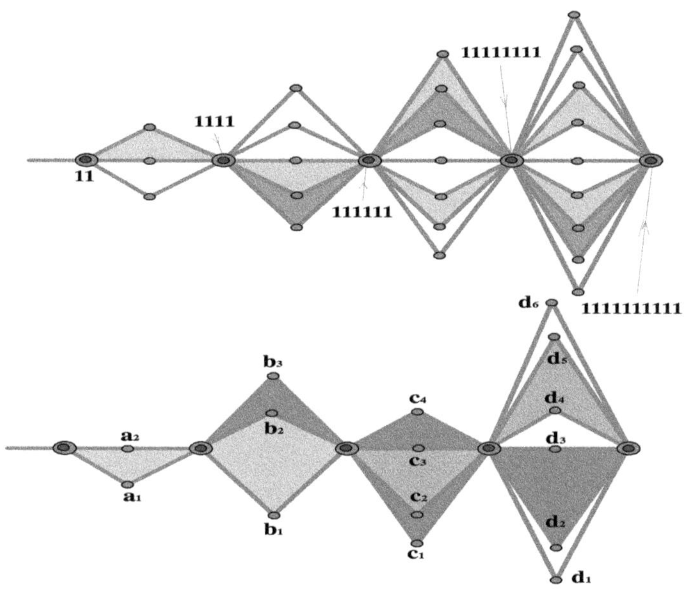

Fig. 6.15 Relations of type **(3)** (the upper diagram) and of type **(22)** (the lower diagram) shown as shadings

Further, we observe that, by the Alexander duality and the Hurewicz Theorem, for any closed $\Theta \subset \mathbf{\Omega}_{\langle d]}$, a minimal generating set of $\pi_1(\mathcal{P}_d^{\mathbf{c}\Theta})$ contains at least $\mathsf{rank}(\bar{H}_{d-2}(\bar{\mathcal{P}}_d^{\Theta}; \mathbb{Z}))$ elements.

Given a closed poset $\Theta \subseteq \mathbf{\Omega}_{\langle d],\, |\sim|' \geq 2}$, we consider two disjoint sets:

$$\Theta_{=2} = \mathbf{\Omega}_{\langle d],\, |\sim|'=2} \cap \Theta, \quad \mathbf{c}\Theta_{=2} = \mathbf{\Omega}_{\langle d],\, |\sim|'=2} \setminus \Theta. \tag{6.40}$$

By definition, $\Theta_{=2}$ and $\mathbf{c}\Theta_{=2}$ both consist of ω's which have some parts 1 and either a *single* entry 3, or *two* 2s.

Now consider a loop γ in $\mathcal{P}_d^{\mathbf{c}\Theta}$. It bounds a 2-disk D in $\mathcal{P}_d$. By a general position argument, we may assume that D avoids all strata R_d^ω with $|\omega|' \geq 3$ and that if γ intersects a stratum $\mathring{\mathsf{R}}_d^\omega$ for some $|\omega|' = 2$, then this intersection is transversal.

Consider a fixed ω with $|\omega|' = 2$ such that γ intersects $\mathring{\mathsf{R}}_d^\omega$. For $x \in \mathring{\mathsf{R}}_d^\omega$, consider a small 2-disk $D_x \subseteq \mathcal{P}_d^{\mathbf{c\Omega}_{\langle d],\, |\sim|' \geq 3}}$ that is centered on x and is transversal to $\mathring{\mathsf{R}}_d^\omega$, so that D_x intersects only the cells $\mathring{\mathsf{R}}_d^{\omega'}$ for $\omega' \succeq \omega$. Let

$$\kappa_x =_{\mathsf{def}} \partial D_x \subset \mathcal{P}_d^{\mathbf{c\Omega}_{\langle d],\, |\sim|' \geq 2}}$$

be the loop bounding D_x (see Fig. 6.14). Since $\mathring{\mathsf{R}}_d^\omega$ is contractible, for each $x \in D \cap \mathring{\mathsf{R}}_d^\omega$, the homotopy class of κ_x depends only on ω (up to inversion). Hence we may speak of κ_ω.

We choose a path $\beta \subset \mathcal{P}_d^{\mathbf{c\Omega}_{\langle d],\, |\sim|' \geq 2}}$ that connects our base point in $\mathring{\mathsf{R}}_d^{()}$ or $\mathring{\mathsf{R}}_d^{(1)}$ with κ_ω. By a general position argument we can again assume that β intersects all codimension 1 strata transversally. Let $\kappa_\omega^\bullet =_{\mathsf{def}} \beta^{-1} \circ \kappa_\omega \circ \beta$ be the loop that starts at the base point, follows a path β, then traverses κ_ω once, and returns to the base point following β^{-1}. By recording the intersections of $\kappa_\omega^\bullet$ with the codimension 1 strata, indexed by the ω_{ij}, together with the direction of the intersections, we obtain an admissible word in $\mathsf{A}^\pm$. By Lemma 6.13, the homotopy class of $\kappa_\omega^\bullet$ in $\mathcal{P}_d^{\mathbf{c\Omega}_{\langle d],\, |\sim|' \geq 2}}$ depends only on this word. In particular, we can express this word as a product of words γ_{ij} or their inverses for the ω_{ij} intersected by the loop κ_ω.

Evidently, if $\omega \in \mathbf{c}\Theta_{=2}$, the loop κ_ω and thus the loop $\kappa_\omega^\bullet$ are contractible in $\mathcal{P}_d^{\mathbf{c}\Theta}$. The following proposition gives a precise analysis of the contractible $\kappa_\omega^\bullet$ and shows that the triviality of the corresponding words provides a presentation of $\pi_1(\mathcal{P}_d^{\mathbf{c}\Theta})$, generalizing Theorem 6.13 (see Fig. 6.14 and Fig. 6.15).

Proposition 6.9. *For any closed poset $\Theta \subset \mathbf{\Omega}_{\langle d],\, |\sim|' \geq 2}$, the fundamental group $\pi_1(\mathcal{P}_d^{\mathbf{c}\Theta})$ is a quotient of the free group $\mathcal{G}_d$ by the normal subgroup, generated by the following relations, one relation for every $\omega \in \mathbf{c}\Theta_{=2} = \mathbf{\Omega}_{\langle d],\, |\sim|'=2} \setminus \Theta$.*

(3) *For each $\omega = (\underbrace{1 \cdots 1}_{i} 3 \underbrace{1 \cdots 1}_{j}) \in \mathbf{c}\Theta_{=2}$ the corresponding relation is*

$$\gamma_{i,j+1}\, \gamma_{i+1,j}^{-1} = 1$$

(22) *For each* $\omega = (\underbrace{1\cdots 1}_{i}2\underbrace{1\cdots 1}_{j}2\underbrace{1\cdots 1}_{\ell}) \in \mathbf{c}\Theta_{=2}$ *the corresponding relation is*

$$\gamma_{i+j,\ell}\,\gamma_{i+j+2,\ell}\,\gamma^{-1}_{i,j+\ell+2}\,\gamma^{-1}_{i,j+\ell} = 1.$$

Remark 6.4. Recall, that whenever a $\gamma_{0,j}$ appears in one of the relations that it can be omitted by $\gamma_{0j} = 1$. Note also that, in case (22), when $j = 0$, the relation $\gamma_{i+j,\ell}\,\gamma_{i+j+2,\ell}\,\gamma^{-1}_{i,j+\ell+2}\,\gamma^{-1}_{i,j+\ell}$ is conjugate to the relation $\gamma_{i+j+2,\ell}\,\gamma^{-1}_{i,j+\ell+2}$. ◇

Proof. First we show that the relations (3) and (22) lie in the kernel of the homomorphism

$$\pi_1(\mathcal{P}_d^{\mathbf{c\Omega}_{\langle d],\,|\sim|'\leq 2}}) \to \pi_1(\mathcal{P}_d^{\mathbf{c}\Theta}), \tag{6.41}$$

induced by the inclusion of spaces $\mathcal{P}_d^{\mathbf{c\Omega}_{\langle d],\,|\sim|'\leq 2}} \subset \mathcal{P}_d^{\mathbf{c}\Theta}$. By the arguments preceding Proposition 6.9, it suffices to show that each relation corresponds to an admissible word representing $\kappa^\bullet_\omega$ for some $\omega \in \mathbf{\Omega}_{\langle d],\,|\sim|'=2}$.

(3) For each $\omega = (\underbrace{1,\cdots,1}_{i},3,\underbrace{1,\cdots,1}_{j}) \in \mathbf{c}\Theta_{=2}$, we notice that $\mathring{\mathsf{R}}^\omega_d$ lies only in the boundaries of the codimension one cells $\mathring{\mathsf{R}}_d^{\omega_{i+1,j}}$ and $\mathring{\mathsf{R}}_d^{\omega_{i,j+1}}$, and of the codimension 0 cells corresponding to $(\underbrace{1,\ldots,1}_{i+j+3})$ and to $(\underbrace{1,\ldots,1}_{i+j+1})$. As we traverse the loop κ_ω, the left diagram in Fig. 6.14 will guide us. When κ_ω enters the cell, labelled by $(\underbrace{1,\ldots,1}_{i+j+3})$, from a point on $\mathring{\mathsf{R}}_d^{\omega_{i,j+1}}$ the $(i+1)^{\text{st}}$ largest real root splits into 2 distinct real roots. On its way towards $\mathring{\mathsf{R}}_d^{\omega_{i+1,j}}$, the smaller of these two roots approaches the $(i+1)^{\text{st}}$ largest real root and at the end merges with it. Then, when entering the cell corresponding to $(\underbrace{1,\ldots,1}_{i+j+1})$, the real double root splits into two conjugate complex roots with small imaginary part and with real part between the $(i+1)^{\text{st}}$ and $(i+2)^{\text{nd}}$ largest real roots. While traversing κ_ω, the value of the real part passes the $(i+1)^{\text{st}}$ real root and then, when the imaginary parts vanish, κ_ω is back in $\mathring{\mathsf{R}}^\omega_{i,j+1}$. Thus $\kappa^\bullet_\omega$ corresponds to the relation $\gamma_{i,j+1}\,\gamma^{-1}_{i+1,j}$.

(22) For $\omega = (\underbrace{1,\cdots,1}_{i},2,\underbrace{1,\cdots,1}_{j},2,\underbrace{1\cdots 1}_{\ell}) \in \mathbf{c}\Theta_{=2}$, the codimension 2 cell $\mathring{\mathsf{R}}^\omega_d$ lies only in the boundary of the codimension 1 cells $\mathring{\mathsf{R}}_d^{\omega_{i+2+j,\ell}}$,

$\mathring{\mathsf{R}}_d^{\omega_{i,j+\ell+2}}$, $\mathring{\mathsf{R}}_d^{\omega_{i+j,\ell}}$, and $\mathring{\mathsf{R}}_d^{\omega_{i,j+\ell}}$. (Note that the last two coincide when $j = 0$). It also lies in the boundary of the codimension 0-cells $(\underbrace{1, \dots, 1}_{k})$ for

$$k = i + j + \ell,\; i + j + \ell + 2,\; i + j + \ell + 4.$$

Again, let us traverse the loop κ_ω, guided by the middle and the right diagrams in Fig. 6.14. We start in $\mathring{\mathsf{R}}_d^{\omega_{i+j+2,\ell}}$ and, when entering the cell, labelled by $(\underbrace{1, \dots, 1}_{i+j+\ell+4})$, we split the $(i + j + 2)^{\mathrm{nd}}$ root into two distinct real roots. Then, on the way through this cell, the $(i + 1)^{\mathrm{st}}$ and $(i + 2)^{\mathrm{nd}}$ largest real roots approach each other and finally merge at a point belonging to the wall $\mathring{\mathsf{R}}_d^{\omega_{i,j+\ell+2}}$. After passing $\mathring{\mathsf{R}}_d^{\omega_{i,j+\ell+2}}$, the $(i + 1)^{\mathrm{st}}$ largest real root splits into two complex conjugate roots. Next, we move towards $\mathring{\mathsf{R}}_d^{\omega_{i,j+\ell}}$ by bringing the ith and $(i+1)^{\mathrm{st}}$ largest real roots together. Then we split this double root into two complex conjugate roots inside the d-cell, labelled by $(\underbrace{1, \dots, 1}_{i+j+\ell})$; further we move the real part of these roots to a value between the $(i + j)^{\mathrm{th}}$ and the $(i + j + 1)^{\mathrm{st}}$ real roots, while letting the imaginary part to approach 0. This leads us to $\mathring{\mathsf{R}}^{\omega_{i+j,\ell}}$, from where we follow the analogous route back to $\mathring{\mathsf{R}}_d^{\omega_{i+j+2,\ell}}$. Note that along the way, we made certain choices that determine which codimension 1 cell we want to approach next. The arguments, preceding the lemma, show that all such choices lead to homotopic paths/loops.

As a result, the loop $\kappa_\omega^\bullet$ translates into the relation

$$\gamma_{i+j,\ell}\, \gamma_{i+j+2,\ell}\, \gamma_{i,j+\ell+2}^{-1}\, \gamma_{i,j+\ell+2}^{-1} = 1.$$

Next we show that the relations (3) and (22) generate the kernel of the homomorphism in (6.41). For each ω with $|\omega|' = 2$, consider a closed regular neighborhood U_ω of the cell $\mathring{\mathsf{R}}_d^\omega$ in the space $\mathcal{P}_d^{\mathbf{c\Omega}\langle d],\, |\sim|' \geq 2} \cup \mathring{\mathsf{R}}_d^\omega$. We may assume that, for distinct ω's, these neighborhoods are disjoint. The space U_ω fibers over $\mathring{\mathsf{R}}_d^\omega$ with fibers being 2-disks. Since the base $\mathring{\mathsf{R}}_d^\omega$ is contractible, the fibration $p_\omega : U_\omega \to \mathring{\mathsf{R}}_d^\omega$ is trivial.

Set $X =_{\mathsf{def}} \mathcal{P}_d^{\mathbf{c\Omega}\langle d],\, |\sim|' \geq 2}$. Adding to X a cell $\mathring{\mathsf{R}}_d^\omega$, where $\omega \in \mathbf{c}\Theta_{=2}$, produces a new space Y. Its homotopy type is the result of attaching a 2-handle to X along its boundary. The spaces X, U_ω, and $X \cap U_\omega$ are path-connected. Thus, by the Seifert–van Kampen Theorem,

$$\pi_1(Y) \approx \pi_1(X) *_{\pi_1(X \cap U_\omega)} \pi_1(U_\omega),$$

which is the free product $*$ of the groups $\pi_1(X)$ and $\pi_1(U_\omega)$, amalgamated over $\pi_1(X \cap U_\omega)$. Since U_ω is homotopy equivalent to the cell $\mathring{\mathsf{R}}_d^\omega$, we get $\pi_1(U_\omega) = 0$. At the same time, using the triviality of p_ω, $X \cap U_\omega = U_\omega \setminus \mathring{\mathsf{R}}_d^\omega$ is homotopy equivalent to a loop κ_ω, which bounds a small 2-disk normal to the stratum $\mathring{\mathsf{R}}_d^\omega$. Therefore, $\pi_1(Y) \approx \pi_1(X)/[\kappa_\omega^\bullet]$. Thus $\pi_1(Y)$ arises from $\pi_1(X)$ by factoring modulo one of the relation of type (3) or (22).

Since the U_ω's were chosen mutually disjoint, it follows that we can repeat this argument for all $\omega \in \mathbf{c}\Theta_{=2}$ and conclude that $\pi_1(\mathcal{P}_d^{\mathbf{c}\Theta})$ arises as a quotient of $\pi_1(\mathcal{P}_d^{\mathbf{c}\mathbf{\Omega}_{\langle d],\,|\sim|'\geq 2}})$ by the relations (3) and (22) for $\omega \in \mathbf{c}\Theta_{=2}$. □

The next theorem shows that in many cases the relations from Proposition 6.9 yield a presentation of a *free* group.

Theorem 6.14. *Let $\Theta \subset \mathbf{\Omega}_{\langle d],\,|\sim|'\geq 2}$ be a closed poset such that either*

(i) Θ contains all $\omega \in \mathbf{\Omega}_{\langle d]}$ such that $\omega = (\underbrace{1\cdots 1}_{i}2\underbrace{1\cdots 1}_{j}2\underbrace{1\cdots 1}_{\ell})$ for some $j > 0$;

or

(ii) Θ is such that for $\omega \in \mathbf{\Omega}_{\langle d]}$ with $|\omega|' = 2$ we have

$$\omega \in \Theta \Leftrightarrow \omega = (\underbrace{1\cdots 1}_{i}3\underbrace{1\cdots 1}_{j})$$

for some $i, j \geq 0$.

Then the fundamental group $\pi_1(\mathcal{P}_d^{\mathbf{c}\Theta})$ is a free group in $\mathsf{rank}(\tilde{H}_{d-2}(\bar{\mathcal{P}}_d^{\Theta};\mathbb{Z}))$ generators.

In particular, in case (ii), we have $\pi_1(\mathcal{P}_d^{\mathbf{c}\Theta}) = \mathbb{Z}$ for $d \geq 4$.

Proof.

(i) By Proposition 6.9 and the assumptions on Θ, the fundamental group $\pi_1(\mathcal{P}_d^{\mathbf{c}\Theta})$ is a quotient of the free group by arbitrary relations of type (3) and some specific relations of type (22).
Any relation of type (3) identifies some $\gamma_{i,j+1}$ with $\gamma_{i+1,j}$. Any relation of type (22) that can appear identifies some γ_{i+j+2} with $\gamma_{i,j+\ell+2}$.
In either case, the relation and one generator can be dropped without changing the isomorphism type of the group.
Note that the number of generators of the free group $\pi_1(\mathcal{P}_d^{\mathbf{c}\Theta})$ is the rank of $H_1(\mathcal{P}_d^{\mathbf{c}\Theta};\mathbb{Z})$. By the Alexander duality, it is equal the

rank of $H^{d-2}(\bar{\mathcal{P}}_d^{\Theta}; \mathbb{Z})$. Thanks to Corollary 6.5, we can compute $\mathsf{rank}(\bar{H}_{d-2}(\bar{\mathcal{P}}_d^{\Theta}; \mathbb{Z}))$ in pure combinatorial terms.
The assertion follows.

(ii) By Proposition 6.9 and the assumptions on Θ, the fundamental group $\pi_1(\mathcal{P}_d^{\mathbf{c}\Theta})$ is a quotient of the free group by *all* possible relations of type (22). First consider compositions of type $(\underbrace{1, \ldots, 1}_{i}, 2, 2, \underbrace{1, \ldots, 1}_{\ell})$ of $d' = i + 4 + \ell \leq d$ where d and d' are of the same parity. The corresponding relation $\gamma_{i,\ell}\, \gamma_{i+2,\ell}\, \gamma_{i,\ell+2}^{-1}\, \gamma_{i,\ell}^{-1} = 1$ identifies $\gamma_{i+2,\ell}$ and $\gamma_{i,\ell+2}$. As a consequence this relation, one generator can be dropped. By iterating this procedure, we identify all $\gamma_{i+2,\ell}$ and $\gamma_{i,\ell+2}$ for $i = 0, \ldots, d' - 4$ and even $d - d' \geq 0$. As a consequence, for a fixed d', we have identified $\gamma_{i,\ell}$ and $\gamma_{i',\ell'}$, where $i + \ell = d' - 2 = i' + \ell'$ if both i and i' are even or both i and i' are odd. After this identification, the group is generated by $\gamma_{1,d'-3}$ for $3 \leq d' \leq d$.
Let $d' \geq 5$ and consider $\omega = (2, 1, 2, \underbrace{1 \cdots 1}_{d'-5})$. Then the relation

$$\gamma_{1,d'-5}\, \gamma_{3,d'-5}\, \gamma_{0,d'-2}^{-1}\, \gamma_{0,d'-4}^{-1} = 1$$

implies $\gamma_{1,d'-5}\, \gamma_{0,d'-4}^{-1} = \gamma_{3,d'-5}\, \gamma_{0,d'-2}^{-1}$, which identifies $\gamma_{1,d'-5}$ and $\gamma_{3,d'-5}$ and, by transitivity, $\gamma_{1,d'-5}$ and $\gamma_{1,d'-3}$. Making this identification and removing the corresponding relations leaves a set of trivial relations on the unique remaining generator $\gamma_{1,1}$ for d even or $\gamma_{1,2}$ for d odd. Thus we have arrived at a group with one generator and no relation. The assertion now follows.

□

As an immediate consequence, we obtain the following result about certain natural spaces of real polynomials.

Corollary 6.10. *The fundamental group of the following spaces is a free group.*

(i) The space of all degree d polynomials with no two distinct roots of multiplicity ≥ 2.
(ii) The space of all degree d polynomials with no root of multiplicity ≥ 3. In this case, the fundamental group is $\mathbb{Z}$.

Proof. The assertions follow immediately from Theorem 6.14. □

The following illuminating example shows however that, in general, $\mathcal{P}_d^{\mathbf{c}\Theta}$ is not free.

Example 6.9. Let $d = 6$ and consider the smallest closed $\Theta \subseteq \mathbf{\Omega}_{\langle 6]}$ containing

$$\{(31), (13), (1311), (1131), (2211), (1221), (1122), (2112)\}. \tag{6.42}$$

By Proposition 6.9, we need to consider the group freely generated by

$$\gamma_{1,1}, \gamma_{2,0}, \gamma_{1,3}, \gamma_{2,2}, \gamma_{3,1}, \gamma_{4,0}$$

as well as by the dummy generators $\gamma_{0,2} = \gamma_{0,4} = 1$. We then have to impose the relations of type (3) for the elements (3111), (1113) and relations of type (22) for (22), (2121), (1212). Fig. 6.16 sketches the 2-complex representing these relations. The type (3) relations identify $\gamma_{0,4}$ and $\gamma_{1,3}$ as well as $\gamma_{3,1}$ and $\gamma_{4,0}$. The type (22) relation, corresponding to the pattern (22), then identifies $\gamma_{0,2}$ and $\gamma_{2,0}$. After these identifications, the type (22) relations that correspond to (2121) and (1212) can be transformed as follows:

$$\gamma_{1,1}\,\gamma_{3,1}\,\gamma_{0,4}^{-1}\,\gamma_{0,2}^{-1} \longrightarrow \gamma_{1,1}\,\gamma_{3,1},$$

$$\gamma_{2,0}\,\gamma_{4,0}\,\gamma_{1,3}^{-1}\,\gamma_{1,1}^{-1} \longrightarrow \gamma_{3,1}\,\gamma_{1,1}^{-1}.$$

Thus we are left with the generators $\gamma_{1,1}$, $\gamma_{2,2}$ and $\gamma_{3,1}$, subject to the relations $\gamma_{1,1}\,\gamma_{3,1} = 1 = \gamma_{3,1}\,\gamma_{1,1}^{-1}$. After eliminating $\gamma_{3,1}$, we are left with the generators $\gamma_{1,1}$, $\gamma_{2,2}$ subject to $\gamma_{1,1}^2 = 1$ which is a presentation of $\mathbb{Z}/2\mathbb{Z}$. Hence, we have shown that

$$\pi_1(\mathcal{P}_6^{\mathbf{c}\Theta}) \approx \mathbb{Z} * \mathbb{Z}/2\mathbb{Z}$$

for the poset Θ introduced in Example 6.9. If one wants to get rid of the free factor $\mathbb{Z}$ as well, one can throw (1131) into Θ, which yields an identification of $\gamma_{2,2}$ and $\gamma_{3,1} = 1$ and leaves the other relation untouched. ◇

Using the same Θ as in Example 6.9, formula (6.42), but for $d = 8$, one can check that $\pi_1(\mathcal{P}_8^{\mathbf{c}\Theta})$ is a *free* group. Guided by this example, one is tempted to speculate that, for any closed Θ and d large enough, $\pi_1(\mathcal{P}_d^{\mathbf{c}\Theta})$ is free.

We will see that this is, in general, not the case, but first we will show how $\pi_1(\mathcal{P}_d^{\mathbf{c}\Theta})$ stabilizes for d large. This stabilization of the fundamental group follows almost immediately from the next lemma, a consequence of Proposition 6.9.

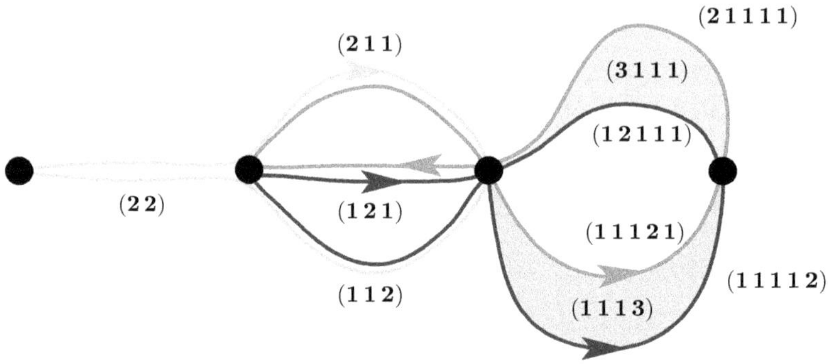

Fig. 6.16 The assembly instructions for the 2-dimensional CW-complex that realizes the fundamental group $\pi_1(\mathcal{P}_6^{\mathbf{c}\Theta})$ for Θ in (6.42). The 2-cells that realize the relations of type (3) are shaded. The three loops, to which the 2-disks that realize relations of type (22) are attached, are shown in three colors (gold for (22), green for (2121), blue for (2112)).

Proposition 6.10. *Let $\Theta \subset \mathbf{\Omega}_{\langle d], |\sim|' \geq 2}$ be a closed poset and $\Theta' = \Theta \cap \mathbf{\Omega}_{\langle d-2]}$. Assume that no ω with $|\omega| \in \{d, d-2\}$ resides in $\Theta_{=2} =_{\mathsf{def}} \mathbf{\Omega}_{\langle d], |\sim|'=2} \cap \Theta$.*

Then $\pi_1(\mathcal{P}_d^{\mathbf{c}\Theta}) \cong \pi_1(\mathcal{P}_{d-2}^{\mathbf{c}\Theta'})$.

Proof. By Proposition 6.9 and the structure of Θ, all the relations of type (3) and (22), that correspond to $\omega \in \mathbf{\Omega}_{\langle d], |\sim|'=2}$ with $|\omega| \in \{d, d-2\}$ and define $\pi_1(\mathcal{P}_d^{\mathbf{c}\Theta})$, are part of a presentation of $\pi_1(\mathcal{P}_{d-2}^{\mathbf{c}\Theta'})$. Indeed, the generators γ_{ij} for $i + j + 2 = d$ and the relations of type (3) and (22), corresponding to $\omega \in \mathbf{c}\Theta_{=2}$ and $|\omega| = d$, are the only difference to the presentations of $\pi_1(\mathcal{P}_{d-2}^{\mathbf{c}\Theta'})$ and $\pi_1(\mathcal{P}_d^{\mathbf{c}\Theta})$.

The relations of type (3) identify all γ_{ij} for $i+j+2 \in \{d, d-2\}$. Since $\gamma_{0,d-2} = \gamma_{0,d-4} = 1$, this implies that $\gamma_{ij} = 1$ for $i + j + 2 \in \{d, d-2\}$. The relations of type (22), corresponding to $\omega \in \mathbf{c}\Theta_{=2}$ with $|\omega| = 2$, then turn into identifications of some γ_{ij} for $i + j + 2 = d - 2$. But since these elements are already identified with 1, the relations of type (22) imply no independent identities, and hence can be removed.

As a result, we have obtained the generators and relations for $\pi_1(\mathcal{P}_{d-2}^{\mathbf{c}\Theta'})$, and the assertion follows. □

Now we can deduce a stabilization result for the fundamental group.

Theorem 6.15. *Let $\Theta \subseteq \mathbf{\Omega}_{\langle d], |\sim|' \geq 2}$ be a closed poset. For $d' \geq d+2$ such that $d' = d \equiv 2$, let $\Theta_{d'}$ be the smallest closed poset in $\mathbf{\Omega}_{\langle d']}$ containing Θ.*

Then for $d' \geq d+2$, we have an isomorphism $\pi_1(\mathcal{P}_{d'}^{\mathbf{c}\Theta_{d'}}) \cong \pi_1(\mathcal{P}_{d+2}^{\mathbf{c}\Theta_{d+2}})$.

Proof. We prove the assertion by induction on d'. For $d' = d+2$, the assertion is trivial, since the relevant polynomial spaces are identical. For $d' \geq d+4$, we have that $\Theta_{=2}$ contains no ω such that $|\omega| \in \{d', d'-2\}$. This implies that $(\Theta_{d'})_{=2}$ contains no ω such that $|\omega| \in \{d', d'-2\}$, and $\Theta_{d'-2} = \Theta_{d'} \cap \mathbf{\Omega}_{\langle d'-2]}$. Hence by Proposition 6.10, $\pi_1(\mathcal{P}_{d'}^{\mathbf{c}\Theta_{d'}}) \cong \pi_1(\mathcal{P}_{d'-2}^{\mathbf{c}\Theta_{d'-2}})$. The assertion now follows from the induction hypothesis. □

Before we can provide a negative answer to the question whether $\pi_1(\mathcal{P}_d^{\mathbf{c}\Theta})$ is free for large d, we need the following simple lemma, which is an immediate consequence of Proposition 6.9 and the definition of the free product of finitely presented groups.

Lemma 6.14. *Let $\Theta \subset \mathbf{\Omega}_{\langle d], |\sim|' \geq 2}$ be closed. Assume that for some $d' < d$ such that $d' \equiv d \mod 2$, the poset Θ contains all $\omega \in \mathbf{\Omega}_{\langle d], |\sim|' = 2}$ with $|\omega| = d'$.*

Then $\pi_1(\mathcal{P}_d^{\mathbf{c}\Theta})$ is the free product of $\pi_1(\mathcal{P}_{d'-2}^{\mathbf{c}\Theta})$ and a group presented by relations of type (3) *and* (22) *for $\omega \notin \Theta$, such that $|\omega| > d'$ and $|\omega|' = 2$.* ◇

Let $\Theta \subset \mathbf{\Omega}_{\langle 6], |\sim|' = 2}$ be the poset from (6.42) (see Example 6.9). For $d \geq 10$, let Θ_d be the smallest closed poset in $\mathbf{\Omega}_{\langle d]}$ containing Θ and all $\omega \in \mathbf{\Omega}_{\langle d], |\sim|' = 2}$ with $|\omega| = 8$. For $d \geq 10$, we have that Θ_d satisfies the conditions of Lemma 6.14 with $d' = 8$. Then, for $d \geq 10$, $\pi_1(\mathcal{P}_d^{\mathbf{c}\Theta_d})$ is the free product of $\pi_1(\mathcal{P}_d^{\mathbf{c}\Theta_6}) \cong \mathbb{Z}/2\mathbb{Z}$ and some group. In particular, $\pi_1(\mathcal{P}_d^{\mathbf{c}\Theta_d})$ is not free stably, as $d \to \infty$.

Next, we formulate some consequences of our results about spaces of polynomials with restrictions on the multiplicities of their critical points. For any closed poset $\Theta \subset \mathbf{\Omega}$ and any $d > 0$, we denote by $\mathcal{P}_{d+1}^{\mathsf{crit}\,\mathbf{c}\Theta}$ the space of polynomials of degree $d+1$, whose derivatives belong to $\mathcal{P}_d^{\mathbf{c}\Theta}$. The relation of the following corollary to Theorem 6.13 and Theorem 6.14 is well known.

Corollary 6.11. *For any closed poset $\Theta \subset \mathbf{\Omega}$ and any $d > 0$, the space $\mathcal{P}_{d+1}^{\mathsf{crit}\,\mathbf{c}\Theta}$ is homotopy equivalent to the space $\mathcal{P}_d^{\mathbf{c}\Theta}$. In particular, $\pi_1(\mathcal{P}_{d+1}^{\mathsf{crit}\,\mathbf{c}\Theta})$ is free for*

(i) $\Theta = \mathbf{\Omega}_{\langle d], |\sim|' \geq 2}$,

(ii) Θ *contains all* $\omega \in \mathbf{\Omega}_{\langle d]}$ *such that* $\omega = (\underbrace{1 \cdots 1}_{i} 2 \underbrace{1 \cdots 1}_{j} 2 \underbrace{1 \cdots 1}_{\ell})$ *for some* $j > 0$.

(iii) Θ *is such that any* $\omega \in \Theta$ *with* $|\omega|' = 2$ *is of the form* $(\underbrace{1 \cdots 1}_{i} 3 \underbrace{1 \cdots 1}_{j})$ *for some* $i, j \geq 0$.

Moreover, in case (iii), we have $\pi_1(\mathcal{P}_{d+1}^{\mathsf{crit}\,\mathbf{c}\Theta}) = \mathbb{Z}$.

Proof. The homotopy equivalence follows from the simple fact that the derivative map $\frac{d}{dx} : \mathcal{P}_{d+1}^{\mathsf{crit}\,\mathbf{c}\Theta} \to \mathcal{P}_{d}^{\mathbf{c}\Theta}$ is a trivial fibration with fiber $\mathbb{R}$. Then (i) follows from Theorem 6.13 and (ii)-(iii) from Theorem 6.14. □

Corollary 6.11, (iii) can be reformulated so that it becomes an analogue of Arnold's Theorem 6.11 in finite degrees d.

Corollary 6.12. *The fundamental group of the space of real monic polynomials of a fixed odd degree* $d > 1$ *with no real critical points of multiplicity higher than* 2 *is isomorphic to the group* $\mathbb{Z}$.

Chapter 7

Spaces of Polynomials with Constrained Divisors as Grassmanians for Immersions & Embeddings

7.1 In this chapter...

"Mathematics is the art of giving the same name to different things."

Henri Poincaré

This chapter, a natural extension of Chapter 5, is also inspired by the works of Arnold [Ar], [Ar1], [Ar2], and Vassiliev [Va], [Va1]–[Va3]. It reflects the content of [K12] and is based heavily on computations from [KSW1], [KSW2], as they are described in Chapter 6.

The research of Ekholm, Szücs, and Terpai [EST], [Sz], [Sz1], [Sz2], [SzTe], [SzTe1], which deals with bordisms of smooth maps whose singularity types are controlled *locally*. Philosophically, our work is close to the investigations in these papers, especially to the work of Lippner [Lip] and [SzTe1]. On one hand, by dealing with immersions of manifolds of a variety of dimensions, the papers [Sz], [Sz1], [Sz2], [SzTe], [SzTe1], and [Lip] contain more general results, while we are concerned here only with immersions of n-dimensional manifolds into $(n+1)$-dimensional targets. On the other hand, this tight squeezing forces us to deal with a more complex combinatorics than the one exhibited by the general Morin type singularities.

The main difference between the forth mentioned research and our efforts here is that we deal with bordisms of smooth maps with not *locally* controlled singularity types, but with bordisms of maps whose singularity types are controlled *semi-locally*, that is, we impose restrictions on the types of singularities and on their *order* along the 1-dimensional fibers of maps.

The classifying spaces for the bordisms we study are fundamentally different from the ones used by the other authors: in our case, they are based on spaces of real polynomials with constrained real divisors.

Let us first describe the results of this chapter informally, in a manner that clarifies their nature, but does not involve their most general forms, which carry the burden of combinatorial decorations.

We study immersions and embeddings $\beta : M \to \mathbb{R}\times Y$ of compact smooth n-manifolds M into products $\mathbb{R} \times Y$, where Y is a given compact smooth n-manifold. We denote by $\mathcal{L}$ the oriented 1-dimensional foliation on $\mathbb{R} \times Y$, formed by the fibers of the obvious projection $\pi : \mathbb{R} \times Y \to Y$. We are interested in the ways the image $\beta(M)$ may be tangent to the fibers of $\mathcal{L}$. More precisely, we impose restrictions Θ on the combinatorial patterns ω of such tangencies and organize $\{\beta : M \to \mathbb{R}\times Y\}$ that do not violate Θ into a set $\mathcal{Q}(Y, \mathbf{c}\Theta)$ of equivalence classes; we call equivalent β's "quasitopic". The quasitopies is a crossover between classical notions of pseudo-isotopies and bordisms (see Fig. 7.2). We will deal with various brands of quasitopies, richly decorated by combinatorial data. Here, to simplify the description of the forthcoming results, we omit the decorations. In particular, we may insist that the cardinality of the fiber of $\pi \circ \beta$ is less than or equal to a given natural number d. In such a case, we are dealing with the quasitopies $\mathcal{Q}_d(Y, \mathbf{c}\Theta)$.

The main observation is that spaces $\mathcal{P}_d^{\mathbf{c}\Theta}$ of real degree d monic polynomials in one variable, whose real divisors do not belong to the forbidden poset Θ, play a role of universal objects (like the Grassmanians) for immersions/embeddings β as above. This is reminiscent to the role played by the Stiefel manifolds of k-frames in $\mathbb{R}^m$ in the Smale theory of immersions [S]. At least for embeddings, we are able to compute $\mathcal{Q}_d^{\mathsf{emb}}(Y, \mathbf{c}\Theta)$ in terms of the homotopy theory of $\mathcal{P}_d^{\mathbf{c}\Theta}$ and Y. The computation uses the results from Chapter 6 about homotopy and homology types of $\mathcal{P}_d^{\mathbf{c}\Theta}$ for many specific Θ's. This information allows us to describe cohomological invariants of $\mathcal{Q}_d^{\mathsf{emb}}(Y, \mathbf{c}\Theta)$ and $\mathcal{Q}_d^{\mathsf{imm}}(Y, \mathbf{c}\Theta)$ (characteristic classes of quasitopies) in purely combinatorial terms, related to Θ. Here the superscript "imm" stands for immersions and "emb" for embeddings. In many special cases, we see subtle connections between $\mathcal{Q}_d^{\mathsf{emb}}(Y, \mathbf{c}\Theta)$ and homotopy groups of spheres. Implicitly, this connection goes back to [Ar], [Ar1] and [Va].

Let D^n denote the standard n-ball. We will see that the sets $\mathcal{Q}_d^{\mathsf{emb}}(D^n, \mathbf{c}\Theta)$ and $\mathcal{Q}_d^{\mathsf{imm}}(D^n, \mathbf{c}\Theta)$ have a group structure (abelian for $n > 1$). We denote these groups by $\mathsf{G}_d^{\mathsf{emb}}(n; \mathbf{c}\Theta)$ and $\mathsf{G}_d^{\mathsf{imm}}(n; \mathbf{c}\Theta)$ and prove that $\mathsf{G}_d^{\mathsf{imm}}(n; \mathbf{c}\Theta)$ is a split extension of $\mathsf{G}_d^{\mathsf{emb}}(n; \mathbf{c}\Theta)$. Then we are able to validate a group isomorphism

$$\mathsf{G}_d^{\mathsf{emb}}(n; \mathbf{c}\Theta) \approx \pi_n(\mathcal{P}_d^{\mathbf{c}\Theta}, pt).$$

With the help of the boundary connected sum operation, the groups $\mathsf{G}_d^{\mathsf{emb}}(n; \mathbf{c}\Theta)$ and $\mathsf{G}_d^{\mathsf{imm}}(n; \mathbf{c}\Theta)$ act naturally on the sets $\mathcal{Q}_d^{\mathsf{emb}}(Y, \mathbf{c}\Theta)$ and $\mathcal{Q}_d^{\mathsf{imm}}(Y, \mathbf{c}\Theta)$, respectively.

For two pairs $A \supset B$, $C \supset D$ of topological spaces, let $[(A, B), (C, D)]$ denote the homotopy classes of continuous maps $f : (A, B) \to (C, D)$, where $f(B) \subset D$.

We prove that the classifying map

$$\Phi^{\mathsf{emb}} : \mathcal{Q}_d^{\mathsf{emb}}(Y, \mathbf{c}\Theta) \to [(Y, \partial Y), (\mathcal{P}_d^{\mathbf{c}\Theta}, pt)]$$

is a bijection, and the classifying map

$$\Phi^{\mathsf{imm}} : \mathcal{Q}_d^{\mathsf{imm}}(Y, \mathbf{c}\Theta) \to [(Y, \partial Y), (\mathcal{P}_d^{\mathbf{c}\Theta}, pt)]$$

is a surjection. Moreover, these maps are equivariant with respect to the $\mathsf{G}_d^{\mathsf{emb}}(n; \mathbf{c}\Theta)$- or $\mathsf{G}_d^{\mathsf{imm}}(n; \mathbf{c}\Theta)$-actions on the quasitopies (the sources of the classifying maps $\Phi^{\mathsf{emb/imm}}$) and the corresponding action of the homotopy groups $\pi_n(\mathcal{P}_d^{\mathbf{c}\Theta}, pt)$ on the sets $[(Y, \partial Y), (\mathcal{P}_d^{\mathbf{c}\Theta}, pt)]$ (the targets of the maps $\Phi^{\mathsf{emb/imm}}$).

We present also a variety of results about the stabilization of $\mathcal{Q}_d^{\mathsf{emb}}(Y, \mathbf{c}\Theta)$ as $d \to \infty$.

In Chapter 8, we will apply and extend the results of this chapter to a study of the traversing smooth flows on manifolds X with boundaries, where the poset Θ controls the combinatorial patterns of tangency of the flow trajectories to the boundary ∂X.

We conclude this introduction with a few examples of computations, performed in Chapter 7. There is nothing exceptional about these propositions-examples, they are exhibited here just to illustrate the intricate flavor of meta-theorems in the chapter. On a positive note, the next four propositions are formulated in terms which do not require familiarity with the notations introduced later. The notion of quasitopies is implicit in these examples.

Consider the space $\mathbb{R}^n$ with the coordinates $u, x_1, \dots, x_{n-1}$ and the projections

$$\pi_n : (u, x_1, \dots, x_{n-1}) \to (x_1, \dots, x_{n-1})$$

We denote by $\mathbb{R}^n_+ \subset \mathbb{R}^n$ the half space $\{x_{n-1} \geq 0\}$.

For any subset $A \subset \mathbb{R}^n$ whose x_1-coordinates belong to an interval $(0, 1)$, we denote by kA the disjoint union of k copies of A, obtained from A by applying $(k-1)$ times the parallel shift

$$\{Sh : (u, x_1, \dots, x_{n-1}) \to (u, x_1 + 1, \dots, x_{n-1})\}.$$

In the following four propositions-examples, we assume that all the submanifolds are in general position with respect to the projections $\{\pi_n\}_n$, i.e., the restrictions of the projections to submanifolds locally are Morin maps [Morin].

Proposition 7.1. *There exists a regularly embedded closed oriented 3-fold $M^3 \subset \mathbb{R}^4$ such that:*

- *the fibers of the projection $\pi_4| : M^3 \to \mathbb{R}^3$ have cardinalities ≤ 9,*
- *every π_4-fiber is tangent to M^3 with order of tangency $\leq \mathbf{3}$,*
- *M^3 is a boundary of an oriented compact smooth 4-manifold $N^4 \subset \mathbb{R}^5_+$.*
- *Any such N^4 must be tangent to* some *π_5-fiber with the order of tangency $\geq \mathbf{4}$, provided that the fibers of the projection $\pi_5| : N^4 \to \mathbb{R}^4$ have cardinalities ≤ 9,*
- *Moreover, for any natural k, the disjoint union $kM^3 \subset \mathbb{R}^4$ of k copies of M^3, has the same properties as M^3.* ◇

Proposition 7.2. *There exists a regularly embedded closed oriented 14-dimensional manifold $M^{14} \subset \mathbb{R}^{15}$ such that:*

- *the fibers of the projection $\pi_{15} : M^{14} \to \mathbb{R}^{14}$ have cardinalities ≤ 9,*
- *every π_{15}-fiber is tangent to M^{14} with order of tangency $\leq \mathbf{3}$,*
- *M^{14} is a boundary of a compact smooth 15-manifold $N^{15} \subset \mathbb{R}^{16}_+$.*
- *Any such N^{15} must be tangent to* some *π_{16}-fiber with the order of tangency $\geq \mathbf{4}$, provided that the fibers of the projection $\pi_{16} : N^{15} \to \mathbb{R}^{15}_+$ have cardinalities ≤ 9.*
- *Moreover, 84 disjoint copies of M^{14} form the boundary of some 15-dimensional $\tilde{N}^{15} \subset \mathbb{R}^{16}_+$, tangent to all π_{16}-fibers with the order of tangency $\leq \mathbf{3}$, provided that the fibers of the projection $\pi_{16} : N^{15} \to \mathbb{R}^{15}_+$ have cardinalities ≤ 9; less than **84** copies of M^{14} do not bound such N^{15}.* ◇

Proposition 7.3. *There exists a regularly embedded closed oriented 7-dimensional manifold $M^7 \subset \mathbb{R}^8$ such that:*

- *the fibers of the projection $\pi_8 : M^7 \to \mathbb{R}^7$ have cardinalities ≤ 8,*
- *every π_8-fiber is tangent to M^7 with order of tangency $\leq \mathbf{3}$,*
- *M^7 is a boundary of a compact orientable smooth 8-manifold $N^8 \subset \mathbb{R}^9_+$.*

- *Any such* N^8 *must be tangent to* some π_8*-fiber with the order of tangency* $\geq \mathbf{4}$*, provided that the fibers of the projection* $\pi_9 : N^8 \to \mathbb{R}^8_+$ *have cardinalities that do not exceed* 8.
- *Moreover,* **12** *disjoint copies of* M^7 *form the boundary of some 8-dimensional* $\tilde{N}^8 \subset \mathbb{R}^9_+$*, tangent to all* π_9*-fibers with the order of tangency* $\leq \mathbf{3}$*, provided that the fibers of the projection* $\pi_9 : N^8 \to \mathbb{R}^8_+$ *have cardinalities that do not exceed* 8*; less than* 12 *copies of* M^7 *do not bound such* $\tilde{N}^8$. ◊

According to Example 7.3, the number of copies of M in these assertions reflect the computations of the homotopy groups $\pi_n(S^2)$ for $n = 3$ and $n = 14$ in case of the first two propositions, and of the homotopy group $\pi_7(S^4)$ in case of the last one.

Evidently, these three propositions depend very much on the delicate interplay between the maximal cardinalities d of the fibers of the maps $\pi_n : M^{n-1} = \partial N^n \to \mathbb{R}^{n-1}$ and $\pi_{n+1} : N^n \to \mathbb{R}^n_+$ and the orders k of tangency of the manifolds to the fibers. For fixed k's, as d increases, these assertions may fail.

In contrast, the next proposition, based on Example 6.8 and Example 7.7, exhibits a stabilization of the cardinalities of the $(\pi_{n+1} : N^n \to \mathbb{R}^n_+)$-fibers for the fixed boundary $M = \partial N^n \subset \mathbb{R}^n$ and the fixed "forbidden" closed poset $\Theta = \langle (4), (33) \rangle$ generated by two tangency patterns, (4) and (33).

Proposition 7.4.

• *Let* $M^{n-1} \subset \mathbb{R}^n$ *be a smooth closed submanifold such that the fibers of the map* $\pi_n : M^{n-1} \subset \mathbb{R}^{n-1}$ *avoid the tangency patterns from the poset* $\langle (4), (33) \rangle$ *and are of the maximal cardinality* $\leq 2n - 6$.

If no smooth compact submanifold $(N^n, \partial N^n) \subset (\mathbb{R}^{n+1}_+, \mathbb{R}^n)$*, subject to the following three properties*

(1) *the boundary* $\partial N^n = M^{n-1}$*,*

(2) *the fibers of* $\pi_{n+1} : N^n \to \mathbb{R}^n_+$ *avoid the tangency patterns from the poset* $\langle (4), (33) \rangle$*,*

(3) *the maximal cardinality of the of* π_{n+1}*-fibers does not exceed* $2n - 4$*, exists, then no such* N^n *that possesses only properties* (1) *and* (2) *exists as well.*

• *Let* $(N^n, \partial N^n) \subset (\mathbb{R}^{n+1}_+, \mathbb{R}^n)$ *be a smooth compact submanifold whose* π_{n+1}*-fibers avoid tangency patterns from the poset* $\langle (4), (33) \rangle$*. Assume that* *the fibers of the map* $\pi_{n+1}| : \partial N^n \to \mathbb{R}^n$ *have the maximal cardinality* $\leq 2n - 6$.

Then there exists another smooth compact submanifold manifold $(\tilde{N}^n, \partial\tilde{N}^n) \subset (\mathbb{R}^{n+1}_+, \mathbb{R}^n)$ *such that:*

(1) $\partial\tilde{N}^n = \partial N^n$,

(2) *the tangency patterns to the fibers* $\pi_{n+1} : \tilde{N}^n \to \mathbb{R}^n_+$ *still avoid* $\langle(4),(33)\rangle$,

(3) *the cardinality of the fibers of the map* $\pi_{n+1}| : \tilde{N}^n \to \mathbb{R}^n$ *is* $\leq 2n-4$. ◇

7.2 Immersions $\{\beta : M^n \to \mathbb{R} \times Y^n\}$ and their self-intersections loci

We start with a description of the main theorem of immersion theory, known as the Smale-Hirsh Theorem [Sm2], [H1], [Gr4], [Ph].

Let M^m, N^n be smooth manifolds, $m \leq n$. We assume that the boundary $\partial N = \emptyset$. If $m = n$, we assume that $M \setminus \partial M$ has no compact components (this assumption is essential!). Let $\mathsf{imm}(M, N)$ be the space of smooth immersions $f : M \to N$, and let $\mathsf{fimm}(M, N)$ be the space of formal immersions $f : M \to N$, i.e., the space of pairs $(f, \delta f)$, where $f : M \to N$ is a continuous (not necessarily smooth) map and $\delta f : TM \to f^*(TN)$ some vector bundle map (over M) which is injective on each fiber. In other words, $\delta f : TM \to f^*(TN)$ is a choice of a linear injection $T_xM \to T_{f(x)}N$ for each $x \in M$, depending continuously on x. We think of δf as a "formal derivative" of f.

Theorem 7.1. *([Sm2], [H1]) Under these assumptions, the map*

$$\mathcal{D} : \mathsf{imm}(M, N) \to \mathsf{fimm}(M, N),$$

given by $f \to (f, df)$, *is a weak homotopy equivalence.*

If M *is compact, then* $\mathcal{D}$ *is a genuine homotopy equivalence.* ◇

In this chapter, we investigate a very special case of immersions $f : M \to N$, where $\dim(M) + 1 = \dim(N)$. However, in our setting the target manifold N carries a simple foliation $\mathcal{L}$ of the product type. We concentrate on the way the image of M interacts with the leaves of $\mathcal{L}$.

We adopt the following definition of a *proper immersion* of manifolds.

Definition 7.1. Let M be a compact smooth n-manifold and W a compact smooth $(n+1)$-manifold. We call an immersion $\beta : M \to W$ proper if:

(1) $\beta(\partial M) \subset \partial W$,
(2) β is transversal to ∂W (thus the kernel of the differential $D(\beta|_{\partial M})$ is also trivial). ◊

When M is a closed manifold, requirements (1) and (2) are vacuous. When Y is a closed manifold, then M must be a closed manifold as well.

Remark 7.1. In this chapter, for a *given* compact smooth n-manifold Y, we are only interested in proper immersions/embeddings $\beta : M \to \mathbb{R} \times Y$ of compact smooth n-manifolds M into the product $\mathbb{R} \times Y$. The existence of such β puts substantial restrictions on the *topological nature* of M. For example, the tangent n-bundle τ_M must be a subbundle of the $(n+1)$-bundle $\beta^*(\tau_Y) \oplus \underline{\mathbb{R}}$, where tau_Y denotes the tangent bundle of Y. By the classical Hirsch Theorem [Hi], the existence of such an immersion β in the homotopy class of a given map $\tilde{\beta} : M \to \mathbb{R} \times Y$ is equivalent to the property of the stable bundle $\tilde{\beta}^*(\tau_{\mathbb{R}\times Y}) \oplus \nu_M$ to be of geometric dimension ≤ 1. Here ν_M is the stable normal bundle of M. The geometric dimension of a stable bundle η is defined to be the minimal dimension of a bundle $\eta^\sharp$ in the stable equivalence class of η. ◊

Definition 7.2 below rests on the following notion. Let G, H be two smooth hypersurfaces in a smooth manifold X. We fix a point $a \in G \cap H$ and consider its neighborhood $U_a \subset X$. Let $g : U_a \to \mathbb{R}$, $h : U_a \to \mathbb{R}$ be two smooth functions such that 0 is a regular value for both, and $G \cap U_a = g^{-1}(0)$, $H \cap U_a = h^{-1}(0)$.

Let q be a natural number. We say that two hypersurfaces G, H are q-jet equivalent at $a \in G \cap H$, if there exists smooth functions $\phi, \chi : U_a \to \mathbb{R}$ such that $\phi(a) \neq 0$, $\chi(a) \neq 0$ and the q-jets are equal: $jet^q_a(\chi \cdot g) = jet^q_a(\phi \cdot h)$. In other words, $jet^q_a(\chi) \cdot jet^q_a(g) = jet^q_a(\phi) \cdot jet^q_a(h)$ modulo the $(q+1)^{st}$ power of the maximal ideal m_0 in the polynomial ring in $\dim X$ variables.

We denote by $\mathsf{j}^q_a(H)$ the q-jet equivalence class of $H \subset X$ at a, and by J^q_a the set of all such q-jet equivalence classes of hypersurfaces through a. The space J^q_a depends on finitely many parameters, the number of which is a function of $\dim X$ and q. Since J^1_a is the Grassmanian of hyperplanes, we view J^q_a as a *generalized Grassmanian.* We denote by $\mathsf{J}^q(X) \to X$ the fibration over X, whose fiber is J^q_a.

Definition 7.2. We call an immersion $\beta : M \to X$ of a smooth compact n-manifold M into a smooth compact $(n+1)$-manifold X q-separable if:

- for any point $a \in \beta(M)$, the set $\beta^{-1}(a)$ is finite,

- there is a natural number q such that, for any point $a \in \beta(M)$, the q-jet equivalence classes of the local branches of $\beta(M)$, labeled by points $\beta^{-1}(a)$, are all distinct. ◇

For the rest of the section, we fix a compact connected smooth n-manifold Y and consider proper immersions $\beta : M \to \mathbb{R} \times Y$ of compact smooth n-manifolds M. By a small perturbation, we may assume that different local branches of $\beta(M)$ are in general position at their mutual intersections (see [GG], Theorem 4.13). In particular, we may assume that β is 1-separable in the sense of Definition 7.2.

Recall the following definition from [LS] and [Th], adjusted to our setting.

Definition 7.3. A smooth proper immersion $\beta : M \to \mathbb{R} \times Y$ is called k-normal if, for each k-tuple of distinct points $x_1, \dots, x_k \in M$ with $\beta(x_1) = \dots = \beta(x_k)$, the images of the tangent spaces $T_{x_1}M, \dots, T_{x_k}M$ under the differential $D\beta$ are in general position in the tangent space $T_y(\mathbb{R} \times Y)$, where $y = \beta(x_i)$. Also, for such a k-tuple, the images of the tangent spaces $T_{x_1}(\partial M), \dots, T_{x_k}(\partial M)$ under the differential $D(\beta|_{\partial M})$ are in general position in the tangent space $T_y(\mathbb{R} \times \partial Y)$.

◇

For a given proper immersion $\beta : (M, \partial M) \to (\mathbb{R} \times Y, \mathbb{R} \times \partial Y)$, we consider the k-fold product map $(\beta)^k : (M)^k \to (\mathbb{R} \times Y)^k$. Let Σ_k be the preimage of the diagonal $\Delta \subset (\mathbb{R} \times Y)^k$ under the map $(\beta)^k$. For a k-normal immersion β, Σ_k is a smooth manifold of dimension $n - k + 1$ [LS]. Following [LS], we call Σ_k the k self-intersection manifold of β. Note that the same construction applies to $\beta| : \partial M \to \mathbb{R} \times \partial Y$, producing the boundary $\partial \Sigma_k$.

Let $p_1 : (M)^k \to M$ be the obvious projection on the first factor of the product. Then $p_1 : \Sigma_k \to M$ is an *immersion* [LS]. Its image $p_1(\Sigma_k)$ is the set of points $x_1 \in M$ such that there exist distinct $x_2, \dots, x_k \in M$ with the property $\beta(x_2) = \beta(x_1), \dots, \beta(x_k) = \beta(x_1)$.

If both M and Y are orientable, then so is Σ_k. In such a case, we denote by $[\Sigma_k]^\beta_\star \in H_{n-k+1}(Y, \partial Y; \mathbb{Z})$ the image of the relative fundamental class $[\Sigma_k]$ under the composition $\pi \circ \beta \circ p_1$, where $\pi : \mathbb{R} \times Y \to Y$ is the obvious projection. Without orientability assumptions, only the classes $[\Sigma_k]^\beta_\star \in H_{n-k+1}(Y, \partial Y; \mathbb{Z}_2)$ are available.

Let $\mathcal{D}_M : H_*(M, \partial M) \to H^{n-*}(M)$ be the Poincaré duality isomorphism. (Here, depending on the orientability of M, the coefficients are $\mathbb{Z}$

or $\mathbb{Z}_2$.) The Poincaré duality isomorphism

$$\mathcal{D}_{\mathbb{R}\times Y} : H_*(\mathbb{R}\times Y,\, \mathbb{R}\times \partial Y) \to H_c^{n+1-*}(\mathbb{R}\times Y)$$

is also available, where the subscript c indicates the cohomology with compact support, and the choice of coefficients $\mathbb{Z}$ or $\mathbb{Z}/2\mathbb{Z}$ depends on the orientability of Y.

Let $\nu^\beta =_{\mathsf{def}} \beta^*(\tau_{\mathbb{R}\times Y})/\tau_M$ be the 1-bundle over M "normal" to $\beta(M)$, and let $W^1(\nu^\beta)$ denote its first Stiefel-Whitney characteristic class. Put $\tilde{\Sigma}_k =_{\mathsf{def}} p_1(\Sigma_k) \subset M$.

Then applying the main result of [LS] to our setting, for $k > 1$, we get a pleasing formula for the cocycle dual to the cycle $\tilde{\Sigma}_k$:

$$\mathcal{D}_M([\tilde{\Sigma}_k]) = \pm\Big(\beta^* \circ \mathcal{D}_{\mathbb{R}\times Y} \circ \beta_*([\tilde{\Sigma}_k]) \;-\; W^1(\nu^\beta)\Big)^{k-1}. \tag{7.1}$$

Applying $\pi \circ \beta \circ p_1$ to Σ_k, these constructions lead directly to

Lemma 7.1. *For compact oriented n-manifolds M and Y and $k \in [2, n+1]$, a proper k-normal immersion $\beta : M \to \mathbb{R}\times Y$ generates the homology class $[\Sigma_k]^\beta_\star \in H_{n-k+1}(Y, \partial Y; \mathbb{Z})$ and its Poincaré dual $[\Sigma_k]_\beta^\star \in H^{k-1}(Y;\mathbb{Z})$. For non-orientable M or Y, a proper k-normal immersion β produces similar classes $[\Sigma_k]^\beta_\star$ and $[\Sigma_k]_\beta^\star$ in the homology/cohomology with coefficients in $\mathbb{Z}_2$.* ♢

Lemma 7.2. *Let two proper k-normal immersions*

$$\beta_0 : (M_0, \partial M_0) \to (\mathbb{R}\times Y \times \{0\},\, \mathbb{R}\times \partial Y \times \{0\})$$

$$\beta_1 : (M_1, \partial M_1) \to (\mathbb{R}\times Y \times \{1\}, \mathbb{R}\times \partial Y \times \{1\})$$

be cobordant with the help of a proper k-normal immersion

$$B : (N, \delta N) \to (\mathbb{R}\times Y \times [0,1],\, \mathbb{R}\times \partial Y \times [0,1]),$$

where the boundary $\partial N = (M_0 \sqcup -M_1) \cup_{(\partial M_0 \sqcup -\partial M_1)} \delta N$, $B|_{M_0} = \beta_0$, $B|_{M_1} = \beta_1$, and $B(\delta N) \subset \mathbb{R}\times \partial Y \times [0,1]$.

Then the k-intersection manifolds

$$\beta_0 \circ (p_0)_1 : (\Sigma_k^{\beta_0}, \Sigma_k^{\beta_0^\partial}) \to (\mathbb{R}\times Y \times \{0\},\, \mathbb{R}\times \partial Y \times \{0\}),$$

$$\beta_1 \circ (p_1)_1 : (\Sigma_k^{\beta_1}, \Sigma_k^{\beta_1^\partial}) \to (\mathbb{R}\times Y \times \{1\},\, \mathbb{R}\times \partial Y \times \{1\})$$

are cobordant with the help of the map

$$B \circ P_1 : (\Sigma_k^B, \Sigma_k^{B^\delta}) \to (\mathbb{R}\times Y \times [0,1],\, \mathbb{R}\times \partial Y \times [0,1]).$$

Here $(p_0)_1 : \Sigma_k^{\beta_0} \to M_0$, $(p_1)_1 : \Sigma_k^{\beta_1} \to M_1$, *and* $P_1 : \Sigma_k^B \to N$ *denote the projections on the first factors of the* k*-th powers* $(M_0)^k$, $(M_1)^k$, *and* $(N)^k$, *respectively. The maps,* $(p_0)_1$, $(p_1)_1$, *and* P_1, *are immersions.*[1]

If the manifolds M_0, M_1, N, *and* Y *are orientable, then so are the manifolds* $\Sigma_k^{\beta_0}$, $\Sigma_k^{\beta_1}$, *and* Σ_k^B.

Proof. By the definition of k-normality, the k-th power B^k of B is transversal to the diagonal $\tilde{\Delta} \subset (\mathbb{R} \times Y \times [0,1])^k$—a manifold with corners $\partial Y \times \partial[0,1]$. Thus, the preimage $\Xi_k^B =_{\mathsf{def}} (B^k)^{-1}(\tilde{\Delta}) \subset (N)^k$ is a compact smooth manifold of dimension $n - k + 2$. Its projection via $P : (N)^k \to N$ on the first factor N is an immersion [LS]. Thus the composition of $B \circ P : \Xi_k \to \mathbb{R} \times Y \times [0,1]$ delivers a cobordism between the pair of maps

$$\beta_0 \circ (p_0)_1 : (\Sigma_k(\beta_0),\, \partial\Sigma_k(\beta_0)) \to (\mathbb{R} \times Y \times \{0\},\, \mathbb{R} \times \partial Y \times \{0\})$$

$$\beta_1 \circ (p_1)_1 : (\Sigma_k(\beta_1),\, \partial\Sigma_k(\beta_1)) \to (\mathbb{R} \times Y \times \{1\},\, \mathbb{R} \times \partial Y \times \{1\}).$$

By [LS], if M_0, M_1, N, and Y are orientable, then so are $\Sigma_k^{\beta_0}$, $\Sigma_k^{\beta_1}$, and Ξ_k^B. □

Corollary 7.1. *Let* $\pi : \mathbb{R} \times Y \to Y$ *be the obvious projection. Let proper* k-normal *immersions* $\beta_0 : (M_0, \partial M_0) \to (\mathbb{R} \times Y, \mathbb{R} \times \partial Y)$ *and* $\beta_1 : (M_1, \partial M_1) \to (\mathbb{R} \times Y, \mathbb{R} \times \partial Y)$ *be cobordant with the help of a proper immersion*[2] $B : (N, \delta N) \to (\mathbb{R} \times Y \times [0,1], \mathbb{R} \times \partial Y \times [0,1])$, *as in Lemma 7.1. Then both maps*

$$\pi\circ\beta_0\circ(p_0)_1 : (\Sigma_k^{\beta_0}, \Sigma_k^{\beta_0^\partial}) \to (Y, \partial Y) \quad \textit{and} \quad \pi\circ\beta_1\circ(p_1)_1 : (\Sigma_k^{\beta_1}, \Sigma_k^{\beta_1^\partial}) \to (Y, \partial Y)$$

define the same element $[(\Sigma_k^\beta, \partial\Sigma_k^\beta) \to (Y, \partial Y)]$ *of the relative bordism group* $\mathbf{B}_{n+1-k}(Y, \partial Y)$.

If M_0, M_1, N, *and* Y *are oriented, then the two immersions* β_0, β_1 *define the same element of the oriented bordism group* $\mathbf{OB}_{n+1-k}(Y, \partial Y)$.

Proof. By the Thom Multijet Transversality Theorem [Th], we may smoothly perturb the immersion map B within the space of immersions, without changing it on $\partial N \setminus \delta N = M_0 \sqcup M_1$, so that its k-th power B^k will become transversal to the diagonal $\tilde{\Delta} \subset (\mathbb{R} \times Y \times [0,1])^k$. For such a perturbation, the cobordism B becomes k-normal, so that $\Xi_k = (B^k)^{-1}(\tilde{\Delta}) \subset (N)^k$ is a compact manifold of dimension $n - k + 2$. Now the claim follows from Lemma 7.2. □

[1] Note that two immersions may be cobordant as smooth maps, but not by a cobordism which is an immersion.

[2] not necessary k-normal!

7.3 Constrained patterns of tangency to the 1-foliation $\mathcal{L}$ and the classifying maps to the polynomial spaces $\mathcal{P}_d^{\mathbf{c}\Theta}$

Let X be a compact topological space. We denote by $C^0(X)$ the algebra of real continuous functions on X. Let

$$\{P(u) = u^d + a_{d-1}u^{d-1} + \ldots + a_1 u + a_0\} \tag{7.2}$$

be a family of real monic polynomials on X, where the functional coefficients $a_i \in C^0(X)$. We fix a closed subposet $\Theta \subset \mathbf{\Omega}_{\leq d}$. Assume that, for each point $x \in X$, the polynomial

$$P(u, x) = u^d + a_{d-1}(x)u^{d-1} + \ldots + a_1(x)u + a_0(x)$$

has a real zero divisor whose combinatorial pattern $\omega(x) \in \mathbf{c}\Theta =_{\mathsf{def}} \mathbf{\Omega} \setminus \Theta$.

Evidently, such polynomial family $P(u, x)$ gives rise to a continuous map $\Phi_P : X \to \mathcal{P}_d^{\mathbf{c}\Theta}$, defined by the formula $x \to P(\sim, x)$. Here the target space $\mathcal{P}_d^{\mathbf{c}\Theta}$ is an open subset of $\mathcal{P}_d \approx \mathbb{R}^d$. The preimage of $\mathbf{c}\Theta$-stratification of $\mathcal{P}_d^{\mathbf{c}\Theta}$ under Φ_P produces an interesting stratification $\mathcal{S}(X, P)$ of the space X.

When X is a smooth compact manifold, it is possible to perturb the functional coefficients of P to form a new polynomial family $\tilde{P}$ so that the perturbed map $\Phi_{\tilde{P}}$ will be transversal to each pure stratum $\{\mathcal{P}_d^{\omega}\}_{\omega \in \mathbf{c}\Theta}$ of $\mathcal{P}_d^{\mathbf{c}\Theta}$. If X has a boundary ∂X, then we may assume also that $\Phi_{\tilde{P}}\big|_{\partial X}$ is transversal to each pure stratum $\mathcal{P}_d^{\omega} \subset \mathcal{P}_d^{\mathbf{c}\Theta}$.

For such polynomial family $\{\tilde{P}(u, x)\}_{x \in X}$, the $\mathbf{c}\Theta$-labeled stratification $\mathcal{S}(X, \tilde{P})$ of X reflects the geometry of the $\mathbf{c}\Theta$-stratification of the target space $\mathcal{P}_d^{\mathbf{c}\Theta}$ and gives rise to a variety of topological invariants of X, generated by $\{\tilde{P}(u, x)\}_{x \in X}$.

The polynomial family $P = \{P(u, x)\}_{x \in X}$ as above generates two loci:

$$\begin{aligned} \partial\mathcal{E}_P &=_{\mathsf{def}} \{(u, x)|\ P(u, x) = 0\} \subset \mathbb{R} \times X, \\ \mathcal{E}_P &=_{\mathsf{def}} \{(u, x)|\ P(u, x) \leq 0\} \subset \mathbb{R} \times X, \end{aligned} \tag{7.3}$$

which are maped in X by the obvious projection $\pi : \mathbb{R} \times X \to X$. The fiber over $x \in X$ of the projection $\pi^{\partial} : \partial\mathcal{E}_P \to X$ is the support of a real zero divisor $D_{\mathbb{R}}(x)$ of $P(u, x)$ of a degree $\leq d$. For some $x \in X$, the x-fiber may be empty; then the divisor $D_{\mathbb{R}}(x)$ is equal to zero.

This setting leads to a natural question:

"*What are the maximal sets $\{U_\alpha \subset X\}_\alpha$ (open, closed, any ...) which admit a* continuous section $\sigma_\alpha : U_\alpha \to \partial\mathcal{E}_P$ *of the projection* $\pi^{\partial} : \partial\mathcal{E}_P \to X$?"

In other words, given a particular root $u_\star$ of $P(u, x_\star)$, what is the maximal set $U_{u_\star, x_\star} \subset X$ over which a global solution of the equation $\{P(u, x) = 0\}$, which extends $u_\star$, exists?

The question resembles the question about the maximal domain of an analytic continuation of a given analytic function f (leading to the Riemannian surface associated with f). Perhaps, tackling this question deserves a different investigation, likely dealing with the interplay between the fundamental groups $\pi_1(X)$ and $\pi_1(\mathcal{P}_d^{\mathsf{c}\Theta})$...

Let $C^k(\sim)$ denote the space of k times continuously differentiable functions. When X is a C^k-differentiable manifold and all the coefficients $a_i \in C^k(X)$ of polynomials like in (7.2), then $\partial\mathcal{E}_P$ is a C^k-differentiable hypersurface in $\mathbb{R} \times X$. Indeed, $\partial\mathcal{E}_P$ admits the C^k-differentiable parametrization

$$(u, a_{d-1}, \dots, a_1) \to (u, a_{d-1}, \dots, a_1,\ P(u) - P(0)).$$

Now, given an appropriate hypersurface $\partial\mathcal{E} \subset \mathbb{R} \times X$, we aim *to reverse* the correspondence $P \rightsquigarrow (\mathcal{E}_P, \partial\mathcal{E}_P)$ by constructing a polynomial family $P = P_{\partial\mathcal{E}}$ on X that generates $\partial\mathcal{E}$.

For a given d, let us consider the crucial for our quest $(d+1)$-dimensional domain

$$\begin{aligned} \mathcal{E}_d &=_{\mathsf{def}} \big\{(u, P) \in \mathbb{R} \times \mathcal{P}_d |\ P(u) \leq 0\big\} \quad \text{and its boundary} \\ \partial\mathcal{E}_d &=_{\mathsf{def}} \big\{(u, P) \in \mathbb{R} \times \mathcal{P}_d |\ P(u) = 0\big\}. \end{aligned} \tag{7.4}$$

Since $\partial\mathcal{E}_d$ is a graph of the map

$$(u, a_1, \dots, a_{d-1}) \longrightarrow a_0 = -\sum_{i=1}^{d-1} a_i u^i,$$

it follows that $\mathcal{E}_d$ is diffeomorphic to the half-space $\mathbb{R}^{d+1}_+$, and $\partial\mathcal{E}_d$ to the space $\mathbb{R}^d$.

We denote by $\pi_d : \mathcal{E}_d \to \mathcal{P}_d$ the obvious projection map. For $d \equiv 0 \mod 2$, its fibers are compact (are finite unions of closed intervals and singletons).

Let Y be a smooth compact manifold, and $\beta : (M, \partial M) \to (\mathbb{R} \times Y, \mathbb{R} \times \partial Y)$ an immersion. We denote by $\mathcal{L}$ the one-dimensional oriented foliation generated by the fibers of the projection map $\pi : \mathbb{R} \times Y \to Y$, and by $\mathcal{L}_y$ the fiber $\pi^{-1}(y)$.

If $\partial Y \neq \emptyset$, we add a collar to Y and denote the resulting manifold by $\hat{Y}$. Similarly, if $\partial M \neq \emptyset$, we add a collar to M, which results in a new manifold

$\hat{M}$. We still assume that $\beta(\partial M) \subset \mathbb{R}\times\partial Y$ and that $\hat{\beta}(\hat{M}\setminus M) \subset \mathbb{R}\times(\hat{Y}\setminus Y)$. Using that β is transversal to $\mathbb{R} \times \partial Y$ and adjusting the collar, we may assume that the immersion $\beta : M \to \mathbb{R} \times Y$ extends to an immersion $\hat{\beta} : \hat{M} \to \mathbb{R} \times \hat{Y}$.

Next, for each point $b \in M$, we introduce a natural number $\mu^\beta(b)$, the multiplicity of tangency between the b-labeled local branch of $\hat{\beta}(\hat{M})$ and the leaf of $\mathcal{L}$ through the point $a =_{\mathsf{def}} \beta(b)$. If a local branch $\tilde{M}$ of $\hat{\beta}(\hat{M})$ is given as the zero set of a locally defined smooth function $z : \mathbb{R} \times \hat{Y} \to \mathbb{R}$ for which 0 is its regular value, then the multiplicity/order of tangency of the π-fiber $\mathcal{L}_y$ with $\tilde{M}$ at a point $a = (u, y) \in \tilde{M} \cap \mathcal{L}_y$ is defined as the natural number $\mu =_{\mathsf{def}} \mu^\beta(b)$ such that the jet $\mathsf{jet}_a^{\mu-1}(z|_{\mathcal{L}_y}) = 0$, but $\mathsf{jet}_a^{\mu}(z|_{\mathcal{L}_y}) \neq 0$. In particular, if the branch is transversal to the leaf, then $\mu^\beta(b) = 1$.

We fix a natural number d and assume that an immersion $\beta : M \to \mathbb{R}\times Y$ is such that each leaf $\{\mathcal{L}_y\}_{y\in Y}$ hits the image $\beta(M)$ so that:

$$\mu^\beta(y) =_{\mathsf{def}} \sum_{\{a \in \mathcal{L}_y \cap \beta(M)\}} \Big(\sum_{\{b \in \beta^{-1}(a)\}} \mu^\beta(b)\Big) \leq d. \tag{7.5}$$

Note that this assumption rules out the infinite multiplicity tangencies $\mu^\beta(y)$.

We order the points $\{a_i\}$ of $\mathcal{L}_y \cap \beta(M)$ by the values of their projections on $\mathbb{R}$ and introduce the combinatorial pattern $\omega^\beta(y)$ of a point $y \in Y$ as the ordered sequence of multiplicities

$$\{\omega_i^\beta(y) =_{\mathsf{def}} \sum_{\{b \in \beta^{-1}(a_i)\}} \mu^\beta(b)\}_i.$$

We denote by $D^\beta(y)$ the real divisor on the leaf $\mathcal{L}_y$, whose support is $\mathcal{L}_y \cap \beta(M)$ and whose multiplicities are the $\{\omega_i^\beta(y)\}_i$.

In particular, the combinatorial type of $D^\beta(y)$ is sensitive to the points $y \in Y$ over which the multiple self-intersections of β reside. Note also that the definition of the divisor $D^\beta(y)$ depends only on the image $\beta(M) \subset \mathbb{R}\times Y$ of M.

A special case of the following theorem can be found in [KSW1], Proposition 3.1.

Theorem 7.2. *Let M and Y be smooth compact n-manifolds.*

For any proper (in the sense of Definition 7.1) immersion $\beta : (M, \partial M) \to (\mathbb{R}\times Y, \mathbb{R}\times\partial Y)$, which satisfies inequality (7.5) and the parity condition $\mu^\beta(y) \equiv d \mod 2$, there exists a smooth map $\Phi^\beta : Y \to \mathcal{P}_d$ such that the locus

$$\big\{(u, y) \in \mathbb{R} \times Y |\;\; \Phi^\beta(y)(u) = 0\big\} = \beta(\partial X).$$

Assume that, for a given pair of closed posets $\Theta \subset \Lambda \subset \mathbf{\Omega}_{\langle d]}$, the immersion β is such that, for each $y \in Y$, no $\omega^\beta(y)$ belongs to Θ and, for each $y \in \partial Y$, no $\omega^\beta(y)$ belongs to Λ.

Then Φ^β maps Y to the open subset $\mathcal{P}_d^{\mathbf{c}\Theta} \subset \mathcal{P}_d$ and ∂Y to the open subset $\mathcal{P}_d^{\mathbf{c}\Lambda} \subset \mathcal{P}_d$. Moreover, any two such maps $\Phi_0^\beta, \Phi_1^\beta : (Y, \partial Y) \to (\mathcal{P}_d^{\mathbf{c}\Theta}, \mathcal{P}_d^{\mathbf{c}\Lambda})$ are homotopic as maps of pairs.

Proof. Let us denote by $\hat{M}^\circ$ the interior of $\hat{M} \supset M$ and by $\hat{Y}^\circ$ the interior of $\hat{Y} \supset Y$.

We claim that the set $\hat{\beta}(\hat{M}^\circ)$ may be viewed as the solution set of the equations

$$\Big\{u^d + \sum_{j=0}^{d-1} a_j(y)\, u^j = 0\Big\}_{y \in \hat{Y}^\circ},$$

where $\{a_j : \hat{Y}^\circ \to \mathbb{R}\}_j$ are some smooth functions.

Let us justify this claim. By Lemma 4.1 from [K2] and Morin's Theorem [Morin] (both based on the Malgrange Preparation Lemma), if a particular branch $\hat{\beta}(\hat{M}^\circ)_\kappa$ of $\hat{\beta}(\hat{M}^\circ)$ is tangent to the leaf $\mathcal{L}_{y_\star}$ at a point $a_\star = (\alpha, y_\star) \in \mathbb{R} \times Y$ with the order of tangency $j = j(a_\star, \kappa)$, then there is a system of local coordinates $(\tilde{u}, \tilde{x}, \tilde{z})$ in the vicinity of $a_\star \in \mathbb{R} \times \hat{Y}^\circ$ such that:

(1) Each branch $\hat{\beta}(\hat{M}^\circ)_\kappa$ is given by the equation $\big\{\tilde{u}^j + \sum_{k=0}^{j-1} \tilde{x}_k\, \tilde{u}^k = 0\big\}$, where $\tilde{x}_k(y_\star) = 0$ and $\tilde{u}(\alpha) = 0$,
(2) each nearby leaf $\mathcal{L}_y$ is locally given by the equations $\{\tilde{x} = \overrightarrow{const},\ \tilde{z} = \overrightarrow{const'}\}$.

Letting $\tilde{u} = u - \alpha$ and writing $\tilde{x}_k$'s as smooth functions of $y \in \hat{Y}^\circ$, the same locus $\hat{\beta}(\hat{M}^\circ)_\kappa$ can be given by the equation

$$\Big\{P_{\alpha,\kappa}(u, y) =_{\mathsf{def}} (u - \alpha)^j + \sum_{j=0}^{j-1} a_{\kappa,k}(y)\, (u - \alpha)^k = 0\Big\}, \tag{7.6}$$

where $a_{\kappa,k} : \hat{Y}^\circ \to \mathbb{R}$ are smooth functions, vanishing at $y_\star$. Therefore, there exists an open neighborhood $U_{y_\star}$ of $y_\star$ in $\hat{Y}^\circ$ such that, in $\mathbb{R} \times U_{y_\star}$, the locus $\hat{\beta}(\hat{M}^\circ)$ is given by the monic polynomial equation

$$\Big\{P_{y_\star}(u, y) =_{\mathsf{def}} \prod_{(\alpha,\, y_\star) \in \mathcal{L}_{y_\star} \cap \hat{\beta}(\hat{M}^\circ)} \Big(\prod_{\kappa \in A_\alpha} P_{\alpha,\kappa}(u, y)\Big) = 0\Big\}, \tag{7.7}$$

of degree $\mu^\beta(y_\star) \leq d$ in u. Here the finite set A_α labels the local branches of $\beta(M)$ that contain the point $(\alpha, y_\star) \in \mathcal{L}_{y_\star} \cap \beta(M)$.

By multiplying $P_{y_\star}(u,y)$ with $(u^2+1)^{\frac{d-\mu^\beta(y_\star)}{2}}$, we get a polynomial $\tilde{P}_{y_\star}(u,y)$ of degree d. For each $y \in U_{y_\star}$, $\tilde{P}_{y_\star}(u,y)$ shares with $P_{y_\star}(u,y)$ the zero set $\hat{\beta}(\hat{M}^\circ) \cap (\mathbb{R} \times U_{y_\star})$, as well as the divisors $D^{\hat{\beta}}(y)$.

For each $y \in \hat{Y}^\circ$, we consider the space $\mathcal{X}_{\hat{\beta}}(y)$ of monic polynomials $\tilde{P}(u)$ of degree d such that their real divisors coincide with the $\hat{\beta}$-induced divisor $D^{\hat{\beta}}(y)$. We view $\mathcal{X}_{\hat{\beta}} =_{\mathsf{def}} \coprod_{y \in \hat{Y}^\circ} \mathcal{X}_{\hat{\beta}}(y)$ as a subspace of $\hat{Y}^\circ \times \mathcal{P}_d$. It is equipped with the obvious projection $p : \mathcal{X}_{\hat{\beta}} \to \hat{Y}^\circ$. The smooth sections of the map p are exactly the smooth functions $\tilde{P}(u,y)$ that interest us. By the previous argument, p admits a *local* smooth section over the vicinity of each $y_\star \in \hat{Y}^\circ$.

Evidently, each p-fiber $\mathcal{X}_{\hat{\beta}}(y)$ is a convex set. Thus, given finitely many smooth sections $\{\sigma_i\}_i$ of p, we conclude that $\sum_i \phi_i \cdot \sigma_i$ is again a section of p, provided that the smooth functions $\phi_i : Y \to [0,1]$ have the property $\sum_i \phi_i \equiv \mathbf{1}$. Note that an individual term $\phi_i \cdot \sigma_i$ may not belong to $\mathcal{P}_d$ due to the failure of the polynomials to be monic.

Since Y is compact, it admits a finite cover by the open sets $\{U_{y_i} \subset \hat{Y}^\circ\}_i$ as above (see formulas (7.6), (7.7)). Let $\{\phi_i : Y \to [0,1]\}_i$ be a smooth partition of unity, subordinated to this finite cover. Then the monic u-polynomial

$$\tilde{P}(u,y) =_{\mathsf{def}} \sum_i \phi_i(y) \cdot \tilde{P}_{y_i}(u,y)$$

of degree d has the desired properties. In particular, its divisor $\tilde{D}^\beta(y) = D^\beta(y)$ for each $y \in Y$. Thus, using $\tilde{P}(u,y)$, any immersion $\beta : (M, \partial M) \to (\mathbb{R} \times Y,\, \mathbb{R} \times \partial Y)$, such that:

(i) β is transversal to $\mathbb{R} \times \partial Y$,
(ii) for $y \in Y$, no $\omega^\beta(y)$ belongs to Θ,
(iii) for $y \in \partial Y$, no $\omega^\beta(y)$ belongs to Λ,

is realized by a smooth map $\Phi^\beta : (Y, \partial Y) \to (\mathcal{P}_d^{\mathbf{c}\Theta}, \mathcal{P}_d^{\mathbf{c}\Lambda})$ for which the image $\beta(M) = \{\Phi^\beta(y)(u) = 0\}$.

If two such maps $\Phi_0^\beta, \Phi_1^\beta : (Y, \partial Y) \to (\mathcal{P}_d^{\mathbf{c}\Theta}, \mathcal{P}_d^{\mathbf{c}\Lambda})$ share the same divisor $D^\beta(y)$ for each $y \in Y$, then the quotients $\Phi_0^\beta(y)/\Phi_1^\beta(y)$ and $\Phi_1^\beta(y)/\Phi_0^\beta(y)$ are *strictly positive* rational functions in u, and their numerators and denominators are monic non-vanishing polynomials of degree d. Such rational functions form a convex set $\mathcal{Q}_+(d)$ that retracts to the point-function $\mathbf{1}$. Therefore, $\Phi_0^\beta, \Phi_1^\beta$ produce a continuous map $\Phi_0^\beta/\Phi_1^\beta$ into a contractible space $\mathcal{Q}_+(d)$. As a result, by the linear homotopy $\{(1-t)\Phi_0^\beta + t\,\Phi_1^\beta\}_{t\in[0,1]}$, the maps $\Phi_0^\beta, \Phi_1^\beta$ are homotopic. □

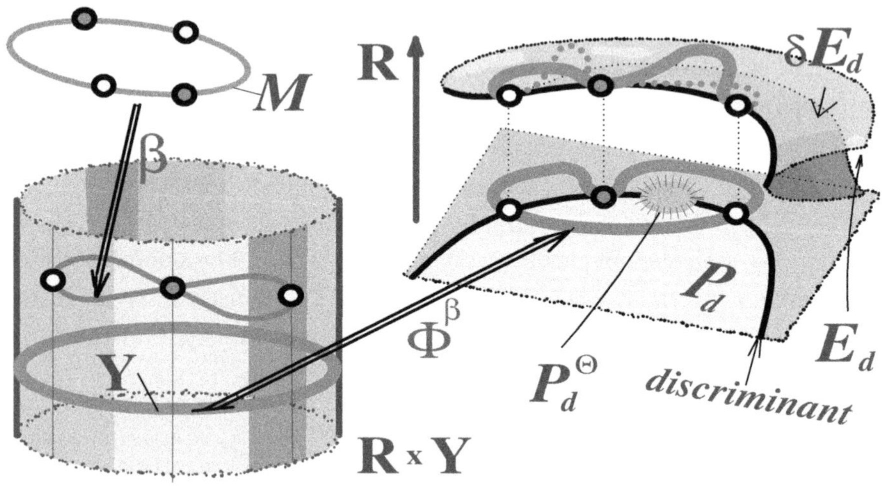

Fig. 7.1 An immersion $\beta : M \to \mathbb{R}\times Y$, where the circle M being immersed as the figure ∞ in the cylinder $\mathbb{R} \times S^1$ and $Y = S^1$. The classifying map $\Phi^\beta : Y \to \mathcal{P}_d^{\mathbf{c}\Theta}$ that β generates (according to Theorem 7.2) is shown as a slanted arrow. To get the figure realistic in $3D$, we have chosen the simplest possible case: $d = 2$, $\Theta = \emptyset$. Note that $\beta(M) = (\mathsf{id} \times \Phi^\beta)^{-1}\big((\mathbb{R} \times \Phi^\beta(Y)) \cap \partial\mathcal{E}_d\big)$, where the surface $\partial\mathcal{E}_d$ is introduced in (7.4).

Definition 7.4. Let Y be a smooth compact n-manifold and the domain $\mathcal{E}_d \subset \mathbb{R}\times\mathcal{P}_d$ be as in (7.4). A smooth map $\Phi : Y \to \mathcal{P}_d$ is called $(\partial\mathcal{E}_d)$-regular if the maps $\mathbb{R} \times Y \stackrel{\mathsf{id}\times\Phi}{\longrightarrow} \mathbb{R} \times \mathcal{P}_d$ and $\mathbb{R} \times \partial Y \stackrel{\mathsf{id}\times\Phi^\partial}{\longrightarrow} \mathbb{R} \times \mathcal{P}_d$ are transversal to the hypersurface $\partial\mathcal{E}_d \subset \mathbb{R} \times \mathcal{P}_d$. $\diamondsuit$

Thus, for a $(\partial\mathcal{E}_d)$-regular Φ, the preimage $(\mathsf{id} \times \Phi)^{-1}(\partial\mathcal{E}_d)$ is a compact smooth n-manifold $(M, \partial M) \subset (\mathbb{R} \times Y, \mathbb{R} \times \partial Y)$. Of course, if $\beta : (M, \partial M) \to (\mathbb{R} \times Y,$
$\mathbb{R} \times \partial Y)$ is an immersion and not an embedding, then any associated map $\Phi^\beta : (Y, \partial Y) \to (\mathcal{P}_d^{\mathbf{c}\Theta}, \mathcal{P}_d^{\mathbf{c}\Lambda})$ is not $(\partial\mathcal{E}_d)$-regular.

Fig. 7.1, which illustrates the construction of the classifying map Φ^β, reveals an interesting duality: if $(u_\star, y_\star) \in \mathbb{R}\times Y$ is the point of *transversal* self-intersection of $\beta(M) =$ "∞", then $y_\star$ is mapped by Φ^β to a point where the curve $\Phi^\beta(Y)$ is *tangent* to the discriminant variety $\mathcal{D}_2$; in contrast, if $(u_*, y_*) \in \mathbb{R}\times Y$ is the point of *tangency* of $\beta(M)$ to $\mathcal{L}$, then y_* is mapped by Φ^β to a point where $\Phi^\beta(Y)$ is *transversal* to $\mathcal{D}_2$. Since the presence of self-intersection point of $\beta(M) =$ "∞" is a stable phenomenon with respect to perturbations of β within immersions, the tangency of $\beta(M)$ to $\mathcal{D}_2$ must be *stable*, despite the fact that a perturbation of the loop $\Phi^\beta(Y)$ can eliminate its tangency to $\mathcal{D}_2$.

Note that, in Fig. 7.1, we can perturb the map $\Phi^\beta : Y \to \mathcal{P}_d$ so that it will become transversal to the discriminant variety. Such a perturbation $\tilde{\Phi}^\beta$ may be assumed to be $(\partial\mathcal{E}_d)$-regular. It will resolve the image $\beta(M)$ into a nonsingular manifold $N \subset \mathbb{R} \times Y$. In general, the resolution N may be topologically different from the original M. These observations are generalized in the next couple of lemmas.

Lemma 7.3. *For a compact smooth manifold* Y*, the* $(\partial\mathcal{E}_d)$*-regular maps* $\{\Phi : Y \to \mathcal{P}_d\}$ *form an open and dense set in the space* $C^\infty(Y, \mathcal{P}_d)$ *of all smooth maps.*

Proof. A smooth map $\Phi : Y \to \mathcal{P}_d$, given by d functions-coefficients $x_{d-1}(y), \dots, x_1(y), x_0(y)$ on Y, is $(\partial\mathcal{E}_d)$-regular if and only if, in any local coordinate system $\vec{y} = \{y_1, \dots y_n\}$ on Y, the system

$$
\begin{aligned}
u^d + x_{d-1}u^{d-1} + \ldots + x_1 u + x_0 &= 0 \\
d\,u^{d-1} + (d-1)x_{d-1}u^{d-2} + \ldots + x_1 &= 0 \\
\Big\{ \frac{\partial x_{d-1}}{\partial y_j} u^{d-1} + \ldots + \frac{\partial x_1}{\partial y_j} u + \frac{\partial x_0}{\partial y_j} &= 0 \Big\}_{j \in [1,n]}
\end{aligned}
\tag{7.8}
$$

of $(n+2)$ equations has no solutions in $\vec{y}$ for all u, and a similar property holds for ∂Y. Indeed, $\partial\mathcal{E}_d$ is given by the equation

$$\wp(u, \vec{x}) =_{\mathsf{def}} u^d + x_{d-1}u^{d-1} + \ldots x_1 u + x_0 = 0.$$

The pull-back $\Psi^*(\wp)$ of $\wp$ under the map $\Psi = (\mathsf{id}, \Phi) : \mathbb{R} \times Y \to \mathbb{R} \times \mathcal{P}_d$ is the function

$$u^d + x_{d-1}(y)u^{d-1} + \ldots x_1(y)u + x_0(y)$$

on $\mathbb{R} \times Y$. Thus the first equation in (7.8) defines the preimage of $\partial\mathcal{E}$ under Ψ. The transversality of Ψ to $\partial\mathcal{E}_d$ can be expressed as the non-vanishing of the 1-jet of $\Psi^*(\wp)$ along the locus $\{\Psi^*(\wp) = 0\}$. In local coordinates on $\mathbb{R} \times Y$, the vanishing of the jet $\mathsf{jet}^1(\Psi^*(\wp))$ is exactly expressed by the constraints imposed in (7.8).

Note that, for each $u \in \mathbb{R}$, the system (7.8) imposes $(n+2)$ *affine* constraints on the functions $\{x_k : Y \to \mathbb{R}\}_{k \in [1,d]}$ and their first derivatives $\{\frac{\partial x_k}{\partial y_j}\}$. Therefore, for any u, (7.8) defines an affine subbundle $\mathcal{W}(u)$ of the jet bundle $\{\mathsf{J}^1(Y, \mathcal{P}_d) \to Y\}$. Thus, the union $\mathcal{W} = \bigcup_{u \in \mathbb{R}} \mathcal{W}(u)$ is a *ruled variety*, residing in $\mathsf{J}^1(Y, \mathcal{P}_d)$. Since $\mathrm{codim}(\mathcal{W}(u)) = n+2$, the codimension of $\mathcal{W}$ in $\mathsf{J}^1(Y, \mathcal{P}_d)$ is $n+1$.

Consider the jet map $\mathsf{jet}^1(\Phi) : Y \to \mathsf{J}^1(Y, \mathcal{P}_d)$. By the Thom Transversality Theorem (see [GG], Theorem 4.13), the space of $\{\Phi\}$ for which $\mathsf{jet}^1(\Phi)$ is transversal to the subvariety $\mathcal{W}$ is open and dense (recall that Y is compact). Since Y is n-dimensional, this transversality implies that $(\mathsf{jet}^1(\Phi))(Y) \cap \mathcal{W} = \emptyset$ for an open and dense set of maps Φ.

A similar arguments apply to the smooth maps $\Phi^\partial : \partial Y \to \mathcal{P}_d$; so we may first perturb a given map $\Phi : Y \to \mathcal{P}_d$ to insure the $(\partial\mathcal{E}_d)$-regularity of $\Phi^\partial = \Phi|_{\partial Y}$ and then perturb Φ to insure its $(\partial\mathcal{E}_d)$-regularity, while keeping the regularity of Φ^∂. Therefore, the set of $(\partial\mathcal{E}_d)$-regular maps Φ is open and dense in $C^\infty(Y, \mathcal{P}_d)$. □

Corollary 7.2. *Let $\Theta \subset \Lambda \subset \mathbf{\Omega}_{\langle d]}$ be closed subposets. For a compact smooth manifold Y, the $(\partial\mathcal{E}_d)$-regular maps $\{\Phi : (Y, \partial Y) \to (P_d^{\mathbf{c}\Theta}, P_d^{\mathbf{c}\Lambda})\}$ form an open and dense set in the space of all smooth maps $C^\infty\big((Y, \partial Y), (P_d^{\mathbf{c}\Theta}, P_d^{\mathbf{c}\Lambda})\big)$.*

Proof. Since Y is compact and the posets Θ, Λ are closed, the target spaces $\mathcal{P}_d^{\mathbf{c}\Theta} \supset \mathcal{P}_d^{\mathbf{c}\Lambda}$ are both open in $\mathcal{P}_d$. Therefore, the claim follows from Lemma 7.3. □

Lemma 7.4. *For a compact n-dimensional manifold Y, any $\partial\mathcal{E}_d$-regular smooth map $h : Y \to \mathcal{P}_d^{\mathbf{c}\Theta}$ is realized by an* embedding $\beta_h : (M, \partial M) \hookrightarrow (\mathbb{R} \times Y, \mathbb{R} \times \partial Y)$, *where the smooth manifold M is n-dimensional and $\mathsf{rk}(D\beta_h) = n$. The manifold M is orientable if Y is.*

If $\partial Y = \emptyset$ and $d \equiv 0 \mod 2$, then M is a boundary of a compact (orientable, if Y is orientable) $(n+1)$-manifold $L \subset \mathbb{R} \times Y$.

Proof. A $\partial\mathcal{E}_d$-regular smooth map $h : Y \to \mathcal{P}_d^{\mathbf{c}\Theta}$ produces the locus $M = (\mathsf{id} \times h)^{-1}(\partial\mathcal{E}_d)$. It is a smooth manifold by the transversality of $\mathsf{id} \times h$ to $\partial\mathcal{E}_d$. If Y is orientable, so is $\mathbb{R} \times Y$. Thus, the pull-back of the normal vector field to $\partial\mathcal{E}_d$ in $\mathcal{E}_d$ helps to orient the tangent bundle of M.

Given an embedding $\beta : M \hookrightarrow \mathbb{R} \times Y$, $\dim M = n$, whose tangent patterns belong to $\mathbf{c}\Theta$, and using that $d \equiv 0 \mod 2$, we conclude that when Y is closed, then $\beta(M)$ bounds a compact $(n+1)$-manifold $X \subset \mathbb{R} \times Y$. Indeed, for each $y \in Y$, we consider the even degree d real monic polynomial $P_y(u)$ whose real divisor is $\mathcal{L}_y \cap \beta(M)$, counted with the multiplicities. Then we define $L \cap \mathcal{L}_y$ as the compact set of $u \in \mathbb{R}$ that satisfies $\{P_y(u) \leq 0\}$. The Y-family of such inequalities determines L. □

As a result, any closed n-manifold M, which is *not* (orientably, if Y is orientable) cobordant to $\emptyset$, does not arise as the preimage $(\mathsf{id} \times \Phi)^{-1}(\partial \mathcal{E}_d)$ via $\partial \mathcal{E}_d$-regular maps $h : Y \to \mathcal{P}_d$ for any closed Y and $d \equiv 0 \mod 2$.

Consider the tautological function $\wp : \mathbb{R} \times \mathcal{P}_d \to \mathbb{R}$ whose value at the point (u, P) is $P(u)$.

Definition 7.5. Let Y be a smooth compact n-manifold. Let $k \leq d+1$.

We call a smooth map $\Phi : Y \to \mathcal{P}_d$ "$(\partial \mathcal{E}_d, k)$-regular" if the map $\mathbb{R} \times Y \xrightarrow{\mathsf{id} \times \Phi} \mathbb{R} \times \mathcal{P}_d$ has the following properties:

- the pull-back $(\mathsf{id} \times \Phi)^*(\wp)$ of the function $\wp$ to the vicinity U_a of every point
$$a = (u, y) \in \mathcal{M}_\Phi =_{\mathsf{def}} \big\{(\mathsf{id} \times \Phi)^*(\wp) = 0\big\} \subset \mathbb{R} \times Y$$
is locally a product of at most $\ell \leq k$ smooth functions $\{z_i : U_a \to \mathbb{R}\}_{i \in [1,\ell]}$ for each of which 0 is a regular value,
- there is a natural number q such that the q-jet equivalence classes of the local branches $\{z_i = 0\}_i$ of $\mathcal{M}_\Phi$ at a are distinct for all points a,[3]
- if $y \in \partial Y$, the restrictions $\{z_i|_{\mathbb{R} \times \partial Y}\}_i$ have 0 as a regular value, and the q-jet equivalence classes of local branches $\{z_i|_{\mathbb{R} \times \partial Y} = 0\}_i$ are distinct as well. ♢

Evidently, $(\partial \mathcal{E}_d, 1)$-regular map is $(\partial \mathcal{E}_d)$-regular.

Informally, we would like to think of $(\partial \mathcal{E}_d, k)$-regular maps Φ as hypersurfaces that divide the space of $(\partial \mathcal{E}_d, k-1)$-regular maps into chambers. Presently, this interpretation is just a wishful thinking... What is clear that, if a map $\Phi : Y \to \mathcal{P}_d$ is $(\partial \mathcal{E}_d, k)$-regular, then there exists its open neighborhood in the space $C^\infty(Y, \mathcal{P}_d)$ that does not contain any $(\partial \mathcal{E}_d, l)$-maps, where $l > k$.

Lemma 7.5. *If a map $\Phi : Y \to \mathcal{P}_d$ is $(\partial \mathcal{E}_d, k)$-regular, then the locus*
$$\mathcal{M}_\Phi =_{\mathsf{def}} \{(\mathsf{id} \times \Phi)^*(\wp) = 0\} \subset \mathbb{R} \times Y$$
is a compact n-dimensional set with singularities of the local types $\{\prod_{i=1}^{\ell} z_i = 0\}$, where $2 \leq \ell \leq k \leq d+1$, and each z_i has 0 as its regular value.

Moreover, $\mathcal{M}_\Phi$ is the image of a smooth compact n-manifold M under an immersion $\beta : M \to \mathbb{R} \times Y$.

[3]The branches are q-separable in the sense of Definition 7.2.

Proof. The validation of the first claim is on the level of definitions.

Recall that we denote by $\mathsf{J}^q(X) \to X$ the fibration over a smooth manifold X, whose fiber over a point $x \in X$ is J^q_x, the space of q-equivalence classes of germs of smooth hypersurfaces in X at x.

It remains to show that, if a map $\Phi : Y \to \mathcal{P}_d$ is $(\partial\mathcal{E}_d, k)$-regular and, for some $q \geq 1$, the branches of the locus $\mathcal{M}_\Phi \subset \mathbb{R} \times Y$ at their mutual intersections are q-separable, then $\mathcal{M}_\Phi$ is the immersed image of some n-manifold M.

In fact, there is an obvious canonical resolution $M \subset \mathsf{J}^q(\mathbb{R}\times Y)$ of $\mathcal{M}_\Phi$, so that the projection $\beta : M \subset \mathsf{J}^q(\mathbb{R}\times Y) \xrightarrow{\pi} \mathbb{R}\times Y$ is the desired immersion. We just associate with each point $a \in \{\prod_{i=1}^{\ell} z_i = 0\}$, $\ell \leq k$, the (unordered) set of ℓ distinct points $\{\mathsf{J}^q_a(\{z_i = 0\})\}_{i\in[1,\ell]} \subset \mathsf{J}^q_a(\mathbb{R} \times Y)$. In this way, each local branch $\mathcal{M}_{\Phi,i} =_{\mathsf{def}} \{z_i = 0\}$ of $\mathcal{M}_\Phi$ defines a smooth section $\sigma_i : \mathcal{M}_{\Phi,i} \to \mathsf{J}^q(\mathbb{R} \times Y)|_{\mathcal{M}_{\Phi,i}}$. Over the vicinity of a, thanks to the q-separability, all the sections $\{\sigma_i\}_i$ are disjoint. Since $\pi\circ\sigma_i = \mathsf{id}$, we conclude that $\beta : M \to \mathcal{M}_\Phi$ is an immersion. □

Definition 7.6. We say that an immersion $\beta : M \to \mathbb{R} \times Y$ is k-flat, if for all $y \in Y$, the reduced multiplicity

$$(\mu^\beta)'(y) =_{\mathsf{def}} \sum_{\{a \in \mathcal{L}_y \cap \beta(M)\}} \Big(\sum_{\{b \in \beta^{-1}(a)\}} (\mu^\beta(b) - 1)\Big) < k. \tag{7.9}$$

In particular, if β is k-flat and k-normal, then the k-intersection manifold $\Sigma_k^\beta = \emptyset$. ◇

7.4 Quasitopies of immersions & embeddings with constrained tangencies against the background of product 1-foliations: the case of general combinatorics

We use the abbreviation "imm" for immersions and "emb" for regular embeddings. When the arguments work equally well for both types, we use the abbreviation "i/e".

Let us recall few facts from Chapter 6, needed now. We fix two natural numbers $d \leq d'$, $d \equiv d' \mod 2$ and consider the embedding $\epsilon_{d,d'} : \mathcal{P}_d \hookrightarrow \mathcal{P}_{d'}$, defined by the formula (6.32). Recall that it preserves the $\mathbf{\Omega}$-stratifications of the two spaces, $\mathcal{P}_d$ and $\mathcal{P}_{d'}$, by the combinatorial types ω of real divisors $D_\mathbb{R}(P)$, $D_\mathbb{R}(\epsilon_{d,d'}(P))$.

For a closed profinite (see Definition 6.5) poset $\Theta \subset \mathbf{\Omega}$, the embeddings $\epsilon_{d,d'}$ make it possible to talk about the stabilization in the homology of spaces $\mathcal{P}_d^{\mathbf{c}\Theta}$, as $d \to \infty$. With the help of maps $\{\epsilon_{d,d'}\}$, it makes sense also to introduce the limit spaces $\bar{\mathcal{P}}_\infty^{\Theta}$ and $\mathcal{P}_\infty^{\mathbf{c}\Theta}$ as in [KSW2].

For a smooth compact connected n-manifold Y, put $Z =_{\mathsf{def}} Y \times [0,1]$ and $Z^\delta =_{\mathsf{def}} \partial Y \times [0,1]$. We denote by $\mathcal{L}^\bullet$ the 1-dimensional oriented foliation on $\mathbb{R} \times Z$, produced by the fibers of the obvious projection $\Pi : \mathbb{R} \times Z \to Z$.

In the Definition 7.7 below, central to our investigations in this chapter, we start with a quite general set of combinatorial input data: $\{n, d, d', \Theta, \Theta', \Lambda\}$. We will gradually restrict them, as we develop the theory. Fig. 7.2 may help the reader to follow our unavoidably cumbersome notations. As in Chapter 5, dealing with oriented and non-oriented doodles and blobs, each definition below comes in two flavors: oriented "$\mathcal{OQ}$" and non-oriented "$\mathcal{Q}$".

Definition 7.7. We fix natural numbers $d \leq d'$, $d \equiv d' \mod 2$, and a triple of closed subposets $\Theta' \subset \Theta \subset \Lambda$ of the universal poset $\mathbf{\Omega}$. Let Y be a fixed smooth compact n-manifold, and M_0, M_1 two (oriented) compact smooth n-manifolds. Let $Z = Y \times [0,1]$ and $Z^\delta = \partial Y \times [0,1]$.

We say that two proper immersions/embeddings,

$$\beta_0 : (M_0, \partial M_0) \to (\mathbb{R}\times Y,\ \mathbb{R}\times\partial Y) \quad \text{and} \quad \beta_1 : (M_1, \partial M_1) \to (\mathbb{R}\times Y,\ \mathbb{R}\times\partial Y),$$

are $(d, d'; \mathbf{c}\Theta, \mathbf{c}\Lambda; \mathbf{c}\Theta')$-quasitopic, if there exists a (oriented) compact smooth
$(n+1)$-manifold N with the corners $\partial M_0 \coprod \partial M_1$ and the boundary

$$\partial N = (M_0 \coprod -M_1) \cup \delta N,$$

where δN is the closure of the complimentary to $M_0 \coprod M_1$ portion of the boundary ∂N, and a smooth proper immersion/embedding $B : N \to \mathbb{R} \times Z$ such that:

- $B|_{M_0} = \beta_0$, $B|_{M_1} = \beta_1$, and $B(\delta N) \subset \mathbb{R} \times \partial Y \times [0,1] =_{\mathsf{def}} \mathbb{R} \times Z^\delta$,
- for each $z \in Z$, the total multiplicity $m_B(z)$ (see (7.5)) of $B(N)$ with respect to the generic fiber $\mathcal{L}_z^\bullet$ is such that: $m_B(z) \leq d'$, $m_B(z) \equiv d' \mod 2$, and the combinatorial tangency pattern $\omega^B(z)$ of $B(N)$ with respect to $\mathcal{L}_z^\bullet$ belongs to the poset $\mathbf{c}\Theta'$,
- for each $z \in Y \times \partial[0,1]$, the total multiplicity of $B(N)$ with respect to $\mathcal{L}_z$ is such that: $m_B(z) \leq d$, $m_B(z) \equiv d \mod 2$, and the combinatorial tangency pattern $\omega^B(z)$ of $B(N)$ with respect to $\mathcal{L}_z$ belongs to the poset $\mathbf{c}\Theta$,

- for each $z \in Z^\delta$, the multiplicity of $B(N)$ with respect to $\mathcal{L}^\bullet_z$ is such that: $m_B(z) \leq d'$, $m_B(z) \equiv d' \mod 2$, and the combinatorial tangency pattern $\omega^B(z)$ of $B(\delta N)$ with respect to $\mathcal{L}^\bullet_z$ belongs to the poset $\mathbf{c}\Lambda$.

We denote by $\mathcal{Q}^{\mathsf{imm}}_{d,d'}(Y, \partial Y; \mathbf{c}\Theta, \mathbf{c}\Lambda; \mathbf{c}\Theta')$ the set of quasitopy classes of such immersions $\beta : (M, \partial M) \to \mathbb{R} \times (Y, \partial Y)$.

We use the notation $\mathcal{Q}^{\mathsf{emb}}_{d,d'}(Y, \partial Y; \mathbf{c}\Theta, \mathbf{c}\Lambda; \mathbf{c}\Theta')$ for the set of quasitopy classes of embeddings $\beta : (M, \partial M) \hookrightarrow \mathbb{R} \times (Y, \partial Y)$. Finally, we use the neutral notation $\mathcal{Q}^{\mathsf{i/e}}_{d,d'}(Y, \partial Y; \mathbf{c}\Theta, \mathbf{c}\Lambda; \mathbf{c}\Theta')$ for both.

It is possible to build a parallel notion of quasitopies for oriented M's by insisting that the cobordisms N are oriented as well. Assuming that Y is oriented, this notion is based on immersions $\beta : (M, \partial M) \to (Y, \partial Y)$ with a trivialized normal bundle $\nu(\beta) =_{\mathsf{def}} \beta^*(T(\mathbb{R} \times Y))/\beta^*(\beta_*(TM))$ and their cobordisms with trivialized normal bundles as well.

We denote by $\mathcal{OQ}^{\mathsf{i/e}}_{d,d'}(Y, \partial Y; \mathbf{c}\Theta, \mathbf{c}\Lambda; \mathbf{c}\Theta')$ the set of these oriented quasitopy classes. ♢

Definition 7.8. We may add an additional assumption to the previous bulleted list in Definition 7.7 by insisting that

- all the immersions β_0, β_1, and B must have *transversal* self-intersections; i.e., β_0, β_1 are k-normal in the sense of Definition 7.3 for $k \leq \min\{n+1, d\}$ and B is k-normal for $k \leq \min\{n+2, d'\}$.

We denote by $\mathcal{Q}^{\times\mathsf{imm}}_{d,d'}(Y, \partial Y; \mathbf{c}\Theta, \mathbf{c}\Lambda; \mathbf{c}\Theta')$ the set of quasitopy classes of such normal immersions. ♢

Thus there is an obvious map

$$\mathcal{B}^\times : \mathcal{Q}^{\times\mathsf{imm}}_{d,d'}(Y, \partial Y; \mathbf{c}\Theta, \mathbf{c}\Lambda; \mathbf{c}\Theta') \to \mathcal{Q}^{\mathsf{imm}}_{d,d'}(Y, \partial Y; \mathbf{c}\Theta, \mathbf{c}\Lambda; \mathbf{c}\Theta'). \quad (7.10)$$

When Y is a closed manifold, we use the following notational simplification: "$\mathcal{Q}^{\mathsf{i/e}}_{d,d'}(Y; \mathbf{c}\Theta; \mathbf{c}\Theta')$". When $\Theta = \Theta'$, we get another simplification: "$\mathcal{Q}^{\mathsf{i/e}}_{d,d'}(Y, \partial Y; \mathbf{c}\Theta, \mathbf{c}\Lambda; \mathbf{c}\Theta)$".

The special cases $d' = d$ and $d' = d + 2$ have significant applications.

Remark 7.2. If in Definition 7.7 we would require that the pair $(N, \delta N)$ is diffeomorphic to the pair $(M_0 \times [0, 1],\ \partial M_0 \times [0, 1])$, then a more recognizable definition of pseudo-isotopy would emerge. Thus, the notion of quasitopy is more flexible than the one of pseudo-isotopy; it is closer to the notion of bordisms. ♢

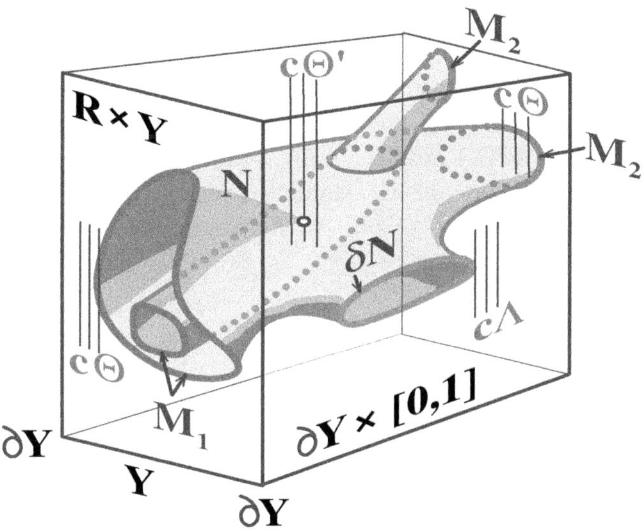

Fig. 7.2 The ingredients of Definition 7.7, which illustrate the notion of quasitopy between two immersions $\beta_1 : M_1 \to \mathbb{R}\times Y$ and $\beta_2 : M_2 \to \mathbb{R}\times Y$. The quasitopy is realized by an immersion-cobordism $B : N \to \mathbb{R}\times Y\times[0,1]$. The vertical lines are the fibers of the projection $p : \mathbb{R}\times Y\times[0,1] \to Y\times[0,1]$. Different combinatorial restrictions are imposed on different parts of $B(N)$: the combinatorial types of the fibers of the map $N \overset{B}{\to} \mathbb{R}\times Y\times[0,1] \overset{p}{\to} Y\times[0,1]$ avoid $\Theta' \subseteq \Theta$, the combinatorial types of the fibers of the map $M_1 \coprod M_2 \overset{B|}{\to} \mathbb{R}\times Y\times(\{0\}\cup\{1\}) \overset{p|}{\to} Y\times(\{0\}\cup\{1\})$ avoid Θ, while the fibers through $B(\delta(N) \subset \mathbb{R}\times\partial Y\times[0,1]$ avoid $\Lambda \supset \Theta$.

Remark 7.3. The quasitopy $\mathcal{Q}^{\mathsf{i/e}}_{d,d'}(\sim,\ \partial\sim; \mathbf{c}\Theta, \mathbf{c}\Lambda; \mathbf{c}\Theta')$ is covariently functorial under the *regular embeddings* $f : (Y, \partial Y) \hookrightarrow (Y', \partial Y')$ of *equidimentional* manifolds Y, Y'.

At the same time, $\mathcal{Q}^{\mathsf{i/e}}_{d,d'}(\sim, \partial\sim; \mathbf{c}\Theta, \mathbf{c}\Lambda; \mathbf{c}\Theta')$ is contravariantly functorial under *submersions* $g : (Y', \partial Y') \to (Y, \partial Y)$, provided that the fibers of g are closed manifolds. The contravariance is delivered via the pull back construction which involves $\mathsf{id}_{\mathbb{R}} \times g$ and $\beta : (M, \partial M) \to (\mathbb{R}\times Y, \mathbb{R}\times\partial Y)$. ♢

Consider the group $\mathsf{Diff}(\mathcal{L})$ of smooth diffeomorphisms of $\mathbb{R}\times Y$ that preserve the 1-foliation $\mathcal{L}$ and its orientation. We denote by $\mathsf{Diff}_0(\mathcal{L})$ its subgroup, generated by the diffeomorphisms that are isotopic to the identity. Similar groups, $\mathsf{Diff}(\mathcal{L}^\bullet)$ and $\mathsf{Diff}_0(\mathcal{L}^\bullet)$ are available for the foliation $\mathcal{L}^\bullet$ on $\mathbb{R}\times Y\times[0,1]$.

Note that the group $\mathsf{Diff}(Y)$ is, in the obvious way, a subgroup of $\mathsf{Diff}(\mathcal{L})$, and the group $\mathsf{Diff}_0(Y)$ is a subgroup of $\mathsf{Diff}_0(\mathcal{L})$.

If $\beta : (M, \partial M) \to (\mathbb{R} \times Y, \mathbb{R} \times \partial Y)$ is a proper immersion/embedding as in Definition 7.7, then, for any $h \in \mathsf{Diff}_0(\mathcal{L})$, the immersion/embedding $h \circ \beta$ is $(d, d'; \mathbf{c}\Theta, \mathbf{c}\Lambda; \mathbf{c}\Theta')$-quasitopic to β. In a similar way, $\mathsf{Diff}_0(\mathcal{L}^\bullet)$ acts on quasitopies of immersion/embedding. Therefore, in what follows, we may ignore the dependence of our constructions on the isotopies of the base manifold Y.

Lemma 7.6. *The map $\mathcal{B}^\times$ in (7.10) is* 1*-to-*1*.*

Proof. The proof is based on the following observation: for any *closed* poset $\Theta \subset \mathbf{\Omega}$, if an immersion $\beta : M \to \mathbb{R} \times Y$ generates a tangency pattern $\omega \in \mathbf{c}\Theta$ with respect to a given fiber $\mathcal{L}_y$ $(y \in Y)$, then a sufficiently C^∞-small perturbation $\tilde{\beta}$ of β, will have a tangency pattern $\tilde{\omega} \in \mathbf{c}\Theta$ with respect to the same fiber $\mathcal{L}_y$. This claim uses the polynomial model (2.10) of each branch $\beta(M)_\kappa$ of $\beta(M)$ in the vicinity of each intersection $\beta(M)_\kappa \cap \mathcal{L}_y$. Since M and Y are compact, it follows that the set of immersions β having tangency patterns from $\mathbf{c}\Theta$ form an open set in the space of all immersions. On the other hand, by [LS], the set of k-normal immersions is dense in the space of all immersions for $k \leq n+1$. Thus, any immersion β with the tangency patterns in $\mathbf{c}\Theta$ is smoothly homotopic to a k-normal immersion $\tilde{\beta}$ with the tangency patterns in $\mathbf{c}\Theta$. Moreover, the homotopy between β and $\tilde{\beta}$ is realized within immersions with tangency patterns in $\mathbf{c}\Theta$. Therefore, the map $\mathcal{B}^\times$ is surjective.

A similar argument, being applied to an immersion-cobordism $B : N \to \mathbb{R} \times Y \times [0,1]$ with B-tangency patterns in $\mathbf{c}\Theta'$ between $\tilde{\beta}$ and the empty immersion, insures that B can be homotoped to a k-normal immersion $\tilde{B}$ (where $k \leq n+2$), the homotopy being equal to $\tilde{\beta}$ on ∂N and is realized within the space of immersions with tangency patterns in $\mathbf{c}\Theta'$. Therefore, the map $\mathcal{B}^\times$ is injective. □

Definition 7.9. For a given closed poset $\Theta_{\langle d]} \subset \mathbf{\Omega}_{\langle d]}$, we denote by $\hat{m}(\mathbf{c}\Theta_{\langle d]})$ the maximum of the entries ω_i for all $\omega = (\omega_1, \dots, \omega_i, \dots, \omega_\ell) \in \mathbf{c}\Theta_{\langle d]}$. ◇

Proposition 7.5. *We adopt the notations of Corollary 7.1. Any proper immersion $\beta : (M, \partial M) \to (\mathbb{R} \times Y, \mathbb{R} \times \partial Y)$ (as in Definition 7.7), which is k-normal for all k in the interval $[2, \min\{n+1, \hat{m}(\mathbf{c}\Theta_{\langle d]})\}]$, generates* *canonically smooth maps*

$$\{\pi \circ \beta \circ p_1 : (\Sigma_k^\beta, \Sigma_k^{\beta^\partial}) \to (Y, \partial Y)\}_k,$$

where Σ_k^β is the kth self-intersection manifold of β. The relative non-oriented bordism classes $[\Sigma_k^\beta] \in \mathbf{B}_{n-k+1}(Y, \partial Y)$ of these maps $\pi \circ \beta \circ p_1$ are invariants of the $(d, d'; \mathbf{c}\Theta, \mathbf{c}\Lambda; \mathbf{c}\Theta')$-quasitopy class of β.

If M, Y are oriented, then oriented bordism classes $[\Sigma_k^\beta] \in \mathbf{OB}_{n-k+1}(Y, \partial Y)$ are invariants of the oriented $(d, d'; \mathbf{c}\Theta, \mathbf{c}\Lambda; \mathbf{c}\Theta')$-quasitopy class of β. In particular, if Y is closed, then the Pontryagin numbers of the manifold Σ_k^β are invariants of the quasitopy class of β.

Proof. If β is an immersion as in Definition 7.7, then its sufficiently small perturbation still has combinatorial tangency to the fibration $\mathcal{L}$ patterns which belong to $\mathbf{c}\Theta$, since Θ is a *closed* poset in $\mathbf{\Omega}_{\langle d]}$. This claim is based on the behavior of real divisors of real polynomials under their perturbations [K3]. Similarly, a sufficiently small perturbation of any cobordism B between such immersions, will still have combinatorial tangency patterns from $\mathbf{c}\Theta'$, since Θ' is a closed poset in $\mathbf{\Omega}_{\langle d']}$. Therefore, by [LS], we may assume that β, within its $(d, d'; \mathbf{c}\Theta, \mathbf{c}\Lambda; \mathbf{c}\Theta')$-quasitopy class, is k-normal for all $k \leq n+1$ and B is k-normal for all $k \leq n+2$. Revisiting Definition 7.9, we notice that also $k \leq \hat{m}(\mathbf{c}\Theta_{\langle d]})$ by the very definition of quasitopies. Now the claim of the proposition follows directly from Corollary 7.1. □

We will focus in two *special cases* of Definition 7.7 to be referred in what follows as the "<u>Λ-condition</u>":

$$\begin{aligned} &(1) \quad d' \equiv d \equiv 0 \mod 2 \ \text{ and } \ \mathbf{c}\Lambda = (\emptyset), \\ &(2) \quad d' \equiv d \equiv 1 \mod 2 \ \text{ and } \ \mathbf{c}\Lambda = (1). \end{aligned} \tag{7.11}$$

The case (1) forces $\delta N = \emptyset$; so M_0, M_1 must be closed and $\partial N = M_0 \coprod M_1$. This is evidently the case when $\partial Y = \emptyset$. The case (2) forces $B|_{\delta N}$ to be transversal to the foliation $\mathcal{L}^\bullet|_{\mathbb{R}\times\partial Y\times[0,1]}$; thus we get diffeomorphisms $B(\delta N) \approx \partial Y \times [0, 1]$ and $\beta_i(\partial M_i) \approx \partial Y$.

Given two compact connected n-manifolds with boundaries, let $Y_1 \# Y_2$ denote their connected sum and $Y_1 \#_\partial Y_2$ their boundary connected sum. The 1-handle $H \approx D^{n-1} \times [0, 1]$ that participates in the connected sum operation is attached to $Y_1 \coprod Y_2$. In the special case $\dim Y_1 = \dim Y_2 = 1$, the manifolds Y_1, Y_2 are closed segments and $Y_1 \#_\partial Y_2$ is understood as a new segment, obtained by attaching one end of Y_1 to an end of Y_2.

If the boundaries of Y_1 and Y_2 are connected, the smooth topological type of $Y_1 \#_\partial Y_2$ does not depend on how the 1-handle H is attached to $Y_1 \coprod Y_2$ in order to form $Y_1 \#_\partial Y_2$. In general, to avoid ambiguity of the operation $\#_\partial$, we pick some elements (connected components) $\kappa_1 \in \pi_0(\partial Y_1)$ and $\kappa_2 \in \pi_0(\partial Y_2)$.

Definition 7.10.

- Under the Λ-condition (7.11), we simplify our former notations as follows:

$$\mathsf{Q}^{\mathsf{i/e}}_{d,d'}(Y; \mathbf{c}\Theta, \mathbf{c}\Theta') =_{\mathsf{def}} \mathcal{Q}^{\mathsf{i/e}}_{d,d'}(Y, \partial Y; \mathbf{c}\Theta, (\emptyset); \mathbf{c}\Theta'), \quad \text{for } d' \equiv d \equiv 0 \mod 2, \tag{7.12}$$

$$\mathsf{Q}^{\mathsf{i/e}}_{d,d'}(Y; \mathbf{c}\Theta, \mathbf{c}\Theta') =_{\mathsf{def}} \mathcal{Q}^{\mathsf{i/e}}_{d,d'}(Y, \partial Y; \mathbf{c}\Theta, (1); \mathbf{c}\Theta'), \quad \text{for } d' \equiv d \equiv 1 \mod 2, \tag{7.13}$$

$$\mathsf{G}^{\mathsf{i/e}}_{d,d'}(n; \mathbf{c}\Theta, \mathbf{c}\Theta') =_{\mathsf{def}} \mathcal{Q}^{\mathsf{i/e}}_{d,d'}(D^n, \partial D^n; \mathbf{c}\Theta, (\emptyset); \mathbf{c}\Theta'), \quad \text{for } d' \equiv d \equiv 0 \mod 2, \tag{7.14}$$

$$\mathsf{G}^{\mathsf{i/e}}_{d,d'}(n; \mathbf{c}\Theta, \mathbf{c}\Theta') =_{\mathsf{def}} \mathcal{Q}^{\mathsf{i/e}}_{d,d'}(D^n, \partial D^n; \mathbf{c}\Theta, (1); \mathbf{c}\Theta'), \quad \text{for } d' \equiv d \equiv 1 \mod 2, \tag{7.15}$$

where D^n stands for the standard n-ball. ◇

Assuming the Λ-condition (7.11), for any the choice of $\kappa_1 \in \pi_0(\partial Y_1)$ and $\kappa_2 \in \pi_0(\partial Y_2)$, let us introduce an operation

$$\uplus : \mathsf{Q}^{\mathsf{i/e}}_{d,d'}(Y_1; \mathbf{c}\Theta, \mathbf{c}\Theta') \times \mathsf{Q}^{\mathsf{i/e}}_{d,d'}(Y_2; \mathbf{c}\Theta, \mathbf{c}\Theta') \to \mathsf{Q}^{\mathsf{i/e}}_{d,d'}(Y_1 \#_\partial Y_2; \mathbf{c}\Theta, \mathbf{c}\Theta'). \tag{7.16}$$

Let $H \approx D^{n-1} \times D^1$ be a 1-handle attached to $Y_1 \coprod Y_2$ using the preferred connected components of ∂Y_1 and ∂Y_2, which are indexed by κ_1 and κ_2. The result is the boundary connected sum $Y_1 \#_\partial Y_2$. From now and on, we assume that the handles are attached so that the corners are smoothened and the resulting manifold has a smooth boundary. Different attachments of H produce diffeomorphc connected sums, provided that the disks $D^{n-1} \times \{0\} \subset H$ and $D^{n-1} \times \{1\} \subset H$ are placed in the preferred connected components $(\partial Y_1)_{\kappa_1}$ and $(\partial Y_2)_{\kappa_2}$ of the boundaries ∂Y_1 and ∂Y_2.

In the case (1) from (7.11), $\mathbf{c}\Lambda = (\emptyset)$, thus forcing the two immersions $\beta_i : M_i \to \mathbb{R} \times Y_i$ $(i = 1, 2)$ to be immersions of *closed* manifolds. With $(\partial Y_1)_{\kappa_1}$ and $(\partial Y_2)_{\kappa_2}$ being fixed, the map $\beta_1 \coprod \beta_2 : M_1 \coprod M_2 \to \mathbb{R} \times (Y_1 \#_\partial Y_2)$ is a well-defined immersion within its quasitopy class. Note that by (1) from (7.11), $(\beta_1 \coprod \beta_2)(M_1 \coprod M_2)$ is disjoint from $\mathbb{R}\times\partial(Y_1\#_\partial Y_2)$. Thus we put $\beta_1 \uplus \beta_2 =_{\mathsf{def}} \beta_1 \coprod \beta_2$.

In the case (2) from (7.11), $\mathbf{c}\Lambda = (1)$, by an action of a diffeomorphism from $\mathsf{Diff}_0(\mathcal{L}_i)$, we insure that $\beta_i(\partial M_i) = \{0\} \times \partial Y_i$, where $i = 1, 2$ and $\{0\} \in \mathbb{R}$. The $\mathsf{Diff}_0(\mathcal{L}_i)$-action does not change the quasitopy classes of $\{\beta_i\}$.

As we attach a 1-handle H to $Y_1 \coprod Y_2$ to form $Y_1\#_\partial Y_2$, we simultaneously attach the 1-handle $\tilde{H} = \{0\} \times H$ to $M_1 \coprod M_2$ to form $M_1\#_\partial M_2$ and extend $\beta_1 \coprod \beta_2$ across $\tilde{H}$ to a new map $\beta : M_1\#_\partial M_2 \to \mathbb{R} \times (Y_1\#_\partial Y_2)$. It is important that the handle $\tilde{H}$ is attached so that the map β is a regular embedding $\tilde{H} \hookrightarrow \mathbb{R} \times H$ and the obvious projection $\beta(\tilde{H}) \to H$ is an immersion.

The transversality of β_i to $\mathbb{R} \times \partial Y_i$, together with the property (2) from (7.11), allows us to smoothen the immersed manifold $\beta(M_1\#_\partial M_2) = \beta_1(M_1) \bigcup \beta(\tilde{H}) \bigcup \beta_2(M_2)$ in the neighborhood of $\tilde{H}$. This smoothing construction is again well-defined within the relative quasitopy classes of β_1, β_2. Thus, we put $\beta_1 \uplus \beta_2 =_{\mathsf{def}} \beta$.

However, in the case (2) from (7.11), when dealing with the *oriented* quasitopies, we face a problem. If M_1 and M_2 are oriented, the handle $\tilde{H}$ must be attached so that the preferred orientations extend across the handle. In "half" of the cases (when the orientations of $\beta_1(M_1)$ and $\beta_2(M_2)$) is "incoherent"), the the obvious projection $\beta(\tilde{H}) \to H$ will fail to have tangency patterns of finite multiplicity, which violates our basic requirement in (7.5).

To summarize, the operation $\uplus$ is well-defined for the elements of the sets $\mathcal{Q}^{\mathsf{i/e}}_d(Y, \partial Y; \mathbf{c}\Theta, (\emptyset); \mathbf{c}\Theta')$ and $\mathcal{OQ}^{\mathsf{i/e}}_d(Y, \partial Y; \mathbf{c}\Theta, (\emptyset); \mathbf{c}\Theta')$, where $d \equiv 0 \mod 2$, and for the elements of the set $\mathcal{Q}^{\mathsf{i/e}}_d(Y, \partial Y; \mathbf{c}\Theta, (1); \mathbf{c}\Theta')$, where $d \equiv 1 \mod 2$.

In particular, by fixing a diffeomorphism $\phi : D^n\#_\partial D^n \approx D^n$, we get two interesting special cases of the operation $\uplus$:

$$\uplus : \mathsf{G}^{\mathsf{i/e}}_{d,d'}(n; \mathbf{c}\Theta, \mathbf{c}\Theta') \times \mathsf{G}^{\mathsf{i/e}}_{d,d'}(n; \mathbf{c}\Theta, \mathbf{c}\Theta') \to \mathsf{G}^{\mathsf{i/e}}_{d,d'}(n; \mathbf{c}\Theta, \mathbf{c}\Theta'), \quad (7.17)$$

$$\uplus : \mathsf{OG}^{\mathsf{i/e}}_{d,d'}(n; \mathbf{c}\Theta, \mathbf{c}\Theta') \times \mathsf{OG}^{\mathsf{i/e}}_{d,d'}(n; \mathbf{c}\Theta, \mathbf{c}\Theta') \to \mathsf{OG}^{\mathsf{i/e}}_{d,d'}(n; \mathbf{c}\Theta, \mathbf{c}\Theta'),$$

provided that, in the last formula, $d \equiv 0 \mod 2$.

Proposition 7.6. *Consider two closed subposets $\Theta' \subset \Theta \subset \mathbf{\Omega}$, where $(\emptyset) \notin \Theta$ for $d \equiv 0 \mod 2$, and $(1) \notin \Theta$ for $d \equiv 1 \mod 2$.*

For $d \equiv d' \equiv 0 \mod 2$, *the operation* $\uplus$ *introduces a* group structure *to the sets* $\mathsf{G}^{\mathsf{i/e}}_{d,d'}(n; \mathbf{c}\Theta, \mathbf{c}\Theta')$ *and* $\mathsf{OG}^{\mathsf{i/e}}_{d,d'}(n; \mathbf{c}\Theta, \mathbf{c}\Theta')$ *(see (7.14) and (7.15)).*

For $d \equiv d' \equiv 1 \mod 2$, *the operation* $\uplus$ *introduces a* group structure *to the sets* $\mathsf{G}^{\mathsf{i/e}}_{d,d'}(n; \mathbf{c}\Theta, \mathbf{c}\Theta')$.

For $n > 1$, *all these groups are abelian.*

Proof. Let $\alpha : X \to X$ be a map of a topological space X. In this proof, we denote by $\tilde{\alpha}$ the product map $\mathsf{id} \times \alpha : \mathbb{R} \times X \to \mathbb{R} \times X$.

In the case $d \equiv d' \equiv 0 \mod 2$, the following arguments work equally well for $\mathsf{OG}^{\mathsf{i/e}}_{d,d'}(n; \mathbf{c}\Theta; \mathbf{c}\Theta')$ and for $\mathsf{G}^{\mathsf{i/e}}_{d,d'}(n; \mathbf{c}\Theta; \mathbf{c}\Theta')$, the oriented and non-oriented quasitopies over the n-ball D^n.

Let $B^n = D^n \#_\partial D^n$ be a n-ball, represented as a connected sum of two standard balls D^n. Let $H \approx D^{n-1} \times I$ be the 1-handle that participates in the construction of the ball B^n as a connected sum: $B^n = D^n \cup H \cup D^n$. We fix a diffeomorphism $\chi : D^n \subset B^n$, which identifies the ball D^n with the first ball in the connected sum B^n, and a diffeomorphism $\phi : B^n \approx D^n$.

Consider an isotopy $\{\psi_t : B^n \to B^n\}_{t \in [0,1]}$ that starts with the identity map and terminates with the diffeomorphism ψ_1 whose image is $\chi(D^n) \subset B^n$. With the help of ϕ^{-1}, we transfer the isotopy ψ_t to an isotopy $\{\eta_t : D^n \to D^n\}_{t \in [0,1]}$. The isotopy ψ_t lifts to the isotopy $\tilde{\psi}_t = \mathsf{id} \times \psi_t$ of $\mathbb{R} \times B^n$, and the isotopy η_t lifts to the isotopy $\tilde{\eta}_t = \mathsf{id} \times \eta_t$ of $\mathbb{R} \times D^n$.

For $d \equiv 0 \mod 2$, the neutral element $\beta_\star$ is represented by $\beta_\star : \emptyset \to \mathbb{R} \times D^n$ (by the "empty" cylinder $\mathbb{R} \times D^n$). For $d \equiv 1 \mod 2$, the neutral element $\beta_\star$ is represented by the obvious embedding $\beta_\star : D^n \to \{0\} \times D^n \subset \mathbb{R} \times D^n$.

Let $d \equiv 0 \mod 2$. We use the lift $\tilde{\eta}_t$ to show that, for any immersion $\beta : M \to \mathbb{R} \times D^n$, the immersion $\tilde{\phi} \circ (\beta \uplus \beta_\star) : M \#_\partial \emptyset \to \mathbb{R} \times D^n$ is isotopic (and thus quasitopic) to $\beta : M \to \mathbb{R} \times D^n$. We identify $M \#_\partial D^n$ with M with the help of a diffeomorphism ρ.

Let $d \equiv 1 \mod 2$. For any immersion $\beta : M \to \mathbb{R} \times D^n$, with the help of $\tilde{\eta}_t$, the immersion $\tilde{\phi} \circ (\beta \uplus \beta_\star) \circ \rho^{-1} : M \to \mathbb{R} \times D^n$ is isotopic to β.

This validates the existence of the neutral element $\beta_\star$ for the operation $\uplus$ within the $(d, d; \mathbf{c}\Theta, \mathbf{c}\Lambda; \mathbf{c}\Theta)$-quasitopy classes. Of course, any $(d, d; \mathbf{c}\Theta, \mathbf{c}\Lambda; \mathbf{c}\Theta)$-quasitopy is automatically a $(d, d'; \mathbf{c}\Theta, \mathbf{c}\Lambda; \mathbf{c}\Theta')$-quasitopy.

Consider an involution $\tau : B^n \to B^n$ that flips the two copies of D^n in the ball B^n and has the ball B^{n-1} in the middle of 1-handle $H \approx B^{n-1} \times I$ as its fixed point set. Then, for any immersion $\beta : M \to \mathbb{R} \times D^n$, the immersion $\tilde{\phi} \circ \tilde{\tau} \circ \tilde{\chi} \circ \beta$ plays the role of the inverse element β^{-1} with respect to $\uplus$. Indeed, consider the unit half-disk D^2_+ and the unit half-cylinder $C^{n+1}_+ =_{\mathsf{def}} D^2_+ \times B^{n-1}$, inscribed in the cylinder $B^n \times [0,\, 1+\epsilon]$, $\epsilon > 0$. We will use the rotations $\tau(\theta)$ of $D^2 \times B^{n-1} \supset C^{n+1}_+$ around the axis B^{n-1} at the angles $\theta \in [0, \pi]$, so that $\tau(\pi) = \tau$. Let $N = M \times [0, \pi]$. In the case $d \equiv 0 \mod 2$, we form an immersion

$$A : N \to \mathbb{R} \times C^{n+1}_+ \subset \mathbb{R} \times B^n \times [0, 1+\epsilon],$$

defined by the formula $A(m, \theta) = \tilde{\tau}(\theta) \circ \beta(m)$.

Let $\{\underline{0}\} \in \mathbb{R}$ denote the origin. In the case $d \equiv 1 \mod 2$, to satisfy Definition 7.7 for $\mathbf{c}\Lambda = (1)$, we need to insure that

$$A(N) \cap \Big\{\mathbb{R} \times \big(\partial(B^n \times [0, 1+\epsilon]) \setminus (B^n \times \{0\})\big)\Big\} = \{\underline{0}\} \times \Big\{\partial(B^n \times [0, 1+\epsilon]) \setminus (B^n \times \{0\})\Big\}.$$

Therefore, we further isotop A *radially* in each of the multipliers D^2_+ onto the rectangle $D^1 \times [0, 1+\epsilon]$. The result is an isotopy of A inside the cylinder $\{\underline{0}\} \times B^n \times [0, 1+\epsilon]$.

Since $\{\tilde{\tau}(\theta)\}$ preserve the combinatorial $\mathcal{L}$-tangency patterns of $\beta(M)$ within the ambient foliation $\mathcal{L}^\bullet$ on $\mathbb{R} \times B^n \times [0, 1+\epsilon]$, we conclude that the cobordism $A : N \to \mathbb{R} \times B^n \times [0, 1+\epsilon]$ delivers $(d, d; \mathbf{c}\Theta, \mathbf{c}\Lambda; \mathbf{c}\Theta)$-quasitopy between $\beta \uplus \beta^{-1}$ and the neutral element $\beta_\star$, provided that $\mathbf{c}\Lambda$ is as in the hypotheses of the proposition.

Next, we need to verify the associativity of the operation $\uplus$. The argument is similar to the one that has validated that $\beta \uplus \beta_\star$ is $(d, d; \mathbf{c}\Theta, \mathbf{c}\Lambda; \mathbf{c}\Theta)$-quasitopic to β. It uses diffeomorphisms $\phi_{21} : D^n \#_\partial D^n \to D^n$ and $\phi_{32} : D^n \#_\partial D^n \#_\partial D^n \to D^n \#_\partial D^n$, embeddings $\chi_1, \chi_2 : D^n \to D^n \#_\partial D^n$ whose images are the first and the second balls, and the embeddings $\chi_{12}, \chi_{23} : D^n \#_\partial D^n \to D^n \#_\partial D^n \#_\partial D^n$ whose images are the union of first and the second ball, and the union of the second and third balls in $D^n \#_\partial D^n \#_\partial D^n$, respectively. We leave the rest of the argument to the reader.

The validation of the fact that $\uplus$ is commutative for $n > 1$ is a bit more involved and similar to the classical proof of the fact that the homotopy groups $\pi_n(\sim)$ of spaces are commutative for $n > 1$ (see Fig. 7.3). Let us sketch this validation.

As before, we put $B^n = D^n_1 \cup H \cup D^n_2$. We start with the case $d \equiv 0 \mod 2$ and two immersions $\beta_i : M_i \to \mathbb{R} \times B^n$ $(i = 1, 2)$ such that the image of β_i belongs to the cylinder $\mathbb{R} \times D^n_i$. Consider an equatorial hyperplane $K^{n-1} \subset B^n$ (the horizontal punctured line in Fig. 7.3), transversal to the

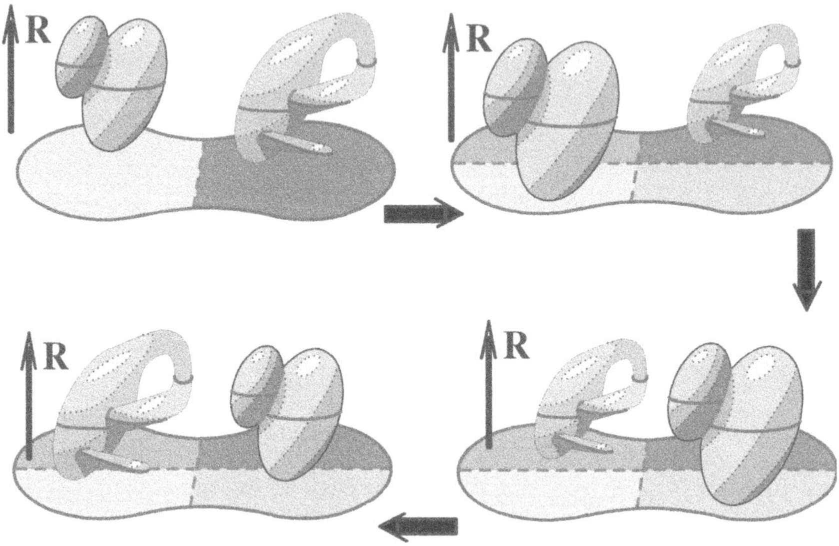

Fig. 7.3 Proving the commutativity of the operation $\uplus$ in the group $\mathsf{G}^{\mathsf{i/e}}_{d,d'}(n; \mathbf{c}\Theta, \mathbf{c}\Theta')$ for $n \geq 2$. The three bold arrows show the effects of three isotopies in $\mathbb{R} \times B^n$.

ball $B^{n-1} \subset H$ and such that K^{n-1} is the fixed point set of another involution λ on B^n. Thus the pair of $(n-1)$-balls K^{n-1} and B^{n-1} divide B^n in the four regions ("quadrants"): $B^n_I, B^n_{II}, B^n_{III}, B^n_{IV}$, ordered counterclockwise around the axle $K^{n-1} \cap B^{n-1}$. Each of these regions is homeomorphic to a n-ball.

First, we choose an isotopy η^t_{III} to compress $D^n_1 = B^n_{II} \cup B^n_{III}$ in the interior of the region B^n_{III} and an isotopy η^t_I to compress $D^n_2 = B^n_I \cup B^n_{IV}$ in the interior of the region B^n_I. We may assume that these isotopies are smooth in the interior of the relevant regions. We lift η^t_{III} to an isotopy $\tilde{\eta}^t_{III}$ and compose it with β_1. We lift η^t_I to an isotopy $\tilde{\eta}^t_I$ and compose it with β_2. Let us denote these compositions by $\beta_{1,III}$ and $\beta_{2,I}$, respectively. Next, we choose an isotopy ζ^t_{II} to compress $B^n_I \cup B^n_{II}$ in the interior of B^n_{II}, and an isotopy ζ^t_{IV} to compress $B^n_{III} \cup B^n_{IV}$ in the interior of B^n_{IV}. The composition $\tilde{\zeta}^t_{II} \circ \beta_{2,I}$ places the image of M_2 over the quadrant B^n_{II} and composition $\tilde{\zeta}^t_{IV} \circ \beta_{1,III}$ places the image M_1 over the quadrant B^n_{IV}. To complete the cycle that "switches" β_1 and β_2, it remains to expand the image $\tilde{\zeta}^t_{II} \circ \beta_{2,I}(M_2) \subset \mathbb{R} \times B^n_{II}$ into $\tilde{\tau}(M_2)$ and the image $\tilde{\zeta}^t_{IV} \circ \beta_{1,III}(M_1) \subset \mathbb{R} \times B^n_{IV}$ into $\tilde{\tau}(M_1)$.

In the case $d \equiv 1 \mod 2$, more careful arguments are needed to insure that $\beta_1 \uplus \beta_2$ is $(d, d; \mathbf{c}\Theta, (1); \mathbf{c}\Theta)$-quasitopic to $(\tilde{\tau} \circ \beta_2) \uplus (\tilde{\tau} \circ \beta_1)$. They are similar to the ones we used to prove that $\beta \uplus (\tilde{\tau} \circ \beta)$ is $(d, d; \mathbf{c}\Theta, (1); \mathbf{c}\Theta)$-quasitopic to the neutral element $\beta_\star$. □

Remark 7.4.
We stress that typically the "1-dimensional" groups $\mathsf{G}^{\mathsf{i/e}}_{d,d'}(1; \mathbf{c}\Theta; \mathbf{c}\Theta')$ are *not* commutative. There are quite a few examples (see Corollary 7.12 and Fig. 7.4) where, for $\Theta' = \Theta$ and $d' = d$, they are free groups with the number of free generators being bounded from above by some quadratic functions in d. In general, these groups form an interesting class, whose presentations are well-understood [KSW1] and described in Chapter 5, and whose group theoretical properties deserve a study. ◇

For a given group G, we denote by $(\mathsf{G})^r$ its r-fold direct product.

Corollary 7.3. *Let $\Theta' \subset \Theta \subset \mathbf{\Omega}$ be two closed subposets, where $(\emptyset) \notin \Theta$ for $d \equiv 0 \mod 2$ and $(1) \notin \Theta$ for $d \equiv 1 \mod 2$. Let $d \equiv d' \mod 2$.*

For any compact connected smooth n-dimensional manifold Y with boundary that has r connected components, the group $\big(\mathsf{G}^{\mathsf{i/e}}_{d,d'}(n; \mathbf{c}\Theta, \mathbf{c}\Theta')\big)^r$ acts, with the help of the operation $\uplus$ from (7.16), on the set $\mathsf{Q}^{\mathsf{i/e}}_{d,d'}(Y; \mathbf{c}\Theta, \mathbf{c}\Theta')$.

Similarly, the group $(\mathsf{OG}^{\mathsf{i/e}}_{d,d'}(n; \mathbf{c}\Theta, \mathbf{c}\Theta'))^r$ acts on $\mathsf{OQ}^{\mathsf{i/e}}_{d,d'}(Y; \mathbf{c}\Theta, \mathbf{c}\Theta')$, provided $d \equiv 0 \mod 2$.

Proof. Temporarily, let $\mathcal{G} =_{\mathsf{def}} \mathsf{G}^{\mathsf{i/e}}_{d,d'}(n; \mathbf{c}\Theta, \mathbf{c}\Theta')$. By Proposition 7.6, $\mathcal{G}$ is a group with respect to the $\uplus$ operation. By the arguments that follow (7.16), for each choice of a connected component $\partial_\kappa Y$ of ∂Y, the group $\mathcal{G}$ acts via formulae (7.16) and (7.17) on the set $\mathsf{Q}^{\mathsf{i/e}}_{d,d'}(Y; \mathbf{c}\Theta, \mathbf{c}\Theta')$. In the case of an even d and $\mathbf{c}\Lambda = (\emptyset)$, for a pair $\beta : M \to \mathbb{R} \times Y$ and $\beta' : M' \to \mathbb{R} \times D^n$, the action is defined by the formula $\beta \coprod \beta' : M \coprod M' \to \mathbb{R} \times (Y \#_{\partial_\kappa} D^n) \approx \mathbb{R} \times Y$, where $\#_{\partial_\kappa}$ is the boundary connected sum that employs the connected component $\partial_\kappa Y$.

In the case of an odd d and $\mathbf{c}\Lambda = (1)$, for a given pair $\beta : M \to \mathbb{R} \times Y$ and $\beta' : M' \to \mathbb{R} \times D^n$, the action is defined by the formula $\beta \#_{\partial_\kappa} \beta' : M \#_{\partial^\dagger_\kappa} M' \to \mathbb{R} \times (Y \#_{\partial_\kappa} D^n)$, where $\#_{\partial^\dagger_\kappa}$ is a boundary connected sum, using the unique component of ∂M embedded by β in $\mathbb{R} \times \partial_\kappa Y$ and the unique component of $\partial M'$ embedded by β' in $\mathbb{R} \times \partial D^n$.

For different choices of elements $\kappa \in \pi_0(\partial Y)$, these $\mathcal{G}_\kappa$-actions evidently commute. Thus $\mathcal{G}^r$ acts on $\mathsf{Q}^{\mathsf{i/e}}_{d,d'}(Y; \mathbf{c}\Theta, \mathbf{c}\Theta')$, where $r = \#(\pi_0(\partial Y))$. Similar arguments work for the oriented case, provided $d \equiv 0 \mod 2$. □

Remark 7.5. We will see soon that the groups $\mathsf{G}^{\mathsf{i/e}}_{d,d'}(n; \mathbf{c}\Theta, \mathbf{c}\Theta')$ may have elements of infinite order, as well as some torsion, as complex as the torsion of the homotopy groups of spheres. As a result, the $\mathsf{G}^{\mathsf{i/e}}_{d,d'}(n; \mathbf{c}\Theta, \mathbf{c}\Theta')$-orbits in the set $\mathsf{Q}^{\mathsf{i/e}}_{d,d'}(Y; \mathbf{c}\Theta, \mathbf{c}\Theta')$ may have complex and diverse periods. We have a limited understanding of the orbit-space of this action; however, for embeddings, this problem will be reduced to the homotopy theory of the spaces of real polynomials with constrained real divisors. In some cases (see Proposition 7.15 and Example 7.5), we can estimate the "size" of the orbit-space $\mathsf{Q}^{\mathsf{i/e}}_{d,d'}(Y; \mathbf{c}\Theta, \mathbf{c}\Theta')/\mathsf{G}^{\mathsf{i/e}}_{d,d'}(n; \mathbf{c}\Theta, \mathbf{c}\Theta')$. ◇

Definition 7.11. • Given two pairs of topological spaces $X_1 \supset A_1$ and $X_2 \supset A_2$, we denote by $[(X_1, A_1), (X_2, A_2)]$ the set of homotopy classes of continuous maps $g : X_1 \to X_2$ such that $g(A_1) \subset A_2$.

• Given three pairs of topological spaces $X_1 \supset A_1$, $X_2 \supset A_2$, and $X_3 \supset A_3$ and a fixed continuous map $\epsilon : (X_2, A_2) \to (X_3, A_3)$, we denote by

$$[[(X_1, A_1),\ \epsilon : (X_2, A_2) \to (X_3, A_3)]] \tag{7.18}$$

the set of homotopy classes $[g]$ of continuous maps $g : (X_1, A_1) \to (X_2, A_2)$, modulo the following equivalence relation: $[g_0] \sim [g_1]$, where $g_0 : (X_1, A_1) \to (X_2, A_2)$ and $g_1 : (X_1, A_1) \to (X_2, A_2)$, if the compositions $\epsilon \circ g_0$ and $\epsilon \circ g_1$ are homotopic as maps from (X_1, A_1) to (X_3, A_3). ◇

If all the spaces above are locally compact CW-complexes, then the set in (7.18) is countable or finite.

The next two propositions deliver new characteristic homotopy classes of immersions/embeddings against a fixed background of 1-foliations of the product type.

Proposition 7.7. *Let $\Theta' \subset \Theta \subset \Lambda \subset \mathbf{\Omega}$ be closed subposets. Let M, Y be smooth compact n-manifolds.*

- *Any proper immersion/embedding $\beta : (M, \partial M) \to (\mathbb{R} \times Y, \mathbb{R} \times \partial Y)$, such that:*
  - *(1) $m^\beta(y) \leq d$ (see (7.5)) and $m^\beta(y) \equiv d \mod 2$ for all $y \in Y$,*
  - *(2) the combinatorial patterns $\{\omega^\beta(y)\}_{y \in Y}$ belong to $\mathbf{c}\Theta$,*

(3) the combinatorial patterns $\{\omega^\beta(y)\}_{y\in\partial Y}$, *belong to* $\mathbf{c}\Lambda$,

generates a continuous map $\Phi_d^\beta : (Y, \partial Y) \to (\mathcal{P}_d^{\mathbf{c}\Theta}, \mathcal{P}_d^{\mathbf{c}\Lambda})$, *whose homotopy class* $[\Phi^\beta]$ *depends on* β *only.*

- *Any such pair*

  $\beta_0 : (M_0, \partial M_0) \to (\mathbb{R}\times Y,\ \mathbb{R}\times\partial Y)$ *and* $\beta_1 : (M_1, \partial M_1) \to (\mathbb{R}\times Y,\ \mathbb{R}\times\partial Y)$

  of $(d, d'; \mathbf{c}\Theta, \mathbf{c}\Lambda; \mathbf{c}\Theta')$*-quasitopic immersions/embeddings generates homotopic composit maps*

  $\epsilon_{d,d'}\circ\Phi^{\beta_0} : (Y, \partial Y) \to (\mathcal{P}_{d'}^{\mathbf{c}\Theta'}, \mathcal{P}_{d'}^{\mathbf{c}\Lambda})$ *and* $\epsilon_{d,d'}\circ\Phi^{\beta_1} : (Y, \partial Y) \to (\mathcal{P}_{d'}^{\mathbf{c}\Theta'}, \mathcal{P}_{d'}^{\mathbf{c}\Lambda})$.

  Thus, we get a well-defined surjective map

$$\Phi_{d,d'}(Y, \partial Y; \mathbf{c}\Theta, \mathbf{c}\Lambda; \mathbf{c}\Theta') :\ \mathcal{Q}^{\mathsf{i/e}}_{d,d'}(Y, \partial Y; \mathbf{c}\Theta, \mathbf{c}\Lambda; \mathbf{c}\Theta') \to$$
$$\to [[(Y, \partial Y),\ \epsilon_{d,d'} : (\mathcal{P}_d^{\mathbf{c}\Theta}, \mathcal{P}_d^{\mathbf{c}\Lambda}) \to (\mathcal{P}_{d'}^{\mathbf{c}\Theta'}, \mathcal{P}_{d'}^{\mathbf{c}\Lambda})]] \quad (7.19)$$

  from the (oriented) quasitopies of immersions/embeddings to the homotopy classes of triples of spaces, as introduced in Definition 7.11.

Proof. The two claims follow from Theorem 7.2, being applied first to the proper immersions $\beta_0 : M_0 \to \mathbb{R} \times Y$ and $\beta_1 : \beta_1 : M_1 \to \mathbb{R} \times Y$, and then to the proper immersion $B : (N, \delta N) \to (\mathbb{R} \times Y \times [0,1],\ \mathbb{R} \times \partial Y \times [0,1])$ of a relative cobordism N between M_0 and $M_1 \cup \delta N$. There is only one technical wrinkle in this application: N is a compact manifold *with corners* $\partial M_0 \coprod \partial M_1$. So, Theorem 7.2 must be adjusted to such a cituation. Note that B is transversal to both hypersurfaces, $\mathbb{R}\times Y\times\partial[0,1]$ and $\mathbb{R}\times\partial Y\times[0,1]$. This fact allows to add a collar to $Y \times [0,1]$ and to extend the arguments of Theorem 7.2 to a new enlarged $(n+1)$-manifold $\hat{Z}$, whose interior contains $Y \times [0,1]$.

Thus the map $\Phi^B : (Y \times [0,1],\ \partial Y \times [0,1]) \to (\mathcal{P}_{d'}^{\mathbf{c}\Theta'}, \mathcal{P}_{d'}^{\mathbf{c}\Lambda})$ delivers a homotopy between the maps $\epsilon_{d,d'} \circ \Phi^{\beta_0}$ and $\epsilon_{d,d'} \circ \Phi^{\beta_1}$. Revisiting Definition 7.11, this implies that (7.19) is a well-defined map. In fact, a similar map may be defined for the oriented quasitopies.

It will follow from formula (7.27) below that the map in (7.19) is *surjective.* □

Corollary 7.4. *Under the hypotheses of Proposition 7.7 and assuming the* Λ*-condition (7.11), we get a well-defined map (see (7.12) for the notations)*

$$\Phi_{d,d'}(Y; \mathbf{c}\Theta, \mathbf{c}\Theta') :\ \mathsf{Q}^{\mathsf{i/e}}_{d,d'}(Y; \mathbf{c}\Theta, \mathbf{c}\Theta') \to$$
$$\to [[(Y, \partial Y),\ \epsilon_{d,d'} : (\mathcal{P}_d^{\mathbf{c}\Theta}, pt) \to (\mathcal{P}_{d'}^{\mathbf{c}\Theta'}, pt')]], \quad (7.20)$$

where the base point $pt \in \mathcal{P}_d^{\mathbf{c}\Lambda}$ *and* $pt' = \epsilon_{d,d'}(pt)$. *Similarly, a map from the oriented quasitopies* $\mathsf{OQ}_{d,d'}^{\mathsf{i/e}}(Y; \mathbf{c}\Theta, \mathbf{c}\Theta')$ *to the same target is available.*

Proof. If $\mathbf{c}\Lambda = (\emptyset)$ or $\mathbf{c}\Lambda = (1)$, then the spaces $\mathcal{P}_d^{\mathbf{c}\Lambda}$ and $\mathcal{P}_{d'}^{\mathbf{c}\Lambda}$ are contractible. Therefore, the corollary follows from Proposition 7.7. □

Let Y be a connected n-dimensional manifold with boundary. Let us pick a preferred component ∂Y_i of ∂Y. Then the group $\pi_n(\mathcal{P}_d^{\mathbf{c}\Theta}, pt)/\mathsf{ker}((\epsilon_{d,d'})_*)$ acts naturally on the set $[[(Y, \partial Y),\ \epsilon_{d,d'}]]$ by forming the connected sum $g \#_{\partial Y_i} f$ of two maps $f : (D^n, \partial D^n) \to (\mathcal{P}_d^{\mathbf{c}\Theta}, pt)$ and $g : (Y, \partial Y) \to (\mathcal{P}_d^{\mathbf{c}\Theta}, pt)$. As a result, we get an action of the group $\big(\pi_n(\mathcal{P}_d^{\mathbf{c}\Theta}, pt)/\mathsf{ker}((\epsilon_{d,d'})_*)\big)^r$ on the set $[[(Y, \partial Y),\ \epsilon_{d,d'}]]$, where $r = \#(\pi_0(\partial Y))$. Let us denote this action by $h \#_{\partial_1} f_1 \#_{\partial_2} \dots \#_{\partial_r} f_r$, where $h \in [[(Y, \partial Y),\ \epsilon_{d,d'}]]$, $f_i \in \pi_n(\mathcal{P}_d^{\mathbf{c}\Theta}, pt)$, and $\#_{\partial_i} f_i$ stands for forming a boundary connected sum map $h \#_\partial f_i$ with h by using the ith component of ∂Y.

• Assuming the Λ-condition (7.11), in what follows, we will use the abbreviation

$$\Phi_{d,d'}(n; \mathbf{c}\Theta, \mathbf{c}\Theta') =_{\mathsf{def}} \Phi_{d,d'}(D^n, \partial D^n; \mathbf{c}\Theta, \mathbf{c}\Lambda; \mathbf{c}\Theta')$$

for the map on the RHS of this equality/definition.

Combining Proposition 7.6 with Proposition 7.7, leads to the following result.

Proposition 7.8. *Under the* Λ*-condition (7.11),*

- *the map in (7.19) generates a group homomorphism*

$$\Phi_{d,d'}(n; \mathbf{c}\Theta, \mathbf{c}\Theta') : \ \mathsf{G}_{d,d'}^{\mathsf{i/e}}(n; \mathbf{c}\Theta, \mathbf{c}\Theta') \to$$
$$\to \pi_n(\mathcal{P}_d^{\mathbf{c}\Theta}, pt)\Big/\mathsf{ker}\Big\{(\epsilon_{d,d'})_* : \pi_n(\mathcal{P}_d^{\mathbf{c}\Theta}, pt) \to \pi_n(\mathcal{P}_{d'}^{\mathbf{c}\Theta'}, pt')\Big\}, \quad (7.21)$$

  where the group operation in $\mathsf{G}_{d,d'}^{\mathsf{i/e}}(n; \mathbf{c}\Theta, \mathbf{c}\Theta')$ *is "*$\uplus$*".*[4]

  For $d \equiv 0 \mod 2$, *the base points* $pt \in \mathcal{P}_d^{\mathbf{c}\Theta}$ *resides in the chamber* $\mathsf{R}_d^{(\emptyset)}$ *of positive monic polynomials, and for* $d \equiv 1 \mod 2$, *in the chamber* $\mathsf{R}_d^{(1)}$ *of monic polynomials with a single simple real root. Similar choices of the base point are made for* $pt' \in \mathcal{P}_{d'}^{\mathbf{c}\Theta'}$.

[4]In Corollary 7.8 we will prove that $\Phi_{d,d'}^{\mathsf{emb}}(n; \mathbf{c}\Theta; \mathbf{c}\Lambda; \mathbf{c}\Theta')$ is an isomorphism and $\Phi_{d,d'}^{\mathsf{imm}}(n; \mathbf{c}\Theta; \mathbf{c}\Lambda; \mathbf{c}\Theta')$ is an epimorphism.

- *Let* $r = \#(\pi_0(\partial Y))$. *The* r^{th} *power of the homomorphism* $\Phi_{d,d'}(n; \mathbf{c}\Theta, \mathbf{c}\Theta')$ *from (7.21) generates a representation of the group* $\big(\mathsf{G}^{\mathsf{i/e}}_{d,d'}(n; \mathbf{c}\Theta, \mathbf{c}\Theta')\big)^r$ *in the group*

$$\Big(\pi_n(\mathcal{P}^{\mathbf{c}\Theta}_d, pt)\Big/\mathsf{ker}\Big\{(\epsilon_{d,d'})_* : \pi_n(\mathcal{P}^{\mathbf{c}\Theta}_d, pt) \to \pi_n(\mathcal{P}^{\mathbf{c}\Theta'}_{d'}, pt')\Big\}\Big)^r.$$

With the help of this representation, the map in (7.19) is equivariant *with respect to the* $\big(\mathsf{G}^{\mathsf{i/e}}_{d,d'}(n; \mathbf{c}\Theta, \mathbf{c}\Theta')\big)^r$*-action on the set* $\mathsf{Q}^{\mathsf{i/e}}_{d,d'}(Y; \mathbf{c}\Theta, \mathbf{c}\Theta')$ *and with respect to the*

$$\big(\pi_n(\mathcal{P}^{\mathbf{c}\Theta}_d, pt)\big/\mathsf{ker}\{(\epsilon_{d,d'})_* : \pi_n(\mathcal{P}^{\mathbf{c}\Theta}_d, pt) \to \pi_n(\mathcal{P}^{\mathbf{c}\Theta'}_{d'}, pt')\}\big)^r\text{-}action$$

on the set $[[(Y, \partial Y),\ \epsilon_{d,d'} : (\mathcal{P}^{\mathbf{c}\Theta}_d, \mathcal{P}^{\mathbf{c}\Lambda}_d) \to (\mathcal{P}^{\mathbf{c}\Theta'}_{d'}, \mathcal{P}^{\mathbf{c}\Lambda}_{d'})]]$.

- *Similar claims hold for the oriented quasitopies* $\mathsf{OQ}^{\mathsf{i/e}}_{d,d'}(Y; \mathbf{c}\Theta, \mathbf{c}\Theta')$ *and for the* $\big(\mathsf{OG}^{\mathsf{i/e}}_{d,d'}(n; \mathbf{c}\Theta, \mathbf{c}\Theta')\big)^r$*-action on them, provided* $d \equiv d' \equiv 0$ mod 2.

Proof. By Proposition 7.7, we get a group representation $\Phi_{d,d'}(n; \mathbf{c}\Theta, \mathbf{c}\Lambda; \mathbf{c}\Theta')$ from (7.21). By Corollary 7.3, the group $\big(\mathsf{G}^{\mathsf{i/e}}_{d,d'}(n; \mathbf{c}\Theta, \mathbf{c}\Theta')\big)^r$ acts on the set $\mathsf{Q}^{\mathsf{i/e}}_{d,d'}(Y; \mathbf{c}\Theta, \mathbf{c}\Theta')$. Employing Corollary 7.4, we get a well-defined map $\Phi(Y) =_{\mathsf{def}} \Phi_{d,d'}(Y, \partial Y; \mathbf{c}\Theta, \mathbf{c}\Lambda; \mathbf{c}\Theta')$ from (7.19).

If a map $F : (Y, \partial Y) \to (\mathcal{P}^{\mathbf{c}\Theta}_d, pt)$, being composed with $\epsilon_{d,d'}$, becomes null-homotopic in $(\mathcal{P}^{\mathbf{c}\Theta'}_d, pt')$, then for any map $G : (D^n, \partial D^n) \to (\mathcal{P}^{\mathbf{c}\Theta}_d, pt)$ such that the composition $\epsilon_{d,d'} \circ G$ is null-homotopic in $(\mathcal{P}^{\mathbf{c}\Theta'}_d, pt')$, the composed map $\epsilon_{d,d'} \circ (G \#_\partial F)$ is null-homotopic in $(\mathcal{P}^{\mathbf{c}\Theta'}_d, pt')$ as well. Thus the action of the group $\pi_n((\mathcal{P}^{\mathbf{c}\Theta}_d, pt))/\mathsf{ker}\{\epsilon_{d,d'}\}$ on the set

$$[[(Y, \partial Y),\ \epsilon_{d,d'} : (\mathcal{P}^{\mathbf{c}\Theta}_d, \mathcal{P}^{\mathbf{c}\Lambda}_d) \to (\mathcal{P}^{\mathbf{c}\Theta'}_{d'}, \mathcal{P}^{\mathbf{c}\Lambda}_{d'})]]$$

is well-defined. Now, thanks to Theorem 7.2, it is on the level of definitions to verify the equivariant property of $\Phi(Y)$, namely, that

$$\Phi(Y)(\beta \uplus_1 \gamma_1 \uplus_2 \gamma_2 \dots \uplus_r \gamma_r) = \Phi(Y)(\beta)\, \#_{\partial_1} \Phi(D^n)(\gamma_1) \dots \#_{\partial_r} \Phi(D^n)(\gamma_r),$$

where $\beta \in \mathsf{Q}^{\mathsf{i/e}}_{d,d'}(Y; \mathbf{c}\Theta, \mathbf{c}\Theta')$ and $\gamma_i \in \mathsf{G}^{\mathsf{i/e}}_{d,d'}(n; \mathbf{c}\Theta, \mathbf{c}\Theta')$, or $\beta \in \mathsf{OQ}^{\mathsf{i/e}}_{d,d'}(Y; \mathbf{c}\Theta, \mathbf{c}\Theta')$ and $\gamma_i \in \mathsf{OG}^{\mathsf{i/e}}_{d,d'}(n; \mathbf{c}\Theta, \mathbf{c}\Theta')$. □

Lemma 7.7. *Let* $\Theta' \subset \Theta \subset \Lambda \subset \mathbf{\Omega}$ *be closed subposets, and let* Y *be a compact smooth manifold. We pick a pair of natural numbers* $d \leq d'$ *of the same parity.*

- *Any smooth $(\partial\mathcal{E}_d)$-regular map $h : (Y, \partial Y) \to (\mathcal{P}_d^{\mathbf{c}\Theta}, \mathcal{P}_d^{\mathbf{c}\Lambda})$ produces a proper regular* embedding $\beta_h : (M, \partial M) \hookrightarrow (\mathbb{R}\times Y, \mathbb{R}\times\partial Y)$ *of some compact n-manifold M. The combinatorial patterns of β_h, relative to the foliation $\mathcal{L}$ on $\mathbb{R}\times Y$, belong to the poset* $\mathbf{c}\Theta$ *and of its restriction to ∂M belong to the poset* $\mathbf{c}\Lambda$. *Moreover, $\Phi^{\beta_h} = h$.*
- *Any pair $h_0, h_1 : (Y, \partial Y) \to (\mathcal{P}_d^{\mathbf{c}\Theta}, \mathcal{P}_d^{\mathbf{c}\Lambda})$ of $(\partial\mathcal{E}_d)$-regular maps such that $\epsilon_{d,d'} \circ h_0$ and $\epsilon_{d,d'} \circ h_1$ are linked by a smooth $(\partial\mathcal{E}_{d'})$-regular homotopy*
$$H : (Y\times[0,1], \partial Y\times[0,1]) \to (\mathcal{P}_{d'}^{\mathbf{c}\Theta'}, \mathcal{P}_{d'}^{\mathbf{c}\Lambda})$$
gives rise to a $(d, d'; \mathbf{c}\Theta, \mathbf{c}\Lambda; \mathbf{c}\Theta')$-quasitopy $B_H : N \hookrightarrow \mathbb{R}\times Y\times[0,1]$ between β_{h_0} and β_{h_1}. Moreover, $\Phi^{B_H} = H$.
- *If Y is orientable, then so are M and N.*

Proof. The first claim of the lemma follows from Definition 7.4 since, for a $\partial\mathcal{E}_d$-regular map, the preimage of the hypersurface $\partial\mathcal{E}_d \subset \mathbb{R}\times\mathcal{P}_d$ under $\mathsf{id}\times h$ is a regularly embedded codimension 1 submanifold M of $\mathbb{R}\times Y$, and the preimage of $\partial\mathcal{E}_d$ under the map $\mathsf{id}\times h|_{\partial Y}$, where $h|_{\partial Y}$ is a $\partial\mathcal{E}_d$-regular map, is the boundary ∂M, regularly embedded in $\mathbb{R}\times\partial Y$. Recall that the map $\mathsf{id}\times h$ pulls back the universal function $\wp(u,x) : \mathbb{R}\times\mathbb{R}^d \to \mathbb{R}$ to the function $(\mathsf{id}\times h)^*(\wp)$ on $\mathbb{R}\times Y$, for which 0 is a regular value. Using this function $(\mathsf{id}\times h)^*(\wp)$, the arguments from Theorem 7.2 imply that $\Phi^{\beta_h} = h$.

Let us validate the second claim. Recall that $\epsilon_{d,d'}(\mathcal{P}_d^{\mathbf{c}\Lambda}) \subset \mathcal{P}_{d'}^{\mathbf{c}\Lambda}$, since $\epsilon_{d,d'}$ preserves the combinatorial types ω of real divisors of real polynomials.

By definition of the map $\epsilon_{d,d'}$ in (6.32), we get
$$\epsilon_{d,d'}^*(\wp_{d'}) = (u^2+1)^{(d'-d)/2}\cdot\wp_d.$$
Thus,
$$\big(\mathsf{id}\times(\epsilon_{d,d'}\circ h)\big)^*(\wp_{d'}) = (\mathsf{id}\times h)^*(\wp_d)\cdot Q,$$
where $Q : \mathbb{R}\times Y \to \mathbb{R}_+$ is a smooth positive function. Thus, the zero loci
$$\{\big(\mathsf{id}\times(\epsilon_{d,d'}\circ h)\big)^*(\wp_{d'}) = 0\} \text{ and } \{\big(\mathsf{id}\times h\big)^*(\wp_d) = 0\}$$
coincide. Moreover, using that $Q > 0$, if the 1-jet of $\big(\mathsf{id}\times h\big)^*(\wp_d)$ does not vanish along $\{\big(\mathsf{id}\times h\big)^*(\wp_d) = 0\}$, then the 1-jet of $(\mathsf{id}\times h)^*(\wp_d)\cdot Q$ does not vanish along the shared locus. Therefore, if $(\mathsf{id}\times h)^*(\wp_d)$ has 0 is a regular value, then $\big(\mathsf{id}\times(\epsilon_{d,d'}\circ h)\big)^*(\wp_{d'})$ has 0 as a regular value as well.

Now, given two $\partial\mathcal{E}_d$-regular maps $h_0, h_1 : (Y, \partial Y) \to (\mathcal{P}_d^{\mathbf{c}\Theta}, \mathcal{P}_d^{\mathbf{c}\Lambda})$ such that $\epsilon_{d,d'}\circ h_0$ and $\epsilon_{d,d'}\circ h_1$ are linked by a smooth $\partial\mathcal{E}_{d'}$-regular homotopy
$$H : (Y\times[0,1], \partial Y\times[0,1]) \to (\mathcal{P}_{d'}^{\mathbf{c}\Theta'}, \mathcal{P}_{d'}^{\mathbf{c}\Lambda}),$$

we conclude that the preimage of $\partial\mathcal{E}_{d'}$ under $\mathsf{id}\times H$ is a regularly embedded codimension 1 submanifold $N \subset \mathbb{R}\times Y\times[0,1]$ with corners $\partial M_0 \coprod \partial M_1$. Its boundary $\partial N = (\partial M_0 \coprod \partial M_1)\bigcup \delta N$, where

$$\delta N = H^{-1}(\partial\mathcal{E}_{d'}) \cap (\mathbb{R}\times\partial Y\times[0,1]).$$

Moreover, for $i = 0,1$, the transversality of $\mathsf{id}\times H$ to $\partial\mathcal{E}_{d'}$ implies the transversality of

$$\beta_i : M_i = (\epsilon_{d,d'}\circ h_i)^{-1}(\partial\mathcal{E}_{d'}) \hookrightarrow \mathbb{R}\times Y$$

to $\mathbb{R}\times\partial Y$ and the transversality of $B : N = H^{-1}(\partial\mathcal{E}_{d'}) \hookrightarrow \mathbb{R}\times Y\times[0,1]$ to $\mathbb{R}\times\partial Y\times[0,1]$. Using the pull-back $(\mathsf{id}\times H)^*(\wp_{d'})$, again, by the arguments from Theorem 7.2, we get that $\Phi^{B_H} = H$.

Finally, if Y is oriented, then so are $\mathbb{R}\times Y$ and $\mathbb{R}\times Y\times[0,1]$. Then the pull-backs under $\mathsf{id}\times h$ or $\mathsf{id}\times H$ of the normal fields $n(\partial\mathcal{E}_d, \mathcal{E}_d)$ or $n(\partial\mathcal{E}_{d'}, \mathcal{E}_{d'})$ help to orient M or N. □

Lemma 7.8 below is a natural generalization of Lemma 7.7 from the case $k = 1$ to the case of a general k. It is based on Lemma 7.5. We skip the proof of Lemma 7.8, since we will not rely on it. However, its formulation may give the reader a slightly wider perspective on desirable efforts.

Lemma 7.8. *Let $\Theta' \subset \Theta \subset \Lambda \subset \mathbf{\Omega}$ be closed sub-posets, and Y a compact manifold. Let $k \leq d$, $k' \leq d'$, $d \leq d'$, $k \leq k'$, and $d \equiv d' \mod 2$.*

- *Any smooth $(\partial\mathcal{E}_d, k)$-regular map $h : (Y, \partial Y) \to (\mathcal{P}_d^{\mathbf{c}\Theta}, \mathcal{P}_d^{\mathbf{c}\Lambda})$ produces a proper immersion $\beta_h : (M, \partial M) \hookrightarrow (\mathbb{R}\times Y, \mathbb{R}\times\partial Y)$ of some compact n-manifold M. The combinatorial patterns of β_h, relative to the foliation $\mathcal{L}$ on $\mathbb{R}\times Y$, belong to $\mathbf{c}\Theta$ and of its restriction to ∂M to $\mathbf{c}\Lambda$. Moreover, $\Phi^{\beta_h} = h$.*
- *Any pair $h_0, h_1 : (Y, \partial Y) \to (\mathcal{P}_d^{\mathbf{c}\Theta}, \mathcal{P}_d^{\mathbf{c}\Lambda})$ of $(\partial\mathcal{E}_d, k)$-regular maps such that $\epsilon_{d,d'}\circ h_0$ and $\epsilon_{d,d'}\circ h_1$ are linked by a smooth $(\partial\mathcal{E}_{d'}, k')$-regular homotopy*

  $$H : (Y\times[0,1],\, \partial Y\times[0,1]) \to (\mathcal{P}_{d'}^{\mathbf{c}\Theta'},\, \mathcal{P}_{d'}^{\mathbf{c}\Lambda}),$$

  where $k \leq k' \leq d'$, gives rise to a $(d, d'; \mathbf{c}\Theta, \mathbf{c}\Lambda; \mathbf{c}\Theta')$-quasitopy $B_H : N \to \mathbb{R}\times Y\times[0,1]$ between the immersions β_{h_0} and β_{h_1}. Moreover, $\Phi^{B_H} = H$.
- *If Y is orientable, then so are M and N.* ◇

Definition 7.12. Let $\Theta' \subset \Theta \subset \Lambda \subset \mathbf{\Omega}$ be closed subposets (which are not necessarily satisfying the Λ-condition). Consider the obvious relaxation map

$$\mathcal{Q}^{\mathsf{i/e}}_{d,d'}(Y, \partial Y; \mathbf{c}\Lambda, \mathbf{c}\Lambda; \mathbf{c}\Theta') \to \mathcal{Q}^{\mathsf{i/e}}_{d,d'}(Y, \partial Y; \mathbf{c}\Theta, \mathbf{c}\Lambda; \mathbf{c}\Theta').$$

We denote by $^\dagger\mathcal{Q}^{\mathsf{i/e}}_{d,d'}(Y, \partial Y; \mathbf{c}\Theta, \mathbf{c}\Lambda; \mathbf{c}\Theta')$ its image. It is represented by immersions/ embeddings $\beta : (M, \partial M) \to (\mathbb{R} \times Y, \mathbb{R} \times \partial Y)$, whose combinatorial tangency patterns to the foliation $\mathcal{L}$ belong to the open subposet $\mathbf{c}\Lambda \subset \mathbf{c}\Theta$. We call such elements combinatorially trivial.

Then we introduce the quotient set

$$\mathbf{Q}^{\mathsf{i/e}}_{d,d'}(Y, \partial Y; \mathbf{c}\Theta, \mathbf{c}\Lambda; \mathbf{c}\Theta') =_{\mathsf{def}} \tag{7.22}$$
$$\mathcal{Q}^{\mathsf{i/e}}_{d,d'}(Y, \partial Y; \mathbf{c}\Theta, \mathbf{c}\Lambda; \mathbf{c}\Theta') /\ ^\dagger\mathcal{Q}^{\mathsf{i/e}}_{d,d'}(Y, \partial Y; \mathbf{c}\Lambda, \mathbf{c}\Lambda; \mathbf{c}\Theta'). \qquad \diamondsuit$$

Under the Λ-condition (7.11), we get bijections

$$\mathbf{Q}^{\mathsf{i/e}}_{d,d'}(Y, \partial Y; \mathbf{c}\Theta, \mathbf{c}\Lambda; \mathbf{c}\Theta') = \tag{7.23}$$
$$= \mathcal{Q}^{\mathsf{i/e}}_{d,d'}(Y, \partial Y; \mathbf{c}\Theta, \mathbf{c}\Lambda; \mathbf{c}\Theta') =_{\mathsf{def}} \mathsf{Q}^{\mathsf{i/e}}_{d,d'}(Y; \mathbf{c}\Theta, \mathbf{c}\Theta'),$$

since $^\dagger\mathcal{Q}^{\mathsf{i/e}}_{d,d'}(Y, \partial Y; \mathbf{c}\Lambda, \mathbf{c}\Lambda; \mathbf{c}\Theta')$ is represented by a single element, an empty embedding (when $d \equiv 0 \mod 2$) or the embedding $\mathsf{id} : Y \to \{0\} \times Y \subset \mathbb{R} \times Y$ (when $d \equiv 1 \mod 2$). In particular, if $\partial Y = \emptyset$, then we get the bijection (7.23).

When Y is *orientable*, similar definitions/constructions are available for the oriented quasitopies $\mathbf{OQ}^{\mathsf{i/e}}_{d,d'}(Y, \partial Y; \mathbf{c}\Theta, \mathbf{c}\Lambda; \mathbf{c}\Theta')$, where $d \equiv 0 \mod 2$.

The next central to our efforts theorem provides homotopy-theoretical invariants of the quasitopy classes of immersions and embeddings into $\mathbb{R} \times Y$. For *embeddings*, assuming the Λ-condition, it actually computes the quasitopies.

Theorem 7.3. *Let $\Theta' \subset \Theta \subset \Lambda \subset \mathbf{\Omega}$ be closed subposets of the universal poset $\mathbf{\Omega}$, and Y be a smooth compact manifold. Assume that $d' \geq d$ and $d' \equiv d \mod 2$.*

- *Then there is a canonical bijection*

$$\mathbf{\Phi}^{\mathsf{emb}} : \mathbf{Q}^{\mathsf{emb}}_{d,d'}(Y, \partial Y; \mathbf{c}\Theta, \mathbf{c}\Lambda; \mathbf{c}\Theta') \xrightarrow{\approx}$$
$$\xrightarrow{\approx} \big[\big[(Y, \partial Y),\ \epsilon_{d,d'} : (\mathcal{P}^{\mathbf{c}\Theta}_d, \mathcal{P}^{\mathbf{c}\Lambda}_d) \to (\mathcal{P}^{\mathbf{c}\Theta'}_{d'}, \mathcal{P}^{\mathbf{c}\Lambda}_{d'})\big]\big]. \tag{7.24}$$

and a canonical surjection

$$\mathbf{\Phi}^{\mathsf{imm}} : \mathbf{Q}^{\mathsf{imm}}_{d,d'}(Y, \partial Y; \mathbf{c}\Theta, \mathbf{c}\Lambda; \mathbf{c}\Theta') \xrightarrow{epi}$$
$$\xrightarrow{epi} \big[\big[(Y, \partial Y),\ \epsilon_{d,d'} : (\mathcal{P}^{\mathbf{c}\Theta}_d, \mathcal{P}^{\mathbf{c}\Lambda}_d) \to (\mathcal{P}^{\mathbf{c}\Theta'}_{d'}, \mathcal{P}^{\mathbf{c}\Lambda}_{d'})\big]\big]. \tag{7.25}$$

• *Assuming the Λ-condition (7.11), we get a bijection*

$$\Phi^{\mathsf{emb}} : \mathsf{Q}^{\mathsf{emb}}_{d,d'}(Y; \mathbf{c}\Theta, \mathbf{c}\Theta') \xrightarrow{\approx}$$
$$\xrightarrow{\approx} \big[\big[(Y, \partial Y),\ \epsilon_{d,d'} : (\mathcal{P}^{\mathbf{c}\Theta}_d, \mathcal{P}^{\mathbf{c}\Lambda}_d) \to (\mathcal{P}^{\mathbf{c}\Theta'}_{d'}, \mathcal{P}^{\mathbf{c}\Lambda}_{d'})\big]\big] \tag{7.26}$$

and a surjection

$$\Phi^{\mathsf{imm}} : \mathsf{Q}^{\mathsf{imm}}_{d,d'}(Y; \mathbf{c}\Theta, \mathbf{c}\Theta') \longrightarrow$$
$$\longrightarrow \big[\big[(Y, \partial Y),\ \epsilon_{d,d'} : (\mathcal{P}^{\mathbf{c}\Theta}_d, \mathcal{P}^{\mathbf{c}\Lambda}_d) \to (\mathcal{P}^{\mathbf{c}\Theta'}_{d'}, \mathcal{P}^{\mathbf{c}\Lambda}_{d'})\big]\big]. \tag{7.27}$$

Moreover, by Proposition 7.9 below, the map Φ^{imm} *admits a* right inverse.

• *If* Y *is oriented, similar claims of surjectivity of maps are valid for the oriented quasitopies* $\mathsf{OQ}^{\mathsf{imm}/\mathsf{emb}}_{d,d'}(Y; \mathbf{c}\Theta, \mathbf{c}\Theta')$*, based on immersions* $\beta : (M, \partial M) \to (Y, \partial Y)$ *with trivialized normal bundles* $\nu(\beta) =_{\mathsf{def}} \beta^*(T(\mathbb{R} \times Y))/T(M)$ *and their cobordisms also with trivialized normal bundles.*

Proof. Recall that the sets $\mathbf{Q}^{\mathsf{emb}/\mathsf{imm}}_{d,d'}(Y, \partial Y; \mathbf{c}\Theta, \mathbf{c}\Lambda; \mathbf{c}\Theta')$ have been defined in (7.22). By Proposition 7.7, the map $\Phi^{\mathsf{imm}/\mathsf{emb}}$ from (7.19) is well-defined. Note that the $\Phi^{\mathsf{imm}/\mathsf{emb}}$-image of the set ${}^\dagger\mathcal{Q}^{\mathsf{imm}/\mathsf{emb}}_{d,d'}(Y, \partial Y; \mathbf{c}\Lambda, \mathbf{c}\Lambda; \mathbf{c}\Theta')$ belongs to $\mathcal{P}^{\mathbf{c}\Lambda}_d$ and thus is trivial in $[[(Y, \partial Y),\ \epsilon_{d,d'}]]$. As a result, the maps $\mathbf{\Phi}^{\mathsf{imm}/\mathsf{emb}}$ in (7.24), (7.25) are well defined.

First, we are going to show that the map $\mathbf{\Phi}^{\mathsf{emb}}$ in (7.24) is surjective. Recall that $\mathcal{P}^{\mathbf{c}\Lambda}_d \subset \mathcal{P}^{\mathbf{c}\Theta}_d$ both are open subspaces of $\mathcal{P}_d \approx \mathbb{R}^d$, and $\mathcal{P}^{\mathbf{c}\Lambda}_{d'} \subset \mathcal{P}^{\mathbf{c}\Theta}_{d'}$ are open subspaces of $\mathcal{P}_{d'} \approx \mathbb{R}^{d'}$. Thus any continuous map $h : (Y, \partial Y) \to (\mathcal{P}^{\mathbf{c}\Theta}_d, \mathcal{P}^{\mathbf{c}\Lambda}_d)$ can be approximated by a smooth map $\tilde{h}$ in the relative homotopy class of h. Similarly, any continuous map $H : (Y \times [0,1], \partial Y \times [0,1]) \to (\mathcal{P}^{\mathbf{c}\Theta'}_{d'}, \mathcal{P}^{\mathbf{c}\Lambda}_{d'})$ can be approximated by a smooth map $\tilde{H}$ which shares with H its relative homotopy class. By Corollary 7.2, $\tilde{h}$ may be approximated further by a $(\partial\mathcal{E}_d)$-regular map $h^\dagger$ that is in the relative homotopy class of h. By Lemma 7.7, $h^\dagger$ is realized by a proper embedding of some compact n-manifold $\beta : (M, \partial M) \subset (\mathbb{R} \times Y, \mathbb{R} \times \partial Y)$ whose tangency to $\mathcal{L}$ patterns belong to $\mathbf{c}\Theta$, while the tangency patterns of $\beta(\partial M)$ belong to $\mathbf{c}\Lambda$. Using that the target set of the map $\mathbf{\Phi}^{\mathsf{emb}}$ is a quotient of the set $[(Y, \partial Y), (\mathcal{P}^{\mathbf{c}\Theta}_d, \mathcal{P}^{\mathbf{c}\Lambda}_d)]$ by the equivalence relation from Definition 7.11, we conclude that $\mathbf{\Phi}^{\mathsf{emb}}$ is surjective.

Since Φ^{emb} factors through Φ^{imm}, it follows that $\mathbf{\Phi}^{\mathsf{emb}}$ factors through $\mathbf{\Phi}^{\mathsf{imm}}$. Thus the map $\mathbf{\Phi}^{\mathsf{imm}}$ is also surjective.

Next, we show that the map $\mathbf{\Phi}^{\mathsf{emb}}$ from (7.24) is injective. By Corollary 7.2, we may assume that any element of $\mathcal{Q}^{\mathsf{emb}}_{d,d'}(Y, \partial Y; \mathbf{c}\Theta, \mathbf{c}\Lambda; \mathbf{c}\Theta')$ can be represented by a $(\partial\mathcal{E}_d)$-regular embedding. Hence, consider a proper $(\partial\mathcal{E}_d)$-regular embedding $\beta_0 : (M, \partial M) \subset (\mathbb{R} \times Y, \mathbb{R} \times \partial Y)$ such that the map

$\psi_0 =_{\mathsf{def}} \epsilon_{d,d'} \circ \Phi^{\mathsf{emb}}(\beta_0)$ is homotopic to a map $\psi_1 : (Y, \partial Y) \to (\mathcal{P}_{d'}^{\mathbf{c}\Theta'}, \mathcal{P}_{d'}^{\mathbf{c}\Lambda})$ whose image is contained in $\mathcal{P}_{d'}^{\mathbf{c}\Lambda}$. By Corollary 7.2, this homotopy $H : (Y, \partial Y) \times [0,1] \to (\mathcal{P}_{d'}^{\mathbf{c}\Theta'}, \mathcal{P}_{d'}^{\mathbf{c}\Lambda})$ can be approximated by a $(\partial\mathcal{E}_{d'})$-regular homotopy $H^\dagger$ which coincides with ψ_0 on $Y \times \{0\}$. Let $N =_{\mathsf{def}} (\mathsf{id} \times H^\dagger)^{-1}(\partial\mathcal{E}_{d'})$. By Lemma 7.7, $H^\dagger$ is produced by a cobordism

$$B : (N, \delta N) \subset (\mathbb{R} \times Y \times [0,1], \mathbb{R} \times \partial Y \times [0,1])$$

whose other end $\beta_1 : (M, \partial M) \subset (\mathbb{R} \times Y \times \{1\}, \mathbb{R} \times \partial Y \times \{1\})$ has tangency patters (to the fibration $\mathcal{L}^\bullet$ on $\mathbb{R} \times Y \times [0,1]$) that belong $\mathbf{c}\Lambda$. In other words, $\beta_0 \in {}^\dagger\mathcal{Q}_{d,d'}^{\mathsf{emb}}(Y, \partial Y; \mathbf{c}\Lambda, \mathbf{c}\Lambda; \mathbf{c}\Theta')$, which represents the trivial element of the quotient $\mathbf{Q}_{d,d'}^{\mathsf{emb}}(Y, \partial Y; \mathbf{c}\Theta, \mathbf{c}\Lambda; \mathbf{c}\Theta')$. Thus we get the bijection (7.24).

When $\mathbf{c}\Lambda = (\emptyset)$ or (1), the set ${}^\dagger\mathcal{Q}_{d,d'}^{\mathsf{emb}}(Y, \partial Y; \mathbf{c}\Lambda, \mathbf{c}\Lambda; \mathbf{c}\Theta')$ consists of a single element. Hence we get the bijection in (7.26) and a surjection (7.27).

Finally, by Proposition 7.9 below, Φ^{imm} admits a right inverse.

Assuming that Y is oriented, the preimages of the loci $\partial\mathcal{E}_d$ and $\partial\mathcal{E}_{d'}$ under regular maps are oriented submanifolds $M \subset \mathbb{R} \times Y$. In the case of oriented quasitopies, these orientations of various components of M may agree or disagree with their original orientations. This remark validates the last claim. □

Corollary 7.5. *Under the hypotheses of Theorem 7.3, including the* Λ*-condition (7.11), the quasitopy set* $\mathsf{Q}_{d,d'}^{\mathsf{emb}}(Y, \partial Y; \mathbf{c}\Theta, \mathbf{c}\Theta')$ *depends only on the* homotopy type *of the pair* $(Y, \partial Y)$.

If Y *is orientable, then the orientation forgetting map*

$$\mathsf{OQ}_{d,d'}^{\mathsf{emb}}(Y, \partial Y; \mathbf{c}\Theta, \mathbf{c}\Theta') \to \mathsf{Q}_{d,d'}^{\mathsf{emb}}(Y, \partial Y; \mathbf{c}\Theta, \mathbf{c}\Theta')$$

is a surjection. In particular, $\mathsf{OG}_{d,d'}^{\mathsf{emb}}(n; \mathbf{c}\Theta; \mathbf{c}\Theta') \to \mathsf{G}_{d,d'}^{\mathsf{emb}}(n; \mathbf{c}\Theta; \mathbf{c}\Theta')$ *is a group epimorphism.*

Proof. The first claim of the corollary follows instantly from the bijection in formula (7.26) and its analogue for the oriented quasitopies.

The rest of the claims can be derived from the following already familiar observation. Since any element $\beta \in \mathsf{Q}_{d,d'}^{\mathsf{emb}}(Y, \partial Y; \mathbf{c}\Theta, \mathbf{c}\Theta')$ is represented by a map $\mathsf{id} \times \Phi^\beta : \mathbb{R} \times (Y, \partial Y) \to \mathbb{R} \times (\mathcal{P}_d^{\mathbf{c}\Theta}, pt)$ which is transversal to $\partial\mathcal{E}_d$, the manifold $M =_{\mathsf{def}} (\mathsf{id} \times \Phi^\beta)^{-1}(\partial\mathcal{E}_d)$ has a trivial normal bundle in $\mathbb{R} \times Y$. Thus M is orientable, when Y is. The same argument applies to the cobordisms that connect quasitopic β's. Therefore, when Y is orientable, the orientation forgetful map from the second claim is onto. □

Remark 7.6. The first claim of Corollary 7.5 is a bit surprising, since not any homotopy equivalence $h : (Y, \partial Y) \to (Y', \partial Y')$ of smooth n-dimensional pairs may be homotoped to a diffeomorphism. Thus it is not clear how to transfer directly a regular embedding $\beta : (M, \partial M) \to (\mathbb{R} \times Y, \mathbb{R} \times \partial Y)$ to a regular embedding $\beta' : (M', \partial M') \to (\mathbb{R} \times Y', \mathbb{R} \times \partial Y')$ without using the bijection Φ^{emb}. Similarly, it is unclear whether $\mathsf{Q}^{\mathsf{imm}}_{d,d'}(Y; \mathbf{c}\Theta, \mathbf{c}\Theta')$ depends only on the homotopy type of the pair $(Y, \partial Y)$. Conjecturally, it does. $\diamondsuit$

Corollary 7.6. *Assume that* $d' \geq d$, $d' \equiv d \mod 2$, *and the* Λ*-condition (7.11) is in place.*

Then the group homomorphism in (7.21) is a group isomorphism *for embeddings and a* split epimorphism *for the immersions.*

With the help of this group isomorphism (and by Theorem 7.8), the map in (7.26) is equivariant, and so is the map in (7.27) with the help of the epimorhism of the corresponding groups.

Proof. By Theorem 7.3, the map in (7.26) is bijective and the map in (7.27) is split surjective for any Y; in particular, this is the case for $Y = D^n$. By Theorem 7.8, all these maps are group homomorphisms. By Proposition 7.9 below, the map in (7.27) is a split epimorphism. $\square$

Letting $d' = d$ and $\Theta' = \Theta$ in Theorem 7.3, we obtain the following two corollaries.

Corollary 7.7. *Let* $\Theta \subset \Lambda \subset \mathbf{\Omega}$ *be a pair of closed subposets, and* Y *be a compact manifold.*

- *Then there is a canonical bijection*

$$\mathbf{\Phi}^{\mathsf{emb}} : \mathbf{Q}^{\mathsf{emb}}_{d,d}(Y, \partial Y; \mathbf{c}\Theta, \mathbf{c}\Lambda; \mathbf{c}\Theta) \stackrel{\approx}{\longrightarrow} [(Y, \partial Y), (\mathcal{P}^{\mathbf{c}\Theta}_d, \mathcal{P}^{\mathbf{c}\Lambda}_d)]. \quad (7.28)$$

and a canonical surjection

$$\mathbf{\Phi}^{\mathsf{imm}} : \mathbf{Q}^{\mathsf{imm}}_{d,d}(Y, \partial Y; \mathbf{c}\Theta, \mathbf{c}\Lambda; \mathbf{c}\Theta) \stackrel{\approx}{\longrightarrow} [(Y, \partial Y), (\mathcal{P}^{\mathbf{c}\Theta}_d, \mathcal{P}^{\mathbf{c}\Lambda}_d)]. \quad (7.29)$$

- *Assuming the* Λ*-condition (7.11), we get a bijection*

$$\Phi^{\mathsf{emb}} : \mathsf{Q}^{\mathsf{emb}}_{d,d}(Y; \mathbf{c}\Theta; \mathbf{c}\Theta) \stackrel{\approx}{\longrightarrow} [(Y, \partial Y), (\mathcal{P}^{\mathbf{c}\Theta}_d, pt)] \quad (7.30)$$

and a surjection

$$\Phi^{\mathsf{imm}} : \mathsf{Q}^{\mathsf{imm}}_{d,d}(Y; \mathbf{c}\Theta; \mathbf{c}\Theta) \longrightarrow [(Y, \partial Y), (\mathcal{P}^{\mathbf{c}\Theta}_d, pt)]. \quad \diamondsuit \quad (7.31)$$

Corollary 7.8. *Let* $\Theta \subset \Lambda \subset \mathbf{\Omega}_{\langle d]}$ *be a pair of closed sub-posets. There exists a group isomorphism (abelian for* $n > 1$*)*

$$\mathbf{\Phi}^{\mathsf{emb}} : \mathbf{Q}^{\mathsf{emb}}_{d,d}(D^n, \partial D^n; \mathbf{c}\Theta, \mathbf{c}\Lambda; \mathbf{c}\Theta) \stackrel{\approx}{\longrightarrow} \pi_n(\mathcal{P}^{\mathbf{c}\Theta}_d, \mathcal{P}^{\mathbf{c}\Lambda}_d). \quad (7.32)$$

Assuming the Λ*-condition (7.11), we get a group epimorphism*

$$\Phi^{\mathsf{emb}} : \mathsf{OG}^{\mathsf{emb}}_{d,d}(n; \mathbf{c}\Theta, \mathbf{c}\Theta) \to \mathsf{G}^{\mathsf{emb}}_{d,d}(n; \mathbf{c}\Theta, \mathbf{c}\Theta) \xrightarrow{\approx} \pi_n(\mathcal{P}^{\mathbf{c}\Theta}_d, pt). \quad (7.33)$$

Proof. The claim follows instantly from Theorem 7.3 and Proposition 7.6. □

Proposition 7.9. *Let* $\Theta' \subset \Theta \subset \Lambda \subset \mathbf{\Omega}$ *be closed subposets, and* Y *be a smooth compact manifold. Assume that* $d' \geq d$ *and* $d' \equiv d \mod 2$.

Then the obvious map

$$\mathcal{A}: \; \mathcal{Q}^{\mathsf{emb}}_{d,d'}(Y, \partial Y; \mathbf{c}\Theta, \mathbf{c}\Lambda; \mathbf{c}\Theta') \to \mathcal{Q}^{\mathsf{imm}}_{d,d'}(Y, \partial Y; \mathbf{c}\Theta, \mathbf{c}\Lambda; \mathbf{c}\Theta') \quad (7.34)$$

is injective. That is, if two embeddings $\beta_1 : (M_1, \partial M_1) \hookrightarrow (\mathbb{R} \times Y, \mathbb{R} \times \partial Y)$ *and* $\beta_2 : (M_2, \partial M_2) \hookrightarrow (\mathbb{R} \times Y, \mathbb{R} \times \partial Y)$ *are quaitopic via immersions, then they are quaitopic via embeddings.*

Moreover, there exists a surjective "resolution map"

$$\mathcal{R}: \; \mathcal{Q}^{\mathsf{imm}}_{d,d'}(Y, \partial Y; \mathbf{c}\Theta, \mathbf{c}\Lambda; \mathbf{c}\Theta') \to \mathcal{Q}^{\mathsf{emb}}_{d,d'}(Y, \partial Y; \mathbf{c}\Theta, \mathbf{c}\Lambda; \mathbf{c}\Theta') \quad (7.35)$$

that serves as the right inverse *of* $\mathcal{A}$. *In other words, there exists a* canonical resolution *of any such immersion into an embedding, well-defined within the relevant quasitopy classes.*

Proof. In the proof, to simplify the notations further, put:

$$\mathcal{Q}^{\mathsf{imm/emb}}(Y) =_{\mathsf{def}} \; \mathcal{Q}^{\mathsf{imm/emb}}_{d,d'}(Y, \partial Y; \mathbf{c}\Theta, \mathbf{c}\Lambda; \mathbf{c}\Theta').$$

Using the map (7.19), we see that the map $\Phi^{\mathsf{emb}}(Y)$ is a composition of the map $\mathcal{A}$ and the map

$$\Phi^{\mathsf{imm}}(Y) : \; \mathcal{Q}^{\mathsf{imm}}(Y) \to \big[\big[(Y, \partial Y), \; \epsilon_{d,d'} : (\mathcal{P}^{\mathbf{c}\Theta}_d, \mathcal{P}^{\mathbf{c}\Lambda}_d) \to (\mathcal{P}^{\mathbf{c}\Theta'}_{d'}, \mathcal{P}^{\mathbf{c}\Lambda}_{d'})\big]\big].$$

However, by formula (7.26) from Theorem 7.3, we have a bijective map

$$\Phi^{\mathsf{emb}}(Y) : \mathcal{Q}^{\mathsf{emb}}(Y) \approx \big[\big[(Y, \partial Y), \; \epsilon_{d,d'} : (\mathcal{P}^{\mathbf{c}\Theta}_d, \mathcal{P}^{\mathbf{c}\Lambda}_d) \to (\mathcal{P}^{\mathbf{c}\Theta'}_{d'}, \mathcal{P}^{\mathbf{c}\Lambda}_{d'})\big]\big].$$

Hence we define the desired surjective map $\mathcal{R}$ as a composition $\mathcal{R} = \Phi^{\mathsf{emb}}(Y)^{-1} \circ \Phi^{\mathsf{imm}}(Y)$. It is on the level of definitions to check that $\mathcal{R} \circ \mathcal{A} = \mathsf{id}$. □

Remark 7.7. Applying the resolution $\mathcal{R}$ to an immersion $\beta : M \to \mathbb{R} \times Y$ produces an embedding $\beta' : M' \to \mathbb{R} \times Y$; we stress that the topology of M' is quite different from the topology of M. ◇

Corollary 7.9. *Each fiber of the resolution map $\mathcal{R}$ consists of the elements that share the same Φ^{imm}-value in $[[(Y, \partial Y),\ \epsilon_{d,d'} : (\mathcal{P}_d^{\mathbf{c}\Theta}, \mathcal{P}_d^{\mathbf{c}\Lambda}) \to (\mathcal{P}_{d'}^{\mathbf{c}\Theta'}, \mathcal{P}_{d'}^{\mathbf{c}\Lambda})]]$.*

In particular, for $Y = D^n$, we get the splittable extension of groups

$$1 \to \mathsf{K} \to \mathsf{Q}_{d,d'}^{\mathsf{imm}}(n; \mathbf{c}\Theta; \mathbf{c}\Theta') \stackrel{\mathcal{R}}{\to} \mathsf{Q}_{d,d'}^{\mathsf{emb}}(n; \mathbf{c}\Theta; \mathbf{c}\Theta') \to 1, \tag{7.36}$$

whose kernel $\mathsf{K} \subset \mathsf{Q}_{d,d'}^{\mathsf{imm}}(n; \mathbf{c}\Theta; \mathbf{c}\Theta')$ consists of quasitopies of immersions that belong to the kernel of the map in (7.20) with $Y = D^n$. Via the operation $\uplus$, the subgroup K acts on the fibers of the map $\mathcal{R}$ from (7.35).

Proof. We use the notations from the previous proof. By Proposition 7.9, each fiber of $\mathcal{R}$ consists of the elements that share the same Φ^{imm}-value in the homotopy set $[[(Y, \partial Y),\ \epsilon_{d,d'} : (\mathcal{P}_d^{\mathbf{c}\Theta}, \mathcal{P}_d^{\mathbf{c}\Lambda}) \to (\mathcal{P}_{d'}^{\mathbf{c}\Theta'}, \mathcal{P}_{d'}^{\mathbf{c}\Lambda})]]$. For $Y = D^n$, this implies that the group $\mathcal{Q}^{\mathsf{imm}}(D^n)$ is a semidirect product $\mathsf{K} \bowtie \mathcal{Q}^{\mathsf{emb}}(D^n)$ of the groups K and $\mathcal{Q}^{\mathsf{emb}}(D^n)$. Again, K is characterized by the property of having trivial Φ^{imm}-values.

Since, for $\beta_1 \in \mathcal{Q}^{\mathsf{imm}}(Y)$ and $\beta_2 \in \mathcal{Q}^{\mathsf{imm}}(D^n)$, we have $\Phi^{\mathsf{imm}}(\beta_1 \uplus \beta_2) = \Phi^{\mathsf{imm}}(\beta_1) \uplus \Phi^{\mathsf{imm}}(\beta_2)$, it follows that $\Phi^{\mathsf{imm}}(\beta_1 \uplus \mathsf{K}) = \Phi^{\mathsf{imm}}(\beta_1) \uplus 0 = \Phi^{\mathsf{imm}}(\beta_1)$. Therefore, K acts on fibers of the map $\mathcal{R}$. □

Since $\beta \in \mathcal{Q}_{d,d'}^{\mathsf{imm}}(Y, \partial Y; \mathbf{c}\Theta, \mathbf{c}\Lambda; \mathbf{c}\Theta')$ and $\mathcal{R}(\beta) \in \mathcal{Q}_{d,d'}^{\mathsf{emb}}(Y, \partial Y; \mathbf{c}\Theta, \mathbf{c}\Lambda; \mathbf{c}\Theta')$ share the same invariants

$$\Phi^{\mathsf{imm}}(\beta) = \Phi^{\mathsf{emb}}(\mathcal{R}(\beta)) \in \big[\big[(Y, \partial Y),\ \epsilon_{d,d'} : (\mathcal{P}_d^{\mathbf{c}\Theta}, \mathcal{P}_d^{\mathbf{c}\Lambda}) \to (\mathcal{P}_{d'}^{\mathbf{c}\Theta'}, \mathcal{P}_{d'}^{\mathbf{c}\Lambda})\big]\big],$$

we should be looking for new invariants that distinguish between β and $\mathcal{R}(\beta)$. We devoted a significant portion of Chapter 5, Section 5.3, to special cases of such invariants.

Relying on Lemma 7.6 and Proposition 7.5, we may use the bordism classes $[\Sigma_k^\beta] \in \mathbf{B}_{n-k+1}(Y, \partial Y)$ (or $[\Sigma_k^\beta] \in \mathbf{OB}_{n-k+1}(Y, \partial Y)$) of the k-self-intersection manifolds $\{\Sigma_k^\beta\}_{k \geq 2}$ of β to produce the desired distinguishing invariants. These quasitopy invariants to some extend "ignore" the foliations $\mathcal{L}, \mathcal{L}^\bullet$. Evidently, they vanish on $\mathcal{Q}_{d,d'}^{\mathsf{emb}}(Y, \partial Y; \mathbf{c}\Theta, \mathbf{c}\Lambda; \mathbf{c}\Theta')$.

Therefore, for $\dim(Y) = n$, assuming that $\min\{n+1,\, \hat{m}(\mathbf{c}\Theta_{\langle d]})\} \geq 2$ (see Definition 7.9 introducing the number $\hat{m}(\mathbf{c}\Theta_{\langle d]})$), the correspondence

$$\beta \rightsquigarrow \{\pi \circ \beta \circ p_1 : (\Sigma_k^\beta, \partial \Sigma_k^\beta) \to (Y, \partial Y)\}_k$$

produces a map

$$\mathcal{BB} : \mathcal{Q}^{\mathsf{imm}}_{d,d'}(Y, \partial Y; \mathbf{c}\Theta, \mathbf{c}\Lambda; \mathbf{c}\Theta')/\mathcal{Q}^{\mathsf{emb}}_{d,d'}(Y, \partial Y; \mathbf{c}\Theta, \mathbf{c}\Lambda; \mathbf{c}\Theta') \to \tag{7.37}$$
$$\to \bigoplus_{k\in \left[2,\, \min\{n+1,\, \hat{m}(\mathbf{c}\Theta_{\langle d]})\}\right]} \mathbf{B}_{n-k+1}(Y, \partial Y),$$

which has a potential to discriminate between $\mathcal{Q}^{\mathsf{imm}}_{d,d'}(Y, \partial Y; \mathbf{c}\Theta, \mathbf{c}\Lambda; \mathbf{c}\Theta')$ and $\mathcal{Q}^{\mathsf{emb}}_{d,d'}(Y, \partial Y; \mathbf{c}\Theta, \mathbf{c}\Lambda; \mathbf{c}\Theta')$. Lemma 7.9 below provides the simplest example of a distinguishing map $\mathcal{BB}$ (c.f. Section 5.3).

Next, we "mix" the tangency patterns of immersions β to $\mathcal{L}$ with the self-intersections of β (recall that the self-intersections of β also influence the $\mathcal{L}$-tangency patterns). For simplicity, we assume that the Λ-condition is valid and Y is oriented.

Let $\beta : (M, \partial M) \to (\mathbb{R} \times Y, \mathbb{R} \times \partial Y)$ be an immersion whose quasitopy class belongs to $\mathcal{Q}^{\mathsf{imm}}_{d,d'}(Y, \partial Y; \mathbf{c}\Theta, \mathbf{c}\Theta')$. Let A be an abelian group. We pick a cohomology class $\theta \in H^{n-k+1}(\mathcal{P}^{\mathbf{c}\Theta}_d, pt; \mathsf{A})$ and evaluate its pull-back $(\Phi^\beta \circ \pi \circ \beta \circ p_1)^*(\theta)$ on the relative fundamental class $[\Sigma^\beta_k, \partial\Sigma^\beta_k]$. This construction leads to the following assertion.

Proposition 7.10. *Let $\Theta' \subset \Theta \subset \mathbf{\Omega}$ be closed posets. We assume the validity of the Λ-condition. Let Y be an oriented smooth compact manifold. Pick a positive integer $k \leq \min\{\hat{m}(\mathbf{c}\Theta_{\langle d]}), n+1\}$. We denote by A an abelian group.*

For any cohomology class $\theta \in H^{n-k+1}(\mathcal{P}^{\mathbf{c}\Theta}_d, pt; \mathsf{A})$, the evaluation

$$\chi_k(\theta, \beta) := \big\langle (\Phi^\beta \circ \pi \circ \beta \circ p_1)^*(\theta),\ [\Sigma^\beta_k, \partial\Sigma^\beta_k] \big\rangle \in \mathsf{A}$$

is an invariant of the quasitopy class $[\beta] \in \mathcal{Q}^{\mathsf{imm}}_{d,d'}(Y, \partial Y; \mathbf{c}\Theta, \mathbf{c}\Theta')$.

If β is an embedding, then evidently $\chi_k(\theta, \beta) = 0$ for all $k \geq 2$.

Proof. By Corollary 7.1, the bordism class of $\pi \circ \beta \circ p_1 : (\Sigma^\beta_k, \partial\Sigma^\beta_k) \to (Y, \partial Y)$ is invariant within the quasitopy class of β. By Theorem 7.2, the cocycle $(\Phi^\beta)^*(\theta)$ is an invariant of the quasitopy class of β.

By the topological Stokes' Theorem, the cocycle $(\Phi^\beta)^*(\theta)$, being evaluated on the cycle $(\pi \circ \beta \circ p_1)(\Sigma^\beta_k, \partial\Sigma^\beta_k)$ is an invariant of $[\beta] \in \mathcal{Q}^{\mathsf{imm}}_{d,d'}(Y, \partial Y; \mathbf{c}\Theta, \mathbf{c}\Theta')$. □

Here is the simplest example of an invariant

$$\rho_\beta =_{\mathsf{def}} \frac{1}{n+1}\big[(\pi \circ \beta \circ p_1) : \Sigma^\beta_{n+1} \to Y\big],$$

delivered by the map $\mathcal{BB}$ from (7.37), which does discriminate between the quasitopies of embeddings and of immersions.

Lemma 7.9. *Let* $\dim M = n = \dim Y$ *and let* ρ_β *be the number of points where exactly* $n+1$ *branches of* $\beta(M)$ *meet transversally in* $\mathbb{R} \times Y$. *If* $\rho_\beta \equiv 1 \mod 2$, *then* $\beta \in \mathcal{Q}^{\mathsf{imm}}_{d,d'}(Y, \partial Y; \mathbf{c}\Theta, \mathbf{c}\Lambda; \mathbf{c}\Theta')$ *is nontrivial, provided* $\mathbf{c}\Lambda = (\emptyset), (1)$.

Proof. Let F_β denote the finite set of points, where exactly $n+1$ branches of $\beta(M)$ meet transversally in $\mathbb{R} \times Y$. Assume that β be quasitopic to the trivial immersion with the help of a cobordism $B : N \to \mathbb{R} \times Y \times [0,1]$, where $\partial N = M$. Then, in general, $B(N)$ may have points where exactly $n+2$ branches of $B(N)$ meet transversally. Let E_B denote their set and let $\eta_B =_{\mathsf{def}} \#(E_B)$. The points, where exactly $n+1$ branches of $B(N)$ meet transversally, form a graph $\Gamma_B \subset \mathbb{R} \times Y \times [0,1]$. If $\mathbf{c}\Lambda = (\emptyset), (1)$, then Γ_B has ρ_β univalent vertices and η_B vertices of valency $n+2$. Every edge of Γ_B that does not terminate at a valency one vertex from Γ_β is attached to E_B. Thus, counting the edges of Γ_B, we get

$$2(n+2)\eta_B - \rho_\beta \equiv 0 \mod 2.$$

Therefore, when $\rho_\beta \equiv 1 \mod 2$, we get a contradiction with the assumption that β is quasitopic to the trivial immersion. As a result, any immersion β with an odd number ρ_β is nontrivial in $\mathcal{Q}^{\mathsf{imm}}_{d,d'}(Y, \partial Y; \mathbf{c}\Theta, \mathbf{c}\Lambda; \mathbf{c}\Theta')$, provided $\mathbf{c}\Lambda = (\emptyset), (1)$. □

Example 7.1. One should compare this example with the claim of Theorem 5.1 from Section 5.3. Recall that the latter dealt with $\Theta = \{\omega \in \mathbf{\Omega} | \max \omega \geq 3\}$ and *unlimited d*.

Pick $n = 1$, $d' = d = 6$. Let $\mathbf{c}\Theta' = \mathbf{c}\Theta$ consist of $\omega \in \mathbf{\Omega}_{\langle 6]}$ such that all their entries are 1's and 2's and a single 2 is present at most. Put $Y = D^1$.

Then $\mathsf{Q}^{\mathsf{emb}}_{6,6}(1; \mathbf{c}\Theta; \mathbf{c}\Theta) \approx \pi_1(\mathcal{P}^{\mathbf{c}\Theta}_6, pt)$. By Theorem 6.13 and Corollary 7.8 (see also [KSW1], Theorem 2.4), the latter group is a free group F_6 in 6 generators (see Fig. 7.4). The group extension in (7.36) reduces to

$$1 \to \mathsf{K} \to \mathsf{Q}^{\mathsf{imm}}_{6,6}(1; \mathbf{c}\Theta; \mathbf{c}\Theta) \stackrel{\mathcal{R}}{\to} \mathsf{F}_6 \to 1.$$

The figure ∞, placed "horizontally" in $\mathbb{R} \times D^1$ (D^1 being the horizontal direction) represents an immersion $\beta : S^1 \to \mathbb{R} \times D^1$. We claim that β belongs to the kernel K from (7.36). Indeed, consider a flip $\tau : \mathbb{R} \times D^1 \to \mathbb{R} \times D^1$ whose fixed point set is the vertical line through the singularity of the figure ∞. Due to the symmetry of ∞ with respect to τ, the path $\Phi^{\mathsf{imm}}(D^1, \partial D^1) \subset (\mathcal{P}^{\mathbf{c}\Theta}_6, pt)$ is relatively contractible. Thus ∞ is indeed in the kernel K. By Lemma 7.9, β is nontrivial in $\mathsf{Q}^{\mathsf{imm}}_{6,6}(1; \mathbf{c}\Theta; \mathbf{c}\Theta)$. By the

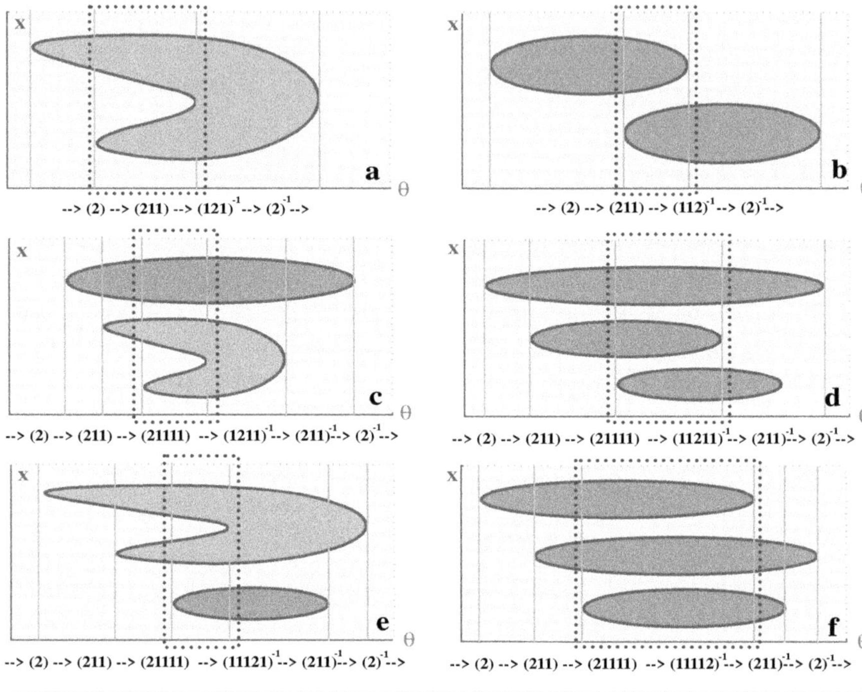

Fig. 7.4 Six generators a, b, c, d, e, f of the free group $\mathsf{Q}^{\mathsf{emb}}_{6,6}(1; \mathbf{c}\Theta, \mathbf{c}\Theta)$, where $\Theta \subset \mathbf{\Omega}_{\langle 6]}$ is the closed poset generated by the elements ω with two 2's or one 3. The vertical lines mark the crossings of the walls of the 6-dimensional chambers in $\mathcal{P}^{\mathbf{c}\Theta}_6$ by a path $\Phi^{\mathsf{emb}}([0,1])$.

same lemma, the quasitopy class of any collection of smooth curves with an *odd* number of crossings is nontrivial in $\mathsf{Q}^{\mathsf{imm}}_{6,6}(1; \mathbf{c}\Theta; \mathbf{c}\Theta)$. At the same time, the immersion $\infty \uplus \infty \subset \mathbb{R} \times D^1$ is trivial in $\mathsf{Q}^{\mathsf{imm}}_{6,6}(1; \mathbf{c}\Theta; \mathbf{c}\Theta)$ since it is the boundary of a horseshoe surface $N \approx \infty \times I \subset \mathbb{R} \times D^1 \times [0,1]$ with the $\mathcal{L}^\bullet$-tangency patterns that belong to $\mathbf{c}\Theta$. Thus the immersion β, such that $\beta(S^1) = \infty$, is an element of order two in the kernel K. However, if we orient S^1, then ∞ becomes an element of infinite order in $\mathsf{ker}(\mathsf{OQ}^{\mathsf{imm}}_{6,6}(1; \mathbf{c}\Theta; \mathbf{c}\Theta) \to \mathsf{OQ}^{\mathsf{emb}}_{6,6}(1; \mathbf{c}\Theta; \mathbf{c}\Theta))$.

In fact, any collection of curves $\beta(M)$ in $\mathbb{R} \times D^1$ with the tangency patterns in $\mathbf{c}\Theta$ and with an odd number of transversal crossings, and which is symmetric under an involution $\tau : \mathbb{R} \times D^1 \to \mathbb{R} \times D^1$ which preserves the oriented foliation $\mathcal{L}$, belongs to K and has order 2 there.

Based on a sparse evidence, we conjecture that $\mathsf{K} \approx \mathbb{Z}_2$. If this is true, then $\mathsf{Q}^{\mathsf{imm}}_{6,6}(1; \mathbf{c}\Theta; \mathbf{c}\Theta)$ is a semi-direct product $\mathbb{Z}_2 \bowtie \mathsf{F}_6$. $\diamondsuit$

The next theorem describes one very general mechanism for generating characteristic cohomology classes for immersions/embeddings $\beta : M \to \mathbb{R}\times Y$ with Θ-restricted tangency patterns to $\mathcal{L}$.

Theorem 7.4. *Let $d' = d$, and $\Theta' = \Theta$. For a compact manifold Y, assuming the Λ-condition, any immersion/embedding $\beta : (M, \partial M) \to (\mathbb{R} \times Y, \mathbb{R} \times \partial Y)$ as in Proposition 7.7 induces a characteristic homomorphism $(\Phi^\beta)^*$ from the (co)homology of the differential complex $\{\partial^\# : \mathbb{Z}[\Theta^\#_{\langle d]}] \to \mathbb{Z}[\Theta^\#_{\langle d]}]\}$ in (6.12) to the cohomology $H^*(Y, \partial Y; \mathbb{Z})$.*

Quasitopic immersions/embeddings induce the same characteristic homomorphisms.

Proof. For any closed poset $\Theta \subset \mathbf{\Omega}_{\langle d]}$, by Corollary 6.6, the homology of the differential complex in (6.12) is isomorphic to $H^*(\mathcal{P}^{\mathbf{c}\Theta}_d, pt; \mathbb{Z})$. By Proposition 7.7, any immersion $\beta : (M, \partial M) \to (\mathbb{R} \times Y, \mathbb{R} \times \partial Y)$, whose tangency patterns belong to $\mathbf{c}\Theta$ and the ones of $\beta|_{\partial M}$ to $\mathbf{c}\Lambda = (\emptyset)$ or (1), produces a map $\Phi^\beta : (Y, \partial Y) \to (\mathcal{P}^{\mathbf{c}\Theta}_d, pt)$ whose homotopy class is determined by β. This β induces a natural map $\beta^* : H^*(\mathcal{P}^{\mathbf{c}\Theta}_d, pt; \mathbb{Z}) \to H^*(Y, \partial Y; \mathbb{Z})$ in cohomology, and therefore, a characteristic homomorphism $(\tilde{\Phi}^\beta)^* : H^*(\mathbb{Z}[\Theta^\#], \partial^\#) \to H^*(Y, \partial Y; \mathbb{Z})$.

By Proposition 7.7, $(d, d; \mathbf{c}\Theta, (\emptyset); \mathbf{c}\Theta)$-quasitopic immersions/embeddings β induce homotopic maps $\tilde{\Phi}^\beta$ and hence the same characteristic homomorphisms $(\tilde{\Phi}^\beta)^*$. $\square$

Thus various homomorphisms $(\tilde{\Phi}^\beta)^*$ may distinguish between different quasitopy classes of immersions or embeddings β. We will soon exhibit multiple examples, where they do.

7.5 Quasitopies of immersions & embeddings with special forbidden combinatorics Θ and the stabilization by d

In this section, we take advantage of a few results of [Va], [Va1], and [KSW2], as formulated in Chapter 5, to advance computations of quasitopies of embeddings with special combinatorial patterns $\mathbf{c}\Theta$ and $\mathbf{c}\Theta'$ of tangency to the foliations $\mathcal{L}$ and $\mathcal{L}^\bullet$, respectively.

We start with the Arnold-Vassiliev case of real polynomials with, so called, moderate singularities [Ar], [Va1].

Let $\Theta_{\max \geq k} \subset \mathbf{\Omega}_{\langle d]}$ be the closed poset consisting of ω's with the maximal entry $\geq k$. For $k \geq 3$, the cohomology $H^j(\mathcal{P}^{\mathbf{c}\Theta_{\max\geq k}}, pt; \mathbb{Z})$ is isomorphic to $\mathbb{Z}$ in each dimension j of the form $(k-2)m$, where the integer $m \leq d/k$, and vanishes otherwise [Ar].

Based on Vassiliev's computation of the cohomology ring $H^*(\mathcal{P}^{\mathbf{c}\Theta_{\max\geq k}}, pt; \mathbb{Z})$ (see [Va], Theorem 1 on page 87), consider the graded ring $\mathcal{V}as_{d,k}$, multiplicatively generated over $\mathbb{Z}$ by the elements $\{e_m\}_{m \leq d/k}$ of the degrees $\deg(e_m) = m(k-2)$, subject to the relations

$$
\begin{aligned}
e_l \cdot e_m &= \frac{(l+m)!}{l! \cdot m!}\, e_{l+m} \text{ for } k \equiv 0 \mod 2, \text{ and} \\
e_1 \cdot e_1 &= 0, \quad e_1 \cdot e_{2m} = e_{2m+1}, \\
e_{2l} \cdot e_{2m} &= \frac{(l+m)!}{l! \cdot m!}\, e_{2l+2m} \text{ for } k \equiv 1 \mod 2.
\end{aligned} \tag{7.38}
$$

Guided by Theorem 7.4 and employing Theorem 7.2, we get the following claim.

Proposition 7.11. *Let $k \geq 3$. Consider an immersion $\beta : (M, \partial M) \to (\mathbb{R} \times Y, \mathbb{R} \times \partial Y)$ whose tangency patterns to the foliation $\mathcal{L}$ belong to $\mathbf{c}\Theta_{\max\geq k} \subset \mathbf{\Omega}_{\langle d]}$ and the ones of $\beta|_{\partial M}$ either form an empty set ($\emptyset$) when $d \equiv 0 \mod 2$, or the set (1) when $d \equiv 1 \mod 2$.*

Then β generates a characteristic ring homomorphism

$$(\Phi^\beta)^* : \mathcal{V}as_{d,k} \to H^*(Y, \partial Y; \mathbb{Z}).$$

The homomorphism $(\Phi^\beta)^*$ is an invariant of the quasitopy class of β. In other words, we get a map

$$\Phi^*_{d,k} : \mathsf{Q}^{\mathsf{imm}}_{d,d}(Y; \mathbf{c}\Theta_{\max\geq k}; \mathbf{c}\Theta_{\max\geq k}) \to \mathsf{Hom}_{\mathsf{ring}}\big(\mathcal{V}as_{d,k},\ H^*(Y, \partial Y; \mathbb{Z})\big). \quad \diamondsuit$$

Remark 7.8. Note that if some generator $e_\ell \in \mathcal{V}as_{d,k}$ is mapped by $(\Phi^\beta)^*$ to zero in $H^{(k-2)\ell}(Y, \partial Y; \mathbb{Z})$, then the images of all e_q with $q > \ell$ must be torsion elements. Thus, some cohomology rings $H^*(Y, \partial Y; \mathbb{Z})$ may not be able to accommodate nontrivial images of the Vassiliev ring $\mathcal{V}as_{d,k}$ in positive degrees.

However, here is a case when the accommodation seems possible: take $Y = \mathbb{CP}^3$, $k = 4$, and $d = 12$. Consider 9-dimensional variety $\mathcal{P}_{12}^{\Theta_{\max\geq 4}}$, its compliment $\mathcal{P}_{12}^{\mathbf{c}\Theta_{\max\geq 4}}$, and the cohomology ring $\mathcal{V}as_{12,4} \approx H^*(\mathcal{P}_{12}^{\mathbf{c}\Theta_{\max\geq 4}}; \mathbb{Z})$. Then the ring homomorphism

$$\Phi^* : \mathcal{V}as_{12,4} \to H^*(\mathbb{CP}^3; \mathbb{Z}) \approx \mathbb{Z}[a]/\{a^4 = 0\}$$

that sends $\Phi^*(e_1) := 6a$, $\Phi^*(e_2) := 18a^2$, and $\Phi^*(e_3) := 36a^3$ embeds $\mathcal{V}as_{12,4}$ as a subring of $\mathbb{Z}[a]/\{a^4 = 0\}$. We speculate that there is an embedding $\Phi : \mathbb{CP}^3 \to \mathcal{P}_{12}^{\mathbf{c}\Theta_{\max\geq 4}}$ that induces that Φ^*. $\diamondsuit$

Combining Propositions 7.10 and 7.11, leads to the following claim.

Corollary 7.10. *Let M, Y be compact oriented smooth n-manifolds. For any generator $e_\ell \in \mathcal{V}as_{d,k}$ of degree $\ell(k-2)$ and an immersion $\beta : (M, \partial M) \to (\mathbb{R} \times Y, \mathbb{R} \times \partial Y)$ with k-moderate tangency patterns to $\mathcal{L}$, the integer*

$$\chi_k(e_\ell, \beta) = \big\langle (\Phi^\beta \circ \pi \circ \beta \circ p_1)^*(e_\ell),\ [\Sigma^\beta_{\ell(k-2)}, \partial\Sigma^\beta_{\ell(k-2)}]\big\rangle$$

is an invariant of the quasitopy class $[\beta] \in \mathcal{Q}^{\mathsf{imm}}_{d,d'}(Y, \partial Y; \mathbf{c}\Theta_{\max\geq k}, \mathbf{c}\Theta_{\max\geq k})$.

The invariant $\chi_k(e_\ell, \beta) = 0$ when β is an embedding and $k > 1$, or when $H^{\ell(k-2)}(Y, \partial Y; \mathbb{Z}) = 0$. $\diamondsuit$

The next proposition is a stabilization result by the increasing $d' \geq d$ for the embeddings with moderate tangencies to the foliation $\mathcal{L}$ on $\mathbb{R} \times Y$.

Proposition 7.12. *Let $k \geq 4$. If $\dim Y \leq (k-2)(\lceil d/k \rceil + 1) - 2$, then the classifying map*

$$\Phi^{\mathsf{emb}} : \mathsf{Q}^{\mathsf{emb}}_{d,d'}(Y; \mathbf{c}\Theta_{\max\geq k}, \mathbf{c}\Theta_{\max\geq k}) \xrightarrow{\approx} \big[(Y, \partial Y), (\mathcal{P}_d^{\mathbf{c}\Theta_{\max\geq k}}, pt)\big] \quad (7.39)$$

is a bijection, and the classifying map

$$\Phi^{\mathsf{imm}} : \mathsf{Q}^{\mathsf{imm}}_{d,d'}(Y; \mathbf{c}\Theta_{\max\geq k}, \mathbf{c}\Theta_{\max\geq k}) \xrightarrow{\approx} \big[(Y, \partial Y), (\mathcal{P}_d^{\mathbf{c}\Theta_{\max\geq k}}, pt)\big] \quad (7.40)$$

is a surjection for any $d' \geq d$, $d' \equiv d \mod 2$.

In particular, for a given Y, $\mathsf{Q}^{\mathsf{emb}}_{d,d'}(Y; \mathbf{c}\Theta_{\max\geq k}, \mathbf{c}\Theta_{\max\geq k})$ stabilizes for all $d' \geq d \geq \frac{k}{k-2}(\dim Y + 4 - k)$, a linear function in $\dim Y$.

Proof. Let $q =_{\mathsf{def}} (k-2)(\lceil d/k \rceil + 1) - 2$. By [Va3], Theorem 3 on page 88, the $\epsilon_{d,d'}$-induced homomorhpism $(\epsilon_{d,d'})_* : \pi_i(\mathcal{P}_d^{\mathbf{c}\Theta_{\max\geq k}}) \to \pi_i(\mathcal{P}_{d'}^{\mathbf{c}\Theta_{\max\geq k}})$ of the homotopy groups is an isomorphism for all $i \leq q$. Thus the two spaces q-connected. Therefore, if $\dim Y \leq q$, then no element of $\big[(Y, \partial Y), (\mathcal{P}_d^{\mathbf{c}\Theta_{\max\geq k}}, pt)\big]$ becomes, under the composition with $\epsilon_{d,d'}$, nill-homotopic. Hence,

$$\big[\big[(Y, \partial Y),\ \epsilon_{d,d'} : \mathcal{P}_d^{\mathbf{c}\Theta_{\max\geq k}} \to \mathcal{P}_{d'}^{\mathbf{c}\Theta_{\max\geq k}}\big]\big] = \big[(Y, \partial Y), (\mathcal{P}_d^{\mathbf{c}\Theta_{\max\geq k}}, pt)\big].$$

In particular, if $\dim Y \leq q$, then $\big[(Y, \partial Y), (\mathcal{P}_d^{\mathbf{c}\Theta_{\max\geq k}}, pt)\big]$ is stable in d, provided $d \geq \frac{k}{k-2}(\dim Y + 4 - k)$. Therefore, by Theorem 7.3, all the claims of the proposition are validated. □

Example 7.2. Let $k = 4$. Assuming the Λ-condition (7.11), by Proposition 7.12 and formula (7.39), for any compact connected surface Y, we get bijections:

$$\mathsf{Q}^{\mathsf{emb}}_{4,4}(Y; \mathbf{c}\Theta_{\max\geq 4}, \mathbf{c}\Theta_{\max\geq 4}) \xrightarrow{\approx} \pi^2(Y/\partial Y),$$

$$\mathsf{Q}^{\mathsf{emb}}_{4,6}(Y; \mathbf{c}\Theta_{\max\geq 4}, \mathbf{c}\Theta_{\max\geq 4}) \xrightarrow{\approx} \pi^2(Y/\partial Y),$$

$$\mathsf{Q}^{\mathsf{emb}}_{6,6}(Y; \mathbf{c}\Theta_{\max\geq 4}, \mathbf{c}\Theta_{\max\geq 4}) \xrightarrow{\approx} \pi^2(Y/\partial Y),$$

whose target is the second cohomotopy group $\pi^2(Y/\partial Y) \approx \mathbb{Z}$, the latter isomorphism being generated by the degree of maps to S^2.

Let $\mathcal{Z}$ be a simply-connected CW-complex whose cohomology ring, truncated in dimensions ≥ 5, is isomorphic to $\mathbb{Z}[e_1, e_2]/\{e_1^2 - 2e_2\}$, where $\deg e_1 = 2, \deg e_2 = 4$. Then by Proposition 7.12, for any compact smooth manifold Y of dimension ≤ 4, we get the bijections:

$$\mathsf{Q}^{\mathsf{emb}}_{8,8}(Y; \mathbf{c}\Theta_{\max\geq 4}, \mathbf{c}\Theta_{\max\geq 4}) \xrightarrow{\approx} [Y/\partial Y,\ \mathcal{Z}],$$

$$\mathsf{Q}^{\mathsf{emb}}_{8,10}(Y; \mathbf{c}\Theta_{\max\geq 4}, \mathbf{c}\Theta_{\max\geq 4}) \xrightarrow{\approx} [Y/\partial Y,\ \mathcal{Z}],$$

$$\mathsf{Q}^{\mathsf{emb}}_{10,10}(Y; \mathbf{c}\Theta_{\max\geq 4}, \mathbf{c}\Theta_{\max\geq 4}) \xrightarrow{\approx} [Y/\partial Y,\ \mathcal{Z}].$$

◇

Corollary 7.11. *Let $\Theta_{\max\geq k} \subset \mathbf{\Omega}_{\langle d]}$ be the closed poset of ω's with at least one entry $\geq k$. Assume that $k \leq d \leq 2k - 1$.*

Then there is a group epimorphism

$$\mathsf{OG}^{\mathsf{emb}}_{d,d}(n; \mathbf{c}\Theta_{\max\geq k}, \mathbf{c}\Theta_{\max\geq k}) \to \mathsf{G}^{\mathsf{emb}}_{d,d}(n; \mathbf{c}\Theta_{\max\geq k}, \mathbf{c}\Theta_{\max\geq k}) \approx \pi_n(S^{k-2}).$$

Thus, the quasitopy group of non-oriented embeddings $\beta : M \hookrightarrow \mathbb{R} \times D^n$ into the cylinder $\mathbb{R} \times D^n$ with no vertical tangencies of $\beta(M)$ to $\mathcal{L}$ of the order $\geq k$ and with the total multiplicities $\{\mu_\beta(y) \leq d\}_{y \in D^n}$ is isomorphic to the group $\pi_n(S^{k-2})$, provided that $d \in [k, 2k-1]$.

As a result, in this range of d's and for $n > 1$, all the groups

$$\mathsf{G}^{\mathsf{emb}}_{d,d}(n; \mathbf{c}\Theta_{\max\geq k}, \mathbf{c}\Theta_{\max\geq k})$$

are finite *abelian, except for $n = k - 2$ and $n = 2k - 5$, where they are infinite of rank one.*

Proof. By Arnold's Theorem 6.1 ([Ar1]), for $d \in [k, 2k+1]$, $\pi_n(\mathcal{P}_d^{\mathbf{c}\Theta_{\max\geq k}}) \approx \pi_n(S^{k-2})$. Applying Corollary 7.8, the claim follows by examining the tables of homotopy groups of spheres. □

Example 7.3. By Corollary 7.11, for $k = 4$, $d \in [4, 9]$, and all n, we get a group isomorphism

$$\mathsf{G}^{\mathsf{emb}}_{d,d}(n; \mathbf{c}\Theta_{\max\geq 4}, \mathbf{c}\Theta_{\max\geq 4}) \approx \pi_n(S^2).$$

In particular, for $d \in [4, 9]$, visiting the table of homotopy groups of spheres, we get:

$$\mathsf{G}^{\mathsf{emb}}_{d,d}(3; \mathbf{c}\Theta_{\max\geq 4}, \mathbf{c}\Theta_{\max\geq 4}) \approx \mathbb{Z}, \quad \mathsf{G}^{\mathsf{emb}}_{d,d}(4; \mathbf{c}\Theta_{\max\geq 4}, \mathbf{c}\Theta_{\max\geq 4}) \approx \mathbb{Z}_2,$$
$$\mathsf{G}^{\mathsf{emb}}_{d,d}(14; \mathbf{c}\Theta_{\max\geq 4}, \mathbf{c}\Theta_{\max\geq 4}) \approx \mathbb{Z}_{84} \times (\mathbb{Z}_2)^2.$$

It would be very interesting to understand and describe, in the sprit of the Arnold's "kidneys" in Fig. 7.4 (see [Ar1]), *the shapes* of embeddings $\beta : M \hookrightarrow \mathbb{R} \times D^n$ that produce such mysterious periods... In principle, Theorem 7.3 contains the instructions for such attempts. Perhaps, at least for $n = 1, 2$, one has a fighting chance... ◊

Let H, G be two groups, and let $\mathsf{Hom}(H, G)$ be their group of homomorphisms. Then G acts on $\mathsf{Hom}(H, G)$ by the conjugation: for any $\phi : H \to G$, $h \in H$, and $g \in G$, we define $(\mathsf{Ad}_g \phi)(h)$ by the formula $g^{-1}\phi(h)g$. We denote by $\mathsf{Hom}^\bullet(H, G)$ the quotient $\mathsf{Hom}(H, G)/\mathsf{Ad}_G$.

The next corollary deals with special Θ's for which $\mathcal{P}_d^{\mathbf{c}\Theta}$ is a $K(\pi, 1)$-space as in Theorem 6.13 (see [KSW1]).

Corollary 7.12. *Let $\mathbf{c}\Theta$ consist of all ω's with entries* 1 *and* 2 *only and no more than a single entry* 2. *Put $\kappa(d) =_{\mathsf{def}} \frac{d(d-2)}{4}$ for $d \equiv 0 \mod 2$, and $\kappa(d) =_{\mathsf{def}} \frac{(d-1)^2}{4}$ for $d \equiv 1 \mod 2$. Let $\mathsf{F}_{\kappa(d)}$ be the* free *group in $\kappa(d)$ generators.*

Assume that either Y is a closed manifold, or $\partial Y \neq \emptyset$ and the Λ-condition is in place. If $\partial Y \neq \emptyset$, then there is a bijection

$$\Phi^{\mathsf{emb}} : \mathsf{Q}^{\mathsf{emb}}_{d,d}(Y; \mathbf{c}\Theta, \mathbf{c}\Theta) \xrightarrow{\approx} \mathsf{Hom}(\pi_1(Y), \mathsf{F}_{\kappa(d)})$$

and a surjection

$$\Phi^{\mathsf{imm}} : \mathsf{Q}^{\mathsf{imm}}_{d,d}(Y; \mathbf{c}\Theta, \mathbf{c}\Theta) \longrightarrow \mathsf{Hom}(\pi_1(Y), \mathsf{F}_{\kappa(d)}).$$

When Y is closed, then similar claims hold with the targets of the maps Φ^{emb} and Φ^{imm} being replaced by the set $\mathsf{Hom}^\bullet(\pi_1(Y), \mathsf{F}_{\kappa(d)})$.

In particular, $\Phi^{\mathsf{emb}} : \mathsf{Q}^{\mathsf{emb}}_{d,d}(S^1; \mathbf{c}\Theta, \mathbf{c}\Theta) \xrightarrow{\approx} \mathsf{F}_{\kappa(d)}/\mathsf{Ad}_{\mathsf{F}_{\kappa(d)}}$*, the set of free cyclic words in* $\kappa(d)$ *letters.*

Thus, if $\pi_1(Y)$ *has no nontrivial free images, then the group* $\mathsf{Q}^{\mathsf{emb}}_{d,d}(Y; \mathbf{c}\Theta, \mathbf{c}\Theta)$ *is trivial.*

Proof. By Theorem 6.13, for such a $\mathbf{c}\Theta$, the space $\mathcal{P}^{\mathbf{c}\Theta}_d$ is of the homotopy type of $K(\mathsf{F}_{\kappa(d)}, 1) = \bigvee_{\sigma=1}^{\kappa(d)} S^1_\sigma$, a the bouquet of circles. Thus, when Y is closed, by the obstruction theory, we get $[Y, \bigvee_{\sigma=1}^{\kappa(d)} S^1_\sigma] \approx \mathsf{Hom}^\bullet(\pi_1(Y), \mathsf{F}_{\kappa(d)})$ (see [Hu], Section VI, F). When $\partial Y \neq \emptyset$, the boundary of Y is mapped by Φ^β to the preferred point $pt \in \mathcal{P}^{\mathbf{c}\Theta}_d$. This makes it possible to consider the based loops in Y, the base point b being chosen in ∂Y, and the based loops in $\bigvee_{\sigma=1}^{\kappa(d)} S^1_\sigma$, the base point pt being the apex $\star$ of the bouquet. Again, by the standard obstruction theory, we get $[(Y, \partial Y), (\bigvee_{\sigma=1}^{\kappa(d)} S^1_\sigma, \star)] \approx \mathsf{Hom}(\pi_1(Y, b), \mathsf{F}_{\kappa(d)})$.

Now, by Corollary 7.8, the claim follows. In particular, if $\pi_1(Y)$ has no nontrivial free images (say, $\pi_1(Y)$ is a finite group), then the group $\mathsf{Q}^{\mathsf{emb}}_{d,d}(Y; \mathbf{c}\Theta, \mathbf{c}\Theta)$ is trivial. □

Proposition 7.13. *Let* $\Theta' = \Theta =_{\mathsf{def}} \{\omega : \; |\omega| \geq q, \; |\omega|' \geq k\} \subset \mathbf{\Omega}$. *Pick the natural numbers* $d' \geq d$, $d' \equiv d \mod 2$, *and* $k \in [3, d)$, $k < q$.

Then, assuming the Λ*-condition (7.11), we get a group homomorphism*

$$\Phi^{\mathsf{i/e}}_{d,d'}(n; \mathbf{c}\Theta, \mathbf{c}\Theta) : \; \mathsf{G}^{\mathsf{i/e}}_{d,d'}(n; \mathbf{c}\Theta, \mathbf{c}\Theta) \to \pi_n(\mathcal{P}^{\mathbf{c}\Theta}_d)/\mathsf{ker}\big((\epsilon_{d,d'})_*\big) \approx$$

$$\pi_n\Big(\bigvee_{i=1}^{A(d,k,q)} S^{k-1}_i\Big)\Big/\mathsf{ker}\Big\{\pi_n\Big(\bigvee_{i=1}^{A(d,k,q)} S^{k-1}_i\Big) \xrightarrow{(\tilde{\epsilon}_{d,d'})_*} \pi_n\Big(\bigvee_{j=1}^{A(d',k,q)} S^{k-1}_j\Big)\Big\}. \tag{7.41}$$

It is an isomorphism for the embeddings and an epimorphism for the immersions. The homomorphism $(\tilde{\epsilon}_{d,d'})_*$ *in (7.41) is induced by the embedding* $\epsilon_{d,d'} : \mathcal{P}^{\mathbf{c}\Theta}_d \hookrightarrow \mathcal{P}^{\mathbf{c}\Theta}_{d'}$. *The number* $A(d, k, q)$ *has been introduced in (6.39).*

For $n < k - 1$, *the group* $\mathsf{G}^{\mathsf{emb}}_{d,d'}(n; \mathbf{c}\Theta, \mathbf{c}\Lambda; \mathbf{c}\Theta)$ *is trivial.*

Proof. By formula (7.21) in Theorem 7.8, we produce the map $\Phi^{\mathsf{i/e}}_{d,d'}(n; \mathbf{c}\Theta, \mathbf{c}\Theta)$. By Corollary 7.8, it is a group epimomorphism for immersions and an isomorphism for the embeddings.

For $k \geq 3$, by Proposition 6.6, the spaces $\mathcal{P}^{\mathbf{c}\Theta}_d$ and $\mathcal{P}^{\mathbf{c}\Theta}_{d'}$ have the homology types of bouquets of $(k-1)$-spheres. Recall that a simply-connected

CW-complex whose torsion-free $\mathbb{Z}$-homology is concentrated in a single dimension r is homotopy equivalent to a bouquet of r-spheres (see [Ha], Theorem 4C.1). Therefore, the spaces $\mathcal{P}_d^{\mathbf{c}\Theta}$ and $\mathcal{P}_{d'}^{\mathbf{c}\Theta}$ have the homotopy types of bouquets of $(k-1)$-spheres, provided $k > 2$. With the help of these homotopy equivalences, the homomorphism $(\tilde{\epsilon}_{d,d'})_* : \pi_n\big(\bigvee_{i=1}^{A(d,\,k,\,q)} S_i^{k-1}\big) \to \pi_n\big(\bigvee_{j=1}^{A(d',\,k,\,q)} S_j^{k-1}\big)$ is induced by the embedding $\epsilon_{d,d'} : \mathcal{P}_d^{\mathbf{c}\Theta} \to \mathcal{P}_{d'}^{\mathbf{c}\Theta}$. This validates the first claim.

The last claim follows from the cellular approximation of a continuous map in its homotopy class. □

For $d' = d$, we are able to compute the quasitopy classes of k-flat (see Definition 7.6) embeddings into $\mathbb{R}\times Y$. They correspond to the case $q = 0$ in the following claim. Corollary 7.13 below follows instantly from Proposition 7.13.

Corollary 7.13. *For $k \in [3, d)$ and $q \in [0, d]$, consider the closed subposet $\Theta^{(q)}_{|\sim|' \geq k} =_{\mathsf{def}} \big\{\omega :\ |\omega|' \geq k,\ |\omega| \in [q, d]\big\}$. Assume the Λ-condition (7.11).*

Then, for a compact connected smooth n-manifold Y with boundary, the quasitopy set of embeddings is described by the formula

$$\mathsf{Q}^{\mathsf{emb}}_{d,d}\big(Y; \mathbf{c}\Theta^{(q)}_{|\sim|' \geq k}, \mathbf{c}\Theta^{(q)}_{|\sim|' \geq k}\big) \approx \Big[(Y, \partial Y),\ \big(\bigvee_{i=1}^{A(d,\,k,\,q)} S_i^{k-1},\ \star\big)\Big],$$

where $\star$ is the apex of the bouquet, and the integer $A(d,\,k,\,q)$ is introduced in (6.39).

In particular, the quasitopic classes of embeddings in the cylinder $\mathbb{R}\times D^n$ with the combinatorial patterns ω, such that $|\omega|' < k$ and $|\omega| \geq q$, are described by the group isomorphism

$$\mathsf{Q}^{\mathsf{emb}}_{d,d}\big(D^n; \mathbf{c}\Theta^{(q)}_{|\sim|' \geq k}, \mathbf{c}\Theta^{(q)}_{|\sim|' \geq k}\big) \approx \pi_n\big(\bigvee_{i=1}^{A(d,\,k,\,q)} S_i^{k-1},\ \star\big).$$

◇

Theorem 7.5. *Let Y be either a closed smooth n-manifold or a manifold with boundary, in which case, we assume the Λ-condition (7.11). We denote by $\langle\omega\rangle$ the closed subposet of $\mathbf{\Omega}_{\langle d]}$, consisting of elements $\omega' \preceq \omega$.*

Then for all $d \leq 13$ and all ω such that $|\omega|' > 2$, the quasitopy $\mathsf{Q}^{\mathsf{emb}}_{d,d}(Y; \mathbf{c}\langle\omega\rangle, \mathbf{c}\langle\omega\rangle)$ either is trivial (consists of single element), or is isomorphic to the cohomotopy set $\pi^k(Y)$, where $k = k(\omega) \in [|\omega|' - 1,\ d-1]$.

For $d \leq 13$, the table (6.36) lists all ω and the corresponding $k = k(\omega)$ for which $\mathcal{P}_d^{\mathbf{c}\langle\omega\rangle}$ is non-contractible (in fact, a homology and even a homotopy k-sphere).

Proof. The claim follows by combining Theorem 7.3 with the table (6.36) from Chapter 5. $\square$

Example 7.4. Let $d = 8$ and $\omega = (4)$. Then, for any closed Y, using the list (6.36), by Theorem 7.5, we get a bijection $\mathsf{Q}^{\mathsf{emb}}_{8,8}(Y; \mathbf{c}\langle(4)\rangle, \mathbf{c}\langle(4)\rangle) \approx \pi^4(Y)$, the 4-cohomotopy set of Y. In particular, the we get the following group isomorphism:

$$\mathsf{Q}^{\mathsf{emb}}_{8,8}(S^7; \mathbf{c}\langle(4)\rangle, \mathbf{c}\langle(4)\rangle) \approx \pi_7(S^4) \approx \mathbb{Z} \times \mathbb{Z}_{12}.$$

Let $d = 12$ and $\omega = (11213)$. Then, for any closed Y, by (6.36), we get a bijection

$$\mathsf{Q}^{\mathsf{emb}}_{12,12}(Y; \mathbf{c}\langle(11213)\rangle, \mathbf{c}\langle(11213)\rangle) \approx \pi^6(Y).$$

In particular, we get an isomorphism $\mathsf{Q}^{\mathsf{emb}}_{12,12}(S^9; \mathbf{c}\langle(11213)\rangle, \mathbf{c}\langle(11213)\rangle) \approx \pi_9(S^6) \approx \mathbb{Z}_{24}$. $\diamondsuit$

7.6 Relation between quasitopies and inner framed cobordisms

For a compact smooth manifold Y of dimension $n \geq k-1$, Proposition 7.14 below points to a somewhat peculiar connection between the quasitopies of embeddings $\mathsf{Q}^{\mathsf{emb}}_{d,d}(Y; \mathbf{c}\Theta^{(q)}_{|\sim|' \geq k}, \mathbf{c}\Theta^{(q)}_{|\sim|' \geq k})$ and the inner framed cobordisms $\mathcal{FB}^{k-1}_{\Xi}(Y)$ of Y, as introduced in the definition below.

Definition 7.13. Let Y be a compact smooth manifold. The inner framed cobordisms $\mathcal{FB}^{k-1}_{\Xi}(Y)$ are based on codimension $k-1$ smooth closed submanifolds Z of Y of the form $Z = \coprod_{\sigma=1}^{\ell} Z_\sigma$. The normal bundle $\nu(Z, Y)$ is framed. The disjoint "components" $\{Z_\sigma\}_\sigma$ of Z are marked with different "colors" σ from a given pallet Ξ of cardinality ℓ.

Two such submanifolds $Z_0, Z_1 \subset Y$ are said to be inner cobordant if there exist codimension $k-1$ submanifold $W = \coprod_{\sigma=1}^{\ell} W_\sigma$ of $Y \times [0,1]$, such that
$\partial W = (Z_0 \times \{0\}) \cup (Z_1 \times \{1\})$; moreover, $\partial W_\sigma = \big((Z_0)_\sigma \times \{0\}\big) \coprod \big((Z_1)_\sigma \times \{1\}\big)$ for each color $\sigma \in \Xi$. The normal bundle $\nu = \nu(W, Y \times [0,1])$ is framed. The bundle restrictions give rise to bundle isomorphisms $\nu|_{Y\times\{0\}} \approx \nu(Z_0, Y)$, $\nu|_{Y\times\{1\}} \approx \nu(Z_1, Y)$, and the framing of ν extends the framings of $\nu(Z_0, Y)$ and $\nu(Z_1, Y)$. $\diamondsuit$

As with quasitopies of immersions (see (7.16)), for any pair Y_1, Y_2 of compact smooth manifolds with boundary and any choice of components

$\kappa_1 \in \pi_0(\partial Y_1)$, $\kappa_2 \in \pi_0(\partial Y_2)$, a connected sum operation

$$\uplus_{\mathcal{F}} : \mathcal{FB}^{k-1}_{\Xi}(Y_1) \times \mathcal{FB}^{k-1}_{\Xi}(Y_2) \to \mathcal{FB}^{k-1}_{\Xi}(Y_1 \#_\partial Y_2) \tag{7.42}$$

may be defined in the obvious way as $Z_1 \coprod Z_2 \hookrightarrow Y_1 \#_\partial Y_2$. This operation $\uplus_{\mathcal{F}}$ converts the inner framed cobordisms $\mathcal{FB}^{k-1}_{\Xi}(D^n)$ of a n-ball into a group. Choosing $\kappa \in \pi_0(\partial Y)$, we get a group action

$$\uplus_{\mathcal{F}} : \mathcal{FB}^{k-1}_{\Xi}(D^n) \times \mathcal{FB}^{k-1}_{\Xi}(Y) \to \mathcal{FB}^{k-1}_{\Xi}(Y). \tag{7.43}$$

Proposition 7.14. *Let Ξ be a pallet of the cardinality $A(d,k,q)$, where the number $A(d,k,q)$ is introduced in (6.39). For a compact smooth manifold Y of dimension $n \geq k-1$, $k \in [3, d-1]$, there is a bijection*

$$Th: \; \mathcal{FB}^{k-1}_{\Xi}(Y) \approx \mathsf{Q}^{\mathsf{emb}}_{d,d}\big(Y; \mathbf{c}\Theta^{(q)}_{|\sim|' \geq k}, \mathbf{c}\Theta^{(q)}_{|\sim|' \geq k}\big),$$

delivered by the Thom construction on the trivialized normal bundle $\nu(Z,Y)$, where $[Z \hookrightarrow Y] \in \mathcal{FB}^{k-1}_{\Xi}(Y)$. The bijection Th respects the operations $\uplus_{\mathcal{F}}$ and $\uplus$ in framed inner bordisms and in the quasitopies:

$$Th(Z_1 \uplus_{\mathcal{F}} Z_2) = Th(Z_1) \uplus Th(Z_2). \tag{7.44}$$

Proof. Given a framed regular embedding $Z =_{\mathsf{def}} \coprod_{\sigma=1}^{A(d,k,q)} Z_\sigma \hookrightarrow Y$ of a closed smooth manifold Z of codimension $k-1$, the Thom construction on the bundle $\nu(Z,Y)$ produces a continuous map $T_Z : Y \to \bigvee_{\sigma=1}^{A(d,k,q)} S^{k-1}_\sigma$ to the bouquet. Since Z is closed, ∂Y is mapped to the base point $\star$ of the bouquet. The homotopy class of T_Z depends only on the inner cobordism class of $Z \subset Y$ in $\mathcal{FB}^{k-1}_{\Xi}(Y)$.

Let us fix a homotopy equivalence $\tau : (\mathcal{P}^{\mathbf{c}\Theta^{(q)}_{|\sim|' \geq k}}, pt) \sim (\bigvee_{\sigma=1}^{A(d,k,q)} S^{k-1}_\sigma, \star)$. By Corollary 7.7, the homotopy class of the map $\tilde{T}_Z =_{\mathsf{def}} \tau^{-1} \circ T_Z$ determines the quasitopy class of some embedding $\beta_{\tilde{T}_Z} : M \subset \mathbb{R} \times Y$, whose combinatorial tangency patterns to the foliation $\mathcal{L}$ belong to the poset $\mathbf{c}\Theta^{(q)}_{|\sim|' \geq k}$, and the tangency patterns of $\beta_{\tilde{T}_Z}| : \partial M \to \mathbb{R} \times \partial Y$ belong to the poset $\mathbf{c}\Lambda = (\emptyset), (1)$. Thus the map $Th : Z \rightsquigarrow \beta_{\tilde{T}_Z}$ is well-defined.

To show that Th is onto, we take any embedding $\beta : M \subset \mathbb{R} \times Y$ of a n-manifold M, whose tangency patterns reside in the poset $\mathbf{c}\Theta^{(q)}_{|\sim|' \geq k}$ and such that the tangency patterns of $\beta| : \partial M \subset \mathbb{R} \times \partial Y$ reside in $\mathbf{c}\Lambda = (\emptyset), (1)$. We use β to produce a map $\Phi^\beta : (Y, \partial Y) \to (\mathcal{P}^{\mathbf{c}\Theta^{(q)}_{|\sim|' \geq k}}, pt)$. Composing Φ^β with the homotopy equivalence τ, we get a map $\tilde{\Phi}^\beta := \tau \circ \Phi^\beta$ from Y to the bouquet, so that ∂Y is mapped to the base point $\star$ of the bouquet. For each sphere S^{k-1}_σ from the bouquet, we fix a point $b_\sigma \in S^{k-1}_\sigma$, different from the apex $\star$, and a $(k-1)$-frame at b_σ. Perturbing $\tilde{\Phi}^\beta$ within its homotopy class, we may assume that $\tilde{\Phi}^\beta$ is transversal to each b_σ. Then the closed framed submanifold $Z_\beta =_{\mathsf{def}} (\tilde{\Phi}^\beta)^{-1}(\coprod_{\sigma\in\Xi} b_\sigma) \subset Y$ is an element of $\mathcal{FB}^{k-1}_\Xi(Y)$. Moreover, for a given β, the map $\tau^{-1} \circ Th(Z_\beta)$ is in the homotopy class of Φ^β.

The injectivity of Th has a similar validation, using the $(n+1)$-dimensional cobordism $B : N \to \mathbb{R} \times Y \times [0,1]$ that bounds an embedding $\beta : M \to \mathbb{R} \times Y$ which is quasitopic to the trivial embedding. Again, we perturb $\tau \circ \Phi^B : Y \times [0,1] \to \bigvee_{\sigma=1}^{A(d,k,q)} S^{k-1}_\sigma$ to make it transversal to $\coprod_{\sigma\in\Xi} b_\sigma$, thus producing a framed cobordism W in $Y \times [0,1]$ which bounds Z_β.

Finally, the validation of the property $Th(Z_1 \uplus_{\mathcal{F}} Z_2) = Th(Z_1) \uplus Th(Z_2)$ is on the level of definitions. □

For $k > 2$ and an oriented n-dimensional Y, where $\partial Y \neq \emptyset$, Proposition 7.14 gives a limited insight into the actions of the (abelian when $n \geq 2$) group $\mathsf{G}^{\mathsf{emb}}_{d,d}(n; \mathbf{c}\Theta^{(q)}_{|\sim|' \geq k}, \mathbf{c}\Theta^{(q)}_{|\sim|' \geq k})$ on the set $\mathsf{Q}^{\mathsf{emb}}_{d,d}\big(Y; \mathbf{c}\Theta^{(q)}_{|\sim|' \geq k}; \mathbf{c}\Theta^{(q)}_{|\sim|' \geq k}\big)$.

It is based on one useful observation: if $\beta : M \hookrightarrow \mathbb{R} \times Y$ is represented, via the inverse Th^{-1} of the Thom map, by a closed normally framed submanifold $Z_\beta = \coprod_{\sigma\in\Xi} Z_{\beta,\sigma} \subset Y$, then the fundamental homology classes $\{[Z_{\beta,\sigma}] \in H_{n-k+1}(Y;\mathbb{Z})\}_{\sigma\in\Xi}$ are *constant* along the $\mathsf{G}^{\mathsf{emb}}_{d,d}(n; \mathbf{c}\Theta^{(q)}_{|\sim|' \geq k}; \mathbf{c}\Theta^{(q)}_{|\sim|' \geq k})$-orbit containing the quasitopy class $Th(Z_\beta)$. Indeed, using the property $Th(Z_1 \uplus_{\mathcal{F}} Z_2) = Th(Z_1) \uplus Th(Z_2)$, all the changes $Z_\beta \rightsquigarrow Z_\beta \uplus_{\mathcal{F}} W$, produced by the $\uplus_{\mathcal{F}}$-action, are such that W is incapsulated in a ball $B^n \subset Y$. Therefore, the fundamental classes of Z_β and $Z_\beta \uplus_{\mathcal{F}} W$ are homologous in Y.

Keeping the previous observation in mind, for a pallet Ξ of cardinality $A(d,k,q)$, consider the subset $H^{[\Xi\sim fr]}_{n-k+1}(Y;\mathbb{Z})$ of the group $\big(H_{n-k+1}(Y;\mathbb{Z})\big)^{A(d,k,q)}$ that is realized by at most $A(d,k,q)$ *disjoint* distinctly colored and normally framed closed submanifolds $\{Z_\sigma\}_{\sigma\in\Xi}$ of Y, $\dim Z_\sigma = n-k+1$.

For $2k-2 > n$, by a general position argument, $H^{[\Xi\sim fr]}_{n-k+1}(Y;\mathbb{Z})$ is actually a *subgroup* of $\big(H_{n-k+1}(Y;\mathbb{Z})\big)^{A(d,k,q)}$.

Proposition 7.15. *Let Y be a compact connected oriented smooth n-manifold Y, where $n \geq k-1$ and $k \in [3, d-1]$. Assume that $\partial Y \neq \emptyset$. Then the orbit-space of the $\mathsf{G}^{\mathsf{emb}}_{d,d}(n; \mathbf{c}\Theta^{(q)}_{|\sim|' \geq k}, \mathbf{c}\Theta^{(q)}_{|\sim|' \geq k})$-action on the set $\mathsf{Q}^{\mathsf{emb}}_{d,d}\big(Y; \mathbf{c}\Theta^{(q)}_{|\sim|' \geq k}, \mathbf{c}\Theta^{(q)}_{|\sim|' \geq k}\big)$ admits a surjection on the set $H^{[\Xi\sim fr]}_{n-k+1}(Y;\mathbb{Z})$.*

Proof. Using Proposition 7.14 and tracing the correspondence $[\beta] \rightsquigarrow Th^{-1}([\beta]) \rightsquigarrow \{[Z_{\beta,\sigma}]\}_\sigma$ validates the claim instantly. □

Example 7.5. Take $k = 3$, $d = 6$, $q = 0$, and $n = 3$. Then $A(6,3,0) = 10$ and $\#\Xi = 10$. Thus we get a homotopy equivalence $\tau : \mathcal{P}_6^{\mathbf{c}\Theta_{|\sim|'\geq 3}} \sim \bigvee_{\sigma=1}^{10} S^2_\sigma$. For a compact oriented 3-manifold Y, any normally framed 1-dimensional submanifold Z acquires an orientation. Using the isomorphism

$$Th : \mathcal{FB}^2_\Xi(Y) \approx \mathsf{Q}^{\mathsf{emb}}_{6,6}(Y; \mathbf{c}\Theta_{|\sim|'\geq 3}, \mathbf{c}\Theta_{|\sim|'\geq 3})$$

from Proposition 7.14, any collection of at most ten disjoint framed closed 1-dimensional submanifolds $Z_1, \dots Z_{10} \subset Y$ (that is, any *framed link* $Z \subset Y$, colored with 10 distinct colors at most) produces a quasitopy class of an embedding $\beta : M^3 \to \mathbb{R} \times Y$, where M is closed. Its tangency patterns ω, with the exception of $(\emptyset)$, have only entries from the list $\{1, 2, 3\}$ so that no more than one 3 and no more than two 2's are present in ω, while $|\omega| \leq 6$.

For an orientable Y, the orbit-space of the $\mathsf{G}^{\mathsf{emb}}_{6,6}(3; \mathbf{c}\Theta_{|\sim|'\geq 3}, \mathbf{c}\Theta_{|\sim|'\geq 3})$-action on $\mathsf{Q}^{\mathsf{emb}}_{6,6}(Y; \mathbf{c}\Theta_{|\sim|'\geq 3}, \mathbf{c}\Theta_{|\sim|'\geq 3})$ admits a surjection onto the group $(H_1(Y;\mathbb{Z}))^{10}$. For example, for $Y = T^3 \setminus \mathrm{int}(D^3)$, where T^3 is the 3-torus, the orbit-space is mapped onto the lattice $\mathbb{Z}^{30}$. ♢

Corollary 7.14. *We adopt the hypotheses of Corollary 7.13.*

For $(k-1)$-dimensional manifold Y and any choice of $\kappa \in \pi_0(\partial Y)$, the group $\mathsf{G}^{\mathsf{emb}}_{d,d}(k-1; \mathbf{c}\Theta^{(q)}_{|\sim|'\geq k}, \mathbf{c}\Theta^{(q)}_{|\sim|'\geq k})$ acts freely and transitively on the set $\mathsf{Q}^{\mathsf{emb}}_{d,d}(Y; \mathbf{c}\Theta^{(q)}_{|\sim|'\geq k}, \mathbf{c}\Theta^{(q)}_{|\sim|'\geq k})$. Thus both sets are in a 1-to-1 correspondence.

For a simply-connected Y of dimension $k > 4$, the group $\mathsf{G}^{\mathsf{emb}}_{d,d}(k; \mathbf{c}\Theta^{(q)}_{|\sim|'\geq k}, \mathbf{c}\Theta^{(q)}_{|\sim|'\geq k})$ acts freely and transitively on the set $\mathsf{Q}^{\mathsf{emb}}_{d,d}(Y; \mathbf{c}\Theta^{(q)}_{|\sim|'\geq k}, \mathbf{c}\Theta^{(q)}_{|\sim|'\geq k})$. Again, both sets are in a 1-to-1 correspondence.

Proof. By Proposition 7.15, there exists a bijection

$$Th : \mathcal{FB}^{k-1}_\Xi(Y) \overset{\approx}{\to} \mathsf{Q}^{\mathsf{emb}}_{d,d}(Y; \mathbf{c}\Theta^{(q)}_{|\sim|'\geq k}, \mathbf{c}\Theta^{(q)}_{|\sim|'\geq k}).$$

When $\dim Y = k-1$, the elements of $\mathcal{FB}^{k-1}_\Xi(Y)$ are represented by finite sets Z of framed oriented singletons, marked with $\#\Xi$ colors at most (in fact, we may assume that no two singletons of the same color have opposite orientations). Evidently, any such finite set Z can be incapsulated into a standard smooth ball $D^{k-1} \subset Y$ such that $\partial D^{k-1} \cap \partial Y = D^{k-2}$, a standard $(k-2)$-ball. Therefore, Z belongs to the $\mathcal{FB}^{k-1}_\Xi(D^{k-1})$-orbit of the empty collection $Z_\emptyset$. By (7.44), the image $Th(Z)$ belongs to the $\mathsf{G}^{\mathsf{emb}}_{d,d}(k-1; \mathbf{c}\Theta^{(q)}_{|\sim|'\geq k}, \mathbf{c}\Theta^{(q)}_{|\sim|'\geq k})$-orbit of the trivial embedding (the empty one for $d \equiv 0 \mod 2$ and the embedding $\beta_\star : Y \to \{0\} \times Y \subset \mathbb{R} \times Y$ for $d \equiv 1 \mod 2$). This validates the first claim.

For a simply-connected Y, $\dim Y = k > 4$, again by Proposition 7.15, any element of $\mathsf{Q}^{\mathsf{emb}}_{d,d}(Y; \mathbf{c}\Theta^{(q)}_{|\sim|' \geq k}; \mathbf{c}\Theta^{(q)}_{|\sim|' \geq k})$ is represented by a framed collection Z of disjoint loops in Y, colored with $\#\Xi$ colors at most. Using that $\pi_1(Y) = 1$ and $\dim Y = k > 4$, we can bound each loop Z_σ by a regularly embedded smooth 2-disk $W_\sigma \subset \mathrm{int}(Y)$ so that all W_σ's are disjoint. Therefore, each loop Z_σ is contained in a standard ball D^k_σ, the regular neighborhood of W_σ. By performing 1-surgeries on $\coprod_\sigma D^k_\sigma$, we construct a k-ball $\tilde{B}^k$ that contains all Z_σ's. Adding a 1-handle $H \subset Y \setminus \tilde{B}^k$ that connects $\partial\tilde{B}^k$ with ∂Y, we exhibit a ball $B^k =_{\mathsf{def}} \tilde{B}^k \cup H$ which contains all the loops and such that $\partial B^k \cap \partial Y$ is a $(k-1)$-dimensional ball. Therefore, the loop union Z is, once more, in the orbit of the empty collection $Z_\emptyset$ of loops. By (7.44), $Th(Z)$ belongs to the $\mathsf{G}^{\mathsf{emb}}_{d,d}(k; \mathbf{c}\Theta^{(q)}_{|\sim|' \geq k}, \mathbf{c}\Theta^{(q)}_{|\sim|' \geq k})$-orbit of the trivial embedding.

In fact, this conclusion is also valid for 3-manifolds Y with a *spherical* boundary, but its validation is not as direct as for the case $k > 4$. If $\pi_1(Y) = 1$, by the Perelman's solution of the Poincaré Conjecture [P1]-[P3], we conclude that Y is the standard ball D^3. Thus, any Ξ-colored and framed link $Z \subset Y$ is obviously contained in D^3.

As usual, the case $k = 4$ seems to be challenging. □

7.7 More about the stabilization of $\mathcal{Q}^{\mathsf{emb}}_{d,d'}(Y, \partial Y; \mathbf{c}\Theta, \mathbf{c}\Lambda; \mathbf{c}\Theta')$ by the increasing degrees d, d'

We start with the embeddings of 1-manifolds in the product $\mathbb{R} \times D^1$ and the forbidden poset Θ that consists of ω's with the reduced norms $|\omega|' \geq 2$. Recall that a similar setting, under the name of "doodles", was discussed in Chapter 5.

Proposition 7.16. *Let* $d' \geq d + 2$, $d' \equiv d \mod 2$. *For a closed subposet* $\Theta \subset \mathbf{\Omega}_{\langle d],\, |\sim|' \geq 2}$, *let* $\hat{\Theta}_{\{d'\}}$ *denote the smallest subposet in* $\mathbf{\Omega}_{\langle d']}$ *that contains* Θ.

Then we get a stabilization by d':

$$\mathsf{Q}^{\mathsf{emb}}_{d+2,\, d'}(D^1; \mathbf{c}\hat{\Theta}_{\{d+2\}}, \mathbf{c}\hat{\Theta}_{\{d'\}}) \approx \pi_1(\mathcal{P}^{\mathbf{c}\hat{\Theta}_{\{d+2\}}}_{d+2}, pt).$$

Proof. The claim follows from Theorem 7.3 by combining it with Theorem 6.15 (see also [KSW1], Theorem 2.15), which claims that

$$(\epsilon_{d+2,d'})_* : \pi_1(\mathcal{P}^{\mathbf{c}\hat{\Theta}_{\{d+2\}}}_{d+2}) \approx \pi_1(\mathcal{P}^{\mathbf{c}\Theta_{\{d'\}}}_{d'})$$

is an isomorphism, induced by the embedding $\epsilon_{d+2,d'}$ of the two spaces. Recall that Chapter 5 contains an explicit description of presentations of such fundamental groups $\pi_1(\mathcal{P}_{d+2}^{\mathbf{c}\hat{\Theta}_{\{d+2\}}})$. $\square$

Theorem 7.6. *Let Θ be a closed poset such that $|\omega|' \geq 3$ for any $\omega \in \Theta$. Let $d' \geq d$ and $d' \equiv d \mod 2$.*

Assuming the Λ-condition (7.11), for any compact smooth Y such that $\dim Y \leq d+2-\psi_\Theta(d+2)$ (see (6.23) for the definition of the ψ-function) we get the classifying map

$$\Phi^{\mathsf{i/e}}_{d,d+2}(Y;\mathbf{c}\Theta,\mathbf{c}\Theta):\ \mathsf{Q}^{\mathsf{i/e}}_{d,d+2}(Y;\mathbf{c}\Theta,\mathbf{c}\Theta) \to [(Y,\partial Y),(\mathcal{P}_d^{\mathbf{c}\Theta},pt)], \quad (7.45)$$

which is bijective for embeddings and surjective for immersions.

For any profinite (in the sense of Definition 6.5) poset Θ and any Y, the set $\mathsf{Q}^{\mathsf{emb}}_{d,d'}(Y;\mathbf{c}\Theta,\mathbf{c}\Theta)$ stabilizes for all sufficiently big d.

Proof. The argument is a bit similar to the one we used in the proof of Proposition 7.13.

Under the hypotheses, by Theorem 6.10, the homomorphism

$$(\epsilon_{d,d'})_* : H_j(\mathcal{P}_d^{\mathbf{c}\Theta};\mathbb{Z}) \to H_j(\mathcal{P}_{d+2}^{\mathbf{c}\Theta};\mathbb{Z})$$

is an isomorphism for all $j < d+2-\psi_\Theta(d+2)$. Since $|\omega|' \geq 3$ for any $\omega \in \Theta$, all the spaces $\{\mathcal{P}_{d'}^{\mathbf{c}\Theta}\}_{d' \geq d}$ are simply connected. Hence, $\epsilon_{d,d+2}$ induces an isomorphism in the integral j-homology of $\mathcal{P}_d^{\mathbf{c}\Theta}$ and of $\mathcal{P}_{d+2}^{\mathbf{c}\Theta}$ for all $j < d+2-\psi_\Theta(d+2)$. By the Whitehead Theorem (the inverse Hurewicz Theorem) (see [Wh], [Wh1]), we conclude that $\epsilon_{d,d+2}$ induces isomorphisms of the homotopy groups of $\mathcal{P}_d^{\mathbf{c}\Theta}$ and of $\mathcal{P}_{d+2}^{\mathbf{c}\Theta}$ in all dimensions $\leq d+2-\psi_\Theta(d+2)$. By standard application of the obstruction theory, if $\dim Y \leq d+2-\psi_\Theta(d+2)$, no nontrivial element of $[(Y,\partial Y),(\mathcal{P}_d^{\mathbf{c}\Theta},pt)]$ becomes, via $\epsilon_{d,d+2}$, null-homotopic in $[(Y,\partial Y),(\mathcal{P}_{d+2}^{\mathbf{c}\Theta},pt')]$. Thus, for such Y, we get

$$[[(Y,\partial Y),\ \epsilon_{d,d+2} : (\mathcal{P}_d^{\mathbf{c}\Theta},pt) \to (\mathcal{P}_{d+2}^{\mathbf{c}\Theta'},pt')]] = [(Y,\partial Y),(\mathcal{P}_d^{\mathbf{c}\Theta},pt)].$$

By Lemma 6.6, for any closed profinite Θ, $\lim_{d'\to\infty}(d'-\psi_\Theta(d')) = +\infty$. Hence, repeating the previous arguments, we get a similar conclusion for all $d' \geq d \gg \dim Y$.

Therefore, by Theorem 7.3, the all the claims of Theorem 7.6 follow. $\square$

Example 7.6. Tracing these lines of arguments and using Vassiliev's computation [V] of the cohomology ring $H^*(\mathcal{P}^{\mathbf{c}\Theta_{\max\geq 4}},pt;\mathbb{Z})$ (see (7.38)), we get the following claim. For all $d \geq 4$ and for any compact connected orientable surface Y, the quasitopies $\mathsf{G}^{\mathsf{emb}}_{d,d+2}(Y; \mathbf{c}\Theta_{\max\geq 4}; \mathbf{c}\Theta_{\max\geq 4})$ of embeddings $\beta : M \to \mathbb{R} \times Y$ (where M is a surface) are stable and isomorphic to $\mathbb{Z}$. $\diamondsuit$

Example 7.7. Let $\Theta = \langle (4), (33) \rangle$. Then by Theorem 7.6 and Example 6.8, we get a bijection

$$\Phi^{\mathsf{emb}}_{d,d+2}(Y; \mathbf{c}\Theta; \mathbf{c}\Theta) : \mathsf{Q}^{\mathsf{emb}}_{d,d+2}(Y; \mathbf{c}\Theta; \mathbf{c}\Theta) \to [(Y, \partial Y), (\mathcal{P}^{\mathbf{c}\Theta}_d, pt)] \quad (7.46)$$

for all $d \geq 2\dim(Y) - 4$. Thus, $\mathsf{Q}^{\mathsf{emb}}_{d,d+2}(Y; \mathbf{c}\Theta; \mathbf{c}\Theta)$ is stable for all $d \geq 2\dim(Y) - 4$.

$\diamondsuit$

7.8 Detour: on cobordant immersions with trivial normal line bundles

We take now a short detour away from immersions against the background of product 1-foliations and towards the immersions at large of n-manifolds into $(n+1)$-manifolds with trivial normal line bundles. With the target manifold being fixed, we consider such immersions up to the natural cobordism relation. Most of the observations below that link such bordisms with the homotopy theory are well-known.

Let Z be a smooth $(n+1)$-dimensional compact manifold with boundary, and M a smooth compact n-dimensional manifold with boundary. We consider proper immersions $\beta : (M, \partial M) \to (Z, \partial Z)$, whose normal 1-bundle $\nu^\beta =_{\mathsf{def}} \beta^*(TZ)/TM$ is *trivialized.*

We denote by $\bar{\beta}$ the immersion β in which the trivialization of ν^β is flipped to the opposite.

As a default, we assume that $\beta(M)$ has transversal self-intersections in Z of the multiplicities less or equal than $\dim(Z)$.

Definition 7.14. We say that two such immersions β_0 and β_1 are cobordant, if there exists a smooth compact $(n+1)$-dimensional manifold N, where

$$\partial N = (-M_0 \sqcup M_1) \bigcup_{\{-\partial M_0 \sqcup \partial M_1\}} \delta N,$$

and a proper immersion

$$B : (N, \delta N) \to (Z \times [0,1],\ \partial Z \times [0,1])$$

so that:

- $B|_{M_0} = \beta_0$, $B|_{M_1} = \beta_1$,
- the trivialized normal bundle

$$\nu^B =_{\mathsf{def}} B^*(T(Z \times [0,1]))/TN$$

of $B(N)$ in $Z \times [0,1]$ is an extension of trivialized bundles $\nu^{\beta_0} \sqcup \nu^{\beta_1}$, and

- the self-intersections of $B(N)$ in $Z \times [0,1]$ are transversal and of the multiplicities less or equal than $\dim(Z) + 1$.

We denote by $\mathsf{IMM}^{\mathsf{fr}}_{(-1)}(Z, \partial Z)$ the set of bordism classes of such *framed* immersions. The index (-1) reflects the codimension of $\beta(M)$ in Z.

If we replace immersions β_0, β_1, B in this setting with framed regular *embeddings*, we get a similar notion of bordism. We denote by $\mathsf{EMB}^{\mathsf{fr}}_{(-1)}(Z, \partial Z)$ the set of embeddings, considered up to the cobordisms which are realized by embeddings in $Z \times [0,1]$ with trivialized normal bundles.

We use the notations $\mathsf{IMM}^{\mathsf{fr}}_{(-1)}(Z)$, $\mathsf{EMB}^{\mathsf{fr}}_{(-1)}(Z)$ for the similar constructions, based on immersions/embeddings of *closed* manifolds M in Z. ◇

Note that if Z is orientable, then the triviality of the normal bundles ν^β and of ν^B implies that M and N are orientable as well.

Szüks' Theorem 2 [Sz2]—a much more general result that the one needed for the setting described here—implies that both sets $\mathsf{EMB}^{\mathsf{fr}}_{(-1)}(Z, \partial Z)$ and $\mathsf{IMM}^{\mathsf{fr}}_{(-1)}(Z, \partial Z)$ have a homotopy-theoretical description:

Theorem 7.7. *Let* $\dim(Z) = n + 1$. *There exist bijections*

$$\mathsf{EMB}^{\mathsf{fr}}_{(-1)}(Z, \partial Z) \approx [Z,\, S^1] \approx H^1(Z; \mathbb{Z}),$$

$$\mathsf{EMB}^{\mathsf{fr}}_{(-1)}(Z) \approx [Z/\partial Z,\, S^1] \approx H^1(Z/\partial Z, ; \mathbb{Z}), \tag{7.47}$$

$$\mathsf{IMM}^{\mathsf{fr}}_{(-1)}(Z, \partial Z) \approx \mathsf{EMB}^{\mathsf{fr}}_{(-1-k)}(Z \times \mathbb{R}^k, \partial Z \times \mathbb{R}^k) \approx [Z,\, \Omega^k(S^{k+1})],$$

$$\mathsf{IMM}^{\mathsf{fr}}_{(-1)}(Z) \approx \mathsf{EMB}^{\mathsf{fr}}_{(-1-k)}(Z \times \mathbb{R}^k) \approx [Z/\partial Z,\, \Omega^k(S^{k+1})], \tag{7.48}$$

for a sufficiently big k ($k \geq 2\dim(Z) + 2$ will do). Here $\Omega^k(S^{k+1})$ denotes the k-fold loop space on the sphere S^{k+1}. ◇

The main idea behind these assertions is to convert a given framed codimension one immersion $\beta : M \to Z$ into a framed embedding $\tilde{\beta} : M \to Z \times \mathbb{R}^k$ by picking any embedding $j : M \hookrightarrow \mathbb{R}^k$ ($k \geq 2\dim(M) + 1$ will do) and using j to embed M by the product map $\tilde{\beta} = \beta \times j$. In this construction, we multiply the product framing of the normal line bundle of β and the constant k-frame on $\mathbb{R}^k$. To invert the construction $\beta \rightsquigarrow \beta \times j$, we need to start with normally framed embedding $\tilde{\beta} : M \hookrightarrow Z \times \mathbb{R}^k$ and to construct, as it is done in [Sz2], a projection $p : Z \times \mathbb{R}^k \to Z$ whose differential Dp is of the maximal rank n on the tangent space $T(\tilde{\beta}(M))$. This step, delivered by an isotopy of $\tilde{\beta}(M)$ in $Z \times \mathbb{R}^k$, is nontrivial (see [Sz2], Theorem 1 on page 23).

By letting $Z = D^{n+1}$, Theorem 7.7 directly leads to the following claim.

Corollary 7.15. *The set* $\mathsf{EMB}^{\mathsf{fr}}_{(-1)}(D^{n+1}, \partial D^{n+1})$ *is trivial, and the set* $\mathsf{EMB}^{\mathsf{fr}}_{(-1)}(D^{n+1})$ *is trivial for* $n > 0$. *The set* $\mathsf{EMB}^{\mathsf{fr}}_{(-1)}(D^1)$ *is in a 1-to-1 correspondence with the set of integers.*

The set $\mathsf{IMM}^{\mathsf{fr}}_{(-1)}(D^{n+1}, \partial D^{n+1})$ *is trivial, and the set* $\mathsf{IMM}^{\mathsf{fr}}_{(-1)}(D^{n+1}) \approx \pi_n^{\mathsf{st}}$, *the stable homotopy group. In particular,* $\mathsf{IMM}^{\mathsf{fr}}_{(-1)}(D^{n+1})$ *is a finite set.* ♢

Proof. By Theorem 7.7, the set $\mathsf{IMM}^{\mathsf{fr}}_{(-1)}(D^{n+1}) \approx \pi_{n+k+1}(S^{k+1})$ at least for $k > 2n + 2$. However, in the range $k > n$, $\pi_{n+k+1}(S^{k+1}) \approx \pi_n^{\mathsf{st}}$, the stable homotopy group. Since the stable homotopy groups of spheres are finite [Ser], $\mathsf{IMM}^{\mathsf{fr}}_{(-1)}(D^{n+1})$ is a finite set. □

By examining the tables of the homotopy groups $\{\pi_n^{\mathsf{st}}\}_{0<n\leq 15}$, and applying Corollary 7.15, we get the following assertion.

Corollary 7.16. *Let* $\mathbb{R}^{n+2}_+$ *denotes the half-space in* $\mathbb{R}^{n+2}$, *bounded by the hyperplane* $\mathbb{R}^{n+1}$.

For any smooth immersion $\beta : M^n \to \mathbb{R}^{n+1}$ *of a closed smooth manifold* M^n *with a trivialized normal line bundle* ν_β, *there is a* β*-independent number* $\mu(n)$ *such that the* $\mu(n)$ *disjoint copies of* β *bound a smooth immersion* $B : W^{n+1} \to \mathbb{R}^{n+2}_+$ *with a trivialized normal line bundle* ν_B. *Moreover, for each dimension* n, *there exists an immersion* β_n *for which less than* $\mu(n)$ *copies of* β_n *do not bound such* B.

In fact, for $1 \leq n \leq 14$, *we get:*

$$\mu(1) = 2,\ \mu(2) = 2,\ \mu(3) = 24,\ \mu(4) = 0,\ \mu(5) = 0,\ \mu(6) = 2,\ \mu(7) = 240,$$

$$\mu(8) = 2,\ \mu(9) = 2,\ \mu(10) = 6,\ \mu(11) = 504,\ \mu(13) = 3,\ \mu(14) = 2,\ \mu(15) = 480.$$

♢

Some of the assertions below about the sets $\mathsf{EMB}^{\mathsf{fr}}_{(-1)}(Z, \partial Z)$, $\mathsf{IMM}^{\mathsf{fr}}_{(-1)}(Z, \partial Z)$ and $\mathsf{EMB}^{\mathsf{fr}}_{(-1)}(Z)$, $\mathsf{IMM}^{\mathsf{fr}}_{(-1)}(Z)$ can be easily derived from Theorem 7.7. When possible, we prefer to present more elementary geometric arguments that do not rely on that theorem. They center on the *resolutions* of self-intersection sets of immersions. As a byproduct, we will get some claims about volume minimizing immersions $M \to Z$ (see Corollary 7.17).

Consider the obvious maps

$$\mathsf{A} : \mathsf{EMB}^{\mathsf{fr}}_{(-1)}(Z, \partial Z) \to \mathsf{IMM}^{\mathsf{fr}}_{(-1)}(Z, \partial Z) \quad \text{and} \quad \mathsf{A} : \mathsf{EMB}^{\mathsf{fr}}_{(-1)}(Z) \to \mathsf{IMM}^{\mathsf{fr}}_{(-1)}(Z).$$

Proposition 7.17.

- *The maps*

$$\mathsf{A} : \mathsf{EMB}^{\mathsf{fr}}_{(-1)}(Z, \partial Z) \to \mathsf{IMM}^{\mathsf{fr}}_{(-1)}(Z, \partial Z),$$

$$\mathsf{A} : \mathsf{EMB}^{\mathsf{fr}}_{(-1)}(Z) \to \mathsf{IMM}^{\mathsf{fr}}_{(-1)}(Z)$$

are injective: if two embeddings, β_0 and β_1, with trivialized normal line bundles are cobordant by a cobordism that is an immersion with a trivialized normal line bundle, then they are cobordant by a cobordism that is an embedding with a trivialized normal line bundle.

- *There exists surjective resolution maps*

$$\mathsf{R} : \mathsf{IMM}^{\mathsf{fr}}_{(-1)}(Z, \partial Z) \to \mathsf{EMB}^{\mathsf{fr}}_{(-1)}(Z, \partial Z),$$

$$\mathsf{R} : \mathsf{IMM}^{\mathsf{fr}}_{(-1)}(Z) \to \mathsf{EMB}^{\mathsf{fr}}_{(-1)}(Z)$$

such that the compositions $\mathsf{R} \circ \mathsf{A} = \mathsf{id}$. Moreover, any $\beta \in \mathsf{IMM}^{\mathsf{fr}}_{(-1)}(Z, \partial Z)$ or $\beta \in \mathsf{IMM}^{\mathsf{fr}}_{(-1)}(Z)$, the embedding $\mathsf{R}(\beta)$ defines the same homology class in $H_n(Z, \partial Z; \mathbb{Z})$ or in $H_n(Z; \mathbb{Z})$, respectively, provided that Z is orientable.

Proof. Consider two embedding

$$\beta_0 : (M_0, \partial M_0) \to (Z, \partial Z) \text{ and } \beta_1 : (M_1, \partial M_1) \to (Z, \partial Z)$$

with trivialized normal bundles that are connected by a cobordism

$$B : (N, \delta N) \to (Z \times [0,1], \partial Z \times [0,1]),$$

which is an immersion with its normal 1-bundle ν^B being trivialized so that the trivializations of ν^{β_0} and of ν^{β_1} extend to the trivialization of ν^B. Then we can perturb B so that it becomes q-normal (see [LS], [Th]) for all $q \leq n+2$. The perturbation B' of B may be chosen so small that the normal 1-bundle $\nu^{B'}$ is still trivial and $B'|_{M_0 \sqcup M_1} = B|_{M_0 \sqcup M_1}$.

We aim to resolve the immersion B' into an embedding $\tilde{B}$ of a new manifold $\tilde{N}$.

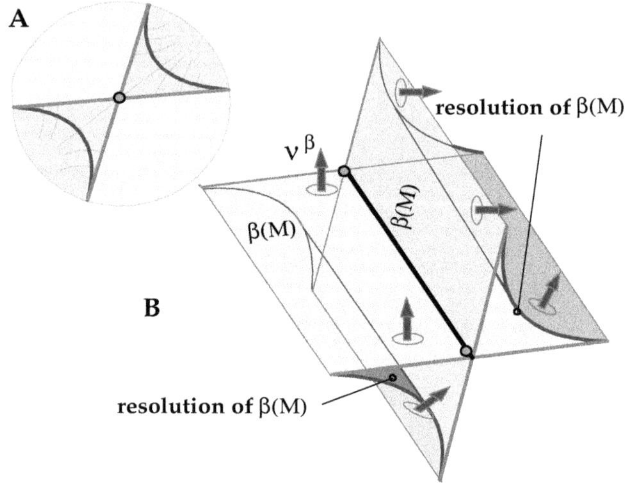

Fig. 7.5 A resolution of an immersion $\beta : M \to Z$ into an embedding. Note that the orientation of the normal line bundle ν^{β} helps to orient the normal line bundle of the resolution. The same picture may represent a resolution of an immersion $B : N \to Z \times [0, 1]$ into an embedding.

Let us start with a local model of such a resolution (see Fig. 7.5, A). Let D^q be an open unit ball in $\mathbb{R}^q$, centered at the origin. We fix a diffeomorphism $\phi : D^q \to \mathbb{R}^q$, where $\phi(\vec{0}) = \vec{0}$ and the radii of D^q are mapped onto the infinite rays emanating from the origin. Let $x_1, \dots, x_q$ be the canonical coordinates on $\mathbb{R}^q$. Consider the hypersurface $H_\epsilon =_{\mathsf{def}} \{x_1 \cdot \ldots \cdot x_q = \epsilon\}$, $\epsilon > 0$. Replacing H_0 with H_ϵ, or rather, $\tilde{H}_0 =_{\mathsf{def}} \phi^{-1}(H_0)$ with $\tilde{H}_\epsilon =_{\mathsf{def}} \phi^{-1}(H_\epsilon)$ mimics the resolution we are after.

To simplify notations, put $B =_{\mathsf{def}} B'$. The resolution of B is carried inductively: we assume that B already has no self-intersections of multiplicities $> q$ and consider the locus $Y =_{\mathsf{def}} Y^B_q \subset B(N)$ where exactly q branches of $B(N)$ intersect transversally. Thus, Y is a compact manifold. Note that $\partial Y \subset \partial Z \times [0, 1]$ since $B|_{M_0 \sqcup M_1}$ is an embedding.

We consider a tubular neighborhood U of Y in $Z \times [0, 1]$ such that $U \cap (\partial Z \times [0, 1])$ is a tubular neighborhood of ∂Y in $\partial Z \times [0, 1]$. Let $\pi : U \to Y$ be the normal fibration whose fibers are open q-balls. For a sufficiently narrow tube U, the intersection of $B(N)$ with U is a fibration over Y with the π-fiber $\tilde{H}_0$ (see Fig. 7.5, B).

Over a small neighborhood V_y of each point $y \in Y$, $B(N)$ is given by a smooth map $\Psi : \pi^{-1}(V_y) \to B^q$ such that Ψ is transversal to each branch of

$\tilde{H}_0 \subset D^q$ and $\Psi^{-1}(\vec{0}) = B(N) \cap \pi^{-1}(V_y)$. Thus, the structure group of the fibration $\pi : U \to Y$ is the group of diffeomorphisms $\mathsf{Diff}(D^q, \tilde{H}_0)$ of D^q that fix $\vec{0}$ and preserve $\tilde{H}_0$ invariant. In general, $\pi_0(\mathsf{Diff}(D^q, \tilde{H}_0)) \neq 1$. However, due to the global triviality of ν^B, we have a consistent way of picking the preferred normals to the branches of $B(N) \cap U$. In particular, we have a consistent way of picking the chambers in the fibers $\pi^{-1}(y) \setminus (\pi^{-1}(y) \cap B(N)) \approx D^q \setminus \tilde{H}_0$, where the product $x_1 \cdot \ldots \cdot x_q > 0$. We call such chambers positive. Hence, the structure group of $\pi : U \to Y$ reduces to the subgroup $\mathsf{Diff}_+(D^q, \tilde{H}_0) \subset \mathsf{Diff}(D^q, \tilde{H}_0)$ whose elements map negative chambers to negative chambers. For $\epsilon > 0$, the group $\mathsf{Diff}_+(D^q, \tilde{H}_0)$ admits a further reduction to the subgroup $\mathsf{Diff}_+(D^q, \{\tilde{H}_\epsilon\}_{\epsilon \geq 0})$ whose diffeomorphisms preserve each hypersurface $\tilde{H}_\epsilon$, $\epsilon \geq 0$, invariant.

Let us show that $\mathsf{Diff}_+(D^q, \{\tilde{H}_\epsilon\}_{\epsilon \geq 0})$ is a deformation retract of $\mathsf{Diff}_+(D^q, \tilde{H}_0)$. Indeed, consider the open set $D^q_+ \subset D^q \setminus \tilde{H}_0$, the union of positive open chambers, and, for each $\epsilon > 0$, its radial retraction $p_\epsilon : D^q_+ \to \tilde{H}_\epsilon$. For every $\psi \in \mathsf{Diff}_+(D^q, \tilde{H}_0)$, consider the map χ_ψ, given by $\chi_\psi(a) =_{\mathsf{def}} p_{\epsilon(a)} \circ \psi(a)$, where $a \in \bar{D}^q_+$ and $\epsilon(a)$ is the non-negative value such that the hypersurface $\tilde{H}_{\epsilon(a)}$ contains a. If $a \in \tilde{H}_0$, then put $\chi_\psi(a) =_{\mathsf{def}} \psi(a)$. Evidently, each $\chi_\psi \in \mathsf{Diff}_+(D^q, \{\tilde{H}_\epsilon\}_{\epsilon \geq 0})$. The t-family $\{p^t_{\epsilon(a)} \circ \psi(a)\}_{t \in [0,1]}$ delivers the deformation of $\mathsf{Diff}_+(D^q, \tilde{H}_0)$ onto $\mathsf{Diff}_+(D^q, \{\tilde{H}_\epsilon\}_{\epsilon \geq 0})$. Here $\{p^t_\epsilon : D^q_+ \to D^q_+\}$ denotes the family of maps that interpolate radially between p_ϵ and the identity map id.

Therefore, we can reduce the structure group of the fibration $\pi : U \to Y$ to the subgroup $\mathsf{Diff}_+(D^q, \{\tilde{H}_\epsilon\}_{\epsilon \geq 0})$. As a result, the hypersurface $\tilde{H}_1$ (where $\epsilon = 1$) in each π-fiber gives rise to a globally defined smooth hypersurface $\mathcal{H} \subset U$. Now, we resolve $B(N)$ into

$$B^\dagger(N^\dagger) =_{\mathsf{def}} \big[B(N) \setminus (B(N) \cap U)\big] \cup \mathcal{H}$$

with no self-intersections of the multiplicity q. Here $N^\dagger =_{\mathsf{def}} (N \setminus B^{-1}(U)) \cup_B \mathcal{H}$.

Critically, for this resolution the trivialization of the normal 1-bundle $\nu^B|_{[B(N) \setminus (B(N) \cap U)]}$ extends over $B^\dagger(N^\dagger)$: just consider the gradient of the function $x_1 \cdot \ldots \cdot x_q$.

Finally, the basis of induction by q is the case "$q = n + 2$". For this case, Y is a finite set, and the resolution $B^\dagger$ of B works instantly.

Let us validate the second claim of the lemma by using an induction by q, whose base is now $q = n + 1$. First, we observe that any immersion $\beta : (M, \partial M) \to (Z, \partial Z)$ is cobordant in $\mathsf{IMM}^{\mathsf{fr}}_{(-1)}(Z, \partial Z)$ to an immersion $\tilde{\beta} : (M, \partial M) \to (Z, \partial Z)$ which is k-normal for all $k \leq n + 1$. Indeed, such

k-normal immersions are dense in C^∞-topology [Th]. Any two C^∞-close immersions are pseudo-isotopic and thus cobordant in $\mathsf{IMM}^{\mathsf{fr}}_{(-1)}(Z, \partial Z)$, the cobordism having a trivial normal bundle in $Z \times [0,1]$.

Now the resolution map $\mathsf{R} : \mathsf{IMM}^{\mathsf{fr}}_{(-1)}(Z, \partial Z) \to \mathsf{EMB}^{\mathsf{fr}}_{(-1)}(Z, \partial Z)$ is defined by applying the previous resolution recipe "$B^\dagger$" to $\tilde{\beta}$ (and not to the cobordism B, as we just did). Of course, the resolution $\beta^\dagger = \mathsf{R}(\beta)$ changes the topology of the manifold M. We denote by $M^\dagger$ the resulting manifold.

Note that the $(n+1)$-chain $K \subset Z$, locally defined as $\bigcup_{\epsilon \in [0,1]} \tilde{H}_\epsilon$, has the boundary $-\tilde{\beta}(M) \cup \beta^\dagger(M^\dagger)$, which validates the equality $[\beta(M)] = [\beta^\dagger(M^\dagger)] \in H_n(Z, \partial Z; \mathbb{Z})$.

In the absolute case, similar claims for the injectivity of $\mathsf{A} : \mathsf{EMB}^{\mathsf{fr}}_{(-1)}(Z) \to \mathsf{IMM}^{\mathsf{fr}}_{(-1)}(Z)$ and its left inverse $\mathsf{R} : \mathsf{IMM}^{\mathsf{fr}}_{(-1)}(Z) \to \mathsf{EMB}^{\mathsf{fr}}_{(-1)}(Z)$ can be validated in exactly the same way. □

Let g be a Riemannian metric on Z, and M a closed manifold. For any immersion $\beta : M \to Z$, we denote by $\beta^*(g)$ the pull-back to M of the Riemannian metric g on Z. Let $vol(\beta, g)$ be the volume of M in the Riemannian metric $\beta^*(g)$.

Corollary 7.17. *Let M be a closed n-manifold, and let g be a Riemannian metric on Z. If an immersion $\beta : M \to Z$ with a trivialized normal line bundle ν^β minimizes the volume $vol(\beta, g)$ in the homology class $[\beta(M)] \in H_n(Z; \mathbb{Z})$, then β is a regular embedding.*

Proof. Again, we argue by an induction in the multiplicity q, the base of induction is $q = n+1$. We assume that β already has no self-intersections of multiplicities $> q$ and consider the locus Y_q where exactly q branches of $\beta(M)$ intersect transversally. As before, we pick a sufficiently narrow tubular neighborhood U of Y_q in Z in which the resolution $\beta^\dagger$ of the minimal β is supported.

We claim that $vol(\beta, g) > vol(\beta^\dagger, g)$ for a sufficiently narrow U. Indeed, outside U, $\beta(M) = \beta^\dagger(M^\dagger)$; thus it is sufficient to compare the g-induced volumes of $\beta(M) \cap U$ and $\beta^\dagger(M^\dagger) \cap U$. We may assume that U has a $(n+1-q)$-disk bundle structure over a q-manifold Y, where Y is the locus of $(n+1-q)$-fold self-intersection of $\beta(M)$, and the disk fibers $\{U_y\}$ of the bundle $U \to Y$ are orthogonal to Y in the metric g. By the Fubini Theorem,

$$vol_g(\beta(M) \cap U) = \int_Y vol_g(\beta(M) \cap U_y)\, dg|_Y, \text{ and}$$

$$vol_g(\beta^\dagger(M^\dagger) \cap U) = \int_Y vol_g(\beta^\dagger(M^\dagger) \cap U_y)\, dg|_Y.$$

However, for each $y \in Y$, (thanks to the hyperbolic-like geometry of the resolution $\beta^\dagger(M^\dagger) \cap U_y$ and the near-euclidean geometry of the fibers U_y, depicted in Fig. 7.5, A) the gradient flow of the product function $\phi^*(x_1 \cdot \ldots \cdot x_q)$, in the positive chambers, decreases the n-volume of compact domains in the constant level loci $\tilde{H}_\epsilon$ of $\phi^*(x_1 \cdot \ldots \cdot x_q)$. Therefore,

$$vol_g(\beta(M) \cap U_y) > vol_g(\beta^\dagger(M^\dagger) \cap U_y).$$

By Proposition 7.17, $[\beta(M)] = [\beta^\dagger(M^\dagger)]$. Thus, the inequality $vol(\beta, g) > vol(\beta^\dagger, g)$ contradicts the hypotheses that β is the volume-minimizing immersion in its homology class $[\beta(M)] \in H_n(Z;\mathbb{Z})$. Therefore, by induction, the the volume-minimizing β does not have self-intersections. So, it must be an embedding. □

Remark 7.9. It is well-known that any integral homology class $h \in H_n(Z;\mathbb{Z})$ is realizable as the image of the fundamental class $[M]$ under an embedding $\beta : M \hookrightarrow Z$ of some closed smooth n-manifold M (see the proof of Lemma 7.10). (In general, when $\dim Z - \dim M \geq 2$, even not any $\mathbb{Z}_2$-homology class has a realization by an immersion $\beta : M \to Z$!)

Definition 7.15. We call a Riemannian metric g on a compact $(n+1)$-manifold Z n-ample, if there is a $\mathbb{Z}$-basis in $H_n(Z;\mathbb{Z})$ whose elements $\{h_i\}$ are realizable by minimal hypersurfaces with trivial normal line bundles, i.e., by *embeddings* $\{\beta_i : M_i \hookrightarrow Z\}$ with trivial normal line bundles $\{\nu^{\beta_i}\}$ and with the *minimal n-volume* $vol(\beta_i, g)$ among all hypersurfaces $H_i \subset Z$ in the homology class of h_i.

◇

For example, the flat metric on a torus T^{n+1} is n-ample, and so is the hyperbolic metric on any closed orientable surface of genus ≥ 2.

For a n-ample metric g, Corollary 7.17 implies that $vol(\beta, g)$ gives rise to a norm $\|h\|_g$ on the space $H_n(Z;\mathbb{R})$, similar to the Thurston semi-norm for 3-folds [Thu].

Note that, for an n-ample g, it is sufficient to define the semi-norm

$$\|h\|_g =_{\mathsf{def}} \inf_{\{\text{immersions } \beta:\, M \hookrightarrow Z,\ \ \beta([M]) = h\}} \{vol(\beta, g)\}$$

on the integral lattice $H_n(Z;\mathbb{Z}) \subset H_n(Z;\mathbb{R})$ so that the triangle inequality $\|h_1 + h_2\|_g \leq \|h_1\|_g + \|h_2\|_g$ holds for all vectors h_1, h_2 of the lattice.

Let us validate the triangle inequality on the lattice $H_n(Z;\mathbb{Z})$. Take any pair of n-volume minimizing immersions $\beta_1 : M_1 \hookrightarrow Z$ and $\beta_2 : M_2 \hookrightarrow Z$, such that $\beta_1([M_1]) = h_1$ and $\beta_2([M_2]) = h_2$. By Corollary 7.17, we may assume that β_1, β_2 are embeddings. Then the immersion $\beta_1 \sqcup \beta_2 : M_1 \sqcup M_2 \to Z$ realizes the homology class $h_1 + h_2$. By the resolution construction R of Corollary 7.17, this immersion has the property

$$\begin{aligned}\|h_1 + h_2\|_g &\leq vol(\mathsf{R}(\beta_1 \sqcup \beta_2), g) \leq vol(\beta_1 \sqcup \beta_2, g)\\ &= vol(\beta_1, g) + vol(\beta_2, g) = \|h_1\|_g + \|h_2\|_g.\end{aligned}$$

$\diamondsuit$

Question 7.1. *For a n-ample metric g on a compact manifold Z, is the unit ball*

$$\{h \in H_n(Z;\mathbb{R}) |\ \|h\|_g \leq 1\}$$

a convex polyhedron*?* $\diamondsuit$

Recall that, by Theorem 7.7, the sets $\mathsf{IMM}^{\mathsf{fr}}_{(-1)}(Z, \partial Z)$ and $\mathsf{IMM}^{\mathsf{fr}}_{(-1)}(Z)$ depend only on the homotopy type of Z and of $(Z, \partial Z)$, respectively.

In the case of $\mathsf{EMB}^{\mathsf{fr}}_{(-1)}(Z)$ or $\mathsf{EMB}^{\mathsf{fr}}_{(-1)}(Z, \partial Z)$, their descriptions reduce simply to the Thom construction, as the next standard lemma testifies.

Lemma 7.10. *There are bijective maps $\mathcal{G} : \mathsf{EMB}^{\mathsf{fr}}_{(-1)}(Z) \stackrel{\approx}{\to} H^1(Z, \partial Z; \mathbb{Z})$ and $\mathcal{G} : \mathsf{EMB}^{\mathsf{fr}}_{(-1)}(Z, \partial Z) \stackrel{\approx}{\to} H^1(Z; \mathbb{Z})$. As a result, $\mathsf{EMB}^{\mathsf{fr}}_{(-1)}(Z)$ depends only on the homology type of $(Z, \partial Z)$, and $\mathsf{EMB}^{\mathsf{fr}}_{(-1)}(Z, \partial Z)$ depends only on the homology type of Z.*

Proof. Let M be a closed manifold. If $\beta : M \to Z$ is a regular embedding whose normal bundle ν^β is trivialized. Then, applying the Thom construction to ν^β, we get a map

$$G(\beta) : (Z/\partial Z) \to Th(\nu^\beta) \to Th(\mathbf{1}) \approx S^1.$$

Let $\star$ be the base point in S^1. Note that $G(\beta)(\partial Z) = \star$. Using Definition 7.14, we conclude that cobordant regular embeddings β_0 and β_1 lead to the homotopic maps $G(\beta_0)$ and $G(\beta_1)$. Therefore, the map

$$\mathcal{G} : \mathsf{EMB}^{\mathsf{fr}}_{(-1)}(Z) \to [(Z, \partial Z), (S^1, \star)],$$

induced by $G(\beta)$, is well-defined.

On the other hand, if a map $G : (Z, \partial Z) \to (S^1, \star)$ is transversal to a point $a \in S^1$, $a \neq \star$, then the preimage $G^{-1}(a)$ (a closed submanifold) delivers a regular embedding $\beta_G : G^{-1}(a) \subset Z$ so that $G(\beta_G) = G$. Thus, $\mathcal{G}$ is surjective.

If $G(\beta) : (Z, \partial Z) \to (S^1, \star)$ is nill-homotopic via a homotopy

$$H : (Z \times [0,1], \partial Z \times [0,1]) \to (S^1, \star),$$

transversal to a, then $(G(\beta))^{-1}(a) \subset Z$ is cobordant to the empty embedding via the cobordism $H^{-1}(a)$.

A similar argument, being applied to a regular embedding $\beta : (M, \partial M) \to (Z, \partial Z)$, produces a map

$$G(\beta) : Z \to Th(\nu^\beta) \to Th(\mathbf{1}) \approx S^1$$

and, therefore, an element of $H^1(Z; \mathbb{Z})$, which depends only on the bordism class of β.

The fact that $\mathcal{G} : \mathsf{EMB}^{\mathsf{fr}}_{(-1)}(Z, \partial Z) \to H^1(Z; \mathbb{Z})$ is a bijection has a similar validation. □

Remark 7.10. Note that flipping the orientation of the normal bundle ν^β leads to the identity $G(\beta) = -G(\bar{\beta})$ in $H^1(Z, \partial Z; \mathbb{Z})$. Thus, when $H^1(Z, \partial Z; \mathbb{Z}) = 0$ or $H^1(Z, \partial Z; \mathbb{Z}) = \mathbb{Z}_2$, inverting a "hairy n-cycle" $\beta(M) \subset Z$ (viewed as an element of $\mathsf{EMB}^{\mathsf{fr}}_{(-1)}(Z)$) inside $Z \times [0,1]$ is easy.

Of course, to invert β, via a path in the space of immersions, is much harder [S], [Hi]. The possibility of such an inversion "$\beta = \bar{\beta}$" for the standard embedding $\beta : S^2 \to D^3$ of the hairy sphere follows from the striking result of Smale [S]. Its visual realization is very intricate indeed. ◊

Proposition 7.17 leads directly to Corollary 7.18 below (c.f. Corollary 7.1). The latter corollary uses the notations from (7.37). In particular, $\mathbf{B}_j(\sim)$ stands for the j-dimensional smooth bordisms and $\mathbf{OB}_j(\sim)$ for the j-dimensional oriented smooth bordisms of the relevant space [St]. Recall that if Z is oriented, then each M acquires an orientation with the help of the oriented line bundle ν^β.

Corollary 7.18. *The relative bordism classes of the k-fold normal self-intersection manifolds $\{[\beta \circ p_1 : \Sigma^\beta_k \to (Z, \partial Z)]\}_{2 \leq k \leq n+1}$ of immersions $\beta : (M, \partial M) \to (Z, \partial Z)$ with trivialized normal bundle ν^β give rise to a map*

$$\mathcal{B}\Sigma : \mathsf{IMM}^{\mathsf{fr}}_{(-1)}(Z, \partial Z) \longrightarrow \bigoplus_{k \in [2, n+1]} \mathbf{B}_{n+1-k}(Z, \partial Z) \tag{7.49}$$

whose "kernel" $\mathcal{B}\Sigma^{-1}(\mathbf{0})$ contains $\mathsf{EMB}^{\mathsf{fr}}_{(-1)}(Z, \partial Z) \approx H^1(Z; \mathbb{Z})$.

The bordism classes of the k-fold normal self-intersection manifolds

$$\{[\beta \circ p_1 : \Sigma_k^\beta \to Z]\}_{2 \leq k \leq n+1}$$

of immersions $\beta : M \to Z$ (M being closed) with trivial normal bundle ν^β give rise to a map

$$\mathcal{B}\Sigma : \; \mathsf{IMM}^{\mathsf{fr}}_{(-1)}(Z) \longrightarrow \bigoplus_{k \in [2, n+1]} \mathbf{B}_{n+1-k}(Z) \tag{7.50}$$

whose "kernel" $\mathcal{B}\Sigma^{-1}(\mathbf{0})$ contains $\mathsf{EMB}^{\mathsf{fr}}_{(-1)}(Z) \approx H^1(Z, \partial Z; \mathbb{Z})$.

If Z is oriented, then the target of $\mathcal{B}\Sigma$ is $\bigoplus_{k \in [2,n+1]} \mathbf{OB}_{n+1-k}(Z, \partial Z)$, the relative oriented bordisms, or $\bigoplus_{k \in [2,n+1]} \mathbf{OB}_{n+1-k}(Z)$, the oriented bordisms. ♢

Recall the familiar example of an invariant

$$\rho(\beta) =_{\mathsf{def}} \left[(\beta \circ p_1) : \Sigma_{n+1}^\beta \to Z\right],$$

delivered by $\mathcal{B}\Sigma$ from (7.50), which does discriminate between the quasitopies of embeddings and immersions. By the definition of Σ_{n+1}^β, the number of points in Σ_{n+1}^β is divisible by $n+1$. Let $\rho_\beta = \frac{1}{n+1}\#(\Sigma_{n+1}^\beta)$ be the number of points in Z where exactly $n+1$ branches of $\beta(M)$ meet transversally. If Z is oriented, then $\rho_\beta \in \mathbb{Z}$; otherwise, $\rho_\beta \in \mathbb{Z}_2$.

Lemma 7.11. *Let $\dim M = n$ and let ρ_β be the number of points in Z where exactly $n+1$ branches of $\beta(M)$ meet transversally. If $\rho_\beta \equiv 1 \mod 2$, then $\beta \in \mathsf{IMM}^{\mathsf{fr}}_{(-1)}(Z)/\mathsf{A}(\mathsf{EMB}^{\mathsf{fr}}_{(-1)}(Z))$ is nontrivial.*

Proof. We recycle the argument in the proof of Lemma 7.9. □

Example 7.8. Let Z be the $3D$-torus or the $3D$-torus from which an open $3D$-ball B is deleted. Let $M_\star = T_1^2 \coprod T_2^2 \coprod T_3^2$ be the disjoint union of three $2D$-tori, and let $\beta_\star : M_\star \to Z$ be the immersion of $M_\star$, whose image is the triple of coordinate subtori in Z which miss B and have a single triple self-intersection. Thus, $\rho_{\beta_\star} = 1$.

Since the torus Z is oriented and $H^1(Z, \partial Z; \mathbb{Z}) \approx \mathbb{Z}^3$, by Corollary 7.18, we get a sequence of maps:

$$0 \to \mathbb{Z}^3 \xrightarrow{\mathsf{A}} \mathsf{IMM}^{\mathsf{fr}}_{(-1)}(Z) \xrightarrow{\tilde{\mathcal{B}\Sigma}} \mathbb{Z}^3 \to 0,$$

where A is injective, $\tilde{\mathcal{B}\Sigma}$ is surjective, and their composition is trivial. The map $\tilde{\mathcal{B}\Sigma}$ is a truncation of the map $\mathcal{B}\Sigma$ from (7.50), or rather of its oriented version. For (co)homological reasons, in the target of $\mathcal{B}\Sigma$, the oriented 1-bordism classes $\{\beta \circ p_1 : \Sigma_2^\beta \to Z\}$ of double intersections determine the oriented 0-bordism class of triple intersections $\beta \circ p_1 : \Sigma_3^\beta \to Z$. Therefore, we dropped the 0-bordism component of the map $\mathcal{B}\Sigma$ and used its truncation $\tilde{\mathcal{B}}\Sigma$. This tells us that $\mathcal{B}\Sigma$ is not an onto map. ◇

Given two immersions $\beta_1 : M_1 \to Z_1$ and $\beta_2 : M_2 \to Z_2$, where M_1, M_2 are closed n-manifolds and Z_1 and Z_2 are compact connected $(n+1)$-manifolds with connected boundaries, we can form a boundary connected sum $Z_1 \#_\partial Z_2$. Consider the obvious map $\beta_1 \sqcup \beta_2 : M_1 \sqcup M_2 \to Z_1 \#_\partial Z_2$ that leaves the 1-handle in $Z_1 \#_\partial Z_2$ untouched. This construction gives rise to a coupling

$$\uplus : \mathsf{IMM}^{\mathsf{fr}}_{(-1)}(Z_1) \times \mathsf{IMM}^{\mathsf{fr}}_{(-1)}(Z_2) \to \mathsf{IMM}^{\mathsf{fr}}_{(-1)}(Z_1 \#_\partial Z_2). \tag{7.51}$$

Let D^{n+1} be a smooth $(n+1)$-dimensional ball. Then, for any connected $(n+1)$-manifold Z with a connected boundary, we get the coupling

$$\uplus : \mathsf{IMM}^{\mathsf{fr}}_{(-1)}(D^{n+1}) \times \mathsf{IMM}^{\mathsf{fr}}_{(-1)}(Z) \to \mathsf{IMM}^{\mathsf{fr}}_{(-1)}(Z) \tag{7.52}$$

since $D^{n+1} \#_\partial Z \approx Z$.

Arguing as in the proof of Proposition 7.6, both sets, $\mathsf{IMM}^{\mathsf{fr}}_{(-1)}(D^{n+1})$ and $\mathsf{EMB}^{\mathsf{fr}}_{(-1)}(D^{n+1})$, are groups, where the group operation $\uplus$ is induced by a connected sum operation as in (7.51), but the group $\mathsf{EMB}^{\mathsf{fr}}_{(-1)}(D^{n+1})$ is trivial for $n \geq 1$.

Proposition 7.18. *Let Z be a connected compact smooth $(n+1)$-manifold with a connected boundary.*

- *For $n \geq 1$ and $k \geq 2n+2$, the finite abelian group*

$$\mathsf{IMM}^{\mathsf{fr}}_{(-1)}(D^{n+1}) \approx \pi_{n+k+1}(S^{k+1}) \approx \pi_n^{\mathsf{st}}$$

  acts on the set $\mathsf{IMM}^{\mathsf{fr}}_{(-1)}(Z)$ via the connected sum operation $\uplus$ as in (7.51).
- *For $n \geq 1$, this π_n^{st}-action preserves the fibers of the surjective resolution map*

$$\mathsf{R} : \mathsf{IMM}^{\mathsf{fr}}_{(-1)}(Z) \to \mathsf{EMB}^{\mathsf{fr}}_{(-1)}(Z) \approx H^1(Z, \partial Z; \mathbb{Z}).$$

Proof. Let $n \geq 1$. By Lemma 7.10, $\mathsf{EMB}^{\mathsf{fr}}_{(-1)}(D^{n+1}) = 0$; so the actions of this group on $\mathsf{IMM}^{\mathsf{fr}}_{(-1)}(Z)$ are trivial as well.

By the arguments as in the proof of Proposition 7.6, the group $\mathsf{IMM}^{\mathsf{fr}}_{(-1)}(D^{n+1})$ is abelian. By Corollary 7.15, $\mathsf{IMM}^{\mathsf{fr}}_{(-1)}(D^{n+1}) \approx \pi_{n+k+1}(S^{k+1}) \approx \pi_n^{\mathsf{st}}$.

Tracing the arguments from the proof of Proposition 7.17, the surjective resolution map $\mathsf{R} : \mathsf{IMM}^{\mathsf{fr}}_{(-1)}(Z) \to \mathsf{EMB}^{\mathsf{fr}}_{(-1)}(Z) \approx H^1(Z, \partial Z; \mathbb{Z})$ has the property $\mathsf{R}(\beta_1 \uplus \beta_2) = \mathsf{R}(\beta_1) \uplus \mathsf{R}(\beta_2)$ for any $\beta_1 \in \mathsf{IMM}^{\mathsf{fr}}_{(-1)}(D^{n+1})$ and $\beta_2 \in \mathsf{IMM}^{\mathsf{fr}}_{(-1)}(Z)$.

Since $\mathsf{R}(\beta_1) \in \mathsf{EMB}^{\mathsf{fr}}_{(-1)}(D^{n+1}) = 0$ for $n \geq 1$, the claim follows. $\square$

7.9 Submersions & embeddings of equidimensional manifolds whose boundaries have constrained tangency patterns to the product 1-foliations

This section forms a bridge between our results from Chapter 7 and the main results from Chapter 8. Its content is based on [K13]. Using the terminology from Chapter 5, we are dealing here with "blobs", not "doodles".

The reader may choose to surf Section 7.9 or to proceed directly to Chapter 8. Since all the results of Section 7.9 are instant derivations of the similar results from the preceding sections of this chapter, we provide just an outline of their validations. Although we will not rely directly on the results of Section 7.9 in Chapter 8, the present section may promote the "right mindset" for the readers.

For a general reference about spaces of submersions see [Ph].

Let Y be a smooth compact n-manifold. Having in mind applications to traversing vector flows, we move away from immersions and embeddings $\beta : M \to \mathbb{R} \times Y$ of n-manifolds M to submersions and regular embeddings $\alpha : X \to \mathbb{R} \times Y$ of compact smooth $(n+1)$-manifolds X with boundary into the equidimensional product $\mathbb{R} \times Y$. When $\dim(X) = \dim(Y) + 1$, the submersions and immersions are the same notion and the restriction of a submersion α to the boundary ∂X is an immersion. Of cause, if $\alpha : X \to \mathbb{R} \times Y$ is an embedding, so is the map $\alpha^\partial = \alpha| : \partial X \to \mathbb{R} \times Y$.

Therefore, many previous constructions and notions, with the help of the correspondence $\alpha \rightsquigarrow \alpha^\partial$, apply instantly to submersions $\alpha : X \to \mathbb{R} \times Y$ such that $\dim X = \dim Y + 1$.

Remark 7.11. The main difference between immersions $\beta : M \to \mathbb{R} \times Y$ and submersions $\alpha : X \to \mathbb{R} \times Y$, where $\dim X = \dim M + 1$, is that not any β is a boundary α^∂ of some α. For example, the figure ∞ in the plane does not bound a submersion of a 2-manifold. See the work of Pappas [Pa] for the comprehensive theory of possible extensions of a given immersion β to a submersion α. Section 7.9 contains some ideas from [Pa]. $\diamondsuit$

Let us introduce the central to this section notion of quasitopy for submersions $\alpha : X \to \mathbb{R} \times Y$, an analogue of Definition 7.7.

We fix a natural number d and consider a closed subposet $\Theta \subset \mathbf{\Omega}$ such that $(\emptyset) \notin \Theta$ (cf. Λ-condition (7.11)). For topological reasons, we will consider only the case $d \equiv 0 \mod 2$.

Let $\mathcal{L}$ be the 1-foliation of $\mathbb{R} \times Y$ by the fibers of the obvious projection $\mathbb{R} \times Y \to Y$, and $\mathcal{L}^\bullet$ be the 1-foliation of $\mathbb{R} \times Y \times [0,1]$ by the fibers of the projection $\mathbb{R} \times Y \times [0,1] \to Y \times [0,1]$.

Let X be a compact smooth $(n+1)$-dimensional manifold with boundary. Consider a smooth map $\alpha : X \to \mathbb{R} \times Y$ with the following properties:

(7.53)

- α is a submersion,
- for each $y \in Y$, the total multiplicity $m_\beta(y)$ of $\alpha(\partial X)$ with respect to the foliation $\mathcal{L}$ is less than or equal to d and $m_\beta(y) \equiv d \mod 2$,
- for each $y \in Y$, the combinatorial tangency pattern $\omega^\alpha(y) \in \mathbf{\Omega}_{\langle d]}$ of $\alpha(\partial X)$ with respect to $\mathcal{L}_y$ does not belong to Θ,
- for each $y \in \partial Y$, $\omega^\alpha(y) = (\emptyset)$.

Note that the normal bundle ν^α to $\alpha(\partial X)$ in $\mathbb{R} \times Y$ is trivial.

Let X_0, X_1 be two compact smooth (oriented) $(n+1)$-dimensional manifolds with boundary. We consider a compact smooth (oriented) $(n+2)$-manifold W with conners $\partial X_0 \coprod \partial X_1$ such that

$$\partial W = (X_0 \coprod X_1) \bigcup_{\{\partial X_0 \coprod \partial X_1\}} \delta W,$$

where δW is a smooth (oriented) cobordism between ∂X_0 and ∂X_1.

As before, let $Z =_{\mathsf{def}} \mathbb{R} \times Y$ and $Z^\partial =_{\mathsf{def}} \mathbb{R} \times \partial Y$.

Let $A : W \to Z \times [0,1]$ be a submersion of equidimensional manifolds. In particular, $A : \delta W \to Z \times [0,1]$, $A|_{\partial X_0} \to \mathbb{R} \times Y \times \{0\}$, and $A|_{\partial X_1} \to \mathbb{R} \times Y \times \{1\}$ are immersions.

The next two definitions lay down the foundation for notions of quasitopy of traversing vector fields, the main subject of Chapter 8 (see Fig. 8.2).

Definition 7.16. Let us fix natural numbers $d' \geq d$, $d' \equiv d \equiv 0 \mod 2$. Consider closed subposets $\Theta' \subset \Theta \subset \mathbf{\Omega}$ such that $(\emptyset) \notin \Theta$.

We say that a two submersions $\alpha_0 : X_0 \to \mathbb{R} \times Y$ and $\alpha_1 : X_1 \to \mathbb{R} \times Y$ are $(d, d'; \mathbf{c}\Theta, \mathbf{c}\Theta')$-quasitopic, if there exists a compact smooth $(n+2)$-manifold W as above and a smooth submersion $A : W \to Z \times [0,1]$ so that:

- $A|_{X_0} = \alpha_0$ and $A|_{X_1} = \alpha_1$;
- for each $z \in Y \times [0,1]$, the total multiplicity $m_A(z)$ of $A(\delta W)$ with respect to the fiber $\mathcal{L}_z^\bullet$ is such that $m_A(z) \leq d'$, $m_A(z) \equiv d \mod 2$, and the combinatorial tangency pattern $\omega^A(z)$ of $A(\delta W)$ with respect to $\mathcal{L}_z^\bullet$ belongs to $\mathbf{c}\Theta'$;
- for each $z \in Y \times (\{0\} \cup \{1\})$, the total multiplicity $m_A(z)$ of $A(\delta W)$ with respect to the fiber $\mathcal{L}_z$ is such that $m_A(z) \leq d$, $m_A(z) \equiv d \mod 2$, and the combinatorial tangency pattern $\omega^A(z)$ of $A(\delta W)$ with respect to $\mathcal{L}$ belongs to $\mathbf{c}\Theta$.
- for each $z \in Z^\partial$, $\omega^A(z) = (\emptyset)$.

We denote by $\overline{\underline{\mathsf{Q}}}^{\mathsf{sub/emb}}_{d,d'}(Y; \mathbf{c}\Theta, \mathbf{c}\Theta')$ the set of quasitopy classes of such submersions/embeddings $\alpha : X \to \mathbb{R} \times Y$. $\diamondsuit$

It is easy to check that the quasitopy of submersions is an equivalence relation.

In Definition 7.7, we have introduced a similar notion of quasitopy for immersions/embeddings $\beta : (M, \partial M) \to (\mathbb{R} \times Y, \mathbb{R} \times \partial Y)$. Recall that there we used the abbreviated notation $\mathsf{Q}^{\mathsf{imm/emb}}_{d,d'}(Y; \mathbf{c}\Theta, \mathbf{c}\Theta') =_{\mathsf{def}} \mathcal{Q}^{\mathsf{imm/emb}}_{d,d'}(Y, \partial Y; \mathbf{c}\Theta, (\emptyset); \mathbf{c}\Theta')$ for the set of equivalence classes of immersions/embeddings β under the quasitopy relation.

As for immersions $\beta : M \to \mathbb{R} \times Y$, for any choice of connected components $\kappa_1 \in \pi_0(\partial Y_1), \kappa_2 \in \pi_0(\partial Y_2)$, the connected sum operation (cf. formula (7.16))

$$\begin{aligned} \uplus : \overline{\underline{\mathsf{Q}}}^{\mathsf{sub/emb}}_{d,d'}(Y_1; \mathbf{c}\Theta, \mathbf{c}\Theta') \times \overline{\underline{\mathsf{Q}}}^{\mathsf{sub/emb}}_{d,d'}(Y_2; \mathbf{c}\Theta, \mathbf{c}\Theta') \to \\ \to \overline{\underline{\mathsf{Q}}}^{\mathsf{sub/emb}}_{d,d'}(Y_1 \#_\partial Y_2; \mathbf{c}\Theta, \mathbf{c}\Theta'), \qquad (7.54) \end{aligned}$$

is well-defined for the quasitopies of submersions (see Fig. 5.2, the lower diagram).

It converts the set $\overline{\underline{\mathsf{Q}}}^{\mathsf{sub/emb}}_{d,d'}(D^n; \mathbf{c}\Theta; \mathbf{c}\Theta')$ into a group $\overline{\underline{\mathsf{H}}}^{\mathsf{sub/emb}}_{d,d'}(n; \mathbf{c}\Theta; \mathbf{c}\Theta')$ (abelian for $n > 1$). This group acts, via the connected sum operation $\uplus$, on the set $\overline{\underline{\mathsf{Q}}}^{\mathsf{sub/emb}}_{d,d'}(Y; \mathbf{c}\Theta; \mathbf{c}\Theta')$, provided that a connected component of ∂Y is chosen.

Following the proof of Proposition 7.7 and Theorem 7.3, we get Theorem 7.8 below, their analogue for submersions. It is crucial that here $d \equiv 0$ mod 2, which implies that any regular embedding $\beta : M \to \mathbb{R} \times Y$, such that all the multiplicities $\{m_\beta(y)\}_{y\in Y}$ are even, bounds a regular embedding $\alpha : X \to \mathbb{R} \times Y$, where $\partial X = M$ and $\alpha|_{\partial X} = \beta$.

For an n-dimensional Y, the next lemma reduces the computation of the quasitopies $\overline{\underline{\mathsf{Q}}}^{\mathsf{emb}}_{d,d'}(Y; \mathbf{c}\Theta; \mathbf{c}\Theta')$, based on regular embeddings $\alpha : X \to \mathbb{R}\times Y$ of $(n+1)$-dimensional manifolds X, to the computation of quasitopies $\mathsf{Q}^{\mathsf{emb}}_{d,d'}(Y; \mathbf{c}\Theta, \mathbf{c}\Theta')$, based on the regular embeddings $\beta : M \to \mathbb{R} \times Y$ of n-dimensional manifolds M.

Lemma 7.12. *For closed posets $\Theta' \subset \Theta \subset \mathbf{\Omega}$ such that $(\emptyset) \notin \Theta$ and $d \leq d'$, $d \equiv d' \equiv 0 \mod 2$, the map*

$$\Delta : \overline{\underline{\mathsf{Q}}}^{\mathsf{emb}}_{d,d'}(Y; \mathbf{c}\Theta; \mathbf{c}\Theta') \to \mathsf{Q}^{\mathsf{emb}}_{d,d'}(Y; \mathbf{c}\Theta; \mathbf{c}\Theta')$$

that takes an embedding $\alpha : X \to \mathbb{R}\times Y$ to the embedding $\alpha^\partial : \partial X \to \mathbb{R}\times Y$ is a bijection.

As a result, the obvious map $\mathcal{A}^\bullet$: $\overline{\underline{\mathsf{Q}}}^{\mathsf{emb}}_{d,d'}(Y; \mathbf{c}\Theta, \mathbf{c}\Theta') \to \overline{\underline{\mathsf{Q}}}^{\mathsf{imm}}_{d,d'}(Y; \mathbf{c}\Theta, \mathbf{c}\Theta')$ is injective. Thus, if two embedding are quasitopic via a submersion, they are quasitopic via an embedding.

Proof. The main step is contained in the proof of Lemma 7.8. Let us describe its flavor. Since, for $d \equiv 0 \mod 2$ and a closed n-manifold M, any $(\partial\mathcal{E}_d)$-$\mathsf{regular}$ embedding $\beta : M \subset \mathbb{R} \times \mathrm{int}(Y)$ bounds a (orientable when Y is orientable) $(n+1)$-manifold $\alpha : X_\beta \subset \mathbb{R} \times Y$, the map Δ is onto. Evidently, the combinatorial tangency types to $\mathcal{L}$ are determined by $\beta(M)$. Thus, every embedding $\alpha \in \overline{\underline{\mathsf{Q}}}^{\mathsf{emb}}_{d,d'}(Y; \mathbf{c}\Theta, \mathbf{c}\Theta')$ produces an element α^∂ in $\mathsf{Q}^{\mathsf{emb}}_{d,d'}(Y; \mathbf{c}\Theta, \mathbf{c}\Theta')$. By the same token, if a $(\partial\mathcal{E}_d)$-regular embedding $\beta : M \subset \mathbb{R} \times \mathrm{int}(Y)$ bounds a $(\partial\mathcal{E}_{d'})$-regular embedding $B : N \subset \mathbb{R} \times Z$ (where $Z =_{\mathsf{def}} Y\times[0, 1]$), whose tangency to $\mathcal{L}^\bullet$ patterns belong to $\mathbf{c}\Theta'$, then $X_\beta \cup_{\partial M} N$ bounds (an orientable when Y is orientable) $(n + 2)$-manifold $W \subset \mathbb{R} \times Z$, provided $d' \equiv 0 \mod 2$. Here $N \subset \mathbb{R} \times Z$ is a compact $(n + 1)$-manifold such that $\partial N = M$ and $B|_M = \beta$. Thus, the map Δ is injective. By the previous argument, it is bijective.

By Proposition 7.9, the map $\mathcal{A} : \mathsf{Q}^{\mathsf{emb}}_{d,d'}(Y; \mathbf{c}\Theta, \mathbf{c}\Theta') \to \mathsf{Q}^{\mathsf{imm}}_{d,d'}(Y; \mathbf{c}\Theta, \mathbf{c}\Theta')$ (see formula (7.34)) is injective. By the argument above, Δ is injective. Chasing the obvious square diagram, formed by the sources and targets of the maps $\mathcal{A}, \mathcal{A}^\bullet$, we conclude that $\mathcal{A}^\bullet$ is injective as well. □

Combining Lemma 7.12 with Theorem 7.3, we get the following theorem.

Theorem 7.8. *We fix natural numbers* $d' \geq d$, $d' \equiv d \equiv 0 \mod 2$, *and closed subposets* $\Theta' \subset \Theta \subset \mathbf{\Omega}_{\langle d]}$ *such that* $(\emptyset) \notin \Theta$. *Let* Y *be a smooth compact* n*-manifold.*

Then any submersion $\alpha : X \to \mathbb{R} \times Y$ *as in (7.53) gives rise (not in a canonical fashion) to a map* $\Psi(\alpha) : (Y, \partial Y) \to (\mathcal{P}_d^{\mathbf{c}\Theta}, \mathcal{P}_d^{(\emptyset)})$. *Moreover,* $(d, d'; \mathbf{c}\Theta, \mathbf{c}\Theta')$*-quasitopic submersions/embeddings* $\alpha_0 : X_0 \to \mathbb{R} \times Y$ *and* $\alpha_1 : X_1 \to \mathbb{R} \times Y$ *produce homotopic maps* $\Psi(\alpha_0)$ *and* $\Psi(\alpha_1)$.

Thus, we get a well-defined map

$$\Psi^{\mathsf{sub/emb}}_{d,d'} : \overline{\underline{\mathsf{Q}}}^{\mathsf{sub/emb}}_{d,d'}(Y; \mathbf{c}\Theta, \mathbf{c}\Theta') \to \big[\big[(Y, \partial Y),\ \epsilon_{d,d'} : (\mathcal{P}_d^{\mathbf{c}\Theta}, \mathcal{P}_d^{(\emptyset)}) \to (\mathcal{P}_{d'}^{\mathbf{c}\Theta'}, \mathcal{P}_{d'}^{(\emptyset)})\big]\big],$$

Conversely, the homotopy class of any continuous map $G : (Y, \partial Y) \to (\mathcal{P}_d^{\mathbf{c}\Theta}, \mathcal{P}_d^{(\emptyset)})$ *is realized by a smooth regular* embedding $\alpha : X \hookrightarrow \mathbb{R} \times Y$ *which has the properties listed in (7.53), that is,* $G = \Psi^{\mathsf{emb}}_{d,d'}(\alpha)$.

Moreover, $\Psi^{\mathsf{emb}}_{d,d'}$ *is a bijection, and* $\Psi^{\mathsf{sub}}_{d,d'}$ *is a surjection, admitting a right inverse.* ◇

See formula (5.16) and Theorem 5.6 for some invariants that distinguish between

$$\overline{\underline{\mathsf{Q}}}^{\mathsf{sub}}_{d,d'}(Y; \mathbf{c}\Theta, \mathbf{c}\Theta) \text{ and } \overline{\underline{\mathsf{Q}}}^{\mathsf{emb}}_{d,d'}(Y; \mathbf{c}\Theta, \mathbf{c}\Theta)$$

when $Y = [0, 1]$ or $Y = S^1$, $d' \geq d \geq 6$ and $d' \equiv d \equiv 0 \mod 2$, and $\Theta = \mathbf{\Omega}^{\max \geq 3}$.

Chapter 8

Spaces of Polynomials with Constrained Divisors as Grassmanians for Traversing Flows and their Convex Envelops

"If you are out to describe the truth, leave elegance to the tailor."

Albert Einstein

8.1 In this chapter...

We continue our study of traversing vector flows v (see Definition 2.1) on smooth compact manifolds X with boundary. The content of Chapter 8 mirrors [K13]. We rely also on [K12], where many ideas which we describe in the chapter are present. Here we apply these ideas to traversing flows. As Chapter 7, this chapter relies heavily on computations from [KSW1], [KSW2].

We deal here with special traversing flows v on compact $(n+1)$-dimensional manifolds X, with the flows that admit a "virtual global section". The existence of such sections allows to establish a transparent correspondence between the universe of immersions into products $\mathbb{R} \times Y$ (the topic of Chapter 7) and the universe of traversing flows with virtual sections. In fact, with this correspondence in place, many results to follow are just clones of the corresponding results from Chapter 7, thus motivating and justifying them. However, we prove also many propositions with no analogues in Chapter 7 (for example, Proposition 8.1, Theorem 8.1, Corollary 8.4, Proposition 8.2, and Theorem 8.8).

Let us now describe the content of this chapter in some detail, but informally and in a manner that does not involve their most general forms, which carry a combinatorial burden.

As before, we are interested in the combinatorial patterns ω that describe how the v-trajectories on X are tangent to the boundary $\partial_1 X =_{\mathsf{def}} \partial X$. We are concerned with the traversing vector flows whose tangency patterns ω do not belong to a given closed poset Θ. Depending on Θ, the very existence of such flows puts strong restrictions on the topology of X (see [K], [K1]-[K10]). For example, in Fig. 1.2 and Fig. 1.3, the forbidden Θ includes the cubic and the double tangencies of $\partial_1 X$ to the family of u-directed vertical lines. In other words, in this figure, the tangency patterns $\omega = (3)$ and $\omega = (22)$ are forbidden.

In order to simplify the investigation of such flows, we "envelop" them into, so called, convex pseudo-envelops $(\hat{X}, \hat{v})$. A convex pseudo-envelop consists of:

(i) a compact smooth $(n+1)$-manifold $\hat{X}$,

(ii) a traversing vector field $\hat{v}$ on it such that the boundary $\partial_1 \hat{X}$ is convex (see Definition 2.3) with respect to the $\hat{v}$-flow,

(iii) a submersion $\alpha : X \to \hat{X}$ such that $\alpha(\partial_1 X)$ avoids a forbidden tangency patterns Θ to the $\hat{v}$-trajectories, and

(iv) the pull-back $\alpha^\dagger(\hat{v})$ of $\hat{v}$, coinciding with the given vector field v (in Fig. 5.2, $\hat{X}$ is the cylinder $\mathbb{R} \times S^1$).

The existence of a convex pseudo-envelop provides the v-flow on X with a "virtual global section", a significantly effective tool. Its existence allows us to transfer the results from Chapter 7 about immersions/embeddings $\{\beta : M \to \mathbb{R} \times Y\}$, where M and Y are compact smooth n-manifolds, to the parallel results about convex envelops of traversing flows.

In the spirit of Chapter 7 and especially of Section 7.8, we organize convex pseudo-envelops $\{\alpha : (X, v) \to (\hat{X}, \hat{v})\}$ with the fixed target $(\hat{X}, \hat{v})$ into the quasitopy equivalence classes $\mathcal{QT}_d(\hat{X}, \hat{v}; \mathbf{c}\Theta)$, where d is a given even natural number. For a typical $\hat{v}$-trajectory $\hat{\gamma} \subset \hat{X}$, d provides an upper bound of the cardinality of the set $\hat{\gamma} \cap \alpha(\partial_1 X)$. Fig. 8.2 may give the reader some impression of the nature of this quasitopy; it depicts a pair of submersions α_0, α_1, linked by a kind of cobordism with a strict combinatorial control of its tangent patterns.

The quasitopies come in two flavors: $\mathcal{QT}_d^{\mathsf{emb}}(\hat{X}, \hat{v}; \mathbf{c}\Theta)$, formed by regular embeddings α, and $\mathcal{QT}_d^{\mathsf{sub}}(\hat{X}, \hat{v}; \mathbf{c}\Theta)$, formed by more general submersions α.

Consider the locus $\partial_1^+ \hat{X}(\hat{v}) \subset \partial_1 \hat{X}$, where $\hat{v}$ points inside of $\hat{X}$, and the locus $\partial_2^- \hat{X}(\hat{v}) \subset \partial_1^+ \hat{X}(\hat{v})$, where $\hat{v}$ is tangent to $\partial_1 \hat{X}$. Then we will construct two classifying maps

$$\begin{aligned}\Phi^{\mathsf{emb}} &: \mathcal{QT}^{\mathsf{emb}}_d(\hat{X}, \hat{v}; \mathbf{c}\Theta) \to [(\partial_1^+\hat{X}(\hat{v}), \partial_2^-\hat{X}(\hat{v})),\, (\mathcal{P}^{\mathbf{c}\Theta}_d, pt)], \\ \Phi^{\mathsf{sub}} &: \mathcal{QT}^{\mathsf{sub}}_d(\hat{X}, \hat{v}; \mathbf{c}\Theta) \to [(\partial_1^+\hat{X}(\hat{v}), \partial_2^-\hat{X}(\hat{v})),\, (\mathcal{P}^{\mathbf{c}\Theta}_d, pt)],\end{aligned} \tag{8.1}$$

whose targets are sets of homotopy classes from the quotient space $\partial_1^+\hat{X}(\hat{v})/\partial_2^-\hat{X}(\hat{v})$ to the universal space $\mathcal{P}^{\mathbf{c}\Theta}_d$, formed by real univariate polynomials of degree d whose real divisors avoid the given poset Θ.

Then we prove (see Theorem 8.2) that Φ^{emb} is a bijection and Φ^{sub} is a split surjection.

For the constant vector field $v^{\|}$ on the standard $(n+1)$-ball D^{n+1}, the sets $\mathcal{QT}^{\mathsf{sub}/\mathsf{emb}}_d(D^{n+1}, v^{\|}; \mathbf{c}\Theta)$ are groups; the group operation is induced by the boundary connected sum of pseudo-envelops/envelops. These groups are abelian for $n > 1$. The group $\mathcal{QT}^{\mathsf{sub}}_d(D^{n+1}, v^{\|}; \mathbf{c}\Theta)$ is a split extension of $\mathcal{QT}^{\mathsf{emb}}_d(D^{n+1}, v^{\|}; \mathbf{c}\Theta)$ (see Theorem 8.4). We prove that

$$\mathcal{QT}^{\mathsf{emb}}_d(D^{n+1}, v^{\|}; \mathbf{c}\Theta) \approx \pi_n(\mathcal{P}^{\mathbf{c}\Theta}_d, pt).$$

In Section 8.3, we will describe an interesting relation between the connected components of the of the space of convex traversing flows on the ball D^{n+1} and the monoid $\mathbf{HS}_{n-1}$ of smooth types of integral homology $(n-1)$-spheres, being considered up to connected sums with smooth homotopy $(n-1)$-spheres (see Proposition 8.1).

The groups $\mathcal{QT}^{\mathsf{sub}/\mathsf{emb}}_d(D^{n+1}, v^{\|}; \mathbf{c}\Theta)$ act on the quasitopies of convex pseudo-envelops $\mathcal{QT}^{\mathsf{sub}/\mathsf{emb}}_d(\hat{X}, \hat{v}; \mathbf{c}\Theta)$, and the classifying maps $\Phi^{\mathsf{sub}/\mathsf{emb}}$ in (8.1) are equivariant.

These classifying maps deliver a variety of (co)homotopy and (co)homology invariants of traversing flows which admit convex pseudo-envelops (Theorem 8.7, Proposition 8.6, and Theorem 8.8).

We also will establish several results (Theorem 8.3, Corollary 8.6) about the stabilization of the sets $\mathcal{QT}^{\mathsf{emb}}_d(\hat{X}, \hat{v}; \mathbf{c}\Theta)$ as $d \to \infty$.

Finally, we compute $\mathcal{QT}^{\mathsf{emb}}_d(\hat{X}, \hat{v}; \mathbf{c}\Theta)$ for many special cases of posets Θ and convex envelops $(\hat{X}, \hat{v})$.

8.2 Enveloped locally generic and traversally generic flows

For enveloped vector fields, we are going to introduce several notions that go hand in hand with the notions of traversing, boundary generic, and traversally generic vector fields, presented in Section 2.2. While in Section 2.2 these notions were defined *intrinsically*, here they will be defined *extrinsically*, via the properties of the ambient convex envelopes. Moreover, in this new context, the boundary $\partial_1 X$ is mapped by an immersion (not necessarily an imbedding) into the ambient manifold $\hat{X}$ that carries a convex flow.

Let v be a traversing and boundary generic (see Definition 2.2 and Definition 8.3 below) vector field on a compact smooth $(n+1)$-manifold X with boundary. Recall that every trajectory γ of such vector field v generates its tangency divisor D_γ, an ordered sequence of points in γ, together with their tangential multiplicities.

We aim to "go around" the fundamental discontinuity of the map $x \to \gamma_x \to D_{\gamma_x}$, where γ_x stands for the v-trajectory through $x \in X$. This requires "to envelop" a given pair (X, v) in a convex envelop/pseudo-envelop $(\hat{X}, \hat{v})$ (see Definition 8.7). The convex pseudo-envelops, when available, will greatly simplify our analysis of traversing flows. In the spirit of Chapter 7, we will apply our results about immersions and submersions (against the background of product 1-foliations) from [K12] to the new environment of convex envelops of traversing flows.

Let us revisit briefly the Morse stratification $\{\partial_j^\pm X(v)\}_{j\in[1,\dim X]}$ for a boundary generic vector field v on a smooth compact $(n+1)$-dimensional manifold X. Recall that this stratification may be described in terms of an auxiliary function $z : \hat{X} \to \mathbb{R}$ that satisfies the three properties, listed in (2.7):

- 0 is a regular value of z,
- $z^{-1}(0) = \partial X$,
- $z^{-1}((-\infty, 0]) = X$.

In terms of z, the locus $\partial_j X =_{\mathsf{def}} \partial_j X(v)$ is defined by the equations:

$$\{z = 0,\ \mathcal{L}_v z = 0,\ \dots,\ \mathcal{L}_v^{(j-1)} z = 0\},$$

where $\mathcal{L}_v^{(k)}$ stands for the k-th iteration of the Lie derivative operator $\mathcal{L}_v$ in the direction of v (see (2.8) and [K2]). The pure stratum $\partial_j X^\circ \subset \partial_j X$ is defined by the additional constraint $\mathcal{L}_v^{(j)} z \neq 0$. The locus $\partial_j X$ is the union of two loci:

(1) $\partial_j^+ X$, defined by the constraint $\mathcal{L}_v^{(j)} z \geq 0$, and

(2) $\partial_j^- X$, defined by the constraint $\mathcal{L}_v^{(j)} z \leq 0$.

The two loci, $\partial_j^+ X$ and $\partial_j^- X$, share a common boundary $\partial_{j+1} X$.

For a boundary generic v, all the strata $\partial_j X$ are smooth $(n+1-j)$-manifolds. The requirement of v being generic with respect to $\partial_1 X$ may be expressed as the property of the differential j-form

$$dz \wedge d(\mathcal{L}_v z) \wedge \ \ldots \ \wedge \ d(\mathcal{L}_v^{(j-1)} z) \tag{8.2}$$

being a nonvanishing section of the bundle $\bigwedge^j T_* X$ along the locus $\partial_j X$ for all $j \in [1, n+1]$.

If v on X is boundary generic, then each point $x \in \partial_1 X$ belongs to a unique minimal stratum $\partial_j X \subset \partial_1 X$ with the maximal $j = j(x) \leq n+1$. In the boundary generic case, at each point $b \in \partial_1 X$, a flag

$$\mathsf{Flag}_b(v) =_{\mathsf{def}} \{T_b(\partial_1 X) = F^n \supset F^{n-1} \supset \ldots \supset F^{n-j(b)+1}\}$$

is generated by the tangent spaces at b to all the Morse strata $\{\partial_j X\}_{j \leq j(b)}$ that contain b.

Let $\hat{v}$ be a traversing vector field on a compact smooth $(n+1)$-dimensional manifold $\hat{X}$. Consider a submersion $\alpha : X \to \mathsf{int}(\hat{X})$, $\dim X = \dim \hat{X}$, such that the self-intersections of $\alpha(\partial_1 X)$ of multiplicities $k \leq \dim(\hat{X})$ are mutually transversal. Let $v = \alpha^\dagger(\hat{v})$ be the transfer of the vector field $\hat{v}$ to X.

For general submersions α, which are not necessarily embeddings, the situation with boundary generic flows is a bit more complex: not only one gets multiple self-intersections $\{\Sigma_k^{\alpha^\partial} \subset \partial_1 X\}_{k \in [2, n+1]}$ of various branches of $\alpha(\partial_1 X)$, but such self-intersections may be tangent to the $\hat{v}$-flow in a variety of ways that produce similar Morse-type stratifications of these loci $\Sigma_k^{\alpha^\partial}$ $(k \geq 2)$, as well. Let us discuss this complication.

Recall that several vector subspaces $\{V_i \subseteq W\}_i$ of a given vector space W are in general position, if the obvious map $W \to \oplus_i (W/V_i)$ is onto. Note that this definition allows for any number of V_i's to coincide with the ambient W.

We associate several flags $\{\mathsf{Flag}_b(v)\}_{b \in \alpha^{-1}(a)}$ (indexed by points $b \in \alpha^{-1}(a)$) with each point $a \in \alpha(\partial_1 X)$. Let α_*^∂ denote the differential of the immersion $\alpha^\partial : \partial_1 X \to \hat{X}$.

Definition 8.1. We say that the flags $\{(\alpha_*^\partial)(\mathsf{Flag}_b(v))\}_{b \in \alpha^{-1}(a)}$ are in general position in the ambient space $T_a \hat{X}$, if the α_*-images of the *minimal* strata $\{F^{n-j(b)+1}(v)\}_{b \in \alpha^{-1}(a)}$ of the flags $\{\mathsf{Flag}_b(v)\}_{b \in \alpha^{-1}(a)}$ are in general position in the tangent space $T_a \hat{X}$. ◇

Example 8.1. Consider Definition 8.1 in the case $n = 2$, depicted in Fig. 8.1.

- If $\#(\alpha^\partial)^{-1}(a) = 1$ (cases (a), (d), (f) in Fig. 8.1), then each flag $(\alpha_*^\partial)\{F^2 \supset F^1 \supset F^0\}$, or $(\alpha_*^\partial)\{F^2 \supset F^1\}$, or $(\alpha_*^\partial)\{F^2\}$ is in general position at a.
- If $\#(\alpha^\partial)^{-1}(a) = 2$ (cases (b), (e) in Fig. 8.1), a pair of flags F, G is in general position, if and only if, it is of the form $(\alpha_*^\partial)\{F^2 \supset F^1\}$ and $(\alpha_*^\partial)\{G^2\}$, or of the form $(\alpha_*^\partial)\{F^2\}$ and $(\alpha_*^\partial)\{G^2\}$.
- If $\#(\alpha^\partial)^{-1}(a) = 3$ (case (c) in Fig. 8.1), a triple of flags $\mathsf{F}, \mathsf{G}, \mathsf{H}$ is in general position only if it is of the form $(\alpha_*^\partial)\{F^2\}$, $(\alpha_*^\partial)\{G^2\}$, $(\alpha_*^\partial)\{H^2\}$. The rest of combinations fail to be generic.

$\diamondsuit$

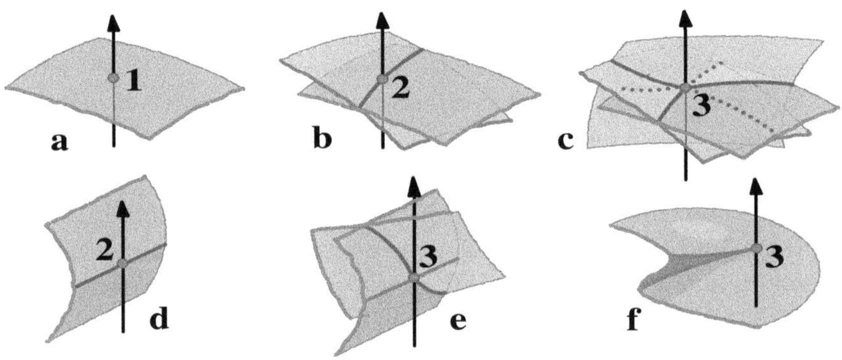

Fig. 8.1 Six locally generic configurations of $\alpha^\partial(\partial_1 X)$ in 3D. The numbers 1, 2, 3 reflect the local multiplicity of the marked point a on the vertical trajectory $\hat{\gamma}_a$.

Definition 8.2. Let $X, \hat{X}$ be smooth compact $(n+1)$-manifolds with boundary and $\hat{v}$ a traversing vector field on $\hat{X}$. We assume that a submersion $\alpha : X \to \hat{X}$ has the following properties: for each point $a \in \alpha(\partial_1 X)$, there exist a natural number $k = k(a) \leq n+1$, an open neighborhood U_a of a in $\hat{X}$, and smooth functions $\{z_1, \dots, z_k : U_a \to \mathbb{R}\}$ such that:

(1) 0 is a regular value for each z_i,
(2) in U_a, the locus $\alpha(\partial_1 X)$ is given by the equation $\{z_1 \cdot \ldots \cdot z_k = 0\}$,
(3) the differential k-form $dz_1 \wedge \ldots \wedge dz_k \,|_{U_a} \neq 0$.[1]

Let $\hat{\gamma}_a$ denote the $\hat{v}$-trajectory through a.

[1]Thus α^∂ is k-normal in the sense of Definition 7.3.

We say that a point $a \in \alpha(\partial_1 X)$ has a multiplicity $j = j(a)$ with respect to $\hat{v}$, if the jet $\mathsf{jet}_a^{j-1}(z_1 \cdot \ldots \cdot z_k\big|_{\hat{\gamma}_a}) = 0$, but $\mathsf{jet}_a^{j}(z_1 \cdot \ldots \cdot z_k\big|_{\hat{\gamma}_a}) \neq 0$. ♢

Definition 8.3. Let $X, \hat{X}$ be smooth compact $(n+1)$-manifolds with boundary and $\hat{v}$ a traversing vector field on $\hat{X}$. Let U be an open $\hat{v}$-flow adjusted neighborhood of a in $\hat{X}$. For each point $a \in \mathsf{int}(\hat{X})$, consider a smooth transversal section S of the $\hat{v}$-flow at a and the flow-generated local projections $\pi : U \to S$.

We say that a submersion $\alpha : X \to \mathsf{int}(\hat{X})$ is locally generic relative to $\hat{v}$ if, for each point $a \in \alpha(\partial_1 X)$,

- the images of the flags[2] $\{\mathsf{Flag}_b(v)\}_{b \in \alpha^{-1}(a)}$, under the differentials $(\alpha^\partial)_*$, are in general position in the tangent space $T_a(\hat{X})$, and
- the images of the flags $\{\mathsf{Flag}_b(v)\}_{b \in \alpha^{-1}(a)}$, under the differentials $(\pi \circ \alpha^\partial)_*$, are in general position in the tangent space $T_a S$. ♢

The next definition, which utilizes the notion of flow convexity, should be compared with Definition 8.5. The latter introduces the more general notion of flow k-convexity.

Definition 8.4. Let $\hat{X}$ be a compact connected smooth manifold with boundary, equipped with a vector field $\hat{v}$. We say that the pair $(\hat{X}, \hat{v})$ is convex if:

- $\hat{v}$ admits a Lyapunov function $\hat{f} : \hat{X} \to \mathbb{R}$ (equivalently, $\hat{v}$ is traversing),
- $\hat{v}$ is boundary generic with respect to $\partial_1 \hat{X}$,
- $\partial_2^+ \hat{X}(\hat{v}) = \emptyset$ (equivalently, $\hat{v}$ is convex as in Definition 2.3). ♢

Remark 8.1. Recall that any compact connected manifold $\hat{X}$ with boundary has a traversing vector field [K1]. By [K1], Lemma 4.1, any traversing vector field $\hat{v}$, admits a Lyapunov function. However, not any compact manifold with boundary admits a traversing *convex* vector field! For example, consider any surface $\hat{X}$, obtained from a closed oriented connected surface, different from the 2-sphere, by removing an open disk. Such an $\hat{X}$ does not admit a traversing convex vector field [K4] (see Lemma 1.1).

If Y is a *closed* manifold, then any vector field $\hat{v} \neq 0$ that is tangent to the fibers of the obvious projection $[0,1] \times Y \to Y$ is evidently convex with respect to the boundary $\partial[0,1] \times Y$. The obvious function $\hat{f} : [0,1] \times Y \to [0,1]$ has the desired Lyapunov property $d\hat{f}(\hat{v}) > 0$. ♢

[2]equivalently, of their minimal strata

Let $\mathsf{conv}(\hat{X})$ denote the space of traversing vector fields $\hat{v}$ on $\hat{X}$ such that $(\hat{X}, \hat{v})$ is a convex pair (as in Definition 8.4). We consider the space $\mathsf{conv}(\hat{X})$ in the C^∞-topology.

Lemma 8.1. *For any pair of vector fields $\hat{v}_0, \hat{v}_1 \in \mathsf{conv}(\hat{X})$ that belong to the same path-connected component of the space $\mathsf{conv}(\hat{X})$, there exists a smooth diffeomorphism $\phi : \hat{X} \to \hat{X}$ such that ϕ maps $\hat{v}_0$-trajectories to $\hat{v}_1$-trajectories, while preserving their orientations.*

If, for a pair $\hat{v}_0, \hat{v}_1 \in \mathsf{conv}(\hat{X})$, there exists a smooth isotopy $\{\psi^t : \partial_1\hat{X} \to \partial_1\hat{X}\}_{t\in[0,1]}$ such that $\psi^1(\partial_1^+\hat{X}(\hat{v}_0)) = \partial_1^+\hat{X}(\hat{v}_1)$, then exists a smooth diffeomorphism $\phi : \hat{X} \to \hat{X}$, an extension of ψ^1, that maps the $\hat{v}_0$-trajectories to the $\hat{v}_1$-trajectories, while preserving their orientations.

Proof. If $\hat{v} \in \mathsf{conv}(\hat{X})$ is a convex boundary generic vector field, then by Theorem 6.6 from [K7], the stratification $\hat{X} \supset \partial_1^+\hat{X}(\hat{v}) \supset \partial_2^-\hat{X}(\hat{v})$ is *stable* under sufficiently small perturbations of $\hat{v}$, up to an isotopy of $\hat{X}$. As a result, within a path connected component of $\hat{v}$ in the space $\mathsf{conv}(\hat{X})$, the smooth topological type of this Morse stratification is stable via an isotopy of $\hat{X}$. In particular, $\partial_1^+\hat{X}(\hat{v}_0)$ and $\partial_1^+\hat{X}(\hat{v}_1)$ are isotopic in $\hat{X}$, provided the $\hat{v}_0$ and $\hat{v}_1$ are connected by a path in $\mathsf{conv}(\hat{X})$.

Let us denote by $\{\hat{\psi}^t : \hat{X} \to \hat{X}\}_{t\in[0,1]}$ the isotopy that transforms the $\hat{v}_0$-induced Morse stratification of $\partial_1\hat{X}$ to the $\hat{v}_1$-induced Morse stratification of $\partial_1\hat{X}$. Let us compare the vector fields $\tilde{v}_0 =_{\mathsf{def}} (\hat{\psi}^1)_*(\hat{v}_0)$ and $\hat{v}_1$. Both vector fields point inside of $\hat{X}$ exactly along $\partial_1^+\hat{X}(\hat{v}_1)$. Since $\hat{v}_0, \hat{v}_1$ are traversing, they admit some Lyapunov functions $f_0, f_1 : \hat{X} \to \mathbb{R}$. Put $\tilde{f}_0 =_{\mathsf{def}} ((\hat{\psi}^1)^{-1})^*(f_0)$. It serves as Lyapunov's function for $\tilde{v}_0$. For $x \in \partial_1^+\hat{X}(\hat{v}_1)$, we denote by $\gamma_x^{\{0\}}$ and $\gamma_x^{\{1\}}$ the $\tilde{v}_0$- and $\hat{v}_1$-trajectories through x. Let $var_0(x)$ stands for the variation of the function $\tilde{f}_0$ along $\gamma_x^{\{0\}}$ and $var_1(x)$ for the variation of the function $\hat{f}_1$ along $\gamma_x^{\{1\}}$.

For each point $y \in \gamma_x^{\{0\}}$, consider the unique point $\tilde{\phi}(y) \in \gamma_x^{\{1\}}$ such that $f_1(\tilde{\phi}(y))/var_1(x) = \tilde{f}_0(y)/var_0(x)$. Now the diffeomorphism $\phi =_{\mathsf{def}} \tilde{\phi} \circ \hat{\psi}^1$ takes $\hat{v}_0$-trajectories to $\hat{v}_0$-trajectories, while preserving their orientations. Hence, we have shown that the isotopy class of $\partial_1^+\hat{X}(\hat{v})$ in $\partial_1\hat{X}$ determines the smooth topological type of a convex pair $(\hat{X}, \mathcal{L}(\hat{v}))$, where $\mathcal{L}(\hat{v})$ denotes the oriented 1-foliation, determined by a convex traversing vector field $\hat{v}$. □

Lemma 8.2. *Let $\hat{X}$ is a connected compact smooth manifold with boundary. If $\#(\pi_0(\partial_1\hat{X})) \geq 3$, then the space $\mathsf{conv}(\hat{X}) = \emptyset$.*

Proof. For a convex $\hat{v}$, $\partial_1^+\hat{X}(\hat{v})$ and $\partial_1^-\hat{X}(\hat{v})$, each is a deformation retract of $\hat{X}$. Therefore, each of these two loci must be connected. Thus each of the loci $\partial_1^+\hat{X}(\hat{v})$, $\partial_1^-\hat{X}(\hat{v})$ must be contained in some connected component of $\partial_1\hat{X}$. When $\#(\pi_0(\partial_1\hat{X})) \geq 3$, this argument forces at least one component of the boundary to be free from both loci. However, the union of the two loci is the entire boundary. This contradiction proves the claim. □

8.3 Homology spheres and convex flows on the standard balls D^{n+1}

The results from this section rely on the pioneering works of Kervaire and Milnor [Ke], [KM] about smooth homotopy spheares. We assume that the reader if familiar with the notions used in these papers.

Let $\mathbf{HS}_{n-1}$ denote the monoid of smooth types of integral homology $(n-1)$-spheres, being considered up to connected sums with smooth homotopy $(n-1)$-spheres.

For $n \neq 4$, let $\mathbf{\Theta}_n$ denote the group of h-cobordism classes of smooth homotopy n-spheres. The group operations in $\mathbf{HS}_{n-1}$ and in $\mathbf{\Theta}_n$ are the connected sums of spheres. We denote by $|\mathbf{\Theta}_n|$ the order of the group $\mathbf{\Theta}_n$.

Proposition 8.1. *Any convex traversing vector field $\hat{v}$ on the standard ball D^{n+1} defines a smooth involution $\tau_{\hat{v}}$ on the sphere $S^n = \partial_1 D^{n+1}$, whose fixed point set $\partial_2^- D^{n+1}(\hat{v})$ is an integral homology $(n-1)$-sphere.*

For $n \geq 6$, any element $[\Sigma]$ of the set $\mathbf{HS}_{n-1}$ arises as the locus $\partial_2^-\hat{X}(\hat{v})$ for a convex traversing vector field $\hat{v}$ on a smooth compact contractible $(n+1)$-dimensional manifold $\hat{X}$, whose boundary $\partial_1\hat{X}$ is a smooth homotopy sphere. Moreover, $\partial_1\hat{X}$ admits a smooth involution $\tau_{\hat{v}}$ such that $(\partial_1\hat{X})^{\tau_{\hat{v}}} = \partial_2^-\hat{X}(\hat{v})$.

For $n \geq 6$, the multiple $|\mathbf{\Theta}_n| \cdot [\Sigma]$ of any given element $[\Sigma] \in \mathbf{HS}_{n-1}$ arises as the locus $\partial_2^- D^{n+1}(\hat{v})$ for a convex traversing vector field $\hat{v}$ on the ball D^{n+1}. The sphere $\partial_1 D^{n+1}$ admits a smooth involution $\tau_{\hat{v}}$ such that

$$(\partial_1 D^{n+1})^{\tau_{\hat{v}}} = \partial_2^- D^{n+1}(\hat{v}) \approx \underbrace{\Sigma \# \ldots \# \Sigma}_{|\mathbf{\Theta}_n|}.$$

Proof. Using a convex $\hat{v}$-flow, $\partial_1^+ D^{n+1}(\hat{v})$ is a deformation retract of D^{n+1} and thus is a contractible manifold with boundary. By the Poincaré duality, $\partial_2^- D^{n+1}(\hat{v})$, the boundary of $\partial_1^+ D^{n+1}(\hat{v})$, is a smooth homology sphere.

By [Ke], Theorem 3, for $n-1 \geq 5$, any smooth homology sphere Σ^{n-1}, after a connected sum $\Sigma^{n-1} \# \Sigma_H^{n-1}$ with a *unique* smooth homotopy $(n-1)$-sphere Σ_H^{n-1}, bounds a contractible smooth manifold W^n.

Consider a smooth metric g on this W^n. We denote by $d_g(x, \partial W^n)$ the smooth distance function to ∂W^n, being restricted to a collar U of ∂W^n in W^n. Let $F : W^n \to \mathbb{R}_+$ be a function that is strictly positive and smooth in the interior of W^n and coincides with the function $\tilde{F}(x) = \sqrt{d_g(x, \partial W^n)}$ in the collar U.

Let us form a smooth manifold $\hat{X}^{n+1} \subset \mathbb{R} \times W^n$, given by the inequality $\{(t, x) : |t| \leq F(x)\}$. It comes equipped with the vector field $\hat{v}$ that is tangent to the fibers of the obvious projection $p : \mathbb{R} \times W^n \to W^n$.

Since W^n is contractible, so is $\hat{X}^{n+1}$. The boundary of $\hat{X}^{n+1}$, the double $D(W^n)$ of W^n (simply-connected for $n > 1$ by the Van Kampen Theorem), using the Poincaré duality, is a smooth homology n-sphere. Therefore, $\partial_1 \hat{X}^{n+1}$ is a smooth homotopy sphere. The vector field $\hat{v}$ on $\hat{X}^{n+1}$ is convex and defines an involution $\tau_{\hat{v}} : \partial_1 \hat{X}^{n+1} \to \partial_1 \hat{X}^{n+1}$, whose fixed point set

$$(\partial_1 \hat{X}^{n+1})^{\tau_{\hat{v}}} = \partial_2^- \hat{X}^{n+1}(\hat{v}) = \partial W = \Sigma^{n-1} \# \Sigma_H^{n-1},$$

a homology sphere. This reasoning validates the second claim.

The connected sum of $|\boldsymbol{\Theta}_n|$ copies of the homotopy sphere $\partial_1 \hat{X}^{n+1}$ is a standard n-sphere. Consider the boundary connected sum $(\hat{Y}^{n+1}, \tilde{v})$ of the $|\boldsymbol{\Theta}_n|$ copies of the pair $(\hat{X}^{n+1}, \hat{v})$. Here the 1-handles $H = D^n \times [0, 1]$ are attached at pairs of points that belong to different copies of ∂W^n's so that $|\boldsymbol{\Theta}_n|$ copies of W^n are connected by the chain of $|\boldsymbol{\Theta}_n|$ many 1-handles $D_+^{n-1} \times [0, 1]$, where $D_+^{n-1} \subset \partial D^n$ is a hemisphere. The fields $\tilde{v}$ in the different copies extend concavely across the 1-handles. Then $\partial_1 \hat{Y}^{n+1}$ is the standard sphere S^n which bounds a contractible manifold $\hat{Y}^{n+1}$. Since $n \geq 6$, by the h-cobordism theorem (see [Mi1]), applied to $\hat{Y}^{n+1} \setminus D^{n+1}$, we conclude that $\hat{Y}^{n+1}$ is the standard ball. Thus we managed to build a convex vector field $v^\#$ on D^{n+1} whose locus $\partial_2^- D^{n+1}(v^\#)$ is a homology sphere $\tilde{\Sigma}^{n-1}$, the $|\boldsymbol{\Theta}_n|$-multiple of the given class $\Sigma^{n-1} \in \mathbf{HS}_{n-1}$. In particular, $\partial_1 D^{n+1}$ admits a smooth involution $\tau_{v^\#}$ whose fixed point set represents $|\boldsymbol{\Theta}_n| \cdot \Sigma^{n-1}$. The third claim has been validated. □

Example 8.2. Consider a free action of the icosahedral group I_{120} on the sphere S^3. Then I_{120} acts freely on $S^7 = \mathsf{join}(S^3, S^3)$. Hence the orbit space $\Sigma^7 =_{\mathsf{def}} S^7/\mathsf{I}_{120}$ is a homology sphere. Recall that $|\boldsymbol{\Theta}_7| = 28$ [KM]. By Proposition 8.1, there is a convex traversing vector field $\hat{v}$ on the ball D^9, such that its locus $\partial_2^- D^9(\hat{v})$ is a connected sum of 28 copies of $[\Sigma^7] \in \mathbf{HS}_7$, and its locus $\partial_1^+ D^9(\hat{v})$ is contractible. Note that $\pi_1(\partial_2^- D^9(\hat{v}))$ is a free product of 28 copies of I_{120}. ◇

Corollary 8.1. *For $n > 6$, $\pi_0(\mathsf{conv}(D^{n+1}))$ admits a surjection onto a subgroup $\mathbf{G}_{n-1}$ of $\mathbf{HS}_{n-1}$ of index $|\mathbf{\Theta}_n|$ at most.*

In particular, here are a few "clean" surjections:

$$\pi_0(\mathsf{conv}(D^7)) \to \mathbf{HS}_5,\ \pi_0(\mathsf{conv}(D^{13})) \to \mathbf{HS}_{11},\ \pi_0(\mathsf{conv}(D^{62})) \to \mathbf{HS}_{60}.$$

Proof. For $n - 1 \geq 5$, let Σ^{n-1} be a given smooth homology sphere, and let Σ_H^{n-1} be the unique homotopy sphere such that $\Sigma^{n-1} \# \Sigma_H^{n-1}$ bounds a smooth contractible manifold [Ke]. By Proposition 8.1, any convex traversing vector field $\hat{v}$ produces a homology sphere $\partial_2^- D^{n+1}(\hat{v})$, and the $|\mathbf{\Theta}_n|$-multiple of any $\Sigma^{n-1} \# \Sigma_H^{n-1}$ is produced this way. On the other hand, by Lemma 8.1, deforming $\hat{v}$ within the space of boundary generic vector fields does not change the smooth isotopy type of the pair $\partial_1^+ D^{n+1}(\hat{v}) \supset \partial_2^+ D^{n+1}(\hat{v})$. In particular, the smooth topological type of this pair is preserved along a path in the space $\mathsf{conv}(D^{n+1})$. Therefore, $\pi_0(\mathsf{conv}(D^{n+1}))$ admits a surjection onto a subgroup $\mathbf{G}_{n-1}$ of $\mathbf{HS}_{n-1}$ of index $|\mathbf{\Theta}_n|$ at most.

The three examples of surjections in the corollary are based on the computations of $|\mathbf{\Theta}_n|$ in [KM], [WX]; we just use some n's for which $|\mathbf{\Theta}_n| = 1$. □

8.4 Convex pseudo-envelops of traversing flows

The next key lemma incapsulates a given convex traversing flow $(\hat{X}, \hat{v})$ into the obvious traversing flow $\tilde{v}$ in a "box" $[0,1] \times Y$ for an appropriate choice of a compact manifold Y, $\dim Y = \dim \partial_1 \hat{X}$. The construction that realizes the embedding $(\hat{X}, \hat{v}) \subset ([0,1] \times Y, \tilde{v})$ delivers a global section $\{0\} \times Y$ of $\hat{v}$ and thus a "global virtual section" of the v-flow on X—the pull-back of $\hat{v}$ under a submertion $\alpha : X \to \hat{X}$. In turn, this section enables us to apply the key Theorem 7.2 to any convex traversing flow $(\hat{X}, \hat{v})$. Therefore, we will be able to transfer many results from Chapter 7 to the environment of convex envelops and pseudo-envelops (see Definition 8.7 below) of traversing boundary generic vector fields.

Lemma 8.3. *Let $\dim(\hat{X}) = n + 1$. If a pair $(\hat{X}, \hat{v})$ is convex, then there exists a compact smooth n-manifold Y such that:*

- $\hat{X} \subset [0,1] \times Y$,

- *$\hat{v}$ is tangent to the fibers of the projection $p : [0,1] \times Y \to Y$,*
- *the obvious function $h : [0,1] \times Y \to [0,1]$ has the property $dh(\hat{v}) > 0$,*
- *with the help of p, the loci $\partial_1^+\hat{X}(\hat{v})$ and $\partial_1^-\hat{X}(\hat{v})$ each is homeomorphic to Y.*

Proof. Since v is a traversing field, by Lemma 2.1, it admits a Lyapunov function $\hat{f} : \hat{X} \to \mathbb{R}$ so that $d\hat{f}(v) > 0$ in $\hat{X}$. We add a collar U to $\hat{X}$ along its boundary $\partial_1\hat{X}$ and denote $\hat{X} \cup_{\partial_1\hat{X}} U$ by $\tilde{X}$. Then, we smoothly extend $\hat{v}$ and $\hat{f}$ in $\tilde{X}$ and denote these extensions by $\tilde{v}$ and $\tilde{f}$. We adjust U so that $\tilde{v} \neq 0$ and $\tilde{f}(\tilde{v}) > 0$ there.

Let $F : \tilde{X} \to \mathbb{R}$ be a smooth function such that 0 is its regular value and $F^{-1}(0) = \partial\hat{X}$, $F^{-1}((-\infty, 0]) = \hat{X}$. Let $\tilde{X}(F,\epsilon) =_{\mathsf{def}} \{x \in \tilde{X} |\; F(x) \leq \epsilon\}$. For a sufficiently small $\epsilon > 0$, $\tilde{X}(F,\epsilon)$ is a smooth compact manifold, contained in $\tilde{X}$.

By definition, $\partial_2\hat{X}(\hat{v}) = \{x \in \partial_1\hat{X} |\; F(x) = 0, \text{ and } (\mathcal{L}_{\tilde{v}}F)(x) = 0\}$. The convexity of $\hat{v}$ in $\hat{X}$ means that $\partial_2^+\hat{X}(\hat{v}) = \emptyset$, thus $\partial_2^-\hat{X}(\hat{v}) = \partial_2\hat{X}(\hat{v})$. By Morin's Theorem [Morin] (see [K2] or [K7] for details), in the vicinity of each point $x \in \partial_2^-\hat{X}(\tilde{v})$ there is a system of smooth coordinates $(u, w, y_1, \dots, y_{n-1})$ in which $F(u, w, y_1, \dots, y_{n-1}) = u^2 + w$, so that $\partial_1\hat{X}$ is given by the equation $u^2 + w = 0$, $\hat{X}$ by the inequality $u^2 + w \leq 0$, and each $\tilde{v}$-trajectory is produced by freezing the coordinates w and $y_1, \dots, y_{n-1}$.

Since $d\tilde{f}(\tilde{v}) > 0$ in $\tilde{X}(F,\epsilon)$, the field $\tilde{v}$ is traversing in $\tilde{X}(F,\epsilon)$. Hence each $\tilde{v}$-trajectory $\gamma \subset \tilde{X}(F,\epsilon)$ is either transversal to $\partial_1\hat{X}$ at a pair of points, or it is simply tangent to $\partial_1\hat{X}$ at a singleton, or does not intersect $\hat{X}$.

Consider the set $Z \subset \tilde{X}(F,\epsilon)$ of $\tilde{v}$-trajectories through the points of $\hat{X}$ (equivalently, the set of $\tilde{v}$-trajectories through $\partial_1^+\hat{X}(\hat{v})$). Every such trajectory γ is a closed oriented segment $[a(\gamma), b(\gamma)]$, where $a(\gamma) \neq b(\gamma) \in \partial Z \subset \partial\tilde{X}(F,\epsilon)$. Thus, the variation $var_\gamma(\tilde{f})$ of the function $\tilde{f}$ along γ is strictly positive. Using compactness of $\hat{X}$, we get that $\inf_{\gamma\subset Z}\{var_\gamma(\tilde{f})\} > 0$.

Let $Y =_{\mathsf{def}} \bigcup_{\gamma\subset Z} a(\gamma) \subset \partial Z$. Using the local model $\{u^2 + w \leq \epsilon\}$, we see that $\tilde{v}$ is transversal to Y for all sufficiently small $\epsilon > 0$.

Next, let us consider a new function on γ:

$$h_\gamma(x) =_{\mathsf{def}} (var_{[a(\gamma),\, x]}\tilde{f})/(var_{[a(\gamma),\, b(\gamma)]}\tilde{f}).$$

The function $h : Z \to \mathbb{R}$, defined as a collection of functions $\{h_\gamma : \gamma \to \mathbb{R}\}_{\gamma\subset Z}$, is evidently a new Lyapunov function for $\tilde{v}$ on Z. It is smooth thanks to the transversality of $\tilde{v}$ to Y and the smooth dependence of solutions of ODEs on initial data. In fact, h gives the product structure $[0,1] \times Y$ to Z. Indeed, any point $x \in Z$ is determined by the unique trajectory γ_x through x and by the value at x of the Lyapunov function h of the $\tilde{v}$-flow.

Finally, the $(-\tilde{v})$-flow defines a smooth map $p : \partial_1^+\hat{X}(\hat{v}) \to Y$ which is a homeomorphism. In fact, $\partial_1^+\hat{X}(\hat{v})$ is diffeomorphic to Y by a small perturbation of p. □

Let $\alpha : X \to \hat{X}$ be a smooth submersion of a compact smooth manifold X with boundary into the interior of a compact connected smooth manifold $\hat{X}$ of the same dimension, where $\partial_1\hat{X} \neq \emptyset$. Assume that $\hat{X}$ is equipped with a traversing vector field $\hat{v}$ so that the pair $(\hat{X}, \hat{v})$ is convex in the sense of Definition 8.4. We denote by $v = \alpha^\dagger(\hat{v})$ the pull-back of $\hat{v}$ under α and by f the pull-back $\alpha^*(\hat{f})$ of the Lyapunov function $\hat{f}$. When α is a regular embedding, to simplify the notations, we identify X and $\alpha(X)$. Under this identification, $v = \hat{v}|_X$ and $f = \hat{f}|_X$.

Definition 8.5. Let $(\hat{X}, \hat{v})$ be a convex pair and let $\alpha : X \to \hat{X}$ be a submersion for which α^∂ is locally generic with respect to $\hat{v}$ in the sense of Definition 8.3.

We say that α is k-convex if $\partial_k^+X(\alpha^\dagger(\hat{v})) = \emptyset$, k-concave if $\partial_k^-X(\alpha^\dagger(\hat{v})) = \emptyset$, and k-flat if $\partial_k X(\alpha^\dagger(\hat{v})) = \emptyset$. ◇

Let an immersion $\alpha^\partial : \partial_1X \to \hat{X}$ be locally generic relative to $\hat{v}$ and k-normal in the sense of Definition 7.3 for all $k \leq n+1$.

For each $\hat{v}$-trajectory $\hat{\gamma} \subset \hat{X}$, we pick a point $a \in \hat{\gamma} \cap \alpha(\partial_1X)$. Such a belongs to the intersection of $k = k(a) \in [1, n+1]$ local branches of $\alpha^\partial(\partial_1X)$, where $k(a) = \#((\alpha^\partial)^{-1}(a))$. In particular, as we remarked on many occasions, a belongs to a unique collection of the α-images of the Morse strata $\{\alpha(\partial_{j(b)}X(v))\}_{b\in\alpha^{-1}(a)}$ with the *maximal possible* indexes $j(b)$ (recall that $v = \alpha^\dagger(\hat{v})$ is the α-transfer of $\hat{v}$). Thanks to α^∂ being locally generic, the tangent spaces of these strata are in general position at a. This setting includes the cases where some or all the strata $\{\alpha(\partial_{j(b)}X(v))\}_{b\in\alpha^{-1}(a)}$ are n-dimensional, i.e., $j(b) = 1$.

$$\text{Let } \; j(a) =_{\mathsf{def}} \sum_{b\in\alpha^{-1}(a)} j(b) \;\geq\; k(a) \quad \text{and}$$

$$\partial_{j(a)}\big(\alpha(X)\big)(\hat{v}) =_{\mathsf{def}} \bigcap_{b\in\alpha^{-1}(a)} \alpha\big(\partial_{j(b)}X(v)\big), \tag{8.3}$$

the latter equation being understood as an identity of the two germs at a of the LHS and RHS loci. By Definition 8.3, the germ of $\partial_{j(a)}(\alpha(X))(\hat{v})$ at a is a smooth submanifold of $\hat{X}$, transversal to the trajectory $\hat{\gamma}_a$.

Let $\mathsf{T}_a \subset T_a\hat{X}$ be the tangent space to the minimal stratum $\partial_{j(a)}(\alpha(X))$ at a, as defined in (8.3). With the help of the $\hat{v}$-flow, the subspace $\mathsf{T}_a \subset T_a\hat{X}$ spreads to form a $(\dim(\hat{X})-j(a))$-dimensional subbundle $\mathsf{T}_a^\bullet$ of the tangent bundle $T\hat{X}\big|_{\hat{\gamma}}$ along the trajectory $\hat{\gamma}$. We denote by $\mathsf{T}_a^\clubsuit$ the image of $\mathsf{T}_a^\bullet$ under the quotient map $T\hat{X}|_{\hat{\gamma}} \to (T\hat{X}|_{\hat{\gamma}})/T\hat{\gamma}$, where $T\hat{\gamma}$ stands for the 1-bundle, tangent to the curve $\hat{\gamma}$. Thus $(T\hat{X}|_{\hat{\gamma}})/T\hat{\gamma}$ serves as the normal bundle to $\hat{\gamma}$.

We introduce a sightly modified version of Definition 3.2 from [K2], a modification that applies to submersions $\alpha : X \to \hat{X}$.

Definition 8.6. Let $\hat{v}$ be a traversing vector field on a compact connected smooth manifold on $\hat{X}$ with boundary. We say that a submersion $\alpha : X \to \mathrm{int}(\hat{X})$ is traversally generic relative to $\hat{v}$, if:

- α^∂ is locally generic in the sense of Definition 8.3 with respect to $\hat{v}$,
- for each $\hat{v}$-trajectory $\hat{\gamma} \subset \hat{X}$, the subbundles $\{\mathsf{T}_a^\clubsuit\}_{a \in \hat{\gamma} \cap \alpha(\partial_1 X)}$ are in general position in the normal to $\hat{\gamma}$ (trivial) n-bundle $(T\hat{X}|_{\hat{\gamma}})/T\hat{\gamma}$. $\diamondsuit$

Example 8.3. The 3-dimensional patterns in Fig. 8.1 may be stacked vertically along a trajectory $\hat{\gamma}$. To get a traversally generic piles in the vicinity of $\hat{\gamma}$, we obey the following rules:

(1) to any stack, we may add any number of configuration of type a from Fig. 8.1, as long as the prescribed parity of $m(\hat{\gamma}) \equiv 0 \mod 2$ is not violated;

(2) no two configurations of multiplicity 3 (of the types c, e, f) reside on $\hat{\gamma}$;

(3) at most two configurations of multiplicity 2 (of the types b, d) reside on $\hat{\gamma}$; moreover, the $(-\hat{v})$-directed projections on the transversal section S to $\hat{\gamma}$ of the fold loci (as in d) or/and of the simple self-intersections (as in b) must be transversal in S. $\diamondsuit$

Let α be a submersion traversally generic relative to $\hat{v}$. Then, for any $\hat{v}$-trajectory $\hat{\gamma}$, by counting the dimensions of the bundles $\{\mathsf{T}_a^\clubsuit\}_{a \in \hat{\gamma} \cap \alpha(\partial_1 X)}$ and using that they are in general position in the n-dimensional bundle, normal to $\hat{\gamma}$, we get that the reduced multiplicity

$$m'(\hat{\gamma}) =_{\mathsf{def}} \sum_{a \in \hat{\gamma} \cap \alpha(\partial_1 X)} (j(a)-1) \leq n. \tag{8.4}$$

For a traversally generic α and a $\hat{v}$-trajectory $\hat{\gamma}$, let $\gamma \subset X$ be any segment in $\alpha^{-1}(\hat{\gamma}) \subset X$ which is bounded by a pair of points in $\partial_1 X$. Then, by Theorem 3.5 from [K2], the total multiplicity $m(\gamma) \leq 2n+2$. Although, for a traversally generic α, there is no α-independent constraint on the total multiplicity

$$m(\hat{\gamma}) =_{\mathsf{def}} \sum_{a \,\in\, \hat{\gamma} \,\cap\, \alpha(\partial_1 X)} j(a)$$

of trajectories $\hat{\gamma}$, there is an universal constraint on the cardinality of the subset $\hat{\gamma}_{\{\geq 2\}} \subset \hat{\gamma} \cap \alpha(\partial_1 X)$ that consists of points a whose multiplicity $m(a) \geq 2$. Namely, $\#(\hat{\gamma}_{\{\geq 2\}}) \leq n$, since no more than n proper vector subspaces may be in general position in an ambient n-dimensional vector space.

Now we are ready to introduce the central notion of a convex pseudo-envelop. Its $2D$-version is familiar from Chapter 1.

Definition 8.7. Let $\hat{v}$ be a traversing vector field on a compact connected smooth manifold $\hat{X}$ with boundary. We assume that $(\hat{X}, \hat{v})$ is convex in the sense of Definition 8.4.

We call such a pair $(\hat{X}, \hat{v})$ a convex pseudo-envelop of a submersion $\alpha : X \to \mathrm{int}(\hat{X})$, if α^∂ is locally generic relative to $\hat{v}$ (in the sense of Definition 8.3).

We think of X as being equipped with the pull-back vector field $v = \alpha^\dagger(\hat{v})$, so that (X, v) is "enveloped" by $(\hat{X}, \hat{v})$.

When $\hat{X}$ is oriented, we may assume that its orientation pull-backs via α to an orientation of X.

If α is a locally generic *regular embedding*, then we call $(\hat{X}, \hat{v})$ a convex envelop of α. ◇

Remark 8.2. Not all traversally generic pairs (X, v) admit convex envelops $(\hat{X}, \hat{v})$. For example, if X has two closed submanifolds (or singular cycles), M and N, of complementary dimensions with a nonzero algebraic intersection number $M \circ N$, then no convex envelop $(\hat{X}, \hat{v})$ of (X, v) exists. Indeed, with the help of the $(-\hat{v})$-flow, M is cobordant in $\hat{X}$ to a cycle M' which resides in $\partial_1^+ \hat{X}(\hat{v})$. Similarly, with the help of the $\hat{v}$-flow, N is cobordant in $\hat{X}$ to a cycle N' which resides in $\partial_1^- \hat{X}(\hat{v})$. Note that if $M \circ N \neq 0$ in X, then evidently the same property holds in any $\hat{X} \supset X$. Since M' and N' are disjoint cycles, their intersection number $M' \circ N' = 0$, which contradicts to the assumption $M \circ N \neq 0$. In particular, the non-triviality of $\mathsf{sign}(X)$, the Wall relative signature of X [Wal], obstructs the existence of a convex envelop for any traversing v on X. As Fig. 5.2 testifies, this argument does not rule out the existence of a convex *pseudo*-envelop for (X, v). In fact, Fig. 5.2 shows that any compact oriented surface X can be pseudo-enveloped. $\diamondsuit$

We denote by $\mathsf{sub}(X, \hat{X}) = \mathsf{imm}(X, \hat{X})$ the space of smooth submersions $\alpha : X \to \mathsf{int}(\hat{X})$.

When α is an embedding, by Theorem 2.1 (or Theorem 3.5 from [K2]), in the space $\mathcal{V}_{\mathsf{trav}}(\hat{X}, \alpha)$ of traversing vector fields $\hat{v}$ on $\hat{X}$, there is an open and dense subset $\mathcal{V}^{\ddagger}(\hat{X}, \alpha)$ such that α is traversally generic (see Definition 8.6) with respect to $\hat{v} \in \mathcal{V}^{\ddagger}(\hat{X}, \alpha)$.

On the other hand, the property of a submersion α to be traversally generic with respect to a *given* traversing vector field $\hat{v}$ (see Definition 8.6) is an "open" property in the C^∞-topology on $\mathsf{sub}(X, \hat{X})$, since it may be expressed in terms of mutual transversality of the relevant strata in the appropriate jet spaces.

We conjecture that the transversal generality of submersions α with respect to a fixed convex pair $(\hat{X}, \hat{v})$ is also a "dense" property. Among other things, the following Theorem 8.1 shows that this conjecture is valid for the regular *embeddings* α into a convex envelop. However, for general submersions α, we are able only to prove that a somewhat *weaker* property "is dense". That property is described in the second claim of Theorem 8.1. Speaking informally, by α-perturbations we can insure the general positions of the singularities of each of the two maps, $\alpha^\partial : \partial_1 X \to \hat{X}$ and $\pi \circ \alpha^\partial : \partial_1 X \to \partial_1^+ \hat{X}(\hat{v})$, but not in relation to each other. Here the map $\pi : \hat{X} \to \partial_1^+ \hat{X}(\hat{v})$ is defined by the $(-\hat{v})$-directed convex flow.

In the proof of Theorem 8.1, we will rely on the following fact.

Lemma 8.4. *Let $\theta : Z \to W$ be a smooth immersion of a compact manifold Z into an open manifold W. Then, for any open neighborhood $\mathcal{U} \subset \mathsf{imm}(Z, W)$ of θ in the C^∞-topology, there exists an open θ-neighborhood $\mathcal{O} \subset \mathcal{U}$ such that any $\theta' \in \mathcal{O}$ is isotopic to θ via a "$\mathcal{U}$-small" isotopy $\{\theta_t : Z \to W\}_{t \in [0,1]} \in \mathcal{U}$.* $\diamondsuit$

Theorem 8.1. *Let $(\hat{X}, \hat{v})$ be a convex pair in the sense of Definition 8.4, and X a compact smooth manifold with boundary, $\dim(X) = \dim(\hat{X}) = n + 1$. Assume that $\mathsf{sub}(X, \hat{X}) \neq \emptyset$. There is an open and dense subset $\mathcal{O} \subset \mathsf{sub}(X, \hat{X})$ such that:*

- *for any $\alpha \in \mathcal{O}$, the local branches of $\alpha(\partial_1 X)$ are in general position in $T_a \hat{X}$ at every point $a \in \alpha(\partial_1 X)$, and*

- *for any $\alpha \in \mathcal{O}$ and each $\hat{v}$-trajectory $\hat{\gamma} \subset \hat{X}$, the $\hat{v}$-invariant subbundles $\{\mathsf{T}^{\clubsuit}_a\}_{a \in \hat{\gamma} \cap \alpha(\partial X)}$, generated by the intersections*
$$\bigcap_{b \in (\alpha^{\partial})^{-1}(a),\; j(b) \geq 2} \alpha_*\big(T_b\big(\partial_{j(b)} X(v)\big)\big)$$
of the tangent spaces to the Morse strata
$$\big\{\partial_{j(b)} X(v)\big\}_{b \in (\alpha^{\partial})^{-1}(a),\; j(b) \geq 2},$$
are in general position in the normal to $\hat{\gamma}$ (trivial) n-bundle $(T\hat{X}|_{\hat{\gamma}})/T\hat{\gamma}$. Here v is the pull-back of $\hat{v}$ under α.

Proof. We assume that the openness of $\mathcal{O}$ in $\mathsf{sub}(X, \hat{X}) = \mathsf{imm}(X, \hat{X})$ is clear, due to general transversality arguments and the compactness of X, and will present the arguments that validate the density of $\mathcal{O}$ in $\mathsf{sub}(X, \hat{X})$.

We divide the proof into three steps: (i), (ii), and (iii).

(i) The set $\mathcal{N} \subset C^\infty(\partial_1 X, \mathsf{int}(\hat{X}))$ of smooth maps $\beta : \partial_1 X \to \mathsf{int}(\hat{X})$ with normal crossings is open and dense (see [GG], Proposition 3.2). Similarly, by [LS], the set of β's that are k-normal immersions (see Definition 7.3) for all $k \leq n+1$ is open and dense.

First we aim to show that, there is an open and dense set $\mathcal{O}_{\mathcal{N}} \subset \mathsf{sub}(X, \mathsf{int}(\hat{X}))$ such that for any $\tilde{\alpha} \in \mathcal{O}_{\mathcal{N}}$, the submersion $\tilde{\alpha}^{\partial} \in \mathcal{N}$. Since the restriction map
$$\mathsf{R}_{\partial_1 X} : \mathsf{imm}(X, \mathsf{int}(\hat{X})) \to \mathsf{imm}(\partial_1 X, \mathsf{int}(\hat{X}))$$
is continuous in the C^∞-topology and the subset $\mathcal{N} \subset C^\infty(\partial_1 X, \mathsf{int}(\hat{X}))$ is open, it follows that $\mathcal{O}_{\mathcal{N}} =_{\mathsf{def}} \mathsf{R}^{-1}_{\partial_1 X}(\mathcal{N})$ is open. Thus it remains to show that this set is also dense.

Let $C \subset X$ be a collar of $\partial_1 X$ inside X, a closed subset equipped with a smooth product structure $\partial_1 X \times [-1, 1]$. We denote by $\partial_1' X =_{\mathsf{def}} \partial_1 X \times \{0\} \subset C$.

Then any immersion (equivalently, submersion) $\alpha : X \to \mathsf{int}(\hat{X})$ restricts to an immersion of the collar C. Thus, we have a map
$$\mathsf{R}_C : \mathsf{imm}(X, \mathsf{int}(\hat{X})) \to \mathsf{imm}(C, \mathsf{int}(\hat{X}))$$
which, according to Hirsh-Smale [H1], is a Serre micro-fibration. The word "micro" means here that the Serre lifting property (see [Hat]) is available for any *sufficiently small* (i.e., localized) path in the base of R_C (see also an excellent survey [Wei] for details). Moreover, using the product structure in C, the further restriction
$$\mathsf{R}_{\partial_1 X} : \mathsf{imm}(C, \mathsf{int}(\hat{X})) \to \mathsf{imm}(\partial_1' X, \mathsf{int}(\hat{X})) \approx \mathsf{imm}(\partial_1 X, \mathsf{int}(\hat{X}))$$

is also a Serre's micro-fibration. The last homeomorphism "$\approx$" of spaces is induced by the diffeomorphism $\epsilon : \partial_1 X \to \partial'_1 X$, defined by the product structure in C.

This implies that any small homotopy of immersions $\{\alpha_t^\partial : \partial'_1 X \to \mathsf{int}(\hat{X})\}_{t\in[0,1]}$, which starts with α^∂, lifts to a homotopy of immersions $\{\alpha_t : X \to \mathsf{int}(\hat{X})\}_{t\in[0,1]}$ which starts with α.

Since the immersions $\beta : \partial_1 X \to \mathsf{int}(\hat{X})$ with normal crossings form a dense and open set in the space $\mathsf{imm}(\partial_1 X, \mathsf{int}(\hat{X}))$ (see [GG], Theorem 5.2, and [LS]), we can find an immersion $\beta \in \mathcal{N}$ that approximates α^∂ arbitrary well in the C^∞-topology. By Lemma 8.4, such approximation β is homotopic to α^∂ via regular immersions $\{\alpha_t^\partial\}_t$, where $\alpha_0^\partial = \alpha^\partial$, $\alpha_1^\partial = \beta$, and the isotopy $\{\alpha_t^\partial\}_t$ is chosen to be small for the Serre micro-fibration $\mathsf{R}_{\partial_1 X}$. By the previous arguments, we can lift α_t^∂ to a small isotopy $\{\alpha_t : X \to \mathsf{int}(\hat{X})\}_{t\in[0,1]}$ which starts with α and terminates with $\beta \in \mathcal{N}$. Therefore, the subset $\mathcal{O}_\mathcal{N} \subset \mathsf{imm}(X, \mathsf{int}(\hat{X}))$ whose boundaries have normal crossings is open and dense.

(ii) For a given smooth map $\rho : \partial_1 X \to \mathsf{int}(\partial_1^+ \hat{X}(\hat{v}))$, let ρ_* denote the differential of ρ. For $\dim(X) = n+1$, consider the locus

$$\mathcal{S}^{(1)}(\rho) =_{\mathsf{def}} \big\{x \in \partial_1 X \big|\ \mathsf{rk}(\rho_*) \leq n-1\big\}.$$

If $S^{(1)}(\rho)$ is a smooth manifold, then the locus

$$\mathcal{S}^{(1,1)}(\rho) =_{\mathsf{def}} \big\{x \in \mathcal{S}^{(1)}(\rho) \big|\ \mathsf{rk}(\rho_*|_{\mathcal{S}^{(1)}(\rho)}) \leq n-2\big\}$$

is well-defined. Continuing this way, the filtration of $\partial_1 X$ by the loci $\big\{\mathcal{S}_{[k]}(\rho) =_{\mathsf{def}} \mathcal{S}^\omega(\rho)\big\}_k$, where $k \leq n+1$ and $\omega = \underbrace{(1,1,\dots,1)}_{k}$, is introduced.

Applying Boardman's Theorems 15.1-15.3, [Bo], the subset $\mathcal{B}$ of maps $\rho \in C^\infty(\partial_1 X, \partial_1^+ \hat{X}(\hat{v}))$, for which all the strata $\{\mathcal{S}_{[k]} =_{\mathsf{def}} \mathcal{S}_{[k]}(\rho)\}$ are smooth manifolds and all the maps

$$\big\{\rho| : \mathcal{S}_{[k]}^\circ =_{\mathsf{def}} \mathcal{S}_{[k]} \setminus \mathcal{S}_{[k+1]} \to \mathsf{int}(\partial_1^+ \hat{X}(\hat{v}))\big\}_k$$

are immersions, is dense and open (since $\partial_1 X$ is compact). Note that each $x \in \partial_1 X$ belongs to a unique pure stratum $\mathcal{S}_{[k(x)]}^\circ$. Moreover, using the Thom Multijet Transversality Theorem (see [GG], Theorem 4.13), the subspace of $\mathcal{B}$, formed maps ρ for which all the spaces $\big\{\rho_*(T_x \mathcal{S}_{[k(x)]})\big\}_{x\in(\pi\circ\rho)^{-1}(y)}$ are in general position in $T_y \partial_1^+ \hat{X}(\hat{v})$ for all $k(x) \geq 2$ and all $y \in \mathsf{int}(\partial_1^+ \hat{X}(\hat{v}))$, form an open and dense subset $\mathcal{B}_{\mathsf{NC}} \subset \mathcal{B}$

(see [GG], Theorem 5.2). Here "NC" abbreviates the condition known as the "normal crossings". Note that if $k(x) = 1$, then $\rho_*(T_x S_{[k(x)]})$ coincides with $T_{\rho(x)}\partial_1^+\hat{X}(\hat{v})$.

We stress that, applying the previous arguments to the map $\rho_\alpha =_{\mathsf{def}} \pi \circ \alpha^\partial$, the strata $\bigcap_{b \in \alpha^{-1}(a)} \alpha\big(\partial_{j(b)}X(v)\big)$ that involve points b's with $j(b) = 1$ become "ρ_α-invisible", thanks to the "erasing" action of π_* on all n-dimensional spaces $\alpha\big(T_a(\partial_1 X(v))\big)$. Because of this short coming, we cannot claim that the traversally generic α to $\hat{v}$ (see Definition 8.6) are dense in $\mathsf{sub}(X, \hat{X})$, although we believe it is.

(iii) Note that the map $\pi : \hat{X} \to \mathsf{int}(\partial_1^+\hat{X}(\hat{v}))$, defined by the $(-\hat{v})$-flow, is smooth due to $\hat{v}$ being convex; moreover, its restriction to $\mathsf{int}(\hat{X})$ is a submersion.

By (i), we can approximate a given submersion $\alpha : X \to \mathsf{int}(\hat{X})$ by a new submersion α_1 such that $\alpha_1^\partial \in \mathcal{N}$.

In turn, by (ii), we can approximate the map $\rho_{\alpha_1} =_{\mathsf{def}} \pi \circ \alpha_1^\partial$ by a smooth map $\rho_1^\dagger : \partial_1 X \to \mathsf{int}(\partial_1^+\hat{X}(\hat{v}))$ such that $\rho_1^\dagger \in \mathcal{B}_{\mathsf{NC}}$.

Let us fix a Lyapunov function $\hat{f} : \hat{X} \to \mathbb{R}$ for $\hat{v}$. In the spirit of Lemma 8.3, the Lyapunov function $\hat{f}$ for $\hat{v}$ and the projection $\pi : \hat{X} \to \partial_1^+\hat{X}(\hat{v})$ define global smooth "coordinates" in the interior of $\hat{X}$. That is, each pair (t, y), where $y \in \partial_1^+\hat{X}(\hat{v})$ and $t \in \hat{f}(\pi^{-1}(y))$, determines a unique point $x \in \hat{X}$ such that $\hat{f}(x) = t$ and $\pi(x) = y$. Let $f_1 =_{\mathsf{def}} \alpha_1^*(\hat{f})$.

For the map $\rho_1^\dagger \in \mathcal{B}_{\mathsf{NC}}$, we define a submersion $\alpha^\dagger_{(\rho_1^\dagger, f_1)} : X \to \hat{X}$ by the formula $\alpha^\dagger_{(\rho_1^\dagger, f_1)}(x) = y$, where $x \in \partial_1 X$ and y is the unique point on the $\hat{v}$-trajectory through $\rho_1^\dagger(x) \in \partial_1^+\hat{X}(\hat{v})$ such that $\hat{f}(y) = f_1(x)$.

By its construction, the submersion $\alpha_2 =_{\mathsf{def}} \alpha^\dagger_{(\rho_1^\dagger, f_1)} : X \to \hat{X}$ is close to the submersion α_1, which approximates the original α. Also, by the construction of α_2, we have $\rho_{\alpha_2} = \rho_1^\dagger \in \mathcal{B}_{\mathsf{NC}}$ and $\alpha_2^\partial \in \mathcal{N}$.

Therefore, we have shown that any given submersion $\alpha \in \mathsf{sub}(X, \hat{X})$ admits an approximation by some $\alpha_2 \in \mathsf{sub}(X, \hat{X})$ such that $\alpha_2^\partial \in \mathcal{N}$ and $\rho_{\alpha_2} \in \mathcal{B}_{\mathsf{NC}}$. These are exactly the two properties that describe the space $\mathcal{O} \subset \mathsf{sub}(X, \hat{X})$ in the theorem. □

Corollary 8.2. *Let $(\hat{X}, \hat{v})$ be a convex pair as in Definition 8.4, and X a compact smooth manifold with boundary,* $\dim(X) = \dim(\hat{X}) = n + 1$.

The regular embeddings *$\alpha : X \subset \hat{X}$ that are traversally generic with respect to $\hat{v}$ (see Definition 8.6) form an open and dense set in the space of all regular smooth embeddings.*

Proof. Since, for a regular embedding $\alpha : X \hookrightarrow \hat{X}$, α^∂ is an embedding, the first claim of Theorem 8.1 is vacuous, and the second claim insures that α is traversally generic. $\square$

Remark 8.3. Consider the k-multiple self-intersection manifolds $\{\Sigma_k^{\alpha^\partial}\}_k$ of k-normal (see [LS]) immersions $\alpha^\partial : \partial_1 X \to \hat{X}$. By definition, $\Sigma_k^{\alpha^\partial}$ is a submanifold of the k-fold product $(\partial_1 X)^k$, the preimage of the diagonal $\Delta \subset (\hat{X})^k$ under the transversal to it map $(\alpha^\partial)^k$. The projection p_1 of $\Sigma_k^{\alpha^\partial}$ on the first factor $\partial_1 X$ of the product $(\partial_1 X)^k$ is an immersion [LS]. By composing p_1 with α^∂, we get an immersion of $\alpha^\partial \circ p_1 : \Sigma_k^{\alpha^\partial} \to \hat{X}$. By using the convex $(-\hat{v})$-flow, we get a map $\pi : \hat{X} \to \partial_1^+ \hat{X}(\hat{v})$. Finally, we obtain a smooth composite map $\pi \circ \alpha^\partial \circ p_1 : \Sigma_k^{\alpha^\partial} \to \partial_1^+ \hat{X}(\hat{v})$.

We notice that, under the hypotheses and notations of Theorem 8.1, if $\hat{v}$ is tangent at a point a to the intersection $\bigcap_{b \in (\alpha^\partial)^{-1}(a)} \alpha_*\big(\partial_{j(b)} X(v)\big)$, then, evidently, each local branch $\alpha_*(\partial_{j(b_\star)} X(v))$, where $b_\star \in (\alpha^\partial)^{-1}(a)$, is tangent to $\hat{v}$ at a. Therefore, $j(b_\star) \geq 2$. In other words, for $k \geq 2$, the singular locus of the map $\pi \circ \alpha^\partial \circ p_1 : \Sigma_k^{\alpha^\partial} \to \partial_1^+ \hat{X}(\hat{v})$ is always *contained* in the singular locus of the map $\pi \circ \alpha^\partial \circ p_1 : (\partial_1 X)^k \to \partial_1^+ \hat{X}(\hat{v})$. $\diamondsuit$

8.5 Quasitopies of convex envelops and pseudo-envelops

Now, let us modify Definition 7.7 so that it applies to convex pseudo-envelops of traversing flows (see Fig. 8.2). This modification is central to our efforts.

Let $\hat{X}$ be a $(n+1)$-dimensional compact smooth manifold and $\hat{v}$ a convex traversing vector field on it. Let $\hat{Z} =_{\mathsf{def}} \hat{X} \times [0,1]$. We denote by $\hat{v}^\bullet$ the vector field on $\hat{Z}$ that is tangent to each slice $\hat{X} \times \{t\}$, $t \in [0,1]$, and is equal to $\hat{v}$ there.

Definition 8.8. Fix natural even numbers $d \leq d'$ and consider closed subposets $\Theta' \subset \Theta$ of the universal poset $\mathbf{\Omega}$ from Section 2, such that $(\emptyset) \notin \Theta$.

We say that a two convex pseudo-envelops, $\alpha_0 : X_0 \to \hat{X}$ and $\alpha_1 : X_1 \to \hat{X}$, are $(d, d'; \mathbf{c}\Theta, \mathbf{c}\Theta')$-$\mathsf{quasitopic}$ in $\hat{X}$, if there exists a compact smooth (orientable when $\hat{X}$ is orientable) $(n+2)$-manifold W (with corners $\partial X_0 \coprod \partial X_1$), whose boundary

$$\partial W = (X_0 \coprod X_1) \bigcup_{\{\partial X_0 \coprod \partial X_1\}} \delta W,$$

and a smooth submersion $A : W \to \hat{Z}$ so that:

- $A|_{X_0} = \alpha_0$ and $A|_{X_1} = \alpha_1$;
- for each $z \in \hat{X} \times \partial[0,1]$, the total multiplicity $m_A(\hat{\gamma}_z)$ of the $\hat{v}^\bullet$-trajectory $\hat{\gamma}_z$ through z, relatively to $A(\partial X_0 \coprod \partial X_1)$, satisfies the constraints

$$\{m_A(\hat{\gamma}_z) \leq d,\ m_A(\hat{\gamma}_z) \equiv 0 \mod 2\},$$

  and the combinatorial tangency pattern $\omega^A(\hat{\gamma}_z)$ of $\hat{\gamma}_z$ with respect to $A(\partial X_0 \coprod \partial X_1)$ does not belong to Θ;
- for each $z \in \hat{Z}$, the total multiplicity $m_A(\hat{\gamma}_z)$ of $\hat{\gamma}_z$ with respect to $A(\delta W)$ satisfies the constraints

$$\{m_A(\hat{\gamma}_z) \leq d',\ m_A(\hat{\gamma}_z) \equiv 0 \mod 2\},$$

  and the combinatorial tangency pattern $\omega^A(\hat{\gamma}_z)$ of $\hat{\gamma}_z$ with respect to $A(\delta W)$ does not belong to Θ';

We denote by $\mathcal{QT}^{\mathsf{sub}}_{d,d'}(\hat{X}, \hat{v};\, \mathbf{c}\Theta; \mathbf{c}\Theta')$ the set of quasitopy classes of such convex pseudo-envelops $\alpha : (X, \alpha^\dagger(\hat{v})) \to (\hat{X}, \hat{v})$.

If we insist that α_0, α_1, and A are regular *embeddings*, then we get $\mathcal{QT}^{\mathsf{emb}}_{d,d'}(\hat{X}, \hat{v};\, \mathbf{c}\Theta; \mathbf{c}\Theta')$, the set of quasitopy classes of convex envelops. $\diamondsuit$

It is easy to check that the quasitopy of convex pseudo-envelops (convex envelops) $\alpha : (X, \alpha^\dagger(\hat{v})) \to (\hat{X}, \hat{v})$ is an equivalence relation.

We are in position to state one of the main results of this chapter.

Theorem 8.2. *Let $\Theta' \subset \Theta \subset \mathbf{\Omega}$ be closed subposets that do not contain the element $(\emptyset)$. Assume that $d' \geq d$ and $d' \equiv d \equiv 0 \mod 2$. Let $\hat{X}$ be a smooth compact connected $(n+1)$-dimensional manifold, equipped with a convex traversing vector field $\hat{v}$.*

Then, under the notations of Definition 7.18, there is a canonical bijection

$$\Phi^{\mathsf{emb}} : \mathcal{QT}^{\mathsf{emb}}_{d,d'}(\hat{X}, \hat{v}; \mathbf{c}\Theta; \mathbf{c}\Theta') \stackrel{\approx}{\longrightarrow}$$
$$\stackrel{\approx}{\longrightarrow} \big[\big[(\partial_1^+ \hat{X}(\hat{v}),\, \partial_2^- \hat{X}(\hat{v})),\ \epsilon_{d,d'} : (\mathcal{P}_d^{\mathbf{c}\Theta}, pt) \to (\mathcal{P}_{d'}^{\mathbf{c}\Theta'}, pt')\big]\big] \tag{8.5}$$

and a canonical surjection

$$\Phi^{\mathsf{sub}} : \mathcal{QT}^{\mathsf{sub}}_{d,d'}(\hat{X}, \hat{v}; \mathbf{c}\Theta; \mathbf{c}\Theta') \stackrel{epi}{\longrightarrow}$$
$$\stackrel{epi}{\longrightarrow} \big[\big[(\partial_1^+ \hat{X}(\hat{v}),\, \partial_2^- \hat{X}(\hat{v})),\ \epsilon_{d,d'} : (\mathcal{P}_d^{\mathbf{c}\Theta}, pt) \to (\mathcal{P}_{d'}^{\mathbf{c}\Theta'}, pt')\big]\big]. \tag{8.6}$$

The map Φ^{sub} admits a right inverse.

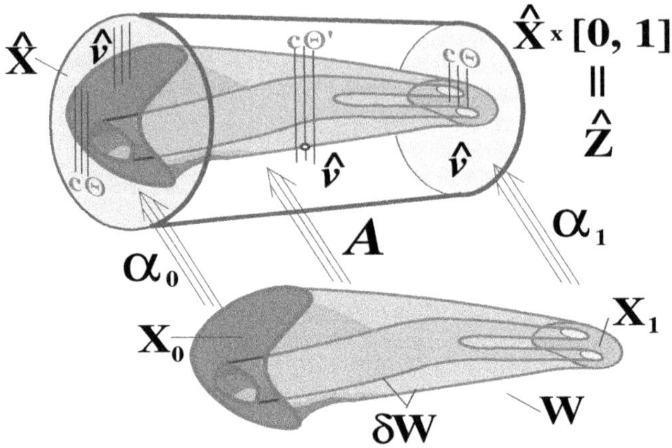

Fig. 8.2 The ingredients of Definition 8.8. For simplicity of the depiction, the maps α_0, α_1, A are shown as embeddings. The vertical lines depict the $\hat{v}^\bullet$-trajectories.

Proof. By Lemma 8.3, any convex pseudo-envelop $\alpha : X \to \hat{X}$ may be incapsulated into a convex pseudo-envelop $\tilde{\alpha} : X \to \hat{X} \subset [0,1] \times Y$ of the product type (we called such a capsule $[0,1] \times Y$ a "box"), where Y is homeomorphic to $\partial_1^+ \hat{X}(\hat{v})$, and its boundary $\partial_1 Y$ to $\partial_2^- \hat{X}(\hat{v})$.

Conversely, any submersion $\tilde{\alpha} : X \to [0,1] \times Y$, where Y is a smooth compact n-manifold, by rounding the corners $\partial[0,1] \times \partial_1 Y$ of $[0,1] \times Y$ produces a convex pseudo-envelop $(\hat{X}, \hat{v}) \subset ([0,1] \times Y, \tilde{v})$, where $\tilde{v} \neq 0$ is tangent to the fibers of the projection $[0,1] \times Y \to Y$. Similarly, the cobordisms A between pairs of quasitopies α_0, α_1 can be incapsulated in boxes of the form $([0,1] \times Y) \times [0,1]$. Therefore, all the results from Chapter 7 apply to $\tilde{\alpha}^\partial$; in particular, the pivotal Theorem 7.3 applies. With its help, the maps Φ^{sub} from (8.6) and Φ^{emb} from (8.5) are generated as compositions of the maps

$$p(\tilde{v}) : (\partial_1^+ \hat{X}(\hat{v}),\, \partial_2^- \hat{X}(\hat{v})) \to (Y, \partial_1 Y)$$

with the maps $\{\Phi^{\tilde{\alpha}^\partial} : (Y, \partial_1 Y) \to (\mathcal{P}_d^{\mathbf{c}\Theta}, \mathcal{P}_d^{(\emptyset)})\}$ from Theorem 7.3; the latter map being generated by the locus $\tilde{\alpha}(\partial_1 X) \subset \mathbb{R} \times Y$. Note that the cell $\mathcal{P}_d^{(\emptyset)} \subset \mathcal{P}_d$ contracts to a singleton $pt \in \mathcal{P}_d^{(\emptyset)}$.

Similarly, for any quasitopy $A : W \to \hat{X} \times [0,1]$ as in Definition 8.8, the product $Y \times [0,1]$ that encapsulates $\hat{X} \times [0,1]$ is mapped to $\mathcal{P}_{d'}^{\mathbf{c}\Theta'}$ with the help of $A(\delta W)$, while $\partial_1 Y \times [0,1]$ are mapped to the cell $\mathcal{P}_{d'}^{(\emptyset)}$, which contracts to the singleton $\epsilon_{d,d'}(pt) \in \mathcal{P}_{d'}^{(\emptyset)}$.

As in Theorem 7.3 and by similar transversality arguments, Φ^{emb} is a 1-to-1 map. Here we need to use Lemma 7.4 to conclude that, under a $\partial\mathcal{E}_d$-regular map (see Definition 7.4), the preimage of the hypersurface $\partial\mathcal{E}_d \subset \mathbb{R} \times \mathcal{P}_d$ (see formula (7.4)) bounds a compact manifold X in $\hat{X}$, provided $d \equiv 0 \mod 2$. Again, as in Proposition 7.9, the bijective map in (8.5) helps to prove that the map in (8.6) is a split surjective one. □

Corollary 8.3. *The sets* $\mathcal{QT}^{\mathsf{emb}}_{d,d'}(\hat{X}, \hat{v}; \mathbf{c}\Theta; \mathbf{c}\Theta')$ *are trivial when* $\dim(\hat{X}) < \max_{\omega\in\Theta}\{|\omega|'\}$.

Proof. Since $\mathrm{codim}(\mathcal{P}_d^{\Theta}, \mathcal{P}_d) = \max_{\omega\in\Theta}\{|\omega|'\}$ and $\mathcal{P}_d$ is contractible, by Theorem 8.2, the claim follows by a general position argument, applied to homotopies of maps from the pair $(\partial_1^+\hat{X}(\hat{v}),\, \partial_2^-\hat{X}(\hat{v}))$ to the pair $(\mathcal{P}_d^{\mathbf{c}\Theta}, pt)$. □

Corollary 8.4. *The sets* $\mathcal{QT}^{\mathsf{emb}}_{d,d'}(\hat{X}, \hat{v}; \mathbf{c}\Theta; \mathbf{c}\Theta')$, $\mathcal{QT}^{\mathsf{sub}}_{d,d'}(\hat{X}, \hat{v}; \mathbf{c}\Theta; \mathbf{c}\Theta')$ *are invariants within the path-connected component of the vector field* $\hat{v}$ *in the space* $\mathsf{conv}(\hat{X})$ *of convex traversing vector fields.*

Proof. By their definitions, the sets $\mathcal{QT}^{\mathsf{emb}}_{d,d'}(\hat{X}, \hat{v}; \mathbf{c}\Theta; \mathbf{c}\Theta')$ and $\mathcal{QT}^{\mathsf{sub}}_{d,d'}(\hat{X}, \hat{v}; \mathbf{c}\Theta; \mathbf{c}\Theta')$ depend only of the smooth topological types of the oriented 1-dimensional foliations $\mathcal{L}(\hat{v})$, and $\mathcal{L}^\bullet(\hat{v}^\bullet)$ on $\hat{X}$ and $\hat{X} \times [0,1]$, respectively. By Lemma 8.1, the smooth topological types of these foliations do not change along any path in the space $\mathsf{conv}(\hat{X})$ that contains the point $\hat{v}$. □

Corollary 8.5. *The set* $\mathcal{QT}^{\mathsf{emb}}_{d,d'}(\hat{X}, \hat{v}; \mathbf{c}\Theta; \mathbf{c}\Theta')$ *depends only on the homotopy type of the pair* $(\partial_1^+\hat{X}(\hat{v}),\, \partial_2^-\hat{X}(\hat{v}))$.

Proof. The claim follows instantly from the bijection in formula (8.5). □

The next theorem contains our main result about stability of the set $\mathcal{QT}^{\mathsf{emb}}_{d,\,d+2}(\hat{X}, \hat{v}; \mathbf{c}\Theta; \mathbf{c}\Theta)$ as a function of d in terms of the function $\psi_\Theta(d+2)$ from formula (6.24).

Theorem 8.3. (short stabilization: $\{\mathbf{d} \Rightarrow \mathbf{d+2}\}$) *Let* Θ *be a closed subposet of* $\mathbf{\Omega}$. *If* $\dim(\hat{X}) < d+2-\psi_\Theta(d+2)$ *and* $|\omega|' > 2$ *for all* $\omega \in \Theta_{\langle d+2]}$, *then there exists a bijection*

$$\Phi^{\mathsf{emb}} : \mathcal{QT}^{\mathsf{emb}}_{d,\,d+2}(\hat{X}, \hat{v}; \mathbf{c}\Theta; \mathbf{c}\Theta) \xrightarrow{\approx} \big[(\partial_1^+\hat{X}(\hat{v}),\, \partial_2^-\hat{X}(\hat{v})),\, (\mathcal{P}_d^{\mathbf{c}\Theta}, pt)\big]$$

Proof. Let $k =_{\mathsf{def}} d+2-\psi_\Theta(d+2)$. If $|\omega|' > 2$ for all $\omega \in \Theta_{\langle d+2]}$, then both $\mathcal{P}_d^{\mathbf{c}\Theta}$ and $\mathcal{P}_{d+2}^{\mathbf{c}\Theta}$ are simply-connected, since $\mathrm{codim}(\bar{\mathcal{P}}_{d+2}^{\Theta}, \bar{\mathcal{P}}_{d+2}) > 2$ and $\mathrm{codim}(\bar{\mathcal{P}}_d^{\Theta}, \bar{\mathcal{P}}_d) > 2$. By Theorem 6.10, we get an isomorphism

$$\epsilon_* : H_j(\mathcal{P}_d^{\mathbf{c}\Theta}; \mathbb{Z}) \approx H_j(\mathcal{P}_{d+2}^{\mathbf{c}\Theta}; \mathbb{Z})$$

for all $j \leq k$. By the Whitehead Theorem (the inverse Hurewicz Theorem) (see Theorem 7.13 in [Wh], page 181, or Theorem 10.1 in [Hu]), the map $\epsilon : \mathcal{P}_d^{\mathbf{c}\Theta} \to \mathcal{P}_{d+2}^{\mathbf{c}\Theta}$ is $(k-1)$-connected. Thus, if $n = \dim(\partial_1^+\hat{X}(\hat{v})) \leq k-1$, then, by the standard application of the obstruction theory, any map

$$\Phi : (\partial_1^+\hat{X}(\hat{v}),\ \partial_2^-\hat{X}(\hat{v})) \to (\mathcal{P}_d^{\mathbf{c}\Theta}, pt)$$

that becomes nill-homotopic in $(\mathcal{P}_{d+2}^{\mathbf{c}\Theta}, pt')$ is nill-homotopic in $(\mathcal{P}_d^{\mathbf{c}\Theta}, pt)$ as well. Now the claim follows from Theorem 8.2. □

Theorem 8.3, being combined with Corollary 6.9, leads to the following important implication.

Corollary 8.6. (long stabilization d $\Rightarrow \infty$) *Let Θ be a closed profinite (see Definition 6.5) poset such that $|\omega|' > 2$ for all $\omega \in \Theta$.*

Then, given a convex pair $(\hat{X}, \hat{v})$, the quasiptopy set $\mathcal{QT}_{d,\,d'}^{\mathsf{emb}}(\hat{X}, \hat{v};$ $\mathbf{c}\Theta; \mathbf{c}\Theta)$ stabilizes towards the homotopy set $\big[(\partial_1^+\hat{X}(\hat{v}),\ \partial_2^-\hat{X}(\hat{v})),\ (\mathcal{P}_\infty^{\mathbf{c}\Theta}, pt)\big]$ for all sufficiently big d relative to $\dim(\hat{X})$. ◇

8.6 Group structure on the ball-shaped convex pseudo-envelops

The connected sum $\alpha_1 \uplus \alpha_2$ of convex pseudo-envelops $\alpha_1 : X_1 \to \hat{X}_1, \alpha_2 : X_2 \to \hat{X}_2$ can be introduced in a fashion, similar to the operation $\uplus$ in formula (7.16) (see also [K12], formulae (3.14) and (3.15)).

Let D_+^{n-1} denote the Southern hemisphere in the sphere ∂D^n. As with the connected sums of submersions and immersions from Chapter 7, there is an ambiguity about how to attach a 1-handle $D_+^{n-1} \times [0,1]$ to $\partial_1^+\hat{X}_1(\hat{v}_1) \coprod \partial_1^+\hat{X}_2(\hat{v}_2)$ and a 1-handle $H = D^n \times [0,1]$ to $\hat{X}_1 \coprod \hat{X}_2$ to form "coordinated" connected sums $\partial_1^+\hat{X}_1(\hat{v}_1) \#_\partial \partial_1^+\hat{X}_2(\hat{v}_2)$ and $\hat{X}_1 \#_\partial \hat{X}_2$. The ambiguity arises if $\partial_2^-\hat{X}_1(\hat{v}_1)$ or $\partial_2^-\hat{X}_2(\hat{v}_2)$ has more than a single connected component. To avoid it, as in formula (7.16), we need to pick a preferred connected component of $\partial_2^-\hat{X}_1(\hat{v}_1)$ and of $\partial_2^-\hat{X}_2(\hat{v}_2)$. Using the local models of convex vector fields (as in the proof of Lemma 8.3), the vector fields $\hat{v}_1$ and $\hat{v}_2$ extend across the 1-handle H so that the convexity of the extended traversing vector field $\hat{v}_{\#_\partial}$ is enforced. For example, the lower diagram in Fig. 5.2 is the connected sum of the upper diagram with itself.

Theorem 8.4. *Let $\Theta' \subset \Theta \subset \mathbf{\Omega}$ be closed subposets which do not contain the element $(\emptyset)$. Assume that $d' \geq d$ and $d' \equiv d \equiv 0 \mod 2$. Let $\hat{v}$ be a convex traversing vector field on the standard ball D^{n+1} such that $\partial_1^+ D^{n+1}(\hat{v})$ is diffeomorphic to the standard ball D^n.*[3]

The group operation in the sources of the maps (8.7) and (8.8) below is the connected sum $\uplus$ of convex envelops/convex pseudo-envelops.

- *There is a group isomorphism*

$$\Phi^{\mathsf{emb}} : \mathcal{QT}^{\mathsf{emb}}_{d,\,d'}(D^{n+1}, \hat{v};\, \mathbf{c}\Theta; \mathbf{c}\Theta') \stackrel{\approx}{\longrightarrow}$$
$$\stackrel{\approx}{\longrightarrow} \big[\big[(D^n,\, S^{n-1}),\ \epsilon_{d,\,d'} : (\mathcal{P}^{\mathbf{c}\Theta}_d, pt) \to (\mathcal{P}^{\mathbf{c}\Theta'}_{d'}, pt')\big]\big]. \tag{8.7}$$

The group homomorphism

$$\Phi^{\mathsf{sub}} : \mathcal{QT}^{\mathsf{sub}}_{d,\,d'}(D^{n+1}, \hat{v};\, \mathbf{c}\Theta; \mathbf{c}\Theta') \longrightarrow$$
$$\stackrel{epi}{\longrightarrow} \big[\big[(D^n,\, S^{n-1}),\ \epsilon_{d,\,d'} : (\mathcal{P}^{\mathbf{c}\Theta}_d, pt) \to (\mathcal{P}^{\mathbf{c}\Theta'}_{d'}, pt')\big]\big]. \tag{8.8}$$

is an split epimorphism. Moreover, we have a split group extension

$$1 \to \mathsf{ker}(\Phi^{\mathsf{sub}}) \to \mathcal{QT}^{\mathsf{sub}}_{d,\,d'}(D^{n+1}, \hat{v};\, \mathbf{c}\Theta; \mathbf{c}\Theta') \stackrel{\mathcal{R}}{\to}$$
$$\stackrel{\mathcal{R}}{\to} \mathcal{QT}^{\mathsf{emb}}_{d,\,d'}(D^{n+1}, \hat{v};\, \mathbf{c}\Theta; \mathbf{c}\Theta') \to 1. \tag{8.9}$$

- *For $n > 1$, all these groups are abelian.*
- *For $n < \max_{\omega \in \Theta_{\langle d]}}\{|\omega|'\}$, the groups $\mathcal{QT}^{\mathsf{emb}}_{d,d'}(D^{n+1}, \hat{v};\, \mathbf{c}\Theta; \mathbf{c}\Theta')$ are trivial.*

Proof. The main observation is that the construction of the boxes $(\hat{X}, \hat{v}) \subset ([0,1] \times Y, \tilde{v})$ in Lemma 8.3 is amenable to the connected sum operation for convex pseudo-envelops of traversing flows. That is, given two boxes $([0,1] \times Y_1, \tilde{v}_1) \supset (\hat{X}_1, \hat{v}_1)$ and $([0,1] \times Y_2, \tilde{v}_2) \supset (\hat{X}_2, \hat{v}_2)$ as in Lemma 8.3, we get that

$$(\hat{X}_1 \#_\partial \hat{X}_2, \hat{v}_1 \#_\partial \hat{v}_2) \subset ([0,1] \times (Y_1 \#_\partial Y_2), \tilde{v}_1 \#_\partial \tilde{v}_2)$$

is also a box as in that lemma. Therefore, all the constructions and arguments from Chapter 7 (like Proposition 7.6, Corollary 7.3, and Theorem 7.3) apply to the convex pseudo-envelops/envelops $\alpha : (X, \alpha^\dagger(\hat{v})) \to (D^{n+1}, \hat{v})$, where the traversing vector field $\hat{v}$ is convex with respect to ∂D^{n+1}. Thus, as in formula (7.54) above (see also formula (7.17) and Proposition 7.6), for a convex vector field $\hat{v}$ with $\partial_1^+ D^{n+1}(\hat{v})$ being a smooth n-ball,

[3]The constant vector field on D^{n+1} will do.

the quasitopies $\mathcal{QT}^{\mathsf{emb}}_{d,d'}(D^{n+1}, \hat{v};\, \mathbf{c}\Theta; \mathbf{c}\Theta')$ and $\mathcal{QT}^{\mathsf{sub}}_{d,d'}(D^{n+1}, \hat{v};\, \mathbf{c}\Theta; \mathbf{c}\Theta')$ are groups. For $n > 1$ they are commutative by arguments as in Proposition 7.6.

The last claim of the theorem follows by the general position argument. □

Recall that, by Corollary 8.5, the set $\mathcal{QT}^{\mathsf{emb}}_{d,d'}(\hat{X}, \hat{v}; \mathbf{c}\Theta; \mathbf{c}\Theta')$ depends only on the homotopy type of the pair $(\partial_1^+\hat{X}(\hat{v}),\ \partial_2^-\hat{X}(\hat{v}))$; however, the corollary does not make any claims about $\mathcal{QT}^{\mathsf{sub}}_{d,d'}(\hat{X}, \hat{v}; \mathbf{c}\Theta; \mathbf{c}\Theta')$. The next proposition, in line with Proposition 8.1 (which deals with the homology n-spheres, $n \geq 6$), is a hint that $\mathcal{QT}^{\mathsf{sub}}_{d,d'}(\hat{X}, \hat{v}; \mathbf{c}\Theta; \mathbf{c}\Theta')$ may also depend only on the homotopy type of the pair $(\partial_1^+\hat{X}(\hat{v}),\ \partial_2^-\hat{X}(\hat{v}))$.

Proposition 8.2. *For $n \geq 5$, assuming that the locus $\partial_2^- D^{n+1}(\hat{v})$ is simply-connected, the groups $\mathcal{QT}^{\mathsf{sub/emb}}_{d,d'}(D^{n+1}, \hat{v};\, \mathbf{c}\Theta; \mathbf{c}\Theta')$ do not depend on the choice of convex traversing vector field $\hat{v}$ on D^{n+1}.*

For $n \leq 3$, the groups $\mathcal{QT}^{\mathsf{sub/emb}}_{d,d'}(D^{n+1}, \hat{v};\, \mathbf{c}\Theta; \mathbf{c}\Theta')$ also do not depend on the choice of convex $\hat{v}$.

Proof. For a convex $\hat{v}$, the locus $\partial_1^+\hat{X}(\hat{v})$ is a deformation retract of D^{n+1} and thus has the homotopy type of a point. By the Poincaré duality, $\partial_2^- D^{n+1}(\hat{v}) = \partial(\partial_1^+\hat{X}(\hat{v}))$ is a homology $(n-1)$-sphere $\Sigma^{n-1} \subset \partial D^{n+1}$. Let us delete a small smooth ball B^n from the interior of $\partial_1^+\hat{X}(\hat{v})$. We denote its complement by W^n.

For $n \geq 5$, assuming that $\pi_1(\partial_2^- D^{n+1}(\hat{v})) = 1$, we may apply the smooth h-cobordism theorem (see [Mi1]) to W to conclude that W is diffeomorphic to the product $S^{n-1} \times [0, 1]$. Therefore, $\partial_1^+ D^{n+1}(\hat{v})$ is a smooth ball $D^n \subset S^n$. Any two regular embeddings $D^n \hookrightarrow S^n$ are diffeotopic. By the proof of Lemma 8.1, any diffeotopy of S^n which maps $\partial_1^+ D^{n+1}(\hat{v}_1)$ to $\partial_1^+ D^{n+1}(\hat{v}_2)$ extends to a diffeotopy of D^{n+1} that maps the $\hat{v}_1$-trajectories to $\hat{v}_2$-trajectories, while preserving their orientation. As a result, for $n \geq 5$, the group $\mathcal{QT}^{\mathsf{emb}}_{d,d'}(D^{n+1}, \hat{v};\, \mathbf{c}\Theta; \mathbf{c}\Theta')$ does not depend on the choice of $\hat{v}$, as long as $\pi_1(\partial_2^- D^{n+1}(\hat{v})) = 1$.

In small dimensions $n \leq 3$, the contractibility of $\partial_1^+ D^{n+1}(\hat{v})$ implies that it is the standard smooth n-ball. The case $n = 1$ is obvious. For $n = 2$, if the domain $\partial_1^+\hat{X}(\hat{v}) \subset S^2$ has a boundary that is a homology 1-sphere, then the domain is the 2-ball, and any 2-balls in S^2 are isotopic. For $n = 3$, if the domain $\partial_1^+ D^4(\hat{v}) \subset S^3$ has a smooth boundary that is a homology 2-sphere. By the classification of 2-surfaces, it follows that $\partial_2^- D^4(\hat{v})$ is the standard 2-sphere. By the solution of the 3-dimensional

Poincaré Conjecture [P1]-[P3], the contractible domain $\partial_1^+ D^4(\hat{v})$ is with the spherical boundary is the 3-ball, and any two 3-balls in S^3 are isotopic. Thus, by Lemma 8.1, the smooth isotopy type of the locus $\partial_1^+ D^{n+1}(\hat{v}) \subset S^n$ determines the smooth topological type of the foliation $\mathcal{L}(\hat{v})$. Since, any two standard n-balls $\partial_1^+ \hat{X}(\hat{v}_1)$ and $\partial_1^+ \hat{X}(\hat{v}_2)$ are isotopic in S^n, the two statements of the proposition are validated.

The difficult case $n = 4$ is wide open. □

Combining Theorem 8.2 and Theorem 8.4, we get the following claim.

Corollary 8.7. *Let $\hat{v}^{||}$ be a constant vector field on the ball $D^{n+1} \subset \mathbb{R}^{n+1}$, and let $\hat{v}$ be a convex vector field on $\hat{X}$. Put $r = \#(\pi_0(\partial_2^- \hat{X}(\hat{v})))$. With the help of the maps from (8.7) and (8.8), the groups*

$$\mathcal{G}^{\mathsf{emb}} =_{\mathsf{def}} \left(\mathcal{QT}^{\mathsf{emb}}_{d,d'}(D^{n+1}, \hat{v}^{||}; \mathbf{c}\Theta; \mathbf{c}\Theta')\right)^r \text{ and}$$

$$\mathcal{G}^{\mathsf{sub}} =_{\mathsf{def}} \left(\mathcal{QT}^{\mathsf{sub}}_{d,d'}(D^{n+1}, \hat{v}^{||}; \mathbf{c}\Theta; \mathbf{c}\Theta')\right)^r$$

are represented in the group

$$\left(\pi_n(\mathcal{P}_d^{\mathbf{c}\Theta}, pt) \big/ \mathsf{ker}\{(\epsilon_{d,d'})_* : \pi_n(\mathcal{P}_d^{\mathbf{c}\Theta}, pt) \to \pi_n(\mathcal{P}_{d'}^{\mathbf{c}\Theta'}, pt')\}\right)^r.$$

We denote by Ψ^{emb} and Ψ^{sub} these two representations (the first one is an isomorphism).

Then the maps $\Phi^{\mathsf{sub}}(\hat{X}, \hat{v})$ in (8.6) and $\Phi^{\mathsf{emb}}(\hat{X}, \hat{v})$ in (8.5) from Theorem 8.2 are equivariant *with respect to the $\mathcal{G}^{\mathsf{sub}}$- and $\mathcal{G}^{\mathsf{emb}}$-actions on their source sets and the $\Psi^{\mathsf{sub}}(\mathcal{G}^{\mathsf{sub}})$- and $\Psi^{\mathsf{emb}}(\mathcal{G}^{\mathsf{emb}})$-actions on their target sets.* ◇

8.7 Quasitopies of convex envelops with generic combinatorics Θ and $d' = d$

From now and until Section 8.9, each result about quasitopies $\mathcal{QT}^{\mathsf{sub/emb}}_{d,d'}(\hat{X}, \hat{v}; \mathbf{c}\Theta; \mathbf{c}\Theta')$ of convex envelops is an easily recognizable image of a similar result from Chapter 7 about the quasitopies $\mathcal{QT}^{\mathsf{imm/emb}}_{d,d'}(Y, \partial Y; \mathbf{c}\Theta, (\emptyset), \mathbf{c}\Theta')$ of immersions/embedding into the products $\mathbb{R} \times Y$.

Theorem 8.5. *Let $(\hat{X}, \hat{v})$ be a convex pair. Let d, k be natural numbers such that $2 < k < d$ and $d \equiv 0 \mod 2$. Put $\mathbf{\Omega}_{|\sim|' \leq k-1} =_{\mathsf{def}} \mathbf{c\Omega}_{|\sim|' \geq k}$.*

Then we get a bijection[4]

$$\Phi : \mathcal{QT}^{\mathsf{emb}}_{d,d}(\hat{X}, \hat{v};\, \mathbf{\Omega}_{|\sim|' \leq k-1}, \mathbf{\Omega}_{|\sim|' \leq k-1}) \approx$$

$$\approx \Big[(\partial_1^+ \hat{X}(\hat{v}),\, \partial_2^- \hat{X}(\hat{v})),\, \big(\bigvee_{\ell=1}^{A(d,\,k)} S^{k-1}_\ell,\, \star\big)\Big].$$

Proof. For $d > k > 2$, Proposition 6.6 describes the homotopy type of the space $\mathcal{P}_d^{\mathbf{\Omega}_{|\sim|' \leq k-1}} = \mathcal{P}_d^{\mathbf{c\Omega}_{|\sim|' \geq k}}$ as a bouquet $\bigvee_{\ell=1}^{A(d,\,k)} S^{k-1}_\ell$ of $(k-1)$-spheres. The space $\mathcal{P}_d^{(\emptyset)}$ is contractible to a point $\star \in \mathcal{P}_d^{(\emptyset)}$. Thus, by Theorem 8.2, the claim follows. □

Now we consider the "combinatorially generic" case of convex pseudo-envelops $\alpha : (X, \alpha^\dagger(\hat{v})) \to (\hat{X}, \hat{v})$ for which $\hat{v}$ is traversally generic with respect to $\alpha(\partial_1 X)$. By Definition 8.6, any such α has tangency patterns that belong to the poset $\mathbf{\Omega}_{|\sim|' \leq n}$, where $n = \dim \hat{X} - 1$.

Corollary 8.8. *Let* $(\hat{X}, \hat{v})$ *be a convex pair,* $\dim \hat{X} = n+1$. *For* $d > n > 2$, *we get a bijection*

$$\Phi : \mathcal{QT}^{\mathsf{emb}}_{d,d}(\hat{X}, \hat{v};\, \mathbf{\Omega}_{|\sim|' \leq n}, \mathbf{\Omega}_{|\sim|' \leq n}) \approx \mathbb{Z}^{A(d,\,n+1)}.$$

These $\mathbb{Z}$*-valued invariants of convex envelops are delivered by the degrees of the maps*

$$\big\{\Phi_\ell : \partial_1^+ \hat{X}(\hat{v}) / \partial_2^- \hat{X}(\hat{v}) \to S^n_\ell\big\}_{\ell \in [1,\, A(d,n+1)]},$$

to the individual spheres, induced by the map Φ *from Theorem 8.5.*

In particular, the $(d, d; \mathbf{\Omega}_{|\sim|' \leq n}, \mathbf{\Omega}_{|\sim|' \leq n})$*-quasitopy class of any* traversally generic *convex envelop* $\alpha : (X, \alpha^\dagger(\hat{v})) \to (\hat{X}, \hat{v})$ *is determined by the collection of such degrees.*

Proof. Since $\hat{X}$ is connected and $\hat{v}$ is convex, the locus $\partial_1^+ \hat{X}$ is connected as well. By Theorem 8.5, the homotopy classes of the classifying maps

$$\Phi^\alpha : (\partial_1^+ \hat{X}(\hat{v}),\, \partial_2^- \hat{X}(\hat{v})) \to \big(\mathcal{P}_d^{\mathbf{\Omega}_{|\sim|' \leq n}},\, \mathcal{P}_d^{(\emptyset)}\big)$$

are in 1-to-1 correspondence with homotopy classes of the corresponding maps

$$\tilde{\Phi}^\alpha : (\partial_1^+ \hat{X}(\hat{v}),\, \partial_2^- \hat{X}(\hat{v})) \to \big(\bigvee_{\ell=1}^{A(d,\,n+1)} S^n_\ell,\, \star\big).$$

[4] see (6.39) for the definition of the number $A(d, k) =_{\mathsf{def}} A(d, k, 0)$

Since $\partial_1^+ \hat{X}(\hat{v})/\partial_2^- \hat{X}(\hat{v})$ is a connected n-dimensional CW-complex, for $n \geq 2$, the latter ones are detected by the degrees of the maps $\{\Phi_\ell^\alpha\}$ to the individual n-spheres $\{S_\ell^n\}$. □

Example 8.4. Take $d = 6$ and $n = 3$. Then $\bar{\mathcal{P}}_6^{\mathbf{\Omega}_{|\sim|' \geq 4}}$ consist of a single 0-dimensional cell (the point "∞" $\in \bar{\mathcal{P}}_6$), one 1-dimensional cell, labelled by $\omega = (6)$, and five 2-dimensional cells, labelled by $\omega = (51), (15), (42), (24), (33)$. Hence, $A(6,4) = \chi(\bar{\mathcal{P}}_6^{\mathbf{\Omega}_{|\sim|' \geq 4}}) - 1 = 4$. The space $\bar{\mathcal{P}}_6^{\mathbf{\Omega}_{|\sim|' \geq 4}}$ has a homotopy type of a bouquet $\bigvee_{\ell=1}^4 S^2$ of four 2-spheres. By the Alexander duality, $\mathcal{P}_6^{\mathbf{c\Omega}_{|\sim|' \geq 4}} = \mathcal{P}_6^{\mathbf{\Omega}_{|\sim|' \leq 3}}$ has a homotopy type of a bouquet $\bigvee_{\ell=1}^4 S^3$ of four 3-spheres.

We pick a convex and traversing vector field $\hat{v}$ on D^4. Recall that $\partial_1^+ D^4(\hat{v})$ is contractible and thus $\partial_2^- D^4(\hat{v})$ a homology 2-sphere, which implies that $\partial_2^- D^4(\hat{v})$ is diffeomorphic to $S^2 \subset S^3$. Thus, by [P1], [P2], $\partial_1^+ D^4(\hat{v})$ is diffeomorphic to the ball D^3. Therefore, for *any* convex $\hat{v}$, we get the group isomorphism

$$\mathcal{QT}_{6,6}^{\mathsf{emb}}(D^4, \hat{v};\, \mathbf{\Omega}_{|\sim|' \leq 3}, \mathbf{\Omega}_{|\sim|' \leq 3}) \approx \pi_3\Big(\bigvee_{\ell=1}^{4} S_\ell^3\Big) \approx \mathbb{Z}^4.$$

As a result, any element $[\alpha] \in \mathcal{QT}_{6,6}^{\mathsf{sub}}(D^4, \hat{v};\, \mathbf{\Omega}_{|\sim|' \leq 3}, \mathbf{\Omega}_{|\sim|' \leq 3})$ generates four integer-valued characteristic invariants. For embeddings α, they determine $[\alpha]$.

At the same time, by Corollary 8.9, $\mathcal{QT}_{6,8}^{\mathsf{emb}}(D^4, \hat{v};\, \mathbf{\Omega}_{|\sim|' \leq 3}, \mathbf{\Omega}_{|\sim|' \leq 4}) = 0$. ◊

The following claim contrasts with Corollary 8.8.

Corollary 8.9. *Under the hypotheses of Corollary 8.8, including $d > n > 2$, the set $\mathcal{QT}_{d,\, d+2}^{\mathsf{emb}}(\hat{X}, \hat{v};\, \mathbf{\Omega}_{|\sim|' \leq n}, \mathbf{\Omega}_{|\sim|' \leq n+1})$ consists of a single element, represented by $X = \emptyset$.*

In particular, the $(d, d+2;\, \mathbf{\Omega}_{|\sim|' \leq n}, \mathbf{\Omega}_{|\sim|' \leq n+1})$-quasitopy class of any traversally generic convex envelop $\alpha : (X, \alpha^\dagger(\hat{v})) \to (\hat{X}, \hat{v})$ is trivial.

Proof. Consider a convex envelop $\alpha : (X, \alpha^\dagger(\hat{v})) \to (\hat{X}, \hat{v})$ as in Theorem 8.5. The corollary claim is based on the observation that any map from the n-dimensional CW-complex $\partial_1^+ \hat{X}(\hat{v})/\partial_2^- \hat{X}(\hat{v})$ to $\mathcal{P}_d^{\mathbf{c\Omega}_{|\sim|' \geq n}}$—homotopically a bouquet of n-spheres—is null-homptopic in $\mathcal{P}_{d+2}^{\mathbf{c\Omega}_{|\sim|' \geq n+1}}$ since the latter space has the homotopy type of a bouquet of $(n+1)$-spheres. Again, by Theorem 8.2, the envelop α is nill-quasitopic. □

8.8 From inner framed cobordisms of $\partial_1^+\hat{X}(\hat{v})$ to quasitopies of k-flat convex envelops

This section is a mirror image of Section 7.6. In what follows, we use the notations from that section. Theorem 8.5 uncovers a connection between quasitopies of certain convex envelops $\alpha : (X, v) \to (\hat{X}, \hat{v})$ and inner framed cobordisms of the boundary $\partial_1\hat{X}$. This connection is described in Proposition 8.3.

In Proposition 7.14, we have seen that the inner framed Ξ-colored codimension $k-1$ bordisms of the n-dimensional manifold Y produce, via the Thom construction, quite special k-flat embeddings $\beta : M \to \mathbb{R} \times Y$, where $\dim M = \dim Y$.

An analogous mechanism, with the help of Lemma 8.3 and Theorem 8.5, generates special k-flat envelops (see Definition 8.5).

Proposition 8.3. *Let $\hat{X}$ be a smooth compact connected $(n+1)$-dimensional manifold, equipped with a convex traversing vector field $\hat{v}$. Let $k \in [3, n+1]$, $k < d$, and $d \equiv 0 \mod 2$. Then the Thom construction delivers a bijection*

$$Th : \mathcal{FB}_{\Xi}^{k-1}(\partial_1^+\hat{X}(\hat{v})) \approx \mathcal{QT}_{d,d}^{\mathsf{emb}}(\hat{X}, \hat{v};\, \mathbf{c\Omega}_{|\sim|' \geq k}, \mathbf{c\Omega}_{|\sim|' \geq k}),$$

where $\mathcal{FB}_{\Xi}^{k-1}(\partial_1^+\hat{X}(\hat{v}))$ denotes the set of inner framed Ξ-colored $(n-k+1)$-dimensional bordisms of the space $\partial_1^+\hat{X}(\hat{v})$. Here the cardinality of the pallet Ξ is $A(d, k)$. $\diamondsuit$

Example 8.5. Let us recycle Example 7.5: take $k = 3$, $d = 6$, $q = 0$, and $n = 3$. Then $\#\Xi = 10$. Thus we get a homotopy equivalence $\tau : \mathcal{P}_6^{\mathbf{c}\Theta_{|\sim|' \geq 3}} \sim \bigvee_{\sigma=1}^{10} S_\sigma^2$.

Let $(\hat{X}, \hat{v})$ be a convex pair, where $\dim \hat{X} = 4$, and let $Z \subset \partial_1^+\hat{X}(\hat{v})$ be a normally framed 1-dimensional closed submanifold, each loop in Z being colored with a color from the pallet Ξ. Using the isomorphism

$$Th : \mathcal{FB}_{\Xi}^2(\partial_1^+\hat{X}(\hat{v})) \approx \mathsf{QT}_{6,6}^{\mathsf{emb}}(\hat{X}, \hat{v}; \mathbf{c}\Theta_{|\sim|' \geq 3}, \mathbf{c}\Theta_{|\sim|' \geq 3}),$$

any *framed link* $Z \subset \partial_1^+\hat{X}(\hat{v})$, colored with 10 distinct colors at most, produces a quasitopy class of a convex envelop $\alpha : (X, \alpha^\dagger(\hat{v})) \to (\hat{X}, \hat{v})$. Its tangency patterns ω, except for $\omega = (\emptyset)$, have only entries from the list $\{1, 2, 3\}$ so that no more than one 3 is present in ω, and no more than two 2's are present, while $|\omega| \leq 6$.

Now, let us assume that D^4 carries a constant vector field $v^{\|}$ (so that $\partial_2^- \hat{X}(\hat{v}) \neq \emptyset$). Note that any element of $H_1(\partial_1^+ \hat{X}(\hat{v})); \mathbb{Z})$ may be realized by a disjoint union of framed oriented loops. Then, like in Example 7.5, the orbit-space of the $\mathsf{G}^{\mathsf{emb}}_{6,6}(D^4,^{\|}; \mathbf{c}\Theta_{|\sim|' \geq 3}; \mathbf{c}\Theta_{|\sim|' \geq 3})$-action on $\mathsf{QT}^{\mathsf{emb}}_{6,6}(\hat{X}, \hat{v}; \mathbf{c}\Theta_{|\sim|' \geq 3}, \mathbf{c}\Theta_{|\sim|' \geq 3})$ admits a surjection on the abelian group $(H_1(\partial_1^+ \hat{X}(\hat{v})); \mathbb{Z}))^{10} \approx (H_1(\hat{X}; \mathbb{Z}))^{10}$, provided that $\partial_1 \hat{X}$ is orientable. Let us give a couple of specific examples of this fact.

Consider the box $T^3_\circ \times [0,1]$, where $T^3_\circ$ is the compliment in the 3-torus T^3 to an open ball D^3. By rounding the corners of the box, we get a convex pair $(\hat{X}, \hat{v})$, where $\hat{X}$ is homeomorpfic to $T^3_\circ \times [0,1]$. Then the orbit-space of the $\mathsf{G}^{\mathsf{emb}}_{6,6}(D^4, v^{\|}; \mathbf{c}\Theta_{|\sim|' \geq 3}, \mathbf{c}\Theta_{|\sim|' \geq 3})$-action on $\mathsf{QT}^{\mathsf{emb}}_{6,6}(\hat{X}, \hat{v}; \mathbf{c}\Theta_{|\sim|' \geq 3}, \mathbf{c}\Theta_{|\sim|' \geq 3})$ is mapped onto the lattice $\mathbb{Z}^{30}$.

Let $M = \mathbb{H}^3/\Gamma$ be a compact hyperbolic 3-manifold, and let $M_\circ =_{\mathsf{def}} M \setminus D^3$. By rounding the corners of the box $M_\circ \times [0,1]$, we get a convex pair $(\hat{X}, \hat{v})$. Then the orbit-space of the $\mathsf{G}^{\mathsf{emb}}_{6,6}(D^4, v^{\|}; \mathbf{c}\Theta_{|\sim|' \geq 3}, \mathbf{c}\Theta_{|\sim|' \geq 3})$-action on $\mathsf{QT}^{\mathsf{emb}}_{6,6}(\hat{X}, \hat{v}; \mathbf{c}\Theta_{|\sim|' \geq 3}, \mathbf{c}\Theta_{|\sim|' \geq 3})$ is mapped onto the abelian group $(\Gamma/[\Gamma, \Gamma])^{10}$, where $[\Gamma, \Gamma]$ denotes the commutator. $\diamondsuit$

By Proposition 8.3 and repeating the (obviously modified) arguments in the proof of Corollary 7.14 , we get the following claim.

Corollary 8.10. *Let $d > k$, and $d \equiv 0 \mod 2$. Let $(\hat{X}, \hat{v})$ be a convex pair.*

- *For $\dim \hat{X} = k \geq 3$ and any choice of $\kappa \in \pi_0(\partial_2^- \hat{X}(\hat{v}))$, the group $\mathcal{QT}^{\mathsf{emb}}_{d,d}(D^k, \hat{v}^{\|}; \mathbf{c\Omega}_{|\sim|' \geq k}, \mathbf{c\Omega}_{|\sim|' \geq k})$ acts freely and transitively on the set $\mathcal{QT}^{\mathsf{emb}}_{d,d}(\hat{X}, \hat{v}; \mathbf{c\Omega}_{|\sim|' \geq k}, \mathbf{c\Omega}_{|\sim|' \geq k})$. Thus, both sets are in a 1-to-1 correspondence and the first one is $(\hat{X}, \hat{v})$-independent.*
- *For $\dim \hat{X} = k + 1 > 5$, a simply-connected $\hat{X}$, and any choice of $\kappa \in \pi_0(\partial_2^- \hat{X}(\hat{v}))$, the group $\mathcal{QT}^{\mathsf{emb}}_{d,d}(D^{k+1}, \hat{v}^{\|}; \mathbf{c\Omega}_{|\sim|' \geq k}, \mathbf{c\Omega}_{|\sim|' \geq k})$ acts freely and transitively on the set $\mathcal{QT}^{\mathsf{emb}}_{d,d}(\hat{X}, \hat{v}; \mathbf{c\Omega}_{|\sim|' \geq k}, \mathbf{c\Omega}_{|\sim|' \geq k})$. Again, both sets are in a 1-to-1 correspondence and the first one is $(\hat{X}, \hat{v})$-independent.* $\diamondsuit$

Example 8.6. Let $\hat{X} = \mathbb{CP}^2 \times D^2$ with $\hat{v}$ on $\mathbb{CP}^2 \times D^2$ being generated by a constant vector field on D^2. Then $\mathcal{QT}^{\mathsf{emb}}_{d,d}(D^6, \hat{v}^{\|}; \mathbf{c\Omega}_{|\sim|' \geq 5}, \mathbf{c\Omega}_{|\sim|' \geq 5})$ acts freely and transitively on the set $\mathcal{QT}^{\mathsf{emb}}_{d,d}(\mathbb{CP}^2 \times D^2, \hat{v}; \mathbf{c\Omega}_{|\sim|' \geq 5}, \mathbf{c\Omega}_{|\sim|' \geq 5})$ for all even $d \geq 6$. Thus both sets are in 1-to-1 correspondence. As a result, $\mathcal{QT}^{\mathsf{emb}}_{d,d}(\mathbb{CP}^2 \times D^2, \hat{v}; \mathbf{c\Omega}_{|\sim|' \geq 5}, \mathbf{c\Omega}_{|\sim|' \geq 5})$ acquires the structure of the abelian group $\pi_5\big(\bigvee_{i=1}^{A(d,5)} S^4_i, \star\big)$. $\diamondsuit$

8.9 Convex envelops with special combinatorics $\mathbf{c\Theta}$

Recall that, for a given $\omega \in \mathbf{\Omega}$, we denote by $\langle\omega\rangle$ the minimal closed poset, generated by ω. Combining Theorem 8.2 and Theorem 7.5 with Proposition 8.2, we get the following result, in which the constraint $d \leq 12$ reflects only the scope of the numerical experiments in [KSW2].

Theorem 8.6. *Assume that $d \leq 12$ and $d \equiv 0 \mod 2$. Let $\omega \in \mathbf{\Omega}_{\leq d}$ be such that $|\omega|' > 2$. Let $\hat{X}$ be a smooth compact connected $(n+1)$-dimensional manifold, equipped with a convex traversing vector field $\hat{v}$.*

Then the quasitopy set $\mathcal{QT}^{\mathsf{emb}}_{d,d}(\hat{X}, \hat{v}; \mathbf{c}\langle\omega\rangle, \mathbf{c}\langle\omega\rangle)$ either is trivial (i.e., consists of a single element), or is isomorphic to the cohomotopy set $\pi^k(\partial_1^+\hat{X}(\hat{v})/\partial_2^-\hat{X}(\hat{v}))$, where $k = k(\omega) \in [|\omega|' - 1,\, d-1]$.

In particular, for a constant vector field $\hat{v} = v^{\|}$, $d \leq 12$, ω and $k = k(\omega)$ from the table (6.36), the group $\mathcal{QT}^{\mathsf{emb}}_{d,d}(D^{n+1}, \hat{v}; \mathbf{c}\langle\omega\rangle, \mathbf{c}\langle\omega\rangle) \approx \pi_n(S^k)$.

A similar claim is valid for any convex vector field $\hat{v}$ on D^{n+1}, provided that either $n \leq 3$ or $n \geq 5$ and $\partial_2^-\hat{X}(\hat{v})$ is simply-connected. $\diamondsuit$

For $d \leq 13$, the table (6.36) lists all ω's and the corresponding $k = k(\omega)$'s for which $\mathcal{P}_d^{\mathbf{c}\langle\omega\rangle}$ is homologically nontrivial. Recall that $\mathcal{P}_d^{\mathbf{c}\langle\omega\rangle}$ is a homology k-sphere and even a homotopy k-sphere, at least when $|\omega|' > 2$.

Let us recycle Example 7.4, while adapting it to the convex envelops setting.

Example 8.7. Let $d = 8$ and $\omega = (4)$. Then, for any convex pair $(\hat{X}, \hat{v})$, using the list in (6.36), we get a bijection

$$\mathcal{QT}^{\mathsf{emb}}_{8,8}(\hat{X}, \hat{v}; \mathbf{c}\langle(4)\rangle, \mathbf{c}\langle(4)\rangle) \approx \pi^4\big(\partial_1^+\hat{X}(\hat{v})/\partial_2^-\hat{X}(\hat{v})\big),$$

where $\pi^4(\sim)$ stands for the 4-cohomotopy set. In particular, using Proposition 8.2, for any convex $\hat{v}$ on D^8 such that the locus $\partial_2^- D^8(\hat{v})$ is simply-connected (the constant $\hat{v}$ will do), we get a group isomorphism:

$$\mathcal{QT}^{\mathsf{emb}}_{8,8}(D^8, \hat{v}; \mathbf{c}\langle(4)\rangle, \mathbf{c}\langle(4)\rangle) \approx \pi_7(S^4) \approx \mathbb{Z} \times \mathbb{Z}_{12}.$$

Let $d = 12$ and $\omega = (11213)$. Then, for any for any convex pair $(\hat{X}, \hat{v})$, by the same list (6.36), we get a bijection

$$\mathcal{QT}^{\mathsf{emb}}_{12,12}(\hat{X}, \hat{v}; \mathbf{c}\langle(11213)\rangle, \mathbf{c}\langle(11213)\rangle) \approx \pi^6\big(\partial_1^+\hat{X}(\hat{v})/\partial_2^-\hat{X}(\hat{v})\big).$$

Again, using Proposition 8.2, for any convex $\hat{v}$ on D^{10}, such that the locus $\partial_2^- D^{10}(\hat{v})$ is simply-connected (again, the constant $\hat{v}$ will do), we get a group isomorphism

$$\mathsf{QT}^{\mathsf{emb}}_{12,12}(D^{10}, \hat{v}; \mathbf{c}\langle(11213)\rangle, \mathbf{c}\langle(11213)\rangle) \approx \pi_9(S^6) \approx \mathbb{Z}_{24}.$$

It would be interesting to produce explicit analytic realizations of embeddings $\alpha : (X, \alpha^\dagger(v^{\|})) \hookrightarrow (D^8, v^{\|})$ or $\alpha' : (X', (\alpha')^\dagger(v^{\|})) \hookrightarrow (D^{10}, v^{\|})$ that would realize the generators of the groups $\mathbb{Z} \times \mathbb{Z}_{12}$ or $\mathbb{Z}_{24}$, respectively. Speaking informally, we are asking: "What are the shapes of $\alpha(\partial_1 X)$ or $\alpha'(\partial_1 X')$ that realize the generators?" ♢

We are adopting the notations we had used prior to Corollary 7.12. Given two groups H, G, we consider $\mathsf{Hom}(H, G)$, the group of their homomorphisms, and $\mathsf{Hom}^\bullet(H, G)$, the quotient $\mathsf{Hom}(H, G)/\mathsf{Ad}_G$.

Proposition 8.4 below deals with special Θ's for which $\mathcal{P}_d^{\mathbf{c}\Theta}$ are $K(\pi, 1)$-spaces and π is a free group.

Proposition 8.4. *Let* $\mathbf{c}\Theta$ *consist of all* ω*'s with entries* 1 *and* 2 *only and no more than a single entry* 2. *Put* $\kappa(d) =_{\mathsf{def}} \frac{d(d-2)}{4}$ *for* $d \equiv 0 \mod 2$. *We denote by* $\mathsf{F}_{\kappa(d)}$ *the* free *group in* $\kappa(d)$ *generators.*

Consider a convex pair $(\hat{X}, \hat{v})$*, where* $\hat{X}$ *is connected. If* $\partial_2^- \hat{X}(\hat{v}) \neq \emptyset$*, then there is a bijection*

$$\Phi^{\mathsf{emb}} : \mathcal{QT}^{\mathsf{emb}}_{d,d}(\hat{X}, \hat{v}; \mathbf{c}\Theta, \mathbf{c}\Theta) \stackrel{\approx}{\longrightarrow} \mathsf{Hom}(\pi_1(\hat{X}), \mathsf{F}_{\kappa(d)})$$

and a surjection

$$\Phi^{\mathsf{sub}} : \mathcal{QT}^{\mathsf{sub}}_{d,d}(\hat{X}, \hat{v}; \mathbf{c}\Theta; \mathbf{c}\Theta) \longrightarrow \mathsf{Hom}(\pi_1(\hat{X}), \mathsf{F}_{\kappa(d)}).$$

When $\partial_2^- \hat{X}(\hat{v}) = \emptyset$*, then similar claims hold, with the target of the maps* Φ^{emb} *and* Φ^{sub} *being replaced by the set* $\mathsf{Hom}^\bullet(\pi_1(\hat{X}), \mathsf{F}_{\kappa(d)})$.

In particular,

$$\Phi^{\mathsf{emb}} : \mathcal{QT}^{\mathsf{emb}}_{d,d}(S^1 \times [0,1], \hat{v}; \mathbf{c}\Theta, \mathbf{c}\Theta) \stackrel{\approx}{\longrightarrow} \mathsf{F}_{\kappa(d)}/\mathsf{Ad}_{\mathsf{F}_{\kappa(d)}},$$

the free group of cyclic *words in* $\kappa(d)$ *letters (see Fig. 7.4). Here* $\hat{v}$ *is tangent to the fibers of the obvious projection* $S^1 \times [0,1] \to S^1$.

If $\pi_1(\hat{X})$ *has no free images, then the group* $\mathcal{QT}^{\mathsf{emb}}_{d,d}(\hat{X}, \hat{v}; \mathbf{c}\Theta, \mathbf{c}\Theta)$ *is trivial.*

Proof. We observe that $\pi_1(\partial_1^+ \hat{X}(\hat{v})) \approx \pi_1(\hat{X})$ since the two spaces are homotopy equivalent due to the convexity of $\hat{v}$. By Lemma 8.3, we can incapsulate $(\hat{X}, \hat{v})$ in a box $\mathbb{R} \times Y$, where Y is homeomorphic to $\partial_1^+ \hat{X}(\hat{v})$. Using this Y, the claim follows from Corollary 7.12. □

The next proposition is a stabilization result, by the increasing $d' \geq d$, for the convex envelops with (using the terminology of [Ar]), k-moderate tangent patterns $\omega \in \mathbf{c}\Theta_{\max \geq k}$. The entries of such ω's are all less than k. Proposition 8.5 below follows instantly from Proposition 7.12 by combining it with Lemma 8.3.

Proposition 8.5. *Let $k \geq 4$. If $\dim Y \leq (k-2)(\lceil d/k \rceil + 1) - 2$, then the classifying map*

$$\Phi^{\mathsf{emb}} : \mathcal{QT}^{\mathsf{emb}}_{d,d'}(\hat{X}, \hat{v}; \mathbf{c}\Theta_{\max \geq k}, \mathbf{c}\Theta_{\max \geq k}) \xrightarrow{\approx} [(\partial_1^+ \hat{X}(\hat{v}), \partial_2^- \hat{X}(\hat{v})), (\mathcal{P}_d^{\mathbf{c}\Theta_{\max \geq k}}, pt)]$$

is a bijection, and the classifying map

$$\Phi^{\mathsf{sub}} : \mathcal{QT}^{\mathsf{imm}}_{d,d'}(\hat{X}, \hat{v}; \mathbf{c}\Theta_{\max \geq k}, \mathbf{c}\Theta_{\max \geq k}) \xrightarrow{\approx} [(\partial_1^+ \hat{X}(\hat{v}), \partial_2^- \hat{X}(\hat{v})), (\mathcal{P}_d^{\mathbf{c}\Theta_{\max \geq k}}, pt)]$$

is a surjection for any $d' \geq d$, $d' \equiv d \mod 2$.

In particular, for a given convex pair $(\hat{X}, \hat{v})$, the quasitopy

$$\mathsf{QT}^{\mathsf{emb}}_{d,d'}(\hat{X}, \hat{v}; \mathbf{c}\Theta_{\max \geq k}, \mathbf{c}\Theta_{\max \geq k})$$

stabilizes for all

$$d' \geq d \geq \frac{k}{k-2}(\dim \hat{X} + 3 - k)$$

—a linear function in $\dim \hat{X}$. ◇

Example 8.8. Let $k = 4$. By Proposition 8.5, for any compact connected 3-*dimensional* convex pair $(\hat{X}, \hat{v})$, we get bijections:

$$\mathcal{QT}^{\mathsf{emb}}_{4,4}(\hat{X}, \hat{v}; \mathbf{c}\Theta_{\max \geq 4}, \mathbf{c}\Theta_{\max \geq 4}) \xrightarrow{\approx} \pi^2(\partial_1^+ \hat{X}(\hat{v}) / \partial_2^- \hat{X}(\hat{v})),$$
$$\mathcal{QT}^{\mathsf{emb}}_{4,6}(\hat{X}, \hat{v}; \mathbf{c}\Theta_{\max \geq 4}, \mathbf{c}\Theta_{\max \geq 4}) \xrightarrow{\approx} \pi^2(\partial_1^+ \hat{X}(\hat{v}) / \partial_2^- \hat{X}(\hat{v})),$$
$$\mathcal{QT}^{\mathsf{emb}}_{6,6}(\hat{X}, \hat{v}; \mathbf{c}\Theta_{\max \geq 4}, \mathbf{c}\Theta_{\max \geq 4}) \xrightarrow{\approx} \pi^2(\partial_1^+ \hat{X}(\hat{v}) / \partial_2^- \hat{X}(\hat{v})),$$

whose targets are the second cohomotopy group $\pi^2(\partial_1^+ \hat{X}(\hat{v}) / \partial_2^- \hat{X}(\hat{v}))$ of the singular connected surface $\partial_1^+ \hat{X}(\hat{v}) / \partial_2^- \hat{X}(\hat{v})$. This cohomotopy group is isomorphic to $\mathbb{Z}$ via the degree invariant. Thus, all the three types of quasitopy classes are determined by this degree. When $\partial_2^- \hat{X}(\hat{v}) = \emptyset$, we may replace $\pi^2(\partial_1^+ \hat{X}(\hat{v}) / \partial_2^- \hat{X}(\hat{v}))$ by $\pi^2(\hat{X})$. ◇

8.10 Characteristic classes of convex pseudo-envelops

In this section, we will use the cohomology $H^*(\mathcal{P}_d^{\mathbf{c}\Theta}; \mathbb{Z})$ of the classifying space $\mathcal{P}_d^{\mathbf{c}\Theta}$ (see [KSW2] and Chapter 6) to produce a variety of characteristic classes (see [MiS]) of convex pseudo-envelops.

Since any convex pseudo-envelop $\alpha : (X, \alpha^\dagger(\hat{v})) \to (\hat{X}, \hat{v})$ produces an immersion $\alpha^\partial : \partial_1 X \to \hat{X}$, the following assertion follows directly from Theorem 7.4.

Theorem 8.7. *Let $(\hat{X}, \hat{v})$ be a convex pair. Let $d' = d$, $d \equiv 0 \mod 2$, and $\Theta' = \Theta$.*

Then any convex pseudo-envelop $\alpha : (X, \alpha^\dagger(\hat{v})) \to (\hat{X}, \hat{v})$ induces a characteristic homomorphism $(\Phi^\alpha)^*$ from the $*$-homology of the differential complex

$$\{(\partial^\#)^* : \mathbb{Z}[\Theta^\#_{\langle d]}]^* \to \mathbb{Z}[\Theta^\#_{\langle d]}]^*\},$$

dual to the differential complex in (6.12), to the cohomology $H^*(\hat{X}; \mathbb{Z})$, and, via α^*, further to the cohomology $H^*(X; \mathbb{Z})$.

The $(d, d; \mathbf{c}\Theta, \mathbf{c}\Theta)$-quasitopic pseudo-envelops/envelops induce the same characteristic homomorphisms. ◇

Let us return the Arnold-Vassiliev case [Ar1], [Va3] of real polynomials with moderate singularities. Let $\Theta_{\mathsf{max} \geq k} \subset \mathbf{\Omega}_{\langle d]}$ be the closed poset, consisting of ω's with the maximal entry $\geq k$. For $k \geq 3$, the cohomology ring

$$\mathcal{V}ass_{d,k} =_{\mathsf{def}} H^*(\mathcal{P}^{\mathbf{c}\Theta_{\max \geq k}}, pt; \mathbb{Z})$$

is multiplicatively generated over $\mathbb{Z}$ by the elements $\{e_m\}_{m \leq d/k}$ of the degrees $\deg(e_m) = m(k-2)$, subject to the relations (7.38) (see [Va3], Theorem 1 on page 87).

Combining Lemma 8.3 with Proposition 7.11, we get the following assertion.

Proposition 8.6. *Let $k \geq 3$. Consider a convex pseudo-envelop $\alpha : (X, v) \to (\hat{X}, \hat{v})$ whose tangency patterns to the $\hat{v}$-flow belong to $\mathbf{c}\Theta_{\mathsf{max} \geq k} \subset \mathbf{\Omega}_{\langle d]}$ (i.e., all the tangencies are k-moderate).*

Then α generates a characteristic ring *homomorphism*

$$(\Phi^\alpha)^* : \mathcal{V}ass_{d,k} \to H^*(\partial_1^+ \hat{X}(\hat{v}), \partial_2^- \hat{X}(\hat{v}); \mathbb{Z}),$$

which is an invariant of the quasitopy class of α. In other words, we get a map

$$\Phi^*_{d,k} : \mathcal{QT}^{\mathsf{imm}}_{d,d}(\hat{X}, \hat{v}; \mathbf{c}\Theta_{\max \geq k}, \mathbf{c}\Theta_{\max \geq k}) \to$$

$$\to \mathsf{Hom}_{\mathsf{ring}}\big(\mathcal{V}ass_{d,k},\ H^*(\partial_1^+ \hat{X}(\hat{v}), \partial_2^- \hat{X}(\hat{v}); \mathbb{Z})\big).$$

In particular, for any generator $e_l \in \mathcal{V}ass_{d,k}$*, we get a characteristic element*

$$(\Phi^\alpha)^*(e_l) \in H^{l(k-2)}(\partial_1^+\hat{X}(\hat{v}), \partial_2^-\hat{X}(\hat{v}); \mathbb{Z}),$$

which is an invariant of the quasitopy class of α.

If $\partial_2^-\hat{X}(\hat{v}) = \emptyset$*, then* $(\Phi^\alpha)^*(e_l)$ *lives in* $H^{l(k-2)}(\hat{X}; \mathbb{Z}) \approx H^{l(k-2)}(\partial_1^+\hat{X}(\hat{v}); \mathbb{Z})$.

If $\hat{X}$ *is oriented and* $(n+1)$*-dimensional, using the Poincaré duality* $\mathcal{D}$ *in* $\partial_1^+\hat{X}(\hat{v})$*, we produce a homology class*

$$\mathcal{D}((\Phi^\alpha)^*(e_l)) \in H_{n-l(k-2)}(\partial_1^+\hat{X}(\hat{v}); \mathbb{Z}) \approx H_{n-l(k-2)}(\hat{X}; \mathbb{Z}),$$

which is again an invariant of the quasitopy class of α. ◇

Remark 8.4. If $(\Phi^\alpha)^*(e_l) = 0$, then, by (7.38), all $\{(\Phi^\alpha)^*(e_q)\}_{q \geq l}$ must be torsion elements in $H^{q(k-2)}(\partial_1^+\hat{X}(\hat{v}), \partial_2^-\hat{X}(\hat{v}); \mathbb{Z})$. For an orientable $\hat{X}$, they may be viewed as elements of $H_{n-q(k-2)}(\hat{X}; \mathbb{Z})$. ◇

Consider the quasitopy invariants $\{[\Sigma_k^\beta] \in \mathbf{OB}_{n-k+1}(Y, \partial Y)\}_k$ from Corollary 7.1. They are defined for oriented M and Y and the k-normal (see Definition 7.3) immersions $\beta : M \to \mathbb{R} \times Y$.

In the spirit of studying blobs in Section 5, for an orientable $\hat{X}$ and convex $\hat{v}$ on it, we have a chance to distinguish between the quasitopies of oriented convex envelops $\mathcal{OQT}^{\mathsf{emb}}_{d,d'}(\hat{X}, \hat{v}; \mathbf{c}\Theta; \mathbf{c}\Theta')$ and oriented convex pseudo-envelops $\mathcal{OQT}^{\mathsf{sub}}_{d,d'}(\hat{X}, \hat{v}; \mathbf{c}\Theta; \mathbf{c}\Theta')$.

By Theorem 8.1, for a given submersion $\alpha : (X, \alpha^\dagger(\hat{v})) \to (\hat{X}, \hat{v})$ with tangency patterns $\omega \in \mathbf{c}\Theta$, there is C^∞-perturbation $\tilde{\alpha}$ of α such that the immersion $\tilde{\alpha}^\partial$ is k-normal within quasitopy class of α. Abusing notations, put $\alpha =_{\mathsf{def}} \tilde{\alpha}$.

Note that, for $\beta = \alpha^\partial$, the normal 1-bundle $\nu^\beta = \nu(\alpha^\partial(\partial_1 X), \hat{X})$ is trivial. Thus, the first Stiefel-Whitney class $W^1(\nu_\beta) = 0$. Assuming that $\hat{X}$ is oriented, the Lashof-Smale formula (7.1) (see [LS], Theorem 1), which lives in $H^{k-1}(\partial_1 X; \mathbb{Z})$, simplifies to

$$\mathcal{D}_{\partial_1 X}([\tilde{\Sigma}_k]) = \pm\big((\alpha^\partial)^* \circ \mathcal{D}_{\hat{X}} \circ \alpha^\partial([\tilde{\Sigma}_k])\big)^{k-1}, \tag{8.10}$$

where $\mathcal{D}_{\partial_1 X}$ and $\mathcal{D}_{\hat{X}}$ are the Poincaré duality isomorphisms in the corresponding manifolds.

Employing Corollary 7.1 (see also Proposition 3.1 from [K12]) and Corollary 7.18 above, we get invariants that, when nontrivial, distinguish between $\mathcal{OQT}^{\mathsf{sub}}_{d,d'}(\hat{X}, \hat{v}; \mathbf{c}\Theta; \mathbf{c}\Theta')$ and $\mathcal{OQT}^{\mathsf{emb}}_{d,d'}(\hat{X}, \hat{v}; \mathbf{c}\Theta; \mathbf{c}\Theta')$:

Proposition 8.7. *Let $\hat{m}(\mathbf{c}\Theta_{\langle d]})$ denote the maximal entry of the collection $\{\omega \in \mathbf{c}\Theta_{\langle d]}\}$. For an orientable $(n+1)$-dimensional $\hat{X}$ and a convex $\hat{v}$, there is a map*

$$\Sigma: \mathcal{OQT}^{\mathsf{sub}}_{d,d'}(\hat{X}, \hat{v}; \mathbf{c}\Theta; \mathbf{c}\Theta')/\mathcal{OQT}^{\mathsf{emb}}_{d,d'}(\hat{X}, \hat{v}; \mathbf{c}\Theta; \mathbf{c}\Theta') \longrightarrow \tag{8.11}$$

$$\stackrel{\mathcal{F}^\partial}{\longrightarrow} \bigoplus_{k \in [2, \min\{n+1, \hat{m}(\mathbf{c}\Theta_{\langle d]})\}]} \mathbf{OB}_{n-k+1}(\hat{X}),$$

delivered by the k-intersection classes

$$\big\{[\Sigma_k^{\alpha^\partial}] \in \mathbf{OB}_{n-k+1}(\hat{X}) \approx \mathbf{OB}_{n-k+1}(\partial_1^+ \hat{X}(\hat{v}))\big\}_k$$

of immersions α^∂. The forgetful map $\mathcal{F}^\partial$ takes each pseudo-envelop α to the immersion α^∂, followed by a map that takes α^∂ to the collection of its k-intersection manifolds.

In particular, if $\partial_2^- \hat{X}(\hat{v}) = \emptyset$ [5] *and $n - k + 1 \equiv 0 \mod 4$, then the Pontryagin numbers and signatures of closed manifolds $\{\Sigma_k^{\alpha^\partial}\}_k$ represent invariants that may distinguish between $\mathcal{OQT}^{\mathsf{sub}}_{d,d'}(\hat{X}, \hat{v}; \mathbf{c}\Theta; \mathbf{c}\Theta')$ and $\mathcal{OQT}^{\mathsf{emb}}_{d,d'}(\hat{X}, \hat{v}; \mathbf{c}\Theta; \mathbf{c}\Theta')$.* $\diamondsuit$

The reader might compare these discriminating invariants with similar but more refined invariants from Section 5.3; the latter ones were based on very special combinatorial settings that take in consideration the position of the vector field $\hat{v}$ in relation to various chambers in which $\hat{X}$ is divided by $\alpha(\partial_1 X)$.

8.11 How to manufacture convex envelops with desired combinatorial tangency patterns

Let us recall one classical construction, leading to the Alexander duality. Let $K \subset S^d$ be a CW-subcomplex of the d-sphere. For an element $a \in H_p(K; \mathbb{Z})$, we denote by $c_a \in H_{p+1}(S^d, K; \mathbb{Z})$ the unique element such that $\partial_*(c_a) = a$. The Alexander duality operator $\mathcal{A}\ell$ is defined by the formula $\mathcal{A}\ell(a) =_{\mathsf{def}} \mathcal{D}(c_a) \in H^{d-p-1}(S^d \setminus K; \mathbb{Z})$, where $\mathcal{D}$ is the Poincaré duality operator, the inverse of the cap product operator $[S^d]\cap$. Pick an element $b \in H_{d-p-1}(S^d \setminus K; \mathbb{Z})$. Then the linking number of a, b is defined by the formula $\mathsf{lk}(a, b) =_{\mathsf{def}} \langle \mathcal{A}\ell(a), b\rangle$, where $\langle \sim, \sim \rangle$ is the natural pairing between cohomology and homology of the dimension $n - p - 1$.

[5]This assumption implies that $\partial_1^+ \hat{X}(\hat{v})$ is a closed manifold; thus, $\hat{X}$ is the product $[0, 1] \times \partial_1^+ \hat{X}(\hat{v})$.

The next lemma is instrumental in producing examples of convex envelopes with prescribed combinatorial patterns of their trajectories.

Lemma 8.5. *For any element* $\omega \in \mathbf{\Omega}_{\langle d]}$, *with the help of the differential* $\partial(\omega)$, *given by formula* (6.10), *the Alexander duality* $\mathcal{A}\ell$ *produces a cohomology class*

$$\theta_\omega =_{\mathsf{def}} \mathcal{A}\ell([\partial\bar{\mathsf{R}}^\omega]) \in H^{|\omega|'}(\mathcal{P}_d^{\mathbf{c}\{\omega_\succ\}};\mathbb{Z}),$$

where $\omega_\succ$ *denotes the set of elements from the poset* $\mathbf{\Omega}_{\langle d]}$ *that are smaller than* ω *(thus,* $\omega \in \mathbf{c}\{\omega_\succ\}$*), and* $\partial\bar{\mathsf{R}}^\omega$ *denotes the algebraic boundary of the cell* $\bar{\mathsf{R}}^\omega$.

Proof. For any $\omega \in \mathbf{\Omega}_{\langle d]}$, take the closed poset $\omega_\succ =_{\mathsf{def}} \{\omega' \prec \omega\} \subset \mathbf{\Omega}_{\langle d]}$ for the role of Θ in Corollary 6.6. We denote by $\bar{\mathsf{R}}_d^\omega$ the one-point compactification of the closed cell $\mathsf{R}_d^\omega \subset \mathcal{P}_d$ (the interior $(\mathsf{R}_d^\omega)^\circ$ of R_d^ω is an open $(d-|\omega|')$-ball). Then the differential $\partial(\omega)$, given by the formula (6.10), represents the $(d-|\omega|'-1)$-cycle $\partial\bar{\mathsf{R}}_d^\omega$ in the chain complex $\mathcal{C}_*(\bar{\mathcal{P}}_d^{\omega_\succ};\mathbb{Z})$ (note that $\partial\bar{\mathsf{R}}_d^\omega$ is a boundary in $\bar{\mathcal{P}}_d^{\omega_\succeq}$, but not in $\bar{\mathcal{P}}_d^{\omega_\succ}$!) and thus defines a nontrivial element $[\partial\bar{\mathsf{R}}_d^\omega] \in H_{d-|\omega|'-1}(\bar{\mathcal{P}}_d^{\omega_\succ};\mathbb{Z})$. By the Alexander duality, this element produces a cohomology class

$$\theta_\omega =_{\mathsf{def}} \mathcal{A}\ell([\partial\bar{\mathsf{R}}_d^\omega]) \in H^{|\omega|'}(\mathcal{P}_d^{\mathbf{c}\{\omega_\succ\}};\mathbb{Z}).$$

□

Example 8.9. If $d = 6$, we get the following cohomology classes:

- $\theta_{(121)} = \mathcal{A}\ell\big(\bar{\mathsf{R}}_6^{(31)} - \bar{\mathsf{R}}_6^{(13)} - \bar{\mathsf{R}}_6^{(2121)} + \bar{\mathsf{R}}_6^{(1212)}\big) \in H^1(\mathcal{P}_6^{\mathbf{c}\{(121)_\succ\}};\mathbb{Z})$,
- $\theta_{(3111)} = \mathcal{A}\ell\big(\bar{\mathsf{R}}_6^{(411)} - \bar{\mathsf{R}}_6^{(321)} + \bar{\mathsf{R}}_6^{(312)}\big) \in H^2(\mathcal{P}_6^{\mathbf{c}\{(3111)_\succ\}};\mathbb{Z})$,
- $\theta_{(31)} = \mathcal{A}\ell\big(\bar{\mathsf{R}}_6^{(4)} - \bar{\mathsf{R}}_6^{(231)} + \bar{\mathsf{R}}_6^{(321)} - \bar{\mathsf{R}}_6^{(312)}\big) \in H^2(\mathcal{P}_6^{\mathbf{c}\{(31)_\succ\}};\mathbb{Z})$,
- $\theta_{(1221)} = \mathcal{A}\ell\big(\bar{\mathsf{R}}_6^{(321)} - \bar{\mathsf{R}}_6^{(141)} + \bar{\mathsf{R}}_6^{(123)}\big) \in H^2(\mathcal{P}_6^{\mathbf{c}\{(1221)_\succ\}};\mathbb{Z})$. ♢

Theorem 8.8 below gives a simple recipe for manufacturing traversing flows with the desired number of v-trajectories of a given combinatorial type ω. Although the resulting construction is explicit, the topological nature of the pull-back X is opaque due to, paraphrasing René Thom, "the misterious nature of transversality".

Theorem 8.8. *For any* $\omega \in \mathbf{\Omega}_{\langle d]}$, *let* $\theta_\omega \in H^{|\omega|'}(\mathcal{P}_d^{\mathbf{c}\{\omega_\succ\}};\mathbb{Z})$ *be the cocycle that takes the value* $\mathsf{lk}(\partial\bar{\mathsf{R}}^\omega, \zeta)$ *on each* $|\omega|'$*-dimensional cycle* $\zeta \in H_{|\omega|'}(\mathcal{P}_d^{\mathbf{c}\{\omega_\succ\}};\mathbb{Z})$. *Let* Y *be an oriented closed and smooth* $|\omega|'$*-dimensional manifold. We denote by* $\hat{v}$ *a non-vanishing vector field that is tangent to* *the fibers of the obvious projection* $\mathbb{R} \times Y \to Y$. *Let a map* $\Phi : Y \to \mathcal{P}_d^{\mathbf{c}(\omega_\succ)}$ *be* $(\partial\mathcal{E}_d)$*-regular in the sense of Definition 7.4*[6].

- *Then the natural coupling*
$$\langle \Phi^*(\theta_\omega), [Y] \rangle = \langle \Phi^*(\mathcal{A}\ell(\partial\bar{\mathsf{R}}_d^\omega)), [Y] \rangle$$
gives the oriented *count of the* $\hat{v}$*-trajectories of the combinatorial type* ω *in the* $(|\omega|' + 1)$*-dimensional convex envelop* $(\mathbb{R} \times Y, \hat{v})$ *of the pair*
$$\big(X_\Phi =_{\mathsf{def}} \{(u, y)|\ \Phi(y)(u) \leq 0\},\ \hat{v}\big) \subset \big(\mathbb{R} \times Y,\ \hat{v}\big).$$
The combinatorial types of the $\hat{v}$*-trajectories relative to* $\partial_1 X_\Phi$ *belong to the poset* $\mathbf{c}\{\omega_\succ\} = \omega_\preceq =_{\mathsf{def}} \{\omega' \in \mathbf{\Omega}_{\langle d]}|\ \omega' \succeq \omega\}$.
- *If the map* Φ *is transversal to the open cell* $(\mathsf{R}_d^\omega)^\circ$ *in* $\mathcal{P}_d^{\mathbf{c}\{\omega_\succ\}}$, *then the number of* $\hat{v}$*-trajectories in* $\mathbb{R} \times Y$, *whose intersection with* $\partial_1 X_\Phi$ *has the combinatorial type* ω, *equals the cardinality of the intersection* $\Phi(Y) \cap (\mathsf{R}_d^\omega)^\circ$. *Thus the number of such trajectories is greater than or equal to the absolute value* $|\langle \Phi^*(\theta_\omega), [Y] \rangle|$.

Proof. The argument is similar to the proof of Lemma 8.5. By Corollary 6.5, the homology class of the cycle $\partial\bar{\mathsf{R}}_d^\omega$ in $H_{d-|\omega|'-1}(\bar{\mathcal{P}}_d^{\{\omega_\succ\}}; \mathbb{Z})$, represented by formula (6.10), is nontrivial since $d - |\omega|' - 1$ is the top grading of the differential complex $(\mathbb{Z}[\omega_\succ], \partial)$.

Now the argument amounts to spelling out the nature of Alexander duality $\mathcal{A}\ell$. By its definition, $\langle \mathcal{A}\ell(\partial\bar{\mathsf{R}}_d^\omega), \zeta \rangle = \mathsf{lk}(\partial\bar{\mathsf{R}}_d^\omega, \zeta)$ for any $\zeta \in H_{|\omega|'}(\mathcal{P}_d^{\mathbf{c}\{\omega_\succ\}}; \mathbb{Z})$. Thus for any oriented closed smooth $|\omega|'$-dimensional manifold Y and its $(\partial\mathcal{E}_d)$-regular map $\Phi : Y \to \mathcal{P}_d^{\mathbf{c}(\omega_\succ)}$, we get
$$\langle \Phi^*(\theta_\omega), [Y] \rangle = \langle \theta_\omega, \Phi_*([Y]) \rangle = \mathsf{lk}(\partial\bar{\mathsf{R}}_d^\omega,\ \Phi(Y)) = \bar{\mathsf{R}}_d^\omega \circ \Phi(Y).$$

Examining the construction of the space $X_\Phi \subset Y \times \mathbb{R}$, given by $(\mathsf{id} \times \Phi)^{-1}(\mathcal{E}_d)$, where the hypersurface $\mathcal{E}_d$ was introduced in (7.4), the latter intersection number gives an *oriented* count of the $\hat{v}$-trajectories in $\mathbb{R} \times Y$ of the combinatorial type ω relative to the boundary $\partial_1 X_\Phi$.

If $\Phi : Y \to \mathcal{P}_d^{\mathbf{c}(\omega_\succ)}$ is transversal to the open cell $(\mathsf{R}_d^\omega)^\circ$, then the cardinality of the geometric intersection $\Phi(Y) \cap (\mathsf{R}^\omega)^\circ$ is exactly the total number of $\hat{v}$-trajectories of the combinatorial type ω with respect to ∂X_Φ. The intersection points from $\Phi(Y) \cap (\mathsf{R}_d^\omega)^\circ$ (equivalently, the trajectories of the type ω with respect to $\partial_1 X_\Phi$) come in two flavors, $\{\oplus, \ominus\}$, depending on whether the canonicalal normal orientation of the cell $(\mathsf{R}_d^\omega)^\circ$ in $\mathcal{P}_d$ agrees or disagrees with the preferred orientation of the cycle $\Phi(Y)$. □

[6]By Corollary 7.2, such maps form and open and dense set in the space of all smooth maps.

Combining formula (6.10) for $\partial\omega$ with Theorem 8.8 leads to the following claim.

Corollary 8.11. *Take* $\omega = (1, \underbrace{2, \dots, 2}_{n}, 1)$. *Let a* $(\partial\mathcal{E}_d)$*-regular map* $\Phi : Y^n \to \mathcal{P}_{2n+2}^{\mathbf{c}\{\omega_\succ\}}$ *be as in Theorem 8.8.*

Then the oriented *number of trajectories of the combinatorial type* ω *in the pull-back* $X_\Phi^{n+1} \subset \mathbb{R} \times Y^n$ *equals the linking number of the cycle* $\Phi(Y^n)$ *with the* $(n+1)$*-cycle*

$$\partial\bar{\mathsf{R}}_{2n+2}^{(12\dots21)} = \bar{\mathsf{R}}_{2n+2}^{(32\dots21)} - \bar{\mathsf{R}}_{2n+2}^{(142\dots21)} + \dots + (-1)^{n-1}\,\bar{\mathsf{R}}_{2n+2}^{(12\dots241)} + (-1)^n\,\bar{\mathsf{R}}_{2n+2}^{(12\dots23)}.$$

In particular, the number of ω*-tangent trajectories in* X_Φ^{n+1} *is at least*

$$|\,\mathsf{lk}(\Phi(Y^n),\, \partial\bar{\mathsf{R}}_{2n+2}^{(12\dots21)})| = |\langle h^*(\theta_{(12\dots21)}),\, [Y]\rangle|. \qquad \diamondsuit$$

8.12 Questions and possible trails to investigate

Recall that the following basic, however, non-trivial question animates many of our investigations in this book and its prequel [K7]:

"For a given closed poset $\Theta \subset \mathbf{\Omega}_{\langle d]}$, *what are the restrictions on the topology of compact smooth manifolds* X *that admit traversing* v*-flows (or their pseudo-envelops/envelops* $(\hat{X}, \hat{v})$*) whose tangency patterns avoid* Θ*?"*

In particular, what smooth topological types X may arise via classifying maps Φ, say, as in Theorem 8.8 or Corollary 8.11?

For $d \equiv 0 \mod 2$, we do not have any problems with manufacturing $(\partial\mathcal{E}_d)$-regular maps from a given compact n-manifold Y to $\mathcal{P}_d^{\mathbf{c}\Theta}$ and, using such maps, to produce embeddings $\alpha : X \subset \mathbb{R} \times Y$ and traversing flows on X that avoid the Θ-patterns. However, the *topological types* of such X's are beyond our control: we get what we get... The resulting X is subject to many restrictions, whose nature we do not understand conceptually. Let us sketch just a couple examples which indicate that the Θ-imposed restrictions on the topology of X can be severe.

In [K] and [K7], Chapters 2-4, we proposed some answers to the previous question for 3-folds X with boundary. For them, the minimal number $gc(X)$ of v-trajectories of the combinatorial type $\omega = (1221)$ was introduced in [K] and [K7] as a measure of complexity of a 3-fold X. This number, called the gradient complexity, was linked directly to the Matveev's combinatorial complexity $c(X)$ defined by counting the strongest isolated singularities of

2-spines of X [Mat]. Via this link, the gradient complexity $gc(X)$ is related to the classification of 3-folds.

For instance, a connected compact oriented 3-fold with a simply-connected boundary, which admits a traversing flow that avoids the tangency pattern $\Theta = (1221)$, is a connected sum of several 3-balls and products $S^1 \times S^2$ ([K7], Theorem 3.14). Therefore, no other 3-folds with a simply-connected boundary can admit a convex pseudo-envelop or even a traversing flow that avoids the tangency pattern (1221). Moreover, no connected compact oriented 3-fold X with a simply-connected boundary has the gradient complexity $gc(X) = 1$ ([K7], Theorem 3.15).

Switching gears, consider a closed hyperbolic $(n+1)$-manifold Z. Let X be obtained from Z by deleting a number $(n+1)$-balls. Then no such X can support a traversing v-flow that avoids the set of isolated $\hat{v}$-trajectories of combinatorial types $\{\omega \in \mathbf{\Omega}_{|\sim|'=n}\}$. Indeed, the *positivity* of Gromov's simplicial semi-norm $\|Z\|_\Delta$ rules out the possibility of such an avoidance (see [AK], Theorem 1). In fact, the positivity of simplicial semi-norms of various homology classes $h \in H_*(X;\mathbb{R})$ is among few general mechanisms known to us that impose constraints on the combinatorial tangency types of any generic traversing flow on X (see [A], [AK], [K10]).

• We conjecture that more refined the invariants, distinguishing between the quasitopies $\mathcal{QT}^{\mathsf{emb}}_{d,d'}(\hat{X}, \hat{v}; \mathbf{c}\Theta; \mathbf{c}\Theta')$ and $\mathcal{QT}^{\mathsf{sub}}_{d,d'}(\hat{X}, \hat{v}; \mathbf{c}\Theta; \mathbf{c}\Theta')$ of convex envelops and pseudo-envelops, may be constructed along the lines developed in Section 5.3 for blobs on ruled pages. They will take in consideration the position of the vector $\hat{v}$ in relation to model singularities from the list $\mathbf{c}\Theta$.

• It seems very probable that many results of Chapters 7 and 8, with minor adjustments, are valid for immersions and embeddings into products $S^1 \times Y$ (instead of the products $\mathbb{R} \times Y$ studied in Chapters 7 and 8). This anticipation is based on the computations in Chapter 6 of the homology $H_*(\mathcal{B}^{\mathbf{c}\Theta}_d; \mathbb{Z})$ of the spaces $\mathcal{B}^{\mathbf{c}\Theta}_d$ of real binary homogeneous forms with Θ-constrained real divisors on the circle $\mathbb{R} \cup \infty \approx S^1$. We suggest that immersions $\beta : M \to S^1 \times Y$ ($\dim M = \dim Y$) and submersions $\alpha : X \to S^1 \times Y$ ($\dim X = \dim Y + 1$) that avoid cyclic tangency patters Θ to the fibers of the obvious map $S^1 \times Y \to Y$ generate classifying maps $Y \to \mathcal{B}^{\mathbf{c}\Theta}_d$, whose homotopy class is well-defined. Using Theorem 6.4, this observation should lead to computations of the appropriate notions of ciclic quasitopies. We do not anticipate significant obstructions along this path.

• We have seen that the spaces $\mathcal{P}^{\mathbf{c}\Theta}_d$ capture faithfully the Θ-avoiding quasitopies of convex envelops $(X, v) \subset (\hat{X}, \hat{v})$, but do not capture faithfully the quasitopies of similar convex *pseudo*-envelops. That is, $\mathcal{P}_d^{\mathbf{c}\Theta}$ is not the universal classifying space for the quasitopies of convex pseudo-envelops.

We speculate that, for $d \equiv 0 \mod 2$, a resolution $\Pi : \mathcal{Q}_d^{\mathbf{c}\Theta} \to \mathcal{P}_d^{\mathbf{c}\Theta}$ of the space $\mathcal{P}_d^{\mathbf{c}\Theta}$ will do a better job of classifying Θ-avoiding pseudo-envelops.

Let us depict in large strokes a construction of the classifying space $\mathcal{Q}_d^{\mathbf{c}\Theta}$. For each polynomial $P(u) \in \mathcal{P}_d$, consider all the decompositions of $P(u)$ into products $\prod_\kappa Q_\kappa(u)$ of monic real *quadratic* polynomials $Q_\kappa(u)$. The order of these quadratic polynomials in the product is unimportant. Crudely, this decomposition is equivalent to introducing couplings among the roots of $P(u)$. By definition, complex conjugate roots are coupled; a real root of multiplicity 2 may be coupled to itself. The map $\Pi : \mathcal{Q}_d \to \mathcal{P}_d$ takes each quadratic decomposition $\prod_\kappa Q_\kappa(u)$ of $P(u)$ to $P(u)$. In particular, for each cell $(\mathsf{R}_d^\omega)^\circ \subset \mathcal{P}_d$, the preimage $\Pi^{-1}((\mathsf{R}_d^\omega)^\circ)$ consists of as many copies of $(\mathsf{R}_d^\omega)^\circ$ as the number of properly defined couplings among the elements of the combinatorial pattern ω.

We conjecture that $\mathcal{QT}_{d,d}^{\mathsf{sub}}(\hat{X}, \hat{v}; \mathbf{c}\Theta; \mathbf{c}\Theta)$ is isomorphic to the set of homotopy classes $[(\partial_1^+ \hat{X}(\hat{v}), \partial_2^- \hat{X}(\hat{v})), (\mathcal{Q}_d^{\mathbf{c}\Theta}, \star)]$. In particular, the challenge is to compute the (co)homology of the space $\mathcal{Q}_d^{\mathbf{c}\Theta}$. This task is beyond the scope of our book.

For now, the tired author and his imaginary readers have reached the end of their long and challenging journey to the Universe of Traversing Flows.

Bibliography

[A] Alpert, H., *Using Simplicial Volume to Count Maximally Broken Morse Trajectories*, 2015.

[AK] Alpert, H., Katz, G., *Using Simplicial Volume to Count Multi-tangent Trajectories of Traversing Vector Fields*, Geometriae Dedicata 180, 323-338, (2016) (arXiv:1503.02583v1 [math.DG] (9 Mar 2015)).

[Ar] Arnold, V. I., *Topological Invariants of Plane Curves and Caustics*, University Lecture Series 5, Publications AMS, 1994.

[Ar1] Arnold V. I., *Spaces of Functions with Moderate Singularities*, Func. Analysis & Its Applications, 23(3), 1-10 (1989) (Russ).

[Ar2] Arnold, V. I., *Springer numbers and morsification spaces*, J. Algebraic Geom., 1 (2) (1992), 197-214 REF].

[Ba] Barannikov, S. A., *The framed Morse complex and its invariants*, Advances in Soviet Mathematics, Vol. 21, V. I. Arnold Ed., 1994, 93-115.

[Be] Besse, A., *Manifolds all of whose Geodesics are Closed*, Springer-Verlag, Berlin, Heidelberg, New York, 1978.

[Bek] Bekenstein, A., (1972). *Black holes and the second law*, Lettere al Nuovo Cimento, 4 (15): 99-104. doi:10.1007/BF02757029.

[Bi] Birkhoff, G. D., *Proof of the ergodic theorem*, Proc. Natl. Acad. Sci. USA, 17 (12): 656-660, (1931).

[Bj] Björner, A., *Topological methods*, In: Handbook of combinatorics, Vol. 2, 1819-1872, Elsevier Sci. B. V., Amsterdam, 1995.

[Bred] Bredon, G., *Introduction to Compact Transformation Groups*, Academic Press, New York - London, 1972.

[Bries] Brieskorn, E. V., *Examples of singular normal complex spaces which are topological manifolds*, Proceedings of the National Academy of Sciences, 55 (6): 1395-1397 (1966).

[Br] Browder, W., (1969), *The Kervaire invariant of framed manifolds and its generalization*, Annals of Mathematics 90 (1): 157-186.

[Bo] Boardman, J. M., *Singularities of Differential Maps*, Publ. Math. I.H.E.S. 33 (1967), 21-57.

[B] Bott, R., *Lectures on Morse theory, old and new*, Bulletin of the American Mathematical Society (N.S.), 7 (2): 331-358, (1982). doi:10.1090/s0273-0979-1982-15038-8.

[BT] Bott, R., Tu, L., *Differential Forms in Algebraic Topology*, New York: Springer, 1982, ISBN 0-387-90613-4.

[BEM] Borman, M., Eliashberg, Y., Murphy, E., *Existence and classification of overtwisted contact structures in all dimensions*, arXiv:1404.6157v2 [math.SG] 12 Oct 2014.

[BCG] Besson, G., Courtois, G., Gallot, S., *Minimal entropy and Mostov's rigidity theorems*, Ergod. Th. and Dynam. Sys. (1996), 16, 623-649.

[BNR] Borodzik, M., Ne'methi, A., Ranicki, A., *Morse Theory for Manifolds with Boundary*, arXiv:1207.3066v4 [math.GT] 16 Sep 2014.

[BFKK] Bartholomew, A., Fenn, R., Kamada, N., Kamada, S., *Doodles on Surfaces*, J. Knot Theory Ramifications, 27 (2018), no. 12, 1850071, 26 pages, MR 3876348.

[BFK] Burago, D., Ferleger, S., Kononenko, A, *Uniform estimates of the number of collisions in semi-dispersing billiards*, Annals of Math., 147 (1998), 695-708.

[BS] Bunimovich, L. A., Sinai, Ya.,G. *On a fundamental theorem in the theory of dispersing billiards*, Mathematics of the USSR-Sbornik, v.19, issue 3, (1973), 407-432.

[BK] Bunimovich, L., Katz, G., *Detecting intrinsic global geometry of an obstacle via layered scattering*, CHAOS, 2022 Jul; 32 (7): 073112. doi:10.1063/5.0091256.

[Ca] Calabi, E., *An Intrinsic Characterization of Harmonic 1-Forms*, Global Analysis, Papers in Honor of K. Kodaira, D.C. Spenser and S. Iyanaga, Eds., 1969, 101-117.

[Can] Cannas da Silva, A., *Lectures on Symplectic Geometry, Corrected 2nd printing*, Springer-Verlag, 2008. pp. 147-148. ISBN 978-3-540-42195-5.

[CS] Cappell, S., Shaneson J., *Some New Four-manifolds*, Annals of Mathematics, 104 (1976), 61-72.

[Car] Cartan, É., *Sur les espaces de Finsler* (French), C.R. Acad. Sci., Paris 196 (1933), 582-586.

[CTGM] Cappell, S., DeTurk, D., Gluck, H., Miller, E., *Cohomology of Harmonic Forms on Riemannian Manifolds With Boundary*, arXiv: math/0508372v1 [math.DG], 19 Aug 2005.

[CKS] Carter, S. J., Kamada, S., Saito, M., *Doodles on Surfaces*, J. Knot Theory Ramifications, 11 (2002), no. 3, 311-322.

[Cerf] Cerf, J., *Sur les diffeomorphismes de la sphere de dimension trois* ($\Gamma_4 = 0$), Lecture Notes in Mathematics, vol. 53, (1968), Berlin - New York: Springer-Verlag.

[Ch] Chernov, N., *Average Length of a Free Trajectory in the Billiard Systems*, Mathematics Education, 3rd series, issue 5, 2001, pp. 100-105 (Russ.).

[CM] Chernov, N., Markarian, R., *Chaotic Billiards*, Math. Surveys and

Monographs, 127, AMS, Providence RI, 2006.

[ChMa] Chen, L., Mann, K., *Structure Theorems for Actions of Homeomorphism Groups*, Duke Math. J. 172 (5), 915-962, 1 April 2023. (arXiv:1902.05117v3 [math.GT] 9 Aug 2020).

[Cr] Croke, C., *Some isoperimetric inequalities and eigenvalue estimates*, Annales scientifiques de l' É.N.S., 4e série, tome 13, $n°4$ (1980), p. 419-435.

[Cr1] Croke, C., *Rigidity Theorems in Riemannian Geometry*, Chapter in Geometric Methods in Inverse Problems and PDE Control, C. Croke, I. Lasiecka, G. Uhlmann, and M. Vogelius eds., IMA vol. Math Appl., 137, Springer, 2004.

[Cr2] Croke, C., *Rigidity and the distance between boundary points*, J. Diff. Geometry, 33 (1991), 445-464.

[Cr3] Croke, C., *Scattering Rigidity with Trapped Geogesics*, Ergodic Theory and Dynamical Systems, v.34, issue 3, (2014), 826-836.

[CK] Croke C., Kleiner, B., *Conjugacy and rigidity for manifolds with a parallel vector field*, J. Diff. Geom. 39 (3) (1994), 659-680.

[CEK] Croke, C., Eberlein P., Kleiner, B., *Conjugacy and rigidity for nonpositively curved manifolds of higher rank*, Topology 35 (1996), 273-286.

[CF] Conner, P. E., Floyd, E. E., *Differentiable Periodic Maps*, Springer-Verlag, 1964.

[C] Cohen, R. L., *Topics in Morse Theory*, Stanford University, 1991.

[D] Dieudonné, J., *Treatise on Analysis, 3*, New York-London: Academic Press, Inc. (1972).

[Do] Donaldson, S. K., *An application of gauge theory to the topology of* 4-*manifolds*, J. Diff. Geom. 18 (1983), 279-315.

[DS] Dunford, N., Schwartz, J. T., *Linear Operators, Part 1: General Theory*, Interscience, New York, London, (1958).

[E] Eliashberg, Y., *Classification of overtwisted contact structures on 3-manifolds*, Invent. Math. 98 (1989), no. 3, 623-637.

[E1] Eliashberg, Y., *Singularities of Folding Type*, Izv. Akad Nauk, 34 (1970), 1110-1126.

[E2] Eliashberg, Y., *Surgery of Singularities of Smooth Mappings*, Izv. Akad Nauk, 36 (1972), 1321-1347.

[EG] Eliashberg, Y. M., Gromov, M. L., *Construction of a Smooth Map with a Given Jacobian*, Functional Analysis and its Applications, v.7, issue 1, (1973) 33-40 (Russ).

[EM] Eliashberg, Y., Mishachev, N., *Introduction to the h-principle*, Graduate Studies in Mathematics, vol. 48, American Mathematical Society (2002).

[ET] Eliashberg, Y., Thurston, W., *Confoliations*, University Lecture Series, vol. 13, American Mathematical Society, Providence, Rhode Island, 1998.

[EKA] El Kacimi-Alaoui, A., *Opérateurs transversalement elliptiques sur un feuilletage riemannien et applications.* Compos. Math. 73, 57-106 (1990).

[EKH] El Kacimi-Alaoui, A., Hector, G. *Dĕcomposition de Hodge basique pour un feuilletage riemannien*, Ann. Inst. Fourier 36, 207-227 (1987).

[EST] Ekholm, T., Szücs, A., Terpai, T., *Bordism groups of fold maps and maps with prescribed number of cusps*, Kyushu Journal of Mathematics (2007), DOI: 10.2206/kyushujm.61.395.

[F] Farber, M., *Topology of Closed One-Forms*, Mathematical Surveys and Monographs 108, AMS, 2004.

[FKL] Farber, M., Katz, G., Levine, J., *Morse Theory of Harmonic Forms*, Topology 37 (1998), 469-483.

[Fed] Federer, H., *Geometric measure theory*, Die Grundlehren der mathematischen Wissenschaften, 153, New York: Springer-Verlag (1969), pp. xiv+676, ISBN 978-3-540-60656-7, MR 0257325

[FM] Friedman, R., Morgan, J. W., *Gauge theory and the topology of four-manifolds*, (Vol. 4), American Mathematical Soc., 1998.

[Ga] Gamelin, T. W., *Uniform Algebras*, Prentice-Hall Series in Modern Analysis, Prentice-Hall, Inc. Englewood Cliffs, N.J., 1969.

[Gab] Gabai, David, *The Smale conjecture for hyperbolic 3-manifolds:* $\mathrm{Isom}(M^3) \simeq \mathrm{Diff}(M^3)$, J. Differential Geom. 58 (2001), 113-149.

[Ge] Geiges, H., *Contact Geometry*, Handbook of Differential Geometry, vol. 2, (F.J.E. Dillen and L.C.A. Verstraelen, eds.)

[Gray] Gray, J. W., *Some global properties of contact structures*, Ann. of Math. (2) 69 (1959), 421-450.

[Gray1] Gray, A., *Tubes*, Progress in Mathematics, vol. 221 (second edition), Birkhauser-Verlag, Basel-Boston-Berlin, 2004, ISBN 3-7643-6907-8.

[Grot] Grothendieck *Produits tensoriels topologiques et espaces nucléaires* (in French), Providence: American Mathematical Society, (1966). ISBN 0-8218-1216-5. OCLC 1315788.

[God] Godement, R., *Topologie Algébraique et Théorie des Faisceaux*, Publications de l'Institut de mathématique de Universite de Strasbourg XIII, Hermann, Paris, 1958.

[Good] Goodman, S., *Vector Fields with Transversal Foliations*, Topology, vol. 24, no. 3, 333-340 (1985).

[Gor] Goresky, M., *Triangulations of stratified objects*, Proc. Amer. Math. Soc., 72 (1978), 193-200.

[Got] Gottlieb, D. H., *All the Way with Gauss-Bonnet and the Sociology of Mathematics*, Math. Monthly, 103 (1996), 457-469.

[Got1] Gottlieb, D. H., *On the Index of Pullback Vector Fields*, Differential Topology Proceedings, Lecture Notes in Mathematics 1350, 167-170 (1988).

[GRS] Gelfand, I., Raikov, D., Shilov, G., *Commutative Normed Rings* (Russian), State Publishing House of Physics-Mathematics Literature, Moscow 1960.

[GG] Golubitsky, M., Guillemin, V., *Stable Mappings and Their Singularities*, Graduate Texts in Mathematics 14, Springer-Verlag, New York, Heidelberg, Berlin, 1973.

[GL] Gordon, C., Luecke, J., *Knots are determined by their complements*, J. Amer. Math. Soc. 2 (1989), no. 2, 371-415.

[GM] Goresky, M., MacPherson, R., *Stratified Morse Theory*, Proceedings of Symposia in Pure Mathematics, Vol. 40 (1983), Part 1, 517-533.

[GM1] Goresky, M., MacPherson, R., *Morse theory for the intersection homology groups*, Analyse et Topologie sur les Espaces Singulieres, Astérisque #101 (1983), 135-192, Société Mathématique de France.

[GM2] Goresky, M., MacPherson, R., *Stratified Morse Theory*, Springer Verlag, N. Y. (1989), Ergebnisse vol. 14.

[GNS] Gurfinkel, T., Noakes L., Stoyanov, L., *Travelling Times in Scattering by Obstacles in Curved Space*, arXiv:2003.12261v1 [math.DG] 27 March 2020.

[Gr] Gromov, M., *Volume and bounded cohomology* Publ. Math. I.H.E.S., tome 56 (1982), 5-99.

[Gr1] Gromov, M., *Singularities, Expanders and Topology of Maps. Part I: Homology versus Volume in the Spaces of Cycles*, Geom. Funct. Anal. vol 19 (2009), 743-841.

[Gr2] Gromov, M., *Partial Differential Relations*, Springer-Verlag, 1986.

[Gr3] Gromov, M. L., *Pseudo holomorphic curves in symplectic manifolds.* Inventiones Mathematicae, 82 (2): 307-347 (1985).

[Gr4] Gromov, M. L., *A Topological Technique for Construction of Solutions of Differential Equations and Inequalities*, Proceedings of ICM (1970), Nice, vol. 2, 221-225.

[Gu] Guth, L., *Minimal number of self-intersections of the boundary of an immersed surface in the plane*, arXiv:0903.3112v1 [math.DG] 18 Mar 2009.

[Ha] Hardon, D. G. C., *The Morse Complex for a Morse Function on a Manifold with Corners*, arXiv:math.GT/0406486 v1, 2004.

[Har1] Hardt, R., *Topological properties of subanalytic sets*, Trans. Amer. Math. Soc., 211 (1975), 57-70.

[Har2] Hardt, R., *Stratifications of real analytic maps and images*, Invent. Math. 28 (1975), 193-208.

[Hart] Hartshorne, R., *Algebraic Geometry*, Graduate Texts in Mathematics 52, New York: Springer-Verlag, 1977.

[Hat] Hatcher, A., *Algebraic Topology*, Cambridge University Press, 2002.

[Hat1] Hatcher, A., *A 50-Year View of Diffeomorphism Groups*, 2012 Slides from a talk at the 50-th Cornell Topology Festival (available at pi.math.cornell.edu/hatcher/Papers/Diff(M)2012.pdf).

[Hat2] Hatcher, A., *A proof of the Smale conjecture,* $\mathrm{Diff}(S^3) \simeq \mathrm{O}(4)$, Ann. of Math. (2), vol. 117, No. 3 (1983), 553-607.

[Hem] Hempel, J., *3-Manifolds*, Annals of Mathematics Studies, Princeton

University Press,1976.
[Hi] Hironaka, H., *Subanalytic sets*, Number Theory, Algebraic Geometry and Commutative Algebra, volume in honor of A. Akizuki, Kinokunya Tokyo 1973, 453-493.
[H] Hirsch, M. W., *Differential Topology*, Springer-Verlag, New York - Heidelberg - Berlin, 1976.
[H1] Hirsch, M. W., *Immersions of Manifolds*, Trans. Amer. Math. Soc. 93 (1959), 242-276.
[Haef] Haefliger, A., *Homotopy and integrability*, Manifolds-Amsterdam 1970 (Proc. Nuffic Summer School) (1971). Lecture Notes in Mathematics, vol. 197, 133-163, Berlin, New York: Springer-Verlag.
[Her] Hermann, H., *Remarks on the foundations of integral geometry*, Rend. Circ. Math. Palermo, II 9 (1960), 91-96.
[He] He, Z., *Odd dimenisonal symplectic manifolds*, MIT Ph.D thesis, 2010.
[Hawk] Hawking, S. W., *Particle creation by black holes*, (1975), Communications in Mathematical Physics. 43 (3): 199-220. doi:10.1007/BF02345020.
[Hopf] Hopf, H., *Vectorfelder in n-dimensionalen Mannigfaltigkeiten*, Math. Annalen 96 (1937), 225-250.
[Hu] Hu, S. T., *Homotopy Theory*, Academic Press, New York & London, 1959.
[Hud] Hudson, J. F. P., *Piecewise Linear Topology*, W. A. Benjamin, Inc., New York - Amsterdam, 1969.
[Huse] Husemoller, D., *Fiber Bundles*, Springer-Verlag, New York-Heidelberg-Berlin, ISBN 0-387-90103-5.
[HHR] Hill, M., Hopkins, M., Ravenel, D., *On non-existence of elements of Kervaire invariant one*, arXiv:0908.3724 [math.AT].
[HILS] De Hoop, M. V., Ilmavirta, J., Lassas, M., Saksala, T., *Determination of a compact Finsler manifold from its boundary distance map and an inverse problem in elasticity*, Communications in Analysis and Geometry, vol 31, no 7 (2023).
[I] Ishii, I., *Flows and Spines*, Tokyo J. Math. 9, No. 2, 505-525 (1986).
[I1] Ishii, I., *Moves for Flows-Spines and Topological Invariants of 3-Manifolds*, Tokyo J. Math. 15, No. 2, 297-312 (1992).
[J] Jaco, W. H., *Lectures on 3-Manifold Topology*, AMS Regional Conference Series in Mathematics 43 (1980).
[Jo] Johnson, F., *On the triangulation of stratified sets and singular varieties*, Trans. Amer. Math. Soc. 275 (1983), 333-343.
[K] Katz, G., *Convexity of Morse Stratifications and Gradient Spines of 3-Manifolds*, JP Journal of Geometry and Topology, vol. 9, no 1 (2009), 1-119.
[K1] Katz, G., *Stratified Convexity & Concavity of Gradient Flows on Manifolds with Boundary*, Applied Mathematics, 2014, vol. 5, 2823-2848. http://www.scirp.org/journal/am

[K2] Katz, G., *Traversally Generic & Versal Flows: Semi-algebraic Models of Tangency to the Boundary*, Asian J. of Math., vol. 21, No. 1 (2017), 127-168 (arXiv:1407.1345v1 [math.GT] 4 July, 2014)).

[K3] Katz, G., *The Stratified Spaces of Real Polynomials & Trajectory Spaces of Traversing Flows*, JP J. of Geometry and Topology, v. 19, No. 2 (2016), 95-160 (arXiv:1407.2984v3 [math.GT] 6 Aug 2014).

[K4] Katz G., *Flows in Flatland: A Romance of Few Dimensions*, Arnold Math. J., DOI 10.1007/s40598-016-0059-1, Springer (2016).

[K5] Katz, G., *Causal Holography in Application to the Inverse Scattering Problem*, Inverse Problems and Imaging J., June 2019, 13(3), 597-633 (arXiv:1703.08874v1 [Math.GT], 27 Mar 2017).

[K6] Katz, G., *Complexity of Shadows & Traversing Flows in Terms of the Simplicial Volume*, J. of Topology and Analysis, vol. 8, No.3 (2016), 501-543 (arXiv:1503.09131v2 [math.GT] 24 Apr 2015).

[K7] Katz, G., *Morse Theory of Gradient Flows, Concavity and Complexity on Manifolds with Boundary*, World Scientific (2019), New Jersey-London-Singapore-Beijing-Shanghai-Hong Kong-Taipei-Chennai-Tokyo, ISBN 978-981-4368-75-9.

[K8] Katz, G., *Causal Holography of Traversing Flows*, Journal of Dynamics and Differential Equations (2020) https://doi.org/10.1007/s10884-020-09910-y.

[K9] Katz, G., *The ball-based origami theorem and a glimpse of holography for traversing flows*, Qual. Theory Dyn. Syst. 19:41 (2020).

[K10] Katz, G., *Gromov's Amenable Localization and Geodesic Flows*, Qual. Theory Dyn. Syst., 20:19, (2021). https://doi.org/10.1007/s12346-021-00448-y

[K11] Katz, G., *Holography of geodesic flows, harmonizing metrics, and billiards' dynamics*, Inverse Problems and Imaging, doi: 10.3934/ipi.2023009, (arXiv:2003.10501v2 [math.DS] 1 April 2020).

[K12] Katz, G., *Spaces of polynomials with constrained divisors as Grassmanians for immersions and embeddings*, Int. J. Topology, 2025, 2, 9, https://doi.org/10.3390/ijt2030009 arXiv:2201.02744v1 [math.GT] 8 Jan 2022

[K13] Katz, G., *Spaces of polynomials with constrained divisors as Grassmanians for traversing flows*, JP Journal of Geometry and Topology, Vol. 29, No 1, (2023), Pages 47-120 (arXiv:2202.00862v2 [math.GT] 18 Apr 2022).

[K14] Katz G., *Harmonic Forms and Near-minimal Singular Foliations*, Comment. Math. Helv. 77 (2002), 39-77.

[K15] Katz, G., *Algebras of Smooth Functions and Holography of Traversing Flows*, Global Journal of Science Frontier Research: F, Mathematics and Decision Science, vol. 23, issue 2, version 1.0, 2023.

[K16] Katz, G., *Recovering Contact Forms from Boundary Data*, arXiv: 2309.14604v3 [math.SG], 11 Mar 2024.

[KSW1] Katz, G., Shapiro B., Welker, V., *Real Polynomials with Constrained Real Divisors. I. Fundamental Groups*, Journal of Topology and Analysis, Vol. 17, No 4 (2025) 1173-1203, https://doi.org/10.1142/S1793525323500553 (arXiv:1908.07941v3 [math.AT] 18 Sep 2020).

[KSW2] Katz, G., Shapiro B., Welker, V., *Spaces of polynomials and bilinear forms with constrained real divisors, II. (Co)homology & stabilization*, arXiv:2112.15205v1 [math.AT] 30 Dec 2021.

[KMS] Kriegl, A., Michhor, P., Schachermayer, W., *Characters of Algebras of Smooth Functions*, Ann. Global Anal. Geom. vol.7, No. 2 (1989), 85-92.

[Ke] Kervaire, M., *Smooth Homology Spheres and their Fundamental Groups*, Transactions of AMS, 144 : 67-72, doi:10.2307/1995269.

[KM] Kervaire, M. A., Milnor, J. W., *Groups of homotopy spheres: I.* Annals of Mathematics (Princeton University Press) 77 (3): 504-537 (1963).

[Kh] Khovanov, M., *Real $K(\pi, 1)$ arrangements from finite root systems*, Math. Research Letters 3, 261-274 (1996).

[KrMa] Kuglikov, B., Matveev, V. S., The geodesic flow of a generic metric does not admit nontrivial integrals polynomial in momenta, arXiv: 1510.01493v1 [math.DG] 6 Oct 2015.

[KA] El Kacimi-Alaoui, A., *Opérateurs transversalement elliptique sur feuilletage riemannien et applications*, Compositio Mathematica, 73, 57-106 (1990).

[KSS] Krámli, A., Simányi, N., Szász, D., *Dispersing billiards without focal points on surfaces are ergodic*, Comm. Math. Phys. 125 (3) (1989) 439-457.

[KSS1] Krámli, A., Simányi, N., Szász, D., *A "transversal" fundamental theorem for semi-dispersing billiards*, Comm. Math. Phys., 129 (1990), 535-560.

[Kos] Kosinski, A., *Differential Manifolds*, Academic Press, Boston, San Diego, New York, London, Sydney, Tokyo, Toronto, 1992.

[LS] Lashof, R., Smale, S., *Self-intersections of Immersed Manifolds*, Journal of Mathematics and Mechanics, Vol. 8, No. 1 (1959), pp. 143-157.

[La] Laudenbach, F., *Morse Complex on Manifolds with Boundary*, Geom. Dedicata, (2011), 153, 47-57.

[LP] Lax, P., Phillips, R., *Scattering Theory*, Academic Press, Amsterdam, 1967.

[L] Levine, H., *Classifying Immersions into $\mathbb{R}^4$ over Stable Maps of 3-Manifolds into $\mathbb{R}^2$*, Lecture Notes in Mathematics 1157, Springer-Verlag, 1985.

[Lin] Lin, Y., *Lefschetz Contact Manifolds and Odd Dimensional Symplectic Geometry*, arXiv:1311.1431v4 [math.SG] 2 Sep 2016.

[Lip] Lipner, G., Singularities of projected immersions revised, Algebraic & Geometric Topology, 9 (2009), 1623-1635.

[Lo] Lojasiewicz, S., *Triangulations of semi-algebraic sets*, Ann. Scu. Norm. di Pisa (1965), 449-474.

[Mal] Malgrange, B., *The preparation theorem for differential functions*, Differential Analysis (Papers presented at the Bombay Colloquium, 1964), 203-208.

[Mat] Matveev, V.S., *Hyperbolic manifolds are geodesically rigid*, Invent. Math. 151(2003) 579-609.

[Mat1] Matveev, V. S., *Closed Manifolds Admitting Metrics with the same Geodesics*, Symmetry and Perturbation Theory-Proceedings of the International Conference on SPT2004, World Scientific Publishing Co. Pte. Ltd., 2005, (ISBN # 9789812702142), pp. 198-208.

[Mat2] Matveev, V. S., *Geodesically equivalent metrics in general relativity*, arXiv:1101.2069v2 [math.DG] 7 Apr 2011.

[MatT] Matveev, V. S., Topalov, P. J., *Metrics with ergodic geodesic flow is completely determined by unparametrized geodesics*, Electron. Res. Announc. Amer. Math. Soc., 6 (2000) 98-104.

[May] May, P. J., *Weak equivalences and quasifibrations*, Lecture Notes in Mathematics, Vol. 1425, 1990, pp. 91-101.

[Me] Melrose, R. *Geometric Scattering Theory*, Cambridge University Press, Cambridge, 1995.

[Mi1] Milnor, J., *On the existence of a connection with curvature zero*, Comm. Math. Helv., 32 (1958), 215-233.

[Mi2] Milnor, J., *Lectures on the h-cobordism Theorem*, Princeton University Press, Princeton, New Jersey, 1965.

[Mi3] Milnor, J., *Morse Theory*, Annals of Mathematics Studies, No 51, Princeton University Press, Princeton, New Jersey, 1963.

[MiS] Milnor, J. H., Stasheff, J. D., *Characteristic Classes*, Annals of Mathematics Studies 76, Princeton University Press, 1974.

[Mos] Moser, J., *On the volume elements on a manifold*, Trans. Amer. Math. Soc. 120 (1965), 286-294.

[Mo] Morse, M. *Singular points of vector fields under general boundary conditions*, Amer. J. Math. 51 (1929), 165-178.

[MS] Morse, M., Van Schaack, G. B., *The critical point theory under general boundary conditions*, Ann. of Math. (2) vol. 35 (1934), no. 3, pp. 545-571.

[Morin] Morin, B., *Formes canoniques des singularities d'une application differentiable*, Comptes Rendus Acad. Sci., Paris 260, (1965), 5662-5665, 6503-6506.

[Mort] Morton, H. R., *Symmetric products of the circle*, (1967) Proc. Cambridge Philos. Soc., 63 (1967), 349-352.

[Most] Mostow, G. D., *Strong rigidity of locally symmetric spaces*, Annals of mathematics studies 78, Princeton University Press (1973), ISBN 978-0-691-08136-6, MR 0385004.

[Mr] Mrcun, J., *On isomorphisms of algebras of smooth functions*, Proceedings of AMS, vol. 133, no. 10, 3109-3113, (2005).

[Mu] Munkres, J.R., *Elementary Differential Topology*, Princeton University

Press, Princeton, (1966).

[Mu1] Munkres, J.R., *Topology: A First Course*, Prentice-Hall, Inc., Englewood Cliffs, New Jersey, (1975).

[MySt] Myers, S. B., Steenrod, N. E. (1939), *The group of isometries of a Riemannian manifold*, Ann. of Math., 2, 40 (2): 400-416, doi:10.2307/1968928, JSTOR 1968928

[Na] Nachbin, L., *Sur les algèbres denses de fonctions diffèrentiables sur une variètè*, C.R. Acad. Sci. Paris 228 (1949) 1549-1551.

[NS] Noakes, L., Stoyanov, L., *Rigidity of scattering lengths and traveling times for disjoint unions of convex bodies*, Proc. Amer. Math. Soc., 143 (2015), pp. 3879-3893.

[NS1] Noakes, L., Stoyanov, L., *Obstacles with non-trivial trapping sets in higher dimensions*, Arch. Math. 107 (2016) 73-80.

[NS2] Noakes, L., Stoyanov, L., *Lence rigidity in scattering by unions of strictly convex bodies in* $\mathbb{R}^2$, SIAM J. Math. Anal., vol. 52, No. 1, pp. 471-480.

[Pal] Palais, R. S., *On the differentiability of isometries*, Proceedings of the American Mathematical Society, 8 (4): 805-807 (1957), doi: 10.1090/S0002-9939-1957-0088000-X.

[Pa] Pappas, C., *Extensions of codimension one immersions*, Trans. of AMS, v. 348, no. 1, August 1996.

[Pan] Pang, M. Y., *Structure of Legendre Foliations*, Transactions of AMS, vol 320, no 2, 1990, 417-455.

[Pe] Penrose, L., Penrose, R., *Puzzles for Christmas*, New Scientist. 25 December 1958. 1580-1581, 1597.

[Ph] Phillips, A., *Submersions of open manifolds*, Topology 6 (1967), 171-206.

[P1] Perelman, G., *The entropy formula for the Ricci flow and its geometric applications*, (2002). arXiv:math.DG/0211159.

[P2] Perelman, G., *Ricci flow with surgery on three-manifolds*, (2003). arXiv:math.DG/0303109.

[P3] Perelman, G., *Finite extinction time for the solutions to the Ricci flow on certain three-manifolds*, (2003), arXiv:math.DG/0307245.

[Raz] Raźny, P., *Cohomology of Manifolds with Structure Group* $\mathsf{U}(n) \times \mathsf{O}(s)$, Geometriae Dedicata (2023) 217:58, https://doi.org/10.1007/s10711-023-00796-w.

[R] Rejeb, K. B., *Periodic flows with global sections*, Ark. Mat., 58 (2020), 39-56.

[S] Santaló, L., *Integral Geometry and Geometric Probability*, Cambridge Mathematical Library, Cambridge University Press (second edition 2004).

[Sa] Sasaki, S., *On the differential geometry of tangent bundle of Riemannian manifolds*, Tohoku Math. J., 10 (1958), 338-354.

[Ser] Serre, J. P., *Groups d'homotopie et classes de groups abelian*, Ann. of

Math. 58 (1953), 258-294 (jstor: 1969789)

[Sch] Schwartzman, *Asymptotic Cycles*, Ann. of Math., 66 (1957), 270-284.

[Sl] N.J.A. Sloane & Inc., T.O.F., 2020. *The on-line encyclopedia of integer sequences*, Available at: `http://oeis.org/`.

[Si] Sinai, Y. G., *Dynamical Systems with Elastic Reflections*, Russian Mathematical Surveys, 25, (1970) 137-191.

[Si1] Sinai, Y. G., *On the foundations of the ergodic hypothesis for a dynamical system of statistical mechanics*, Doklady Akademii Nauk SSSR (1963) (in Russian), 153 (6), 1261-1264 (in English, Sov. Math Dokl. 4 (1963) pp. 1818-1822).

[SiC] Sinai, Ya.G., Chernov, N., *Ergodic properties of some systems of two-dimensional disks and three-dimensional balls*, Uspekhi mat. nauk., 42 (1987), 153-174.

[Sm] Smale, S., *On the structure of manifolds*, Amer. J. Math., 84 (1962), 387-399.

[Sm1] Smale, S., *Generalized Poincare's conjecture in dimensions greater than four*, Ann. of Math. (2) 74 (1961), 39-406.

[Sm2] Smale, S., *The Classification of Immersions of Spheres in Euclidean Spaces*, Ann. of Math. 69 (1959), 327-344.

[Sp] Spanier, E. H., *Algebraic Topology*, McGraw-Hill Book Company, New York - San Francisco - St. Louis - Toronto - London - Sydney, 1966.

[SW] Shapiro B., Welker, V., *Combinatorics and topology of stratifications of the space of monic polynomials with real coefficients*, Result. Math., 33 (1998) 338-355.

[St] Stoyanov, L., *Santaloos formula and stability of trapping sets of positive measure*, J. Differential Equations 263 (2017) 2991-3008.

[Su] Sullivan, D., *A Cohomological Characterization of Foliations Consisting of Minimal Surfaces*, Comment. Math. Helvetici, 54 (1979), 218-223.

[Su1] Sullivan, D., *A Foliation of Geodesics is Characterized by Having no Tangent Homologies*, Journal of Pure and Applied Algebra, 13 (1978), 101-104.

[SU] Stefanov, P., Uhlmann, G., *Stability estimates for the X-ray transform for tensor fields and boundary rigidity*, Duke Math. J., 123(3):445-467, 2004.

[SU1] Stefanov, P., Uhlmann, G., *Boundary rigidity and stability for generic simple metrics*, J. Amer. Math. Soc., 18(4): 975-1003, 2005.

[SU2] Stefanov, P., Uhlmann, G., *Boundary and lens rigidity, tensor holography, and analytic microlocal analysis*, In Algebraic Analysis of Differential Equations. Springer, 2008.

[SU3] Stefanov, P., Uhlmann, G., *Rigidity for metrics with the same lengths of geodesics*, Math. Res. Lett., 5(1-2): 83-96, 1998.

[SU4] Stefanov, P., Uhlmann, G., *Local lens rigidity with incomplete data for a class of non-simple Riemannian manifolds*, J. Differential Geom., 82(2), 383-409, 2009.

[SUV] Stefanov, P., Uhlmann, G., Vasy, A., *Boundary rigidity with partial data*, J. Amer. Math. Soc., 29, 299-332 (2016), arXiv.1306.2995.

[SUV1] Stefanov, P., Uhlmann, G., Vasy, A., *Local recovery of the compressional and shear speeds from the hyperbolic DN map*, preprint.

[SUV2] Stefanov, P., Uhlmann, G., Vasy, A., *Inverting the local geodesic X-ray transform on tensors*, Journal d'Analyse Mathematique, to appear, arXiv:1410.5145.

[SUV3] Stefanov, P., Uhlmann, G., Vasy, A., *Local and global boundary rigidity and the geodesic X-ray transform in the normal gauge*, arXiv:1702.03638, 2017.

[Sz] Szücs, A., *On the cobordism group of cooriented, codimension one Morin maps*, Journal of Singularities, vol. 4, (2012), 196-205.

[Sz1] Szücs, A., *On the singularities of hyperplane projections of immersions*, Bull. London Math. Soc. 32 (2000), 364-374 MR1750481.

[Sz2] Szücs, A., *Cobordism of Singular Maps*, Geometry & Topology, 12 (2008) 2379-2452.

[SzTe] Szücs, A., Terpai, T., *Bordism groups of fold maps*, Acta Math. Hung. 126, 334-351 (2010). https://doi.org/10.1007/s10474-009-9070-5

[SzTe1] Szücs, A., Terpai, T., *Classifying spaces for projections of immersions with controlled singularities*, Acta Math. Hungar. 160, 273-291 (2020). https://doi.org/10.1007/s10474-020-01028-3

[Tab] Tabachnikov, S., *Geometry of Billiards*, AMS, Mathematics Advanced Study Semesters (2005).

[Th] Thom, R., *La classification des immersions*, Sémin. Bourbaki 157, 1957-58.

[Thu] Thurston, W. P., *A norm for the homology of* 3*-manifolds*, Memours of AMS, vol. 59, no. 339 (1986), 100-130. MR88h:57014.

[TuTo] Tutyshev, S. G., Toth, V. T., *Gravitational Lensing by an Extended Mass Distribution*, arXiv:2106.06696v2 [gr-qc] 5 Aug 2021.

[Va] Vasiliev, V. A., *Complements of Discriminants of Smooth Maps: Topology and Applications*, Translations of Mathematical Monographs, vol. 98, American Math. Society publication, 1994.

[Va1] Vassiliev, V. A., *Topology of spaces of functions without complicated singularities*, Funct. Anal. and its Appl., (1989), 23:4, p. 24-36.

[Va2] Vasiliev, V. A., *Homology of spaces of homogeneous polynomials in* $\mathbb{R}^2$ *without multiple zeros*, Proc. Steklov Inst. Math., 221 (1998), 133-138.

[Va3] Vassiliev V. A., *Homology of spaces of non-resultant polynomial systems in* $\mathbb{R}^2$ *and* $\mathbb{C}^2$, Arnold Math. J., 1:3 (2015), 233-242.

[V] Vigdal Abascal, E., *Measures définies sur les espaces des feuilles d'un feuilletage*, Rend. Circ. Math. Palermo, (2) 15 (1966), 247-256.

[V1] Vigdal Abascal, E., *Sur les feuilletages réguliers et les problèmes qui s'y reportent*, Acta Ci Compostelana 3 (1966), 3-12.

[V2] Vigdal Abascal, E., *On regular foliations*, Ann. Inst. Fourier (Grenoble) 17 (1967), 129-133.

[Ve] Verona, A., *Triangulation of stratified fiber bundles*, Manuscripta Math., 30 (1979), 425-445.

[Walt] Walters, P., *An introduction to ergodic theory*, Graduate Texts in Mathematics, 79, Springer-Verlag (1982).

[Wa] Wall, C. T. C., *Classification of* $(n-1)$*-connected* $2n$*-manifolds*, Ann. of Math. 75 (1962), 163-189.

[Wa1] Wall, C. T. C., *Non-Additivity of the Signature*, Inventiones Math., 7, 269-274 (1969).

[Wa2] Wall, C. T. C., *Determination of the cobordism ring*, The Annals of Mathematics, Vol. 72, No. 2, 292-311 (1960).

[Wa3] Wall, C. T. C., *Surgery on Compact Manifolds*, Academic Press, London & New York (1970), ISBN 0-12-732720-9.

[Wald] Waldhausen, F., *On irreducible* 3*-manifolds which are sufficiently large*, Ann. of Math., (2), 87, 56-88 (1968).

[WX] Wang, G., Xu, Z., *The triviality of the 61-stem in the stable homotopy groups of spheres*, Annals of Mathematics, 186 (2): 501-580, (2017), arXiv:1601.02184

[Wei] Weiss, M. S., *Immersion theory for homotopy theorists*, Graduate J. Math., 5 (2020), 9-27.

[Wen] Wen, H., *Simple Riemannian Surfaces are Scattering Rigid*, arXiv: 1405.1712v1 [math.DG], 7 May, 2014.

[We] Weyl, H., *On the volume of tubes*, American J. Of Mathematics, vol. 61, no 2 (Apr., 1939), pp. 461-472.

[We1] Weyl, H., *Zur Infinitisimalgeometrie: Einordnung der projectiven und der konformen Auffasung*, Nachrichten von der K. Gesellschaft der Wissenschaften zu Göttingen, Mathematisch-Physikalische Klasse; "Selecta Hermann Weyl", Birkhaüser Verlag, Basel und Stuttgart, 1956.

[W1] Whitney, H., *On regular closed curves in the plane*, Comp. Math. 4 (1937) 276-284.

[W2] Whitney, H., *Analytic extensions of functions defined in closed sets*, Transactions of the American Mathematical Society, American Mathematical Society, 36 (1): 63-89, (1934).

[W3] Whitney, H., *Geometric Integration Theory*, Princeton, New Jersey, Princeton University Press, 1957.

[Wh] Whitehead, J. H. C., *Combinatorial homotopy. I.*, Bull. Amer. Math. Soc., 55 (1949), 213-245.

[Wh1] Whitehead, J. H. C., *Combinatorial homotopy. II.*, Bull. Amer. Math. Soc., 55 (1949), 453-496.

[Zh] Zhou, X., *Recovery of the* C^∞*-jet from the boundary distance function*, arXiv:1103.5509v1 [math.DG] 28 March 2011.

www.ingramcontent.com/pod-product-compliance
Ingram Content Group UK Ltd.
Pitfield, Milton Keynes, MK11 3LW, UK
UKHW020405150926
8624IPUK00004B/355